Selected Titles in This Series

Volume

11 **Luc Vinet, Editor**
Advances in mathematical sciences: CRM's 25 years
1997

10 **Donald E. Knuth**
Stable marriage and its relation to other combinatorial problems: An introduction to the mathematical analysis of algorithms
1997

9 **D. Levi, L. Vinet, and P. Winternitz, Editors**
Symmetries and integrability of difference equations
1996

8 **J. Feldman, R. Froese, and L. M. Rosen, Editors**
Mathematical quantum theory II: Schrödinger operators
1995

7 **J. Feldman, R. Froese, and L. M. Rosen, Editors**
Mathematical quantum theory I: Field theory and many-body theory
1994

6 **Guido Mislin, Editor**
The Hilton Symposium 1993
Topics in topology and group theory
1994

5 **D. A. Dawson**
Measure-valued processes, stochastic partial differential equations, and interacting systems
1994

4 **Hershy Kisilevsky and M. Ram Murty, Editors**
Elliptic curves and related topics
1994

3 **Rémi Vaillancourt and Andrei L. Smirnov, Editors**
Asymptotic methods in mechanics
1993

2 **Philip D. Loewen**
Optimal control via nonsmooth analysis
1993

1 **M. Ram Murty, Editor**
Theta functions
1993

Advances in Mathematical Sciences: CRM's 25 Years

Volume 11

C R M

CRM PROCEEDINGS & LECTURE NOTES

Centre de Recherches Mathématiques
Université de Montréal

Advances in Mathematical Sciences: CRM's 25 Years

Luc Vinet, *Editor*

The Centre de Recherches Mathématiques (CRM) of the Université de Montréal was created in 1968 to promote research in pure and applied mathematics and related disciplines. Among its activities are special theme years, summer schools, workshops, postdoctoral programs, and publishing. The CRM is supported by the Université de Montréal, the Province of Québec (FCAR), and the Natural Sciences and Engineering Research Council of Canada. It is affiliated with the Institut des Sciences Mathématiques (ISM) of Montréal, whose constituent members are Concordia University, McGill University, the Université de Montréal, the Université du Québec à Montréal, and the Ecole Polytechnique.

American Mathematical Society
Providence, Rhode Island USA

The production of this volume was supported in part by the Fonds pour la formation de Chercheurs et l'Aide à la Recherche (Fonds FCAR) and the Natural Sciences and Engineering Research Council of Canada (NSERC).

1991 *Mathematics Subject Classification.* Primary 00B15.

Library of Congress Cataloging-in-Publication Data

Advances in mathematical sciences: CRM's 25 years / Luc Vinet, editor.
p. cm. — (CRM proceedings & lecture notes, ISSN 1065-8580 ; v. 11)
In English, with two contributions in French.
ISBN 0-8218-0686-6 (alk. paper)
1. Mathematics. I. Vinet, Luc. II. Université de Montréal, Centre de recherches mathématiques. III. Series.
QA7.A233 1997
510–dc21 97-26729
CIP

∞ The paper used in this book is acid-free and falls within the guidelines
established to ensure permanence and durability.
This volume was typeset using AMS-TeX and AMS-LaTeX,
the American Mathematical Society's TeX macro systems,
and submitted to the American Mathematical Society in camera ready
form by the Centre de Recherches Mathématiques.
Visit the AMS home page at URL: http://www.ams.org/

10 9 8 7 6 5 4 3 2 1 02 01 00 99 98 97

Contents

Preface

The Centre de Recherches Mathématiques (CRM) was founded at the Université de Montréal in 1968 and celebrated in 1994 the 25th anniversary of the beginning of its operations. A symposium entitled *"Advances in Mathematics"* was organized on this occasion and the present volume grew out of that event.

The CRM is by far the oldest of the Canadian institutes for research in the mathematical sciences. Its founders showed great vision at the time, in designing, with the help of the Université de Montréal, an organizational model that is now viewed in many countries as a "must have" resource for mathematical research. These founders include Roger Gaudry, then rector of the Université de Montréal, Maurice l'Abbé, chair of the Department of Mathematics and later vice-rector for research, Jacques St-Pierre, who acted as CRM's first director and became vice-rector of the Université de Montréal after creating its Computer Science Department, and André Aisenstadt, a Montreal philanthropist whose lasting support and involvement have been vital to the CRM.

The activities of the CRM fall into three broad categories: on-going research projects, scientific events, and publishing. The research programs are carried out by regular, associate, and visiting members of the Centre: this makes for a group of about 70 researchers (not counting graduate students) in residence at the CRM producing over 150 technical reports each year. The research areas are quite diversified and include in addition to pure and applied mathematics, parts of theoretical physics, statistics, and computer science. It is fitting here to stress the outstanding stature of the early members of the Centre; these people have contributed much to the reputation of the CRM and have set the high standards which led to its development into a world-class research institute. The CRM has also benefited from the leadership of remarkable directors; especially crucial was the term of Francis Clarke (1984–1993) when the national status of the CRM was firmly established. In keeping with this national mission, the CRM organizes and hosts every year some 15 workshops, conferences, or symposia, many of which take place in the framework of theme programs. Themes since 1993 are: *Dynamical Systems* (1993–1994), *Geometry and Topology* (1994–1995), *Applied and Numerical Analysis* (1995–1996), *Combinatorics and Group Theory* (1996–1997), *Statistics* (1997–1998), *Number Theory and Algebraic Geometry* (1998–1999), *Mathematical Physics* (1999–2000). The theme years involve important visitor programs and incorporate the Aisenstadt Chair which allows one or two distinguished scientists to be in residence at the Centre. In addition, the CRM organizes a yearly summer school in Banff, Alberta and supports a number of scientific events across the country. Altogether, some 1,000 scientists participate in CRM activities each year. With respect to publishing, let us mention that the CRM produces volumes in a number of series both in English and French. In this context our collaboration with the American Mathematical Society is most important.

The CRM is moreover actively pursuing partnerships with business and industry. This activity is at the heart of the recently established Network for Computing and Mathematical Modelling, NCM_2, whose purpose is to enhance research in industrial mathematics in Canada. Five major research centres and close to 20 organizations from business, industry, and government are involved in the NCM_2. The CRM also awards (i) the André-Aisenstadt Prize to outstanding young Canadian mathematicians, (ii) the CRM–CAP Prize for exceptional achievement by a Canadian in theoretical and mathematical physics, and (iii) jointly with the Fields Institute, the CRM-Fields Institute Prize for outstanding accomplishments in mathematics by a Canadian.

The CRM owes its success to the generous collaboration and involvement of numerous people: its regular, associate, and visiting members, its postdoctoral fellows and graduate students, its committee members, the organizers of its scientific events, its staff, the Université de Montréal administration, and its partners. I wish to take this occasion to thank them all. I also wish to acknowledge the sustained support that the CRM has received from federal (CRSNG) and provincial (FCAR) agencies, the Université de Montréal, private foundations, the Canadian Institute for advanced Research (CIAR), and of course the André and Nussia Aisenstadt Foundation.

In preparing this volume we wish we could have invited all of CRM's friends to contribute. This was of course impossible. We have thus chosen to solicit papers from associate members and long-term regular members of the Centre. The list of authors also includes eminent mathematicians from Canada and abroad, most of whom have directly contributed to the life of the CRM be it as directors, organizers, committee members, Aisenstadt chairholders, or in some other capacity. While very distinguished, this group of contributors is but a small subset of those we would have liked to include. Nevertheless, I believe that the present volume illustrates the great vehicle that the CRM has been for research in the mathematical sciences and the promise it holds for the next 25 years!

Luc Vinet

Centre de Recherches Mathématiques
CRM Proceedings and Lecture Notes
Volume **11**, 1997

The Problem of Classifying Automorphic Representations of Classical Groups

James Arthur

In this article we shall give an elementary introduction to an important problem in representation theory. The problem is to relate the automorphic representations of classical groups to those of the general linear group. Thanks to the work of a number of people over the past twenty-five years, the automorphic representation theory of $\mathrm{GL}(n)$ is in pretty good shape. The theory for $\mathrm{GL}(n)$ now includes a good understanding of the analytic properties of Rankin-Selberg L-functions, the classification of the discrete spectrum, and cyclic base change. One would like to establish similar things for classical groups. The goal would be an explicit comparison between the automorphic spectra of classical groups and $\mathrm{GL}(n)$ through the appropriate trace formulas. There are still obstacles to be overcome. However with the progress of recent years, there is also reason to be optimistic.

We shall not discuss the techniques here. Nor will we consider the possible applications. Our modest aim is to introduce the problem itself, in a form that might be accessible to a nonspecialist. In the process we shall review some of the basic constructions and conjectures of Langlands that underlie the theory of automorphic representations.

1. We shall begin with a few of the basic concepts from the theory for the general linear group. For the present, then, we take $G = \mathrm{GL}(n)$. The adèles of $\mathbb{Q}$ form a locally compact ring

$$\mathbb{A} = \mathbb{R} \times \mathbb{Q}_2 \times \mathbb{Q}_3 \times \mathbb{Q}_5 \times \cdots$$

in which $\mathbb{Q}$ embeds diagonally as a discrete subring. Consequently $G(\mathbb{A})$ is a locally compact group which contains $G(\mathbb{Q})$ as a discrete subgroup. One can form the Hilbert space $L^2\big(G(\mathbb{Q})\backslash G(\mathbb{A})\big)$ of functions which are square integrable with respect to the right $G(\mathbb{A})$-invariant measure. The primary object of study is the regular representation

$$\big(R(y)f\big)(x) = f(xy), \quad f \in L^2\big(G(\mathbb{Q})\backslash G(\mathbb{A})\big), \quad x, y \in G(\mathbb{A}),$$

on the Hilbert space.

1991 *Mathematics Subject Classification.* Primary: 22E55.

Supported in part by an NSERC Research Grant.

This is the final form of the paper.

The unitary representation R is highly reducible. For this discussion we shall define an *automorphic representation* informally as an irreducible unitary representation π of $G(\mathbb{A})$ which occurs in the decomposition of R. This notion would be precise certainly if π occurred as a discrete summand of R. However, the irreducible constituents of R depend on several continuous parameters and one wants to include all of these. The proper definition [**14**] in fact includes irreducible representations of $G(\mathbb{A})$ which come from the analytic continuation of these parameters, but there is no need to consider such objects here. It is known [**5**] that any such π has a decomposition

$$\pi = \pi_{\mathbb{R}} \otimes \pi_2 \otimes \pi_3 \otimes \pi_5 \otimes \cdots$$

as a restricted tensor product, with each π_p being an irreducible unitary representation of the group $G(\mathbb{Q}_p)$.

Anyone seeing these definitions for the first time could well ask why automorphic representations are interesting. To get a feeling for the situation, we fix a prime p and recall the construction of the unramified representations of $G(\mathbb{Q}_p)$—the simplest family of irreducible representations $\{\pi_p\}$ of this group.

The representations in the family are determined by elements $u = (u_1, \ldots, u_n)$ in $\mathbb{C}^n$. Such an element defines a character of the Borel subgroup

$$B(\mathbb{Q}_p) = \left\{ b = \begin{pmatrix} b_{11} & & * \\ & \ddots & \\ 0 & & b_{nn} \end{pmatrix} \right\} \subseteq G(\mathbb{Q}_p)$$

of $\mathrm{GL}(n, \mathbb{Q}_p)$ by

$$\chi_u(b) = |b_{11}|_p^{u_1+(n-1)/2} \, |b_{22}|_p^{u_2+(n-3)/2} \cdots |b_{nn}|_p^{u_n-(n-1)/2}.$$

Let $\pi_{p,u}^+$ be the corresponding induced representation of $G(\mathbb{Q}_p)$. It acts on a space of functions $f_p : G(\mathbb{Q}_p) \to \mathbb{C}$ which satisfy

$$f_p(bx) = \chi_u(b) f_p(x), \quad b \in B(\mathbb{Q}_p), \quad x \in G(\mathbb{Q}_p),$$

be right translation—

$$\big(\pi_{p,u}^+(y) f_p\big)(x) = f_p(xy), \quad x, y \in G(\mathbb{Q}_p).$$

The vector $\frac{1}{2}\big(n-1, n-3, \ldots, -(n-1)\big)$ comes from the usual Jacobian factor, and is included so that $\pi_{p,u}^+$ will be unitary if u is purely imaginary. If u is purely imaginary, $\pi_{p,u}^+$ is known to be irreducible as well as unitary. In general, $\pi_{p,u}^+$ can have several irreducible constituents, but there is a canonical one—the irreducible constituent $\pi_{p,u}$ of $\pi_{p,u}^+$ which contains a $G(\mathbb{Z}_p)$-fixed vector. Thus, any u determines an irreducible representation $\pi_{p,u}$ of $G(\mathbb{Q}_p)$. Since the p-adic absolute values in the definition of χ_u are powers of p, it is clear that $\pi_{p,u}$ remains the same if u is translated by a vector in $(2\pi i/\log p)\mathbb{Z}^n$. In fact if u' is any other vector in $\mathbb{C}^n$, it is known that $\pi_{p,u'}$ is equivalent to $\pi_{p,u}$ if and only if

$$(u_1', \ldots, u_n') \equiv (u_{\sigma(1)}, \ldots, u_{\sigma(n)}) \left(\mathrm{mod} \left(\frac{2\pi i}{\log p} \right) \mathbb{Z}^n \right),$$

for some permutation σ in S_n.

By definition, the *unramified* representations of $G(\mathbb{Q}_p)$ are the ones in the family $\{\pi_{p,u} : u \in \mathbb{C}^n\}$. Set

$$t(\pi_{p,u}) = \begin{pmatrix} p^{-u_1} & & 0 \\ & \ddots & \\ 0 & & p^{-u_n} \end{pmatrix},$$

regarded as a semisimple conjugacy class in $\mathrm{GL}(n,\mathbb{C})$. This is a special case of a general construction [**13**] of Langlands. In the present situation it gives a bijection between the unramified representations of $\mathrm{GL}(n,\mathbb{Q}_p)$ and the semisimple conjugacy classes in $\mathrm{GL}(n,\mathbb{C})$.

Now suppose that π is an automorphic representation of $\mathrm{GL}(n,\mathbb{A})$. It is known that the local components π_p of π are unramified for almost all p. In other words, π determines a family

$$t(\pi) = \{t(\pi_p) : p \notin S\}$$

of semisimple conjugacy classes in $\mathrm{GL}(n,\mathbb{C})$. Here $S = S_\pi$ is a finite set of completions of $\mathbb{Q}$ which includes the Archimedean place $\mathbb{R}$. Returning to the original question, automorphic representations are interesting because the corresponding families $t(\pi)$ are believed to carry fundamental arithmetic information. What is important is not the fact that almost all π_p are unramified—this would be true of any irreducible representation of $G(\mathbb{A})$ with some weak continuity hypothesis—but that π is automorphic. It is only then that the semisimple conjugacy classes $\{t(\pi_p)\}$ will be related one to another in a way that is governed by fundamental arithmetic phenomena.

In order to package the data $t(\pi)$ conveniently, one defines the local L-function

$$L(s,\pi_p) = \det\big(I - t(\pi_p)p^{-s}\big)^{-1}, \quad s \in \mathbb{C}, \quad p \in S,$$

as the reciprocal (evaluated at p^{-s}) of the characteristic polynomial of the conjugacy class $t(\pi_p)$. One can then define a global L-function

$$L_S(s,\pi) = \prod_{p \notin S} L(s,\pi_p)$$

as an Euler product which converges in some right half plane. It is known that $L_S(s,\pi)$ has analytic continuation as a meromorphic function of $s \in \mathbb{C}$, and satisfies a functional equation [**9**]. The basic proof is a generalization of the one used by Tate for $\mathrm{GL}(1)$. It exploits the embedding of $\mathrm{GL}(n)$ into the space of $(n \times n)$-matrices. The proof entails defining local L-functions $L(s,\pi_p)$ for every p (including $p = \mathbb{R}$). If one forms the product

$$L(s,\pi) = \prod_p L(s,\pi_p)$$

over all p, the functional equation takes the form

$$L(s,\pi) = \epsilon(s,\pi)L(1-s,\tilde{\pi}),$$

where $\tilde{\pi}$ is the contragredient representation of π, and the ϵ-factor is a simple function of the form

$$\epsilon(s,\pi) = a_\pi(p^{r_\pi})^s, \quad a_\pi \in \mathbb{C}, \quad r_\pi \in \mathbb{Z}.$$

For an elementary example, take $G = \mathrm{GL}(1)$. Then $G(\mathbb{A}) = \mathbb{A}^*$ is the group of idèles, while $G(\mathbb{Q})\backslash G(\mathbb{A}) = \mathbb{Q}^*\backslash\mathbb{A}^*$ is the quotient group of idèle classes. We shall consider an automorphic representation $\pi = \bigotimes_p \pi_p$ with $S_\pi = \{\mathbb{R}, 2\}$. Then if $p \notin$

S, π_p is determined by an unramified character on the group $B(\mathbb{Q}_p) = \mathbb{Q}_p^* = G(\mathbb{Q}_p)$. Any such prime is of course odd. Set

$$t(\pi_p) = p^{-u} = \begin{cases} 1, & \text{if } p \equiv 1 \pmod 4, \\ -1, & \text{if } p \equiv 3 \pmod 4. \end{cases}$$

In other words,

$$\pi_p(x_p) = |x_p|_p^u = \begin{cases} 1, & \text{if } p \equiv 1 \pmod 4, \\ (-1)^{v(x_p)}, & \text{if } p \equiv 3 \pmod 4, \end{cases}$$

where $v(x_p) \in \mathbb{Z}$ is the valuation of a point $x_p \in \mathbb{Q}_p^*$. It is then easy to define characters $\pi_{\mathbb{R}}$ and π_2 on $\mathbb{R}^*$ and $\mathbb{Q}_2^*$ respectively so that $\pi = \bigotimes_p \pi_p$ is trivial on the subgroup $\mathbb{Q}^*$ of $\mathbb{A}^*$, and is hence an automorphic representation of GL(1). Observe that the definition of π_p for $p \notin S$ matches the splitting law of the prime p in the Gaussian integers $\mathbb{Z}\left[\sqrt{-1}\right]$; p is of the form

$$p = \left(a + \sqrt{-1}b\right)\left(a - \sqrt{-1}b\right) = a^2 + b^2, \quad a, b \in \mathbb{Z},$$

if and only if p is congruent to 1 modulo 4. This is no co-incidence. The Kronecker-Weber theorem can be read as the construction of an automorphic representation for any cyclic extension of $\mathbb{Q}$ in terms of how rational primes behave in the extension. The Artin reciprocity law gives a similar construction in the more general case that $\mathbb{Q}$ is replaced by an arbitrary number field F. It can be regarded as a classification of abelian extensions of F in terms of automorphic representations of GL(1) (relative to F).

This is a good point to recall Langlands' nonabelian generalization of the Artin reciprocity law. Suppose that

$$\phi : \mathrm{Gal}(\overline{\mathbb{Q}}/\mathbb{Q}) \longrightarrow \mathrm{GL}(n, \mathbb{C})$$

is an n-dimensional representation of the Galois group of an algebraic closure of $\mathbb{Q}$ which is continuous, that is, which factors through a finite quotient $\mathrm{Gal}(E/\mathbb{Q})$ of $\mathrm{Gal}(\overline{\mathbb{Q}}/\mathbb{Q})$. Then ϕ is unramified outside a finite set $S = S_\phi$ of primes. For any prime $p \notin S$, there is a Frobenius conjugacy class Fr_p in $\mathrm{Gal}(E/\mathbb{Q})$, and hence a conjugacy class $\phi(\mathrm{Fr}_p)$ in $\mathrm{GL}(n, \mathbb{C})$. Langlands conjectured that for any ϕ there is an automorphic representation π of $\mathrm{GL}(n)$ such that

$$t(\pi_p) = \phi(\mathrm{Fr}_p), \quad p \notin S_\pi \cap S_\phi.$$

This conjecture is very difficult, and has been established in only a limited number of cases [**15, 16, 4**]. It is known, however, that there is at most one π with this property [**10**].

We recall also that there is an Artin L-function attached to ϕ which is completely parallel to the construction of an automorphic L-function. It is defined by an Euler product

$$L(s, \phi) = \prod_p L(s, \phi_p)$$

which converges in a right half plane, with local factors given by

$$L(s, \phi_p) = \det\left(I - \phi(\mathrm{Fr}_p)p^{-s}\right)^{-1}$$

if p does not belong to S_ϕ. The function $L(s,\phi)$ has analytic continuation and satisfies a functional equation

$$L(s,\phi) = \epsilon(s,\phi)L(1-s,\tilde{\phi}),$$

with an ϵ-factor of the form

$$\epsilon(s,\phi) = a_\phi(p^{r_\phi})^s, \quad a_\phi \in \mathbb{C}, \quad r_\phi \in \mathbb{Z}.$$

Langlands' conjectural reciprocity law, which is actually a special case of his functoriality principle, was formulated for all places p. It asserts that $L(s,\pi_p) = L(s,\phi_p)$ for all p, or in global form, that

$$L(s,\pi) = L(s,\phi).$$

In other words, every Artin L-function is an automorphic L-function.

2. Suppose now that G belongs to one of the other three families $\mathrm{SO}(2n+1)$, $\mathrm{Sp}(2n)$ and $\mathrm{SO}(2n)$ of classical groups. We shall assume that G is quasi-split. Then G will actually be split if it is of the form $\mathrm{SO}(2n+1)$ or $\mathrm{Sp}(2n)$. In the remaining case, G could be a nonsplit form of $\mathrm{SO}(2n)$ which splits over a quadratic extension E of $\mathbb{Q}$. (We exclude the exceptional quasi-split forms of $\mathrm{SO}(8)$.)

With suitable modifications, the constructions of §1 all carry over to G. (They were introduced by Langlands for any reductive group over any global field F [**13**].) In particular, an automorphic representation of $G(\mathbb{A})$ has a decomposition $\pi = \bigotimes_p \pi_p$, in which π_p is an unramified representation of $G(\mathbb{Q}_p)$ for all p outside a finite set $S = S_\pi$. Each such π_p is a constituent of a representation induced from an unramified quasi-character of a Borel subgroup $B(\mathbb{Q}_p)$ of $G(\mathbb{Q}_p)$. The reader unfamiliar with these things could try at this point to construct a semisimple conjugacy class $t(\pi_p)$, in analogy with $\mathrm{GL}(n)$. He/she will discover that such a conjugacy class exists, but that it occurs naturally in a complex group which is dual to G. If G is split, one can take the dual group $\widehat{G}$ given by the table

G	$\widehat{G}$
$\mathrm{SO}(2n+1)$	$\mathrm{Sp}(2n,\mathbb{C})$
$\mathrm{Sp}(2n)$	$\mathrm{SO}(2n+1,\mathbb{C})$
$\mathrm{SO}(2n)$	$\mathrm{SO}(2n,\mathbb{C})$

If G is not split, one must take a semi-direct product

$$\widehat{G} \rtimes \mathrm{Gal}(E/\mathbb{Q}),$$

in which $\mathrm{Gal}(E/\mathbb{Q})$ acts on $\widehat{G} = \mathrm{SO}(2n,\mathbb{C})$ by conjugation through the isomorphism

$$\mathrm{Gal}(E/\mathbb{Q}) \simeq \mathrm{O}(2n,\mathbb{C})/\mathrm{SO}(2n,\mathbb{C}).$$

The two cases are combined in Langlands' original construction of the L-group

$${}^LG = \widehat{G} \rtimes \mathrm{Gal}(\overline{\mathbb{Q}}/\mathbb{Q}),$$

where $\mathrm{Gal}(\overline{\mathbb{Q}}/\mathbb{Q})$ acts trivially on $\widehat{G}$ in case G is split, and acts on $\widehat{G}$ through its quotient $\mathrm{Gal}(E/\mathbb{Q})$ if G is not split. In the case of the general linear group, one obviously takes ${}^L\big(\mathrm{GL}(n)\big)$ to be the direct product of $\mathrm{GL}(n,\mathbb{C})$ with $\mathrm{Gal}(\overline{\mathbb{Q}}/\mathbb{Q})$.

Thus, to any automorphic representation π of $G(\mathbb{A})$ there is associated a family

$$t(\pi) = \big\{t(\pi_p) : p \notin S_\pi\big\}$$

of semisimple conjugacy classes in the complex reductive group ${}^LG$. We have to remind ourselves that the situation is more concrete than the final notation suggests; if G is split, for example, one can always replace ${}^LG$ by the complex connected group $\widehat{G}$. As with $\mathrm{GL}(n)$, the numerical data which determine these conjugacy classes are believed to carry fundamental arithmetic information. In fact, the data obtained in this way ought to be a subset of the data obtained from general linear groups. This is the essence of the problem we shall presently discuss, and is also a special case of Langlands' functoriality principle. (For an introduction to the functoriality principle, see [**1**].)

If the automorphic representations of classical groups are to be understood in terms of $\mathrm{GL}(n)$, why study them at all? There are compelling reasons to do so. Suppose for example that $G = \mathrm{Sp}(2n)$. One can form the Siegel moduli space

$$S(N) = \Gamma(N)\backslash\mathcal{H},$$

where $\mathcal{H}$ is the Siegel upper half space of genus n, and $\Gamma(N)$ is the congruence subgroup

$$\{\gamma \in \mathrm{Sp}(2n, \mathbb{Z}) : \gamma \equiv I \pmod{N}\}$$

of $\mathrm{Sp}(2n, \mathbb{R})$. Then $S(N)$ is a complex algebraic variety. The L^2-cohomology of $S(N)$, $H^*_{(2)}\big(S(N)\big)$, is a very interesting object which is directly related to certain automorphic representations π of $\mathrm{Sp}(2n, \mathbb{A})$. For such π, the conjugacy classes $t(\pi_p)$ are governed by the eigenvalues of Hecke operators acting on the cohomology. (See [**3**] for an introduction to these and related questions.) In this way one studies quite different properties of π than one could get from the corresponding object on a general linear group.

To attach an L-function to an automorphic representation π of $G(\mathbb{A})$, one has first to embed ${}^LG$ in a general linear group. Suppose that

$${}^Lr : {}^LG \longrightarrow \mathrm{GL}(V)$$

is a complex analytic, finite dimensional representation of ${}^LG$. This determines local L-factors

$$L(s, \pi_p, {}^Lr) = \det\big(1 - {}^Lr(t(\pi_p))p^{-s}\big)^{-1}, \quad p \notin S_\pi,$$

for almost all p. One would like to be able to define L-factors for all p, and to show that the Euler product

$$L(s, \pi, {}^Lr) = \prod_p L(s, \pi_p, {}^Lr)$$

has analytic continuation and functional equation. The case of $G = \mathrm{GL}(n)$ and ${}^Lr$ the standard n-dimensional representation of $\mathrm{GL}(n, \mathbb{C})$ was discussed in §1. Despite considerable progress [**8**], however, the general case is still far from solved.

Finally, we recall that the Langlands reciprocity conjecture applies equally well to L-homomorphisms

$$\phi_G : \mathrm{Gal}(\overline{\mathbb{Q}}/\mathbb{Q}) \longrightarrow {}^LG$$

attached to G. (An *L-homomorphism* is one which is compatible with projections of the domain and co-domain onto $\mathrm{Gal}(\overline{\mathbb{Q}}/\mathbb{Q})$.) For each ϕ_G there should exist an automorphic representation π_G of $G(\mathbb{A})$ with the property that for any ${}^Lr : {}^LG \to \mathrm{GL}(V)$, the Artin L-function $L(s, {}^Lr \circ \phi_G)$ equals the automorphic L-function $L(s, \pi_G, {}^Lr)$.

This completes our discussion of some of the general properties of automorphic representations. We can now formulate the problem we set out to describe.

Observe that for our classical group G there is a canonical embedding

$$r_G : \widehat{G} \hookrightarrow \mathrm{GL}(N, \mathbb{C}),$$

with N equal to either $2n$ or $2n+1$. This can be extended to an L-embedding

$$^L r_G : {}^L G = \widehat{G} \rtimes \mathrm{Gal}(\overline{\mathbb{Q}}/\mathbb{Q}) \hookrightarrow \mathrm{GL}(N, \mathbb{C}) \times \mathrm{Gal}(\overline{\mathbb{Q}}/\mathbb{Q}) = {}^L\big(\mathrm{GL}(N)\big).$$

By composing with $^L r_G$, we obtain a map $\phi_G \to \phi$,

$$\begin{array}{ccc} & \mathrm{Gal}(\overline{\mathbb{Q}}/\mathbb{Q}) & \\ {}^{\phi_g}\swarrow & & \searrow^{\phi} \\ {}^L G & \xrightarrow{\ {}^L r_G\ } & {}^L\big(\mathrm{GL}(N)\big) \end{array}$$

between L-homomorphisms into the two L-groups. We shall identify ϕ with its projection onto $\mathrm{GL}(N, \mathbb{C})$, that is, with an N-dimensional representation of $\mathrm{Gal}(\overline{\mathbb{Q}}/\mathbb{Q})$. As such it is self-contragredient. Conversely suppose that ϕ is an arbitrary self-contragredient N-dimensional representation of $\mathrm{Gal}(\overline{\mathbb{Q}}/\mathbb{Q})$. We assume also that ϕ is irreducible. Then ϕ factors through an orthogonal or a symplectic group. More precisely, there is a unique G, and an L-homomorphism ϕ_G for G, such that ϕ is equivalent to $^L r_G \circ \phi_G$. (This is an easy consequence of the self-contragredience of ϕ—see for example §3 below.)

The problem is to show that there is a similar mapping $\pi_G \to \pi$ between automorphic representations. The mapping should reduce to $\phi_G \to \phi$ for the automorphic representations attached (by Langlands' conjectural reciprocity law) to L-homomorphisms. As in this special case, the general mapping will be defined in terms of the families $t(\pi)$ of conjugacy classes.

Problem.

(i) *If π_G is an automorphic representation of the classical group G, show that there is an automorphic representation π of* $\mathrm{GL}(N, \mathbb{A})$ *such that*

$$^L r_G\big(t(\pi_{G,p})\big) = t(\pi_p)$$

for almost all p.

(ii) *Conversely, suppose that π is a self-contragredient automorphic representation of* $\mathrm{GL}(N, \mathbb{A})$. *If π is cuspidal, show that π is the image of an automorphic representation π_G of $G(\mathbb{A})$, for a unique G as above.*

The problem is analogous to the base change problem, solved originally for $\mathrm{GL}(2)$ by Langlands [**15**]. That a similar question could be posed for the outer automorphism

$$x \to \tilde{x} = {}^t x^{-1}, \quad x \in \mathrm{GL}(N),$$

of $\mathrm{GL}(N)$ was I believe first noticed by Jacquet. However, there are some new phenomena here. The most obvious is the possibility of lifting representations from more than one G to a given $\mathrm{GL}(N)$. If $N = 2n$ is even, $\widehat{G}$ could be either $\mathrm{SO}(2n, \mathbb{C})$ or $\mathrm{Sp}(2n, \mathbb{C})$; that is, G could be either $\mathrm{SO}(2n)$ or $\mathrm{SO}(2n+1)$. It was pointed out by Shalika that one ought to be able to separate these two cases by looking at the symmetric square and alternating square L-functions.

Let S^2 (respectively Λ^2) be the finite dimensional representation

$$g : X \to {}^t gXg, \quad g \in \mathrm{GL}(2n, \mathbb{C}),$$

of $\mathrm{GL}(2n, \mathbb{C})$ on the space of symmetric (resp. skew-symmetric) $(2n \times 2n)$-matrices. Consider a self-contragredient irreducible Galois representation

$$\phi : \mathrm{Gal}(\overline{\mathbb{Q}}/\mathbb{Q}) \longrightarrow \mathrm{GL}(2n, \mathbb{C}).$$

Then ϕ factors through $\mathrm{O}(2n, \mathbb{C})$ (resp. $\mathrm{Sp}(2n, \mathbb{C})$) if and only if the representation $S^2 \circ \phi$ (resp. $\Lambda^2 \circ \phi$) of $\mathrm{Gal}(\overline{\mathbb{Q}}/\mathbb{Q})$ contains the trivial representation. This is the case if and only if the Artin L-function $L(s, S^2 \circ \phi)$ (resp. $L(s, \Lambda^2 \circ \phi)$) has a pole at $s = 1$. This suggests the following supplement to the problem.

(iii) *Suppose that π is a self-contragredient cuspidal automorphic representation of* $\mathrm{GL}(2n)$. *Show that π is the image of an automorphic representation π_G of* $\mathrm{SO}(2n)$ *(respectively* $\mathrm{SO}(2n+1)$*) if and only if the automorphic L-function $L(s, \pi, S^2)$ (resp. $L(s, \pi, \Lambda^2)$) has a pole at $s = 1$.*

We shall state a second supplement to the problem that concerns automorphic ϵ-factors. Suppose that

$$\phi : \mathrm{Gal}(\overline{\mathbb{Q}}/\mathbb{Q}) \longrightarrow \mathrm{GL}(N, \mathbb{C})$$

is an irreducible Galois representation. If we apply the functional equation of the Artin L-function $L(s, \phi)$ twice, we obtain

$$\epsilon(s, \phi)\epsilon(1 - s, \tilde{\phi}) = 1.$$

Assume that ϕ is self-contragredient. Setting $s = \frac{1}{2}$, we see that

$$\epsilon\left(\frac{1}{2}, \phi\right) = \pm 1.$$

The self-contragredience of ϕ means that it factors through an orthogonal or a symplectic group. If ϕ factors through $\mathrm{Sp}(N, \mathbb{C})$, $\epsilon\left(\frac{1}{2}, \phi\right)$ can be either 1 or -1; the actual value of this sign has interesting number theoretic implications [**6**]. If ϕ factors through $\mathrm{O}(N, \mathbb{C})$, however, $\epsilon\left(\frac{1}{2}, \phi\right)$ is known to equal 1 [**7**]. One would like to establish the automorphic version of this property.

(iv) *Suppose that π is a self-contragredient cuspidal automorphic representation of* $\mathrm{GL}(N)$. *If π is the image of an automorphic representation π_G of a group G with $\widehat{G} = \mathrm{SO}(N, \mathbb{C})$, show that $\epsilon\left(\frac{1}{2}, \pi\right) = 1$.*

3. It is known that an automorphic representation π of $\mathrm{GL}(N)$ is uniquely determined by the family $t(\pi)$ of conjugacy classes. In other words, the map

$$\pi \longrightarrow t(\pi),$$

from the automorphic representations of $\mathrm{GL}(N)$ to families of semisimple conjugacy classes in $\mathrm{GL}(N, \mathbb{C})$, is injective. (The objects in the range are to be regarded as equivalence classes, two families being equivalent if they are equal at almost all p.) This is a theorem of Jacquet-Shalika [**10**], which is an extension of the earlier result for cuspidal automorphic representations. (Keep in mind that we have adopted a restrictive definition of automorphic representation. What we are calling an automorphic representation really includes an extra condition, that of being *globally tempered*; it is only with this condition that the injectivity is valid.)

The corresponding assertion for a classical group G is generally false. If

$$t_G = \{t_{G,p} : p \notin S\}$$

is a family of semisimple conjugacy classes in ${}^LG$, the set of automorphic representations π_G of $G(\mathbb{A})$ such that $t(\pi_G) = t_G$ could be an infinite packet. In particular, the mapping $\pi_G \to \pi$ of our problem could have large fibres. An important part of the problem is to determine these fibres. There is a precise conjectural description of the preimage of any π, based on the theory of endoscopy [**12**] and its extension to nontempered representations [**2**]. We shall not repeat it here. It suffices to say that the description is motivated by the case that π is attached to a self-contragredient Galois representation. We shall conclude this article with a few remarks on the structure of such Galois representations.

Consider an L-homomorphism

$$\phi : \mathrm{Gal}(\overline{\mathbb{Q}}/\mathbb{Q}) \longrightarrow {}^L\big(\mathrm{GL}(N)\big).$$

We have agreed not to distinguish between such an object and the corresponding N-dimensional Galois representation. Thus, ϕ has a decomposition

$$\phi = \ell_1\phi_1 \oplus \cdots \oplus \ell_r\phi_r$$

into irreducible Galois representations

$$\phi_i : \mathrm{Gal}(\overline{\mathbb{Q}}/\mathbb{Q}) \longrightarrow {}^L\big(\mathrm{GL}(N_i)\big),$$

which occur with multiplicities ℓ_i. Suppose that ϕ is self-contragredient. Then there is a permutation $i \to \tilde{\imath}$ of period 2 on the set of indices such that $\tilde{\phi}_i = \phi_{\tilde{\imath}}$ and $\ell_i = \ell_{\tilde{\imath}}$.

We are going to confine our attention to a special case. We assume that for every i, $\ell_i = 1$ and $\tilde{\phi}_i = \phi_i$. In particular, the irreducible representation

$$\sigma \longrightarrow \tilde{\phi}_i(\sigma) = {}^t\phi_i(\sigma)^{-1}, \quad \sigma \in \mathrm{Gal}(\overline{\mathbb{Q}}/\mathbb{Q}),$$

is equivalent to ϕ_i. It follows that for each i, there is a matrix $A_i \in \mathrm{GL}(N_i, \mathbb{C})$ such that

$${}^t\phi_i(\sigma)^{-1} = A_i\phi_i(\sigma)A_i^{-1}, \quad \sigma \in \mathrm{Gal}(\overline{\mathbb{Q}}/\mathbb{Q}).$$

Applying this equation twice, we see that ${}^tA_i^{-1}A_i$ is an intertwining operator for the representation ϕ_i. It follows from Schur's lemma that ${}^tA_i = cA_i$ for some $c \in \mathbb{C}^*$. Applying this last identity twice, we find that $c^2 = 1$, so that A_i is either skew-symmetric or symmetric. Therefore ϕ_i is either of symplectic or orthogonal type. More precisely, if we replace ϕ_i by a suitable $\mathrm{GL}(N_i, \mathbb{C})$-conjugate, we can assume that either

$$\mathrm{Image}(\phi_i) \subseteq \mathrm{Sp}(N_i, \mathbb{C}) \subseteq \mathrm{GL}(N_i, \mathbb{C})$$

or

$$\mathrm{Image}(\phi_i) \subseteq \mathrm{O}(N_i, \mathbb{C}) \subseteq \mathrm{GL}(N_i, \mathbb{C}).$$

Separating the indices i into two disjoint sets I^1 and I^2 according to whether ϕ_i is symplectic or orthogonal, we obtain a decomposition

$$\phi = \phi^1 \oplus \phi^2,$$

where

$$\phi^1 = \bigoplus_{j \in I^1} \phi_j : \mathrm{Gal}(\overline{\mathbb{Q}}/\mathbb{Q}) \longrightarrow \prod_j \mathrm{Sp}(N_j, \mathbb{C}) \subseteq \mathrm{Sp}(N^1, \mathbb{C})$$

and

$$\phi^2 = \bigoplus_{k \in I^2} \phi_k : \mathrm{Gal}(\overline{\mathbb{Q}}/\mathbb{Q}) \longrightarrow \prod_k \mathrm{O}(N_k, \mathbb{C}) \subseteq \mathrm{O}(N^2, \mathbb{C}),$$

in which $N^1 = \sum_j N_j$ and $N^2 = \sum_k N_k$.

The maps ϕ^1 and ϕ^2 can be analyzed separately. For the first one, we note that $\mathrm{Sp}(N^1, \mathbb{C})$ is connected and equals $(\widehat{G^1})$, where $G^1 = \mathrm{SO}(N^1+1)$. There is nothing more to say in this case. For the second case, observe that the map

$$\mathrm{Gal}(\overline{\mathbb{Q}}/\mathbb{Q}) \longrightarrow \mathrm{O}(N^2, \mathbb{C}) / \mathrm{SO}(N^2, \mathbb{C}) = \mathbb{Z}/2\mathbb{Z}$$

obtained from ϕ^2 by projection, determines a quadratic character η of $\mathrm{Gal}(\overline{\mathbb{Q}}/\mathbb{Q})$. Suppose first that N^2 is odd. Then $\mathrm{O}(N^2, \mathbb{C})$ is the direct product of $\mathrm{SO}(N^2, \mathbb{C})$ with $\mathbb{Z}/2\mathbb{Z}$. Setting $G^2 = \mathrm{Sp}(N^2 - 1)$, we use η to define an embedding of

$${}^L(G^2) = \mathrm{SO}(N^2, \mathbb{C}) \times \mathrm{Gal}(\overline{\mathbb{Q}}/\mathbb{Q})$$

into

$${}^L\big(\mathrm{GL}(N^2)\big) = \mathrm{GL}(N^2, \mathbb{C}) \times \mathrm{Gal}(\overline{\mathbb{Q}}/\mathbb{Q})$$

so that ϕ^2 factors through ${}^L(G^2)$. Next suppose that N^2 is even. Then $\mathrm{O}(N^2, \mathbb{C})$ is a semi-direct product of $\mathrm{SO}(N^2, \mathbb{C})$ with $\mathbb{Z}/2\mathbb{Z}$. Let G^2 be the quasi-split form of $\mathrm{SO}(N^2)$ obtained from η and the action of the nonidentity component of $\mathrm{O}(N^2)$ on $\mathrm{SO}(N^2)$. Again there is an embedding of

$${}^L(G^2) = \mathrm{SO}(N^2, \mathbb{C}) \underset{\eta}{\rtimes} \mathrm{Gal}(\overline{\mathbb{Q}}/\mathbb{Q})$$

into

$${}^L\big(\mathrm{GL}(N^2)\big) = \mathrm{GL}(N^2, \mathbb{C}) \times \mathrm{Gal}(\overline{\mathbb{Q}}/\mathbb{Q})$$

such that ϕ^2 factors through ${}^L(G^2)$.

We have shown that the original Galois representation factors through ${}^LG$, for a unique classical group $G = G^1 \times G^2$. The groups obtained in this way (taken together with the embeddings ${}^LG \hookrightarrow {}^L(\mathrm{GL}(N))$) are called the twisted endoscopic groups for $\mathrm{GL}(N)$. (See [**11**]). They arise naturally from the twisted trace formula for $\mathrm{GL}(N)$, which of course is where one would begin the study of our problem. If one is interested in the image and fibres of the maps $\pi_G \to \pi$, one should really state the problem in terms of these general endoscopic groups. However, for the study of classical groups, the primitive case that G equals G^1 or G^2 is obviously what is important.

The conjectural description of the contribution of ϕ to the spectrum of G we have alluded to (that is, the preimage in G of the automorphic representation π of $\mathrm{GL}(N)$ attached to ϕ) is given in terms of a group

$$S_\phi = S_\phi(G) = \mathrm{Cent}\big(\mathrm{Image}(\phi), \widehat{G}\big),$$

the centralizer in $\widehat{G}$ of the image of ϕ [**2**, Conjecture 8.1]. For example, ϕ should contribute to the discrete spectrum of G if and only if $S_\phi(G)$ is finite. It is clear that for ϕ as above,

$$S_\phi\big(\mathrm{GL}(N)\big) = (\mathbb{C}^*)^r.$$

One also sees easily that

$$S_\phi(G) = \begin{cases} (\mathbb{Z}/2\mathbb{Z})^r, & \text{if each } N_i \text{ is even,} \\ (\mathbb{Z}/2\mathbb{Z})^{r-1}, & \text{if some } N_i \text{ is odd.} \end{cases}$$

Thus, ϕ contributes to the continuous spectrum of $\mathrm{GL}(N)$, but ought to contribute to the discrete spectrum of G. This property actually characterizes the special case we have been considering. If $\ell_i > 1$ or $\tilde{\phi}_i \neq \phi_i$ for some i, and if ϕ factors through ${}^LG$, the group $S_\phi(G)$ will be infinite. Then ϕ should contribute only to the continuous spectrum of G. In this more general situation, there could also be several different G such that ϕ factors through ${}^L G$.

What is apparent is that one will need some analogue of the group $S_\phi(G)$ to determine the fibres of the map $\pi_G \to \pi$. It is no solution to use $S_\phi(G)$ itself—the Langlands reciprocity law is far from being established, and even if it were, it would not be surjective. What we need instead is the construction of a group $S_\pi(G)$, for any self-contragredient automorphic representation π of $\mathrm{GL}(N)$, which reduces to $S_\phi(G)$ in case π comes from ϕ. Now we can write π formally as

$$\pi = \ell_1\pi_1 \oplus \cdots \oplus \ell_r\pi_r,$$

where each π_i is a (unitary) *cuspidal* automorphic representation of $\mathrm{GL}(N_i)$. The notation means that π is a representation induced from a parabolic subgroup with Levi component

$$\mathrm{GL}(N_1)^{\ell_1} \times \cdots \times \mathrm{GL}(N_r)^{\ell_r},$$

and embedded into $L^2\big(\mathrm{GL}(N,\mathbb{Q})\backslash \mathrm{GL}(N,\mathbb{A})\big)$ by an Eisenstein series. If we would handle the cuspidal components π_i, we could copy the construction above; we would be able to attach twisted endoscopic groups $G = G^1 \times G^2$ to π, and to define the groups $S_\pi(G)$. It is enough to treat the case that π_i is self-contragredient. One would need to show that each such π_i is attached to a unique endoscopic group G_i for $\mathrm{GL}(N_i)$, and that G_i is *primitive* in the sense that $\widehat{G}_i$ equals either $\mathrm{Sp}(N_i,\mathbb{C})$ or $\mathrm{SO}(N_i,\mathbb{C})$. This is essentially part (ii) of the problem stated above.

The remarks of this section have been concerned with setting up the definitions. One needs to define the group $S_\pi(G)$ in order even to state what the image and fibres of the maps $\pi_G \to \pi$ should be. These groups are therefore at the heart of things. The required properties of the cuspidal components π_i will have to be established as part of the full solution of the problem. One can foresee an elaborate inductive argument on the rank N of $\mathrm{GL}(N)$, which is based on the interplay of the stabilized twisted trace formula of $\mathrm{GL}(N)$, and the stabilized trace formulas of the endoscopic groups G.

References

1. J. Arthur, *Automorphic representations and number theory*, 1980 Seminar on Harmonic Analysis (Montreal, Que., 1980), CMS Conf. Proc. 1, Amer. Math. Soc., Providence, R.I., 1981, pp. 3–51.
2. ______, *Unipotent automorphic representations: Conjectures*, Astérisque **171–172** (1989), 13–71.
3. ______, *L^2-cohomology and automorphic representations*, Advances in the Mathematical Sciences—CRM's 25 years (Ed. L. Vinet), CRM Proceedings and Lecture Notes, American Mathematical Society, Providence, 1996, pp.???
4. J. Arthur and L. Clozel, *Simple algebras, base change and the advanced theory of the trace formula*, Ann. of Math. Stud. **120**, Princeton University Press, 1989.
5. D. Flath, *Decomposition of representations into tensor products*, Automorphic Forms, Representations and L-Functions (Proc. Sympos. Pure Math., Oregon State Univ., Corvallis, Ore., 1977), Part I, Proc. Sympos. Pure Math. **XXXIII**, Amer. Math. Soc., Providence, R.I., 1979, 179–184.
6. A. Fröhlich, *Galois module structure of algebraic integers*, Ergeb. Math. Grenzgeb. (3) **1**, Springer-Verlag, Berlin-New York, 1983.
7. A. Fröhlich and J. Queyrut, *On the functional equations of the Artin L-function for characters of real representations*, Invent. Math. **20** (1973), 125–138.
8. S. Gelbart and F. Shahidi, *Analytic properties of automorphic L-functions*, Perspect. Math. **6**, Academic Press, Inc., Boston, MA., 1988.

9. R. Godement and H. Jacquet, *Zeta functions of simple algebras*, Lecture Notes in Math., Vol. 260, Springer-Verlag, Berlin-New York, 1972.
10. H. Jacquet and J. Shalika, *On Euler products and the classification of automorphic representations.* II, Amer. J. Math. **103** (1981), 777–815.
11. R. Kottwitz and D. Shelstad, *Twisted endoscopy* 1: *Definitions, norm mappings and transfer factors*, preprint.
12. J.-P. Labesse and R. Langlands, *L-indistinguishability for* SL(2), Canad. J. Math. **31** (1979), no. 4, 726–785.
13. R. Langlands, *Problems in the theory of automorphic forms*, Lectures in modern analysis and applications, III. Lecture Notes in Math., Vol. 170, Springer, Berlin, 1970, 18–86.
14. ———, *On the notion of an automorphic representation*, Automorphic forms, representations and L-functions (Proc. Sympos. Pure Math., Oregon State Univ., Corvallis, Ore., 1977), Part 1, Amer. Math. Soc., Providence, R.I., 1979, pp. 203–207.
15. ———, *Base change for* GL(2), Ann. of Math. Stud. **96**. Princeton University Press, Princeton, N.J.; University of Tokyo Press, Tokyo, 1980.
16. J. Tunnell, *Artin's conjecture for representations of octahedral type*, Bull. Amer. Math. Soc. (N.S.) **5** (1981), 173–175.

Department of Mathematics, University of Toronto, Toronto, Ontario (Canada) M5S 1A1

E-mail address: arthur@math.toronto.edu

Centre de Recherches Mathématiques
CRM Proceedings and Lecture Notes
Volume **11**, 1997

A Survey of Results Relating to Giuga's Conjecture on Primality

J. M. Borwein and E. Wong

ABSTRACT. This article is an expanded version of the talk given by the first author at the 25th Anniversary Conference of the Centre de Recherches Mathématiques.

In 1950, G. Giuga conjectured that if an integer n satisfies $\sum_{k=1}^{n-1} k^{n-1} \equiv -1 \mod n$, then n must be a prime. In this paper, we survey and complement a recent article (B^3G) on this interesting and now well established conjecture.

Giuga proved that n is a counterexample to his conjecture if and only if each prime divisor p of n satisfies $(p-1) \mid (n/p-1)$ and $p \mid (n/p-1)$. Using this characterization, he proved computationally that any counterexample has at least 1000 digits; equipped with more computing power, E. Bedocchi later raised this bound to 1700 digits. By improving on their method, we determine that any counterexample has at least 13,800 digits.

We also give some new results on the second of the above conditions. This leads to open questions about what we call Giuga numbers and Giuga sequences. Finally we study eight natural variations on our theme. Complete details are provided in this section since the results are entirely new.

1. Our Main Survey

A. Giuga's conjecture and some equivalences. In 1950, G. Giuga [**G**] conjectured:

CONJECTURE. For each integer n it is true that

$$n \text{ is prime} \iff s_n := \sum_{k=1}^{n-1} k^{n-1} \equiv -1 \mod n.$$

Fermat's little theorem is that if p is a prime, then $k^{p-1} \equiv 1 \mod p$ for $k = 1, \ldots, p-1$. Thus, for each prime p, $s_p \equiv -1 \mod p$. Giuga's question becomes:

Does there exist a non-prime n such that $s_n \equiv -1 \mod n$?

1991 *Mathematics Subject Classification.* Primary: 11A41, 11Y11; Secondary: 11Y50.

Key words and phrases. Primality, Carmichael numbers, Computational number theory.

The research of the first author was supported by NSERC and the Shrum Endowment at Simon Fraser University.

The research of the second author was supported by an NSERC graduate fellowship.

This work is based on joint work with David Borwein, Peter Borwein, and Roland Girgensohn.

This is the final form of the paper.

This question has resisted solution for more than forty years. We will see in part why!

The key to dealing with Giuga's conjecture is:

THEOREM 1 ([**G, R**]). *$s_n \equiv -1 \mod n$ if and only if for each prime divisor p of n we have $(p-1) \mid (n/p - 1)$ and $p \mid (n/p-1)$.*

PROOF. Considering residue classes, for a prime p, we have

$$\sum_{k=1}^{p-1} k^{n-1} \equiv \begin{cases} -1 & \mod p \quad \text{if } (p-1) \mid (n-1), \\ 0 & \mod p \quad \text{if } (p-1) \nmid (n-1). \end{cases}$$

Therefore, for each prime divisor p of n with $n = p \cdot q$, we get

$$\sum_{k=1}^{n-1} k^{n-1} \equiv q \sum_{k=1}^{p-1} k^{n-1} \equiv \begin{cases} -q & \mod p \quad \text{if } (p-1) \mid (n-1), \\ 0 & \mod p \quad \text{if } (p-1) \nmid (n-1). \end{cases} \tag{1}$$

First, assume $s_n \equiv -1 \mod n$. Then for each prime divisor p of n, $n = p \cdot q$, we have

$$-1 \equiv \begin{cases} -q & \mod p \quad \text{if } (p-1) \mid (n-1), \\ 0 & \mod p \quad \text{if } (p-1) \nmid (n-1). \end{cases} \tag{2}$$

This is only possible if $(p-1) \mid (n-1) = q(p-1) + (q-1)$. So $(p-1) \mid (q-1)$. It then also follows from (2) that $-1 \equiv -q \mod p$, or $p \mid (q-1)$.

Conversely, assume $p \mid (q-1)$ and $(p-1) \mid (q-1)$. From (1) we have that $s_n \equiv -q \mod p$; since $q \equiv 1 \mod p$, we have that $s_n \equiv -1 \mod p$ for each prime divisor p of n. Now, n must be square-free: else, there would exist a prime divisor p of n with $p \mid q$. This contradicts $p \mid (q-1)$.
Since each of the prime divisors of n divides $s_n + 1$, this is also true for n. In other words, $s_n \equiv -1 \mod n$. □

We return to arguments of this type in the final section.

AGOH'S CONJECTURE. Let B_k denote the k-th Bernoulli number. Then $nB_{n-1} \equiv -1 \mod n$ if and only if n is a prime.

- The denominator of nB_{n-1}, being square-free, is invertible $\mod n$. Indeed one has

  THEOREM 2 (von Staudt-Clausen's Theorem). *The denominator of B_{2k} is given by*

  $$\prod_{\substack{p \text{ prime} \\ (p-1) \mid 2k}} p.$$

  (This allows rapid computation of denominators of large Bernoulli numbers!)
- Takashi Agoh (Tokyo) has shown his recent conjecture is equivalent to Giuga's. This can be seen from the well-known

  $$s_{n-1} = \sum_{i=1}^{n} \binom{n}{i} n^{i-1} B_{n-i}$$

  and analysis involving von Staudt-Clausen's theorem.

EXAMPLE. Bernoulli numbers are generated by

$$\frac{te^t}{e^t-1} = \sum_{k=0}^{\infty} \frac{B_k}{k!} t^k.$$

After B_1 the odd ones are 0, while

$$B_2 = \frac{1}{6}, \quad B_4 = \frac{-1}{30}, \quad B_6 = \frac{1}{42}, \quad B_8 = \frac{-1}{30},$$

and

$$B_{10} = \frac{5}{66}, \quad B_{12} = \frac{-691}{2730}, \quad B_{14} = \frac{7}{6}.$$

Thus

(i) $15B_{14} = \frac{35}{2} \equiv -5 \mod 15$.
(ii) $13B_{12} = \frac{-691}{210} \equiv \frac{-2}{2} \equiv -1 \mod 13$.
(iii) $\text{denom}(B_{10^5}) = 9{,}355{,}235{,}774{,}427{,}510 = 2{\cdot}3{\cdot}5{\cdot}11{\cdot}17{\cdot}41{\cdot}101{\cdot}251{\cdot}401{\cdot}4001$.

B. Carmichael and Giuga numbers.

RECALL. $s_n \equiv -1 \mod n$ iff *for each prime divisor* p *of* n *we have*:
(a) $(p-1) \mid (n/p - 1)$ *and*
(b) $p \mid (n/p - 1)$.

Square-free composite numbers satisfing (a) are the *Carmichael numbers*, introduced by Carmichael in 1912; (a) is equivalent to

$$(p-1) \mid (n-1) \text{ for all prime divisors } p \text{ of } n,$$

and to the strong *pseudo-primality* condition

$$a^{n-1} \equiv 1 \mod n \quad \text{for } (a, n) = 1.$$

- Carmichael numbers are odd. The three smallest are **561** $= 3 \cdot 11 \cdot 17$, **1105** $= 5 \cdot 13 \cdot 17$ and **1729** $= 7 \cdot 13 \cdot 19$. It has only recently been proved that there are infinitely many Carmichael numbers. Indeed their density exceeds $n^{2/7}$ (see [**AGP**], 1994).

Correspondingly, we call a composite number n with $p \mid (n/p - 1)$ for all prime divisors p of n a *Giuga number*. Giuga showed that (b) is equivalent to the condition

$$\sum_{p|n} \frac{1}{p} - \prod_{p|n} \frac{1}{p} \in \mathbb{N}.$$

The three smallest Giuga numbers are:

30: $1/2 + 1/3 + 1/5 - 1/30 = 1$.
858: $1/2 + 1/3 + 1/11 + 1/13 - 1/858 = 1$.
1722: $1/2 + 1/3 + 1/7 + 1/41 - 1/1722 = 1$.

We do not know if there are infinitely many Giuga numbers?

Giuga's theorem can now be restated as:

THEOREM 1*. *A composite integer* n *satisfies* $s_n \equiv -1 \mod n$ *if and only if it is both a Carmichael number and a Giuga number.*

Giuga's conjecture is that such a number cannot exist.

C. Computationally produced bounds. Odd Giuga numbers have at least 10 prime factors, since with fewer prime factors the sum $1/p_1+\cdots+1/p_m-1/n$ is smaller than 1. Also, the number of factors $\equiv 2 \mod 4$.

This lower bound increases dramatically in the presence of the Carmichael condition. Any Carmichael number n satisfies:

If p is a prime factor of n, then for no k is $kp+1$ a prime factor of n.
(If not, then $(kp+1)-1 = kp \mid (n-1)$ and $p \mid n$, a contradiction.)

- Thus no Carmichael number is divisible by both 3 and 7, or by 7 and 43, etc.
- This property allowed Giuga (1950) to prove by hand computation that any counterexample uses more than 360 primes and so has at least 1000 digits.
- E. Bedocchi ([**B**],1985) used the same method to extend this to 1700 digits.
- D. Borwein, J. M. Borwein, P. B. Borwein, and R. Girgensohn ([**B**^{3}**G**, **G**],1994) have improved on this method by dramatically reducing the number of cases to be looked at and so we calculate that any counterexample contains at least 3459 distinct primes and so has at least 13887 digits (the 8th power of the previous estimate).

Sketch of how this works ...

Computing lower bounds for a counterexample. We must look at $n = q_1 \cdots q_k$ such that

(i) $q_i \not\equiv 1 \mod q_j$ for all i, j, and
(ii) $1/q_1 + \cdots + 1/q_k > 1$.

[For $m \in \mathbb{N}$, denote by p_m the m-th odd prime.] A set of odd primes, $\{q_1, \ldots, q_k\}$, is *normal* if condition (i) holds.

(a) Let S_m be the set of all normal sets with maximum element smaller than p_m. For each $S \in S_m$, $S = \{q_1, \ldots, q_k\}$, define

$$T_m(S) := \{q_1, \ldots, q_k,\ q_{k+1}, \ldots, q_r\}$$

to be the smallest set of odd primes which contains S, and satisfies $q_j \geq p_m$, has $S \cup \{q_j\}$ normal for $j > k$, and $\sum_{j=1}^{r} 1/q_j > 1$.

(b) Let $r_m(S)$ be the number of elements of $T_m(S)$.
For example,

$$T_6\big(\{3,5\}\big) = \{3, 5,\ 17, 23, 29, 47, 53, \ldots, 7919, 7937\}$$
$$r_6\big(\{3,5\}\big) = 383$$

and

$$T_1\big(\{\}\big) = \{3, 5, 7, 11, 13, 17, 19, 23, 29\}, \quad r_1\big(\{\}\big) = 9.$$

(c) Now define the (non-decreasing) sequence $(i_m)_{m\in\mathbb{N}}$ by

$$i_m = \min\big\{r_m(S) \mid S \in S_m\big\}.$$

Now, the number of prime factors of a counterexample to Giuga's conjecture exceeds i_m for each $m \in \mathbb{N}$.

- Indeed, the prime factors form a normal set, and the subset S of the factors smaller than p_m is a member of S_m. Since any normal set of primes which contains S and satisfies condition (ii) above must contain at least $r_m(S)$ elements, we have that n has at least $r_m(S) \geq i_m$ prime factors; this is true for any $m \in \mathbb{N}$.

- Thus, any counterexample is bigger than $\prod_{j=1}^{i_m} p_j$, and so has at least as many digits. Giuga estimated $i_9 > 361$, this yields more than 1000 digits; Bedocchi computed $i_9 = 554$, this yields more than 1700 digits.
- To compute i_m, one has to find $r_m(S)$ for all $S \in \mathcal{S}_m$. Since the number of elements of $\mathcal{S}_m$ increases geometrically with m, the time needed to compute i_m gets out of hand quickly. With our R4000 Challenge server and Maple, we were able, with considerable effort, to compute $i_{19} = 825$.
- Fortunately we found a good *branch and bound algorithm*, based on the following observation, that allows us to look at only some sets in $\mathcal{S}_m$, not all of them.

Consider a set $S \in \mathcal{S}_m$ and the associated value $r_m(S)$. Now, S has at most two "successors" in the set $\mathcal{S}_{m+1}$, namely S itself and the set $S' = S \cup \{p_m\}$. We will now show that $r_{m+1}(S) \geq r_m(S)$ and $r_{m+1}(S') \geq r_m(S)$. In fact, there are two cases:

CASE (i). *$S \cup \{p_m\}$ is normal.* Then S has both successors S and S' in $\mathcal{S}_{m+1}$. Also, $p_m \in T_m(S)$. However, $p_m \notin T_{m+1}(S)$, but every other element of $T_m(S)$ is contained in $T_{m+1}(S)$. So, $T_{m+1}(S)$ must contain at least one higher prime to get $\sum_{q \in T_{m+1}(S)} 1/q \geq 1$. Thus, $r_{m+1}(S) \geq r_m(S)$.

As regards S', the set $T_m(S)$ may contain primes congruent to 1 mod p_m. These are missing in $T_{m+1}(S')$, since $p_m \in S'$. For each we add primes to get $\sum_{q \in T_{m+1}(S')} 1/q \geq 1$. Again, $r_{m+1}(S') \geq r_m(S)$.

CASE (ii). *$S \cup \{p_m\}$ is not normal.* Then the only successor of S in $\mathcal{S}_{m+1}$ is S itself. Also, $T_m(S) = T_{m+1}(S)$; the prime p_m is not contained in either set. Therefore, $r_m(S) = r_{m+1}(S)$.

This shows $\{i_m\}$ is indeed nondecreasing. And more: the values $r_{k+1}, r_{k+2}, \ldots$ for all successors in $\mathcal{S}_{k+1}, \mathcal{S}_{k+2}, \ldots$ of a set $S \in \mathcal{S}_k$ do not fall below $r_k(S)$. To compute i_m given upper bound $I \geq i_m$, we do not have to look at any successor in the sets $\mathcal{S}_{k+1}, \ldots, \mathcal{S}_m$ of a set $S \in \mathcal{S}_k$ with $r_k(S) > I$.

Algorithm. Start with $A_1 := \mathcal{S}_1$, and let A_{k+1} consist of the successors in $\mathcal{S}_{k+1}$ of all $S \in A_k$ with $r_k(S) \leq I$. Then $i_m = \min\{r_m(S) \mid S \in A_m\}$.

- The bound $I = I(m)$ can of course be chosen as the value $r_m(S)$ for some $S \in \mathcal{S}_m$. We save considerable time and space if we correctly guess a set with $r_m(I)$ close to i_m.
- Preliminary computational results suggested that the following might hold. Let $L_5 := \{5, 7\}$, and define

$$L_{k+1} := \begin{cases} L_k \cup \{p_k\} & \text{if } L_k \cup \{p_k\} \text{ is normal,} \\ L_k & \text{otherwise.} \end{cases}$$

  Then it appears that for $m \geq 5$,

$$r_m(L_m) = i_m.$$

- We have no proof of this but since the sets L_m yielded "exact" upper bounds for i_m, we employed them to iteratively compute all values i_m first for $m \leq 100$ (in Maple) and later for $m \leq 130$ (in C).
- We *always* observe that $r_m(L_m) = i_m$! The best way to "pack" such primes?

- It was surprisingly hard to translate a fairly straight-forward Maple program into C. While Maple handled the data structures (lists of sets of variable length) easily, it was a non-trivial problem to implement these in C. We did gain a speed-up of a factor of up to 5, though. We thank Jerry Kuch (now a graduate student at Waterloo) for doing this conversion from Maple to C.
- $i_1 = 9$, $i_2 = 27$, $i_3 = 65$, $i_4 = 114$, $i_5 = 127$, $i_6 = 202$, $i_7 = 278$, $i_8 = 323$, $i_9 = i_{10} = i_{11} = 554$, $i_{12} = i_{13} = i_{14} = i_{15} = i_{16} = 704$, $i_{17} = i_{18} = 751$, $i_{19} = i_{20} = 825$, $\dots$, $i_{49} = i_{50} = 2121$, $\dots$, $i_{74} = i_{75} = 2657$, $\dots$, $i_{99} = i_{100} = i_{101} = 3050$, $\dots$, $i_{129} = 3204$.
- $i_{100} = 3050$ implies that any counterexample to Giuga's conjecture has at least 12,055 digits, while $i_{135} = 3459$ implies that any counterexample has at least 13887 digits.
- Each new case now takes several weeks of (single processor) cpu time!!
- As Bedocchi points out, this method is *inherently incapable* of showing that Giuga's conjecture holds for all integers: the set L_{27692} is normal, has 8135 elements and satisfies $\sum_{q \in L_{27692}} 1/q > 1$.

  Therefore, $i_m \leq 8,135$ for $m \geq 27,692$.

The Maple Code

```
# p(i) is the i-th prime.

# Que(m,bound) returns all sets for index m (i.e.,
# subsets of the set of first m-1 odd primes such
# that  no element divides another element-1)
# with estimate  below the bound.

Que := proc(m,bound) local i,n,S,Q,q,Sp;

if m=1 then RETURN({{}}); fi;

# If m=1 then empty set is only "feasible" set.

S:=Que(m-1,bound);

# Collect "feasible" sets for index m-1 in S.
# A "feasible" estimate is one below the bound.

n:=nops(S);

# n is number of sets Que returned for index m-1.

Q:={};

# Now collect all feasible sets for index m in Q.
# by going through all sets in S.

for i from 1 to n do

q:=convert(S[i],'*');

 # There are two cases:
```

```
   if gcd(q,p(m)-1)>1 then
      Q:=Q union {S[i]};

# 1st case: (m-1)st odd prime can't enter S[i].
# Then S[i] is in Q and its estimate doesn't change.
# 2nd case:(m-1)st odd prime can enter S[i].
# Then compute estimate of set S[i].

 else
      if Est(S[i],m,bound+1)<=bound
         then Q:=Q union {S[i]};
      fi;

# If estimate is below bound, put S[i] into Q.

      Sp:=S[i] union {p(m)};
      if Est(Sp,m,bound+1)<=bound
          then Q:=Q union {Sp};
      fi; fi;
od;

# Compute estimate of S[i]+{p(m)}. If estimate is
# below bound, put S[i]+{p(m)} into Q.

Q; end;

 # Return all feasible sets for index m.

# Est(S,m,bound) adds reciprocals of primes in S and
# then continues to add reciprocals of all primes >=
# m-th odd prime that are not congruent to 1
# modulo any prime in S.
# Est stops adding when sum exceeds 1 or when
# the number of primes added exceeds the bound.
# Est returns the number of primes it added.


Est:=proc(S,m,bound) local c,n,q,t,s,k,j,L;
n:=m+1;k:=nops(S);
s:=evalf(sum(1/S[j],j=1..k));
q:=product(S[j],j=1..k);
while k<bound and s<1 do
        c:=char(p(n)-1,q);s:=s+c/p(n); k:=k+c;
        n:=n+1;
        od;
k;
end;


char:=proc(a,b) local s;
s:=1; if gcd(a,b)>1 then s:=0;fi;s;
end;
```

D. Giuga sequences and Carmichael sequences. One promising approach to Giuga's conjecture is to study Giuga numbers in more depth. There is much more structure when we relax the condition that the numbers p in the definition be prime.

DEFINITION. *A finite, increasing sequence of integers,* $[n_1, \dots, n_m]$, *is called a Giuga sequence if*

$$\sum_{i=1}^{m} \frac{1}{n_i} - \prod_{i=1}^{m} \frac{1}{n_i} \in \mathbb{N}.$$

A *proper* Giuga sequence is one whose elements are prime.

THEOREM 3. *A finite, increasing sequence* $[n_1, \dots, n_m]$ *is a Giuga sequence if and only if it satisfies* $n_i \mid (n_1 \cdots n_{i-1} \cdot n_{i+1} \cdots n_m - 1)$ *for* $i = 1, \dots, m$.

- We have found all Giuga sequences up to length 7 and some of length 8.
- There is: 1 sequence of length 3 ($[2,3,5]$); 2 of length 4 ($[2,3,7,41]$ & $[2,3,11,13]$); 3 of length 5 ($[2,3,7,43,1805]$, $[2,3,7,83,85]$ & $[2,3,11,17,59]$); 17 of length 6; 27 of length 7; hundreds of length 8.
- We know only 11 Giuga numbers (proper Giuga sequences).
  - **3 factors:**
    $30 = 2 \cdot 3 \cdot 5$
  - **4 factors:**
    $858 = 2 \cdot 3 \cdot 11 \cdot 13, \quad 1722 = 2 \cdot 3 \cdot 7 \cdot 41$
  - **5 factors:**
    $66198 = 2 \cdot 3 \cdot 11 \cdot 17 \cdot 59$
  - **6 factors:**
    $2214408306 = 2 \cdot 3 \cdot 11 \cdot 23 \cdot 31 \cdot 47057$
    $24423128562 = 2 \cdot 3 \cdot 7 \cdot 43 \cdot 3041 \cdot 4447$
  - **7 factors:**
    $432749205173838 = 2 \cdot 3 \cdot 7 \cdot 59 \cdot 163 \cdot 1381 \cdot 775807$
    $14737133470010574 = 2 \cdot 3 \cdot 7 \cdot 71 \cdot 103 \cdot 67213 \cdot 713863$
    $550843391309130318 = 2 \cdot 3 \cdot 7 \cdot 71 \cdot 103 \cdot 61559 \cdot 29133437$
  - **8 factors:**
    $244197000982499715087866346$
    $= 2 \cdot 3 \cdot 11 \cdot 23 \cdot 31 \cdot 47137 \cdot 28282147 \cdot 3892535183$
    $554079914617070801288578559178$
    $= 2 \cdot 3 \cdot 11 \cdot 23 \cdot 31 \cdot 47059 \cdot 2259696349 \cdot 110725121051$
- For all of these examples, the 'sum minus product' value is 1; to reach value 2, the sequence must have at least 59 factors.
- To find all Giuga sequences of length m, in principle one might check all sequences whose elements are below an a priori bound (the sum of their reciprocals must exceed 1). The number of these grows exponentially; even for length 7 there are too many to check this crudely.

Fortunately, we have the following theorem which yields *all* Giuga sequences of length m with a given initial segment of length $(m-2)$. *It was discovered from our preliminary computations.*

THEOREM 4. (a) *Take an initial sequence of length $m-2$, $[n_1, \dots, n_{m-2}]$. Let*

$$P = n_1 \cdots n_{m-2}, \quad S = \frac{1}{n_1} + \cdots + \frac{1}{n_{m-2}}.$$

Fix an integer $v > s$ (this will be the sum minus product value). Take any integers a, b with $a \cdot b = P(P + S - v)$ and $b > a$. Let

$$n_{m-1} := \frac{(P+a)}{P(v-S)}, \quad n_m := \frac{(P+b)}{P(v-S)}.$$

Then

$$S + \frac{1}{n_{m-1}} + \frac{1}{n_m} - \frac{1}{P n_{m-1} n_m} = v.$$

The sequence $[n_1, \dots, n_{m-1}, n_m]$ is a Giuga sequence if and only if n_{m-1} is an integer.

(b) *Conversely, if $[n_1, \dots, n_{m-1}, n_m]$ is a Giuga sequence with sum minus product value v, and if we define*

$$a := n_{m-1} P(v - S) - P, \quad b := n_m P(v - S) - P$$

(with P and S the product and the sum of the first $m-2$ terms) then a and b are integers and $a \cdot b = P(P + S - v)$.

Application. Odd Giuga sequences (with value $v = 1$) have length $m \equiv 2 \mod 4$. The cases $m = 2$ and $m = 6$ can be excluded as the sum of reciprocals never exceeds 1.

Now, $m = 10$ can be ruled out, with the use of Theorem 4. Again, this is where computational feasibility ends. For $m = 14$, some new insight is needed.

For interest we record:

THEOREM 5. *Take a Giuga sequence of length m, $[n_1, \dots, n_{m-1}, n_m]$, which satisfies*

(3) $$n_m = n_1 \cdots n_{m-1} - 1.$$

Let

$$\tilde{n}_m := n_1 \cdots n_{m-1} + 1, \quad \tilde{n}_{m+1} := n_1 \cdots n_{m-1} \tilde{n}_m - 1.$$

Then $[n_1, \dots, n_{m-1}, \tilde{n}_m, \tilde{n}_{m+1}]$ is a Giuga sequence with the same sum minus product value.

- For each m, new Giuga sequences pop out of thin air, some with property (3) and so leading to infinite families. Two are:
  (a) $n_1 = 2$, $n_k = n_1 \cdots n_{k-1} + 1$ for $k = 2, \dots, m-1$, $n_m = n_1 \cdots n_{m-1} - 1$;
  (b) $n_1 = 2$, $n_2 = 3$, $n_3 = 11$, $n_4 = 23$, $n_5 = 31$, $n_k = n_1 \cdots n_{k-1} + 1$ for $k = 6, \dots, m-1$, $n_m = n_1 \cdots n_{m-1} - 1$.

  Similarly, a *Carmichael sequence* is a finite increasing sequence of integers, $[a_1, \dots, a_n]$, such that

  $$(a_i - 1) \mid \left(\prod_{j=1}^{n} a_j - 1 \right) \text{ for } i = 1, \dots, n.$$

- There are infinitely many Carmichael sequences of length 3. Take two relatively prime integers $c, d \in \mathbb{N}$ and let a_1 be a solution of the two congruences $d a_1 \equiv -1 \mod c$, $c a_1 \equiv -1 \mod d$. (There always exists a solution which is unique modulo $c \cdot d$, by the Chinese remainder theorem.) Define

$a_2 := c(a_1 - 1) + 1$ and $a_3 := d(a_1 - 1) + 1$. Then $[a_1, a_2, a_3]$ is always a Carmichael sequence. (This contains, as a special case, constructions such as Chernick's observation (see [**AGP**]) that $a_1 = 6k + 1$, $a_2 = 12k + 1$, $a_3 = 18k + 1$ is always a Carmichael sequence.)

The following two recursions produce Carmichael sequences of length $m + 1$ and $m + 2$ out of the Carmichael sequence $[a_1, \ldots, a_m]$.

(a) Let $a_{m+1} := \prod_{j=1}^{m} a_j$. Then $[a_1, \ldots, a_m, a_{m+1}]$ is a Carmichael sequence.

(b) Let $a_{m+1} := \prod_{j=1}^{m} a_j$ and $a_{m+2} := d(a_{m+1} - 1) + 1$ where d is a divisor of $a_{m+1} + 1$. Then $[a_1, \ldots, a_m, a_{m+1}, a_{m+2}]$ is a Carmichael sequence.

In particular, if we start with an odd 3-factor Carmichael sequence, then we can extend it to arbitrary length by iterating the step (b) with $d = 2$.

E. Open problems.

1. Giuga's conjecture: *Show that no integer exists which is both a Giuga number and a Carmichael number.*
   More generally: *Show that no Giuga sequence can be a Carmichael sequence.*
2. Does every Giuga sequence contain two factors n_i, n_j with $n_j \mid (n_i - 1)$? If yes, then Giuga's conjecture is proved.
3. Find a Giuga sequence consisting of odd factors (or odd primes) only, or prove that none exist.
4. Are there infinitely many proper Giuga sequences?
5. In the notation above, is $r_m(L_m) = i_m$ for $m = 5, \ldots, 27692$? If so, then a counterexample to Giuga's conjecture has at least 36,069 digits.
6. Find a fast way to generate all Giuga sequences of a given length.
7. Are there Giuga sequences with a sum minus product value higher than 1?
8. Agoh's conjecture: *Let B_k denote the k-th Bernoulli number. Then n is a prime, If and only if $nB_{n-1} \equiv -1 \mod n$.*

2. Variations on the Theme of Giuga

A. The Eightfold way. In this final part we consider Giuga's condition along with seven variants, obtained by replacing $n-1$ with $\phi(n)$ in the exponent and right-hand side of the summation. We also vary the summation by restricting the original sum, which we call *Type* I, to only those k which are relatively prime to n (we call this a sum of *Type* II). Since each of these variants is satisfied by any prime, we are interested in the composite n for which they also hold.

We use $[n]$ to denote the integers between 1 and n, and $[n]^*$ for the subset of $[n]$ of integers relatively prime to n. p will denote a prime, and q a prime power. We will assume that $n > 2$, with prime factorization $p_1^{r_1} p_2^{r_2} \cdots p_l^{r_l}$, and define $q_i = p_i^{r_i}$ for convenience. The notation $O(p^r)$ will be used to denote an arbitrary constant divisible by p^r.

We will use $S(n, m)$ to denote the sum $\sum_{k \in [n]} k^m$ of Type I, and $S^*(n, m)$ for the sum $\sum_{k \in [n]^*} k^m$ of Type II. We are only concerned with the residue classes of $S(n, m)$ and $S^*(n, m)$ modulo n.

We start with a computational theorem in Section B concerning the case where n is a prime power. We apply this theorem in Section C to sums of Type I, and in Section D to sums of Type II, and derive analogues of Giuga's theorem for the variants mentioned above. Finally, in Section E, we discuss what is known about the n which satisfy the eight different conditions.

B. A computational theorem. In this section we examine sums of the form $\sum_{k\in[p^r]^*} k^m$ modulo p^r.

To begin, we need the following simple lemma:

LEMMA. *If $a \equiv 1 \mod p$, then $a^{p^i} \equiv 1 \mod p^{i+1}$, for all $i \geq 0$.*

PROOF. If $a = 1 + O(p^j)$, then the binomial expansion gives

$$a^p = 1 + pO(p^j) + O(p^{2j}) = 1 + O(p^{j+1}),$$

and the result follows by induction.

The multiplicative structure of $[p^r]^*$ is well-known (see [**NZM**]). When p is odd, it is cyclic with some generator α_p. If $p = 2$ and $r > 1$, then it is generated by the following elements: 5 which has order 2^{r-2} and -1 which has order 2. □

We now prove the following:

THEOREM 6. *If $q = p^r$, then $S = \sum_{k\in[q]^*} k^m$ is congruent to either 0 or $\phi(q)$ modulo q, according to the following conditions:*

- If $p \geq 3$, then $S = \phi(q)$ iff $p - 1 \mid m$.
- If $q = 2$, then $S = \phi(2) = 1$.
- If $q = 2^r$, $r \geq 2$, then $S(q, m) = \phi(q)$ iff m is even.

PROOF. In the first case, we have

$$S = \sum_{k=0}^{\phi(q)-1} \beta^k, \tag{4}$$

where $\beta = \alpha_p^m$. The condition $p - 1 \nmid m$ is equivalent to $\beta - 1$ being non-zero modulo p and hence invertible modulo q. Thus if $p - 1 \nmid m$ then S is a geometric series which evaluates to

$$\frac{\beta^{\phi(q)} - 1}{\beta - 1} = 0.$$

Now suppose $p - 1 \mid m$. Then we may assume without loss of generality that $m = (p-1)p^{s-1}$, where $1 \leq s \leq r$, and thus $\beta \equiv 1 \mod p^s$. We then see that the sum (4) consists of m repetitions of the sum

$$T = \sum_{k=0}^{p^{r-s}-1} \beta^k.$$

Since α_p is primitive, these summands are distinct modulo q and all congruent to 1 modulo p^s. Thus they must be a permutation of the arithmetic progression $1 + np^s$, $0 \leq n < p^{r-s}$, which sums to

$$p^{r-s} + p^s p^{r-s} \frac{(p^{r-s} - 1)}{2} = p^{r-s} + O(p^r).$$

This gives $T = p^{r-s}$ hence $S = mT = (p-1)p^{r-1} = \phi(q)$.

The second case is trivial. The third case, for odd m, is also simple: the terms cancel out in pairs. Thus we may assume $m = 2^s$ with $1 \le s \le r-2$. The sum S factors into

$$S = \left(1 + (-1)^m\right) \sum_{k=0}^{q/4-1} \beta^k,$$

where $\beta = 5^m$. Since m is even, the first factor is 2. The second factor consists of m repetitions of

$$T = \sum_{k=0}^{q/4m-1} \beta^k. \tag{5}$$

The summands in (5) are distinct modulo q and congruent to 1 modulo $4m$, thus they form the arithmetic progression $1 + 4mn$, $0 \le n < q/4m$, which sums to

$$\frac{q}{4m} + 4m\frac{(q/4m)(q/4m-1)}{2} = \frac{q}{4m} + O\left(\frac{q}{2}\right).$$

Multiplication by $2m$ yields $S = 2mT = q/2 = \phi(q)$. □

Note in particular that $S^*(p^r, m)$ is always divisible by p^{r-1}, regardless of m. This observation allows us to prove that the results of Theorem 6 also hold for most sums of the form $\sum_{k\in[p^r]} k^m$. More specifically, we have the following theorem:

THEOREM 7. *If $m > 1$, then $S(p^r, m) \equiv S^*(p^r, m) \mod p^r$.*

PROOF. This is trivial for $m > r$ (and hence $r = 1$), which is really enough for the following sections, but the full result has a nice proof by induction: we have, for $r > 1$, $S(p^r, m) = S^*(p^r, m) + p^m S(p^{r-1}, m)$. The last term is divisible by $p^m(p^{r-2})$ and thus 0 modulo p^r. □

C. Sums of Type I. Theorem 7 readily applies to sums of Type I. Recall that n factors into $\prod_{i=1}^{l} p_i^{r_i}$. For a given i, $1 \le i \le l$, the sum $\sum_{k\in[n]} k^m$ modulo q_i simply consists of n/q_i repetitions of $\sum_{k\in[q_i]} k^m$. Therefore $S(n, m) \equiv 0$ or $(n/q_i)\phi(q_i) \mod q_i$, depending on whether the conditions of Theorem 6 are satisfied.

We first consider the case of the right-hand side $n - 1$. Now, $S(n, m)$ equals $n-1$ modulo n iff it equals $n-1$ modulo each of the q_i. This gives us the following theorem.

THEOREM 8. *$\sum_{k\in[n]} k^m \equiv n - 1 \mod n$ if and only if n is square-free, and for all odd p dividing n, $p - 1 \mid m$ and $p \mid (n/p - 1)$.*

PROOF. First, n must be square-free, or else for some i, p_i divides $\sum_{k\in[q_i]} k^m$, so that $\sum_{k\in[n]} k^m$ is not relatively prime to n, hence not congruent to $n-1$. By the same logic, we need $\sum_{k\in[p]} k^m$ to be non-zero modulo p, hence the condition that $p - 1 \mid m$ for odd p. If $p = 2$, we have the second case of Theorem 1, and the congruence holds trivially. Finally, $(n/p)\phi(p) \equiv n - 1 \mod p$ iff $n/p \equiv 1 \mod p$. It is clear, from the proof, that these conditions are also sufficient. □

Theorem 8 has two corollaries, the second of which is Giuga's theorem.

COROLLARY 1. *$S(n, \phi(n)) \equiv n - 1 \mod n$ iff n is square-free, and for each prime p dividing n, $p \mid (n/p - 1)$.*

COROLLARY 2. *$S(n, n-1) \equiv n-1 \mod n$ iff n is odd, square-free, and for each prime p dividing n, $p \mid (n/p - 1)$ and $p - 1 \mid n - 1$.*

Both corollaries are immediate from Theorem 8. We see that the condition in Corollary 1 is exactly that n is a Giuga number. The conditions of Corollary 2 (Giuga's theorem) can thus be restated as before: n is both Carmichael and Giuga.

We now consider the case $S(n,m) \equiv \phi(n) \mod n$. This will hold if and only if for each i, $(n/q_i)S(q_i, m) \equiv \phi(n) \mod q_i$. By examining the two possibilities for the left-hand side of this congruence, we arrive at the following theorem.

THEOREM 9. *$\sum_{k\in[n]} k^m \equiv \phi(n) \mod n$ if and only if, for each i, $1 \leq i \leq l$, one of the following conditions holds:*

- *$S(q_i, m) \equiv 0 \mod q_i$ and $p_i \mid p' - 1$ for some prime p' dividing n,*
- *$S(q_i, m) \equiv \phi(q_i) \mod q_i$ and $n/q_i \equiv \phi(n/q_i) \mod p_i$.*

PROOF. For the first case, we need $\phi(q_i)\phi(n/q_i) \equiv 0 \mod q_i$, and this is equivalent to $p_i \mid \phi(n/q_i)$, which gives us the first condition. For the second case, we must have $(n/q_i)\phi(q_i) \equiv \phi(n) \mod q_i$. Dividing through by $\phi(q_i)$ (which contains a factor of $p_i^{r_i - 1}$), we get $(n/q_i) \equiv \phi(n/q_i) \mod p_i$. □

Theorem 9 has two corollaries analogous to Corollaries 1 and 2, the first of which follows trivially.

COROLLARY 3. *$S(n, \phi(n)) \equiv \phi(n) \mod n$ iff $n/q \equiv \phi(n/q) \mod p$ for all primes p dividing n, where q is defined as the highest power of p which divides n.*

We call n a *co-Giuga number* if it satisfies the conditions of Corollary 3. Note that, while prime powers are not Giuga numbers, they are co-Giuga numbers. The next corollary requires a descent argument.

COROLLARY 4. *$S(n, n-1) \equiv \phi(n) \mod n$ iff n is co-Giuga, and $p - 1 \mid n - 1$ for all primes p dividing n.*

PROOF. We must rule out the first condition in Theorem 9. Suppose that for some p_i, $p_i \nmid n - 1$ so $S(q_i, n-1) \equiv 0 \mod q_i$. We can choose such a p_i to be maximal. Then there must exist p' dividing n such that $p_i \mid p' - 1$. But clearly $p_i \nmid n - 1$ and thus $p' - 1 \nmid n - 1$. Since $p' > p_i$, this is a contradiction. □

If n satisfies the condition $p - 1 \mid n - 1$ above (n does not have to be square-free), we say n is a *generalized Carmichael number*. The square-free generalized Carmichael numbers are exactly the classical Carmichael numbers. Thus we can restate the conditions of Corollary 4 as: n is both co-Giuga and generalized Carmichael.

D. Sums of Type II. We now consider the sums $S^*(n,m) = \sum_{k\in[n]^*} k^m$ of Type II. For a given n, define $[n]_i^*$ as the subset of $[n]^*$ consisting of all elements congruent to 1 modulo n/q_i. Then $[n]_i^*$ has cardinality $\phi(q_i)$, and furthermore is congruent to $[q_i]^*$ modulo q_i. By the Chinese remainder theorem we can factor $[n]^*$ in the following sense:

Every $x \in [n]^*$ can be written uniquely as $x = x_1 x_2 \cdots x_l$, with $x_i \in [n]_i^*$.

This representation allows us to factor $S = S^*(n,m)$ into $S_1 S_2 \cdots S_l$, where

$$S_i = \sum_{k\in[n]_i^*} k^m.$$

Now, S_i is clearly congruent to $\phi(q_j)$ modulo q_j for $i \neq j$; thus we are only interested in S_i modulo q_i, which is equivalent to $\sum_{k\in[q_i]^*} k^m$. Theorem 6 now shows that $S = \prod S_i$ is congruent to 0 or $\phi(n)$ modulo q_i according to whether S_i is 0 or $\phi(q_i)$.

We are now in a position to prove results similar to those in Section C.

THEOREM 10. *$S = \sum_{k\in[n]^*} k^m \equiv n-1 \mod n$ if and only if n is prime and $n-1 \mid m$.*

PROOF. Since $n-1$ is relatively prime to n, we cannot have $S_i \equiv 0 \mod q_i$ for any i. Thus $S_i \equiv \phi(q_i) \mod q_i$ for all i, so that $S = \phi(n)$. But $\phi(n) < n-1$ when n is not prime. The result now follows from Theorem 6. □

The following two corollaries are immediate, but included for completion.

COROLLARY 5. *$S^*(n, \phi(n)) \equiv n-1 \mod n$ iff n is prime.*

COROLLARY 6. *$S^*(n, n-1) \equiv n-1 \mod n$ iff n is prime.*

THEOREM 11. *$S = \sum_{k\in[n]^*} k^m \equiv \phi(n) \mod n$ if and only if, for each i, $1 \leq i \leq l$, either $S^*(q_i, m) \equiv \phi(q_i) \mod q_i$, or $p_i \mid p' - 1$ for some prime p' dividing n.*

PROOF. As in the proof of Theorem 9, if $S^*(q_i, m) \equiv 0 \mod q_i$, then we need $\phi(n) \equiv 0 \mod q_i$, and this is equivalent to the p' condition. If $S^*(q_i, m) \equiv \phi(q_i) \mod q_i$, then we have $S^*(n, m) \equiv \phi(n) \mod q$ by the arguments preceding Theorem 10. □

The final two corollaries complete the Eightfold way. Corollary 7 also follows immediately from Euler's theorem.

COROLLARY 7. *$S^*\big(n, \phi(n)\big) \equiv \phi(n) \mod n$ for all natural numbers n.*

COROLLARY 8. *$S^*(n, n-1) \equiv \phi(n) \mod n$ iff n is odd and a generalized Carmichael number, i.e. $p-1 \mid n-1$ for each prime $p \mid n$.*

PROOF. The descent argument of Corollary 4 applies here to prove that n must be a generalized Carmichael number. Suppose $n > 2$ is even, hence a power of two. Then $S^*(n, n-1) \equiv 0 \mod n$ by Theorem 6, so n must be odd. Finally, Theorem 11 shows that the condition is sufficient. □

We note that Agoh's conjecture also has counterparts within the Eightfold way.

E. Co-Giuga and generalized Carmichael numbers. In this final section, we take a closer look at the classes of co-Giuga and generalized Carmichael numbers introduced in Sections C and D.

The co-Giuga numbers include prime powers, and also must have the normality property that $p \nmid q-1$ for any two primes p, q dividing n. Furthermore, it is not hard to show that the co-Giuga condition is equivalent to

$$\sum_{p|n}\left(1-\frac{1}{p}\right) - \prod_{p|n}\left(1-\frac{1}{p}\right) \in \mathbb{N}.$$

Using this characterization, and the monotonicity of the above expression, we showed that there are no co-Giuga numbers, other than prime powers, up to 10^{30000}.

It is interesting to note that the co-Giuga condition is independent of the multiplicity of prime factors: that is, n is co-Giuga if and only if $\prod_{p|n} p$ is co-Giuga.

The generalized Carmichael numbers include all prime powers, Carmichael numbers, and many other numbers. The smallest five of these "others" are 45, 225, 325, 405, and 637. There are infinitely many such numbers, for instance, $3^{2r}5^s$ for any $r, s \geq 1$. More generally, given any two primes $p < q$ with $p \nmid q-1$, $p^{r\phi(q-1)}q^{s\phi(p-1)}$ is a generalized Carmichael number. In fact, a similar pattern exists for any normal family of primes.

References

[AGP] W. R. Alford, A. Granville, and C. Pomerance, *There are infinitely many Carmichael Numbers*, Ann. of Math. (2) **139** (1994), no. 3, 703–722.

[B] E. Bedocchi, *Note on a conjecture about prime numbers*, Riv. Mat. Univ. Parma (4) **11** (1985), 229–236. (Italian)

[B³G] D. Borwein, J. M. Borwein, P. B. Borwein, and R. Girgensohn, *Giuga's conjecture on primality*, Amer. Math. Monthly **103** (1996), no. 1, 40–50.

[G] G. Giuga, *Su una presumibile proprietà caratteristica dei numeri primi*, Istit. Lombardo Accad. Sci. Lett. Rend. A **83** (1950), 511–528.

[HW] G. H. Hardy, E. M. Wright, *An Introduction to the theory of numbers*, Oxford University Press, Oxford, 1959.

[NZM] I. Niven, H. Zuckermann, and H. L. Montgomery. *An introduction to the theory of numbers*, 5th ed., John Wiley & Sons, Inc., New York, 1991.

[R] P. Ribenboim, *The book of prime number records*, Springer-Verlag, 1989.

Department of Mathematics and Statistics, Simon Fraser University, Burnaby, BC (Canada) V5A 1S6

E-mail address: `jborwein@cecm.sfu.ca`

Department of Mathematics and Statistics, Simon Fraser University, Burnaby, BC (Canada) V5A 1S6

E-mail address: `erick@cs.sfu.ca`

Centre de Recherches Mathématiques
CRM Proceedings and Lecture Notes
Volume **11**, 1997

Solving Equations by the Decrease Principle

F. H. Clarke

1. Introduction

Let $f\colon \mathbb{R}^n \to \mathbb{R}$ be a differentiable function, and suppose that a certain point $x_0 \in X$ is not a critical point of $x_0\colon f'(x_0) \neq 0$. Then for any $\rho > 0$, it is clear that

$$\inf_{x\in B(x_0;\rho)} f(x) < f(x_0),$$

where $B(x_0;\rho)$ signifies the open ball around x_0 of radius ρ. In fact, one can quantify such an assertion more precisely. Suppose that for every $x \in B(x_0;\rho)$ one has $\|f'(x)\| \geq \delta$, where δ is a positive parameter. Then it is the case that

$$\inf_{x\in B(x_0;\rho)} f(x) \leq f(x_0) - \rho\delta.$$

This is an instance of what we call the *Decrease Principle.*

It turns out that when the Decrease Principle is extended to nondifferentiable functions, it leads to a powerful and unified framework for obtaining solvability criteria in a wide variety of contexts. The results include the classical theorems of Graves and Lyusternik as special cases, as well as the inverse and implicit function theorems.

To explain the pertinence of the Decrease Principle to equation solving, consider for example the situation in which we wish to solve an equation

$$F(x) = \alpha$$

for x in terms of α, where $\mathbb{F}\colon \mathbb{R}^n \to \mathbb{R}^n$ is a given smooth function. Let us define $f(x,\alpha) := \|F(x) - \alpha\|$; then the original vector equation $F(x) = \alpha$ is equivalent to the scalar inequality $f(x,\alpha) \leq 0$. Suppose now that $F'(x)$ is locally nonsingular, the usual type of hypothesis that intervenes classically. Then, whenever $F(x) \neq \alpha$, we have

$$f'_x(x,\alpha) = F'(x)^* \frac{(F(x) - \alpha)}{\|F(x) - \alpha\|},$$

so that $\|f'_x(x,\alpha)\| \geq \delta$ for some $\delta > 0$. In view of the Decrease Principle, f can therefore be decreased at a fixed rate δ, until (in a suitable setting) we have $f = 0$; i.e., the equation holds.

1991 *Mathematics Subject Classification.* Primary: 26B05.
This is the final form of the paper.

We remark that f is nondifferentiable when $F(x) = \alpha$. Trying to circumvent this by taking f^2 (for example) will not work in this approach, for a fixed rate of decrease does not result. There are other reasons for which nonsmooth (and even extended-valued) functions intervene in the analysis, as we will see. We proceed first to recall some basic facts from nonsmooth analysis [**1, 2**].

2. Proximal Subgradients

We shall work on a Hilbert space X and consider a function $f\colon X \to (-\infty, \infty]$ which is lower semicontinuous and not identically $+\infty$; the class of such functions is denoted $\mathcal{F}$.

Let $x \in \operatorname{dom} f := \{x \in X : f(x) < \infty\}$. Then an element ζ of X is said to be a *proximal subgradient* of f at x provided that for some $\sigma \geq 0$, for some $\eta > 0$, one has

$$f(y) - f(x) + \sigma\|y - x\|^2 \geq \langle \zeta, y - x\rangle \quad \forall y \in B(x;\eta).$$

The set of proximal subgradients of f at x is denoted $\partial_P f(x)$, a set which may be empty or unbounded.

Surprisingly perhaps, there is a very complete and productive calculus associated with proximal subgradients. One of the cornerstones of this calculus is the Mean Value Inequality of Clarke and Ledyaev. We proceed to state this result now and then to derive a proximal version of the Decrease Principle from it.

In the statement below, Y is a subset of X, and the following notation is used: $B := B(0;1)$,

$$[x, Y] := \{tx + (1-t)y : 0 \leq t \leq 1, y \in Y\},$$
$$\hat{r}(f, Y) := \lim_{\delta \downarrow 0} \inf_{w \in Y + \delta B} \{f(w) - f(x)\}.$$

THEOREM 2.1 ([**3**]). *Let Y be a closed, convex, bounded subset of X, and let $x \in \operatorname{dom} f$, where $f \in \mathcal{F}$. Then, for any $\varepsilon > 0$ and $\bar{r} < \hat{r}$, there exists $z \in [x, Y] + \varepsilon B$ and $\zeta \in \partial_P f(z)$ such that*

$$\bar{r} < \langle \zeta, y - x\rangle \quad \forall y \in Y.$$

Further, we can choose z to satisfy

$$f(z) < \inf_{[x,Y]} + |\bar{r}| + \varepsilon.$$

As a guide to understanding this result, we suggest deriving from it the following conclusion when f is a smooth function and $X = \mathbb{R}^n$: for some $z \in [x, Y]$ one has

$$\min_Y f - f(x) \leq \langle f'(z), y - x\rangle \quad \forall y \in X.$$

This reveals the theorem as a uniform, multidirectional extension of the classical mean value principle.

The uses of Theorem 2.1 are many and varied. Here is the consequence we are interested in at the moment, the proximal Decrease Principle.

THEOREM 2.2. *Let $x_0 \in \operatorname{dom} f$, where $f \in \mathcal{F}$. Suppose that for some $\delta > 0$ and $\rho > 0$ one has the following condition:*

$$z \in x_0 + \rho B, \quad \zeta \in \partial_P f(z) \implies \|\zeta\| \geq \delta.$$

Then

$$\inf_{x \in x_0 + \rho B} f(x) \leq f(x_0) - \rho\delta.$$

PROOF. We shall show that for any $s \in (0, \rho)$, for all $\varepsilon > 0$ sufficiently small, one has

$$\inf_{x_0 + \rho B} f \leq f(x_0) - s\delta + \varepsilon, \tag{2.1}$$

a fact which implies the estimate of the theorem. Given $s \in (0, \rho)$, set $Y := x_0 + s\overline{B}$, and observe that the corresponding quantity $\hat{r}$ define above satisfies (for $x := x_0$)

$$\infty > \hat{r} \geq \inf_{x_0 + \rho B} f - f(x_0).$$

Choose $\varepsilon \in (0, \rho - s)$, and set $\bar{r} := \hat{r} - \varepsilon$. Applying Theorem 2.1 gives

$$\begin{aligned} \inf_{x_0 + \rho B} f - f(x_0) &\leq \inf_{x_0 + s\overline{B}} f - f(x_0) \\ &\leq \hat{r} = \bar{r} + \varepsilon \\ &< \langle \zeta, su \rangle + \varepsilon \quad \forall u \in B \end{aligned}$$

where $\zeta \in \partial_P f(z)$ and $z \in Y + \varepsilon B = x_0 + s\overline{B} + \varepsilon B \subset x_0 + \rho B$. By hypothesis, we have $\|\zeta\| \geq \delta > 0$. Choosing $u = -\zeta/\|\zeta\|$ in the preceding inequality therefore gives precisely (2.1). □

3. Solving Equations

Consider an equation

$$f(x, \alpha) = 0 \tag{3.1}$$

where f is a given function of two variables. It is a familiar consideration in mathematics to seek to solve this equation for x, while viewing α as a parameter. Typically this is done in a neighborhood of a given point (x_0, α_0) for which (3.1) is satisfied, and the important issues are these: For a given α sufficiently near to α_0, does there continue to be at least one value of x for which (3.1) holds? How does this set of solutions vary with α? When can we be sure there is a suitable "nice" function $x(\alpha)$ which "tracks" solutions as α varies; i.e., such that $f\big(x(\alpha), \alpha\big) = 0$, and $x(\alpha_0) = x_0$?

We shall sketch a way to deal with such issues via nonsmooth analysis, and in so doing recover some classical results as well as establish new ones; full details are given in [**4**].

For a given *nonnegative* function f, we are interested, for given α, in the points $x \in X$ (a Hilbert space) for which (3.1) holds, and we wish to limit attention to just those x lying in a prescribed subset Ω of X. We define, then, the feasible set $\Phi(\alpha)$ as follows:

$$\Phi(\alpha) := \big\{x \in \Omega\colon f(x, \alpha) = 0\big\}. \tag{3.2}$$

The following result is the key to our approach; in its statement, $\partial_P f(x, \alpha)$ refers to the proximal subdifferential of the function $x \mapsto f(x, \alpha)$.

THEOREM 3.1 (Solvability Theorem). *Let A be a parameter set, and suppose that for each $\alpha \in A$, the function $x \mapsto f(x, \alpha)$ is nonnegative and belongs to $\mathcal{F}$. Suppose that for some $\delta > 0$ and subset V of X, one has the following nonstationarity condition*:

$$\alpha \in A, \quad x \in V, \quad f(x,\alpha) > 0, \quad \zeta \in \partial_P f(x,\alpha) \implies \|\zeta\| \geq \delta.$$

Then for all $(x, \alpha) \in X \times A$ one has

$$\min\{d(x, \operatorname{comp} V), d(x, \operatorname{comp} \Omega), d(x, \Phi(\alpha))\} \leq \frac{f(x,\alpha)}{\delta}. \tag{3.3}$$

PROOF. If the conclusion is false, then $f(x, \alpha)$ is finite, and there exists $\rho > 0$ such that

$$\min\{d(x, \operatorname{comp} V), d(x, \operatorname{comp} \Omega), d(x, \Phi(\alpha))\} > \rho > \frac{f(x,\alpha)}{\delta}.$$

It follows that

$$x + \rho B \subset V, \quad x + \rho B \subset \Omega, \quad d(x, \Phi(\alpha)) > \rho.$$

Thus we have $f(x', \alpha) > 0$ for all $x' \in x + \rho B$. By hypothesis, then, the proximal subgradients of $f(\cdot, \alpha)$ on the open set $x + \rho B$ have norm bounded below by δ. We invoke the Decrease Principle (Theorem 2.2) to conclude

$$0 \leq \inf_{x' \in x + \rho B} f(x', \alpha) \leq f(x) - \rho\delta < 0.$$

This contradiction proves the theorem. □

We remark that the context in which the theorem is most often used is that in which a given point (x_0, α_0) has been identified at which $f = 0$, and in which V and Ω are neighborhoods of x_0. Then (3.3) asserts that

$$d(x, \Phi(\alpha)) \leq \frac{f(x,\alpha)}{\delta} \tag{3.4}$$

whenever $f(x, \alpha)$ is sufficiently small. In cases where f is continuous, we then derive (3.4) for all (x, α) sufficiently near (x_0, α_0). Of course, part of the interest of (3.4) is that it implies the nonemptiness of $\Phi(\alpha)$ (since $d(x, \Phi(\alpha)) = +\infty$ when $\Phi(\alpha)$ is empty).

The following application of the Solvability Theorem illustrates these remarks, together with the reduction of vector-valued equations to a single scalar one. We consider the equation

$$F(x, \alpha) = 0 \tag{3.5}$$

where F is a mapping from $X \times M$ to Y, with M a metric space and Y another Hilbert space. We assume that some open set U of X exists such that F is continuous on $U \times M$, and such that the partial Gâteaux derivative $F'_x(x, \alpha)$ exists for all $(x, \alpha) \in U \times M$, and is continuous there (jointly in (x, α)).

THEOREM 3.2 (Graves-Lyusternik). *Let $(x_0, \alpha_0) \in U \times M$ be a point satisfying $F(x_0, \alpha_0) = 0$, and suppose that $F'_x(x_0, \alpha_0)$ is onto: $F'_x(x_0, \alpha_0)X = Y$. Let Ω be any neighborhood of x_0. Then for some $\delta > 0$, for all (α, x) sufficiently near (α_0, x_0), one has*

$$d(x, \Phi(\alpha)) \leq \frac{\|F(x,\alpha)\|}{\delta},$$

where $\Phi(\alpha) := \{x' \in \Omega \colon F(x, \alpha) = 0\}$.

PROOF. By the Open Mapping Principle of Banach, there exists $\delta > 0$ such that $F'_x(x_0, \alpha_0)\overline{B}_X \supset 2\delta \overline{B}_Y$. Since $(x, \alpha) \mapsto F'_x(x, \alpha)$ is continuous, there exist neighborhoods V and A of x_0 and α_0 such that

$$F'_x(x, \alpha)\overline{B}_X \supset \delta \overline{B}_Y \quad \forall (x, \alpha) \in V \times A.$$

It follows that for any unit vector y in Y, for any $(x, \alpha) \in V \times A$, one has $\|F'_x(x, \alpha)^* y\| \geq \delta$. Indeed, there exists a unit vector $v \in X$ such that $F'_x(x, \alpha)v = \delta y$. Then

$$\|F'_x(x, \alpha)^* y\| \geq \langle F'_x(x, \alpha)^* y, v\rangle = \langle y, F'_x(x, \alpha)v\rangle = \langle y, \delta y\rangle = \delta$$

as claimed.

We are now ready to apply the Solvability Theorem 3.1, with $f(x, \alpha) := \|F(x, \alpha)\|$. We need to verify the nonstationarity condition of the theorem. If $f(x, \alpha) > 0$, then the only possible proximal subgradient of $f(\cdot, \alpha)$ at x is its derivative there, namely $F'_x(x, \alpha)^* F(x, \alpha)/\|F(x, \alpha)\|$ (by the classical Chain Rule). We have shown above that such a vector has norm at least δ, as required.

Once the Solvability Theorem has been applied, there remains only to observe that (3.3) reduces to the required estimate for all (x, α) sufficiently near (x_0, α_0) since then

$$\min\{d(x, \operatorname{comp} V), d(x, \operatorname{comp} \Omega)\} > \frac{f(x, \alpha)}{\delta}. \quad \square$$

We remark that the conclusion of the theorem subsumes as special cases stability results with respect to either a parameter change or to a change in the variable x. If α alone changes (somewhat), then x_0 is no longer a solution to $F(x, \alpha) = 0$, but is not far from one; if only x changes, then there is a solution for the original equation which approximates x to the same extent to which x is infeasible (i.e., $\|F(x, \alpha_0)\|$). The term $\|F(x, \alpha)\|$ which appears in the estimate given by Theorem 3.2 is nondifferentiable at (x_0, α_0), and this is necessarily the case: no term of the form $\theta(\|F(x, \alpha)\|)$ with $\theta(0) = 0$ and θ differentiable can provide the estimate, as simple examples show.

Solutions constrained to a given set. The following result will be used in the next theorem, which will illustrate the use of *extended-valued* f. N_S^L refers to the "limiting normal cone" of S, which reduces to the usual normal set when S is smooth or convex. We shall omit the details and the proofs.

LEMMA 3.3. *Let F and (x_0, α_0) satisfy all the hypotheses of* Theorem 3.2, *where we take X and Y finite-dimensional. Let S be a closed subset of X containing x_0, and assume in addition that the following "constraint qualification" holds*:

$$(3.6) \qquad 0 \in F'_x(x_0, \alpha_0)^* y + N_S^L(x_0) \implies y = 0.$$

Then there exist $\delta > 0$ and neighborhoods V and A of x_0 and α_0 with the following property: *for all $(x, \alpha) \in V \times A$, for any unit vector y, any vector ζ belonging to*

$$F'_x(x, \alpha)^* y + N_S^L(x)$$

satisfies $\|\zeta\| \geq \delta$.

THEOREM 3.4. *Let (x_0, α_0) be a solution of the equality/constraint system*

$$F(x, \alpha) = 0, \quad x \in S$$

under the hypotheses of Lemma 3.3, *and in particular in the presence of the constraint qualification* (3.6). *Then for some $\delta > 0$, for all (x, α) sufficiently near (x_0, α_0) having $x \in S$, we have*

$$d(x, \Phi_S(\alpha)) \leq \frac{\|F(x, \alpha)\|}{\delta},$$

where $\Phi_S(\alpha) := \{x \in S \colon F(x, \alpha) = 0\}$. In particular, $\Phi_S(\alpha)$ is nonempty for α sufficiently near α_0.

We omit the proof, which is carried out by taking

$$f(x, \alpha) := \begin{cases} \|F(x, \alpha)\| & \text{if } x \in S \\ +\infty & \text{otherwise} \end{cases}$$

in the Solvability Theorem. The condition (3.6) leads to the required nonstationarity hypothesis.

EXAMPLE 3.5. *If the function $F \colon \mathbb{R}^3 \times M \to \mathbb{R}^2$ satisfies the hypothesis of* Theorem 3.2, *with*

$$F'_x(x_0, \alpha_0) = \begin{pmatrix} 1 & 0 & 0 \\ 1 & 1 & -1 \end{pmatrix},$$

then for every α sufficiently near α_0 there exists $x = (x_1, x_2, x_3)$ in $\mathbb{R}^3$ with $x_2 \geq 0$, $x_3 \geq 0$ such that $F(x, \alpha) = 0$. In fact, we can also require that $x_2 x_3 = 0$.

Implicit and inverse functions. Under additional structural assumptions, the equation $F(x, \alpha) = 0$ can be shown to uniquely define x as a function of α (locally), and the implicit function inherits regularity from F. We make the same assumptions as in Theorem 3.2, but now with $X = Y$.

THEOREM 3.6. (a) *Let $F'_x(x_0, \alpha_0)$ be onto and one-to-one. Then there exist neighborhoods Ω of x_0 and W of α_0 and a unique continuous function $\hat{x}(\cdot) \colon W \to \Omega$, with $\hat{x}(\alpha_0) = x_0$ such that $F(\hat{x}(\alpha), \alpha) = 0 \ \forall \alpha \in W$.*

(b) *If in addition F is Lipschitz in a neighborhood of (x_0, α_0), then we can affirm that $\hat{x}$ is Lipschitz.*

(c) *If in addition M is a Hilbert space and $F(x, \alpha)$ is C^1 (in both variables) near (x_0, α_0), then we can affirm that $\hat{x}$ is C^1.*

PROOF. It is a well-known consequence of the Closed Graph Theorem that $F'_x(x_0, \alpha_0)^{-1}$ is a continuous linear operator, and this can be used to show that $F'_x(x_0, \alpha_0)^*$ is onto (with inverse $(F'_x(x_0, \alpha_0)^{-1})^*$). Thus for some $\eta > 0$, we have $F'_x(x_0, \alpha_0)^* \overline{B} \supset 2\eta \overline{B}$. For some convex neighborhood $\Omega \times N \subset U \times M$ of (x_0, α_0), one has

$$\|F'_x(x, \alpha) - F'_x(x_0, \alpha_0)\| < \eta \quad \forall (x, \alpha) \in \Omega \times N$$

Now let x, $x' \in \Omega$ and $\alpha \in N$, $x' \neq x$. Then, for any vector θ in $\overline{B}$ one has, by the Mean Value Theorem, for some $z \in [x, x']$:

$$\begin{aligned}
\big\langle \theta, F(x',\alpha) - F(x,\alpha)\big\rangle &= \big\langle \theta, F'_x(z,\alpha)(x'-x)\big\rangle \\
&= \big\langle \theta, F'_x(x_0,\alpha_0)(x'-x)\big\rangle \\
&\qquad + \big\langle \theta, \big(F'_x(z,\alpha) - F'_x(x_0,\alpha_0)\big)(x'-x)\big\rangle \\
&\geq \big\langle F'_x(x_0,\alpha_0)^*\theta, x'-x\big\rangle - \eta\|x'-x\| \\
&= \eta\|x'-x\|,
\end{aligned}$$

where θ has been chosen in $\overline{B}$ to satisfy

$$F'_x(x_0,\alpha_0)^*\theta = 2\eta\frac{(x'-x)}{\|x'-x\|}.$$

It follows that

$$\big\|F(x',\alpha) - F(x,\alpha)\big\| \geq \eta\|x'-x\| \quad \forall (x,\alpha) \in \Omega \times N. \tag{3.7}$$

We now invoke Theorem 3.2 to deduce the existence of a neighborhood $W \subset N$ of α_0 such that

$$\Phi(\alpha) := \big\{x \in \Omega\colon F(x,\alpha) = 0\big\} \neq \emptyset \quad \text{for all } \alpha \in W.$$

It follows from (3.7) that $\Phi(\alpha)$ is a singleton $\big\{\hat{x}(\alpha)\big\}$ for each $\alpha \in W$, and that $\hat{x}$ is the unique function from W to Ω such that $F\big(\hat{x}(\alpha),\alpha\big) = 0$ $\forall \alpha \in W$. We now show that $\hat{x}(\cdot)$ is continuous.

From (3.7) we have

$$\eta\big\|\hat{x}(\alpha') - \hat{x}(\alpha)\big\| \leq \big\|F\big(\hat{x}(\alpha'),\alpha'\big) - F\big(\hat{x}(\alpha),\alpha'\big)\big\| = \big\|F\big(\hat{x}(\alpha),\alpha'\big)\big\|,$$

which tends to 0 as α' tends to α. This establishes continuity. In addition, if F is Lipschitz near (x_0,α_0), then for α and α' sufficiently near α_0 the last term above satisfies

$$\big\|F\big(\hat{x}(\alpha),\alpha'\big)\big\| = \big\|F\big(\hat{x}(\alpha),\alpha'\big) - F\big(\hat{x}(\alpha),\alpha\big)\big\| \leq K d_M(\alpha,\alpha'),$$

for some suitable Lipschitz constant K, where d_M is the metric on M. We deduce that $\hat{x}(\cdot)$ is Lipschitz near α_0, proving (b).

Finally, let M be a Hilbert space and let F be C^1 in (x,α). The Taylor expansion gives

$$\begin{aligned}
0 &= F\big(\hat{x}(\alpha),\alpha\big) \\
&= F(x_0,\alpha_0) + F'_x(x_0,\alpha_0)\big(\hat{x}(\alpha) - x_0\big) \\
&\qquad + F'_\alpha(x_0,\alpha_0)(\alpha-\alpha_0) + o\big(\hat{x}(\alpha) - x_0, \alpha-\alpha_0\big),
\end{aligned}$$

where $o(x-x_0,\alpha-\alpha_0)/\big\|(x-x_0,\alpha-\alpha_0)\big\| \to 0$ as $(x,\alpha) \to (x_0,\alpha_0)$. Since $\hat{x}(\cdot)$ is Lipschitz near α_0, $o\big(\hat{x}(\alpha)-x_0,\alpha-\alpha_0\big)$ can be written $o(\alpha-\alpha_0)$. Then we can solve the Taylor expansion explicitly for $\hat{x}(\alpha)$ and deduce from the resulting expression that $\hat{x}(\cdot)$ is differentiable at α_0, with derivative $-F'_x(x_0,\alpha_0)^{-1}F'_\alpha(x_0,\alpha_0)$. But this argument holds not just at α_0 but at any α in a neighborhood, whence $\hat{x}(\cdot)$ is C^1 near α_0. □

A special case of the theorem is that of inverse functions.

COROLLARY 3.7. *Let F satisfy the hypotheses of the theorem and have the form $F(x,\alpha) = G(x) - \alpha$ (so that $X = Y = M$). Then there exist neighborhoods W of α_0 and Λ of x_0 and a C^1 function $\hat{x}$ on W such that*

$$G\big(\hat{x}(\alpha)\big) = \alpha \quad \forall \alpha \in W, \qquad \hat{x}\big(G(x)\big) = x \quad \forall x \in \Lambda.$$

PROOF. The function $\hat{x}(\cdot)$ of the theorem is C^1 and satisfies $G\big(\hat{x}(\alpha)\big) = \alpha$ for $\alpha \in W$, the neighborhood of α_0 provided by the theorem. Now pick $\Lambda \subset \Omega$ so that $G(x) \in W \ \forall x \in \Lambda$. Then, as shown in the theorem, $\Phi\big(G(x)\big)$ is a singleton when $x \in \Lambda$; that is, there is a unique $x' \in \Omega$ such that $G(x') = G(x)$. But then this x', namely $\hat{x}\big(G(x)\big)$, must equal x. □

Systems of mixed equalities and inequalities. We consider now the solutions of a parametrized system of the form

$$h(x) = \alpha, \quad g(x) \le \beta \tag{3.8}$$

where the vector $(\alpha, \beta) \in \mathbb{R}^m \times \mathbb{R}^p$ is the parameter. Let x_0 satisfy the system for $(\alpha,\beta) = (0,0)$, and let us suppose that the functions $g\colon X \to \mathbb{R}^p$ and $h\colon X \to \mathbb{R}^m$ are Lipschitz in a neighborhood of x_0. The limiting proximal subdifferential $\partial_L f$ of a locally Lipschitz function f is defined as follows:

$$\partial_L f(x) := \big\{\lim \zeta_i \colon \zeta_i \in \partial_P f(x_i), x_i \to x \text{ as } i \to \infty\big\}.$$

THEOREM 3.8. *Let Ω be a neighborhood of x_0, and set*

$$\Phi(\alpha,\beta) := \big\{x \in \Omega\colon \ \text{system (3.8) is satisfied}\big\}.$$

Suppose that the following constraint qualification is satisfied at x_0:

$$\gamma \ge 0, \quad \big\langle \gamma, g(x_0)\big\rangle = 0, \quad 0 \in \partial_L\big\{\big\langle\gamma, g(\cdot)\big\rangle + \big\langle\lambda, h(\cdot)\big\rangle\big\}(x_0) \implies \gamma = 0 \\ \text{and} \quad \lambda = 0.$$

Then for some $\delta > 0$, for all x sufficiently near x_0, and for all (α,β) sufficiently near $(0,0)$, one has

$$d\big(x, \Phi(\alpha,\beta)\big) \le \frac{\max_{i,j}\{(g_i(x) - \beta_i)_+, |h_j(x) - \alpha_j|\}}{\delta}$$

where $\big(g_i(x) - \beta_i\big)_+$ signifies $\max\big(g_i(x) - \beta_i, 0\big)$. In particular, $\Phi(\alpha,\beta)$ is nonempty for (α,β) near $(0,0)$.

PROOF. We set

$$f(x,\alpha,\beta) := \max_{i,j}\big\{\big(g_i(x) - \beta_i\big)_+, \big|h_j(x) - \alpha_j\big|\big\} \ge 0.$$

Then x is solution of (3.8) iff $f(x,\alpha,\beta) = 0$. Note that f is Lipschitz in a neighborhood of $(x_0, 0, 0)$.

LEMMA. *There exist neighborhoods V of x_0 and A of $(0,0)$ and $\delta > 0$ such that*

$$x \in V, \quad (\alpha,\beta) \in A, \quad f(x,\alpha,\beta) > 0, \quad \zeta \in \partial_P f(x,\alpha,\beta) \implies \|\zeta\| \ge \delta.$$

We omit the proof of this lemma, which uses standard results in nonsmooth analysis. Once we are armed with it, the application of the Solvability Theorem 3.1 is immediate and gives the required conclusion.

□

When the inequalities are absent in (3.8), it is natural to seek nonsmooth analogues of the implicit function assertions of Theorem 3.6. We shall examine for illustrative purposes the issue of an inverse function induced by an equation $G(x) = \alpha$, which was treated in the case of smooth data in Corollary 3.7. The solvability aspect is at hand, as a special case of the preceding theorem:

COROLLARY 3.9. *Let $G\colon X \to \mathbb{R}^n$ be Lipschitz on a neighborhood of the point x_0, where $G(x_0) = \alpha_0$. Suppose that*

$$0 \in \partial_L \langle \zeta, G(\cdot) \rangle (x_0) \implies \zeta = 0. \tag{3.9}$$

Then, for any neighborhood Ω of x_0, there exists $\delta > 0$ such that for all (x, α) sufficiently near (x_0, α_0) one has

$$d\big(x, \Phi(\alpha)\big) \leq \frac{\|G(x)\|}{\delta},$$

where $\Phi(\alpha) := \big\{x \in \Omega\colon G(x) = \alpha\big\}$. (This reduces to a special case of Theorem 3.2 *when G is C^1.)*

What is required now to deduce the existence of an inverse is a condition akin to (3.7) which implies that G is locally one-to-one. A convenient way to formulate such a condition, which will also turn out to be a more readily verifiable criterion implying the constraint qualification (3.9), is in terms of the *generalized Jacobian* ∂G [**1**].

Let $X = \mathbb{R}^m$. We define $\partial G(x)$ as follows:

$$\partial G(x) := \operatorname{co}\big\{\lim_{i\to\infty} G'(x_i)\colon G'(x_i) \text{ exists }, x_i \to x\big\}.$$

(We remark that an arbitrary set of measure 0 can be avoided in calculating $\partial G(x)$, without the results below being affected.) We identify $\partial G(x)$ with a convex set of $n \times m$ matrices, and we say that $\partial G(x)$ is *of maximal rank* provided that every matrix in $\partial G(x)$ is of maximal rank. When $n = m$, it is alternate terminology to say that $\partial G(x)$ is *nonsingular.* Here are some immediate properties of ∂G.

PROPOSITION 3.10. (a) *Let $G\colon \mathbb{R}^m \to \mathbb{R}^n$ be Lipschitz near x. Then $\partial G(x)$ is compact, convex and nonempty, where the space of $n \times m$ matrices is identified with $\mathbb{R}^{nm}$. Further, ∂G is upper semicontinuous: for all $\varepsilon > 0$ there exists $r > 0$ such that*

$$x' \in x + rB \implies \partial G(x') \subset \partial G(x) + \varepsilon B_{n\times m},$$

where B_{nm} denotes the open unit ball in the space of $n \times m$ matrices.

(b) *For any $\zeta \in \mathbb{R}^n$, one has $\partial\langle \zeta, G(\cdot)\rangle(x) = \zeta^* \partial G(x)$.*

PROPOSITION 3.11. *Let $G\colon \mathbb{R}^n \to \mathbb{R}^n$ be Lipschitz near x_0, and suppose that $\partial G(x_0)$ is nonsingular. Then the constraint qualification* (3.9) *holds, and for some $\eta > 0$, for some neighborhood Ω of x_0, one has*

$$\big\|G(x) - G(y)\big\| \geq \eta\|x - y\| \quad \forall x, y \in \Omega.$$

PROOF.

LEMMA. *There are positive numbers r and η with the following property. Given any unit vector $v \in \mathbb{R}^n$, there is a unit vector w in $\mathbb{R}^n$ such that, whenever x lies in $x_0 + rB$ and the matrix M belongs to $\partial G(x)$, then one has*

$$\langle w, Mv\rangle \geq \eta.$$

To see this, let S be the unit sphere in $\mathbb{R}^n$ and note that the subset $\partial G(x_0)S$ of $\mathbb{R}^n$ is compact and does not contain 0. Hence, $d\big(0, GF(x_0)S\big)/2 =: \eta > 0$. For some $\varepsilon > 0$ small enough, we will have $d\big(0, \big(\partial G(x_0) + \varepsilon B_{n\times n}\big)S\big) > \eta$. Now for some $r > 0$, for all $x \in x_0 + rB$, one has $\partial G(x) \subset \partial G(x_0) + \varepsilon B_{n\times n}$, in light of Proposition 3.10, where r is small enough so that G is Lipschitz on $x_0 + r\overline{B}$. Now let any unit vector v be given. It follows from the above that the convex set $\big(\partial G(x_0) + \varepsilon B_{n\times n}\big)v$ is distance at least η from 0. By the separation theorem, there exists a unit vector w such that $\langle \zeta, w\rangle \geq \eta$ for all $\zeta \in \big(\partial G(x_0) + \varepsilon B_{n\times n}\big)v$, and all the more for any ζ of the form Mv, $M \in \partial G(x)$, $x \in x_0 + rB$. This establishes the lemma.

Now let x and y belong to $\Omega := x_0 + rB$. We shall complete the proof of the Proposition by showing that $\big\|G(x) - G(y)\big\| \geq \eta\|x - y\|$. We may suppose $x \neq y$. Set

$$v = \frac{y - x}{\|y - x\|}, \qquad \lambda = \|y - x\|,$$

so that $y = x + \lambda v$.

Let π be the plane perpendicular to v and passing through x. The set P of points x' in $x_0 + rB$ where $G'(x')$ fails to exist is of measure 0, and hence by Fubini's theorem, for almost every x' in π, the ray

$$x' + tv, \quad t \geq 0$$

meets P in a set of 0 one-dimensional measure. Choose an x' with the above property and sufficiently close to x so that $x' + tv$ lies in $x_0 + rB$ for every t in $[0, \lambda]$. Then the function

$$t \to G(x' + tv)$$

is Lipschitzian for t in $[0, \lambda]$ and has a.e. on this interval the derivative $G'(x' + tv)v$. Thus

$$G(x' + \lambda v) - G(x') = \int_0^\lambda G'(x' + tv)v\, dt.$$

Let w correspond to v as in the Lemma above. We deduce

$$w \cdot \big[G(x' + \lambda v) - G(x')\big] = \int_0^\lambda w \cdot \big[G'(x' + tv)v\big]\, dt \geq \int_0^\lambda \eta\, dt = \lambda\eta.$$

Recalling the definition of λ, we arrive at:

$$\big\|G(x' + \lambda v) - G(x')\big\| \geq \eta\|x - y\|.$$

This conclusion has been obtained for almost all x' in a neighborhood of x. Since G is continuous, the Proposition is proved. □

We now possess all the necessary components to prove the Lipschitz Inverse Function Theorem [**1**].

THEOREM 3.12. *If $G\colon \mathbb{R}^n \to \mathbb{R}^n$ is Lipschitz near x_0 and $\partial G(x_0)$ is nonsingular, then there exist neighborhoods W of $G(x_0)$ and Λ of x_0 and a Lipschitz function $\hat{x}$ on W such that*

$$G\big(\hat{x}(\alpha)\big) = \alpha \quad \forall \alpha \in W, \quad \hat{x}\big(G(x)\big) = x \quad \forall x \in \Lambda.$$

We remark that the theorem fails (even for $n = 1$) if the hypothesis that $\partial G(x_0)$ be nonsingular is replaced by the condition that $G'(x)$ be nonsingular whenever it exists.

References

1. F. H. Clarke, *Optimization and nonsmooth analysis*, 2nd ed., Classics Appl. Math. vol. 5, SIAM, Philadelphia, PA, 1990. (Originally published by Wiley, New York, 1983.)
2. ———, *Methods of dynamic and nonsmooth optimization*, CBMS-NSF Regional Conf. Ser. in Appl. Math., vol. 57, SIAM, Philadelphia, 1989.
3. F.H. Clarke and Yu.S. Ledyaev, *Mean value inequalities in Hilbert space*, Trans. Amer. Math. Soc. **344** (1994), no. 1, 307–324.
4. F. H. Clarke, Yu. S. Ledyaev, R. J. Stern, and P. R. Wolenski, *Nonsmooth analysis and control theory* (to appear).

Centre de recherches mathématiques, Université de Montréal, C.P. 6128, succ. Centre-ville, Montréal, Québec H3C 3J7, Canada

E-mail address: clarke@CRM.UMontreal.CA

Centre de Recherches Mathématiques
CRM Proceedings and Lecture Notes
Volume **11**, 1997

Differential Equations for Linear Shells: Comparison Between Intrinsic and Classical Models

Michel C. Delfour and Jean-Paul Zolésio

ABSTRACT. The oriented boundary (signed, algebraic) distance function of a set remarkably describes its geometric properties and those of its boundary. When combined with tangential differential operators defined through an extension in a neighborhood of a submanifold they yield a very powerful and simple intrinsic tangential differential calculus on $C^{1,1}$ submanifolds of $\mathbb{R}^N$. To illustrate this point a new intrinsic approach has been developed to the modeling and analysis of thin shells. We feel that it is more natural and tractable for the optimization, control and sensitivity analysis of shells. In this paper we give the relevant basic intrinsic differential geometry and provide the correspondence between intrinsic tangential derivatives and classical covariant derivatives using local bases and Christoffel symbols. We extend previous results on the $P(1,2)$ model to the simpler $P(1,1)$ model. We show how the asymptotic membrane equation model and the classical models of Naghdi and Koiter can be expressed in terms of intrinsic operators and make comparison with the $P(1,0)$ model and approximations of the $P(1,1)$ model.

1. Introduction

The topic of shells is a classical one which is fairly well covered in many books. For instance the reader is referred to [**5, 7, 45, 61, 87, 88**], but this is by no means an exhaustive list. Research in the theory of shells has currently been stimulated by a number of questions arising from applications in control and design of large space structures, flexible robots, materials science, etc, where intrinsic and mathematically more tractable models would be much needed. Substantial work is also going on asymptotic methods by Ph. Ciarlet, V. Lodz, C. Mardare, B. Miara, E. Sanchez-Palencia, ..., and on polynomial approximations by S. M. Alessandrini, D. N. Arnold, R. S. Falk, A. Madureira, M. Dauge, I. Gruais, J.C. Paumier, A. Raoult, ... This is again a very incomplete list.

1991 *Mathematics Subject Classification.* Primary 73K15, 73K12; Secondary 49Q, 73O62.

The research of the first author has been supported in part by National Sciences and Engineering Research Council of Canada operating grant A-8730 and by a FCAR grant from the Ministère de l'Éducation du Québec.

This is the final form of the paper.

The mathematical theory of shells very much follows the theory of elliptic partial differential equations with the difference that everything takes place on a submanifold of $\mathbb{R}^3$. Part of the difficulty resides in the complexity of the expressions. In the literature, there has been independent interest in getting intrinsic equations for shells (cf. for instance the work of [**87, 88**] and [**56**]). For Control and Optimal Design there is a definite interest in working directly with equations involving intrinsic differential operators rather than parameterized differential operators which might hide the natural structure of the equations.

Our interest in shells grew out of work started in 1991 and presented at the IFIP Workshop in June 1992 (Sophia-Antipolis, France) on the the use of the distance functions to generate topologies on equivalence classes of sets in order to make sense of the continuity of shape functionals and characterize compact families of sets in order to obtain new problem formulations and existence results.

It appeared that the oriented boundary (signed) distance function provides a very natural and simple tool to describe the geometric properties of a set and its boundary. To illustrate this on a nontrivial example we presented a first elementary linear model of thin shells at an IMA Workshop in November 1992 (Minneapolis, USA). The models were developed in a completely intrinsic way from first principles without references to established models using covariant derivatives and Christoffel symbols. It became quickly desirable to be able to compare models and find explicit correspondences between intrinsic concepts and differential operators and covariant ones. The object of this paper is precisely to fill up that gap and to throw a bridge between these approaches.

In § 2 we first recall some key properties of the oriented distance function, elementary notions of differential geometry and Federer's decomposition of the measure along the level sets of a function. In § 3 tangential derivatives are introduced via an extension in a neighborhood of the submanifold, and the tangential differential calculus developed in [**35, 36**] is expanded and sharpened. We provide complete correspondence between tangential differential operators and differential operators with respect to covariant/contravariant bases. The fundamental observation is that tangential derivatives on a $C^{1,1}$ submanifold Γ of $\mathbb{R}^N$ can be directly defined as restrictions to Γ of derivatives in $\mathbb{R}^N$ of the composition of the function with the projection onto the submanifold. This was the key to make the tangential calculus on a submanifold *fully operational* and our motivation to try it in the theory of shells. All the computations are carried out in $\mathbb{R}^N$ and the final expressions are obtained by restriction to Γ. It is possible to recover in a systematic way many results in Differential Geometry. This approach to the tangential calculus naturally extends to submanifolds Γ of codimension larger than one for which the square of the distance function to the submanifold Γ is $C^{1,1}$ in a neighborhood. Such sets are the sets with positive reach characterized by H. Federer. § 4 gives the intrinsic version for a submanifold of codimension one of the corresponding theorems in $\mathbb{R}^N$ for Sobolev spaces, Korn's, Poincaré's inequality, etc...

The material on thin shells begins with § 5 where the problem of finding an *asymptotic model* for the deformation of the midsurface of the shell is considered. Developments are made with respect to the normal local coordinate and the *asymptotic $P(1,0)$ model* is characterized. Existence of solutions is established in an appropriate quotient function space. This model yields the accepted *asymptotic membrane equation* and the *Love-Kirchhoff condition.* In § 6 we express Naghdi's and Koiter's models and the *asymptotic membrane equation* in terms of intrinsic tangential operators. In § 7 we give the main results for the $P(1,1)$ model. This model uses a first order development of the displacement vector and the linear strain tensor with respect to the normal variable to the midsurface. It is remarkable that the results are essentially the same as the ones obtained in an earlier paper for a $P(1,2)$ model (second order development for the strain tensor and first order for the displacement vector, cf. [**36**]). This model also preserves the approximate rigid displacements. It turns out that the solution(s) of both models converge to the solution(s) of the asymptotic $P(1,0)$ model. This material is new and more details can be found in [**41**]. Finally in § 8 the $P(1,1)$ model is further approximated by neglecting terms which are relatively small with respect to others in the expression of the energy functional. This approximation does not destroy the structure of the models and does not change the mathematical results. Comparisons are made with Naghdi's model in order to isolate the *occasional mathematical contradictions* in the foundations of the theory of shell.

As a final remark we would like to explain why we have considered shells in the N-dimensional Euclidean space rather than in $\mathbb{R}^3$. Firstly restricting the analysis to $\mathbb{R}^3$ does not simplify any of the arguments. Secondly the constructions and technical results on the various aspects of the theory of thin shells contribute to the fundamental constructions and results in intrinsic differential geometry. They can be used in different contexts and for other applications.

Note added in the proofs. Further results are now available for the $P(2,1)$ and $P(2_n,1)$ models and the asymptotic $P(2,1)$ and $P(2_n,1)$ models in [**40**], [**42**] and [**43**]. They require some slight extensions of the basic material in this paper. One of the very nice result which comes out of the analysis is that the asymptotic versions of these two models yield as a special case the classical asymptotic bending dominated equation.

2. Distance Functions and Differential Geometry

2.1. Notation and definitions. Since the superscript t will often appear in the book, the transposed of a vector v and a matrix A will be respectively denoted $^*v$ and $^*A$. The inverse of A will be denoted by A^{-1} and the inverse of $^*A$ by $^*A^{-1}$. The inner product in $\mathbb{R}^N$ will be written

$$x \cdot y = \sum_{i=1}^{N} x_i y_i, \quad |x| = \sqrt{x \cdot x}.$$

For a linear transformation $A\colon \mathbb{R}^N \to \mathbb{R}^K$

$$|A| = \max_{|x|_{\mathbb{R}^N} \le 1} |Ax|_{\mathbb{R}^K},$$

and if A_{ij} and B_{ij} are the matrix representations of A and B with respect to some bases $\{a_1, \dots, a_N\}$ and $\{b_1, \dots, b_K\}$

$$A \cdot\cdot B = \sum_{i=1}^{N} \sum_{j=1}^{K} A_{ij} B_{ij}.$$

The norm $|A|$ is equivalent to the norm $\sqrt{A \cdot\cdot A}$.

Let $\{e_1, \dots, e_N\}$ be the standard unitary orthonormal basis in $\mathbb{R}^N$, that is $(e_i)_j = \delta_{ij}$, the Kronecker index function. We shall use the notation $\zeta = (\zeta', \zeta_N)$

for a point $\zeta = (\zeta_1, \dots, \zeta_N)$ in $\mathbb{R}^N$, where $\zeta' = (\zeta_1, \dots, \zeta_{N-1})$. Denote by B the open unit ball in $\mathbb{R}^N$ and define the sets

$$B_0 = \{\zeta \in B \colon \zeta_N = 0\}, \quad B_\pm = \{\zeta \in B : \zeta_N \gtrless 0\}. \tag{2.1}$$

DEFINITION 2.1. a) A set Ω in $\mathbb{R}^N$ is said to be ***locally of class*** C^k, $1 \le k \le \infty$, if for each $x \in \partial\Omega$ there exist
(i) a neighborhood $U(x)$ of x,
(ii) and a map $f_x \colon U(x) \to B$ with the following properties:

$$f_x \in C^k\big(U(x), B\big), \quad h_x = f_x^{-1} \in C^k\big(B, U(x)\big), \tag{2.2}$$

$$\operatorname{int}\Omega \cap U(x) = h_x(B_+), \tag{2.3}$$

$$\Gamma_x \overset{\text{def}}{=} \partial\Omega \cap U(x) = h_x(B_0), \quad B_0 = f_x(\Gamma_x). \tag{2.4}$$

b) A set Ω in $\mathbb{R}^N$ is said to be ***locally*** (k,λ)***-Hölderian or of class*** $C^{k,\lambda}$, $1 \le k$, $0 < \lambda \le 1$, if the conditions of part a) are verified with a map $f_x \in C^{k,\lambda}\big(U(x), B\big)$, such that $h_x = f_x^{-1} \in C^{k,\lambda}\big(B, U(x)\big)$.

2.2. Covariant differential calculus on Γ. Consider a locally $C^{1,1}$ domain Ω in $\mathbb{R}^N$. Thus its boundary $\Gamma = \partial\Omega$ is an $(N-1)$-dimensional submanifold of $\mathbb{R}^N$ of class $C^{1,1}$. At each point $x \in \Gamma$ there is a diffeomorphism f_x of class C^2 from a neighborhood $U(x)$ of x onto B. With the notation $h_x = f_x^{-1}$ by definition of B_0, $\{e_1, \dots, e_{N-1}\} \subset B_0$ and the tangent space $T_y\Gamma$, $\Gamma = \partial\Omega$, at y to Γ_x is the vector space spanned by the $N-1$ vectors

$$\big\{Dh_x(\zeta)e_i \colon 1 \le i \le N-1\big\}, \quad \zeta = f_x(y) \in B_0, \tag{2.5}$$

where $Dh_x(\zeta)$ is the Jacobian matrix of h_x at the point ζ

$$(Dh_x)_{\ell m} \overset{\text{def}}{=} \partial_m (h_x)_\ell.$$

The ***inward unit normal*** is given by

$$a_N(\zeta') = \frac{{}^*Dh_x(\zeta', 0)^{-1} e_N}{|{}^*Dh_x(\zeta', 0)^{-1} e_N|}. \tag{2.6}$$

We introduce the map

$$\zeta' \mapsto \Phi(\zeta') \overset{\text{def}}{=} h_x(\zeta', 0) \colon B_0 \subset \mathbb{R}^{N-1} \to \Gamma \cap U(x) \subset \mathbb{R}^N.$$

The ***covariant basis*** at a point $y = \Phi(\zeta') \in U(x) \cap \Gamma$ is defined as

$$a_\alpha(\zeta') = \frac{\partial h_x}{\partial \zeta_\alpha}(\zeta', 0) = \frac{\partial \Phi}{\partial \zeta_\alpha}(\zeta'), \quad \alpha = 1, \dots, N-1, \quad h_x(\zeta', 0) = y,$$

and a_N is chosen as the ***inward unit normal*** (2.6). The standard convention that the Greek indices ranges from 1 to $N-1$ and the and the result follows by definition of the quotient Roman indices from 1 to N will be followed together with Einstein's rule of summation over repeated indices. In general the covariant basis is not even orthogonal and it is natural to introduce the ***contravariant basis*** $\{a^i\} = \{a^i(\zeta')\}$ defined from the covariant one $\{a_i\} = \{a_i(\zeta')\}$ as $a^i \cdot a_j = \delta_{ij}$. In those bases a vector $v \colon \Gamma \to \mathbb{R}^N$ has covariant and contravariant representations

$$v = v^i a_i = v_i a^i, \quad v^i \overset{\text{def}}{=} v \cdot a^i, \quad v_i \overset{\text{def}}{=} v \cdot a_i. \tag{2.7}$$

If the vector v is tangent then $v = v^\alpha a^\alpha = v_\alpha a^\alpha$. The inner product of two vectors v and w becomes

$$v \cdot w = v^i w_i = v_i w^i. \tag{2.8}$$

Similarly a linear transformation A of $\mathbb{R}^N$ can have several representations

$$Av \cdot w = \begin{cases} A^{ij} a_i \cdot a_j, & A^{ij} = Aa^i \cdot a^j \\ A_{ij} a^i \cdot a^j, & A_{ij} = Aa_i \cdot a_j \\ A^i_{\cdot j} a_i \cdot a^j, & A^i_{\cdot j} = Aa^i \cdot a_j \\ A_i^{\cdot j} a^i \cdot a_j, & A_i^{\cdot j} = Aa_i \cdot a^j \end{cases} \tag{2.9}$$

where the dot before j shows that j is the second index and i the first. For the double product it is easy to verify that

$$A \cdot\cdot B = A_{ij} B^{ij} = A^{ij} B_{ij}, \quad \operatorname{tr} A = A \cdot\cdot I = A_{ii} = A^{ii}.$$

Consider a C^1-function $w\colon B_0 \to \mathbb{R}$. The ***partial derivative*** of w will be denoted by

$$w_{,\alpha} \stackrel{\text{def}}{=} \frac{\partial w}{\partial \xi^\alpha}, \quad w_{,N} \stackrel{\text{def}}{=} 0. \tag{2.10}$$

The partial derivative of a C^1 vector functions $v\colon B_0 \to \mathbb{R}^N$ is denoted in a similar way

$$v_{,\alpha} \stackrel{\text{def}}{=} \frac{\partial v}{\partial \xi^\alpha}, \quad v_{,N} \stackrel{\text{def}}{=} 0. \tag{2.11}$$

The ***first, second and third fundamental forms*** a, b, and c are defined as

$$a_{\alpha\beta} \stackrel{\text{def}}{=} a_\alpha \cdot a_\beta, \quad b_{\alpha\beta} \stackrel{\text{def}}{=} -a_\alpha \cdot a_{N,\beta}, \quad c_{\alpha\beta} \stackrel{\text{def}}{=} b^\lambda_\alpha b_{\lambda\beta},$$

where

$$a_{N,\beta} = \frac{\partial a_N}{\partial \zeta_\beta}, \quad b^\lambda_\alpha = a^\lambda \cdot a^\mu b_{\mu\alpha}.$$

The second fundamental form defines a symmetric linear transformation of the ***tangent hyperplane*** $T_x\Gamma$ onto $T_x\Gamma$. Its $N-1$ eigenvalues are the ***principal curvatures*** κ_i at x, $1 \le i \le N-1$, of the submanifold Γ. The ***mean curvature*** H and the ***Gauss curvature*** K are defined as

$$H = \frac{1}{N-1} \sum_{\alpha=1}^{N-1} \kappa_\alpha \quad \text{and} \quad K = \prod_{\alpha=1}^{N-1} \kappa_\alpha.$$

The factor $1/(N-1)$ is used to make the mean curvature of the unit sphere equal to one in all dimensions. However the sum is more fundamental and we shall modify the definition of H as follows

$$\boxed{H \stackrel{\text{def}}{=} \sum_{\alpha=1}^{N-1} \kappa_\alpha} \tag{2.12}$$

and still refer to H as the mean curvature. So H is the trace of $b_{\alpha\beta}$ and K its determinant. Finally the choice of the inner normal for a_N is necessary to make the principal curvatures of the boundary of the ball positive. Note that by the Cayley–Hamilton theorem the transformation b satisfies its $(N-1)$-th order characteristic polynomial.

The ***Christoffel symbols*** are defined as

$$\Gamma^{\alpha}_{\beta\gamma} \stackrel{\text{def}}{=} a^{\alpha} \cdot a_{\beta,\gamma}. \tag{2.13}$$

For a C^2-mapping

$$a_{\gamma,\beta} = \frac{\partial}{\partial \xi^{\beta}} \frac{\partial}{\partial \xi^{\gamma}} \Phi = \frac{\partial}{\partial \xi^{\gamma}} \frac{\partial}{\partial \xi^{\beta}} \Phi = a_{\beta,\gamma}, \tag{2.14}$$

and $\Gamma^{\alpha}_{\beta\gamma} = \Gamma^{\alpha}_{\gamma\beta}$.

Finally the ***covariant and contravariant derivatives*** are defined as

$$v_{\alpha}\big|_{\gamma} \stackrel{\text{def}}{=} v_{\alpha,\gamma} - \Gamma^{\lambda}_{\alpha\gamma} v_{\lambda}, \quad v^{\alpha}\big|_{\gamma} \stackrel{\text{def}}{=} v^{\alpha}_{,\gamma} + \Gamma^{\alpha}_{\lambda\gamma} v^{\lambda}. \tag{2.15}$$

2.3. The oriented distance function. The ***oriented*** (resp. ***algebraic*** or ***signed***) ***distance function*** plays an important role in the description of the geometric properties of a domain and its boundary. Among its many interesting properties its smoothness in a neighborhood of the boundary of a given set is equivalent to the smoothness of the boundary of the set. In addition their gradient and Laplacian coincide with the exterior unit ***normal*** and the ***mean curvature*** on the boundary of the set.

Given a subset Ω of $\mathbb{R}^N$ the ***distance function*** from a point x to a set Ω in $\mathbb{R}^N$ is defined as

$$d_{\Omega}(x) \stackrel{\text{def}}{=} \begin{cases} \inf_{y\in\Omega} |y-x|, & \Omega \neq \varnothing \\ +\infty, & \Omega = \varnothing. \end{cases} \tag{2.16}$$

If d_{Ω} is finite in $\mathbb{R}^N$, then $\overline{\Omega} \neq \varnothing$ and a fortiori $\Omega \neq \varnothing$. For $\Omega \neq \varnothing$ the ***set of projections*** from a point x onto Ω is defined as

$$\Pi_{\Omega}(x) = \big\{p \in \overline{\Omega} \colon |p-x| = d_{\Omega}(x)\big\}.$$

The distance function d_{Ω} provides a good description of a domain Ω from the "outside", but says nothing about its "inside" where it is identically zero. To get a complete description of a subset Ω of $\mathbb{R}^N$, it is natural to take into account its complement $\complement\Omega = \{y \in \mathbb{R}^N : y \notin \Omega\}$.

DEFINITION 2.2. Let Ω be a subset of $\mathbb{R}^N$. The ***oriented distance function*** to the set Ω is defined as

$$b_{\Omega}(x) \stackrel{\text{def}}{=} d_{\Omega}(x) - d_{\complement\Omega}(x), \forall x \in \mathbb{R}^N. \tag{2.17}$$

The set of projections of a point x onto $\partial\Omega$ is defined as

$$\Pi_{\partial\Omega}(x) = \big\{p \in \partial\Omega \colon |p-x| = d_{\partial\Omega}(x)\big\}.$$

Recall that $d_{\Omega} = +\infty$ when $\Omega = \varnothing$. Hence $b_{\Omega} = +\infty$ if $\Omega = \varnothing$ and $b_{\Omega} = -\infty$ if $\Omega = \mathbb{R}^N$. When $\partial\Omega \neq \varnothing$, b_{Ω} coincides with the ***signed or algebraic distance function*** to the boundary

$$b_{\Omega}(x) = \begin{cases} d_{\Omega}(x), & x \in \mathbb{R}^N \setminus \overline{\Omega} = \operatorname{int} \complement\Omega \\ 0, & x \in \overline{\Omega} \cap \overline{\complement\Omega} = \partial\Omega \\ -d_{\complement\Omega}(x), & x \in \mathbb{R}^N \setminus \overline{\complement\Omega} = \operatorname{int} \Omega. \end{cases} \tag{2.18}$$

Noting that $b_{\complement\Omega} = -b_{\Omega}$, it amounts to choose the negative sign for the interior of Ω and the positive sign for the interior of its complement.

The smoothness of a set Ω is equivalent to the smoothness of the function b_Ω in a neighborhood of its boundary $\partial\Omega$ for domains which are are class $C^{1,1}$ or smoother (cf. for instance [**32, 58**], Lem. 14.16, p. 355]).

THEOREM 2.1. *Let $k \geq 1$ be an integer, $0 \leq \lambda \leq 1$, and Ω be a domain in $\mathbb{R}^N$ with a nonempty boundary $\partial\Omega$. If for each x in $\partial\Omega$ there exists a neighborhood $V(x)$ of x where b_Ω is of class $C^{k,\lambda}$, then the boundary $\partial\Omega$ is of class $C^{k,\lambda}$ and for each x in $\partial\Omega$ there exists a neighborhood $U(x)$ of x where*

$$\forall y \in U(x), \quad \Pi_{\partial\Omega}(y) = \{p_{\partial\Omega}(y)\} \text{ is a singleton,} \tag{2.19}$$

$$p_{\partial\Omega} \in C^{k-1,\lambda}(U(x);\mathbb{R}^N), \quad n(y) = \nabla b_\Omega(y)\ \forall y \in \partial\Omega. \tag{2.20}$$

THEOREM 2.2. *Let Ω be a domain in $\mathbb{R}^N$ with a nonempty boundary $\partial\Omega$ of class $C^{k,\lambda}$ (resp. $C^{1,1}$) for $k \geq 2$ and $0 \leq \lambda \leq 1$. Then for each x on $\partial\Omega$, there exists a neighborhood of x where b_Ω is $C^{k,\lambda}$ (resp. $C^{1,1}$). Moreover in all cases*

$$\nabla b_\Omega = n \circ p_{\partial\Omega} = \nabla b_\Omega \circ p_{\partial\Omega}, \tag{2.21}$$

where n is the unit outward external normal to $\partial\Omega$.

2.4. Flow of the gradient of $\mathbf{b}_\Omega$. For any set $A \subset \mathbb{R}^N$ and $h > 0$ define the neighborhood

$$U_h(A) = \{x \in \mathbb{R}^N : d_A(x) < h\},$$

of A. In particular for a set $\Omega \subset \mathbb{R}^N$, $d_{\partial\Omega} = |b_\Omega|$

$$U_h(\partial\Omega) = \{x \in \mathbb{R}^N : |b_\Omega(x)| < h\}.$$

THEOREM 2.3. *Let Ω be a bounded $C^{1,1}$ domain in $\mathbb{R}^N$ and let $h > 0$ be such that $b_\Omega \in C^{1,1}(U_{2h}(\partial\Omega))$.*

(i) *Let ψ be a C^∞-function such that*

$$\psi(x) = \begin{cases} 1, & x \in U_h(\partial\Omega) \\ \in [0,1], & x \in U_{2h}(\partial\Omega) \setminus U_h(\partial\Omega) \\ 0, & x \in \mathbb{R}^N \setminus U_{2h}(\partial\Omega). \end{cases} \tag{2.22}$$

The equation

$$\frac{dx}{dz}(z) = \psi(x(z))\nabla b_\Omega(x(z)), \quad x(0) = X \in \mathbb{R}^N, \tag{2.23}$$

has a unique solution $x(z;X)$. Denote by

$$T_z(X) \stackrel{\text{def}}{=} x(z;X) \tag{2.24}$$

the transformation generated by the solutions of (2.23). Then for all X and z such that $|b(X)| < h$ and $|b(X) + z| < h$

$$T_z(X) = X + z\nabla b_\Omega(X), \quad b_\Omega \circ T_z = b_\Omega + z, \tag{2.25}$$

$$\nabla(b_\Omega \circ T_z) = \nabla b_\Omega = \nabla b_\Omega \circ p_{\partial\Omega}, \tag{2.26}$$

$$p_{\partial\Omega} \circ T_z = p_{\partial\Omega}, \nabla b_\Omega \circ T_z = \nabla b_\Omega \quad \text{and} \quad (T_z)^{-1} = T_{-z}.$$

(ii) *The set* $\Omega_z \overset{\text{def}}{=} T_z(\Omega)$, $|z| < h$, *is of class* $C^{1,1}$

$$\partial\Omega_z = \{x\colon b_\Omega(x) = z\}, \quad \operatorname{int}\Omega_z = \{x\colon b_\Omega(x) < z\},$$

$$b_{\Omega_z} \in C^{1,1}(U_{h-|z|}(\partial\Omega_z)), \tag{2.27}$$

$$b_{\Omega_z} \circ T_z = b_\Omega, \quad p_{\partial\Omega_z} = T_z \circ p_{\partial\Omega}. \tag{2.28}$$

(iii) *For all* $x \in U_h(\partial\Omega)$ *and* z *such that* $|z + b_\Omega(x)| < h$

$$DT_z(x) = I + zD^2 b_\Omega(x), \tag{2.29}$$

$$j_z(x) \overset{\text{def}}{=} \det(DT_z(x)) = \sum_{i=0}^{N-1} \lambda_i(x) z^i, \tag{2.30}$$

since $|\nabla b_\Omega(x)|^2 = 1$, $D^2 b_\Omega(x)\nabla b_\Omega(x) = 0$, *and* $\det D^2 b_\Omega(x) = 0$. *Moreover*

$${}^*DT_z^{-1}(x)\nabla b_\Omega(x) = \nabla b_\Omega(x), \tag{2.31}$$

$$\frac{dDT_z}{dz} = D^2 b_\Omega \circ T_z DT_z = D^2 b_\Omega, \quad DT_0 = I, \tag{2.32}$$

$$\frac{dj_z}{dz} = \Delta b_\Omega \circ T_z j_z, \quad j_0 = 1, \tag{2.33}$$

$$D^2 b_\Omega = D^2 b_\Omega \circ T_z [I + zD^2 b_\Omega], \tag{2.34}$$

$$D^2 b_\Omega \circ T_z = [I - zD^2 b_\Omega \circ T_z] D^2 b_\Omega, \tag{2.35}$$

$$[I + zD^2 b_\Omega]^{-1} = I - zD^2 b_\Omega \circ T_z. \tag{2.36}$$

2.5. Federer's decomposition of the integral along level sets. This theorem can be found in [**49**] or [**90**]. We quote the version given in [**47**]

THEOREM 2.4. *Let* $g\colon \mathbb{R}^N \to \mathbb{R}$ *be a Lipschitz continuous function such that*

$$|\nabla g| > 0 \quad a.e.$$

For any Lebesgue summable function $f\colon \mathbb{R}^N \to \mathbb{R}$, *we have the decomposition of the integral along the level curves of* g

$$\int_{\{g>t\}} f\,dx = \int_t^\infty \left(\int_{\{g=s\}} \frac{f}{|\nabla g|}\, dH_{N-1} \right) ds.$$

One application of the above theorem is to obtain the celebrated Steiner-Minkowski formula for the area of a dilated set. This formula was extended to sets of positive reach in 1959 by [**49**] who introduced the ***curvature measures*** for that purpose. For a set Ω of class $C^{1,1}$ with oriented distance function $b = b_\Omega$, pick $g = b$ and assume that $\mathrm{m}(\partial\Omega) = 0$. Then $|\nabla b| = 1$ almost everywhere and the formula becomes

$$\int_{\{b>t\}} f\,dx = \int_t^\infty \left(\int_{\{b=z\}} f\,dH_{N-1} \right) dz.$$

If $b \in C^{1,1}(U_h(\partial\Omega))$ for some $h > 0$ and $-h < s < t < h$, then $T_z(\partial\Omega) = \{x\colon b(x) = z\}$ and we can use the usual change of variable formula

$$\int_{T_z(\partial\Omega)} f\,dH_{N-1} = \int_{\partial\Omega} f \circ T_z \omega_z\, dH_{N-1},$$

where from the previous theorem the expression of the canonical density ω_z reduces to

$$\omega_z(x) = \left|{}^*(DT_z(x))^{-1}\nabla b(x)\right| \det(DT_z(x)) = j_z(x).$$

As a result for $-h < s < t < h$

$$\int_{s<b<t} f\,dx = \int_s^t \Big(\int_{\partial\Omega} f \circ T_z j_z \, dH_{N-1} \Big)\, dz \tag{2.37}$$
$$= \int_{\partial\Omega} \Big(\int_s^t f \circ T_z j_z \, dz \Big)\, dH_{N-1}.$$

For functions of the form

$$f = q \circ p_{\partial\Omega}, \quad q\colon \partial\Omega \to \mathbb{R},$$

for some H_{N-1}-measurable function q the composition is generally not Lebesgue measurable even when $p_{\partial\Omega}$ is continuous. However we have the following result.

THEOREM 2.5. *Let Ω be a subset of $\mathbb{R}^N$ of class $C^{1,1}$ with bounded boundary $\partial\Omega$ and let $h > 0$ be such that $b_\Omega \in C^{1,1}(U_h(\partial\Omega))$.*

(i) *The map*

$$(X, z) \mapsto T(X, z) \overset{\text{def}}{=} T_z(X) = X + z\nabla b(X)\colon \partial\Omega \times]-h, h[\to U_h(\partial\Omega),$$

and its inverse $T^{-1}(x) = \big(p(x), b(x)\big)$ are both Lipschitz continuous.

(ii) *For $p \geq 1$ the maps*

$$V \mapsto v = V \circ T\colon L^p\big(U_h(\partial\Omega)\big) \to L^p\big(\partial\Omega \times]-h, h[\big),$$
$$V \mapsto v = V \circ T\colon W^{1,p}\big(U_h(\partial\Omega)\big) \to W^{1,p}\big(\partial\Omega \times]-h, h[\big),$$

are continuous linear bijections.

(iii) *For $p \geq 1$*

$$\begin{matrix} h\colon \partial\Omega \to \mathbb{R} \\ H_{N-1} - \text{measurable} \end{matrix} \implies \begin{matrix} h \circ p_{\partial\Omega} \\ \text{Lebesgue measurable in } U_h(\partial\Omega) \end{matrix}$$
$$h \in L^p(\partial\Omega) \implies h \circ p_{\partial\Omega} \in L^p\big(U_h(\partial\Omega)\big).$$

PROOF. (i) Is obvious.

(ii) This can be proved by a density argument or by adapting a classical lemma which can be found for instance in [**80**, Lemme 3.1, p. 65].

(iii) An $(H_{N-1} \times \mathrm{m})$-measurable function $u\colon \partial\Omega \times]-h, h[\to \mathbb{R}$, $u \circ T^{-1}$ is Lebesgue measurable in $\mathbb{R}^N$. In particular for an H_{N-1}-measurable function $\tilde{u}$ on $\partial\Omega$, the map $u(X, z) = \tilde{u}(X)$ is $(H_{N-1} \times \mathrm{m})$-measurable. From part (ii) applied to $(u \circ T^{-1})(x) = \tilde{u}\big(p(x)\big) = (\tilde{u} \circ p)(x)$. □

3. Tangential Differential Calculus

3.1. Tangential derivatives. In this section the set Ω of class $C^{1,1}$ is fixed and we adopt the notation

$$\Gamma = \partial\Omega, \quad b = b_\Omega, \quad p = p_{\partial\Omega}.$$

Assume that Γ is bounded and that there exists $h > 0$ such that $b \in C^{1,1}\big(U_{2h}(\Gamma)\big)$. The ***projection*** of a point x onto Γ is given by

$$p(x) = x - b(x)\nabla b(x),$$

and the ***orthogonal projection operator*** onto the tangent plane $T_{p(x)}\Gamma$ is given by

$$P(x) = I - \nabla b(x)^* \nabla b(x).$$

The notation for the tangent plane should not be confused with the notation T_z introduced in Theorem 2.3 for the transformation of $\mathbb{R}^N$. Similarly D^2b can be considered as a transformation of $T_{p(x)}$ since

$$D^2b(x)\colon T_{p(x)} \to T_{p(x)}.$$

Moreover

$$Dp(x) = I - \nabla b(x)^* \nabla b(x) - bD^2b(x),$$

coincides with $P(x)$ on Γ.

3.1.1. *Definitions of the tangential gradient.* The classical way to define the tangential gradient of a scalar function $f\colon \Gamma \to \mathbb{R}$ is through an extension of f in a neighborhood of Γ and to show that it is independent of the choice of the extension.

DEFINITION 3.1 (Classical). Given an extension $F \in C^1\big(U_h(\Gamma)\big)$ of $f \in C^1(\Gamma)$, the ***tangential gradient*** of f is defined as

$$\nabla_\Gamma f \overset{\text{def}}{=} \nabla F\big|_\Gamma - \frac{\partial F}{\partial n} n.$$

The notation is quite natural and the subscript Γ of $\nabla_\Gamma f$ indicates that the gradient is with respect to the variable in Γ.

THEOREM 3.1. *Assume that $b \in C^{1,1}\big(U_{2h}(\Gamma)\big)$ and that $f \in C^1(\Gamma)$, then for any C^1-extension F of f*

(i) $\nabla_\Gamma f = (P\nabla F)\big|_\Gamma$ *and* $n \cdot \nabla_\Gamma f = \nabla b \cdot \nabla_\Gamma f = 0.$

(ii) $\nabla(f \circ p) = [I - bD^2b]\nabla_\Gamma f \circ p$ *and* $\nabla(f \circ p)\big|_\Gamma = \nabla_\Gamma f.$

In view of the second part of the theorem $f \circ p$ plays the role of a canonical extension and this suggests to start instead with the following definition.

DEFINITION 3.2 (Intrinsic). Given $b \in C^{1,1}\big(U_{2h}(\Gamma)\big)$ and $f \in C^1(\Gamma)$

$$\nabla_\Gamma f \overset{\text{def}}{=} \nabla(f \circ p)\big|_\Gamma. \tag{3.1}$$

In that case we get the analogue of the previous theorem.

THEOREM 3.2. *Given $b \in C^{1,1}\big(U_{2h}(\Gamma)\big)$ and $f \in C^1(\Gamma)$*

(i) $\nabla b \cdot \nabla(f \circ p) = 0$ *in* $U_h(\Gamma)$ *and* $n \cdot \nabla_\Gamma f = 0$ *on* Γ.

(ii) *Given any C^1-extension F of f*

$$\nabla F\big|_\Gamma - \frac{\partial F}{\partial n} n = (P\nabla F)\big|_\Gamma = \nabla_\Gamma f \text{ and } \nabla(f \circ p) = [I - bD^2b]\nabla_\Gamma f \circ p.$$

The extension $f \circ p$ has many useful properties. For instance its gradient is tangent to the level curves of b. We shall refer to $f \circ p$ as the ***canonical extension*** of a map $f\colon \Gamma \to \mathbb{R}$ in a neighborhood $U_h(\Gamma)$ of Γ.

3.1.2. *First order derivatives.* The ***tangential Jacobian matrix*** of a vector function $v \in C^1(\Gamma)^M$, $M \geq 1$, is defined in the same way as the gradient

$$\boxed{D_\Gamma v \overset{\text{def}}{=} D(v \circ p)\big|_\Gamma \text{ or } (D_\Gamma v)_{ij} = (\nabla_\Gamma v_i)_j} \tag{3.2}$$

If ${}^*v = (v_1, \ldots, v_M)$, then ${}^*D_\Gamma v = (\nabla_\Gamma v_1, \ldots, \nabla_\Gamma v_M)$, where $\nabla_\Gamma v_i$ is a column vector. From the previous theorems we can recover the definition from an extension $V \in C^1\big(U_h(\Gamma)\big)^M$ of v

$$\begin{aligned} {}^*D_\Gamma v &= (P\nabla V_1, \ldots, P\nabla V_M)\big|_\Gamma = (I - \nabla b^* \nabla b)^* DV\big|_\Gamma \\ &= {}^*DV\big|_\Gamma - \nabla b^* (DV\, \nabla b)\big|_\Gamma \end{aligned}$$

and

$$D_\Gamma v = DV\big|_\Gamma - DV\, n^* n = (DV\, P)\big|_\Gamma \tag{3.3}$$

Also

$$^*D(v \circ p) = [I - bD^2 b](\nabla_\Gamma v_1, \dots, \nabla_\Gamma v_M) \circ p = [I - bD^2 b]^* D_\Gamma v \circ p$$

and we have for the extension

$$D(v \circ p) = D_\Gamma v \circ p[I - bD^2 b] \tag{3.4}$$

Note that

$$D(v \circ p)\nabla b = 0 \quad \text{and} \quad D_\Gamma(v) n = 0 \tag{3.5}$$

For a vector function $V \in C^1(\Gamma)^N$ the ***tangential divergence*** is

$$\operatorname{div}_\Gamma v \stackrel{\text{def}}{=} \operatorname{div}(v \circ p)\big|_\Gamma \tag{3.6}$$

and it is easy to show that

$$\begin{aligned} \operatorname{div}_\Gamma v &= \operatorname{div}(v \circ p)\big|_\Gamma = \operatorname{tr} D(v \circ p)\big|_\Gamma = \operatorname{tr} D_\Gamma v, \\ \operatorname{div}_\Gamma v &= \operatorname{tr}\big[DV\big|_\Gamma - DV\, n^* n\big] = \operatorname{div} V\big|_\Gamma - DV\, n \cdot n. \end{aligned}$$

The ***tangential linear strain tensor*** of elasticity is given by

$$\varepsilon_\Gamma(v) \stackrel{\text{def}}{=} \frac{1}{2}(D_\Gamma v + {}^*D_\Gamma v) \quad \text{and} \quad \varepsilon_\Gamma(v) = \varepsilon(v \circ p)_{|\Gamma} \tag{3.7}$$

The tangential Jacobian matrix of the normal n is specially interesting since $n \circ p = n = \nabla b = \nabla b \circ p$ and $b \circ p = 0$. As a result

$$D_\Gamma(n) = D_\Gamma(\nabla b) = D^2 b\big|_\Gamma = {}^*D_\Gamma(\nabla b) = {}^*D_\Gamma(n) \tag{3.8}$$

and since it is symmetrical we also have

$$\varepsilon_\Gamma(n) = \varepsilon_\Gamma(\nabla b) = D^2 b\big|_\Gamma \tag{3.9}$$

Finally the ***tangential vectorial divergence*** of a matrix or tensor function A is defined as

$$(\vec{\operatorname{div}}_\Gamma A)_i = \operatorname{div}_\Gamma A_{i,\cdot} \tag{3.10}$$

3.1.3. *Second order derivatives.* The simplest one is the ***Laplace-Beltrami operator*** of a function $f \in C^2(\Gamma)$ which is defined as

$$\Delta_\Gamma f \stackrel{\text{def}}{=} \operatorname{div}_\Gamma(\nabla_\Gamma f) \tag{3.11}$$

Recall from Theorem 3.2 the identity

$$\nabla(f \circ p) = [I - bD^2 b]\nabla_\Gamma f \circ p.$$

Then

$$\begin{aligned} \operatorname{div}(\nabla_\Gamma f \circ p) &= \operatorname{div}\big(\nabla(f \circ p)\big) + \operatorname{div}(bD^2 b \nabla_\Gamma f \circ p), \\ &= \Delta(f \circ p) + b \operatorname{div}(D^2 b \nabla_\Gamma f \circ p) + (D^2 b \nabla_\Gamma f \circ p) \cdot \nabla b, \\ &= \Delta(f \circ p) + b \operatorname{div}(D^2 b \nabla_\Gamma f \circ p), \end{aligned}$$

and by taking restrictions to Γ

$$\boxed{\Delta_\Gamma f = \Delta(f \circ p)\big|_\Gamma}$$

Finally the ***tangential Hessian matrix*** of second order derivatives is defined as

$$\boxed{D^2_\Gamma f \overset{\text{def}}{=} D_\Gamma(\nabla_\Gamma f)} \tag{3.12}$$

The Hessian matrix is not symmetrical and does not coincide with the restriction of the Hessian matrix of the canonical extension. Specifically

$$\begin{aligned} D^2(f\circ p) &= D\big(\nabla(f\circ p)\big) = D\big([I - bD^2b]\nabla_\Gamma f \circ p\big), \\ D^2(f\circ p)\big|_\Gamma &= D_\Gamma(\nabla_\Gamma f) - D^2 b \nabla_\Gamma f^* \nabla b = D^2_\Gamma f - D^2 b \nabla_\Gamma f^* n. \end{aligned}$$

Of course since $D^2(f\circ p)$ is symmetrical we also have

$$\begin{aligned} D^2(f\circ p) &= {}^*D\big(\nabla(f\circ p)\big) = {}^*D\big([I - bD^2b]\nabla_\Gamma f \circ p\big), \\ D^2(f\circ p)\big|_\Gamma &= {}^*D_\Gamma(\nabla_\Gamma f) - \nabla b^*(D^2 b \nabla_\Gamma f) = {}^*D^2_\Gamma f - n^*(D^2 b\nabla_\Gamma f). \end{aligned}$$

As a final result we have the following identity

$$\boxed{D^2_\Gamma f - (D^2 b \nabla_\Gamma f)^* n = D^2(f\circ p)\big|_\Gamma = {}^*D^2_\Gamma f - n^*(D^2 b \nabla_\Gamma f)} \tag{3.13}$$

since by definition ${}^*D^2_\Gamma f = {}^*(D^2_\Gamma f)$. So the Hessian and its transposed differ by first derivative terms as is well-known in Differential Geometry. Note that $D^2 b\nabla_\Gamma f = {}^*D_\Gamma(\nabla_\Gamma f)n = {}^*D^2_\Gamma(f)n$ and that we also can write

$$\boxed{D^2_\Gamma f - {}^*D^2_\Gamma(f)[n^*n] = D^2(f\circ p)\big|_\Gamma = {}^*D^2_\Gamma(f) - [n^*n]D^2_\Gamma(f)} \tag{3.14}$$

3.1.4. *Some useful identities.* The fact that we can go back to the differential calculus in $\mathbb{R}^N$ considerably simplifies the differential calculus on submanifolds. As a last illustration we give a few useful formulae for a function f and vector functions v and w on Γ. By using extensions and by taking restrictions to Γ

$$D_\Gamma(fv) = v^*\nabla_\Gamma f + f D_\Gamma v, \tag{3.15}$$

$$\text{div}_\Gamma(fv) = \nabla_\Gamma f \cdot v + f\text{div}_\Gamma v, \tag{3.16}$$

$$\nabla_\Gamma(v\cdot w) = {}^*D_\Gamma(v)w + {}^*D_\Gamma(w)v. \tag{3.17}$$

In particular we shall often use formula (3.17) with $w = n$

$$\nabla_\Gamma(v\cdot n) = {}^*D_\Gamma(v)n + {}^*D^2 b v, \tag{3.18}$$

or by introducing the notation

$$\boxed{v_n \overset{\text{def}}{=} v\cdot n, \quad v_\Gamma \overset{\text{def}}{=} v - v_n n = Pv}$$

$$\boxed{\nabla_\Gamma(v_n) = {}^*D_\Gamma(v)n + {}^*D^2 b v_\Gamma} \tag{3.19}$$

since $D^2 b\nabla b = 0$. For more details see also [**36, 41**].

3.2. Projected tangential derivatives. In the theory of shells equations are expressed in tensor forms with respect to the first $(N-1)$ tangent vectors of the covariant or the contravariant bases. The projected operators will be used in the decomposition of $D_\Gamma(v)$ and $\varepsilon_\Gamma(v)$ into their normal and tangent images. First recall from (3.5) that by definition of $D_\Gamma(v)$

$$D_\Gamma(v)n = 0 \implies D_\Gamma(v)P = D_\Gamma(v). \tag{3.20}$$

and that $D_\Gamma(v)$ maps $T_x\Gamma$ into all of $\mathbb{R}^N$. Even in the case of a tangent vector function v, the range of $D_\Gamma(v)$ has a normal component. Let $v_n = v \cdot n$ and $v_\Gamma = v - v_n n$ denote the respective normal component and tangential part of v. For any vector V

$$n \cdot D_\Gamma(v)V = {}^*D_\Gamma(v)n \cdot V = [\nabla_\Gamma v_n - D^2 b v_\Gamma] \cdot V$$

using identity (3.19). So even for tangent vectors V and v, the term $-D^2bv_\Gamma$ is generally not zero. It is therefore natural to introduce the ***projected tangential Jacobian matrix*** and the ***projected tangential linear strain tensor***

$$D_\Gamma^P(v) \overset{\text{def}}{=} PD_\Gamma(v)P, \quad \varepsilon_\Gamma^P(v) \overset{\text{def}}{=} P\varepsilon_\Gamma(v)P. \tag{3.21}$$

By definition it is easy to verify that

$$D_\Gamma^P(v) = PD_\Gamma(v), \quad \varepsilon_\Gamma^P(v) = \frac{1}{2}\left(D_\Gamma^P(v) + {}^*D_\Gamma^P(v)\right)$$

and that

$$D_\Gamma^P(v) = [I - n^*n]D_\Gamma(v) = D_\Gamma(v) - n^*\left({}^*D_\Gamma(v)n\right) \tag{3.22}$$

$$\implies \boxed{D_\Gamma(v) = D_\Gamma^P(v) + n^*\left({}^*D_\Gamma(v)n\right)} \tag{3.23}$$

Then

$$\begin{aligned} D_\Gamma^P(v) &= D_\Gamma(v) - n^*(\nabla_\Gamma v_n - D^2 b v_\Gamma) \\ &= D_\Gamma(v_\Gamma) + n^*\nabla_\Gamma v_n + v_n D^2 b - n^*(\nabla_\Gamma v_n - D^2 b v_\Gamma), \end{aligned}$$

and

$$\boxed{\begin{aligned} D_\Gamma^P(v) &= D_\Gamma(v_\Gamma) + n^*(D^2 b v_\Gamma) + v_n D^2 b \\ D_\Gamma^P(v_\Gamma) &= D_\Gamma(v_\Gamma) + n^*(D^2 b v_\Gamma) \\ D_\Gamma^P(v) &= D_\Gamma^P(v_\Gamma) + v_n D^2 b \end{aligned}} \tag{3.24}$$

The same considerations now yield

$$\boxed{\begin{aligned} \varepsilon_\Gamma(v) &= \varepsilon_\Gamma^P(v) + \frac{1}{2}\left[{}^*D_\Gamma(v)n^*n + n^*({}^*D_\Gamma(v)n)\right] \\ &= \varepsilon_\Gamma^P(v) + \left[\varepsilon_\Gamma(v)n^*n + n^*(\varepsilon_\Gamma(v)n)\right] \\ \varepsilon_\Gamma^P(v) &= \varepsilon_\Gamma(v_\Gamma) + \frac{1}{2}\left[(D^2 b v_\Gamma)^*n + n^*(D^2 b v_\Gamma)\right] + v_n D^2 b \\ \varepsilon_\Gamma^P(v_\Gamma) &= \varepsilon_\Gamma(v_\Gamma) + \frac{1}{2}\left[(D^2 b v_\Gamma)^*n + n^*(D^2 b v_\Gamma)\right] \\ \varepsilon_\Gamma^P(v) &= \varepsilon_\Gamma^P(v_\Gamma) + v_n D^2 b \end{aligned}} \tag{3.25}$$

Finally

$$\boxed{D_\Gamma(v) = D_\Gamma^P(v) + n^*(\nabla_\Gamma v_n - D^2 b v_\Gamma)} \tag{3.26}$$

and the following identities

$$\begin{aligned} D_\Gamma(u) \cdot\cdot D_\Gamma(v) &= D_\Gamma^P(u) \cdot\cdot D_\Gamma^P(v) + ({}^*D_\Gamma^P(u)n) \cdot ({}^*D_\Gamma^P(v)n) \\ &= D_\Gamma^P(u) \cdot\cdot D_\Gamma^P(v) + (\nabla_\Gamma u_n - D^2 b u_\Gamma) \cdot (\nabla_\Gamma v_n - D^2 b v_\Gamma) \end{aligned}$$

and

$$\boxed{\begin{aligned} \|D_\Gamma(u)\|^2 &= \|D_\Gamma^P(u)\|^2 + |{}^*D_\Gamma(u)n|^2 \\ \|D_\Gamma(u)\|^2 &= \|D_\Gamma^P(u)\|^2 + |\nabla_\Gamma u_n - D^2 b u_\Gamma|^2 \end{aligned}} \tag{3.27}$$

$$\boxed{\operatorname{div}_\Gamma v = \operatorname{tr} D_\Gamma(v) = \operatorname{tr} D_\Gamma^P(v)} \tag{3.28}$$

Again by direct computation

$$\boxed{\varepsilon_\Gamma(v) = \varepsilon_\Gamma^P(v) + \frac{1}{2}\left[n^*(\nabla_\Gamma v_n - D^2 b v_\Gamma) + (\nabla_\Gamma v_n - D^2 b v_\Gamma)^* n\right]}$$

and we have a series of identities

$$\begin{aligned} \varepsilon_\Gamma(u) \cdot\cdot \varepsilon_\Gamma(v) &= \varepsilon_\Gamma^P(u) \cdot\cdot \varepsilon_\Gamma^P(v) + 2(\varepsilon_\Gamma(u)n) \cdot (\varepsilon_\Gamma(v)n) \\ &= \varepsilon_\Gamma^P(u) \cdot\cdot \varepsilon_\Gamma^P(v) + \frac{1}{2}(\nabla_\Gamma u_n - D^2 b u_\Gamma) \cdot (\nabla_\Gamma v_n - D^2 b v_\Gamma), \end{aligned}$$

and

$$\boxed{\begin{aligned} \|\varepsilon_\Gamma u\|^2 &= \|\varepsilon_\Gamma^P u\|^2 + 2\left|\varepsilon_\Gamma(u)n\right|^2 \\ &= \|\varepsilon_\Gamma^P u\|^2 + \frac{1}{2}|\nabla_\Gamma u_n - D^2 b u_\Gamma|^2 \end{aligned}} \tag{3.29}$$

$$\boxed{\operatorname{div}_\Gamma v = \operatorname{tr} \varepsilon_\Gamma v = \operatorname{tr} \varepsilon_\Gamma^P v} \tag{3.30}$$

In both cases $D_\Gamma^P v$ and $\varepsilon_\Gamma^P v$ map $T_x\Gamma$ into $T_x\Gamma$ and they can be considered as extensions to $\mathbb{R}^N$.

We can do the same thing with the tangential Hessian matrix

$$\begin{aligned} D_\Gamma^2 f &= P D_\Gamma^2 f + n^* n D_\Gamma^2 f \\ &= P D_\Gamma^2 f + n^*({}^*D_\Gamma^2 f n) = P D_\Gamma^2 f - n^*(D^2 b \nabla_\Gamma f), \end{aligned}$$

and finally

$$P D_\Gamma^2 f = D_\Gamma^2 f + n^*(D^2 b \nabla_\Gamma f) = {}^*\left(D_\Gamma^2 f + n^*(D^2 b \nabla_\Gamma f)\right), \tag{3.31}$$

and this operator is symmetrical in view of identity (3.13).

This definition of projected operator can be extended to matrix or tensor valued functions A defined in the neighborhood $S_h(\Gamma)$ of Γ since the projection operator $P = I - \nabla b^* \nabla b$ is defined there. By defining

$$A^P \overset{\text{def}}{=} PAP, \tag{3.32}$$

A can be decomposed as follows

$$A = A^P + (PA\nabla b)^* \nabla b + \nabla b^*(P^* A \nabla b) + (A\nabla b \cdot \nabla b)\nabla b^* \nabla b. \tag{3.33}$$

In particular for two such functions A and B

$$A \cdot\cdot B = A^P \cdot\cdot B^P + (PA\nabla b)\cdot(PB\nabla b) + (P^*A\nabla b)\cdot(P^*B\nabla b) + (A\nabla b\cdot\nabla b)(B\nabla b\cdot\nabla b), \tag{3.34}$$

and

$$\|A\|^2 = \|A^P\|^2 + |PA\nabla b|^2 + |P^*A\nabla b|^2 + |A\nabla b\cdot\nabla b|^2. \tag{3.35}$$

3.3. Covariant and tangential derivatives. The classical objects of Differential Geometry are usually intrinsic even if they are introduced through local bases. So it is not too surprising that the intrinsic tangential differential operators defined in the last section coincide with the ones introduced via covariant or contravariant bases. We follow the notation and definitions given in §§ 2.1 and 2.2 of § 2. Recall that $\{e_1, \dots, e_N\}$ is the standard orthonormal basis in $\mathbb{R}^N$ such that $(e_i)_j = \delta_{ij}$.

3.3.1. *Partial derivative and fundamental forms.* Denote by $F = f \circ p$ the extension of a C^1-function $f\colon \Gamma \to \mathbb{R}$ in a neighborhood of Γ. By definition the *partial derivative* of f is given by

$$\boxed{f_{,\alpha} \overset{\text{def}}{=} \frac{\partial}{\partial \xi^\alpha}(f\circ\Phi), \quad f_{,N} \overset{\text{def}}{=} 0} \tag{3.36}$$

since $f\circ\Phi$ is independent of the normal displacement to the submanifold Γ. Observe that

$$f\circ\Phi = (f\circ p)\circ\Phi = F\circ\Phi,$$

and consequently

$$\frac{\partial}{\partial\xi^\alpha}(f\circ\Phi) = (\nabla F\circ\Phi)\cdot\frac{\partial\Phi}{\partial\xi^\alpha} = \big[(\nabla F\circ p)\circ\Phi\big]\cdot a_\alpha.$$

But

$$\nabla F\circ p = \nabla(f\circ p)\circ p = \nabla_\Gamma f\circ p,$$

and

$$\frac{\partial}{\partial\xi^\alpha}(f\circ\Phi) = [\nabla_\Gamma f\circ p\circ\Phi]\cdot a_\alpha = (\nabla_\Gamma f\circ\Phi)\cdot a_\alpha.$$

In the sequel it will be convenient to use the notation a_i for both a_i and $a_i\circ\Phi^{-1}$ whenever no confusion arises. Hence

$$\boxed{f_{,\alpha}\circ\Phi^{-1} = \nabla_\Gamma f\cdot a_\alpha} \tag{3.37}$$

Moreover

$$\boxed{f_{,N}\circ\Phi^{-1} = 0 = \nabla_\Gamma f\cdot a_N} \tag{3.38}$$

since $\nabla_\Gamma f\cdot n = 0$. Therefore

$$\nabla_\Gamma f = (f_{,\alpha}a^\alpha)\circ\Phi^{-1}, \tag{3.39}$$

can be viewed as an extension of the partial derivative from the tangent space to the ambient space $\mathbb{R}^N$.

This readily extends to C^1 vector functions $v\colon \Gamma\to\mathbb{R}^N$ by using the extension $V = v\circ p$ in the neighborhood $U_h(\Gamma)$ of Γ

$$\boxed{v_{,\alpha} \overset{\text{def}}{=} \frac{\partial}{\partial\xi^\alpha}(v\circ\Phi), \quad v_{,N} \overset{\text{def}}{=} 0} \tag{3.40}$$

By the same technique

$$(3.41) \qquad \boxed{v_{,i} \circ \Phi^{-1} = D_\Gamma(v)a_i \quad \text{and} \quad D_\Gamma(v) = (v_{,\alpha}a^\alpha) \circ \Phi^{-1}}$$

As an illustration of the above identity, consider the *second fundamental form*

$$(3.42) \qquad b_{\alpha\beta} \overset{\text{def}}{=} -a_\beta \cdot a_{N,\alpha}.$$

It is not to be confused with the oriented distance function b. By definition

$$a_{N,\alpha} = D_\Gamma(a_N \circ \Phi^{-1}) \circ \Phi a_\alpha = -\big(D_\Gamma(\nabla b)\big) \circ \Phi a_\alpha.$$

But $\nabla b = \nabla b \circ p = n \circ p$

$$D_\Gamma(n) = D_\Gamma(\nabla b) = D(\nabla b \circ p)\big|_\Gamma = D(\nabla b)\big|_\Gamma = D^2 b\big|_\Gamma,$$

and $a_{N,\alpha} \circ \Phi^{-1} = -D^2 b a_\alpha$. Finally

$$(3.43) \qquad b_{\alpha\beta} \circ \Phi^{-1} = a_\beta \cdot D^2 b a_\alpha.$$

For simplicity we shall write $D^2 b$ for $D^2 b \circ \Phi$ whenever no confusion arises. Note that the definition of $b_{\alpha\beta}$ readily extends from $T_x\Gamma$ to $\mathbb{R}^N$ as follows

$$(3.44) \qquad \boxed{b_{ij} \overset{\text{def}}{=} D^2 b a_i \cdot a_j}$$

and since $D^2 b \nabla b = 0$ $b_{Ni} = b_{iN} = 0$. Therefore the second fundamental form coincides with the bilinear form associated with the matrix $D^2 b$.

In a similar way consider the *third fundamental form*

$$(3.45) \qquad c_{\alpha\beta} \overset{\text{def}}{=} b_\alpha^\lambda b_{\lambda\beta},$$

where

$$(3.46) \qquad b_i^j \overset{\text{def}}{=} a^j \cdot D^2 b a_i.$$

From (3.43)

$$\begin{aligned} a^\lambda \cdot a^\beta b_{\beta\alpha} &= a^\lambda \cdot a^\beta a_\beta \cdot D^2 b a_\alpha \\ &= a^\lambda \cdot [D^2 b a_\alpha - D^2 b a_\alpha \cdot a_N a^N]. \end{aligned}$$

But $D^2 b a_N \circ \Phi^{-1} = -D^2 b \nabla b = 0$ and necessarily

$$(3.47) \qquad b_\alpha^\lambda = a^\lambda \cdot D^2 b a_\alpha,$$

which can also be extended to indices equal to N. Therefore

$$c_{\alpha\beta} = a^\lambda \cdot D^2 b a_\alpha a_\lambda \cdot D^2 b a_\beta = D^2 b a_\alpha \cdot D^2 b a_\beta,$$

and

$$(3.48) \qquad c_{\alpha\beta} = a_\beta \cdot (D^2 b)^2 a_\alpha.$$

By extending $c_{\alpha\beta}$ to $\mathbb{R}^N$

$$(3.49) \qquad \boxed{c_{ij} = a_i \cdot (D^2 b)^2 a_j \quad \text{and} \quad c_{Ni} = c_{iN} = 0}$$

So $D^2 b$ and $(D^2 b)^2$ which map $T_x\Gamma$ onto $T_x\Gamma$ are extension of the second and third fundamental forms to $\mathbb{R}^N$. It is also interesting to observe that the orthogonal projection operator P is the extension to $\mathbb{R}^N$ of the first fundamental form

$$a_{\alpha\beta} = a_\alpha \cdot a_\beta.$$

3.3.2. *Christoffel symbols.* By definition

$$\boxed{\Gamma^{\alpha}_{\beta\gamma} \overset{\text{def}}{=} a^{\alpha} \cdot a_{\beta,\gamma}} \tag{3.50}$$

and if we recall that for any vector function v on Γ, $v_{,N} = 0$, this extends to $\mathbb{R}^N$

$$\Gamma^{i}_{jk} \overset{\text{def}}{=} a^{i} \cdot a_{j,k}. \tag{3.51}$$

Hence from (3.41)

$$\Gamma^{i}_{jk} = a^{i} \cdot D_{\Gamma}(a_j \circ \Phi^{-1}) \circ \Phi a_k \quad \text{and} \quad \Gamma^{i}_{jN} = 0. \tag{3.52}$$

For simplicity we shall denote $D_{\Gamma}(a_j \circ \Phi^{-1}) \circ \Phi$ by $D_{\Gamma} a_j$. Moreover

$$a^{i} \cdot a_j = \delta_{ij} \implies \frac{\partial}{\partial \xi^{\gamma}} (a^{i} \cdot a_j) = 0,$$

and for all γ $D_{\Gamma}(a^{i}) a_{\gamma} \cdot a_j + D_{\Gamma}(a_j) a_{\gamma} \cdot a^{i} = 0$, which yields

$$\boxed{\begin{aligned} \Gamma^{i}_{j\gamma} &= a^{i} \cdot D_{\Gamma}(a_j) a_{\gamma} \\ &= -D_{\Gamma}(a^{i}) a_{\gamma} \cdot a_j = -{}^{*}D_{\Gamma}(a^{i}) a_j \cdot a_{\gamma} \end{aligned}} \tag{3.53}$$

For a C^2-mapping

$$a_{\gamma,\beta} = \frac{\partial}{\partial \xi^{\beta}} \frac{\partial}{\partial \xi^{\gamma}} \Phi = \frac{\partial}{\partial \xi^{\gamma}} \frac{\partial}{\partial \xi^{\beta}} \Phi = a_{\beta,\gamma}, \tag{3.54}$$

and $\Gamma^{i}_{\beta\gamma} = \Gamma^{i}_{\gamma\beta}$. Finally from (3.51), (3.52) and (3.53)

$$\begin{aligned} \Gamma^{i}_{Nj} &= -D_{\Gamma}(a^{i}) a_j \cdot a_N = a_j \cdot {}^{*}D_{\Gamma}(a^{i}) n \circ \Phi \\ &= a_j \cdot \left[-D^2 b a^{i} + \nabla_{\Gamma}(a^{i} \circ \Phi^{-1} \cdot n) \circ \Phi \right]. \end{aligned}$$

But $a^{i} \circ \Phi^{-1} \cdot n = -\delta_{i,N}$ is constant and

$$\Gamma^{i}_{Nj} = -a^{i} \cdot D^2 b a_j = -b^{i}_{j}. \tag{3.55}$$

3.3.3. *Covariant derivatives.* We cover the vector case, but all this readily extends to tensors of higher order. Recall that for $v \colon \Gamma \to \mathbb{R}^N$

$$\boxed{v_{\alpha}\big|_{\gamma} \overset{\text{def}}{=} v_{\alpha,\gamma} - \Gamma^{\lambda}_{\alpha\gamma} v_{\lambda}, \quad v^{\alpha}\big|_{\gamma} \overset{\text{def}}{=} v^{\alpha}_{,\gamma} + \Gamma^{\alpha}_{\lambda\gamma} v^{\lambda}} \tag{3.56}$$

For any two vector functions $u, v \colon \Gamma \to \mathbb{R}^N$

$$\begin{aligned} &\frac{\partial}{\partial \xi^{\gamma}} (u \circ \Phi \cdot v \circ \Phi) = (u \circ \Phi)_{,\gamma} \cdot v \circ \Phi + u \circ \Phi \cdot (v \circ \Phi)_{,\gamma}, \\ &\frac{\partial}{\partial \xi^{\gamma}} (u \circ \Phi \cdot v \circ \Phi) = \left({}^{*}D_{\Gamma}(u) v + {}^{*}D_{\Gamma}(v) u \right) \circ \Phi \cdot a_{\gamma}. \end{aligned} \tag{3.57}$$

Again for simplicity we shall write $D_{\Gamma}(u)$ instead of $D_{\Gamma}(u) \circ \Phi$. Hence

$$v_{\alpha,\gamma} \overset{\text{def}}{=} \frac{\partial}{\partial \xi^{\gamma}} (v \circ \Phi \cdot a_{\alpha}) = D_{\Gamma}(v) a_{\gamma} \cdot a_{\alpha} + v \cdot D_{\Gamma}(a_{\alpha}) a_{\gamma},$$

and from (3.55)

$$\begin{aligned} v_\alpha\big|_\gamma &= D_\Gamma(v)a_\gamma \cdot a_\alpha + v \cdot D_\Gamma(a_\alpha)a_\gamma - a^\lambda \cdot D_\Gamma(a_\alpha)a_\gamma v_\lambda \\ &= D_\Gamma(v)a_\gamma \cdot a_\alpha + (v - a^\lambda v_\lambda) \cdot D_\Gamma(a_\alpha)a_\gamma \\ &= D_\Gamma(v)a_\gamma \cdot a_\alpha + a^N v_N \cdot D_\Gamma(a_\alpha)a_\gamma \\ &= D_\Gamma(v)a_\gamma \cdot a_\alpha - {}^*D_\Gamma(a_\alpha)\nabla b \circ \Phi \cdot a_\gamma v_N. \end{aligned}$$

Recall from identity (3.17) that

$${}^*D_\Gamma(a_\alpha)\nabla b = \nabla_\Gamma(a_\alpha \circ \Phi^{-1} \cdot \nabla b) - D^2 b a_\alpha = -D^2 b a_\alpha.$$

Finally

$$\boxed{\begin{aligned} v_\alpha\big|_\gamma &= D_\Gamma^P(v)a_\gamma \cdot a_\alpha + D^2 b a_\alpha \cdot a_\gamma v_N \\ &= \left[D_\Gamma^P(v) + D^2 b v_N\right] a_\gamma \cdot a_\alpha \end{aligned}} \tag{3.58}$$

But

$$D_\Gamma^P(v) = D_\Gamma^P(v_\Gamma) + D_\Gamma^P(v_n n) = D_\Gamma^P(v_\Gamma) + v_n D^2 b$$

and since $v_N = -v_n$

$$\boxed{v_\alpha\big|_\gamma = D_\Gamma^P(v_\Gamma)a_\gamma \cdot a_\alpha \quad \text{and} \quad v_\alpha\big|_\gamma - v_N b_{\alpha\gamma} = D_\Gamma^P(v)a_\gamma \cdot a_\alpha} \tag{3.59}$$

Similarly we can show that

$$\boxed{v^\alpha\big|_\gamma = D_\Gamma^P(v_\Gamma)a_\gamma \cdot a^\alpha} \tag{3.60}$$

$$\boxed{v^\alpha\big|_\gamma - b_\gamma^\alpha v_N = D_\Gamma^P(v_\Gamma)a_\gamma \cdot a^\alpha} \tag{3.61}$$

From the above identities

$$\boxed{\varepsilon_\Gamma^P(v_\Gamma)a_\alpha \cdot a_\gamma = \frac{1}{2}\left(v_\alpha\big|_\gamma + v_\gamma\big|_\alpha\right)} \tag{3.62}$$

$$\boxed{\varepsilon_\Gamma^P(v)a_\alpha \cdot a_\gamma = \frac{1}{2}\left(v_\alpha\big|_\gamma + v_\gamma\big|_\alpha\right) - b_{\alpha\gamma}v_N} \tag{3.63}$$

$$\boxed{\varepsilon_\Gamma^P(v)a^\alpha \cdot a_\gamma = \frac{1}{2}\left(v^\alpha\big|_\gamma + v^\gamma\big|_\alpha\right) - \frac{1}{2}(b_\gamma^\alpha + b_\alpha^\gamma)v_N} \tag{3.64}$$

3.3.4. *A few useful formulas.* In the theory of shells some identities will repeatedly occur. We summarize them below.

THEOREM 3.3. *For all u and v in $C^1(\Gamma)^N$*

$$(D^2 b D_\Gamma^P(v))a_\alpha \cdot a_\beta = b_\beta^\gamma\left[v_\gamma\big|_\alpha - b_{\alpha\gamma}v_N\right] = b_\beta^\gamma v_\gamma\big|_\alpha - c_{\alpha\beta}v_N, \tag{3.65}$$

$$(D_\Gamma^P(v)D^2 b)a_\alpha \cdot a_\beta = b_\alpha^\gamma\left[v_\beta\big|_\gamma - b_{\beta\gamma}v_N\right] = b_\alpha^\gamma v_\beta\big|_\gamma - c_{\alpha\beta}v_N, \tag{3.66}$$

where $c_\alpha^\beta = c_{\alpha\gamma} a_\beta^\gamma$, $c^{\alpha\beta} = c_{\alpha\gamma} a^{\gamma\beta}$. *Moreover*

(3.67) $$a_\beta \cdot \varepsilon_\Gamma^P(u) a^\beta = \operatorname{tr} \varepsilon_\Gamma^P(u) = \operatorname{tr} \varepsilon_\Gamma(u) = \operatorname{div}_\Gamma u,$$

(3.68) $$\begin{aligned} \varepsilon_\Gamma^P(u) \cdot\cdot\, \varepsilon_\Gamma^P(v) &= a^\alpha \cdot \varepsilon_\Gamma(u) a_\beta \varepsilon_\Gamma(v) a^\beta \cdot a_\alpha \\ &= \varepsilon_\Gamma(u) \cdot\cdot\, \varepsilon_\Gamma(v) - 2\varepsilon_\Gamma(u) n \cdot \varepsilon_\Gamma(v) n \\ &= \varepsilon_\Gamma(u) \cdot\cdot\, \varepsilon_\Gamma(v) - \frac{1}{2}(\nabla_\Gamma u_n - D^2 b u_\Gamma) \cdot (\nabla_\Gamma v_n - D^2 b v_\Gamma). \end{aligned}$$

PROOF. (i)For identity (3.65)

$$\begin{aligned} \big(D^2 b D_\Gamma(v)\big) a_\alpha \cdot a_\beta &= D_\Gamma(v) a_\alpha \cdot D^2 b a_\beta = D_\Gamma(v) a_\alpha \cdot a_\gamma a^\gamma \cdot D^2 b a_\beta \\ &= b_\beta^\gamma \big[v_\gamma\big|_\alpha - b_{\alpha\gamma} v_N \big] = b_\beta^\gamma v_\gamma\big|_\alpha - c_{\alpha\beta} v_N, \\ (D_\Gamma(v) D^2 b) a_\alpha \cdot a_\beta &= D^2 b a_\alpha \cdot {}^* D_\Gamma(v) a_\beta = D^2 b a_\alpha \cdot a^\gamma a_\gamma \cdot {}^* D_\Gamma(v) a_\beta \\ &= b_\alpha^\gamma \big[v_\beta\big|_\gamma - b_{\beta\gamma} v_N \big] = b_\alpha^\gamma v_\beta\big|_\gamma - c_{\alpha\beta} v_N. \end{aligned}$$

(ii) For the other formulas recall that the greek indices are summed from 1 to $N-1$. By definition in § 2

$$\begin{aligned} \operatorname{tr} \varepsilon_\Gamma(u) &= e_i \cdot \varepsilon_\Gamma(u) e_i = e_i \cdot a^j a_j \cdot \varepsilon_\Gamma(u) a^\ell a_\ell \cdot e_i \\ &= a_j \cdot \varepsilon_\Gamma(u) a^\ell a^j \cdot e_i e_i \cdot a_\ell \\ &= a_j \cdot \varepsilon_\Gamma(u) a^j = a_\beta \cdot \varepsilon_\Gamma(u) a^\beta = \operatorname{tr} \varepsilon_\Gamma^P(u), \end{aligned}$$

since $a_N \cdot \varepsilon_\Gamma(u) a^N = n \cdot \varepsilon_\Gamma(u) n = 0$. For (3.68) (cf. (2.7) and (2.9) in § 2)

$$\begin{aligned} \varepsilon_\Gamma(u) \cdot\cdot\, \varepsilon_\Gamma(v) &= \varepsilon_\Gamma(u) e_i \cdot e_j \varepsilon_\Gamma(v) e_i \cdot e_j \\ &= \big[\varepsilon_\Gamma(u) a_k a^k \cdot e_i\big] \cdot (a^\ell a_\ell \cdot e_j) [\varepsilon_\Gamma(v) a^m a_m \cdot e_i] \cdot (a_q a^q \cdot e_j) \\ &= \varepsilon_\Gamma(a_k) \cdot a^\ell \varepsilon_\Gamma(v) a^m \cdot a_q a^k \cdot e_i a_\ell \cdot e_j a_m \cdot e_i a^q \cdot e_j \\ &= \varepsilon_\Gamma(a_k) \cdot a^\ell \varepsilon_\Gamma(v) a^m \cdot a_q a^k \cdot a_m a_\ell \cdot a^q \\ &= \varepsilon_\Gamma(u) a_m \cdot a^\ell \varepsilon_\Gamma(v) a^m \cdot a_\ell. \end{aligned}$$

But

$$\begin{aligned} &\varepsilon_\Gamma(u) a_N \cdot a^\ell \varepsilon_\Gamma(v) a^N \cdot a_\ell + \varepsilon_\Gamma(u) a_m \cdot a^N \varepsilon_\Gamma(v) a^m \cdot a_N \\ &\qquad = \varepsilon_\Gamma(u) n \cdot a^\ell a_\ell \cdot\cdot\, \varepsilon_\Gamma(v) n + \varepsilon_\Gamma(u) n \cdot a_m a^m \cdot \varepsilon_\Gamma(v) n \\ &\qquad\qquad = 2\varepsilon_\Gamma(u) n \cdot \varepsilon_\Gamma(v) n, \end{aligned}$$

and

$$\begin{aligned} \varepsilon_\Gamma(u) \cdot\cdot\, \varepsilon_\Gamma(v) &= \varepsilon_\Gamma(u) a_\beta \cdot a^\alpha \varepsilon_\Gamma(v) a^\beta \cdot a_\alpha + 2\varepsilon_\Gamma(u) n \cdot \varepsilon_\Gamma(v) n \\ &= \varepsilon_\Gamma^P(u) \cdot\cdot\, \varepsilon_\Gamma^P(v) + 2\varepsilon_\Gamma(u) n \cdot \varepsilon_\Gamma(v) n. \end{aligned}$$

□

4. Sobolev Spaces and Some Related Topics

4.1. Introduction. In this section we give some basic results for Sobolev spaces of scalar and vector functions defined in an open domain ω of the submanifold Γ of codimension one which is the boundary of a domain Ω of class $C^{1,1}$ in $\mathbb{R}^N$. This is the minimum smoothness for which there is a complete correspondence between properties described by local maps and global intrinsic ones. For less smooth submanifolds of $\mathbb{R}^N$ or manifolds which are not embedded in $\mathbb{R}^N$ the situation is more complex. For C^∞ manifolds the reader is referred for instance [**6**], [**63**, p. 142], and [**64**, p. 471] and related works.

Many results will be established through local maps by using the diffeomorphism of Theorem 2.1 in of § 2, but direct intrinsic proofs will also be given whenever available. For this diffeomorphism explicit expressions are available for the covariant and contravariant bases as a function of the oriented distance function. Most global intrinsic properties will follow from Theorem 2.5 in § 2 by going to a small neighborhood of Γ and inferring properties on Γ from those of the canonical extension in the neighborhood as was done for the tangential calculus of § 3.

4.2. Spaces of continuous functions. If α is a subset of Γ, the space of continuous functions from α to $\mathbb{R}$ will be denoted $C(\alpha)$. For an open subset ω of Γ we shall consider in addition to the spaces $C(\omega)$ and $C(\overline{\omega})$ the spaces $C^n(\omega)$ and $C^n(\overline{\omega})$ of n-times continuously differentiable functions

$$C^1(\omega) \stackrel{\text{def}}{=} \{f \in C(\omega) \colon \nabla_\Gamma f \in C(\omega)^N\}, \tag{4.1}$$

$$C^n(\omega) \stackrel{\text{def}}{=} \{f \in C^{n-1}(\omega) \colon \nabla_\Gamma f \in C^{n-1}(\omega)^N\}, \quad n \geq 2. \tag{4.2}$$

The spaces $C^n(\overline{\omega})$ are defined in a similar way with $\overline{\omega}$ in place of ω. It is important to observe that the smoothnes of a function in ω is limited by the smoothness of Γ.

In the remainder of this paper we shall make the following assumption.

ASSUMPTION 4.1. Let Γ be the boundary of a domain Ω in $\mathbb{R}^N$ and let ω be a bounded open domain in Γ. Assume that there exists $h > 0$ such that $b_\Omega \in C^{1,1}(S_h(\omega))$, where

$$S_h(\omega) \stackrel{\text{def}}{=} \{x \in \mathbb{R}^N \colon |b_\Omega(x)| < h \text{ and } p_{\partial\Omega}(x) \in \omega\}, \tag{4.3}$$

is the truncated *orthogonal cylinder* generated by ω.

$S_h(\omega)$ corresponds to the union of the images $T_z(\omega)$ of ω for all $|z| < h$. If γ is the (relative) boundary of ω in Γ, the *lateral boundary* of $S_h(\omega)$

$$\Sigma_h(\gamma) \stackrel{\text{def}}{=} \{x \in \mathbb{R}^N \colon |b_\Omega(x)| < h \text{ and } p_{\partial\Omega}(x) \in \gamma\}, \tag{4.4}$$

is a submanifold of $\mathbb{R}^N$ *orthogonal* to Γ.

The next ingredient is the smoothness of the boundary γ of the domain ω in Γ. It is convenient to characterize it by the smoothness of the lateral boundary $\Sigma_h(\gamma)$ of the orthogonal cylinder $S_h(\omega)$ near γ.

DEFINITION 4.1. Under Assumption 4.1 let ω be a bounded open domain with (relative) boundary γ in Γ.

(i) Given an integer $k \geq 2$ and a real number $0 \leq \lambda \leq 1$, the boundary γ is $C^{k,\lambda}$ (resp. $C^{1,1}$) if there exist $h > 0$ and $0 < h' \leq h$ such that the piece $\Sigma_{h'}(\gamma)$ of the lateral boundary of $S_h(\omega)$ is $C^{k,\lambda}$ (resp. $C^{1,1}$).

(ii) The boundary γ is *Lipschitzian* if there exist $h > 0$ and $0 < h' \leq h$ such that the piece $\Sigma_{h'}(\gamma)$ of the lateral boundary of $S_h(\omega)$ is Lipschitzian.

(iii) The boundary γ has the *cone property* if there exist $h > 0$ and $0 < h' \leq h$ such that the piece $\Sigma_{h'}(\gamma)$ of the lateral boundary of $S_h(\omega)$ has the cone property.

(iv) The boundary γ has the *uniform cone property* if there exist $h > 0$, $0 < h' \le h$, $\lambda > 0$, $0 < \beta \le \pi/2$, $\rho > 0$ such that

$$\begin{aligned} &\forall x \in \Sigma_{h'}(\gamma),\ \exists \zeta_x \in \mathbb{R}^N,\ |\zeta_x| = 1 \text{ such that} \\ &\forall y \in B(x,\rho) \cap \overline{S}_h(\omega), \quad y + C(\lambda,\beta,\zeta_x) \subset S_h(\omega). \end{aligned}$$

where $C(\lambda,\beta,\zeta)$ is the open cone of vertex 0, height λ, aperture β, and direction ζ defined as

$$C(\lambda,\beta,\zeta) \stackrel{\text{def}}{=} \left\{y \in \mathbb{R}^N : \frac{1}{\tan\beta}|P(y)| < y\cdot\zeta < \lambda\right\}$$

and P is the orthogonal projection onto the hyperplane orthogonal to ζ.

For the corresponding definitions in $\mathbb{R}^N$ the reader is referred for instance to [**84**]. From previous considerations condition (i) is equivalent to say that the oriented distance function $b_{S_h(\omega)}$ associated with the set $S_h(\omega)$ has the required smoothness in a neighborhood of $\Sigma_{h'}(\gamma)$.

For a bounded open connected domain ω in Γ with a Lipschitzian boundary γ verifying Assumption 4.1, there exists a sufficiently small $h > 0$ such that $S_h(\omega)$ is a bounded open connected domain in $\mathbb{R}^N$ with a Lipschitzian boundary. So for $f \in C^1(\omega)$ such that $\nabla_\Gamma f = 0$ in ω

$$\nabla(f\circ p) = [I - bD^2b]\nabla_\Gamma f \circ p = 0 \text{ in } S_h(\omega),$$

and since $S_h(\omega)$ is a bounded connected open domain with a Lipschitzian boundary, there exists a constant such that

$$f\circ p = c \text{ in } S_h(\omega) \implies f = c \text{ in } \omega.$$

This is similar to what we have in $\mathbb{R}^N$ and the result is obtained by going in a neighborhood of ω.

In the same spirit define the subspaces

$$\begin{aligned} &\mathcal{D}^0(\omega) \stackrel{\text{def}}{=} \left\{f \in C(\Gamma)\colon \operatorname{supp} f \text{ is compact and } \operatorname{supp} f \subset \omega\right\}, \\ &\mathcal{D}^n(\omega) \stackrel{\text{def}}{=} \left\{f \in \mathcal{D}^{n-1}(\omega)\colon \nabla_\Gamma f \in \mathcal{D}^{n-1}(\omega)^N\right\}, \quad n \ge 1, \quad \mathcal{D}(\omega) \stackrel{\text{def}}{=} \mathcal{D}^\infty(\omega). \end{aligned}$$

4.3. Sobolev spaces. For submanifolds Γ which are the boundary of a C^∞ domain Ω in $\mathbb{R}^N$ the situation is simpler (cf. for instance to the book of [**6**]), but more recent results should now be available. However when Ω is only of class $C^{1,1}$, the traces on Γ of C^∞ functions defined in $\mathbb{R}^N$ are not necessarily in $C^\infty(\Gamma)$. The smoothness of functions on Γ is linked to the smoothness of Γ.

One way to define the distribution corresponding to a function f in $L^r(\omega)$, $1 \le r < \infty$, is

$$\langle f, \Phi\rangle \stackrel{\text{def}}{=} \int_\omega f\Phi\, dH_{N-1},$$

for all $\Phi \in \mathcal{D}(\mathbb{R}^N)$. Its distributional gradient would be defined as

$$\langle \nabla f, \vec{\Phi}\rangle \stackrel{\text{def}}{=} -\int_\omega f \operatorname{div}\vec{\Phi}\, dH_{N-1},$$

for all $\vec{\Phi} \in \mathcal{D}(\mathbb{R}^N)^N$. So f and $\nabla_\Gamma f$ become distributions in $\mathbb{R}^N$ with support in ω. If for instance Γ was sufficiently smooth to have $\mathcal{D}(\omega)$ dense in $L^2(\omega)$, then we could use functions Φ of the form $\varphi \circ p$ in a neighborhood of ω and zero outside a larger neighborhood and the above definitions would reduce to

$$\langle f, \varphi \rangle \stackrel{\text{def}}{=} \int_\Gamma f\varphi \, dH_{N-1},$$

for all $\varphi \in \mathcal{D}(\omega)$ and

$$\langle \nabla f, \vec{\varphi} \rangle \stackrel{\text{def}}{=} -\int_\Gamma f \mathrm{div}_\Gamma \vec{\varphi} \, dH_{N-1},$$

for all $\vec{\varphi} \in \mathcal{D}(\omega)^N$. Furthermore

$$\mathrm{div}_\Gamma \vec{\varphi} = \mathrm{div}_\Gamma(\vec{\varphi_\Gamma} + \vec{\varphi} \cdot nn) = \mathrm{div}_\Gamma \vec{\varphi_\Gamma} + H\vec{\varphi} \cdot n,$$

and for $f \in L^2(\omega)$

$$\langle \nabla f, \vec{\varphi} \rangle \stackrel{\text{def}}{=} -\int_\Gamma f \mathrm{div}_\Gamma \vec{\varphi_\Gamma} + Hf\vec{\varphi} \cdot n \, dH_{N-1},$$

and in view of the tangential Green's formula the *distributional tangential gradient* would be given by

$$\langle \nabla_\Gamma f, \vec{\varphi_\Gamma} \rangle \stackrel{\text{def}}{=} -\int_\Gamma f \mathrm{div}_\Gamma \vec{\varphi_\Gamma} \, dH_{N-1}.$$

For the purpose of this section we shall however proceed by density. For a real number $r \geq 1$ we first discuss the definition of $W^{1,r}(\omega)$ for a bounded open domain ω in the locally $C^{1,1}$ submanifold Γ with Lipschitzian boundary γ. The definition will be extended to $W^{k,r}(\omega)$ for a locally $C^{k,1}$ submanifold Γ. Under Assumption 4.1 from Theorem 2.5 in § 2 the map

$$x \mapsto T(X,z) \stackrel{\text{def}}{=} T_z(X) = X + z\nabla b(X) \colon \omega \times]-h, h[\to S_h(\omega),$$

and its inverse $T^{-1}(x) \stackrel{\text{def}}{=} \big(p(x), b(x)\big)$ are both Lipschitzian for $b \in C^{1,1}\big(S_h(\omega)\big)$. The space $L^r(\omega) \stackrel{\text{def}}{=} L^r(\omega; H_{N-1})$ is well-defined for the Hausdorff measure H_{N-1} and $f \mapsto f \circ p \colon L^r(\omega) \to L^r\big(S_h(\omega)\big)$ is well-defined, linear and continuous in view of the properties of T and T^{-1}. Moreover by using Federer's formula

$$\int_{S_h(\omega)} |f \circ p|^r \, dx = \int_{-h}^{h} dz \int_\omega d\Gamma j_z |f|^r = \int_\omega \alpha_0(h) |f|^r \, d\Gamma,$$

where

$$\alpha_0(h) = \int_{-h}^{h} j_z \, dz.$$

But $j_0 = 1$ and there exists $\bar{h} > 0$ such that

$$\forall h, 0 < h \leq \bar{h}, \quad h \leq j_z(X) \leq 4h, (H_{N-1})\text{-a.e. in } \omega,$$

and hence the norms $\|f \circ p\|_{L^r(S_h(\omega))}$ and $\|f\|_{L^r(\omega)}$ are equivalent. Therefore the linear subspace $\{f \circ p \colon f \in L^r(\omega)\}$ is closed in $L^r\big(S_h(\omega)\big)$ and we have the usual density properties

$$L^r\big(S_h(\omega)\big) = \overline{C^\infty\big(\overline{S_h(\omega)}\big)}^{L^r} = \overline{\mathcal{D}\big(S_h(\omega)\big)}^{L^r}.$$

Therefore for each $f \in L^r(\omega)$

$$\exists \{F_k\} \subset C^\infty\big(\overline{S_h(\omega)}\big) \text{ such that } F_k \to f \circ p \text{ in } L^r\big(S_h(\omega)\big),$$

and it can be shown that

(4.5) $$C^\infty(|\overline{\omega}) \stackrel{\text{def}}{=} \left\{F\big|_{\overline{\omega}} : \forall F \in C^\infty\left(\overline{S_h(\omega)}\right)\right\}$$

is dense in $L^r(\omega)$.

Introduce the norm

(4.6) $$\|f\|_{W^{1,r}(\omega)} \stackrel{\text{def}}{=} \|f\|_{L^r(\omega)} + \|\nabla_\Gamma f\|_{L^r(\omega)}.$$

For $f \in C^\infty(|\overline{\omega})$, there exists $F \in C^\infty\left(\overline{S_h(\omega)}\right)$ such that

$$\nabla_\Gamma f = F\big|_{\overline{\omega}} - \frac{\partial F}{\partial n} n,$$

and $\nabla_\Gamma f$ is the restriction to $\overline{\omega}$ of the $C^{0,1}$ vector function

$$g(F) \stackrel{\text{def}}{=} \nabla F - (\nabla F \cdot \nabla b)\nabla b,$$

for which the $W^{1,r}(\omega)$-norm of f is well-defined. It is then natural to introduce the definition

(4.7) $$\boxed{W^{1,r}(\omega) \stackrel{\text{def}}{=} \overline{C^\infty(|\overline{\omega})}^{W^{1,r}}}$$

For $f \in C^\infty(|\overline{\omega})$

$$\begin{aligned}\int_{S_h(\omega)} |\nabla(f \circ p)|^r \, dx &= \int_{S_h(\omega)} \left|[I - bD^2 b]\nabla_\Gamma f \circ p\right|^r dx \\ &= \int_{-h}^{h} dz \int_\omega d\Gamma j_z \left|[I + zD^2 b]^{-1}\nabla_\Gamma f\right|^r,\end{aligned}$$

and there exists $\bar{h}$, $0 < h \le \bar{h}$, such that for all z, $|z| \le \bar{h}$, and (H_{N-1}) almost all $X \in \overline{\omega}$

$$\frac{1}{2} I \le I + zD^2 b(X) \le 2I, \quad \frac{1}{2} \le j_z(X) \le 2.$$

Thus the norms

$$\left\|\nabla(f \circ p)\right\|_{W^{1,r}(S_h(\omega))} \quad \text{and} \quad \|\nabla_\Gamma f\|_{W^{1,r}(\omega)}$$

are equivalent. For $f \in W^{1,r}(\omega)$, $\nabla_\Gamma f$ is defined from the sequence $\{f_n\} \subset C^\infty(|\overline{\omega})$ such that $f_n \to f$ in $W^{1,r}(\omega)$ as the limit

$$\nabla_\Gamma f \stackrel{\text{def}}{=} \lim_{n\to\infty} \nabla_\Gamma f_n \text{ in } L^r(\omega)^N.$$

In view of the equivalence of norms there exists $F \in W^{1,r}(S_h(\omega))$ such that

$$f_n \circ p \to F \text{ in } W^{1,r}(S_h(\omega)),$$

and necessarily $F = f \circ p$. Therefore we have the following convergences in $L^r(S_h(\omega))^N$

$$\begin{gathered}[I - bD^2 b]\nabla_\Gamma f_n \circ p = \nabla(f_n \circ p) \to \nabla(f \circ p), \\ \nabla_\Gamma f_n \circ p = [I - bD^2 b]^{-1}\nabla(f_n \circ p) \to [I - bD^2 b]^{-1}\nabla(f \circ p), \\ \nabla_\Gamma f_n \circ p \to \nabla_\Gamma f \circ p,\end{gathered}$$

and

(4.8) $$\boxed{\nabla(f \circ p) = [I - bD^2 b]\nabla_\Gamma f \circ p}$$

So all the properties of the tangential gradient of $C^1(\omega)$ functions extend to elements of $W^{1,r}(\omega)$.

For an integer $n \geq 1$ such that $b \in C^{n,1}\big(S_h(\omega)\big)$ define the Sobolev spaces

$$W^{n,r}(\omega) \overset{\text{def}}{=} \overline{C^\infty(|\overline{\omega})}^{W^{n,r}}, \quad W_0^{n,r}(\omega) \overset{\text{def}}{=} \overline{\mathcal{D}(|\omega)}^{W^{n,r}} \tag{4.9}$$

where

$$\mathcal{D}(|\omega) \overset{\text{def}}{=} \Big\{F\big|_\omega : F \in \mathcal{D}\big(S_h(\omega)\big)\Big\}. \tag{4.10}$$

For $r = 2$ we use the standard notation $H^n(\omega) \overset{\text{def}}{=} W^{n,2}(\omega)$ and $H_0^n(\omega) \overset{\text{def}}{=} W_0^{n,2}(\omega)$.

4.4. Tangential Poincaré's inequality: Scalar case. As for the tangential calculus many results can be obtained by going to $\mathbb{R}^N$. As an other illustration we give the proof of the Poincaré's inequality on Γ. For the notion of n-capacity the reader is referred to [**47**].

THEOREM 4.1. *Under Assumption 4.1 let ω be a bounded open domain in Γ with a Lipschitzian (relative) boundary γ. Then there exists a constant $c > 0$ such that*

$$\forall f \in H_0^1(\omega), \quad \int_\omega |f|^2\, d\Gamma \leq c \int_\omega |\nabla_\Gamma f|^2\, d\Gamma. \tag{4.11}$$

If, in addition, ω is connected and γ_0 is a subset of γ with non-zero $(N-1)$-capacity, then the result is still true with

$$H_{\gamma_0}^1(\omega) \overset{\text{def}}{=} \{f \in H^1(\omega) : f = 0 \text{ on } \gamma_0\}$$

in place of $H_0^1(\omega)$.

PROOF. Consider the following quotient and its infimum

$$q(f) = \frac{\int_\omega |\nabla_\Gamma f|^2\, d\Gamma}{\int_\omega |f|^2\, d\Gamma}, \quad f \neq 0,$$

$$q \overset{\text{def}}{=} \inf\big\{q(f) : \|f\|_{L^2} = 1, f \in H_{\gamma_0}^1(\omega)\big\},$$

and proceed by contradiction. □

4.5. Laplace–Beltrami operator. Elliptic problems in a bounded open domain ω can be handled in much the same way as their counterparts in $\mathbb{R}^N$. To illustrate this point consider the Laplace–Beltrami operator with variable coefficients. Let

$$a_{ij} \in L^\infty(\omega), \quad 1 \leq i,\ j \leq N, \tag{4.12}$$

and $f \in L^2(\omega)$ be given. Assume that

$$\exists \alpha > 0, \forall \xi \in \mathbb{R}^N, \quad \sum_{i,j=1}^N a_{ij}(x)\xi^j\xi^i \geq \alpha|\xi|^2, \tag{4.13}$$

and that

$$\int_\omega f\, v\, d\Gamma = 0, \quad \forall v \in N, \tag{4.14}$$

where

$$N = \big\{f \in H^1(\omega) : \nabla_\Gamma f = 0\big\}. \tag{4.15}$$

Consider the variational problem: to find $u \in H^1(\omega)$ such that for all $v \in H^1(\omega)$

$$\langle Au, v\rangle = \int_\omega fv\, d\Gamma, \tag{4.16}$$

$$\langle Au, v\rangle \stackrel{\text{def}}{=} \sum_{i,j=1}^{N} \int_\omega a_{ij}(x)(\nabla_\Gamma u)_j(\nabla_\Gamma v)_i\, d\Gamma, \tag{4.17}$$

or if a denotes the matrix $\{a_{ij}\}$

$$\langle Au, v\rangle = \int_\omega a\nabla_\Gamma u \cdot \nabla_\Gamma v\, d\Gamma. \tag{4.18}$$

In view of (4.13)

$$\forall v \in H^1(\omega), \quad \langle Av, v\rangle \geq \alpha\|\nabla_\Gamma v\|_{L^2}^2,$$

and the kernel of the operator A is given by N defined in (4.15). As a result we obtain the "standard" result under condition (4.14)

$$\exists! u \in H^1(\omega)/N,\ \forall v \in H^1(\omega), \qquad \int_\omega a\nabla_\Gamma u \cdot \nabla_\Gamma v\, dx = \int_\omega fv\, d\Gamma, \tag{4.19}$$

or equivalently (4.19) has a solution up to an additive element of N. When ω is connected with a Lipschitzian boundary γ the kernel N coincides with the constants since

$$\nabla(f\circ p) = [I - bD^2b]\nabla_\Gamma f \circ p = 0 \implies \exists c, \quad f\circ p = c \text{ in } S_h(\omega).$$

If the coefficients $a_{ij} \in W^{1,\infty}(\omega)$ and $b \in C^{2,1}(S_h(\omega))$, then

$$\begin{cases} -\operatorname{div}_\Gamma\big(a(x)\nabla_\Gamma u\big) + H(x)a(x)\nabla_\Gamma u \cdot \nabla b = f \text{ in } \omega \\ a(x)\nabla_\Gamma u \cdot \nu = 0 \text{ on } \gamma, \end{cases} \tag{4.20}$$

where ν is the unit exterior normal to the boundary γ of ω in Γ (that is $\nu \cdot \nabla b = 0$) which is given by

$$\nu = \nabla b_{S_h(\omega)}\big|_\gamma, \quad (H_{N-1})\text{-a.e.}$$

and $H = \Delta b$ is the mean curvature. When $a = I$ then we recover the Laplace–Beltrami operator

$$-\Delta_\Gamma u = f \text{ in } \omega, \quad \nabla_\Gamma u \cdot \nu = 0 \text{ on } \gamma.$$

Therefore we have the same picture as in $\mathbb{R}^N$ up to the fact that the familiar differential operators are now tangential.

From Theorem 4.1 the homogeneous Dirichlet tangential boundary value problem: to find $u \in H_0^1(\omega)$ such that for all $v \in H_0^1(\omega)$

$$\int_\omega \nabla_\Gamma u \cdot \nabla_\Gamma v\, d\Gamma = \int_\omega fv\, d\Gamma,$$

or in strong form

$$\Delta_\Gamma u = f \text{ in } \omega, \quad u = 0 \text{ on } \Gamma,$$

has a unique solution in $H_0^1(\omega)$.

4.6. Canonical local basis. In the translation of some classical results into intrinsic form it will often be convenient to choose as local basis in each point x of ω the one generated by the $C^{1,1}$-diffeomorphism introduced in the proof of Theorem 2.1 of § 2 in a neighborhood of $\Gamma = \partial\Omega$ where $b = b_\Omega$ is $C^{1,1}$. With this choice the covariant and the contravariant bases are directly expressed in terms of the function b and its derivatives.

With each $x \in \omega$ was associated a local orthonormal basis $\{\tau_1, \dots, \tau_N\}$ of $\mathbb{R}^N$ such that $\tau_N = -\nabla b_\Omega(x)$, a neighborhood $\mathcal{U}(x)$ of x in $\mathbb{R}^N$, and the open ball $B(0, \varepsilon)$ of radius ε at the origin such that the mapping

$$y \mapsto m_x(y) = \left(\{(y - x) \cdot \tau_\alpha\}_{\alpha=1}^{N-1}, -b_\Omega(y)\right) : \mathcal{U}(x) \to B(0, \varepsilon), \tag{4.21}$$

be invertible. The Jacobian matrix of m_x at y is given by

$$Dm_x(y) = \begin{bmatrix} {}^*\tau_1 \\ {}^*\tau_2 \\ \vdots \\ {}^*\tau_{N-1} \\ -\,{}^*\nabla b_\Omega(y) \end{bmatrix} \tag{4.22}$$

where ${}^*z$ denotes the row vector corresponding to the column vector z. Its inverse is given by

$$(Dm_x)^{-1}(y) = \left[\sigma_1(y), \dots, \sigma_N(y)\right], \tag{4.23}$$

where the $\sigma_i(y)$'s are the following column vectors

$$\begin{aligned} \sigma_\alpha(y) &= \tau_\alpha - \frac{\tau_\alpha \cdot \nabla b(y)}{\nabla b(x) \cdot \nabla b(y)} \nabla b(x)\,, \quad 1 \le \alpha \le N-1, \\ \sigma_N(y) &= -\frac{1}{\nabla b(x) \cdot \nabla b(y)} \nabla b(x). \end{aligned} \tag{4.24}$$

The map $\xi' \mapsto \Phi(\xi') = m_x^{-1}(\xi', 0) \colon V \to \omega_x \stackrel{\text{def}}{=} \mathcal{U}(x) \cap \omega$ defined on

$$V = \{\xi' = (\xi_1, \dots, \xi_{N-1}) \in \mathbb{R}^{N-1} \colon |\xi'| \le \varepsilon\},$$

is also invertible. So the first $N-1$ vectors $\sigma_\alpha(y)$ coincide with the covariant basis

$$\begin{aligned} a_\alpha(\xi') = \frac{\partial \Phi}{\partial \xi'_\alpha}(\xi') &= D(m_x^{-1})_{.\alpha}(\xi', 0) \\ &= \left[(Dm_x)^{-1} \circ m_x^{-1}\right]_{.\alpha}(\xi', 0) \\ &= \left\{[(Dm_x)^{-1}_{.\alpha} \circ \Phi]\right\}(\xi') = [\sigma_\alpha \circ \Phi](\xi'), \end{aligned}$$

or for $y \in \omega_x$

$$\begin{aligned} a_\alpha(y) &= \sigma_\alpha(y) = \tau_\alpha - \frac{\tau_\alpha \cdot n(y)}{n(x) \cdot n(y)} n(x), \quad 1 \le \alpha \le N-1, \\ a_N(y) &= -n(y). \end{aligned} \tag{4.25}$$

It is easy to check that the contravariant basis is given by

$$\begin{aligned} a^\alpha(y) &= \tau_\alpha - \tau_\alpha \cdot n(y) n(y), \quad 1 \le \alpha \le N-1 \\ a^N(y) &= -n(y). \end{aligned} \tag{4.26}$$

4.7. Tangential Korn's inequality. In this section we give a proof of the tangential Korn's inequality via local coordinates. Let ω be a bounded open domain in Γ for which Assumption 4.1 is verified. We first start with a domain ω which is completely described by a single map Φ. Then we extend to the case where Γ is Lipschizian in a neighborhood of ω, that is $\overline{\omega}$ can be described by a finite number of Lipschizian maps. In the scalar case associate with a function $f\colon \omega \to \mathbb{R}$ the function $f \circ \Phi\colon A \subset B_0 \to \mathbb{R}$. First

$$\int_\omega |f|^2 \, d\Gamma = \int_A |f \circ \Phi|^2 \, |\det \Phi| \, d\zeta', \tag{4.27}$$

and the L^2-norms are equivalent. From (3.39)

$$\nabla_\Gamma f \cdot e_i = \nabla_\Gamma f \cdot a_\alpha a^\alpha \cdot e_i = f_{,\alpha} a^\alpha \cdot e_i, \tag{4.28}$$

$$f_{,\alpha} = \nabla_\Gamma f \cdot a_\alpha = \nabla_\Gamma f \cdot e_i e_i \cdot a_\alpha, \tag{4.29}$$

and $\nabla_\Gamma f$ exists if and only if $\{f_{,\alpha}\}$ exists. In addition

$$|\nabla_\Gamma f| \le \Big\{\sum_\alpha |f_{,\alpha}|^2\Big\}^{1/2} \Big\{\sum_\alpha |a^\alpha|^2\Big\}^{1/2}, \tag{4.30}$$

$$\Big\{\sum_\alpha |f_{,\alpha}|^2\Big\}^{1/2} \le |\nabla_\Gamma f| \Big\{\sum_\alpha |a_\alpha|^2\Big\}^{1/2}, \tag{4.31}$$

and if

$$|a^\bullet| \stackrel{\text{def}}{=} \Big\{\sum_\alpha |a^\alpha|^2\Big\}^{1/2} \quad \text{and} \quad |a_\bullet| \stackrel{\text{def}}{=} \Big\{\sum_\alpha |a_\alpha|^2\Big\}^{1/2},$$

are bounded then

$$\nabla_\Gamma f \in L^2(\omega)^N \iff \nabla(f \circ \Phi) \in L^2(A)^{N-1}, \tag{4.32}$$

and again we have the equivalence of norms. When $\overline{\omega}$ is compact it can be described by a finite number of local maps and the above inequalities and equivalences which are verified locally extend globally.

For a vector function $v\colon \omega \to \mathbb{R}^N$ first assume that v is a tangential vector, that is $v_\Gamma = v$ on ω. Associate with v the vector functions $v \circ \Phi\colon A \to \mathbb{R}^N$ and $\vec{v}\colon A \to \mathbb{R}^{N-1}$

$$\vec{v}_\alpha \stackrel{\text{def}}{=} v_\alpha, \quad v_\alpha \stackrel{\text{def}}{=} (v \circ \Phi) \cdot a_\alpha. \tag{4.33}$$

From (3.56), (3.59) and (3.53) since $v = v_\Gamma$

$$D(\vec{v})_{\alpha\beta} = v_{\alpha,\beta} = \big[D^P_\Gamma(v) - D^P_\Gamma(a^\lambda) v_\lambda\big] a_\beta \cdot a_\alpha, \tag{4.34}$$

$$\varepsilon(\vec{v})_{\alpha\beta} = \big[\varepsilon^P_\Gamma(v) - \varepsilon^P_\Gamma(a^\lambda) v_\lambda\big] a_\beta \cdot a_\alpha, \tag{4.35}$$

and by symmetry of $D^P_\Gamma(a^\lambda) a_\alpha \cdot a_\beta = \Gamma^\lambda_{\beta\alpha} = \Gamma^\lambda_{\alpha\beta} = \varepsilon^P_\Gamma(a^\lambda) a_\alpha \cdot a_\beta$

$$\big(D(\vec{v}) - \varepsilon(\vec{v})\big)_{\alpha\beta} = \big[D^P_\Gamma(v) - \varepsilon^P_\Gamma(v)\big] a_\beta \cdot a_\alpha. \tag{4.36}$$

There exist positive constants such that

$$c_1 \big\|D(\vec{v}) - \varepsilon(\vec{v})\big\| \le \big\|D^P_\Gamma(v) - \varepsilon^P_\Gamma(v)\big\| \le c_2 \big\|D(\vec{v}) - \varepsilon(\vec{v})\big\|,$$
$$\big\|\varepsilon(\vec{v})\big\| \le c_3 \big\|\varepsilon^P_\Gamma(v) - v_\lambda \varepsilon^P_\Gamma(a^\lambda)\big\| \le c_4 \big[\|\varepsilon^P_\Gamma(v)\| + \|v\|\big].$$

By Korn's inequality in $\mathbb{R}^{N-1}$

$$\exists c > 0, \forall \vec{v}, \quad \|D\vec{v}\| \le c\big[\|\varepsilon(\vec{v})\| + \|\vec{v}\|\big]. \tag{4.37}$$

Therefore in view of the above inequalities for all v tangential

$$\exists c > 0, \forall v \in H^1(\omega)^N, \quad \|D_\Gamma^P(v)\| \le c\big[\|\varepsilon_\Gamma^P(v)\| + \|v\|\big]. \tag{4.38}$$

This last inequality can now be used locally for the general case of Γ compact and $C^{1,1}$ by considering a finite number of points $\{x_i\}$ of Γ, neighborhoods U_i of x_i, and a C^∞ partition of unity $\{\rho_i\}$ of $U_h(\Gamma)$. Indeed

$$v = \sum_i v_i \rho_i \implies \big\|D_\Gamma^P(v)\big\| \le \sum_i \big\|D_\Gamma^P(\rho_i v)\big\|,$$

from (4.38) for each i

$$\big\|D_\Gamma^P(\rho_i v)\big\| \le c_i \big[|\rho_i v|^2 + \big\|\varepsilon_\Gamma^P(\rho_i v)\big\|^2\big]^{1/2},$$

and

$$\varepsilon_\Gamma^P(\rho_i v) = \rho_i \varepsilon_\Gamma^P(v) + \frac{1}{2}\big[v^* \nabla_\Gamma \rho_i + \nabla_\Gamma \rho_i{}^* v\big] \implies \big\|\varepsilon_\Gamma^P(\rho_i v)\big\| \le c_i\big[\|\varepsilon_\Gamma^P(v)\| + \|v\|\big].$$

Therefore there exists $c > 0$ such that for all tangential $v \in H^1(\omega)^N$

$$\boxed{\big\|D_\Gamma^P(v)\big\| \le c\big[\|\varepsilon_\Gamma^P(v)\| + \|v\|\big]} \tag{4.39}$$

But $D_\Gamma(v)n = 0$, $P^* D_\Gamma(v)n = {}^*D_\Gamma(v)n$, $D_\Gamma(v)n \cdot n = 0$ and from (3.33)

$$D_\Gamma(v) = D_\Gamma^P(v) + n^*\big({}^*D_\Gamma(v)n\big) = D_\Gamma^P(v) - n^*(D^2 b v),$$

and there exists $c > 0$ such that for all tangential $v \in H^1(\omega)^N$

$$\boxed{\big\|D_\Gamma(v)\big\| \le c\big[\|\varepsilon_\Gamma^P(v)\| + \|v\|\big]} \tag{4.40}$$

As for the general case the vector function v is decomposed into its tangential part v_Γ and its normal component v_n, $v = v_\Gamma + v_n n$. Then from identity (3.35) applied to $D_\Gamma(v)$

$$\begin{aligned}
\big\|D_\Gamma(v)\big\|^2 &= \big\|D_\Gamma^P(v)\big\|^2 + \big\|P^* D_\Gamma(v)n\big\|^2 \\
&= \big\|D_\Gamma^P(v)\big\|^2 + \|\nabla v_n - D^2 b v_\Gamma\|^2 \\
&= \big\|D_\Gamma^P(v_\Gamma) + v_n D^2 b\big\|^2 + \|\nabla v_n - D^2 b v_\Gamma\|^2.
\end{aligned}$$

Similarly

$$\begin{aligned}
\big\|\varepsilon_\Gamma(v)\big\|^2 &= \big\|\varepsilon_\Gamma^P(v)\big\|^2 + 2\big\|P\varepsilon_\Gamma(v)n\big\|^2 \\
&= \big\|\varepsilon_\Gamma^P(v_\Gamma) + v_n D^2 b\big\|^2 + \frac{1}{2}\|\nabla v_n - D^2 b v_\Gamma\|^2.
\end{aligned}$$

and

$$\big\|D_\Gamma(v)\big\|^2 - 2\big\|\varepsilon_\Gamma(v)\big\|^2 d = \big\|D_\Gamma^P(v_\Gamma) + v_n D^2 b\big\|^2 - 2\big\|\varepsilon_\Gamma^P(v_\Gamma) + v_n D^2 b\big\|^2. \tag{4.41}$$

But

$$\begin{aligned}
\big\|D_\Gamma^P(v_\Gamma) + v_n D^2 b\big\| &\le \big\|\varepsilon_\Gamma^P(v_\Gamma) + v_n D^2 b\big\| + \big\|D_\Gamma^P(v_\Gamma) - \varepsilon_\Gamma^P(v_\Gamma)\big\|, \\
\big\|D_\Gamma^P(v_\Gamma) + v_n D^2 b\big\|^2 &\le 2\big[\|\varepsilon_\Gamma^P(v_\Gamma) + v_n D^2 b\|^2 + \big\|D_\Gamma^P(v_\Gamma) - \varepsilon_\Gamma^P(v_\Gamma)\big\|^2\big],
\end{aligned}$$

and

$$\boxed{\big\|D_\Gamma(v)\big\|^2 \le 2\big\|\varepsilon_\Gamma(v)\big\|^2 + 2\big\|D_\Gamma^P(v_\Gamma) + v_n D^2 b\big\|^2} \tag{4.42}$$

Since v_Γ is tangential we can use Korn's inequality (4.39) and

$$\left\|\varepsilon_\Gamma^P(v_\Gamma)\right\|^2 \le \left\|\varepsilon_\Gamma^P(v_\Gamma)+v_n D^2 b\right\|^2 + \|v_n D^2 b\|^2 \le c\big[\|\varepsilon_\Gamma(v)\| + \|v\|\big],$$

and necessarily

$$\left\|D_\Gamma^P(v_\Gamma) - \varepsilon_\Gamma^P(v_\Gamma)\right\| \le c\big[\|\varepsilon_\Gamma(v)\| + \|v\|\big].$$

Putting all this together we have the intrinsic Korn's inequality

$$\exists c>0, \forall v \in H^1(\omega)^N, \quad \left\|D_\Gamma(v)\right\| \le c\big[\|\varepsilon_\Gamma(v)\| + \|v\|\big]. \tag{4.43}$$

THEOREM 4.2. *Under Assumption 4.1 let ω a bounded open domain with a Lipschitzian boundary γ. There exists $c>0$ such that for all*

$$v \in E_\Gamma(\omega) = \left\{v \in L^2(\omega)^N : \varepsilon_\Gamma^P(v_\Gamma) \in L^2(\omega)^{N\times N}\right\},$$

$$\int_\omega \left\|D_\Gamma^P(v_\Gamma)\right\|^2 d\Gamma \le c^2 \int_\omega |v_\Gamma|^2 + \left\|\varepsilon_\Gamma^P(v_\Gamma)\right\|^2 d\Gamma, \tag{4.44}$$

and $E_\Gamma(\omega) = \left\{v \in L^2(\omega)^N : v_\Gamma \in H^1(\omega)^N\right\}$. Similarly there exists $c>0$ such that for all

$$v \in E(\omega) = \left\{v \in L^2(\omega)^N : \varepsilon_\Gamma(v) \in L^2(\omega)^{N\times N}\right\},$$

$$\int_\omega \left\|D_\Gamma(v)\right\|^2 d\Gamma \le c^2 \int_\omega |v|^2 + \left\|\varepsilon_\Gamma(v)\right\|^2 d\Gamma, \tag{4.45}$$

and $E(\omega) = H^1(\omega)^N$.

We give another proof of the intrinsic Korn's inequality using an argument developped by R. Drai and J. P. Zolésio. It follows the classical proof in $\mathbb{R}^N$ (cf. for instance [**84**], [**4**]). In their original proof the basic inequality

$$\|u\|_{L^2(\omega)/\mathbb{R}} \le c(\omega)\left\|\nabla_\Gamma(u)\right\|_{H^{-1}(\omega)}, \tag{4.46}$$

is obtained by a simple argument which necessitates the density of $H_0^1(\omega)^N$ in

$$H(\mathrm{div},\omega) \stackrel{\mathrm{def}}{=} \left\{v \in L^2(\omega)^N : \mathrm{div}_\Gamma v \in L^2(\omega)\right\}.$$

The equivalent result in $\mathbb{R}^N$ is true for domains $\Omega \subset \mathbb{R}^N$ with a C^2 boundary (cf. [**86**], [**84**] p. 13). As it stands even with an extension of this proof to a domain ω in Γ with a C^2 boundary γ, the condition that $b \in C^{2,1}(S_h(\omega))$ is stronger than the $C^{1,1}$ assumption of Theorem 4.2 which uses local coordinates. However this proof presents an intrinsic interest. Here inequality (4.46) is proved directly by a different argument in Lemmas 4.1 and 4.2 for a domain ω with Lipschitzian boundary.

LEMMA 4.1. *Let ω be a bounded open domain in Γ with Lipschitzian boundary verifying Assumption 4.1. Further assume that $b \in C^{2,1}(S_h(\omega))$. Given $u \in L^2(\omega)$, $\xi \in H_0^1(\omega)^N$ and*

$$\Xi \stackrel{\mathrm{def}}{=} \frac{h^2-b^2}{h^2}\big(\det[I-bD^2b]\big)[I-bD^2b]^{-1}\xi\circ p,$$

then

$$\int_{S_h(\omega)} u\circ p\, \mathrm{div}(\Xi)\, dx = \frac{4h}{3}\int_\omega u\mathrm{div}_\Gamma(\xi)\, d\Gamma = \frac{4h}{3}\int_\omega u\mathrm{div}_\Gamma(\xi_\Gamma)\, d\Gamma.$$

PROOF. By assumption on ξ,

$$\xi \circ p \in H^1_{\Sigma_h(\gamma)}\big(S_h(\omega)\big)^N = \big\{v \in H^1\big(S_h(\omega)\big)^N : v|_{\Sigma_h(\gamma)} = 0\big\}.$$

Since $b \in C^{2,1}$, $I - bD^2b \in C^{0,1} \subset W^{1,\infty}$ and it is easy to check that $\Xi \in H^1_0\big(S_h(\omega)\big)^N$. For $u \in H^1(\omega)$, $u \circ p \in H^1\big(S_h(\omega)\big)^N$ and

$$\begin{aligned}
-\int_{S_h(\omega)} & u \circ p \operatorname{div} \Xi \, dx \\
&= \int_{S_h(\omega)} \nabla(u \circ p) \cdot \Xi \, dx \\
&= \int_{S_h(\omega)} \frac{h^2 - b^2}{h^2} [I - bD^2b] \nabla_\Gamma(u) \circ p \cdot [I - bD^2b]^{-1} \xi \circ p \det[I - bD^2b] \, dx \\
&= \int_{S_h(\omega)} \frac{h^2 - b^2}{h^2} \big(\nabla_\Gamma(u) \cdot \xi\big) \circ p \det[I - bD^2b] \, dx \\
&= \int_\omega d\Gamma \int_{-h}^{h} dz j_z \frac{h^2 - b^2}{h^2} \nabla_\Gamma(u) \cdot \xi \det[I - bD^2b] \circ T_z \, dx \\
&= \frac{4h}{3} \int_\omega \nabla_\Gamma(u) \cdot \xi \, d\Gamma.
\end{aligned}$$

since

$$[I - bD^2b] \circ T_z = [I + zD^2b]^{-1} \implies \det[I - bD^2b] \circ T_z = \det[I + zD^2b]^{-1} = \frac{1}{j_z}.$$

Finally

$$\int_\omega \nabla_\Gamma(u) \cdot \xi \, d\Gamma = \int_\omega \nabla_\Gamma(u) \cdot \xi_\Gamma \, d\Gamma = -\int_\omega u \operatorname{div}_\Gamma(\xi_\Gamma) \, d\Gamma,$$

and

$$-\int_{S_h(\omega)} u \circ p \operatorname{div}(\Xi) \, dx = -\frac{4h}{3} \int_\omega u \operatorname{div}_\Gamma(\xi_\Gamma) \, d\Gamma.$$

By density the result is true for $u \in L^2(\omega)$. □

LEMMA 4.2. *Let ω be a bounded open domain in Γ with Lipschitzian boundary verifying Assumption 4.1. Further assume that $b \in C^{2,1}(S_h(\omega))$. There exists $c > 0$ such that*

$$\forall u \in L^2(\omega), \quad \|u\|_{L^2(\omega)/\mathbb{R}} \leq c \big\|\nabla_\Gamma(u)\big\|_{H^{-1}(\omega)}.$$

PROOF. In the Lipschitzian domain $S_h(\omega)$ there exists $c(h) > 0$ such that

$$\forall U \in L^2\big(S_h(\omega)\big), \quad \|U\|_{L^2(S_h(\omega))/\mathbb{R}} \leq c(h) \|\nabla U\|_{H^{-1}(S_h(\omega))}.$$

In particular this holds for $U = u \circ p$. By definition of the H^{-1}-norm

$$(4.47) \qquad \|\nabla(u \circ p)\|_{H^{-1}} = \sup_{\Xi \in H_0^1(S_h(\omega))^N} \frac{-\int_{S_h(\omega)} u \circ p \operatorname{div} \Xi \, dx}{\|\Xi\|_{H^1(S_h(\omega))}}$$
$$\geq \sup_{\xi \in H^1(\omega)^N} \frac{-\int_{S_h(\omega)} u \circ p \operatorname{div} \Xi(\xi) \, dx}{\|\Xi(\xi)\|_{H^1(S_h(\omega))}}$$
$$\geq \sup_{\xi_\Gamma \in H^1(\omega)^N} \frac{-\int_{S_h(\omega)} u \circ p \operatorname{div} \Xi(\xi_\Gamma) \, dx}{\|\Xi(\xi_\Gamma)\|_{H^1(S_h(\omega))}},$$

where

$$\Xi(\xi) = \frac{h^2 - b^2}{h^2} \det[I - bD^2 b] \, [I - bD^2 b]^{-1} \xi \circ p.$$

By Lemma 4.1

$$\int_{S_h(\omega)} u \circ p \operatorname{div} \Xi(\xi) \, dx = \frac{4h}{3} \int_\omega u \operatorname{div}_\Gamma(\xi_\Gamma) \, d\Gamma.$$

Moreover

$$\int_{S_h(\omega)} |\Xi(\xi)|^2 \, dx \leq c \int_{S_h(\omega)} |\xi \circ p|^2 \, dx \leq c \int_\omega d\Gamma \int_{-h}^{h} dz j_z |\xi|^2 \, dx$$
$$\leq c4h \int_\omega d\Gamma |\xi|^2 \, d\Gamma,$$

for h sufficiently small since $j_0 = 1$. Similarly

$$\int_{S_h(\omega)} |D\Xi(\xi)|^2 \, dx \leq c \, h \|\xi\|^2_{H^1(\omega)}.$$

Therefore there exists $c > 0$ such that for all $\xi \in H_0^1(\omega)^N$

$$\|\Xi(\xi)\|_{H_0^1(S_h(\omega))} \leq c \, h^{1/2} \|\xi\|^2_{H^1(\omega)}$$

So for $\xi = \xi_\Gamma$ the right-hand side of (4.47) is greater than

$$\frac{4}{3c} h^{1/2} \sup_{\xi_\Gamma \in H_0^1(\omega)^N} \frac{-\int_\omega u \operatorname{div}_\Gamma \xi_\Gamma \, d\Gamma}{\|\xi_\Gamma\|_{H_0^1}} = \frac{4}{3c} h^{1/2} \|\nabla_\Gamma(u)\|_{H^{-1}(\omega)}.$$

Finally

$$\int_{S_h(\omega)} |u \circ p|^2 dx = \int_\omega d\Gamma \int_{-h}^{h} dz j_z |u|^2 \geq h \int_\omega |u|^2 d\Gamma,$$

for h sufficiently small ($j_z \geq 1/2$ for $|z| < h$). Hence

$$h^{1/2} \|u\|_{L^2(\omega)/\mathbb{R}} \leq \|u \circ p\|_{L^2(S_h(\omega))/\mathbb{R}}$$

and

$$\|u\|_{L^2(\omega)/\mathbb{R}} \leq \frac{4}{3c} \|\nabla_\Gamma(u)\|_{H^{-1}(\omega)}.$$

□

THEOREM 4.3. *Let ω be a bounded open domain in Γ with Lipschitzian boundary verifying Assumption 4.1. Further assume that $b \in C^{2,1}(S_h(\omega))$. Then there exists $c > 0$ such that for all*

$$v \in E(\omega) \overset{\text{def}}{=} \{v \in L^2(\omega)^N : \varepsilon_\Gamma(v) \in L^2(\omega)^{N\times N}\},$$

$$\int_\omega \|D_\Gamma(v)\|^2 \, d\Gamma \le c^2 \int_\omega |v|^2 + \|\varepsilon_\Gamma(v)\|^2 \, d\Gamma. \tag{4.48}$$

Moreover $E(\omega) = H^1(\omega)^N$ and the H^1-norm is equivalent to the norm

$$\{\|v\|_{L^2}^2 + \|\varepsilon_\Gamma(v)\|_{L^2}^2\}^{1/2}.$$

PROOF. We follow the classical proof in $\mathbb{R}^N$ which uses the identity

$$\partial_j(\partial_i(u_k\circ p)) = \partial_i\left[\frac{\partial_j(u_k\circ p)+\partial_k(u_j\circ p)}{2}\right] + \partial_j\left[\frac{\partial_k(u_i\circ p)+\partial_i(u_k\circ p)}{2}\right] - \partial_k\left[\frac{\partial_i(u_j\circ p)+\partial_j(u_ji\circ p)}{2}\right].$$

For $u \in H^1(\omega)^N$ recall that

$$D_\Gamma^2(u_k)_{ij} = \left[D_\Gamma(\nabla_\Gamma(u_k))\right]_{ij} = \left[\nabla_\Gamma(D_\Gamma(u))_{ki}\right]_j,$$

and hence

$$\left[\nabla_\Gamma(\varepsilon_\Gamma(u))_{ki}\right]_j = \frac{1}{2}\left[D_\Gamma^2(u_k)_{ij} + D_\Gamma^2(u_i)_{kj}\right]. \tag{4.49}$$

We proceed as in the $\mathbb{R}^N$ case taking into account the fact that the matrix of second tangential derivatives is no longer symmetrical

$$\begin{aligned} 2D_\Gamma^2(u_k)_{ij} = {} & \left[D_\Gamma^2(u_k)_{ji} + D_\Gamma^2(u_j)_{ki}\right] + \left[D_\Gamma^2(u_k)_{ij} - {}^*D_\Gamma^2(u_k)_{ij}\right] \\ & + \left[D_\Gamma^2(u_i)_{kj} + D_\Gamma^2(u_k)_{ij}\right] + \left[D_\Gamma^2(u_i)_{jk} - {}^*D_\Gamma^2(u_i)_{jk}\right] \\ & - \left[D_\Gamma^2(u_j)_{ik} + D_\Gamma^2(u_i)_{jk}\right] - \left[D_\Gamma^2(u_j)_{ki} - {}^*D_\Gamma^2(u_j)_{ki}\right], \end{aligned}$$

and from identity (4.49)

$$\begin{aligned} 2D_\Gamma^2(u_k)_{ij} = {} & 2\left[\nabla_\Gamma(\varepsilon_\Gamma(u))_{kj}\right]_i + 2\left[\nabla_\Gamma(\varepsilon_\Gamma(u))_{ik}\right]_j - 2\left[\nabla_\Gamma(\varepsilon_\Gamma(u))_{ji}\right]_k \\ & + \left[D_\Gamma^2(u_k) - {}^*D_\Gamma^2(u_k)\right]_{ij} + \left[D_\Gamma^2(u_i) - {}^*D_\Gamma^2(u_i)\right]_{jk} \\ & - \left[D_\Gamma^2(u_j) - {}^*D_\Gamma^2(u_j)\right]_{ki}. \end{aligned}$$

But

$$\begin{aligned} D_\Gamma^2(u_k) - {}^*D_\Gamma^2(u_k) &= D^2b\nabla_\Gamma(u_k)^*n - n^*(D^2b\nabla_\Gamma(u_k)), \\ D_\Gamma^2(u_k)_{ij} - {}^*D_\Gamma^2(u_k)_{ij} &= \left[(D^2b)_{i\ell}n_j - n_i(D^2b)_{j\ell}\right]\nabla_\Gamma(u_k)_\ell. \end{aligned}$$

Finally

$$\begin{aligned}\left[\nabla_\Gamma(D_\Gamma(u)_{ki})\right]_j = {}& \left[\nabla_\Gamma(\varepsilon_\Gamma(u)_{kj})\right]_i + \left[\nabla_\Gamma(\varepsilon_\Gamma(u)_{ik})\right]_j - \left[\nabla_\Gamma(\varepsilon_\Gamma(u)_{ji})\right]_k \\ &+ \frac{1}{2}\left[(D^2b)_{i\ell}n_j - (D^2b_{j\ell})n_i\right]\nabla_\Gamma(u_k)_\ell \\ &+ \frac{1}{2}\left[(D^2b)_{j\ell}n_k - (D^2b_{k\ell})n_j\right]\nabla_\Gamma(u_i)_\ell \\ &- \frac{1}{2}\left[(D^2b)_{k\ell}n_i - (D^2b_{i\ell})n_k\right]\nabla_\Gamma(u_j)_\ell.\end{aligned}$$

For a vector function ξ in $H_0^1(\omega)^N$ we have

$$\begin{aligned}\nabla_\Gamma\big(D_\Gamma(u)_{ki}\big)\cdot\xi &= \nabla_\Gamma\big(D_\Gamma(u)_{ki}\big)\cdot\xi_\Gamma \\ &= \sum_j\left[\nabla_\Gamma(\varepsilon_\Gamma(u)_{kj})\right]_i(\xi_\Gamma)_j + \left[\nabla_\Gamma(\varepsilon_\Gamma(u)_{ik})\right]\cdot\xi_\Gamma \\ &\quad - \sum_j\left[\nabla_\Gamma(\varepsilon_\Gamma(u)_{ji})\right]_k(\xi_\Gamma)_j \\ &\quad + \frac{1}{2}\left[(D^2b)_{i\ell}n\cdot\xi_\Gamma - (D^2b\xi_\Gamma)_\ell n_i\right]\nabla_\Gamma(u_k)_\ell \\ &\quad + \frac{1}{2}\left[(D^2b\xi_\Gamma)_\ell n_k - (D^2b)_{k\ell}\xi_\Gamma\cdot n\right]\nabla_\Gamma(u_i)_\ell \\ &\quad - \frac{1}{2}\sum_j\left[(D^2b)_{k\ell}n_i - (D^2b)_{i\ell}n_k\right]\sum_j\nabla_\Gamma(u_j)_\ell(\xi_\Gamma)_j.\end{aligned}$$

The three terms involving a summation can be rearranged

$$\begin{aligned}\sum_j\left[\nabla_\Gamma(\varepsilon_\Gamma(u)_{kj})\right]_i(\xi_\Gamma)_j &= \left[\nabla_\Gamma(\varepsilon_\Gamma(u)\xi_\Gamma)_k)\right]_i - \left[\varepsilon_\Gamma(u)D_\Gamma(\xi_\Gamma)\right]_{ki}, \\ \sum_j\left[\nabla_\Gamma(\varepsilon_\Gamma(u)_{ji})\right]_k(\xi_\Gamma)_j &= \left[\nabla_\Gamma(\varepsilon_\Gamma(u)\xi_\Gamma)_i)\right]_k - \left[\varepsilon_\Gamma(u)D_\Gamma(\xi_\Gamma)\right]_{ik}, \\ \sum_j\nabla_\Gamma(u_j)_\ell(\xi_\Gamma)_j &= \left[\nabla_\Gamma(u\cdot\xi_\Gamma) - {}^*D_\Gamma(\xi_\Gamma)n\right]_\ell.\end{aligned}$$

Finally

$$\begin{aligned}\nabla_\Gamma\big(D_\Gamma(u)_{ki}\big)\cdot\xi_\Gamma = {}& \left[\nabla_\Gamma(\varepsilon_\Gamma(u)\xi_\Gamma)_k)\right]_i - \left[\varepsilon_\Gamma(u)D_\Gamma(\xi_\Gamma)\right]_{ki} \\ &+ \left[\nabla_\Gamma(\varepsilon_\Gamma(u)_{ik})\right]\cdot\xi_\Gamma \\ &- \left[\nabla_\Gamma(\varepsilon_\Gamma(u)\xi_\Gamma)_i)\right]_k + \left[\varepsilon_\Gamma(u)D_\Gamma(\xi_\Gamma)\right]_{ik} \\ &+ \frac{1}{2}\left[\big(D^2b\nabla_\Gamma(u_k)\big)_i n\cdot\xi_\Gamma - (D^2b\xi_\Gamma)\cdot\nabla_\Gamma(u_k)n_i\right] \\ &+ \frac{1}{2}\left[(D^2b\xi_\Gamma)\cdot\nabla_\Gamma(u_i)n_k - \big(D^2b\nabla_\Gamma(u_i)\big)_k\xi_\Gamma\cdot n\right] \\ &- \frac{1}{2}\left[\big(D^2b\nabla_\Gamma(u\cdot\xi_\Gamma)\big)_k n_i - \big(D^2b\nabla_\Gamma(u\cdot\xi_\Gamma)\big)_i n_k\right] \\ &+ \frac{1}{2}\left[(D^2b^*D_\Gamma(\xi)u)_k n_i - \big(D^2b^*D_\Gamma(\xi)u\big)_i n_k\right].\end{aligned}$$

After integration by parts whenever necessary

$$
\begin{aligned}
(4.50)\quad & -\int_\omega D_\Gamma(u)_{ki}\mathrm{div}_\Gamma\xi_\Gamma\, d\Gamma \\
&= -\int_\omega \varepsilon_\Gamma(u)_{ik}\mathrm{div}_\Gamma\xi_\Gamma + \left[\varepsilon_\Gamma(u)D_\Gamma(\xi_\Gamma)\right]_{ki} - \left[\varepsilon_\Gamma(u)D_\Gamma(\xi_\Gamma)\right]_{ik} d\Gamma \\
&\quad + \int_\omega H\{[\varepsilon_\Gamma(u)\xi_\Gamma]_k n_i - [\varepsilon_\Gamma(u)\xi_\Gamma]_i n_k\} \\
&\qquad + \frac{1}{2}\left[(D^2b\xi_\Gamma)\cdot\nabla_\Gamma(u_i)n_k - (D^2b\xi_\Gamma)\cdot\nabla_\Gamma(u_k)n_i\right] \\
&\qquad - \frac{1}{2}\{[\vec{\mathrm{div}}_\Gamma(D^2bn_k)]_i - [\vec{\mathrm{div}}_\Gamma(D^2bn_i)]_k\}u\cdot\xi_\Gamma \\
&\qquad + \frac{1}{2}\{[D^2b^*D_\Gamma(\xi_\Gamma)u]_k n_i - [D^2b^*D_\Gamma(\xi_\Gamma)u]_i n_k\}\, d\Gamma,
\end{aligned}
$$

where we have used

$$
\begin{aligned}
\int_\omega [D^2b\nabla_\Gamma(u\cdot\xi_\Gamma)]_i n_k\, d\Gamma
&= \int_\omega (n_k D^2 b_{\cdot i})\cdot\nabla_\Gamma(u\cdot\xi_\Gamma)\, d\Gamma \\
&= \int_\omega H(n_k D^2 b_{\cdot i})n(u\cdot\xi_\Gamma)\, d\Gamma - \int_\omega \mathrm{div}_\Gamma(n_k D^2 b_{\cdot i})(u\cdot\xi_\Gamma)d\Gamma \\
&= -\int_\omega \left[\vec{\mathrm{div}}_\Gamma(n_k D^2 b)\right]_i (u\cdot\xi_\Gamma)\, d\Gamma.
\end{aligned}
$$

There are three groups of terms in the right-hand side of identity (4.50). When it is divided by the H^1-norm of ξ_Γ and the supremum is taken over ξ_Γ, the first term on the first line is $\|\nabla_\Gamma(\varepsilon_\Gamma(u)_{ik})\|_{H^{-1}}$ and the next two terms and the terms of the second line are bounded by a constant times $\|\varepsilon_\Gamma(u)\|_{L^2}$. The last two lines are bounded by $c\|u\|_{L^2}$ since $b\in C^{2,1}$ and $D^2b\in C^{0,1}$ implies $D^2b\in W^{1,\infty}(\omega)^{N\times N}$. As for the middle lines the terms are all of order zero in ξ_Γ and order one in u. Since $b\in C^{2,1}$ upon integration by parts we get terms bounded by $c\|u\|_{L^2}$. To conclude

$$
\|\nabla_\Gamma(D_\Gamma(u)_{ki})\|_{H^{-1}} \le c\left[\|\nabla_\Gamma(\varepsilon_\Gamma(u)_{ik})\|_{H^{-1}} + \|\varepsilon_\Gamma(u)\|_{L^2} + \|u\|_{L^2}\right].
$$

But $\varepsilon_\Gamma(u)$ and u belong to L^2. So from Lemma 4.2 and by equivalence of norms

$$
\|D_\Gamma(u)_{ki}\|_{L^2(\omega)} \le c\left[\|D_\Gamma(u)_{ki}\|_{H^{-1}(\omega)} + \|\nabla_\Gamma(D_\Gamma(u)_{ki})\|_{H^{-1}(\omega)}\right].
$$

But $D_\Gamma(u)_{ki} = \left[\nabla_\Gamma(u_k)\right]_i$ implies that

$$
\|D_\Gamma(u)_{ki}\|_{H^{-1}(\omega)} \le \|\nabla_\Gamma(u_k)\|_{H^{-1}(\omega)} \le \|u_k\|_{L^2(\omega)}.
$$

Similarly

$$
\|\nabla_\Gamma(\varepsilon_\Gamma(u)_{ik})\|_{H^{-1}} \le c\|\varepsilon_\Gamma(u)_{ik}\|_{L^2(\omega)} \le c\|\varepsilon_\Gamma(u)\|_{L^2(\omega)}.
$$

Finally putting all the above estimates together

$$
\|D_\Gamma(u)_{ki}\|_{L^2(\omega)} \le c\left[\|u\|_{L^2(\omega)} + \|\varepsilon_\Gamma(u)\|_{L^2(\omega)}\right]
$$

and this completes the proof. □

4.8. Tangential Poincaré's inequality: Vectorial case. As in the case of the Korn's inequality we give a proof of the tangential Poincaré's inequality via local coordinates and the corresponding inequality in $\mathbb{R}^{N-1}$. Choose as local basis in each point x of ω the one generated by the $C^{1,1}$-diffeomorphism in a neighborhood of $\Gamma = \partial\Omega$ where $b = b_\Omega$ is of class $C^{1,1}$ (cf. § 4.6). From the explicit expressions of the elements of the covariant and contravariant bases in terms of the function b in § 4.6, we have a direct access to all the quantities appearing in the identity

$$\varepsilon(\vec{v}_\Gamma)_{\alpha\beta} = \left[\varepsilon_\Gamma^P(v_\Gamma) - \varepsilon_\Gamma^P(a^\lambda)v_\lambda\right] a_\beta \cdot a_\alpha.$$

For the plate we can choose the τ_i's independent of x as $\tau_i(x) = e_i$. From the previous formulae $a_\alpha = a^\alpha = e_\alpha$ are independent of both y and x and $\Gamma^\lambda_{\alpha\beta} = 0$. Similarly for the cylinder in $\mathbb{R}^3$ whose axis coincides with the 3-axis, we can choose

$$\tau_1(x) = \frac{(-x_2, x_1, 0)}{\sqrt{x_1^2 + x_2^2}}, \quad \tau_2(x) = (0,0,1) \implies a_\alpha(y) = \tau_\alpha(x),$$

and again $\Gamma^\lambda_{\alpha\beta} = 0$. However in the general case $\Gamma^\lambda_{\alpha\beta} \neq 0$. By direct computation

$$\begin{aligned} D_\Gamma(a^\lambda)(y) &= \left[n(y)^*(D^2b(y)\tau_\lambda) + \tau_\lambda \cdot n(y) D^2b(y)\right], \\ D_\Gamma^P(a^\lambda)(y) &= P D_\Gamma(a^\lambda)(y) = -\tau_\lambda \cdot n(y) D^2 b(y), \end{aligned}$$

and

$$\varepsilon(\vec{v}_\Gamma)_{\alpha\beta} = \left[\varepsilon_\Gamma^P(v_\Gamma) - v_\lambda \tau_\lambda \cdot n D^2 b\right] a_\beta \cdot a_\alpha. \tag{4.51}$$

Since $b \in C^{1,1}$ the norm of the matrix D^2b is uniformly bounded by a constant $c > 0$ in $\overline{\omega}$. Moreover in view of (4.51) there exists a constant $c_x > 0$ such that for all $v \in H^1(\omega_x)^N$,

$$\left\|\varepsilon(\vec{v}_\Gamma)\right\| \leq c_x \left[\|\varepsilon_\Gamma^P(v_\Gamma)\| + c\|v_\Gamma\| \max_{\substack{1 \leq \lambda \leq N-1 \\ y \in \operatorname{supp} v_\Gamma}} \left|\tau_\lambda \cdot n(y)\right|\right],$$

and since for all λ we have the identity $\tau_\lambda \cdot n(x) = 0$

$$\left|\tau_\lambda \cdot n(y)\right| = \left|\tau_\lambda \cdot \big(n(y) - n(x)\big)\right| \leq \left|n(y) - n(x)\right|$$

and

$$\left\|\varepsilon(\vec{v}_\Gamma)\right\| \leq c_x \left[\|\varepsilon_\Gamma^P(v_\Gamma)\| + c\|v_\Gamma\| \max_{y \in \operatorname{supp} v_\Gamma} |n(y) - n(x)|\right].$$

The boundary ∂V of V in $\mathbb{R}^{N-1}$ is $\{\xi' \in \mathbb{R}^{N-1} : |\xi'| = \varepsilon\}$ (cf. § 4.6). It is mapped onto the (relative) boundary γ_x of ω_x in Γ which is connected as the image of ∂V through the diffeomorphism Φ.

Given a subset $\gamma_{x,0}$ of γ_x of non-zero $(N-1)$-capacity, define the spaces

$$\begin{aligned} H^1_{\gamma_{x,0}}(\omega_x)^N &= \left\{v \in H^1(\omega_x)^N : v = 0 \text{ on } \gamma_{x,o}\right\}, \\ H^1_{\Phi^{-1}(\gamma_{x,0})}(V)^{N-1} &= \left\{\xi' \in H^1(V)^{N-1} : \xi' = 0 \text{ on } \Phi^{-1}(\gamma_{x,0})\right\}. \end{aligned}$$

If ω_x is connected and $\Phi^{-1}(\gamma_{x,0})$ has a non-zero $(N-1)$-capacity, we have Poincaré's inequality in the domain $\Phi^{-1}(\gamma_{x,0})$ of $\mathbb{R}^{N-1}$: there exists $c_x > 0$ such that

$$\forall \vec{v} \in H^1_{\Phi^{-1}(\gamma_{x,0})}(V)^{N-1}, \quad \|\vec{v}\| \leq c_x \left\|\varepsilon(\vec{v})\right\|.$$

and finally there exists a constant $c_x > 0$ such that for all $v \in H^1_{\gamma_{x,0}}(\omega_x)^N$

$$\|v_\Gamma\| \leq c_x \left[\|\varepsilon_\Gamma^P(v_\Gamma)\| + c\|v_\Gamma\| \max_{y \in \operatorname{supp} v_\Gamma} |n(y) - n(x)|\right].$$

By assumption $n \in C^{0,1}(\overline{\omega})^N$ and for any $\varepsilon > 0$ there exists $\delta > 0$ such that

$$\forall y, z \in \overline{\omega}, \quad |y - z| < \delta \implies \big|n(y) - n(x)\big| \le \varepsilon,$$

and for all $v \in H^1_{\gamma_x,0}\big(\omega_x \cap B(x,\delta)\big)^N$

$$\|v_\Gamma\| \le c_x \big\|\varepsilon_\Gamma^P(v_\Gamma)\big\| + cc_x \max_{y \in B(x,\delta)} \big|n(y) - n(x)\big|$$

and

$$[1 - \varepsilon c c_x]\, \|v_\Gamma\| \le c_x \big\|\varepsilon_\Gamma^P(v_\Gamma)\big\|.$$

By choosing $\varepsilon \le 1/(2cc_x)$ for each $x \in \omega$ there exists $\delta_x > 0$ such that $B(x, \delta_x) \subset \mathcal{U}(x)$ and

$$\forall v \in H^1_{\gamma_x,0}\big(\omega \cap B(x,\delta_x)\big)^N, \quad \|v_\Gamma\| \le 2c_x \big\|\varepsilon_\Gamma^P(v_\Gamma)\big\|. \tag{4.52}$$

To go from the local inequality we now follow the proof of Theorem 4.1.

THEOREM 4.4. *Under Assumption 4.1 let ω be a bounded open domain in Γ with a Lipschitzian boundary γ. Then there exists a constant $c > 0$ such that for all $v \in H^1_0(\omega)^N$*

$$\int_\omega |v_\Gamma|^2 \, d\Gamma \le c^2 \int_\omega \big\|\varepsilon_\Gamma^P(v_\Gamma)\big\|^2 \, d\Gamma, \tag{4.53}$$

$$\int_\omega |v|^2 \, d\Gamma \le c^2 \int_\omega \big\|\varepsilon_\Gamma^P(v_\Gamma)\big\|^2 + |\nabla_\Gamma v_n|^2 \, d\Gamma. \tag{4.54}$$

The result is also true for

$$H^1_{\gamma_0}(\omega)^N = \big\{v \in H^1(\omega)^N : v = 0 \text{ on } \gamma_0\big\}.$$

if ω is connected and γ_0 is a subset of γ with non-zero $(N-1)$-capacity.

PROOF. Assume that $v = v_\Gamma$ is tangential. Consider the following quotient and its infimum

$$q(v) = \frac{\int_\omega \|\varepsilon_\Gamma^P(v)\|^2 \, d\Gamma}{\int_\omega |v|^2 \, d\Gamma}, \quad v \neq 0,$$

$$q \stackrel{\text{def}}{=} \inf\big\{q(v) \colon \|v\|_{L^2} = 1, v \in H^1_{\gamma_0}(\omega)^N\big\}.$$

We proceed by contradiction. Assume that $q = 0$. Let $\{v_k\}$ be a sequence in $H^1_{\gamma_0}(\omega)^N$ such that

$$\int_\omega |v_k|^2 \, d\Gamma = 1 \quad \text{and} \quad \int_\omega \big\|\varepsilon_\Gamma^P(v_k)\big\|^2 \, d\Gamma \to 0,$$

as k goes to infinity. By Korn's inequality (4.44) the sequence is bounded in $H^1_{\gamma_0}(\omega)^N$ and there exists $v \in H^1_{\gamma_0}(\omega)^N$ and a subsequence, still denoted $\{v_k\}$, such that $v_k \to v$ in $H^1_{\gamma_0}(\omega)^N$-weak. Therefore

$$1 = \int_\omega |v_k|^2 \, d\Gamma \to \int_\omega |v|^2 \, d\Gamma,$$

$$\int_\omega \big\|\varepsilon_\Gamma^P(v)\big\|^2 \, d\Gamma \le \liminf_{k \to \infty} \int_\omega \big\|\varepsilon_\Gamma^P(v_k)\big\|^2 \, d\Gamma = 0,$$

and $\varepsilon_\Gamma^P(v) = 0$.

Now we are going to use Poincaré's inequality locally. Since ω is connected there exists δ'_x, $0 < \delta'_x \le \delta_x$, sufficiently small such that $\omega'_x \stackrel{\text{def}}{=} \omega \cap B(x, \delta'_x)$ is connected. Denote by γ'_x the boundary of ω'_x in Γ. By compactness of ω there exists a finite sequence $\{x_i\}$ of points in ω such that $\omega \subset \cup B(x, \delta_{x'_i})$ and $\omega = \cup \omega'_i$ where $\omega'_i \stackrel{\text{def}}{=} \omega \cap B(x, \delta_{x'_i})$.

There exists i_0 such that $\gamma_0 \cap \gamma'_{x_0}$ has non-zero $(N-1)$-capacity and $v = 0$ on $\gamma \cap \gamma_{x'_0}$. Therefore using the Poincaré's inequality in $\omega_{x'_0}$, $v = 0$ in $\omega_{x'_0}$. Then we go to a $i_1 \neq i_0$ such that $\omega_{x'_0} \cap \omega_{x'_1} \neq \varnothing$ and hence $v = 0$ on a part of the boundary $\gamma_{x'_1}$. So we conclude that $v = 0$ in $\omega_{x'_1}$. Proceeding in that way we visit all the $\omega_{x'_i}$'s and conclude that $v = 0$ in ω. This is a contradiction since the L^2-norm of v is equal to one. Necessarily $q > 0$ and we can pick $c = 1/q$. The second identity is obtained by combining the first with the scalar case in the previous section. □

REMARK 4.1. Recall that

$$\left\|\varepsilon_\Gamma(v)\right\|^2 = \left\|\varepsilon^P_\Gamma(v_\Gamma) + v_n D^2 b\right\|^2 + \frac{1}{2}\|\nabla_\Gamma v_n - D^2 b v_\Gamma\|^2.$$

So for v tangential ($v_n = 0$) the inequalities of the theorem remain true with $\left\|\varepsilon_\Gamma(v_\Gamma)\right\|$ in place of $\left\|\varepsilon^P_\Gamma(v_\Gamma)\right\|$. However when v is not tangential $\left\|\varepsilon_\Gamma(v)\right\| = 0$ is equivalent to the system of equations

$$\varepsilon^P_\Gamma(v_\Gamma) + v_n D^2 b = 0, \quad \nabla_\Gamma v_n - D^2 b v_\Gamma = 0,$$

which does not necessarily have a unique solution in $H^1_0(\omega)^N$.

5. Approximations of Thin Shells

5.1. Some elementary linear elasticity theory. We start with a brief review of N-dimensional linear elasticity for linear isotropic homogeneous materials with small loads and small displacements.

Let B be an open bounded body in $\mathbb{R}^N$. Let $V\colon B \to \mathbb{R}^N$ be the displacement vector field resulting from the action of the applied forces. The ***linear strain tensor*** is defined as

$$\varepsilon(V) = \frac{1}{2}[DV + {}^*DV],$$

and the resulting ***stress tensor*** is denoted by σ. The relationship between σ and ε is the ***constitutive law*** which we assume here to be the simplest possible one

$$\sigma = 2\mu\varepsilon(V) + \lambda \operatorname{tr}\varepsilon(V) I, \tag{5.1}$$

where $\mu > 0$ and $\lambda \geq 0$ are the Lamé coefficients. They can also be expressed in term of the Young modulus E and the Poisson coefficient ν

$$2\mu = \frac{E}{1+\nu} \quad \text{and} \quad \lambda = \frac{E}{1+\nu}\frac{\nu}{1-2\nu}.$$

The strain or deformation energy is given by

$$\begin{aligned} E(V) &= \frac{1}{2}\int_B \sigma(V) \cdot\cdot\, \varepsilon(V)\, dx \\ &= \frac{1}{2}\int_B 2\mu\varepsilon(V) \cdot\cdot\, \varepsilon(V) + \lambda\left|\operatorname{tr}\varepsilon(V)\right|^2 dx. \end{aligned}$$

This strain energy is the result of the work of the body forces F and the boundary forces g

$$\int_B F \cdot V\, dx + \int_{\partial B} g \cdot V\, d(\partial B).$$

The equilibrium is typically given by the deformation vector function which minimizes

$$\inf_{V \text{ and boundary conditions}} E(V) + W(V).$$

The choice of the appropriate function space is given by ***Korn's inequality***: there exists $c > 0$ such that

$$\int_B \|DV\|^2\, dx \le c^2 \int_B \|\varepsilon(V)\|^2 + |V|^2\, dx,$$

which means that

$$H^1(B)^N = \{V \in L^2(D)^N : \varepsilon(V) \in L^2(D)^{N\times N}\}.$$

Moreover the kernel of ε corresponds to the (approximate) rigid displacements

$$\ker \varepsilon(V) = \{V : V(x) = a + Bx, a \in \mathbb{R}^N \quad \text{and} \quad B + {}^*B = 0\},$$

so that the solution of the Neumann boundary value problem is unique up to rigid displacements if the following compatibility condition is verified

$$\int_B F \cdot V\, dx + \int_{\partial B} g \cdot V\, d(\partial B) = 0, \forall V \in \ker \varepsilon.$$

For Dirichlet boundary conditions we have a Poincaré's inequality

$$\int_B \|V\|^2\, dx \le c^2 \int_B \|\varepsilon(V)\|^2\, dx, \forall V \in H_0^1(B)^N,$$

and this gives the uniqueness of solution in $H_0^1(B)^N$.

Another approach is look at the stationary points of the functional

$$G(\tau, v) = \int_B \frac{1}{2} C\tau \cdot\cdot \tau - \tau \cdot\cdot \varepsilon(V) + F \cdot v\, dx + \int_{\partial B} g \cdot V\, d\gamma, \tag{5.2}$$

for an $N \times N$-matrix function τ, a vector function V and a transformation C which maps symmetric $N \times N$-matrices into $N \times N$-matrices. This is often referred to as the *Hellinger–Reissner* principle. The stationary points (σ, U) are formally characterized by the following system of variational equations

$$\int_B C\sigma \cdot\cdot \tau - \tau \cdot\cdot \varepsilon(U)\, dx = 0, \forall \tau,$$

$$\int_B -\sigma \cdot\cdot \varepsilon(v) + F \cdot V\, dx + \int_{\partial B} g \cdot V\, d\gamma = 0, \forall V.$$

When C is invertible and C^{-1} is symmetrical and coercive, the second equation characterizes the minimizing element u of the functional

$$J(U) = \inf_V J(v), \quad J(V) = \int_B \frac{1}{2} C^{-1}\varepsilon(V) \cdot\cdot \varepsilon(V) - F \cdot V\, d\Gamma - \int_{\partial B} g \cdot V\, d\gamma.$$

However the variational equation

$$\int_B -C^{-1}\varepsilon(U) - F \cdot V\, dx - \int_{\partial B} g \cdot V\, d\gamma = 0, \forall V,$$

still makes sense for transformations C which are not necessarily symmetrical but which verify the conditions of the Lax–Milgram Theorem

$$\exists \alpha > 0 \text{ such that } |C^{-1}\tau \cdot\cdot \tau| \ge \alpha \|\tau\|^2.$$

In particular this is true when C^{-1} is coercive and the whole theory applies to more general constitutive laws than (5.1).

5.2. Definition of the shell. A *shell* is a bounded open domain in $\mathbb{R}^3$ around a *midsurface* of (possibly variable) *thickness* $2h$. The midsurface ω is a bounded open subset in a two-dimensional submanifold of $\mathbb{R}^3$ and the (variable) thickness is a strictly positive function $\tilde{h}$ defined on ω.

Of course this construction readily extends to $\mathbb{R}^N$. Let the midsurface ω be a bounded open domain in Γ which satisfies Assumption 4.1. When $\omega = \Gamma$, the shell has no boundary; when $\omega \neq \Gamma$ we denote by γ the boundary of ω in Γ. Since $\overline{\omega}$ is compact, there exists a bounded neighborhood $N(\omega)$ of ω such that $b \in C^{1,1}\big(N(\omega)\big)$. For each $X \in \omega$, we introduce the quantities

$$0 < h^-(X) \stackrel{\text{def}}{=} \sup\big\{-b(x)\colon x \in N(\omega), p(x) = X\big\} < \infty,$$
$$0 < h^+(X) \stackrel{\text{def}}{=} \sup\big\{b(x)\colon x \in N(\omega), p(x) = X\big\} < \infty.$$

The ***thickness of the shell*** is a Lipschitz continuous function $\tilde{h}\colon \omega \to \mathbb{R}^+$ such that

$$\forall X \in \omega, \quad 0 < \tilde{h}(X) \leq \min_{X \in \omega}\big\{h^-(X), h^+(X)\big\}.$$

Given ω and $\tilde{h}$ the ***shell*** is the set

$$S_{\tilde{h}}(\omega) \stackrel{\text{def}}{=} \Big\{x \in \mathbb{R}^N : p(x) \in \omega, \big|b(x)\big| < \tilde{h}\big(p(x)\big)\Big\}.$$

It will be assumed that ω and $\tilde{h}$ are such that the set $S_{\tilde{h}}(\omega)$ is a bounded open domain in $\mathbb{R}^N$ with a Lipschitzian boundary.

When $\omega \neq \Gamma$, $S_{\tilde{h}}(\omega)$ has a ***lateral boundary***

$$\Sigma_{\tilde{h}}(\gamma) = \Big\{x \in \mathbb{R}^N : p(x) \in \gamma, \big|b(x)\big| < \tilde{h}\big(p(x)\big)\Big\},$$

which is an $(N-1)$-dimensional surface normal to the midsurface Γ. It is important to keep in mind that we use the distance function $b = b_\Omega$ and not the distance function to ω.

5.3. Expansion of the linear strain tensor. For convenience assume that the thickness of the shell is constant. As for linear elasticity in $\mathbb{R}^N$ the unknown is a displacement vector defined in $S_h(\omega)$. Since the thickness is the variable which will be small we start with a displacement vector function

$$(X, z) \mapsto v(X, z)\colon \omega \times \,]-h, h[\to \mathbb{R}^N,$$

expressed in the local coordinates variables (X, z) with respect to the submanifold Γ. Assume that it can be developed as an infinite sum

$$v(X, z) = \sum_{i=0}^{\infty} v^i(X) z^i, \quad v^i : \omega \to \mathbb{R}^N.$$

Since we want to compute the corresponding strain tensor in $\mathbb{R}^N$ it is convenient to use the transformation T_z and the mapping

$$T\colon \omega \times \,]-h, h[\to S_h(\omega) = \big\{x \in \mathbb{R}^N : \big|b(x)\big| < h,\, p(x) \in \omega\big\}, \tag{5.3}$$

$$T(X, z) \stackrel{\text{def}}{=} T_z(X) = X + z\nabla b(X), \quad T^{-1}(x) = \big(p(x), b(x)\big). \tag{5.4}$$

When $b \in C^{1,1}(S_h(\omega))$, T is a bijection from $\omega\times\,]-h,h[$ to $S_h(\omega)$ and T and T^{-1} are Lipschitz continuous. Going to $\mathbb{R}^N$ we get

$$V(x) = (v \circ T^{-1})(x) = \sum_{i=0}^{\infty} (v^i \circ p)(x) b(x)^i.$$

This expression is most interesting in view of the tangential calculus developed in § 3 since we have $D(v^i \circ p)$ in terms of $D_\Gamma(v^i)$ and we know the meaning of the derivatives of b.

From (3.4) a direct computation yields the following expressions in $S_h(\omega)$

$$\boxed{DV = \sum_{i=0}^{\infty} b^i \{D_\Gamma(v^i) \circ p[I - bD^2 b] + (i+1)(v^{i+1*} n) \circ p\}} \tag{5.5}$$

$$\boxed{DV = \sum_{i=0}^{\infty} b^i \{D_\Gamma(v^i) + (i+1)(v^{i+1*} n)\} \circ p[I - bD^2 b]} \tag{5.6}$$

since

$$\begin{aligned} v^{i+1} \circ p^* \nabla b[I - bD^2 b] &= v^{i+1} \circ p^* \nabla b - b v^{i+1} \circ p^* (D^2 b \nabla b), \\ &= v^{i+1} \circ p^* \nabla b. \end{aligned}$$

In local coordinates

$$\boxed{DV \circ T_z = \sum_{i=0}^{\infty} z^i \{D_\Gamma(v^i)[I + zD^2 b]^{-1} + (i+1)(v^{i+1*} n)\}}$$

$$\boxed{DV \circ T_z = \sum_{i=0}^{\infty} z^i \{D_\Gamma(v^i) + (i+1)(v^{i+1*} n)\}[I + zD^2 b]^{-1}}$$

For $\|zD^2 b\|$ small, that is

ASSUMPTION 5.1. Let Γ be the boundary of a domain Ω in $\mathbb{R}^N$ and let ω be a bounded open domain in Γ. Assume that there exists $h > 0$ such that $b_\Omega \in C^{1,1}(S_h(\omega))$ and there exists β, $0 < \beta < 1$, such that

$$\forall X \in \omega, \quad h\|D^2 b_\Omega(X)\| \le \beta.$$

With this assumption the matrix $[I + zD^2 b]^{-1}$ can be expressed as

$$[I + zD^2 b]^{-1} = \sum_{i=0}^{\infty} (-D^2 b)^i z^i. \tag{5.7}$$

Notice that we have two levels of approximation: the asymptotic development of v with respect to z and the one of $[I + zD^2 b]^{-1}$ with respect to $zD^2 b$ which means that the shell can be thin or shallow or both. After substitution

$$\boxed{DV \circ T_z = \sum_{i=0}^{\infty} z^i \left\{ (i+1)(v^{i+1*} n) + \sum_{k=0}^{i} D_\Gamma(v^k)(-D^2 b)^{i-k} \right\}}$$

The linear strain tensor $2\varepsilon(V) = D(V) + {}^*DV$ is given by

(5.8)
$$\boxed{\varepsilon(V) \circ T_z = \sum_{i=1}^{\infty} \varepsilon^i z^i}$$

where

(5.9)
$$\boxed{\begin{aligned} 2\varepsilon^i(v) &= (i+1)[v^{i+1*}n + n^* v^{i+1}] \\ &\quad + \sum_{k=0}^{i} [D_\Gamma(v^k)(-D^2 b)^{i-k} + (-D^2 b)^{i-k*} D_\Gamma(v^k)] \end{aligned}}$$

5.4. Expansion of the strain energy. We make the following assumption.

ASSUMPTION 5.2. Let C be a bijective linear mapping which transforms symmetrical $N \times N$ matrices onto symmetrical $N \times N$ matrices. Assume, in addition, that there exists $\nu > 0$ such that $C^{-1}\tau \cdot\cdot \tau \geq \nu \, \|\tau\|^2$ for all τ.

Next we use Federer's formula to compute the strain energy. The first term is

$$\begin{aligned} \int_{S_h(\omega)} C^{-1}\varepsilon \cdot\cdot \overline{\varepsilon} \, dx &= \int_\omega \left[\int_{-h}^{h} j_z C^{-1} \varepsilon \circ T_z \cdot\cdot \overline{\varepsilon} \circ T_z \, dz \right] d\Gamma, \\ &= \sum_{i=0}^{\infty} \int_\omega \alpha_i(h) \sum_{j=0}^{i} C^{-1}\varepsilon^j \cdot\cdot \overline{\varepsilon}^{i-j} \, d\Gamma, \end{aligned}$$

where $\varepsilon = \varepsilon(\{v_i\})$, $\overline{\varepsilon} = \varepsilon(\{\overline{v}_i\})$ and the α_n's result from the integration with respect to z from $-h$ to h

(5.10)
$$\begin{aligned} &\boxed{\alpha_n(h) = \int_{-h}^{h} j_z z^n \, dz, \quad n \geq 0} \\ &= h^{n+1} \sum_{i=0}^{N-1} [1 - (-1)^{n+i+1}] \frac{h^i}{n+i+1} \kappa_i \end{aligned}$$

They are polynomials in odd powers of h. For instance for the first three

(5.11)
$$\begin{aligned} \alpha_0(h) &= 2h\left\{1 + \frac{h^2}{2} \sum_{i=2}^{N-1} [1 - (-1)^{i+1}] \frac{h^{i-2}}{i+1} \kappa_i \right\} \\ &= 2h\big(1 + h^2 c_0(h)\big), \\ \alpha_1(h) &= 2\frac{h^3}{3}\left\{\kappa_1 + 3\frac{h^2}{2} \sum_{i=3}^{N-1} [1 - (-1)^{i+2}] \frac{h^{i-3}}{i+2} \kappa_i \right\}, \\ &= 2\frac{h^3}{3}\big(\kappa_1 + h^2 c_1(h)\big) \\ \alpha_2(h) &= 2\frac{h^3}{3}\left\{1 + 3\frac{h^2}{2} \sum_{i=2}^{N-1} [1 - (-1)^{i+3}] \frac{h^{i-2}}{i+3} \kappa_i \right\}, \\ &= 2\frac{h^3}{3}\big(1 + h^2 c_2(h)\big). \end{aligned}$$

For $N = 3$ we have $j_z = 1 + \kappa_1 z + \kappa_2 z^2$ and for $m \geq 1$

$$\boxed{\alpha_{2m}(h) = 2\frac{h^{2m+1}}{2m+1} + 2\frac{h^{2m+3}}{2m+3}\kappa_2, \quad \alpha_{2m+1}(h) = 2\frac{h^{2m+3}}{2m+3}\kappa_1}$$

and for the first five terms

$$\begin{aligned} {}_0(h) &= 2h + 2\frac{h^3}{3}\kappa_2, & \alpha_1(h) &= 2\frac{h^3}{3}\kappa_1, \\ \alpha_2(h) &= 2\frac{h^3}{3} + 2\frac{h^5}{5}\kappa_2, & \alpha_3(h) &= 2\frac{h^5}{5}\kappa_1, \\ \alpha_4(h) &= 2\frac{h^5}{5} + 2\frac{h^7}{7}\kappa_2, & \alpha_5(h) &= 2\frac{h^7}{7}\kappa_1. \end{aligned}$$

For the linear term we proceed in the same way using Federer's formula. For simplicity let $g = 0$. Assume that the force F is of the form

$$F = f \circ T^{-1}, \quad f = \sum_{i=0}^{\infty} f_i z^i, \ f_i \in L^2(\omega)^N, \ \left\{\sum_{i=0}^{\infty} h^{2i} \|f_i\|^2_{L^2(\omega)}\right\}^{1/2} < \infty. \tag{5.12}$$

Then

$$\begin{aligned} \int_{S_h(\omega)} F \cdot V \, dx &= \int_\omega \left\{ \int_{-h}^{h} j_z \sum_{j=0}^{\infty} f^j z^j \cdot \sum_{i=0}^{\infty} v^i z^i \, dz \right\} d\Gamma, \\ &= \sum_{i=0}^{\infty} \int_\omega \left\{ \alpha_i(h) \sum_{j=0}^{i} f^{i-j} \cdot v^j \, dz \right\} d\Gamma. \end{aligned}$$

Finally

$$\begin{aligned} &\int_{S_h(\omega)} C^{-1}\varepsilon(V) \cdot\cdot\, \varepsilon(\overline{V}) - F \cdot V \, dx \\ &\qquad = \sum_{i=0}^{\infty} \int_\omega \alpha_i(h) \sum_{j=0}^{i} \{C^{-1}\varepsilon^{i-j}(v) \cdot\cdot\, \varepsilon^j(\overline{v}) - f^{i-j} \cdot \overline{v}^j\} \, d\Gamma. \end{aligned} \tag{5.13}$$

Define

$$B_j^i \overset{\text{def}}{=} C^{-1}\varepsilon^{i-j}(v) \cdot\cdot\, \varepsilon^j(\overline{v}) - f^{i-j} \cdot \overline{v}^j.$$

For $N = 3$ we can write explicitly the sequence of powers of h. The sum (5.13) can be split into two sums

$$\begin{aligned} &\int_{S_h(\omega)} C^{-1}\varepsilon(V) \cdot\cdot\, \varepsilon(\overline{V}) - F \cdot V \, dx \\ &\qquad = \sum_{m=0}^{\infty} \int_\omega \left[2\frac{h^{2m+1}}{2m+1} + 2\frac{h^{2m+3}}{2m+3}\kappa_2 \right] \sum_{j=0}^{2m} B_j^{2m} \, d\Gamma \\ &\qquad\qquad + \sum_{m=0}^{\infty} \int_\omega 2\frac{h^{2m+3}}{2m+3}\kappa_1 \sum_{j=0}^{2m+1} B_j^{2m+1} \, d\Gamma, \end{aligned}$$

and after regrouping terms the right-hand side becomes

$$\begin{aligned} &\int_\omega 2hB_0^0 \, d\Gamma \\ &\quad + \sum_{m=0}^{\infty} \int_\omega 2\frac{h^{2m+3}}{2m+3} \left[\kappa_2 \sum_{j=0}^{2m} B_j^{2m} + \kappa_1 \sum_{j=0}^{2m+1} B_j^{2m+1} + \sum_{j=0}^{2m+2} B_j^{2m+2} \right] d\Gamma. \end{aligned} \tag{5.14}$$

This identity must be zero for all $\overline{v}$. It is customary to proceed by polynomial approximation of the displacement variable v, but there are obviously many ways to do the approximation: by direct projection or by mixed approximations in the solution and the test functions. The good approximations are obviously the ones which converge and converge with the best rates.

5.5. Asymptotic membrane equation and Love–Kirchhoff condition. It is easy to formally deal with the term in $2h$ in expression (5.14)

$$\int_\omega \{C^{-1}\varepsilon^0(v) \cdot\cdot\, \varepsilon^0(\overline{v}) - f^0 \cdot \overline{v}^0\}\, d\Gamma = 0,\ \forall \overline{v}^0, \overline{v}^1, \tag{5.15}$$

where

$$\varepsilon^0(v) = \frac{1}{2}\left[v^{1*}n + n^*v^1\right] + \varepsilon_\Gamma(v^0).$$

This characterizes the minimizing elements (v^0, v^1) of the energy functional

$$e_0(v^0, v^1) \stackrel{\text{def}}{=} \int_\omega \frac{1}{2} C^{-1}\varepsilon^0(v^0, v^1) \cdot\cdot\, \varepsilon^0(v^0, v^1) - f^0 \cdot v^0\, d\Gamma. \tag{5.16}$$

The two test functions $\overline{v}^0$ and $\overline{v}^1$ yield the following two variational equations for the vector functions v^0 and v^1

$$\int_\omega \left\{C^{-1}\varepsilon^0(v) \cdot\cdot\, \frac{1}{2}[\overline{v}^{1*}n + n^*\overline{v}^1]\right\} d\Gamma = 0, \forall \overline{v}^1,$$

$$\int_\omega \{C^{-1}\varepsilon^0(v) \cdot\cdot\, \varepsilon_\Gamma(\overline{v}^0) - f^0 \cdot \overline{v}^0\}\, d\Gamma = 0, \forall \overline{v}^0.$$

By using the decomposition (3.34) of § 3

$$\begin{aligned} C^{-1}\varepsilon^0(v) \cdot\cdot\, \varepsilon_\Gamma(\overline{v}^0) = \left[C^{-1}\varepsilon^0(v)\right]^P \cdot\cdot\, \varepsilon_\Gamma^P(\overline{v}^0) &+ 2P\left[C^{-1}\varepsilon^0(v)\right]n \cdot P\varepsilon_\Gamma^P(\overline{v}^0)n \\ &+ \left([C^{-1}\varepsilon^0(v)]n \cdot n\right)\left(\varepsilon_\Gamma^P(\overline{v}^0)n \cdot n\right), \end{aligned}$$

we get

$$\boxed{\begin{gathered} \left[C^{-1}\varepsilon^0(v)\right]n = 0 \\ \int_\omega \{[C^{-1}\varepsilon^0(v)]^P \cdot\cdot\, \varepsilon_\Gamma^P(\overline{v}^0) - f^0 \cdot \overline{v}^0\}\, d\Gamma = 0, \forall \overline{v}^0 \end{gathered}}$$

For the special constitutive law $C^{-1}\varepsilon = 2\mu\varepsilon + \lambda \operatorname{tr} \varepsilon I$

$$\begin{aligned} \left[C^{-1}\varepsilon^0(v)\right]n =& 2\mu\left[\varepsilon_\Gamma(v_0)n + \frac{1}{2}v_\Gamma^1 + v_n^1 n\right] + \lambda\left[\operatorname{tr}\varepsilon_\Gamma(v_0) + v_n^1\right]n \\ =& 2\mu\left[\varepsilon_\Gamma(v_0)n + \frac{1}{2}v_\Gamma^1\right] + \left[\lambda \operatorname{tr}\varepsilon_\Gamma(v_0) + (2\mu + \lambda)v_n^1\right]n, \end{aligned}$$

and the condition $\left[C^{-1}\varepsilon^0(v)\right]n = 0$ yields the following explicit expression for the normal and tangential components of v^1 in term of v^0

$$\boxed{\begin{aligned} v_n^1 = v^1 \cdot n =& -\frac{\lambda}{2\mu + \lambda}\operatorname{div}_\Gamma(v^0) \\ =& -\frac{\lambda}{2\mu + \lambda}\left[Hv_n^0 + \operatorname{div}_\Gamma(v_\Gamma^0)\right] \\ v_\Gamma^1 = -2\varepsilon_\Gamma(v^0)n =& -\nabla_\Gamma(v_n^0) + D^2 b v_\Gamma^0 \end{aligned}} \tag{5.17}$$

The second equation is the *Love-Kirchhoff condition.* The substitution of the value of v^1 into the term

$$\begin{aligned}
\left[C^{-1}\varepsilon^0(v)\right]^P &= P\left[2\mu\varepsilon^0(v) + \lambda \operatorname{tr}\varepsilon^0(v) I\right]P \\
&= 2\mu\varepsilon^{0P}(v) + \lambda \operatorname{tr}\varepsilon^0(v) P \\
&= 2\mu\varepsilon_\Gamma^P(v^0) + \lambda\left[\operatorname{tr}\varepsilon_\Gamma^P(v^0) + v_n^1\right]P \\
&= 2\mu\varepsilon_\Gamma^P(v^0) + \frac{2\mu\lambda}{2\mu+\lambda}\operatorname{tr}\varepsilon_\Gamma^P(v^0) P
\end{aligned}$$

$$\left[C^{-1}\varepsilon^0(v)\right]^P \cdot\cdot\, \varepsilon_\Gamma^P(\overline{v}^0) = 2\mu\varepsilon_\Gamma^P(v^0) \cdot\cdot\, \varepsilon_\Gamma^P(\overline{v}^0) + \frac{2\mu\lambda}{2\mu+\lambda}\operatorname{tr}\varepsilon_\Gamma^P(v^0)\operatorname{tr}\varepsilon_\Gamma^P(\overline{v}^0),$$

yields the second equation for v^0: for all $\overline{v}^0$

$$\boxed{\begin{aligned}\int_\omega & 2\mu\varepsilon_\Gamma^P(v^0) \cdot\cdot\, \varepsilon_\Gamma^P(\overline{v}^0) \\ &+ \frac{2\mu\lambda}{2\mu+\lambda}\operatorname{tr}\varepsilon_\Gamma^P(v^0)\operatorname{tr}\varepsilon_\Gamma^P(\overline{v}^0) - f^0\cdot\overline{v}^0\, d\Gamma = 0\end{aligned}} \tag{5.18}$$

or in term of the *equivalent constitutive law*

$$C_e^{-1}\beta \overset{\text{def}}{=} 2\mu\beta + \frac{2\mu\lambda}{2\mu+\lambda}\operatorname{tr}\beta I, \tag{5.19}$$

the variational equation

$$\boxed{\int_\omega \left[C_e^{-1}\varepsilon_\Gamma^P(v^0)\right] \cdot\cdot\, \varepsilon_\Gamma^P(\overline{v}^0) - f^0\cdot\overline{v}^0\, d\Gamma = 0, \forall \overline{v}^0} \tag{5.20}$$

We shall see in § 6.4 that this is precisely the *asymptotic membrane equation* with the right coefficients without introducing any ad hoc assumption. It is important to notice that we simultaneously get v^0 and v^1.

5.6. Polynomial approximations. It is convenient to first introduce some notation. Always formally for $h > 0$ consider the minimization of the N-dimensional (in practice $N = 3$) energy functional

$$E_h(V_h) = \inf_V E_h(V), \quad E_h(V) \overset{\text{def}}{=} \int_{S_h(\omega)} \frac{1}{2}C^{-1}\varepsilon(V) \cdot\cdot\, \varepsilon(V) - F\cdot V\, dx,$$

where V_h is a minimizing element in a function space $B\big(S_h(\omega)\big)$ which will be specified later along with applicable boundary conditions for a shell with boundary. These definitions have their counterparts in local coordinates by using the transformation $T(x) = \big(p(x), b(x)\big)$ and its inverse $T^{-1}(X,z) = X + z\nabla b(X)$ and introducing the variables $v = V \circ T^{-1}$ and $u = U \circ T^{-1}$

$$\begin{aligned}
e_h(v) &\overset{\text{def}}{=} \frac{E_h(v\circ T^{-1})}{2h} = \frac{1}{2}a_h(v,v) - l_h(v), \\
a_h(u,v) &\overset{\text{def}}{=} \int_{S_h(\omega)} \frac{1}{2h}C^{-1}\varepsilon(u\circ T^{-1}) \cdot\cdot\, \varepsilon(v\circ T^{-1})\, dx, \\
l_h(v) &\overset{\text{def}}{=} \int_{S_h(\omega)} \frac{1}{2h}F\cdot v\circ T^{-1}\, dx.
\end{aligned}$$

The displacement V in $S_h(\omega)$ can be approximated to the k-th order by the vector function V^k specified in terms of a sequence of vector functions $\vec{v}^k = \{v^0, v^1, \dots, v^k\}$ in ω as

$$V^k = \sum_{i=0}^{k} b^i v^i \circ p.$$

For $k = 1$, $v^i = 0$ for $i \geq 2$ and expressions (5.9) in § 5.3 for ε^i simplify

$$\varepsilon^0(v) = \frac{1}{2}(v^{1*}n + n^* v^1) + \varepsilon_\Gamma(v^0), \text{ and for } i \geq 1, \tag{5.21}$$

$$\varepsilon^i(v) = \frac{1}{2}\sum_{k=0}^{1}\left[D_\Gamma(v^k)(-D^2 b)^{i-k} + (-D^2 b)^{i-k*} D_\Gamma(v^k)\right].$$

In the approximation process the energy term $e_h(v^0, v^1)$ can be further approximated by dropping terms multiplied by powers of h higher than a predetermined one. For instance this type of truncation was considered in [**36**] where the linear strain tensor was truncated as

$$\varepsilon(V^1) \circ T_z \simeq \varepsilon^0(v^0, v^1) + z\varepsilon^1(v^0, v^1) + z^2\varepsilon^2(v^0, v^1),$$

and in § 7 we shall also consider the case

$$\varepsilon(V^1) \circ T_z \simeq \varepsilon^0(v^0, v^1) + z\varepsilon^1(v^0, v^1).$$

In the first case terms in the energy expression $e_h(v^0, v^1)$ multiplied by powers of h strictly greater than 2 are dropped (that is, keep ε^0 and ε^1); in the second one those multiplied by powers of h strictly greater than 4 are dropped (that is, keep ε^0, ε^1 and ε^2).

These approximations are closer to the general framework introduced by [**3**] for a mixed formulation of plates where both the tensor σ and hence ε and the vector v are approximated by polynomial in z. Their approximation involves different polynomial orders for tangential and normal components. This construction can also be incorporated in the variational method. What are the most appropriate polynomial orders k and p for v and ε to achieve a given convergence rate. In the sequel this type of approximation will be denoted by $P(k,p)$. For more material and a discussion of polynomial approximations the reader is also referred to [**82**].

The choice of approximations to the 3-D problem and their convergence is quite an active field of research. The reader is referred to the early work of [**76**] in 1959, [**45**], [**15**] and co-workers, and [**3**] on plates.

For plates the central result seems to be that the order of convergence of the solution of the 3-D model to the asymptotic model is $h^{1/2}$. This was mathematically proved by [**44**] in 1980. This rate was also established for mixed formulations of the Hellinger–Reissner type by [**3**]. A slightly better rate can be found in [**12**, pp. 80–86]. For shells recent work by [**74, 75**] indicates orders of $h^{1/6}$ and $h^{1/5}$ for Koiter's model.

6. Some Classical Linear Models of Shells

6.1. Introduction. In § 5 we have discussed linear models of the $P(1,1)$-type characterized by a first order (in z) approximation of the displacement vector and the strain tensor

$$V \circ T_z = e + z\ell, \quad \varepsilon \circ T_z = \varepsilon^0(e,\ell) + z\varepsilon^1(e,\ell). \tag{6.1}$$

In this section we use the notation (e, ℓ) in place of (v^0, v^1). This model will be studied in more details in § 7. In the literature many models are of this type but the surface Jacobian

$$j_z = 1 + Hz + Kz^2, \tag{6.2}$$

is usually approximated by one which means that after integration with respect to z we do not get cross terms $\varepsilon^0 \cdot\cdot\, \varepsilon^1$, $i \neq j$, but only $\|\varepsilon^0\|^2$ and $\|\varepsilon^1\|^2$.

In this section we use the material from § 3.3 in § 3 to rewrite the linear models of Naghdi and Koiter and the asymptotic membrane model in intrinsic form. We use the notation and definitions from the book of [**7**]. Naghdi and Koiter's linear models can be respectively found in § 3 and § 4 of Chapter I. For the asymptotic model we use [**23**].

6.2. Naghdi's linear model. We use the definitions and variational forms from [**7**, Chapter I, § 3]. The details of the transcription are given in § 6.2.2.

6.2.1. *Intrinsic equations.* We have the following variational equation

$$\begin{aligned} \frac{2hE}{1+\nu} \int_\omega \varepsilon^0(e, \ell_\Gamma) \cdot\cdot\, \varepsilon^0(\bar{e}, \bar{\ell}_\Gamma) + \frac{h^2}{3} \hat{\varepsilon}^{1P}(e, \ell_\Gamma) \cdot\cdot\, \hat{\varepsilon}^{1P}(\bar{e}, \bar{\ell}_\Gamma) \\ + \frac{\nu}{1-\nu} \left\{ \operatorname{tr} \varepsilon^0(e, \ell_\Gamma) \operatorname{tr} \varepsilon^0(\bar{e}, \bar{\ell}_\Gamma) + \frac{h^2}{3} \operatorname{tr} \hat{\varepsilon}^{1P}(e, \ell_\Gamma) \operatorname{tr} \hat{\varepsilon}^{1P}(\bar{e}, \bar{\ell}_\Gamma) \right\} d\Gamma \\ = \int_\omega p \cdot \bar{e}\, d\Gamma + \int_\gamma N \cdot \bar{e} - M \cdot \bar{\ell}_\Gamma \, d\gamma, \end{aligned}$$

where ℓ_Γ and $\bar{\ell}_\Gamma$ are tangential and

$$\varepsilon^0(v, \zeta) \stackrel{\text{def}}{=} \varepsilon_\Gamma(v) + \frac{1}{2}[\zeta^* n + n^* \zeta], \quad \varepsilon_\Gamma^P(v) \stackrel{\text{def}}{=} P \varepsilon_\Gamma(v) P, \tag{6.3}$$

$$\hat{\varepsilon}^1(v, \zeta) \stackrel{\text{def}}{=} \varepsilon_\Gamma(\zeta) + \frac{1}{2}[D^2 b D_\Gamma v + {}^* D_\Gamma v D^2 b], \tag{6.4}$$

$$\hat{\varepsilon}^{1P}(v, \zeta) \stackrel{\text{def}}{=} P \hat{\varepsilon}^1(v, \zeta) P. \tag{6.5}$$

The intrinsic expression of the model is obviously not unique. It can also be written in terms of $\varepsilon_\Gamma^P(v) = \varepsilon_\Gamma^{0P}(v, \ell_\Gamma)$ and $\operatorname{div}_\Gamma e = \operatorname{tr} \varepsilon_\Gamma(v) = \operatorname{tr} \varepsilon_\Gamma^P(v) = \operatorname{tr} \varepsilon_\Gamma^{0P}(v, \ell_\Gamma)$ as follows

$$\boxed{\begin{aligned} 2h \int_\omega 2\mu \left\{ \varepsilon_\Gamma^P(e) \cdot\cdot\, \varepsilon_\Gamma^P(\bar{e}) + \frac{1}{2}(2\varepsilon_\Gamma(e) n + \ell_\Gamma) \cdot (2\varepsilon_\Gamma(\bar{e}) n + \bar{\ell}_\Gamma) \right\} \\ + \frac{2\mu\lambda}{2\mu+\lambda} \operatorname{div}_\Gamma e \operatorname{div}_\Gamma \bar{e}\, d\Gamma \\ + \frac{2h^3}{3} \int_\omega 2\mu \hat{\varepsilon}^{1P}(e, \ell_\Gamma) \cdot\cdot\, \hat{\varepsilon}^{1P}(\bar{e}, \bar{\ell}_\Gamma) + \frac{2\mu\lambda}{2\mu+\lambda} \operatorname{tr} \hat{\varepsilon}^{1P}(e, \ell_\Gamma) \operatorname{tr} \hat{\varepsilon}^{1P}(\bar{e}, \bar{\ell}_\Gamma)\, d\Gamma \\ = \int_\omega p \cdot \bar{e}\, d\Gamma + \int_\gamma N \cdot \bar{e} - M \cdot \bar{\ell}_\Gamma \, d\gamma \end{aligned}}$$

The detailed computations are given in the next section with

$$2h = e, \quad \ell_\Gamma = -\beta, \quad e = u, \quad \bar{\ell}_\Gamma = -\delta, \quad \bar{e} = v, \tag{6.6}$$

and the Lamé's coefficients since

$$2\mu = \frac{E}{1+\nu}, \qquad \lambda = \frac{E}{1+\nu}\frac{\nu}{1-2\nu} \implies \frac{\nu}{1-\nu} = \frac{\lambda}{2\mu+\lambda}.$$

Note that $\hat{\varepsilon}^1(v,\zeta)$ is not quite the same as the one in equation (5.21) of § 5: we have the terms $D^2bD_\Gamma v$ instead of $D_\Gamma vD^2b$ and a plus sign in front of them. This results from the way this equation was approximated. Recall that

$$\varepsilon^1(v,\zeta) \stackrel{\text{def}}{=} \varepsilon_\Gamma(\zeta) - \frac{1}{2}[D_\Gamma v^* D^2b + D^2b^* D_\Gamma v]$$
$$\implies \hat{\varepsilon}^{1P}(v,\zeta) - \varepsilon^{1P}(v,\zeta) = \varepsilon_\Gamma^P v^* D^2b + D^2b^* \varepsilon_\Gamma^P v,$$

The square of the norm of the right-hand side of the last expression is multiplied by h^3

$$2\frac{h^3}{3}\|\varepsilon_\Gamma^P v^* D^2b + D^2b^* \varepsilon_\Gamma^P v\|^2,$$

and is negligible compared to $2h\|\varepsilon_\Gamma^P v\|^2$. Therefore changing $\hat{\varepsilon}^1(v,\zeta)$ into $\varepsilon^1(v,\zeta)$ does not change the mathematical structure and the mechanical interpretation of Naghdi's model.

Naghdi's model which uses the assumption $\ell_n = 0$ and $\sigma n \cdot n = 0$ seems to be approximately between cases (3) and (5) once ℓ_n has been eliminated from the equations.

6.2.2. *Computations.* The displacement vector is

$$U = u + \xi^3\beta_\alpha a^\alpha, \tag{6.7}$$

where $-e/2 < \xi^3 < e/2$, e is the thickness of the shell, u and β are maps from ω to $\mathbb{R}^3$ and

$$\beta = \beta_\alpha a^\alpha, \quad \beta_\alpha = \beta \cdot a_\alpha \implies \beta \cdot n = 0. \tag{6.8}$$

We shall use the same notation for a vector $v\colon \omega \longrightarrow \mathbb{R}^3$ and its 2-dimensional representation $v \circ \Phi$ in terms of the $1-2$ coordinates. We now go through several definitions and give their tangential equivalent

$$\varphi_\alpha(u) \stackrel{\text{def}}{=} u_{3,\alpha} + b_\alpha^\lambda u_\lambda, \tag{6.9}$$

$$\gamma_{\alpha\beta}(u) \stackrel{\text{def}}{=} \frac{1}{2}\big(u_\alpha|_\beta + u_\beta|_\alpha\big) - b_{\alpha\beta}u_3, \tag{6.10}$$

$$d_{\lambda\alpha}(u) \stackrel{\text{def}}{=} u_\lambda|_\alpha - b_{\lambda\alpha}u_3, \tag{6.11}$$

$$\chi_{\alpha\mu}(u,\beta) \stackrel{\text{def}}{=} \frac{1}{2}\big[\beta_\alpha|_\mu + \beta_\mu|_\alpha - b_\alpha^\lambda d_{\lambda\mu}(u) - b_\mu^\lambda d_{\lambda\alpha}(u)\big], \tag{6.12}$$

$$\varepsilon_{\alpha\beta}(u,\beta) \stackrel{\text{def}}{=} \gamma_{\alpha\beta}(u) + \xi^3\chi_{\alpha\beta}(u,\beta). \tag{6.13}$$

Then

$$\begin{aligned}\varphi_\alpha(u) &= \nabla_\Gamma(u \cdot a_3) \cdot a_\alpha + D^2ba_\alpha \cdot a^\lambda u_\lambda = \big[D^2bu - \nabla_\Gamma(u \cdot n)\big] \cdot a_\alpha \\ &= -{}^*D_\Gamma(u)n \cdot a_\alpha = -2\varepsilon_\Gamma(u)n \cdot a_\alpha,\end{aligned}$$

and

$$\boxed{\varphi_\alpha(u) = -2\varepsilon_\Gamma(u)n \cdot a_\alpha.} \tag{6.14}$$

Next from (3.63) in § 3

$$\boxed{\gamma_{\alpha\beta}(u) = \varepsilon_\Gamma(u)a_\alpha \cdot a_\beta = \varepsilon_\Gamma^P(u)a_\alpha \cdot a_\beta = \varepsilon^{0P}(u, -\delta)a_\alpha \cdot a_\beta} \tag{6.15}$$

since

$$\varepsilon^{0P}(u, -\delta)a_\alpha \cdot a_\beta = P\varepsilon_\Gamma(u)P - \frac{1}{2}P[\delta^* n + n^*\delta]P = \varepsilon_\Gamma^P(u),$$

and from (3.58) in § 3

$$d_{\lambda\alpha}(u) = D_\Gamma u a_\alpha \cdot a_\lambda + D^2 b a_\lambda \cdot a_\alpha u_3 - D^2 b a_\alpha \cdot a_\lambda u_3,$$

$$\boxed{d_{\lambda\alpha}(u) = D_\Gamma(u)a_\alpha \cdot a_\lambda = D_\Gamma^P(u)a_\alpha \cdot a_\lambda} \tag{6.16}$$

By using $\hat{\varepsilon}^1$ as defined in (6.4)

$$\chi_{\alpha\mu}(u, \beta) = -a_\alpha \cdot \hat{\varepsilon}^1(u, -\beta)a_\mu = -a_\alpha \cdot \hat{\varepsilon}^{1P}(u, -\beta)a_\mu,$$

and

$$\boxed{\varepsilon_{\alpha\beta}(u, \beta) = a_\alpha \cdot \{\varepsilon^{0P}(u, -\beta) - \xi^3 \hat{\varepsilon}^{1P}(u, -\beta)\}a_\beta} \tag{6.17}$$

The underlying variational equation for Naghdi's model consists in finding (u, β) such that for all (v, δ)

$$a^S\big((u, \beta), (v, \delta)\big) + b^S\big((u, \beta), (v, \delta)\big) = f^S(v, \delta), \tag{6.18}$$

where the bilinear forms a^S and b^S and the linear form f^S are given by

$$a^S\big((u, \beta), (v, \delta)\big) \overset{\text{def}}{=} \int_\omega eE^{\alpha\beta\lambda\mu}\left\{\gamma_{\alpha\beta}(u)\gamma_{\lambda\mu}(v) + \frac{e^2}{12}\chi_{\alpha\beta}(u, \beta)\chi_{\lambda\mu}(v, \delta)\right\} d\Gamma, \tag{6.19}$$

$$b^S\big((u, \beta), (v, \delta)\big) \overset{\text{def}}{=} \int_\omega \frac{eEa^{\alpha\beta}}{2(1+\nu)}\big(\varphi_\alpha(u) + \beta_\alpha\big)\big(\varphi_\beta(v) + \delta_\beta\big)\, d\Gamma, \tag{6.20}$$

$$f^S(v, \delta) \overset{\text{def}}{=} \int_\omega p \cdot v\, d\Gamma + \int_{\partial\Gamma} N \cdot v - M \cdot \delta\, d\gamma, \tag{6.21}$$

and

$$E^{\alpha\beta\lambda\mu} = \frac{E}{2(1+\nu)}\left[a^{\alpha\lambda}a^{\beta\mu} + a^{\alpha\mu}a^{\beta\lambda} + \frac{2\nu}{1-\nu}a^{\alpha\beta}a^{\lambda\mu}\right]. \tag{6.22}$$

First consider the term in γ of a^S

$$E^{\alpha\beta\lambda\mu}\gamma_{\alpha\beta}(u)\gamma_{\lambda\mu}(v) = \frac{E}{2(1+\nu)}\left[a^{\alpha\lambda}a^{\beta\mu} + a^{\alpha\mu}a^{\beta\lambda} + \frac{2\nu}{1-\nu}a^{\alpha\beta}a^{\lambda\mu}\right]\gamma_{\alpha\beta}(u)\gamma_{\lambda\mu}(v). \tag{6.23}$$

Observe that since γ is symmetrical

$$A \overset{\text{def}}{=} a^{\alpha\lambda}a^{\beta\mu}\gamma_{\alpha\beta}(u)\gamma_{\lambda\mu}(v) = a^{\alpha\mu}a^{\beta\lambda}\gamma_{\alpha\beta}(u)\gamma_{\lambda\mu}(v).$$

So we only need to compute the first term

$$\begin{aligned}
A &\overset{\text{def}}{=} \varepsilon_\Gamma(u)a_\beta \cdot a_\alpha a^\alpha \cdot a^\lambda \varepsilon_\Gamma(v)a_\lambda \cdot a_\mu a^\mu \cdot a^\beta \\
&= \varepsilon_\Gamma(u)a_\beta \cdot [a^\lambda - a_3 a^3 \cdot a^\lambda]\varepsilon_\Gamma(v)a_\lambda \cdot [a^\beta - a_3 a^3 \cdot a^\beta] \\
&= \varepsilon_\Gamma(u)a_\beta \cdot a^\lambda a_\lambda \cdot \varepsilon_\Gamma(v)a^\beta = \varepsilon_\Gamma^P(u) \cdot\cdot\, \varepsilon_\Gamma^P(v) \\
&= \varepsilon^{0P}(u, -\beta) \cdot\cdot\, \varepsilon^{0P}(v, -\delta),
\end{aligned}$$

from (3.68) in § 3. Now

$$\begin{aligned}B &\stackrel{\text{def}}{=} a^{\alpha\beta}a^{\lambda\mu}\gamma_{\alpha\beta}(u)\gamma_{\lambda\mu}(v)\\&= \varepsilon_\Gamma(u)a_\alpha \cdot a_\beta a^\beta \cdot a^\alpha \varepsilon_\Gamma(v)a_\lambda \cdot a_\mu a^\mu \cdot a^\lambda\\&= \varepsilon_\Gamma(u)a_\alpha \cdot a^\alpha \varepsilon_\Gamma(v)a_\lambda \cdot a^\lambda = \operatorname{tr}\varepsilon_\Gamma(u)\operatorname{tr}\varepsilon_\Gamma(v) = \operatorname{tr}\varepsilon_\Gamma^P(u)\operatorname{tr}\varepsilon_\Gamma^P(v)\\&\qquad= \operatorname{tr}\varepsilon^{0P}(u,-\beta)\operatorname{tr}\varepsilon^{0P}(v,-\delta).\end{aligned}$$

Finally

$$\boxed{\begin{aligned}&E^{\alpha\beta\lambda\mu}\gamma_{\alpha\beta}(u)\gamma_{\lambda\mu}(v)\\&\quad= 2A + \frac{2\nu}{1-\nu}B\\&\quad= 2\varepsilon^{0P}(u,-\beta)\cdot\cdot\,\varepsilon^{0P}(v,-\delta) + \frac{2\nu}{1-\nu}\operatorname{tr}\varepsilon^{0P}(u,-\beta)\operatorname{tr}\varepsilon^{0P}(v,-\delta)\\&\quad= \varepsilon_\Gamma^P(u)\cdot\cdot\,\varepsilon_\Gamma^P(v) + \frac{2\nu}{1-\nu}\operatorname{tr}\varepsilon_\Gamma^P(u)\operatorname{tr}\varepsilon_\Gamma^P(v)\end{aligned}}$$

Next we turn to the term in χ. Again

$$C \stackrel{\text{def}}{=} a^{\alpha\mu}a^{\beta\lambda}\chi_{\alpha\beta}(u,\beta)\chi_{\lambda\mu}(v,\delta) = a^{\alpha\lambda}a^{\beta\mu}\chi_{\alpha\beta}(u,\beta)\chi_{\lambda\mu}(v,\delta),$$

and from definition (6.4) of $\hat\varepsilon^1$

$$\begin{aligned}C &= a_\alpha \cdot \hat\varepsilon^{1P}(u,-\beta)a_\beta a^\beta \cdot a^\lambda a_\lambda \cdot \hat\varepsilon^{1P}(v,-\delta)a_\mu a^\mu \cdot a^\alpha\\&= a_\alpha \cdot \hat\varepsilon^{1P}(u,-\beta)a^\lambda a_\lambda \cdot \hat\varepsilon^{1P}(v,-\delta)a^\alpha\\&= \hat\varepsilon^{1P}(u,-\beta)\cdot\cdot\,\hat\varepsilon^{1P}(v,-\delta).\end{aligned}$$

Next

$$\begin{aligned}D &\stackrel{\text{def}}{=} a^{\alpha\beta}a^{\lambda\mu}\chi_{\alpha\beta}(u,\beta)\chi_{\lambda\mu}(v,\delta)\\&= a_\alpha \cdot \big[\chi_{\alpha\beta}(u,\beta)\big]a_\beta a^\beta \cdot a^\alpha a_\lambda \cdot \big[\chi_{\lambda\mu}(v,\delta)\big]a_\mu a^\mu \cdot a^\lambda\\&= a_\alpha \cdot \big[\chi_{\alpha\beta}(u,\beta)\big]a^\alpha a_\lambda \cdot \big[\chi_{\lambda\mu}(v,\delta)\big]a^\lambda\\&= \operatorname{tr}\hat\varepsilon^{1P}(u,-\beta)\operatorname{tr}\hat\varepsilon^{1P}(v,-\delta).\end{aligned}$$

Finally

$$\begin{aligned}&E^{\alpha\beta\lambda\mu}\chi_{\alpha\beta}(u,\beta)\chi_{\lambda\mu}(v,\delta)\\&\quad= \frac{E}{2(1+\nu)}\left[2\hat\varepsilon^{1P}(u,-\beta)\cdot\cdot\,\hat\varepsilon^{1P}(v,-\delta) + \frac{2\nu}{1-\nu}\operatorname{tr}\hat\varepsilon^{1P}(u,-\beta)\operatorname{tr}\hat\varepsilon^{1P}(u,-\delta)\right].\end{aligned}$$

As for the bilinear form b^S the situation is simpler

$$\begin{aligned}F &\stackrel{\text{def}}{=} a^{\alpha\beta}\big(\varphi_\alpha(u) + \beta_\alpha\big)\big(\varphi_\beta(v) + \delta_\beta\big)\\&= a^\beta \cdot \alpha^\alpha a_\alpha \cdot \big[-2\varepsilon_\Gamma(u)\nabla b + \beta\big]\,\big[-2\varepsilon_\Gamma(v)\nabla b + \delta\big] \cdot a_\beta\\&= \big(-2\varepsilon_\Gamma(u)\nabla b + \beta\big) \cdot a^\beta a_\beta \cdot \big(-2\varepsilon_\Gamma(v)\nabla b + \delta\big)\\&= \big(-2\varepsilon_\Gamma(u)\nabla b + \beta\big) \cdot \big(-2\varepsilon_\Gamma(v)\nabla b + \delta\big)\\&= \big(2\varepsilon_\Gamma(u)\nabla b - \beta\big) \cdot \big(2\varepsilon_\Gamma(v)\nabla b - \delta\big).\end{aligned}$$

The term $2A + F$ can be slightly rearranged. To see this consider the expression

$$G \stackrel{\text{def}}{=} \varepsilon^0(u, -\beta) \cdot\cdot\, \varepsilon^0(v, -\delta).$$

Then

$$\begin{aligned}
G &= \varepsilon^0\big(u, -2\varepsilon_\Gamma(u)n + 2\varepsilon_\Gamma(u)n - \beta\big) \cdot\cdot\, \varepsilon^0\big(v, -2\varepsilon_\Gamma(v)n + 2\varepsilon_\Gamma(v)n - \delta\big) \\
&= \varepsilon^0\big(u, -2\varepsilon_\Gamma(u)n\big) \cdot\cdot\, \varepsilon^0\big(v, -2\varepsilon_\Gamma(v)n\big) \\
&\qquad + \varepsilon^0\big(0, 2\varepsilon_\Gamma(u)n - \beta\big) \cdot\cdot\, \varepsilon^0\big(v, -2\varepsilon_\Gamma(v)n\big) \\
&\qquad + \varepsilon^0\big(u, -2\varepsilon_\Gamma(u)n\big) \cdot\cdot\, \varepsilon^0\big(0, 2\varepsilon_\Gamma(v)n - \delta\big) \\
&\qquad + \varepsilon^0\big(0, 2\varepsilon_\Gamma(u)n - \beta\big) \cdot\cdot\, \varepsilon^0\big(0, 2\varepsilon_\Gamma(v)n - \delta\big).
\end{aligned}$$

But

$$\begin{aligned}
&\varepsilon^0\big(0, -2\varepsilon_\Gamma(u)n + \beta\big) \cdot\cdot\, \varepsilon^0\big(v, -2\varepsilon_\Gamma(v)n\big) \\
&\quad = \frac{1}{2}\big[(-2\varepsilon_\Gamma(u)n + \beta)^* n + n^*(-2\varepsilon_\Gamma(u)n + \beta)\big] \cdot\cdot\, \varepsilon^0\big(v, -2\varepsilon_\Gamma(v)n\big) \\
&\quad = \big(-2\varepsilon_\Gamma(u)n + \beta\big) \cdot \varepsilon^0\big(v, -2\varepsilon_\Gamma(v)n\big)n \\
&\quad = -\big(2\varepsilon_\Gamma(u)n - \beta\big) \cdot \varepsilon^0\big(v, -2\varepsilon_\Gamma(v)n\big)n,
\end{aligned}$$

and

$$\begin{aligned}
\varepsilon^0(v, -2\varepsilon_\Gamma(v)n)n &= \big\{\varepsilon_\Gamma(v) - [\varepsilon_\Gamma(v)n^*n + n^*(\varepsilon_\Gamma(v)n)]\big\}n \\
&= \varepsilon_\Gamma(v)n - \varepsilon_\Gamma(v)n = 0.
\end{aligned}$$

Also

$$\begin{aligned}
\varepsilon^0(0, -2\varepsilon_\Gamma(u)n + \beta) &\cdot\cdot\, \varepsilon^0(0, -2\varepsilon_\Gamma(v)n + \delta) \\
&= \frac{1}{4}\big[(-2\varepsilon_\Gamma(u)n + \beta)^* n + n^*(-2\varepsilon_\Gamma(u)n + \beta)\big] \\
&\qquad\qquad \cdot\cdot\, \big[(-2\varepsilon_\Gamma(v)n + \delta)^* n + n^*(-2\varepsilon_\Gamma(v)n + \delta)\big] \\
&= \frac{1}{2}\big(2\varepsilon_\Gamma(u)n - \beta\big) \cdot \big(2\varepsilon_\Gamma(v)n - \delta\big),
\end{aligned}$$

and

$$\begin{aligned}
\text{(6.24)} \qquad \varepsilon^0(u, -2\varepsilon_\Gamma(u)n) &\\
&= \varepsilon_\Gamma(u) - \frac{1}{2}\big[2\varepsilon_\Gamma(u)n^*n + n^*(2\varepsilon_\Gamma(u)n)\big] \\
&= \frac{1}{2}\big[D_\Gamma(u) - n^*nD_\Gamma(u) + {}^*D_\Gamma(u) - {}^*D_\Gamma(u)n^*n\big] \\
&= \frac{1}{2}\big[PD_\Gamma(u) + {}^*D_\Gamma(u)P\big] = \varepsilon_\Gamma^P(u).
\end{aligned}$$

Moreover

$$\begin{aligned}
\operatorname{tr} \varepsilon^0(u, -\beta) &= \operatorname{tr} \varepsilon_\Gamma(u) = \operatorname{div}_\Gamma u, \\
\varepsilon^{0P}(u, -\beta) &= \varepsilon_\Gamma^P(u) - \frac{1}{2}P\big[\beta^*\nabla b + \nabla b^*\beta\big]P = \varepsilon_\Gamma^P(u), \\
\operatorname{tr} \varepsilon^{0P}(u, -\beta) &= \operatorname{tr} \varepsilon_\Gamma^P(u) = \operatorname{div}_\Gamma u.
\end{aligned}$$

So finally $2G = 2A + F$. Putting the two bilinear forms together

$$a^S\big((u,\beta),(v,\delta)\big) + b^S\big((u,\beta),(v,\delta)\big)$$
$$= \frac{eE}{1+\nu}\int_\omega \varepsilon^0(u,-\beta)\cdot\cdot\,\varepsilon^0(v,-\delta) + \frac{e^2}{12}\hat{\varepsilon}^{1P}(u,-\beta)\cdot\cdot\,\varepsilon^{1P}(v,-\delta)$$
$$+\frac{\nu}{1-\nu}\left\{\operatorname{tr}\varepsilon^0(u,-\beta)\operatorname{tr}\varepsilon^0(v,-\delta) + \frac{e^2}{12}\operatorname{tr}\hat{\varepsilon}^{1P}(u,-\beta)\operatorname{tr}\hat{\varepsilon}^{1P}(v,-\delta)\right\}d\Gamma$$
$$= \frac{eE}{1+\nu}\int_\omega \varepsilon_\Gamma^P(u)\cdot\cdot\,\varepsilon_\Gamma^P(v) + \frac{1}{2}(2\varepsilon_\Gamma(u)n-\beta)\cdot(2\varepsilon_\Gamma(v)n-\delta)$$
$$+\frac{e^2}{12}\hat{\varepsilon}^{1P}(u,-\beta)\cdot\cdot\,\hat{\varepsilon}^{1P}(v,-\delta)$$
$$+\frac{\nu}{1-\nu}\left\{\operatorname{div}_\Gamma u \operatorname{div}_\Gamma v + \frac{e^2}{12}\operatorname{tr}\hat{\varepsilon}^{1P}(u,-\beta)\operatorname{tr}\hat{\varepsilon}^{1P}(v,-\delta)\right\}d\Gamma.$$

Also
$$f^S(v,\delta) = \int_\omega p\cdot v\,d\Gamma + \int_\gamma N\cdot v - M\cdot\delta\,d\gamma.$$

6.3. Koiter's linear model. It is mathematically the same model as Naghdi's model with the condition

(6.25)
$$\ell_\Gamma + 2\varepsilon_\Gamma(e)n = 0.$$

In the variables u and β

(6.26)
$$\beta_\alpha + \varphi_\alpha(u) = 0 \implies a_\alpha\cdot[\beta - {}^*D_\Gamma un] = 0,$$

which can be rewritten as

(6.27)
$$\boxed{\beta - 2\varepsilon_\Gamma(u)n = 0} \text{ or } \boxed{\beta = \nabla_\Gamma(u_n) - D^2bu_\Gamma}$$

Koiter's model which uses the assumption $\ell_n = 0$ and $\sigma n = 0$ seems to be approximately between cases (4) and (6) once ℓ_n has been eliminated from the equations.

6.4. Asymptotic membrane equation.

6.4.1. *Intrinsic equations.* It is specified in the next section with the coefficients λ and μ such that $\lambda \geq 0$ and $\mu > 0$ (cf. [**23**]).

(6.28)
$$\int_\omega 2\mu\varepsilon^0\big(e,-2\varepsilon_\Gamma(e)n\big)\cdot\cdot\,\varepsilon^0\big(\bar{e},-2\varepsilon_\Gamma(\bar{e})n\big)$$
$$+\frac{2\mu\lambda}{\lambda+2\mu}\operatorname{tr}\varepsilon^0\big(e,-2\varepsilon_\Gamma(e)n\big)\operatorname{tr}\varepsilon^0\big(\bar{e},-2\varepsilon_\Gamma(\bar{e})n\big)\,d\Gamma = \int_\omega \frac{1}{2}f\cdot\bar{e}\,d\Gamma.$$

The identification with the variables of § 6.4.2 are

(6.29)
$$e = \eta, \quad \bar{e} = \zeta.$$

Recall from (6.24) that

(6.30)
$$\varepsilon^0\big(e,-2\varepsilon_\Gamma(e)\nabla b\big) = \varepsilon_\Gamma^P(e),$$

and hence

(6.31)
$$\boxed{\int_\omega 2\mu\varepsilon_\Gamma^P(e)\cdot\cdot\,\varepsilon_\Gamma^P(\bar{e}) + \frac{2\mu\lambda}{\lambda+2\mu}\operatorname{tr}\varepsilon_\Gamma^P(e)\operatorname{tr}\varepsilon_\Gamma^P(\bar{e})\,d\Gamma = \int_\omega \frac{1}{2}f\cdot\bar{e}\,d\Gamma}$$

or in terms of E and ν

$$\frac{E}{1+\nu}\int_\omega \varepsilon^{0P}(e,\ell_\Gamma)\cdot\cdot\,\varepsilon^{0P}(\bar e,\bar\ell_\Gamma)+\frac{\nu}{1-\nu}\operatorname{tr}\varepsilon^{0P}(e,\ell_\Gamma)\operatorname{tr}\varepsilon^{0P}(\bar e,\bar\ell_\Gamma)\,d\Gamma = \int_\omega \frac{1}{2}f\cdot\bar e\,d\Gamma, \tag{6.32}$$

since

$$2\mu=\frac{E}{1+\nu}\quad\text{and}\quad\lambda=\frac{E}{1+\nu}\frac{\nu}{1-2\nu}.$$

6.4.2. *Computations.* It is characterized by the bilinear form

$$B(\zeta,\eta)\overset{\text{def}}{=}\int_\omega A^{\alpha\beta\rho\delta}\gamma_{\rho\sigma}(\zeta)\gamma_{\alpha\beta}(\eta)\,d\Gamma, \tag{6.33}$$

$$A^{\alpha\beta\rho\delta}\overset{\text{def}}{=}\frac{4\lambda\mu}{\lambda+2\mu}a^{\alpha\beta}a^{\rho\delta}+2\mu[a^{\alpha\rho}a^{\beta\sigma}+a^{\alpha\sigma}a^{\beta\rho}], \tag{6.34}$$

$$\gamma_{\alpha\beta}(\eta)\overset{\text{def}}{=}\frac{1}{2}(\eta_{\alpha,\beta}+\eta_{\beta,\alpha})-\Gamma^\rho_{\alpha\beta}\eta_\beta-b_{\alpha\beta}\eta_3. \tag{6.35}$$

Clearly

$$\gamma_{\alpha\beta}(\eta)=\frac{1}{2}\left(\eta_\alpha\big|_\beta+\eta_\beta\big|_\alpha\right)-b_{\alpha\beta}\eta_3=a_\alpha\cdot\varepsilon_\Gamma(\eta)a_\beta=a_\alpha\cdot\varepsilon^P_\Gamma(\eta)a_\beta.$$

But this is the same γ as in Naghdi's model. Therefore

$$\begin{aligned}B(\zeta,\eta)=\int_\omega & 4\mu\varepsilon^0\big(\zeta,-2\varepsilon_\Gamma(\zeta)\nabla b\big)\cdot\cdot\,\varepsilon^0\big(\eta,-2\varepsilon_\Gamma(\eta)\nabla b\big)\\ &+\frac{4\lambda\mu}{\lambda+2\mu}\operatorname{tr}\varepsilon^0\big(\zeta,-2\varepsilon_\Gamma(\zeta)\nabla b\big)+\operatorname{tr}\varepsilon^0\big(\eta,-2\varepsilon_\Gamma(\eta)\nabla b\big)\,d\Gamma,\end{aligned}$$

$$\begin{aligned}B(\zeta,\eta)=\int_\omega & 4\mu\varepsilon^0\big(\zeta,-2\varepsilon_\Gamma(\zeta)\nabla b\big)\cdot\cdot\,\varepsilon^0\big(\eta,-2\varepsilon_\Gamma(\eta)\nabla b\big)\\ &+\frac{4\lambda\mu}{\lambda+2\mu}\operatorname{div}_\Gamma\zeta\operatorname{div}_\Gamma\eta\,d\Gamma.\end{aligned}$$

7. P(1, 1) Linear Model of Thin Shells

7.1. The linear P(1, 1) model.

7.1.1. *The linear strain tensor.* For simplicity we shall work with a shell $S_h(\omega)$ of constant thickness, but almost everything we shall say applies to shells $S_{\tilde h}(\omega)$ with variable thickness under appropriate assumptions on the function $\tilde h$. In § 5, we considered the asymptotic membrane model $P(1,0)$ in our terminology. In this section we consider the $P(1,1)$ polynomial model which will now depend on the thickness h. This choice is motivated by the fact that many classical models are of that type. The material for the $P(1,1)$ model is new and more details can be found in [**41**] and the one for the $P(1,2)$ model in [**36**].

ASSUMPTION 7.1. At each point x of the shell the *displacement vector* $V(x)$ is of the form

$$V(x)=e\circ p(x)+b(x)\ell\circ p(x),\quad x\in S_h(\omega), \tag{7.1}$$

for vector-valued mappings e and ℓ defined on ω.

Note that the notation (v^0, v^1) of § 5 has been changed to (e, ℓ). In addition to the *unconstrained model* we shall discuss six special cases which have mechanical significance, but we first introduce the underlying function analytic machinery. Equations (5.5) and (5.6) in § 5 yield the following expressions in $S_h(\omega)$

$$DV = [D_\Gamma e \circ p + bD_\Gamma \ell \circ p]\,[I - bD^2 b] + \ell \circ p^* \nabla b, \tag{7.2}$$

$$DV = [D_\Gamma e \circ p + bD_\Gamma \ell \circ p + \ell \circ p^* \nabla b]\,[I - bD^2 b], \tag{7.3}$$

$$DV\,\nabla b = \ell \circ p. \tag{7.4}$$

Also

$$^*DV\,\nabla b = [I - bD^2 b]\big[(^*D_\Gamma en) \circ p + \ell_\Gamma \circ p + b(\nabla_\Gamma \ell_n) \circ p\big] - \ell \circ p + 2\ell_n \circ p \nabla b.$$

From (7.3) the ***linear strain tensor*** $\varepsilon(V)$ over $S_h(\omega)$ becomes

$$\begin{aligned} 2\varepsilon(V) = {} & [D_\Gamma e \circ p + bD_\Gamma \ell \circ p + \ell \circ p^* \nabla b]\,[I - bD^2 b] \\ & + [I - bD^2 b]^* [D_\Gamma e \circ p + bD_\Gamma \ell \circ p + \ell \circ p^* \nabla b], \end{aligned} \tag{7.5}$$

$$\begin{aligned} & \varepsilon(V)\nabla b \\ & = \frac{1}{2}[I - bD^2 b]\big[(^*D_\Gamma en + \ell_\Gamma) \circ p + b(\nabla_\Gamma \ell_n) \circ p\big] + \ell_n \circ p \nabla b, \end{aligned} \tag{7.6}$$

$$\varepsilon(V)\nabla b \cdot \nabla b = \ell_n \circ p \cdot \nabla b. \tag{7.7}$$

In terms of the local coordinates (X, z), the tensor $\varepsilon(V) \circ T_z$ is almost affine in z, that is the sum of a tensor in X and z times another tensor in X

$$DV \circ T_z = [D_\Gamma e + zD_\Gamma \ell + \ell^* \nabla b]\,[I + zD^2 b]^{-1}. \tag{7.8}$$

$$\begin{aligned} 2\varepsilon(V) \circ T_z = {} & [D_\Gamma e + zD_\Gamma \ell + \ell^* \nabla b]\,[I + zD^2 b]^{-1} \\ & + [I + zD^2 b]^{-1*} [D_\Gamma e + zD_\Gamma \ell + \ell^* \nabla b]. \end{aligned} \tag{7.9}$$

$$\begin{aligned} & \varepsilon(V) \circ T_z \nabla b \\ & = \frac{1}{2}[I + zD^2 b]^{-1}\big[2\varepsilon_\Gamma(e)\nabla b + \ell_\Gamma + b(\nabla_\Gamma \ell_n) \circ p] + \ell_n \circ p \nabla b\big]. \end{aligned} \tag{7.10}$$

The nonlinear part comes from the matrix $[I + zD^2 b]^{-1}$. Under Assumption 5.1

$$\begin{aligned} DV \circ T_z = {} & D_\Gamma e + \ell^* \nabla b + \big[D_\Gamma \ell - D_\Gamma e D^2 b\big] z \\ & + \sum_{i=2}^{\infty} \big[D_\Gamma \ell - D_\Gamma e D^2 b\big] (-D^2 b)^{i-1} z^i. \end{aligned} \tag{7.11}$$

$$\begin{aligned} 2\varepsilon(V) \circ T_z = {} & 2\varepsilon_\Gamma(e) + \ell^* \nabla b + \nabla b^* \ell \\ & + \big\{2\varepsilon_\Gamma(\ell) - [D_\Gamma e D^2 b + D^2 b^* D_\Gamma e]\big\} z \\ & + \sum_{i=2}^{\infty} \big\{[D_\Gamma \ell - D_\Gamma e D^2 b](-D^2 b)^{i-1} \\ & \qquad + (-D^2 b)^{i-1}\big[^*D_\Gamma \ell - D^2 b^* D_\Gamma e]\big\} z^i. \end{aligned} \tag{7.12}$$

At this juncture it is convenient to introduce the following notation which coincides with the one of § 5

(7.13) $$2\varepsilon^0(e,\ell) = D_\Gamma e + \ell^*\nabla b + {}^*D_\Gamma e + \nabla b^*\ell,$$

(7.14) $$2\varepsilon^1(e,\ell) = D_\Gamma \ell - D_\Gamma e D^2 b + {}^*D_\Gamma \ell - D^2 b^* D_\Gamma e,$$

(7.15) $$2\varepsilon^2(e,\ell) = [D_\Gamma \ell - D_\Gamma e D^2 b](-D^2 b), + (-D^2 b)[{}^*D_\Gamma \ell - D^2 b^* D_\Gamma e]$$

(7.16) $$2\varepsilon^n(e,\ell) = [D_\Gamma \ell - D_\Gamma e D^2 b](-D^2 b)^{n-1} + (-D^2 b)^{n-1}[{}^*D_\Gamma \ell - D^2 b^* D_\Gamma e], \quad n \geq 3.$$

7.1.2. *Approximate rigid displacements.* The following result is central to the subsequent development of the model.

THEOREM 7.1 (Approximate rigid displacements). *Under Assumption 5.1 let* ω *be a bounded open domain in* Γ. *For* e *and* ℓ *in* $H^1(\Gamma)^N$ *and* $V = e \circ p + b\,\ell \circ p$

(i)

(7.17) $$\varepsilon(V) = 0 \text{ in } S_h(\omega)$$

if and only if

(7.18) $$\varepsilon^0(e,\ell) = \varepsilon^1(e,\ell) = \varepsilon^2(e,\ell) = 0 \text{ in } \omega,$$

if and only if there exists a vector $a \in \mathbb{R}^N$ *and an* $N \times N$ *matrix* B *such that*

(7.19) $$e(X) = a + BX, \quad \ell(X) = B\nabla b(X),\ \forall X \in \omega,$$

where B *is antisymmetric*

(7.20) $$B + {}^*B = 0.$$

In particular (7.19) *and* (7.20) *imply that* ℓ *is tangential.*

(ii) *For all* z, $|z| < h$, *and* $X \in \omega$,

$$\varepsilon(V) \circ T_z = \varepsilon^0 + [I + zD^2 b]^{-1}\{z[\varepsilon^1 + \varepsilon^1 z D^2 b + z D^2 b \varepsilon^1] + z^2 \varepsilon^2\}[I + zD^2 b]^{-1}.$$

(iii) $\ker \varepsilon^0 \cap \ker \varepsilon^1 = \ker \varepsilon^0 \cap \ker \varepsilon^1 \cap \ker \varepsilon^2$.

REMARK 7.1. When ω is not connected the pair (a, B) is constant in each connected part of ω, but can be different from one connected part to another.

PROOF. (i) and (ii) From [**36**].

(iii) The proof uses identity (3.35) in § 3 applied to the $\varepsilon^i(e,\ell)$'s

$$\|\varepsilon^i(e,\ell)\|^2 = \|\varepsilon^{iP}(e,\ell)\|^2 + 2\|\varepsilon^i(e,\ell)n\|^2 - \|\varepsilon^i(e,\ell)n\cdot n\|^2,$$

for $i = 0, 1\,2$. First

$$\varepsilon^{0P}(e,\ell) = P\left\{\varepsilon_\Gamma(e) + \frac{1}{2}[\ell^* n + n^*\ell]\right\}P = \varepsilon_\Gamma^P(e),$$

$$\varepsilon^0(e,\ell)n = \varepsilon_\Gamma(e)n + \frac{1}{2}\ell_\Gamma + \ell_n n,$$

$$\varepsilon^0(e,\ell)n\cdot n = \ell_n,$$

$$\|\varepsilon^0(e,\ell)\|^2 = \|\varepsilon_\Gamma^P(e)\|^2 + \frac{1}{2}\|2\varepsilon_\Gamma(e)n + \ell_\Gamma\|^2 + \|\ell_n\|^2.$$

Similarly

$$\begin{aligned}\varepsilon^{1P}(e,\ell) &= \frac{1}{2}P\big[D_\Gamma(\ell) - D_\Gamma(e)D^2b + {}^*D_\Gamma(\ell) - D^2b^*D_\Gamma(e)\big]P\\ &= \varepsilon_\Gamma^P(\ell) - \frac{1}{2}\big[D_\Gamma^P(e)D^2b + D^2b^*D_\Gamma^P(e)\big],\\ \varepsilon^1(e,\ell)n &= \frac{1}{2}\big[{}^*D_\Gamma(\ell)n - D^2b^*D_\Gamma(e)n\big], \qquad \varepsilon^1(e,\ell)n\cdot n = 0,\\ \|\varepsilon^1(e,\ell)\|^2 &= \left\|\varepsilon_\Gamma^P(\ell) - \frac{1}{2}\big[D_\Gamma^P(e)D^2b + D^2b^*D_\Gamma^P(e)\big]\right\|^2\\ &\qquad + \frac{1}{2}\big\|{}^*D_\Gamma(\ell)n - D^2b^*D_\Gamma(e)n\big\|^2.\end{aligned}$$

Finally

$$\begin{aligned}\varepsilon^{2P}(e,\ell) &= -\frac{1}{2}P\Big[(D_\Gamma(\ell) - D_\Gamma(e)D^2b)D^2b + D^2b({}^*D_\Gamma(\ell) - D^2b^*D_\Gamma(e))\Big]P\\ &= -\frac{1}{2}\Big[(D_\Gamma^P(\ell) - D_\Gamma^P(e)D^2b)D^2b + D^2b\big({}^*D_\Gamma^P(\ell) - D^2b^*D_\Gamma^P(e)\big)\Big],\\ \varepsilon^2(e,\ell)n &= -\frac{1}{2}D^2b\big[{}^*D_\Gamma(\ell)n - D^2b^*D_\Gamma(e)n\big] = -D^2b\varepsilon^1(e,\ell)n,\\ &\qquad \varepsilon^2(e,\ell)n\cdot n = 0,\\ \|\varepsilon^2(e,\ell)\|^2 &= \left\|\frac{1}{2}\big[(D_\Gamma^P(\ell) - D_\Gamma^P(e)D^2b)D^2b + D^2b({}^*D_\Gamma^P(\ell) - D^2b^*D_\Gamma^P(e))\big]\right\|^2\\ &\qquad + \frac{1}{2}\big\|D^2b[{}^*D_\Gamma(\ell)n - D^2b^*D_\Gamma(e)n]\big\|^2.\end{aligned}$$

If $\varepsilon^1(e,\ell) = 0 = \varepsilon^0(e,\ell)$, then

$$\varepsilon_\Gamma^P(e) = 0, \quad {}^*D_\Gamma(e)n + \ell_\Gamma = 0, \quad \ell_n = 0,$$

$$\varepsilon_\Gamma^P(\ell) - \frac{1}{2}\big[D_\Gamma^P(e)D^2b + D^2b^*D_\Gamma^P(e)\big] = 0, \quad {}^*D_\Gamma(\ell)n - D^2b^*D_\Gamma(e)n = 0.$$

Therefore $\ell = \ell_\Gamma$ and the only independent identities are

$$\boxed{\varepsilon_\Gamma^P(e) = 0, \quad {}^*D_\Gamma(e)n + \ell_\Gamma = 0} \tag{7.21}$$

$$\boxed{\varepsilon_\Gamma^P(\ell_\Gamma) - \frac{1}{2}\big[D_\Gamma^P(e)D^2b + D^2b^*D_\Gamma^P(e)\big] = 0} \tag{7.22}$$

since

$${}^*D_\Gamma(\ell_\Gamma)n - D^2b^*D_\Gamma(e)n = -D^2b\big(\ell_\Gamma + {}^*D_\Gamma(e)n\big) = 0.$$

Recall that for a vector v

$${}^*D_\Gamma(v) = (\nabla_\Gamma v_1, \dots, \nabla_\Gamma v_N), \quad \big(D_\Gamma(v)\big)_{ij} = (\nabla_\Gamma v_i)_j.$$

Differentiate the Love-Kirchhoff term

$$\begin{aligned}\big(\ell_\Gamma + {}^*D_\Gamma(e)n\big)_i &= (\ell_\Gamma)_i + \big({}^*D_\Gamma(e)\big)_{ik}n_k = (\ell_\Gamma)_i + \big(\nabla_\Gamma(e_k)\big)_i n_k,\\ \nabla_\Gamma\big(\ell_\Gamma + {}^*D_\Gamma(e)n\big)_i &= \nabla_\Gamma(\ell_\Gamma)_i + \big(\nabla_\Gamma(e_k)\big)_i\nabla_\Gamma n_k + n_k\nabla_\Gamma\big((\nabla_\Gamma(e_k))_i\big),\end{aligned}$$

$$D_\Gamma\big(\ell_\Gamma + {}^*D_\Gamma(e)n\big) = D_\Gamma(\ell_\Gamma) + {}^*D_\Gamma(e)D^2b + \sum_k n_k D_\Gamma^2(e_k). \tag{7.23}$$

Now

$$D_\Gamma^P(\ell_\Gamma + {}^*D_\Gamma(e)n) = D_\Gamma^P(\ell_\Gamma) + {}^*D_\Gamma^P(e)D^2b + \sum_k n_k PD_\Gamma^2(e_k), \tag{7.24}$$

$$\varepsilon_\Gamma^P(\ell_\Gamma + {}^*D_\Gamma(e)n) = \varepsilon_\Gamma^P(\ell_\Gamma) + \frac{1}{2}\left[{}^*D_\Gamma^P(e)D^2b + D^2bD_\Gamma^P(e)\right] + \sum_k n_k PD_\Gamma^2(e_k).$$

since $PD_\Gamma^2(e_k)$ is symmetrical (see (3.31) of § 3). By using (7.21) and (7.22)

$$\varepsilon_\Gamma^P(e)D^2b + D^2b\varepsilon_\Gamma^P(e) + \sum_k n_k PD_\Gamma^2(e_k) = 0,$$

and since $\varepsilon_\Gamma^P(e) = 0$,

$$\sum_k n_k PD_\Gamma^2(e_k) = 0 \implies D_\Gamma^P(\ell_\Gamma) = -{}^*D_\Gamma^P(e)D^2b,$$

from (7.24). Now

$$D_\Gamma^P(\ell_\Gamma) - D_\Gamma^P(e)D^2b = -2\varepsilon_\Gamma^P(e)D^2b = 0$$

and necessarily $\varepsilon^{2P}(e,\ell) = 0$. This in addition to the fact that

$$\varepsilon^2(e,\ell)n = -D^2b\,\varepsilon^1(e,\ell)n$$

implies that $\varepsilon^2(e,\ell) = 0$. □

COROLLARY 1. *Under the assumptions of the theorem*

$$\ker\varepsilon^0 \cap \ker\varepsilon^1 = \ker\varepsilon^0 \cap \ker\varepsilon^1 \cap \ker\varepsilon^2 \tag{7.25}$$

$$= \left\{(e, -{}^*D_\Gamma(e)n) \colon e \in H^1(\omega)^N, \varepsilon_\Gamma^P(e) = 0, \sum_k n_k PD_\Gamma^2(e_k) = 0\right\}.$$

7.1.3. *Spaces $\mathcal{H}$, $\mathcal{V}$, $\mathcal{N}$.* In view of Theorem 7.1 it is natural to introduce the following spaces

$$\mathcal{H} = L^2(\omega)^N \times L^2(\omega)^N, \tag{7.26}$$

$$\mathcal{V} = \left\{(e,\ell) \in \mathcal{H} \colon \varepsilon^i(e,\ell) \in L^2(\omega)^{N\times N}, 0 \le i \le 2\right\}, \tag{7.27}$$

$$\mathcal{N} = \left\{(e,\ell) \colon \varepsilon^i(e,\ell) = 0 \text{ in } \omega, 0 \le i \le 2\right\}, \tag{7.28}$$

with norms

$$\left|(e,\ell)\right|_{\mathcal{H}}^2 = |e|_{L^2(\omega)}^2 + |\ell|_{L^2(\omega)}^2, \tag{7.29}$$

$$\left\|(e,\ell)\right\|_{\mathcal{V}}^2 = \left|(e,\ell)\right|_{\mathcal{H}}^2 + \sum_{i=0}^2 \left\|\varepsilon^i(e,\ell)\right\|_{L^2(\omega)}^2, \tag{7.30}$$

and from Theorem 7.1 (i) when ω is connected

$$\mathcal{N} = \left\{(e,\ell) \colon \ell(X) = B\nabla b(X)\right\},\ e(X) = a + BX, \forall a \in \mathbb{R}^N, \tag{7.31}$$
$$\forall B \text{ an } N\times N \text{ matrix such that } B + {}^*B = 0\big\}.$$

The subspace $\mathcal{N}$ characterizes the approximate rigid displacements. Note that the seminorm on the space $\mathcal{V}$

$$|||(e,\ell)||| = \left\{\sum_{i=0}^2 \left\|\varepsilon^i(e,\ell)\right\|_{L^2(\omega)}^2\right\}^{1/2}, \tag{7.32}$$

becomes a norm on the quotient space $\mathcal{H}/\mathcal{N}$.

THEOREM 7.2. *Under Assumption 5.1 let ω be a bounded open domain in Γ.*

(i) *The space*

$$\{(e,\ell) \in \mathcal{H}\colon \varepsilon^0(e,\ell) \in L^2(\omega)^{N\times N}\},$$

with norm

$$\|e\|_{L^2} + \|\ell\|_{L^2} + \|\varepsilon^0(e,\ell)\|_{L^2},$$

is equal to $H^1(\omega)^N \times L^2(\omega)^N$ with equivalent norm $\|e\|_{H^1} + \|\ell\|_{L^2}$.

(ii) *The following spaces are equal with equivalent norms:*

a) $H^1(\omega)^N \times H^1(\omega)^N$ *with norm* $\|e\|_{H^1} + \|\ell\|_{H^1}$.

b) $\{(e,\ell) \in \mathcal{H}\colon \varepsilon^i(e,\ell) \in L^2(\omega)^{N\times N}, i = 0,1\}$ *with norm*

$$\|e\|_{L^2} + \|\ell\|_{L^2} + \|\varepsilon^0(e,\ell)\|_{L^2} + \|\varepsilon^1(e,\ell)\|_{L^2}.$$

c) $\mathcal{V} = \{(e,\ell) \in \mathcal{H}\colon \varepsilon^i(e,\ell) \in L^2(\omega)^{N\times N}, i = 0,1,2\}$ *with norm*

$$\|e\|_{L^2} + \|\ell\|_{L^2} + \|\varepsilon^0(e,\ell)\|_{L^2} + \|\varepsilon^1(e,\ell)\|_{L^2} + \|\varepsilon^2(e,\ell)\|_{L^2}.$$

d) $\{(e,\ell) \in \mathcal{H}\colon \varepsilon^0(e,\ell) \in L^2(\omega)^{N\times N}$ *and* $\varepsilon_\Gamma(\ell) \in L^2(\omega)^{N\times N}\}$ *with norm*

$$\|e\|_{L^2} + \|\ell\|_{L^2} + \|\varepsilon^0(e,\ell)\|_{L^2} + \|\varepsilon_\Gamma(\ell)\|_{L^2}.$$

(iii) *The seminorm*

$$\|\varepsilon^0(e,\ell)\|_{L^2} + \|\varepsilon^1(e,\ell)\|_{L^2},$$

is equivalent to the seminorm

$$\|\varepsilon^0(e,\ell_\Gamma)\|_{L^2} + \|\varepsilon^1(e,\ell_\Gamma)\|_{L^2} + \|\ell_n\|_{H^1}.$$

PROOF. (i) By definition of $\varepsilon^0(e,\ell)$

$$\varepsilon^0(e,\ell) = \varepsilon_\Gamma(e) + \frac{1}{2}[\ell^* n + n^*\ell],$$

and

$$\|\varepsilon^0(e,\ell)\|_{L^2} \le \|\varepsilon_\Gamma(e)\|_{L^2} + \|\ell\|_{L^2} \le \|e\|_{H^1} + \|\ell\|_{L^2}.$$

Conversely by Korn's inequality of Theorem 4.2 of § 4

$$\|e\|_{H^1} \le c\big(\|e\|_{L^2} + \|\varepsilon_\Gamma(e)\|_{L^2}\big),$$

and

$$\begin{aligned}\|\varepsilon_\Gamma(e)\|_{L^2} &\le \left\|\varepsilon_\Gamma(e) + \frac{1}{2}[\ell^* n + n^*\ell]\right\|_{L^2} + \left\|\frac{1}{2}[\ell^* n + n^*\ell]\right\|_{L^2}\\ &\le \|\varepsilon^0(e,\ell)\|_{L^2} + \|\ell\|_{L^2}.\end{aligned}$$

Hence the norms $\|\varepsilon^0(e,\ell)\|_{L^2} + \|\ell\|_{L^2} + \|e\|_{L^2}$ and $\|\varepsilon_\Gamma(e)\|_{L^2} + \|\ell\|_{L^2} + \|e\|_{L^2}$ are equivalent.

(ii) By definition of $\varepsilon^1(e,\ell)$

$$\varepsilon^1(e,\ell) = \varepsilon_\Gamma(\ell) - \frac{1}{2}\big[D_\Gamma(e)D^2b + D^2b^* D_\Gamma(e)\big],$$

and

$$\|\varepsilon^1(e,\ell)\| \le \|\varepsilon_\Gamma(\ell)\| + c\|D_\Gamma(e)\| \le c'\big[\|\ell\|_{H^1} + \|e\|_{H^1}\big].$$

Conversely by Korn's inequality

$$\|\ell\|_{H^1} \le c\|\ell\|_{L^2} + \|\varepsilon_\Gamma(\ell)\|_{L^2},$$

and again by Korn's inequality applied to e

$$\begin{aligned}\|\varepsilon_\Gamma(\ell)\|_{L^2} &\le \|\varepsilon^1(e,\ell)\|_{L^2} + c\|D_\Gamma(e)\|_{L^2}\\ &\le \|\varepsilon^1(e,\ell)\|_{L^2} + c\big[\|\varepsilon_\Gamma(e)\|_{L^2} + \|e\|_{L^2}\big],\end{aligned}$$

and from part (i) we have the equivalence of norms for a) and b). For the equivalence of b) and c) it is sufficient to consider the term

$$\varepsilon^2(e,\ell) = -\frac{1}{2}\big[(D_\Gamma(\ell) - D_\Gamma(e)D^2b)D^2b + D^2b^*(D_\Gamma(\ell) - D_\Gamma(e)D^2b)\big],$$

and

$$\|\varepsilon^2(e,\ell)\| \le c\|D_\Gamma(\ell) - D_\Gamma(e)D^2b\| \le c'\big[\|\ell\|_{H^1} + \|e\|_{H^1}\big].$$

This establishes the equivalence.

Finally for the equivalence of a) and d) from part (i)

$$\begin{aligned}\|\varepsilon^0(e,\ell)\| + \|\varepsilon_\Gamma(\ell)\| &\le \|\varepsilon_\Gamma(e)\| + \|\ell\| + \|\varepsilon_\Gamma(\ell)\|,\\ \|\varepsilon_\Gamma(e)\| + \|\varepsilon_\Gamma(\ell)\| &\le \|\varepsilon^0(e,\ell)\| + \|\ell\| + \|\varepsilon_\Gamma(\ell)\|.\end{aligned}$$

So we have the equivalence of the seminorms and the norms and the space is equal to $H^1(\omega)^N \times H^1(\omega)^N = \mathcal{V}$.

(iii) Recall from Theorem 7.1 that

$$\begin{aligned}\|\varepsilon^0(e,\ell)\|^2 &= \|\varepsilon_\Gamma^P(e)\|^2 + \frac{1}{2}\|2\varepsilon_\Gamma(e)n + \ell_\Gamma\|^2 + \|\ell_n\|^2,\\ \|\varepsilon^1(e,\ell)\|^2 &= \left\|\varepsilon_\Gamma^P(\ell) - \frac{1}{2}\big[D_\Gamma^P(e)D^2b + D^2b^*D_\Gamma^P(e)\big]\right\|^2\\ &\qquad + \frac{1}{2}\|{}^*D_\Gamma(\ell)n - D^2b^*D_\Gamma(e)n\|^2,\end{aligned}$$

and $\|\ell_n\| \le \|\varepsilon^0(e,\ell)\|$. But

$$\begin{aligned}D_\Gamma(\ell_n n) = \ell_n D^2b + n^*\nabla_\Gamma \ell_n, &\quad {}^*D_\Gamma(\ell_n n)n = \ell_n D^2b + \nabla_\Gamma \ell_n{}^*n,\\ {}^*D_\Gamma(\ell_n n)n = \nabla_\Gamma \ell_n, &\quad D_\Gamma^P(\ell_n n) = \ell_n D^2b = \varepsilon_\Gamma^P(\ell_n n),\end{aligned}$$

and

$$\begin{aligned}\nabla_\Gamma \ell_n &= {}^*D_\Gamma(\ell_n n)n = {}^*D_\Gamma(\ell)n - {}^*D_\Gamma(\ell_\Gamma)n\\ &= {}^*D_\Gamma(\ell)n - D^2b^*D_\Gamma(e)n + D^2b(\ell_\Gamma + {}^*D_\Gamma(e)n)\\ &= 2\varepsilon^1(e,\ell)n + 2D^2b\varepsilon^0(e,\ell)n\\ &\qquad \Longrightarrow \|\nabla_\Gamma \ell_n\| \le c\big(\|\varepsilon^1(e,\ell)\| + \|\varepsilon^0(e,\ell)\|\big).\end{aligned}$$

In the other direction observe that

$$\|\varepsilon^0(e,\ell)\|^2 = \|\varepsilon^0(e,\ell_\Gamma)\|^2 + \|\ell_n\|^2,$$

$$\begin{aligned}\|\varepsilon^1(e,\ell)\|^2 &= \left\|\varepsilon_\Gamma^P(\ell_\Gamma) + \ell_n D^2b - \frac{1}{2}\big[D_\Gamma^P(e)D^2b + D^2b^*D_\Gamma^P(e)\big]\right\|^2\\ &\quad + \frac{1}{2}\|\nabla_\Gamma \ell_n - D^2b(\ell_\Gamma + {}^*D_\Gamma(e)n)\|^2\\ &\qquad \Longrightarrow \|\varepsilon^1(e,\ell)\|^2 \le 2\big[\|\varepsilon^1(e,\ell_\Gamma)\|^2 + \|\ell_n D^2b\|^2\big]\\ &\qquad\qquad + \|\nabla_\Gamma \ell_n\|^2 + \|2D^2b\varepsilon^0(e,\ell_\Gamma)n\|^2.\end{aligned}$$

and this gives the inequality in the other direction. □

7.1.4. *Spaces $\mathcal{V}_0$, $\mathcal{V}_{\gamma_0}$, and global Poincaré's inequality.* In this section ω is a bounded open domain in Γ with Lipschitzian boundary γ. In preparation to the study of the homogeneous Dirichlet boundary value problem we introduce the spaces

$$\mathcal{V}_0 = \{(e,\ell) \in \mathcal{V} \colon e|_\gamma = 0 \text{ and } \ell_\Gamma|_\gamma = 0\}, \tag{7.33}$$

$$\mathcal{N}_0 = \{(e,\ell) \in \mathcal{V}_0 \colon \varepsilon^0(e,\ell) = \varepsilon^1(e,\ell) = 0\}. \tag{7.34}$$

When ω is connected and γ_0 is a subset of the boundary γ of ω with non-zero $(N-1)$-capacity

$$\mathcal{V}_{\gamma_0} = \{(e,\ell) \in \mathcal{V} \colon e|_{\gamma_0} = 0 \text{ and } \ell_\Gamma|_{\gamma_0} = 0\}, \tag{7.35}$$

$$\mathcal{N}_{\gamma_0} = \{(e,\ell) \in \mathcal{V}_{\gamma_0} \colon \varepsilon^0(e,\ell) = \varepsilon^1(e,\ell) = 0\}. \tag{7.36}$$

In both cases $q(e,\ell)$

$$q(e,\ell)^2 = \|\varepsilon^0(e,\ell)\|^2 + \|\varepsilon^1(e,\ell)\|^2, \tag{7.37}$$

is a seminorm on the above spaces. We shall need the following lemma.

LEMMA 7.1. *Under Assumption 5.1 let ω be a bounded open domain in Γ with a Lipschitzian boundary γ. As h goes to zero, there exists a constant $c = c(h) > 0$ such that for all $(e,\ell) \in \mathcal{V}_0$*

$$\begin{aligned}\int_\omega 2h|e|^2 + 2\frac{h^3}{3}|\ell|^2\, d\Gamma \\ \le c^2 \int_\omega 2h\|\varepsilon^0(e,\ell)\|^2 + 2\frac{h^3}{3}\|\varepsilon^1(e,\ell)\|^2 + 2\frac{h^5}{5}\|\varepsilon^2(e,\ell)\|^2\, d\Gamma.\end{aligned} \tag{7.38}$$

The result is also true for all $(e,\ell) \in \mathcal{V}_{\gamma_0}$ when ω is connected and γ_0 is a subset of γ with a non-zero $(N-1)$-capacity.

PROOF. From Theorem 7.2 (iii) the seminorm

$$\|\varepsilon^0(e,\ell)\|_{L^2} + \|\varepsilon^1(e,\ell)\|_{L^2},$$

is equivalent to the seminorm

$$\|\varepsilon^0(e,\ell_\Gamma)\|_{L^2} + \|\varepsilon^1(e,\ell_\Gamma)\|_{L^2} + \|\ell_n\|_{H^1}.$$

So we have the inequality for ℓ_n and it is sufficient to prove (7.38) for (e,ℓ_Γ) (cf. [**36**]). □

From this we get a global Poincaré's inequality involving only $\varepsilon^0(e,\ell)$ and $\varepsilon^1(e,\ell)$.

THEOREM 7.3 (Global Poincaré's inequality). *Under Assumption 5.1 let ω be a bounded open domain in Γ with Lipschitzian boundary γ (resp. bounded open connected domain in Γ with Lipschitzian boundary γ and γ_0 be a subset of γ with non-zero $(N-1)$-capacity). Then*

(i) $\mathcal{N}_0 = \{(0,0)\}$ *(resp. $\mathcal{N}_{\gamma_0} = \{(0,0)\}$) and*
(ii) *$q(e,\ell)$ is a norm on $\mathcal{V}_0$ (resp. $\mathcal{V}_{\gamma_0}$) and there exists a constant $c > 0$ such that*

$$\|e\|_{L^2(\omega)} + \|\ell\|_{L^2(\omega)} \le c\big[\|\varepsilon^0(e,\ell)\|_{L^2(\omega)} + \|\varepsilon^1(e,\ell)\|_{L^2(\omega)}\big], \tag{7.39}$$

and $\|\varepsilon^0(e,\ell)\|_{L^2(\omega)} + \|\varepsilon^1(e,\ell)\|_{L^2(\omega)}$ is a norm on $\mathcal{V}_0$ (resp. $\mathcal{V}_{\gamma_0}$) which is equivalent to the $H^1 \times H^1$-norm.

PROOF. (i) From Theorem 7.1

$$\varepsilon^0(e,\ell) = 0 = \varepsilon^1(e,\ell) \quad \Rightarrow \varepsilon^0(e,\ell) = \varepsilon^1(e,\ell) = \varepsilon^2(e,\ell) = 0.$$

By Lemma 7.1 this last property implies that $(e,\ell) = (0,0)$ for all $(e,\ell) \in \mathcal{V}_0$ and a fortiori $\mathcal{N}_0 = \{(0,0)\}$.

(ii) Consider the quotient and its infimum

$$q(e,\ell) = \frac{\int_\omega \|\varepsilon^0(e,\ell)\|^2 + \|\varepsilon^1(e,\ell)\|^2 \, d\Gamma}{\int_\omega |e|^2 + |\ell|^2 \, d\Gamma}, \qquad (e,\ell) \neq (0,0),$$

$$q \overset{\text{def}}{=} \inf\{q(e,\ell) \colon \forall (e,\ell) \in \mathcal{V}_0, \|e\|^2_{L^2} + \|\ell\|^2_{L^2} = 1\}.$$

We now proceed by contradiction. Assume that $q = 0$. Let $\{(e_k,\ell_k)\}$ be a sequence in $\mathcal{V}_0$ such that

$$\int_\omega |e_k|^2 + |\ell_k|^2 \, d\Gamma = 1 \quad \text{and} \quad \int_\omega \left\|\varepsilon^0(e_k,\ell_k)\right\|^2 + \left\|\varepsilon^1(e_k,\ell_k)\right\|^2 d\Gamma \to 0,$$

as k goes to infinity. The sequence is bounded in $\mathcal{V}_0$ and there exists $(e,\ell) \in \mathcal{V}_0$ and a subsequence, still denoted $\{(e_k,\ell_k)\}$, such that $(e_k,\ell_k) \to (e,\ell)$ in $H^1(\omega)^N \times H^1(\omega)^N$-weak. Therefore

$$1 = \int_\omega |e_k|^2 + |\ell_k|^2 \, d\Gamma \to \int_\omega |e|^2 + |\ell|^2 \, d\Gamma,$$

$$\int_\omega \left\|\varepsilon^0(e,\ell)\right\|^2 + \left\|\varepsilon^1(e,\ell)\right\|^2 d\Gamma \\ \leq \liminf_{k\to\infty} \int_\omega \left\|\varepsilon^0(e_k,\ell_k)\right\|^2 + \left\|\varepsilon^1(e_k,\ell_k)\right\|^2 d\Gamma = 0.$$

But from part (i) an element $(e,\ell) \in \mathcal{V}_0$ such that $\varepsilon^0(e,\ell) = \varepsilon^1(e,\ell) = 0$ is $(0,0)$. This is a contradiction with the fact that its L^2-norm is one. Necessarily q is both finite and strictly positive and we can pick $c = 1/q$. Hence we have the global Poincaré's inequality on $\mathcal{V}_0$. □

7.2. Discussion of mechanical properties. In the previous section we have just used a standard development of the displacement vector V into e and ℓ without any additional constraints. Yet in the literature the vector function ℓ is often assumed to be tangential, that is, for all $X \in \omega$, $\ell(X) \cdot n(X) = 0$. This assumption combined with the special form of the displacement vector yields $\varepsilon(V)\nabla b \cdot \nabla b = 0$ for the strain tensor. This property seems to be mechanically unacceptable. To get around it Naghdi and Koiter have used the condition $\sigma(V)\nabla b \cdot \nabla b = 0$ to artificially define a non-zero value for $\varepsilon(V)\nabla b \cdot \nabla b$. This *occasional contradiction* which is accepted in the literature will be discussed in § 7.2.4 and in § 8. In this section we consider six constrained cases which do not a priori assume that the vector function ℓ be tangential.

7.2.1. *Constraints on the strain tensor.* We consider two cases:

(1) $\varepsilon\big|_\omega n \cdot n \; \left(= \varepsilon \circ T_0 n \cdot n = \varepsilon \circ T_z n \cdot n\big|_{z=0} =\right) \; 0$ in ω.

(2) $\varepsilon|_\omega n \; \left(= \varepsilon \circ T_0 n = \varepsilon \circ T_z n\big|_{z=0} =\right) \; 0$ in ω.

Recall the identity

$$\varepsilon \circ T_z n = \ell_n n + \frac{1}{2}[I + zD^2 b]^{-1}[{}^*D_\Gamma(e)n + \ell_\Gamma + z\nabla_\Gamma \ell_n].$$

In case (1)

$$\varepsilon|_\omega n\cdot n=0 \text{ in } \omega \iff \ell_n=0 \text{ in } \omega \ (\ell \text{ is tangent to } \omega) \iff \varepsilon\nabla b\cdot\nabla b=0 \text{ in } S_h(\omega).$$

In case (2)

$$\varepsilon|_\omega n=0 \text{ in } \omega \ (\text{plane strain}) \iff \ell_\Gamma+{}^*D_\Gamma(e)n=0 \text{ and } \ell_n=0 \text{ in } \omega \iff \varepsilon\nabla b=0 \text{ in } S_h(\omega) \ (\text{plane strain})$$

In both cases the condition(s) at the level $z=0$ extends to all level sets of b and hence in $S_h(\omega)$. In case (2) the vector ℓ is specified by

$$\ell=-{}^*D_\Gamma(e)n,$$

since ${}^*D_\Gamma(e)n$ is a tangent vector. So ℓ can be completely eliminated.

In the sequel the tangential term ${}^*D_\Gamma(e)n+\ell_\Gamma$ will be referred to as the ***Love-Kirchhoff term*** since the condition $\ell_\Gamma+{}^*D_\Gamma(e)n=0$ is the ***Love-Kirchhoff condition*** which can also be written in any of the following forms

$$\ell_\Gamma=-{}^*D_\Gamma(e)n=-2\varepsilon_\Gamma(e)n=-\nabla_\Gamma e_n+D^2be_\Gamma. \tag{7.40}$$

7.2.2. *Constraints on the stress tensor.* However it is usually argued that $\sigma|_\omega n\cdot n=0$ is the correct mechanical assumption and not $\varepsilon|_\omega n\cdot n=0$. This leads to the following two cases:

(3) $\sigma\big|_\omega n\cdot n=0$ in ω.
(4) $\sigma\big|_\omega n=0$ in ω.

For the simple constitutive law $\sigma=2\mu\varepsilon+\lambda\operatorname{tr}\varepsilon I$ case (3) yields

$$\sigma\big|_\omega n\cdot n=0 \text{ in } \omega \iff \ell_n=-\frac{\lambda}{2\mu+\lambda}\operatorname{div}_\Gamma(e) \text{ in } \omega,$$

since $\operatorname{tr}\varepsilon|_\omega=\operatorname{tr}\varepsilon_\Gamma(e)=\operatorname{div}_\Gamma(e)$, and case (4)

$$\sigma|_\omega n=0 \text{ in } \omega \ (\text{plane stress}) \iff \begin{cases}\ell_n=-\frac{\lambda}{2\mu+\lambda}\operatorname{div}_\Gamma(e) \text{ in } \omega\\ \ell_\Gamma=-{}^*D_\Gamma(e)n \text{ in } \omega.\end{cases}$$

This follows from

$$\varepsilon\big|_\omega=\varepsilon_\Gamma(e)+\frac{1}{2}[\ell^*n+n^*\ell] \implies \varepsilon\big|_\omega n\cdot n=\ell_n \quad\text{and}\quad \operatorname{tr}\varepsilon\big|_\omega=\operatorname{div}_\Gamma(e)+\ell_n$$
$$\implies 0=\sigma\big|_\omega n\cdot n=2\mu\ell_n+\lambda\big(\operatorname{div}_\Gamma(e)+\ell_n\big).$$

For case (4) ℓ is specified and can be eliminated as in case (2).

7.2.3. *Stronger constraints on the stress tensor.* Contrarily to cases (1) and (2), the constraints on the s!! tress tensor at level $z=0$ do not imply that they be verified at all levels $|z|<h$ and that $\sigma\nabla b\cdot\nabla b=0$ in $S_h(\omega)$. Yet in the respective linear models of Naghdi and Koiter we specifically have $\sigma\nabla b\cdot\nabla b=0$ and $\sigma\nabla b=0$ in $S_h(\omega)$. So we consider the following two additional cases

(5) $\sigma\nabla b\cdot\nabla b=0$ in $S_h(\omega)$.
(6) $\sigma\nabla b=0$ in $S_h(\omega)$.

Again for the simple constitutive law $\sigma=2\mu\varepsilon+\lambda\operatorname{tr}\varepsilon I$ case (5) yields

$$\sigma\nabla b\cdot\nabla b=0 \text{ in } S_h(\omega) \iff \sigma\circ T_z\, n\cdot n=0 \text{ in } \omega \text{ and } |z|<h \iff \ell_n=-\frac{\lambda}{2\,\mu+\lambda}\operatorname{tr}\varepsilon^P\circ T_z \text{ in } \omega \text{ and } |z|<h$$

In case (6)

$$\begin{array}{ccccc} \sigma\nabla b = 0 & & \sigma \circ T_z\, n = 0 & & \ell_n = -\dfrac{\lambda}{2\,\mu+\lambda}\, \operatorname{tr} \varepsilon^P \circ T_z \\ \text{in } S_h(\omega) & \Longleftrightarrow & \text{in } \omega \text{ and } |z| < h & \Longleftrightarrow & \text{in } \omega \text{ and } |z| < h \\ \text{(plane stress)} & & \text{(plane stress)} & & \ell_\Gamma = -{}^*D_\Gamma(e)n \text{ in } \omega \\ & & & & \nabla_\Gamma \ell_n = 0, \quad \text{in } \omega. \end{array}$$

Recalling the definition of the orthogonal projection $P = I - \nabla b^* \nabla b$ and the projected operator $\varepsilon^P \stackrel{\text{def}}{=} P\varepsilon P$ in $S_h(\omega)$, we get the following identity

$$\varepsilon = \varepsilon^P + (\varepsilon\nabla b)^*\nabla b + \nabla b^*(\varepsilon\nabla b) - \nabla b\cdot(\varepsilon\nabla b)\nabla b^*\nabla b$$

$$\Longrightarrow \operatorname{tr}\varepsilon = \operatorname{tr}\varepsilon^P + \varepsilon\nabla b\cdot\nabla b$$

$$\Longrightarrow \varepsilon\nabla b\cdot\nabla b = -\frac{\lambda}{2\mu+\lambda}\operatorname{tr}\varepsilon^P.$$

The last equivalence for case (5) comes from the identity

$$0 = \sigma\nabla b\cdot\nabla b = 2\mu\varepsilon\nabla b\cdot\nabla b + \lambda\operatorname{tr}\varepsilon.$$

For case (6) it comes from

$$0 = \sigma\nabla b = 2\mu\varepsilon\nabla b + \lambda\operatorname{tr}\varepsilon\nabla b,$$

which contains the condition for case (5) plus the fact that the tangential component $P(\varepsilon\nabla b) = (\varepsilon\nabla b)_\Gamma$ of $\varepsilon\nabla b$ is zero, that is for all $|z| < h$

$$0 = (\varepsilon\circ T_z n)_\Gamma = \frac{1}{2}[I + zD^2 b]^{-1}\left[{}^*D_\Gamma(e)n + \ell_\Gamma + z\nabla_\Gamma\ell_n\right] \text{ in } \omega$$

$$\Longrightarrow \ell_\Gamma + {}^*D_\Gamma(e)n = 0 = \nabla_\Gamma\ell_n \text{ in } \omega.$$

In case (6) when ω is simply connected,

$$\ell_n = -\frac{\lambda}{2\mu}\operatorname{div}_\Gamma(e) = \text{constant}.$$

This becomes a tangential ***incompressibility/inextensibility*** condition of the midsurface surface when the constant is zero. It is the surface equivalent of the zero divergence volume condition in three-dimension.

7.2.4. *Tangential condition and occasional contradictions.* It is important to emphasize that ℓ was not assumed to be tangent in the initial asymptotic development of the displacement vector. However this assumption is relatively standard in the literature. If we do we get the undesired condition

$$\varepsilon\circ T_z n\cdot n = 0 \text{ in } \omega, \forall |z| < h.$$

The standard assumption in the literature is $\sigma\nabla b\cdot\nabla b = 0$ in $S_h(\omega)$ as in the model of Naghdi. This condition implies that

$$0 = \sigma|_{z=0} n\cdot n = 2\mu\varepsilon|_{z=0} n\cdot n + \lambda\operatorname{tr}\varepsilon|_{z=0}|n|^2 = \lambda\operatorname{tr}\varepsilon|_{z=0} = \lambda\operatorname{div}_\Gamma(e),$$

and we get the tangential *incompressibility/inextensibility* condition

$$\operatorname{div}_\Gamma(e) = 0 \text{ in } \omega.$$

It arises from the combined assumptions that ℓ be tangent and that $\sigma|_{z=0} n\cdot n = 0$. To get around this constraint and the fact that $\varepsilon\nabla b\cdot\nabla b = 0$, it is accepted to introduce a *mathematical compromise* in the foundations of the theory. Even if the tangential condition $\ell_n = 0$ implies that $\varepsilon\nabla b \cdot \nabla b = 0$, this is ignored and the value of $\varepsilon\nabla b \cdot \nabla b$ is extracted from the condition $\sigma\nabla b \cdot \nabla b = 0$:

$$0 = \sigma\nabla b \cdot \nabla b = 2\mu\varepsilon\nabla b \cdot \nabla b + \lambda \operatorname{tr} \varepsilon \implies 0 = 2\mu\varepsilon\nabla b \cdot \nabla b + \lambda \operatorname{tr} \varepsilon$$
$$\implies \varepsilon\nabla b \cdot \nabla b = -\frac{\lambda}{2\mu + \lambda} \operatorname{tr} \varepsilon^P.$$

Therefore since ℓ is assumed tangential $\ell = \ell_\Gamma$, $\varepsilon = \varepsilon(e, \ell_\Gamma)$ but

$$\varepsilon\nabla b \cdot \nabla b = -\frac{\lambda}{2\mu + \lambda} \operatorname{tr} \varepsilon^P(e, \ell_\Gamma),$$

that is from (7.6) and (7.5)

$$\begin{aligned} \varepsilon^P(e, \ell_\Gamma) &= \frac{1}{2}\{[D_\Gamma^P(e) \circ p + bD_\Gamma^P(\ell_\Gamma) \circ p][I - bD^2 b] \\ &\qquad + [I - bD^2 b]({}^*D_\Gamma^P(e) \circ p + b^* D_\Gamma^P(\ell_\Gamma) \circ p)\} \\ \big[\varepsilon(e, \ell_\Gamma)\nabla b\big]_\Gamma &= \frac{1}{2}[I - bD^2 b]\big({}^*D_\Gamma(e)n + \ell_\Gamma\big) \circ p. \end{aligned}$$

The reader is referred to [**7**, p. 19] for comparison.

This mathematical contradiction is well known in the literature

> The occurrence of occasional contradictions is, of course, not an uncommon phenomenon in approximate theories. Nevertheless, it may be considered an unfortunate trait of the classical theory of thin shells that its basic assumptions are already mutually contradictory. We have no doubt that much of the still existing confusion in shell theory is due to this flaw in the foundations. ([**66**, pp. 15–16]).

Thus removing the tangential condition on ℓ removes the mechanically unrealistic zero condition on $\varepsilon\nabla b \cdot \nabla b$ and the inherent *mathematical compromise*. We shall see later in § 8 that by introducing reasonable approximations cases (3)–(5) and (4)–(6) yield models which are respectively very similar to the models of Naghdi and Koiter.

7.3. Strain energy and work of the external forces. The equivalence of the conditions $\varepsilon(V) \circ T_z = 0$ and $\varepsilon^0 = \varepsilon^1 = 0$, and the characterization of the function space $\mathcal{V}$ in terms of those two tensors suggest to use a first order approximation of the strain tensor.

ASSUMPTION 7.2. The strain tensor $\varepsilon(V)$ will be approximated by the following first order (in z) ***approximate strain tensor***

$$\tilde{\varepsilon}(V) \circ T_z \stackrel{\text{def}}{=} \varepsilon^0(e, \ell) + \varepsilon^1(e, \ell) z. \tag{7.41}$$

This confirms the interest in the $P(1,1)$ model. We now compute the total energy made up of the strain energy plus the work of the external forces for the static case using the approximate strain tensor. For the dynamical case of small vibrations we subtract the kinetic energy and characterize a stationary point of this total energy functional. By external forces we mean forces and torques applied to the shell. For simplicity we have assumed that they are *distributed* within $S_h(\omega)$, but more general loadings can be considered.

The *strain energy* $\mathcal{P}$ and the *work of the external forces* $\mathcal{W}$ are given by

$$\mathcal{P} = \frac{1}{2} \int_{S_h(\omega)} \sigma \cdot\cdot\, \tilde{\varepsilon}(V)\, dx, \quad \mathcal{W} = \int_{S_h(\omega)} \big[F \cdot V + (m \circ p) \cdot (\ell \circ p)\big]\, dx. \tag{7.42}$$

Assume that F is of the form

$$F = f^0 \circ p + b f^1 \circ p, \quad f^0, f^1, m \in L^2(\omega)^N. \tag{7.43}$$

and that $S_h(\omega)$ is made up of an elastic material which obeys the *constitutive law*

$$\sigma = C^{-1}\tilde{\varepsilon}(V), \tag{7.44}$$

where C satisfies Assumption 5.2. Hence

$$\sigma \circ T_z \cdot\cdot\, \tilde{\varepsilon}(V) \circ T_z = \big[C^{-1}\tilde{\varepsilon}(V) \circ T_z\big] \cdot\cdot \tilde{\varepsilon}(V) \circ T_z. \tag{7.45}$$

Recall expressions $V \circ T_z = e + z\ell$, $\tilde{\varepsilon}(V) \circ T_z = \varepsilon^0 + z\varepsilon^1$, and the measure decomposition (2.37). If in addition C is symmetrical then from (7.42) and (7.44)

$$\mathcal{P} = \frac{1}{2} \int_{-h}^{h} dz \int_{\omega} d\Gamma \big\{\big[C^{-1}\tilde{\varepsilon}(V) \circ T_z\big] \cdot\cdot \tilde{\varepsilon}(V) \circ T_z\big\} j_z, \tag{7.46}$$

$$\mathcal{W} = \int_{-h}^{h} dz \int_{\omega} d\Gamma \big[F \circ T_z \cdot V \circ T_z + m \cdot (\ell \circ p \circ T_z)\big] j_z. \tag{7.47}$$

Using identity (7.45)

$$\sigma(V) \circ T_z \cdot\cdot \tilde{\varepsilon}(V) \circ T_z\, j_z = \sum_{i,j=0}^{1} z^{i+j} j_z \,[C^{-1}\varepsilon^i \cdot\cdot\, \varepsilon^j] = \sum_{n=0}^{2} z^n j_z A_n,$$

where

$$A_n = \sum_{\substack{i=0 \\ 0 \le n-i \le 1}}^{1} C^{-1}\varepsilon^i \cdot\cdot\, \varepsilon^{n-i}. \tag{7.48}$$

Similarly in the integrand of (7.47)

$$\begin{aligned} (F \cdot V) \circ T_z &= m \cdot \ell + (f^0 + z f^1) \cdot (e + z\ell) \\ &= f^0 \cdot e + m \cdot \ell + z(f^1 \cdot e + f^0 \cdot \ell) + z^2 f^1 \cdot \ell. \end{aligned} \tag{7.49}$$

The final expressions for $\mathcal{P}$ and $\mathcal{W}$ become

$$\mathcal{P} = \frac{1}{2} \int_{\omega} \int_{-h}^{h} \sum_{n=0}^{2} z^n j_z \, dz A_n \, d\Gamma = \frac{1}{2} \int_{\omega} \sum_{n=0}^{2} \alpha_n(h) A_n \, d\Gamma, \tag{7.50}$$

$$\mathcal{W} = \int_{-h}^{h} dz \int_{\omega} d\Gamma \big[m \cdot \ell + (f^0 + z f^1) \cdot (e + z\ell)\big] j_z \tag{7.51}$$

$$= \int_{\omega} \alpha_0(h)(f^0 \cdot e + m \cdot \ell) + \alpha_1(h)(f^1 \cdot e + f^0 \cdot \ell) + \alpha_2(h) f^1 \cdot \ell \, d\Gamma, \tag{7.52}$$

where the α_n's where the α_n's are the polynomials in odd powers of h defined by (5.11) in § 5.

7.4. Static linear unconstrained model.

7.4.1. *Bilinear form.* The spaces $\mathcal{H}$, $\mathcal{V}$, and $\mathcal{N}$, and their associated norms and seminorm have been defined in (7.26) to (7.30). Now introduce the operator $A\colon \mathcal{V} \to \mathcal{V}'$ associated with $\mathcal{P}$: for all (e,ℓ) and $(\overline{e},\overline{\ell})$ in $\mathcal{V}$ and $\varepsilon^i = \varepsilon^i(e,\ell)$, $\overline{\varepsilon}^i = \varepsilon^i(\overline{e},\overline{\ell})$

$$\big\langle A(e,\ell),(\overline{e},\overline{\ell})\big\rangle_{\mathcal{V}} = \sum_{n=0}^{2} \int_\omega \alpha_n(h) a_n\big((e,\ell),(\overline{e},\overline{\ell})\big)\, d\Gamma, \tag{7.53}$$

where

$$a_n\big((e,\ell),(\overline{e},\overline{\ell})\big) = \sum_{\substack{i=0 \\ 0\le n-i\le 1}}^{1} C^{-1}\varepsilon^i \cdot\cdot\, \overline{\varepsilon}^{n-i}. \tag{7.54}$$

Note that $A_n = a_n\big((e,\ell),(e,\ell)\big)$, $0 \le n \le 2$. Similarly introduce the continuous linear control operator associated with $\mathcal{W}$: for all (e,ℓ) and $(\overline{e},\overline{\ell})$ in $\mathcal{V}$

$$\begin{gathered} \mathcal{B}\colon U \overset{\text{def}}{=} L^2(\omega)^N \times L^2(\omega)^N \to \mathcal{H}', \\ \big\langle \mathcal{B}(f,m),(e,\ell)\big\rangle_{\mathcal{H}} \\ = \int_\omega \alpha_0(f^0\cdot e + m\cdot\ell) + \alpha_1(f^1\cdot e + f^0\cdot \ell) + \alpha_2 f^1\cdot \ell\, d\Gamma. \end{gathered} \tag{7.55}$$

LEMMA 7.2. **[36]** *Under Assumptions 5.1 and 5.2 there exist* $\overline{h} > 0$ *and* $\alpha > 0$ *such that for all* $0 < h < \overline{h}$ *and all* $(e,\ell)\in\mathcal{V}$

$$\big\langle A(e,\ell),(e,\ell)\big\rangle_{\mathcal{V}} \ge 2\mu h\alpha \sum_{n=0}^{1} h^{2n}\big\|\varepsilon^n(e,\ell)\big\|^2. \tag{7.56}$$

If, in addition, C *is symmetrical, then* A *is symmetrical*

$$\forall (e,\ell),(\overline{e},\overline{\ell}),\quad \big\langle A(e,\ell),(\overline{e},\overline{\ell})\big\rangle_{\mathcal{V}} = \big\langle A(\overline{e},\overline{\ell}),(e,\ell)\big\rangle_{\mathcal{V}}.$$

Note that for convenience we have assumed that C was symmetrical in the previous section in order to use a minimization problem formulation. If fact this is not necessary and the results in the next sections follow from the Principle of virtual work and the Lax-Milgram theorem.

7.4.2. *Neumann problem and shells without boundary.* If the elements of the dual $\mathcal{H}'$ of $\mathcal{H}$ are identified with those of $\mathcal{H}$, then from the above lemma the operator A is a $\mathcal{V}$-$\mathcal{H}$ coercive operator. If $\omega = \Gamma$, the domain ω has no boundary; if ω has a boundary γ, we have an homogeneous Neumann boundary problem. Both cases are covered by the following theorem.

THEOREM 7.4. *Let Assumptions 5.1 and 5.2 be verified. Given* $\overline{h} > 0$ *as specified in* Lemma 7.2, $0 < h \le \overline{h}$, *and assuming that the following condition is verified for all* $(e,\ell)\in\mathcal{N}$

$$\int_\omega \alpha_0(f^0\cdot e + m\cdot\ell) + \alpha_1(f^1\cdot e + f^0\cdot\ell) + \alpha_2 f^1\cdot\ell\, d\Gamma = 0, \tag{7.57}$$

then for all h, $0 < h \le \overline{h}$, *there exists a unique solution* $(\hat{e},\hat{\ell}) \in \mathcal{V}/\mathcal{N}$ *to the variational equation: for all* $(e,\ell)\in\mathcal{V}$

$$\big\langle A(\hat{e},\hat{\ell}),(e,\ell)\big\rangle_{\mathcal{V}} + \big\langle \mathcal{B}(f,m),(e,\ell)\big\rangle_{\mathcal{H}} = 0. \tag{7.58}$$

7.4.3. *Dirichlet problem.* For a shell with boundary and homogeneous Dirichlet boundary conditions the results are analogous by the same arguments as in the proof of Theorem 7.4 using the global Poincaré inequality Theorem 7.3.

THEOREM 7.5. *Let Assumptions 5.1 and 5.2 be verified. Given $\overline{h} > 0$ as specified in* Lemma 7.2 *and h, $0 < h \leq \overline{h}$, there exists a unique solution $(\hat{e}, \hat{\ell}) \in \mathcal{V}_0$ to the variational equation: for all $(e, \ell) \in \mathcal{V}_0$*

$$\big\langle A(\hat{e}, \hat{\ell}), (e, \ell)\big\rangle_{\mathcal{V}} + \big\langle \mathcal{B}(f, m), (e, \ell)\big\rangle_{\mathcal{H}} = 0, \tag{7.59}$$

where

$$\mathcal{V}_0 = \big\{(e, \ell) \in H^1(\omega)^N \times H^1(\omega)^N : e\big|_\gamma = 0 \text{ and } \ell_\Gamma\big|_\gamma = 0\big\}. \tag{7.60}$$

The same results hold with the space

$$\mathcal{V}_{\gamma_0} = \big\{(e, \ell) \in \mathcal{V}: e|_{\gamma_0} = 0 \text{ and } \ell_\Gamma|_{\gamma_0} = 0\big\}$$

for ω connected and a subset γ_0 of γ with non-zero $(N - 1)$-capacity.

REMARK 7.2. It is important to note that there is no condition on the normal component ℓ_n of ℓ.

7.5. Dynamical linear unconstrained model.

7.5.1. *Kinetic energy.* To get the dynamical linear model it remains to compute the kinetic energy. Assume that $V(x, t)$, $e(x, t)$, and $\ell(x, t)$ depend on the time t and denote by V', e', and ℓ' the derivatives with respect to the time t:

$$\begin{aligned} \mathcal{K} &= \int_{S_h(\omega)} \frac{1}{2}\rho |V'|^2 \, dx = \int_{S_h(\omega)} \frac{1}{2}\rho |e' \circ p + b\ell' \circ p|^2 \, dx \\ &= \frac{1}{2}\rho \int_\omega \alpha_0(h)|e'|^2 \; + \alpha_1(h) 2e' \cdot \ell' + \alpha_2(h)|\ell'|^2 \, d\Gamma, \end{aligned} \tag{7.61}$$

where ρ is the volume density and e' and ℓ' are the time derivatives of e and ℓ. Define the operator $\mathcal{M}\colon \mathcal{H} \to \mathcal{H}'$ as follows:

$$\big\langle \mathcal{M}(e, \ell), (\overline{e}, \overline{\ell})\big\rangle_{\mathcal{H}} = \rho \int_\omega \alpha_0(h) e \cdot \overline{e} + \alpha_1(h)[e \cdot \overline{\ell} + \overline{e} \cdot \ell] + \alpha_2(h) \ell \cdot \overline{\ell} \, d\Gamma. \tag{7.62}$$

For h sufficiently small $\mathcal{M}$ is symmetrical, positive and invertible.

7.5.2. *Dynamical shell equation.* Identify the elements of the dual $\mathcal{H}'$ of $\mathcal{H}$ with those of $\mathcal{H}$ and denote by $\Lambda\colon \mathcal{H}' \to \mathcal{H}$ the corresponding canonical isomorphism. Then the linear operators

$$M = \Lambda\mathcal{M}, \quad B = \Lambda\mathcal{B}, \tag{7.63}$$

are bounded from $\mathcal{H}$ to $\mathcal{H}$ and from U to $\mathcal{H}$. For h sufficiently small, M is a positive symmetrical continuous and invertible linear operator. A stationary point (e, ℓ) of the total energy $\mathcal{P} + \mathcal{W} - \mathcal{K}$ verifies the following second order dynamical equation

$$\frac{d^2}{dt^2}\left(M \begin{bmatrix} e(t) \\ \ell(t) \end{bmatrix}, \begin{bmatrix} \varphi \\ \psi \end{bmatrix}\right)_{\mathcal{H}} + \left\langle A \begin{bmatrix} e(t) \\ \ell(t) \end{bmatrix}, \begin{bmatrix} \varphi \\ \psi \end{bmatrix}\right\rangle_{\mathcal{V}} + \left(B \begin{bmatrix} f(t) \\ m(t) \end{bmatrix}, \begin{bmatrix} \varphi \\ \psi \end{bmatrix}\right)_{\mathcal{H}} = 0, \tag{7.64}$$

for all $(\varphi, \psi) \in \mathcal{V}$. Standard existence and uniqueness theorems (cf. [**71**], Thm 1.1 and Rem. 1.3, p. 294]) apply.

THEOREM 7.6. *Let Assumptions 5.1 and 5.2 be verified and further assume that C is symmetrical. Let $0 < h \leq \overline{h}$ for $\overline{h} > 0$ as specified in* Lemma 7.2.

(i) (Shell without boundary or Neumann boundary conditions). *There exists a unique solution*

$$(e,\ell) \in C^1([0,T];\mathcal{H}) \cap C([0,T];\mathcal{V}). \tag{7.65}$$

to equation (7.64) *verifying the initial conditions*

$$\big(e(0),\ell(0)\big) = (e_0,\ell_0) \text{ in } \mathcal{V}, \quad \big(e'(0),\ell'(0)\big) = (e_1,\ell_1) \text{ in } \mathcal{H}. \tag{7.66}$$

(ii) (Dirichlet boundary conditions). *There exists a unique solution*

$$(e,\ell) \in C^1\big([0,T];\mathcal{H}\big) \cap C\big([0,T];\mathcal{V}_0\big). \tag{7.67}$$

to equation (7.64) *verifying the initial conditions*

$$\big(e(0),\ell(0)\big) = (e_0,\ell_0) \text{ in } \mathcal{V}_0, \quad \big(e'(0),\ell'(0)\big) = (e_1,\ell_1) \text{ in } \mathcal{H}. \tag{7.68}$$

When ω is connected and γ_0 is a subset of γ with non-zero $(N-1)$ capacity, the same results hold with the space

$$\mathcal{V}_{\gamma_0} = \big\{(e,\ell) \in \mathcal{V}\colon e\big|_{\gamma_0} = 0 \text{ and } \ell_\Gamma|_{\gamma_0} = 0\big\}.$$

7.6. Discussion of the six constrained cases. We now show that we the same results for the six constrained cases. They are obtained from the unconstrained static and dynamical models by going to linear subspaces of $\mathcal{V}$ (resp. $\mathcal{V}_0$ or $\mathcal{V}_{\gamma_0}$) and $\mathcal{H}$. The important observation is that the approximation of the strain tensor does not affect the conditions associated with the six cases, since

$$2\tilde{\varepsilon}(V) \circ T_z n\big[I - zD^2b + (zD^2b)^2\big]\big(2\varepsilon_\Gamma(e)n + \ell_\Gamma + z\nabla_\Gamma \ell_n\big) + \ell_n n,$$
$$\tilde{\varepsilon}(V) \circ T_z n \cdot n = \ell_n.$$

Moreover in view of (3.33) in § 3

$$\begin{aligned}
\varepsilon^0(e,\ell) &= \varepsilon_\Gamma^P(e) + \frac{1}{2}\big[(\ell_\Gamma + 2\varepsilon_\Gamma(e)n)^* n + n^*(\ell_\Gamma + 2\varepsilon_\Gamma(e)n)\big] + \ell_n n^* n,\\
\varepsilon^1(e,\ell) &= \varepsilon_\Gamma^P(\ell_\Gamma) + \ell_n D^2 b - \frac{1}{2}\big[D_\Gamma^P(e) D^2 b + D^2 b^* D_\Gamma^P(e)\big]\\
&\quad - \frac{1}{2}\Big[\big(D^2 b(\ell_\Gamma + 2\varepsilon_\Gamma(e)n)\big)^* n + n^*\big(D^2 b(\ell_\Gamma + 2\varepsilon_\Gamma(e)n)\big)\Big]\\
&\quad + \frac{1}{2}[\nabla_\Gamma \ell_n{}^* n + n^* \nabla_\Gamma \ell_n].
\end{aligned}$$

In each of the six cases constraints will be added on ℓ_n, $\nabla_\Gamma \ell_n$, or the Love-Kirchhoff term $\ell_\Gamma + 2\varepsilon_\Gamma(e)n)$.

7.6.1. *Constraints on the strain tensor.* In case (1), it is now sufficient to work in the subspaces

$$\mathcal{H}_1 = \big\{(e,\ell) \in \mathcal{H}\colon \ell_n = 0 \text{ in } \omega\big\}, \quad \mathcal{V}_1 = \big\{(e,\ell) \in \mathcal{V}\colon \ell_n = 0 \text{ in } \omega\big\}, \tag{7.69}$$

$$\mathcal{N}_1 = \big\{(e,\ell) \in \mathcal{V}_1 \colon \varepsilon^0(e,\ell) = \varepsilon^1(e,\ell) = 0\big\}. \tag{7.70}$$

The spaces $\mathcal{H}_1$ and $\mathcal{V}_1$ are closed linear subspaces of $\mathcal{H}$ and $\mathcal{V}$, respectively. Moreover in view of Theorem 7.1 (i) $\mathcal{N}_1 = \mathcal{N}$. Therefore everything we have said before applies with the above subspaces. For the dynamical case we identify the elements of the dual $\mathcal{H}_1'$ of $\mathcal{H}_1$ with those of $\mathcal{H}_1$ instead of $\mathcal{H}'$ and $\mathcal{H}$. Here

$$\varepsilon^0(e,\ell) = \varepsilon_\Gamma^P(e) + \frac{1}{2}\big[(\ell_\Gamma + 2\varepsilon_\Gamma(e)n)^* n + n^*(\ell_\Gamma + 2\varepsilon_\Gamma(e)n)\big],$$

$$\varepsilon^1(e,\ell) = \varepsilon_\Gamma^P(\ell_\Gamma) - \frac{1}{2}\big[D_\Gamma^P(e)D^2b + D^2b^* D_\Gamma^P(e)\big]$$

$$- \frac{1}{2}\Big[\big(D^2b(\ell_\Gamma + 2\varepsilon_\Gamma(e)n)\big)^* n + n^*\big(D^2b(\ell_\Gamma + 2\varepsilon_\Gamma(e)n)\big)\Big].$$

In case (2) it is also sufficient to work in the subspaces

(7.71) $$\mathcal{H}_2 = \big\{(e,\ell) \in \mathcal{H}\colon 2\varepsilon_\Gamma(e)n + \ell = 0 \text{ in } \omega\big\},$$

(7.72) $$\mathcal{V}_2 = \big\{(e,\ell) \in \mathcal{V}\colon 2\varepsilon_\Gamma(e)n + \ell = 0 \text{ in } \omega\big\},$$

(7.73) $$\mathcal{N}_2 = \big\{(e,\ell) \in \mathcal{V}_2\colon \varepsilon^0(e,\ell) = \varepsilon^1(e,\ell) = 0\big\}.$$

The space $\mathcal{H}_2$ is a closed linear subspace of $\mathcal{H}$. Recalling that

$$\ell = -2\varepsilon_\Gamma(e)n = D^2be - \nabla_\Gamma e_n,$$

it is easy to check that

$$\mathcal{H}_2 = \big\{(e, D^2be - \nabla_\Gamma e_n)\colon e \in L^2(\omega)^N \text{ and } e_n \in H^1(\omega)\big\}$$

with equivalent norm

$$\big\{\|e\|_{L^2}^2 + \|\nabla_\Gamma e_n\|_{L^2}^2\big\}^{1/2}.$$

Indeed since $b \in C^{1,1}(\overline{S_h(\omega)})$ for all $(e,\ell) \in \mathcal{H}_2$

$$\|e\|^2 + \|\ell\|^2 = \|e\|^2 + \|\nabla_\Gamma e_n - D^2be\|^2 \le c^2\big[\|e\|^2 + \|\nabla_\Gamma e_n\|^2\big].$$

In the other direction

$$\|\nabla_\Gamma e_n\| \le \|\nabla_\Gamma e_n - D^2be\| + \|D^2be\| \le c\big[\|\ell\| + \|e\|\big].$$

It is also easy to verify that

$$\mathcal{V}_2 = \big\{(e, -2\varepsilon_\Gamma(e)n)\colon e \in L^2(\omega)^N \text{ and } \varepsilon_\Gamma(e)n \in H^1(\omega)\big\},$$

is a closed subspace of $\mathcal{V}$. Here

(7.74) $$\varepsilon^0(e,\ell) = \varepsilon^{0P}(e,\ell_\Gamma) = \varepsilon_\Gamma^P(e),$$

(7.75) $$\varepsilon^1(e,\ell) = \varepsilon^{1P}(e,\ell_\Gamma) = \varepsilon_\Gamma^P(\ell_\Gamma) - \frac{1}{2}\big[D_\Gamma^P(e)D^2b + D^2b^* D_\Gamma^P(e)\big].$$

Recall that in view of (7.23) in the proof of Theorem 7.1

$$D_\Gamma(\ell_\Gamma) = -D_\Gamma({}^*D_\Gamma(e)n) = -\Big[{}^*D_\Gamma(e)D^2b + \sum_k n_k D_\Gamma^2(e_k)\Big],$$

$$D_\Gamma^P(\ell_\Gamma) = -\Big[{}^*D_\Gamma^P(e)D^2b + \sum_k n_k P D_\Gamma^2(e_k)\Big],$$

$$\varepsilon_\Gamma^P(\ell_\Gamma) = -\frac{1}{2}\big[{}^*D_\Gamma^P(e)D^2b + D^2bD_\Gamma^P(e)\big] - \sum_k n_k P D_\Gamma^2(e_k),$$

and

$$\varepsilon^1(e,\ell) = -\big[\varepsilon_\Gamma^P(e)D^2b + D^2b\varepsilon_\Gamma^P\big] - \sum_k n_k P D_\Gamma^2(e_k).$$

Note that the term $\varepsilon_\Gamma(e)n$ is meaningful as an element of $H^1(\omega)^N$ even if b is only $C^{1,1}$. The above spaces are respectively isomorphic to

$$H_2 = \{e \in L^2(\omega)^N : e_n \in H^1(\omega)\}, \tag{7.76}$$

$$V_2 = \{e \in H^1(\omega)^N : \varepsilon_\Gamma(e)n \in H^1(\omega)^N\}, \tag{7.77}$$

with equivalent norms

$$\|e\|_{H_2}^2 = \|e\|_{L_2}^2 + \|\nabla_\Gamma e_n\|_{L_2}^2, \quad \|e\|_{V_2}^2 = \|e\|_{H^1}^2 + \|\varepsilon_\Gamma(e)n\|_{H^1}^2.$$

If we assume that Ω is of class $C^{2,1}(\overline{S_h(\omega)})$, then $\mathcal{V}_2$ and V_2 become:

$$\mathcal{V}_2 = \{(e, D^2 be - \nabla_\Gamma e_n) : e \in H^1(\omega)^N \quad \text{and} \quad e_n \in H^2(\omega)\},$$

$$V_2 = \{e \in H^1(\omega)^N : e_n \in H^2(\omega)^N\}.$$

Such spaces where the normal component of e is one degree smoother than e itself are typical in the Theory of Plates where $\partial\Omega$ is of class C^∞.

Moreover in view of Theorem 7.1 (i) $\mathcal{N}_2 = \mathcal{N}$. By definition $\mathcal{N}_2 \subset \mathcal{N}$. In the other direction recall that for all $(e, \ell) \in \mathcal{N}$

$$e(X) = a + BX \quad \text{and} \quad \ell(X) = B\nabla b(X),$$

with ${}^*B + B = 0$. So by definition

$$(e \circ p)(x) = a + Bp(x) \implies D_\Gamma(e) = BDp\big|_\Gamma = B[I - n{}^*n],$$

$${}^*D_\Gamma(e)n = [I - n{}^*n]{}^*Bn = {}^*Bn - n(Bn \cdot n) = {}^*Bn = -Bn = -\ell,$$

since ${}^*B + B = 0$ implies that $2Bn \cdot n = 0$. Finally

$$0 = \ell + {}^*D_\Gamma(e)n = \ell + 2\varepsilon_\Gamma(e)n,$$

and $\mathcal{N} \subset \mathcal{N}_2$. Note that $\mathcal{N}$ is isomorphic to

$$N = \{e : e(X) = a + BX, \forall a \in \mathbb{R}^N \quad \text{and} \\ \forall B \text{ an } N \times N \text{ matrix such that } B + {}^*B = 0\}.$$

For the dynamical case we identify the elements of the dual $\mathcal{H}_2'$ of $\mathcal{H}_2$ with those of $\mathcal{H}_2$ instead of $\mathcal{H}'$ and $\mathcal{H}$. Therefore everything we have said before applies with the above subspaces.

7.6.2. *Constraints on the stress tensor.* In case (3) the condition on ℓ_n is included in the definitions of the spaces

$$\mathcal{H}_3 = \left\{(e, \ell) \in \mathcal{H} : \ell_n = -\frac{\lambda}{2\mu + \lambda}\operatorname{div}_\Gamma e \text{ in } \omega\right\},$$

$$\mathcal{V}_3 = \left\{(e, \ell) \in \mathcal{V} : \ell_n = -\frac{\lambda}{2\mu + \lambda}\operatorname{div}_\Gamma e \text{ in } \omega\right\},$$

$$\mathcal{N}_3 = \{(e, \ell) \in \mathcal{V}_3 : \varepsilon^0(e, \ell) = \varepsilon^1(e, \ell) = 0\}.$$

The spaces $\mathcal{H}_3$ and $\mathcal{V}_3$ are closed linear subspaces of $\mathcal{H}$ and $\mathcal{V}$, respectively. Moreover in view of Theorem 7.1 (i) $\mathcal{N}_3 = \mathcal{N}$. Clearly $\mathcal{N}_3 \subset \mathcal{N}$. For all $(e, \ell) \in \mathcal{N}$, $\ell_n = Bn \cdot n = 0$ and $\operatorname{div}_\Gamma(e) = \operatorname{tr} D_\Gamma(e) = \operatorname{tr}(B - Bn{}^*n) = \operatorname{tr} B - Bn \cdot n = 0$ since $B + {}^*B = 0$ implies $2 \operatorname{tr} B = 0$. So the condition is verified since $\ell_n = \operatorname{div}_\Gamma(e) = 0$ and $\mathcal{N}_3 = \mathcal{N}$. For the dynamical case we identify the elements of the dual $\mathcal{H}_3'$ of $\mathcal{H}_3$ with those of $\mathcal{H}_3$ instead of $\mathcal{H}'$ and $\mathcal{H}$. Therefore previous results apply in the above subspaces.

Case (4) is case (3) plus the Love-Kirchhoff conditions

$$\mathcal{H}_4 = \left\{(e,\ell) \in \mathcal{H} \colon \ell_n = -\frac{\lambda}{2\mu+\lambda}\operatorname{div}_\Gamma e, \ell_\Gamma + 2\varepsilon_\Gamma(e)n = 0 \text{ in } \omega\right\},$$

$$\mathcal{V}_4 = \left\{(e,\ell) \in \mathcal{V} \colon \ell_n = -\frac{\lambda}{2\mu+\lambda}\operatorname{div}_\Gamma e, \ell_\Gamma + 2\varepsilon_\Gamma(e)n = 0 \text{ in } \omega\right\},$$

$$\mathcal{N}_4 = \left\{(e,\ell) \in \mathcal{V}_4 \colon \varepsilon^0(e,\ell) = \varepsilon^1(e,\ell) = 0\right\}.$$

The spaces $\mathcal{H}_4$ and $\mathcal{V}_4$ are closed linear subspaces of $\mathcal{H}$ and $\mathcal{V}$, respectively. Moreover in view of Theorem 7.1 (i) and the cases (2) and (3) $\mathcal{N}_4 = \mathcal{N}$. For the dynamical case we identify the elements of the dual $\mathcal{H}_4'$ of $\mathcal{H}_4$ with those of $\mathcal{H}_4$. Therefore everything we have said before applies with the above subspaces. Here

$$\begin{aligned}\varepsilon^0(e,\ell) &= \varepsilon_\Gamma^P(e) + \ell_n n^* n,\\ \varepsilon^1(e,\ell) &= \varepsilon_\Gamma^P(\ell_\Gamma) + \ell_n D^2 b - \frac{1}{2}\left[D_\Gamma^P(e) D^2 b + D^2 b^* D_\Gamma^P(e)\right]\\ &\qquad + \frac{1}{2}\left[\nabla_\Gamma \ell_n{}^* n + n^* \nabla_\Gamma \ell_n\right].\end{aligned}$$

7.6.3. *Stronger constraints on the stress tensor.* In case (5) the condition on ℓ_n really contains three conditions

$$\ell_n = -\frac{\lambda}{2\mu+\lambda}\operatorname{div}_\Gamma e \text{ in } \omega, \tag{7.78}$$

$$0 = \operatorname{tr}\varepsilon^{1P} = \operatorname{tr}(D_\Gamma^P \ell - D_\Gamma^P e D^2 b) = \operatorname{div}_\Gamma \ell - D_\Gamma^P e \cdot\cdot D^2 b \text{ in } \omega. \tag{7.79}$$

Again they can be included in the definitions of the spaces

$$\mathcal{H}_5 = \left\{(e,\ell) \in \mathcal{H} \colon \ell_n = -\frac{\lambda}{2\mu+\lambda}\operatorname{div}_\Gamma e, \operatorname{div}_\Gamma \ell = D_\Gamma^P e \cdot\cdot D^2 b \text{ in } \omega\right\},$$

$$\mathcal{V}_5 = \left\{(e,\ell) \in \mathcal{V} \colon \ell_n = -\frac{\lambda}{2\mu+\lambda}\operatorname{div}_\Gamma e, \operatorname{div}_\Gamma \ell = D_\Gamma^P e \cdot\cdot D^2 b \text{ in } \omega\right\},$$

$$\mathcal{N}_5 = \left\{(e,\ell) \in \mathcal{V}_5 \colon \varepsilon^0(e,\ell) = \varepsilon^1(e,\ell) = 0\right\}.$$

The spaces $\mathcal{H}_5$ and $\mathcal{V}_5$ are closed linear subspaces of $\mathcal{H}$ and $\mathcal{V}$, respectively. Moreover in view of Theorem 7.1 (i) $\mathcal{N}_5 = \mathcal{N}$. To see this we only need to check the last two conditions

$$\operatorname{div}_\Gamma \ell = D_\Gamma^P e \cdot\cdot D^2 b, \quad D_\Gamma^P \ell \cdot\cdot D^2 b = D_\Gamma^P e \cdot\cdot (D^2 b)^2.$$

since we have already shown that $\ell_n = \operatorname{div}_\Gamma e = 0$. First

$$D_\Gamma^P \ell = P D^2 b B = D^2 b B = D_\Gamma \ell \quad \Longrightarrow \begin{cases}\operatorname{div}_\Gamma \ell = D^2 b \cdot\cdot B\\ D_\Gamma^P \ell \cdot\cdot D^2 b = D^2 b B \cdot\cdot D^2 b = B \cdot\cdot (D^2 b)^2\end{cases}$$

and

$$\begin{aligned}D_\Gamma^P e \cdot\cdot D^2 b &= PB[I - n^* n] \cdot\cdot D^2 b = B \cdot\cdot D^2 b - B n^* n \cdot\cdot D^2 b\\ &= B \cdot\cdot D^2 b - Bn \cdot D^2 b n = B \cdot\cdot D^2 b.\end{aligned}$$

Putting the above identities together we get $\mathcal{N} \subset \mathcal{N}_5$. For the dynamical case we identify the elements of the dual $\mathcal{H}_5'$ of $\mathcal{H}_5$ with those of $\mathcal{H}_5$ instead of $\mathcal{H}'$ and $\mathcal{H}$. Therefore everything we have said before applies with the above subspaces.

Case (6) is case (5) plus the Love-Kirchhoff conditions

$$\mathcal{H}_6 = \Big\{(e,\ell) \in \mathcal{H} \colon \ell_n = -\frac{\lambda}{2\mu+\lambda}\operatorname{div}_\Gamma e,\ \nabla_\Gamma \ell_n = 0 \\ \operatorname{div}_\Gamma \ell = D_\Gamma^P e \cdot\cdot D^2 b,\ \ell_\Gamma + 2\varepsilon_\Gamma(e) n = 0 \text{ in } \omega\Big\},$$

$$\mathcal{V}_6 = \Big\{(e,\ell) \in \mathcal{V} \colon \ell_n = -\frac{\lambda}{2\mu+\lambda}\operatorname{div}_\Gamma e,\quad \nabla_\Gamma \ell_n = 0 \\ \operatorname{div}_\Gamma \ell = D_\Gamma^P e \cdot\cdot D^2 b,\ \ell_\Gamma + 2\varepsilon_\Gamma(e) n = 0 \text{ in } \omega\Big\},$$

$$\mathcal{N}_6 = \big\{(e,\ell) \in \mathcal{V}_6 \colon \varepsilon^0(e,\ell) = \varepsilon^1(e,\ell) = 0\big\}.$$

The spaces $\mathcal{H}_6$ and $\mathcal{V}_6$ are closed linear subspaces of $\mathcal{H}$ and $\mathcal{V}$, respectively. Moreover in view of Theorem 7.1 (i) and the cases (2) to (5) $\mathcal{N}_6 = \mathcal{N}$. For the dynamical case we identify the elements of the dual $\mathcal{H}_6'$ of $\mathcal{H}_6$ with those of $\mathcal{H}_6$ instead of $\mathcal{H}'$ and $\mathcal{H}$. Therefore everything we have said before applies with the above subspaces. Here

$$\varepsilon^0(e,\ell) = \varepsilon_\Gamma^P(e) + \ell_n n^* n,$$
$$\varepsilon^1(e,\ell) = \varepsilon_\Gamma^P(\ell_\Gamma) + \ell_n D^2 b - \frac{1}{2}\big[D_\Gamma^P(e) D^2 b + D^2 b^* D_\Gamma^P(e)\big].$$

8. Approximate Linear Models of Shells

8.1. Introduction. In § 6 we have seen the two important linear models of Naghdi and Koiter. They correspond to some first order approximation (in z) of the strain tensor

$$\varepsilon \circ T_z = \varepsilon^0 + z\varepsilon^1, \tag{8.1}$$

and the approximation by one of the surface Jacobian

$$j_z = 1 + Hz + Kz^2 \simeq 1 \qquad (\text{for } N = 3). \tag{8.2}$$

As was pointed out by Koiter the term "approximation" is very important.

In this section we explicit the unconstrained model of § 7 and the six constrained cases. We introduce further approximations which preserve the theoretical results of § 7 and discuss the relationship between Naghdi's and Koiter's models and cases (3)–(5) and (4)–(6), respectively.

8.2. Approximate unconstrained linear model. In this section we explicit the unconstrained model of § 7 under the assumption that $j_z \simeq 1$. In that case there is no cross term and the operator A simplifies. However all the results of § 7 remain true. For all (e,ℓ) and $(\bar e,\bar\ell)$ in $\mathcal{V}$ with the notation $\varepsilon^i = \varepsilon^i(e,\ell)$, $\bar\varepsilon^i = \varepsilon^i(\bar e,\bar\ell)$

$$\big\langle A(e,\ell),(\bar e,\bar\ell)\big\rangle_{\mathcal{V}} = \int_\omega 2h(2\mu\varepsilon^0 \cdot\cdot \bar\varepsilon^0 + \lambda \operatorname{tr}\varepsilon^0 \operatorname{tr}\bar\varepsilon^0) \\ + \frac{2h^3}{3}(2\mu\varepsilon^1 \cdot\cdot \bar\varepsilon^1 + \lambda \operatorname{tr}\varepsilon^1 \operatorname{tr}\bar\varepsilon^1)\, d\Gamma. \tag{8.3}$$

For the purpose of the discussion it will be convenient to use the notation (7.48) and (7.54) of § 7

$$A_0 = 2\mu\|\varepsilon^0\|^2 + \lambda|\operatorname{tr}\varepsilon^0|^2,\quad A_1 = 2\mu 2\varepsilon^0 \cdot\cdot \varepsilon^1 + \lambda 2 \operatorname{tr}\varepsilon^0 \operatorname{tr}\varepsilon^1, \\ A_2 = 2\mu\|\varepsilon^1\|^2 + \lambda|\operatorname{tr}\varepsilon^1|^2, \tag{8.4}$$

and

$$
\begin{aligned}
a_0\big((e,\ell),(\bar e,\bar\ell)\big) &= 2\mu\varepsilon^0\cdot\cdot\bar\varepsilon^0 + \lambda\operatorname{tr}\varepsilon^0\operatorname{tr}\bar\varepsilon^0,\\
a_1\big((e,\ell),(\bar e,\bar\ell)\big) &= 2\mu[\varepsilon^0\cdot\cdot\bar\varepsilon^1 + \bar\varepsilon^0\cdot\cdot\varepsilon^1] + \lambda[\operatorname{tr}\varepsilon^0\operatorname{tr}\bar\varepsilon^1 + \operatorname{tr}\bar\varepsilon^0\operatorname{tr}\varepsilon^1],\\
a_2\big((e,\ell),(\bar e,\bar\ell)\big) &= 2\mu\varepsilon^1\cdot\cdot\bar\varepsilon^1 + \lambda\operatorname{tr}\varepsilon^1\operatorname{tr}\bar\varepsilon^1.
\end{aligned}
\tag{8.5}
$$

8.2.1. *Decomposition of ε^0 and ε^1.* In order to compute the terms entering in the definition of A we use identities (3.33) to (3.35) in § 3 applied to the ε^i's

$$\varepsilon^i = \varepsilon^{iP} + (\varepsilon^i n)^* n + n^*(\varepsilon^i n) - (\varepsilon^i n\cdot n)n^* n, \tag{8.6}$$

$$\varepsilon^i\cdot\cdot\bar\varepsilon^i = \varepsilon^{iP}\cdot\cdot\bar\varepsilon^{iP} + 2(\varepsilon^i n)\cdot(\bar\varepsilon^i n) - (\varepsilon^i n\cdot n)(\bar\varepsilon^i n\cdot n), \tag{8.7}$$

$$\operatorname{tr}\varepsilon^i = \operatorname{tr}\varepsilon^{iP} + \varepsilon^i n\cdot n, \tag{8.8}$$

$$\|\varepsilon^i\|^2 = \|\varepsilon^{iP}\|^2 + 2|\varepsilon^i n|^2 - |\varepsilon^i n\cdot n|^2. \tag{8.9}$$

Proceeding term by term we get

$$
\begin{gathered}
\varepsilon^{0P}(e,\ell) = P\Big\{\varepsilon_\Gamma(e) + \frac{1}{2}[\ell^* n + n^*\ell]\Big\}P = \varepsilon_\Gamma^P(e),\\
\varepsilon^0(e,\ell)n = \varepsilon_\Gamma(e)n + \frac{1}{2}\ell_\Gamma + \ell_n n, \quad \varepsilon^0(e,\ell)n\cdot n = \ell_n,\\
\operatorname{tr}\varepsilon^0(e,\ell) = \operatorname{div}_\Gamma(e) + \ell_n,\\
\varepsilon^0\cdot\cdot\bar\varepsilon^0 = \varepsilon_\Gamma^P(e)\cdot\cdot\varepsilon_\Gamma^P(\bar e) + \frac{1}{2}\big(2\varepsilon_\Gamma(e)n + \ell_\Gamma\big)\cdot\big(2\varepsilon_\Gamma(\bar e)n + \bar\ell_\Gamma\big) + \ell_n\bar\ell_n,\\
\big\|\varepsilon^0(e,\ell)\big\|^2 = \big\|\varepsilon_\Gamma^P(e)\big\|^2 + \frac{1}{2}\big\|2\varepsilon_\Gamma(e)n + \ell_\Gamma\big\|^2 + \|\ell_n\|^2.
\end{gathered}
$$

Similarly

$$
\begin{aligned}
\varepsilon^{1P}(e,\ell) &= \frac{1}{2}P\big[D_\Gamma(\ell) - D_\Gamma(e)D^2b + {}^*D_\Gamma(\ell) - D^2b^*D_\Gamma(e)\big]P\\
&= \varepsilon_\Gamma^P(\ell) - \frac{1}{2}\big[D_\Gamma^P(e)D^2b + D^2b^*D_\Gamma^P(e)\big],\\
\varepsilon^1(e,\ell)n &= \frac{1}{2}\big[{}^*D_\Gamma(\ell)n - D^2b^*D_\Gamma(e)n\big]\\
&= \frac{1}{2}\big[\nabla_\Gamma\ell_n - D^2b(2\varepsilon_\Gamma(e)n + \ell_\Gamma)\big] = \frac{1}{2}\big[\nabla_\Gamma\ell_n - D^2b(2\varepsilon_\Gamma(e)n + \ell)\big]\\
&= \frac{1}{2}[\nabla_\Gamma\ell_n - D^2b2\varepsilon^0(e,\ell_\Gamma)n] = \frac{1}{2}\big[\nabla_\Gamma\ell_n - D^2b2\varepsilon^0(e,\ell)n\big],
\end{aligned}
$$

$$
\begin{gathered}
\varepsilon^1(e,\ell)n\cdot n = 0,\\
\operatorname{tr}\varepsilon^1(e,\ell) = \operatorname{tr}\varepsilon^{1P}(e,\ell) = \operatorname{div}_\Gamma(\ell) - D^2b\cdot\cdot D_\Gamma^P(e),
\end{gathered}
$$

$$
\begin{aligned}
\varepsilon^1\cdot\cdot\bar\varepsilon^1 = \varepsilon^{1P}(e,\ell)\cdot\cdot\varepsilon^{1P}(\bar e,\bar\ell)&\\
+\frac{1}{2}\big(\nabla_\Gamma\ell_n - D^2b(2\varepsilon_\Gamma(e)n + \ell_\Gamma)\big)&\cdot\big(\nabla_\Gamma\bar\ell_n - D^2b\big(2\varepsilon_\Gamma(\bar e)n + \bar\ell_\Gamma\big)\big),\\
\big\|\varepsilon^1(e,\ell)\big\|^2 = \bigg\|\varepsilon_\Gamma^P(\ell) - \frac{1}{2}\big[D_\Gamma^P(e)D^2b + D^2b^*D_\Gamma^P(e)\big]\bigg\|^2&\\
+\frac{1}{2}\big|\nabla_\Gamma\ell_n - D^2b(2\varepsilon_\Gamma(e)n + \ell_\Gamma)\big|^2&.
\end{aligned}
$$

8.2.2. *Unconstrained model.* It is useful to isolate the variable ℓ_n and the Love-Kirchhoff term $2\varepsilon^0(e,\ell_\Gamma)n = \ell_\Gamma + 2\varepsilon_\Gamma(e)n$ in the previous unconstrained expressions

$$\varepsilon^{0P}(e,\ell) = \varepsilon_\Gamma^P(e) = \varepsilon^{0P}(e,\ell_\Gamma),$$
$$\varepsilon^0(e,\ell)n = \varepsilon^0(e,\ell_\Gamma)n + \ell_n n, \quad \varepsilon^0(e,\ell_\Gamma)n = \varepsilon_\Gamma(e)n + \frac{1}{2}\ell_\Gamma, \quad \varepsilon^0(e,\ell)n\cdot n = \ell_n,$$
$$\operatorname{tr}\varepsilon^0(e,\ell) = \operatorname{tr}\varepsilon^{0P}(e,\ell_\Gamma) + \ell_n, \quad \operatorname{tr}\varepsilon^{0P}(e,\ell_\Gamma) = \operatorname{div}_\Gamma(e),$$
$$\|\varepsilon^0(e,\ell)\|^2 = \|\varepsilon^{0P}(e,\ell_\Gamma)\|^2 + 2|\varepsilon^0(e,\ell_\Gamma)n|^2 + |\ell_n|^2,$$
$$\varepsilon^0\cdot\cdot\,\overline{\varepsilon}^0 = \varepsilon^{0P}(e,\ell_\Gamma)\cdot\cdot\,\varepsilon^{0P}(\overline{e},\overline{\ell}_\Gamma) + 2\varepsilon^0(e,\ell_\Gamma)n\cdot\varepsilon^0(\overline{e},\overline{\ell}_\Gamma)n + \ell_n\overline{\ell}_n.$$

So we compute the terms a_0 and A_0

$$\begin{aligned} a_0\big((e,\ell),(\overline{e},\overline{\ell})\big) &= 2\mu\varepsilon^0\cdot\cdot\,\overline{\varepsilon}^0 + \lambda\operatorname{tr}\varepsilon^0\operatorname{tr}\overline{\varepsilon}^0 \\ &= 2\mu\big[\varepsilon^{0P}(e,\ell_\Gamma)\cdot\cdot\,\varepsilon^{0P}(\overline{e},\overline{\ell}_\Gamma) + 2\varepsilon^0(e,\ell_\Gamma)n\cdot\varepsilon^0(\overline{e},\overline{\ell}_\Gamma)n + \ell_n\overline{\ell}_n\big] \\ &\qquad + \lambda\big(\operatorname{tr}\varepsilon^{0P}(e,\ell_\Gamma) + \ell_n\big)\big(\operatorname{tr}\varepsilon^{0P}(\overline{e},\overline{\ell}_\Gamma) + \overline{\ell}_n\big), \end{aligned}$$

and

$$\begin{aligned} A_0 &= 2\mu\|\varepsilon^0\|^2 + \lambda|\operatorname{tr}\varepsilon^0|^2 \\ &= 2\mu\big[\|\varepsilon^{0P}(e,\ell_\Gamma)\|^2 + 2|\varepsilon^0(e,\ell_\Gamma)n|^2 + |\ell_n|^2\big] + \lambda\big|\operatorname{tr}\varepsilon^{0P}(e,\ell_\Gamma) + \ell_n\big|^2. \end{aligned}$$

Similarly for ε^1 we separate the terms in ℓ_Γ from those in ℓ_n

$$\begin{aligned} \varepsilon^{1P}(e,\ell) &= \varepsilon_\Gamma^P(\ell) - \frac{1}{2}\big[D_\Gamma^P(e)D^2b + D^2b^*D_\Gamma^P(e)\big], \\ &= \varepsilon^{1P}(e,\ell_\Gamma) + \ell_n D^2b, \end{aligned}$$
$$\varepsilon^1(e,\ell)n = \frac{1}{2}\nabla_\Gamma\ell_n - D^2b\varepsilon^0(e,\ell_\Gamma)n, \quad \varepsilon^1(e,\ell)n\cdot n = 0,$$
$$\operatorname{tr}\varepsilon^1(e,\ell) = \operatorname{tr}\varepsilon^{1P}(e,\ell) = \operatorname{tr}\varepsilon^{1P}(e,\ell_\Gamma) + \ell_n\Delta b,$$
$$\begin{aligned} \varepsilon^1\cdot\cdot\,\overline{\varepsilon}^1 &= \big[\varepsilon^{1P}(e,\ell_\Gamma) + \ell_n D^2b\big]\cdot\cdot\,\big[\varepsilon^{1P}(\overline{e},\overline{\ell}_\Gamma) + \overline{\ell}_n D^2b\big] \\ &\qquad + 2\Big(\frac{1}{2}\nabla_\Gamma\ell_n - D^2b\varepsilon^0(e,\ell_\Gamma)n\Big)\cdot\Big(\frac{1}{2}\nabla_\Gamma\overline{\ell}_n - D^2b\varepsilon^0(\overline{e},\overline{\ell}_\Gamma)n\Big), \end{aligned}$$
$$\|\varepsilon^1(e,\ell)\|^2 = \|\varepsilon^{1P}(e,\ell_\Gamma) + \ell_n D^2b\|^2 + 2\Big|\frac{1}{2}\nabla_\Gamma\ell_n - D^2b\varepsilon^0(e,\ell_\Gamma)n\Big|^2.$$

Next we compute a_2 and A_2

$$\begin{aligned} a_2\big((e,\ell),(\overline{e},\overline{\ell})\big) &= 2\mu\varepsilon^1\cdot\cdot\,\overline{\varepsilon}^1 + \lambda\operatorname{tr}\varepsilon^1\operatorname{tr}\overline{\varepsilon}^1 \\ &= 2\mu\Big\{[\varepsilon^{1P}(e,\ell_\Gamma) + \ell_n D^2b]\cdot\cdot\,[\varepsilon^{1P}(\overline{e},\overline{\ell}_\Gamma) + \overline{\ell}_n D^2b] \\ &\qquad + 2\Big(\frac{1}{2}\nabla_\Gamma\ell_n - D^2b\varepsilon^0(e,\ell_\Gamma)n\Big)\cdot\Big(\frac{1}{2}\nabla_\Gamma\overline{\ell}_n - D^2b\varepsilon^0(\overline{e},\overline{\ell}_\Gamma)n\Big)\Big\} \\ &\qquad + \lambda\big(\operatorname{tr}\varepsilon^{1P}(e,\ell_\Gamma) + \ell_n\Delta\big)\big(\operatorname{tr}\varepsilon^{1P}(\overline{e},\overline{\ell}_\Gamma) + \overline{\ell}_n\Delta b\big). \end{aligned}$$

and

$$\begin{aligned} A_2 &= 2\mu\|\varepsilon^1\|^2 + \lambda|\operatorname{tr}\varepsilon^1|^2 \\ &= 2\mu\left[\left\|\varepsilon^{1P}(e,\ell_\Gamma) + \ell_n D^2 b\right\|^2 + 2\left|\frac{1}{2}\nabla_\Gamma \ell_n - D^2 b\varepsilon^0(e,\ell_\Gamma)n\right|^2\right] \\ &\qquad + \lambda\left|\operatorname{tr}\varepsilon^{1P}(e,\ell_\Gamma) + \ell_n \Delta b\right|^2. \end{aligned}$$

8.2.3. *Approximation of the energy functional.* At this juncture we introduce some additional approximations which do not affect the theoretical results of § 7. The general idea is to do it in such a way that the $\mathcal{V}$-$\mathcal{H}$ coercivity and the continuity of the operator A is not affected. Recall the form of the strain energy term

$$\int_\omega 2hA_0 + 2\frac{h^3}{3}A_2\, d\Gamma.$$

From the previous computations it is reasonable to neglect in the term A_2 the terms

$$2\frac{h^3}{3}\|\ell_n D^2 b\|^2 \quad \text{and} \quad 2\frac{h^3}{3}|\ell_n \Delta b|^2,$$

as compared to $2h|\ell_n|^2$ in A_0 and the term

$$2\frac{h^3}{3}\left|D^2 b\varepsilon^0(e,\ell_\Gamma)n\right|^2,$$

as compared to the term $2h\left|\varepsilon^0(e,\ell_\Gamma)n\right|^2$. Now introduce the following approximations of a_2 and A_2

$$\begin{aligned} \tilde{a}_2\big((e,\ell),(\bar{e},\bar{\ell})\big) &= 2\mu\left[\varepsilon^{1P}(e,\ell_\Gamma)\cdot\cdot\,\varepsilon^{1P}(\bar{e},\bar{\ell}_\Gamma) + \frac{1}{2}\nabla_\Gamma\ell_n\cdot\nabla_\Gamma\bar{\ell}_n\right] \\ &\qquad + \lambda\operatorname{tr}\varepsilon^{1P}(e,\ell_\Gamma)\operatorname{tr}\varepsilon^{1P}(\bar{e},\bar{\ell}_\Gamma), \end{aligned}$$

and

$$\tilde{A}_2 = 2\mu\left[\left\|\varepsilon^{1P}(e,\ell_\Gamma)\right\|^2 + \frac{1}{2}|\nabla_\Gamma\ell_n|^2\right] + \lambda\left|\operatorname{tr}\varepsilon^{1P}(e,\ell_\Gamma)\right|^2,$$

and the approximation $\tilde{A}$ of the operator A

$$\boxed{\begin{aligned} &\langle\tilde{A}(e,\ell),(\bar{e},\bar{\ell})\rangle_{\mathcal{V}} \\ &= \int_\omega 2h\Big\{2\mu\big[\varepsilon^{0P}(e,\ell_\Gamma)\cdot\cdot\,\varepsilon^{0P}(\bar{e},\bar{\ell}_\Gamma) + 2\varepsilon^0(e,\ell_\Gamma)n\cdot\varepsilon^0(\bar{e},\bar{\ell}_\Gamma)n + \ell_n\bar{\ell}_n\big] \\ &\qquad +\lambda\big(\operatorname{tr}\varepsilon^{0P}(e,\ell_\Gamma)+\ell_n\big)\big(\operatorname{tr}\varepsilon^{0P}(\bar{e},\bar{\ell}_\Gamma)+\bar{\ell}_n\big)\Big\} \\ &\quad + 2\frac{h^3}{3}\Big\{2\mu\left[\varepsilon^{1P}(e,\ell_\Gamma)\cdot\cdot\,\varepsilon^{1P}(\bar{e},\bar{\ell}_\Gamma) + \frac{1}{2}\nabla_\Gamma\ell_n\cdot\nabla_\Gamma\bar{\ell}_n\right] \\ &\qquad + \lambda\operatorname{tr}\varepsilon^{1P}(e,\ell_\Gamma)\operatorname{tr}\varepsilon^{1P}(\bar{e},\bar{\ell}_\Gamma)\Big\}d\Gamma \end{aligned}}$$

The term

$$\boxed{\int_\omega 2h\Big\{2\mu\big[\|\varepsilon^{0P}(e,\ell_\Gamma)\|^2 + |\ell_n|^2\big] + \lambda\left|\operatorname{tr}\varepsilon^{0P}(e,\ell_\Gamma)+\ell_n\right|^2\Big\}\,d\Gamma}$$

is the ***membrane energy***, the term

$$\boxed{\int_\omega 2\mu\Big\{h\big|\nabla_\Gamma(e_n) - D^2 b e_\Gamma + \ell_\Gamma\big|^2 + \frac{h^3}{3}|\nabla_\Gamma \ell_n|^2\Big\}\, d\Gamma}$$

is the ***shear energy*** and the term

$$\boxed{\int_\omega 2\frac{h^3}{3}\{2\mu\|\varepsilon^{1P}(e,\ell_\Gamma)\|^2 + \lambda|\operatorname{tr}\varepsilon^{1P}(e,\ell_\Gamma)|^2\}\, d\Gamma}$$

is the ***deflection energy***. They are analogous to the expressions found in the literature except for the presence of the variable ℓ_n and the value of the coefficients.

8.3. Approximate constrained linear models. In the following sections we specialize this approximate model to the six constrained cases.

8.3.1. *Constraints on the strain tensor.* In case (1) $\ell_n = 0$ and

$$\begin{aligned}
\langle \tilde{A}(e,\ell),(\bar{e},\bar{\ell})\rangle_{\mathcal{V}}
&= \int_\omega 2h\{2\mu[\varepsilon^{0P}(e,\ell_\Gamma)\cdot\cdot\,\varepsilon^{0P}(\bar{e},\bar{\ell}_\Gamma) + 2\varepsilon^0(e,\ell_\Gamma)n\cdot\varepsilon^0(\bar{e},\bar{\ell}_\Gamma)n] \\
&\qquad +\lambda \operatorname{tr}\varepsilon^{0P}(e,\ell_\Gamma)\operatorname{tr}\varepsilon^{0P}(\bar{e},\bar{\ell}_\Gamma)\} \\
&\qquad + 2\frac{h^3}{3}\{2\mu[\varepsilon^{1P}(e,\ell_\Gamma)\cdot\cdot\,\varepsilon^{1P}(\bar{e},\bar{\ell}_\Gamma)] + \lambda\operatorname{tr}\varepsilon^{1P}(e,\ell_\Gamma)\operatorname{tr}\varepsilon^{1P}(\bar{e},\bar{\ell}_\Gamma)\}\, d\Gamma.
\end{aligned}$$

In case (2) $\ell_n = 0$ and the Love-Kirchhoff term $\varepsilon^0(e,\ell_\Gamma)n$ is zero

$$\begin{aligned}
\langle \tilde{A}(e,\ell),(\bar{e},\bar{\ell})\rangle_{\mathcal{V}}
&= \int_\omega 2h\{2\mu\varepsilon^{0P}(e,\ell_\Gamma)\cdot\cdot\,\varepsilon^{0P}(\bar{e},\bar{\ell}_\Gamma) + \lambda\operatorname{tr}\varepsilon^{0P}(e,\ell_\Gamma)\operatorname{tr}\varepsilon^{0P}(\bar{e},\bar{\ell}_\Gamma)\} \\
&\qquad + 2\frac{h^3}{3}\{2\mu\varepsilon^{1P}(e,\ell_\Gamma)\cdot\cdot\,\varepsilon^{1P}(\bar{e},\bar{\ell}_\Gamma) + \lambda\operatorname{tr}\varepsilon^{1P}(e,\ell_\Gamma)\operatorname{tr}\varepsilon^{1P}(\bar{e},\bar{\ell}_\Gamma)\}\, d\Gamma.
\end{aligned}$$

8.3.2. *Constraints on the stress tensor.* In case (3)

$$\ell_n = -\frac{\lambda}{\lambda+2\mu}\operatorname{tr}\varepsilon^{0P}(e,\ell_\Gamma) = -\frac{\lambda}{\lambda+2\mu}\operatorname{div}_\Gamma(e), \tag{8.10}$$

and ℓ_n can be eliminated from the terms a_0 and A_0. Recall that

$$\begin{aligned}
a_0\big((e,\ell),(\bar{e},\bar{\ell})\big) &= 2\mu[\varepsilon^{0P}(e,\ell_\Gamma)\cdot\cdot\,\varepsilon^{0P}(\bar{e},\bar{\ell}_\Gamma) + 2\varepsilon^0(e,\ell_\Gamma)n\cdot\varepsilon^0(\bar{e},\bar{\ell}_\Gamma)n + \ell_n\bar{\ell}_n] \\
&\qquad + \lambda\big(\operatorname{tr}\varepsilon^{0P}(e,\ell_\Gamma)+\ell_n\big)\big(\operatorname{tr}\varepsilon^{0P}(\bar{e},\bar{\ell}_\Gamma)+\bar{\ell}_n\big).
\end{aligned}$$

The terms in ℓ_n and $\operatorname{tr}\varepsilon^{0P}$ can easily be isolated

$$\begin{aligned}
2\mu\ell_n\bar{\ell}_n &+ \lambda\big(\operatorname{tr}\varepsilon^{0P}(e,\ell_\Gamma)+\ell_n\big)\big(\operatorname{tr}\varepsilon^{0P}(\bar{e},\bar{\ell}_\Gamma)+\bar{\ell}_n\big) \\
&= 2\mu\Big(\frac{\lambda}{\lambda+2\mu}\Big)^2\operatorname{tr}\varepsilon^{0P}(e,\ell_\Gamma)\operatorname{tr}\varepsilon^{0P}(\bar{e},\bar{\ell}_\Gamma) \\
&\qquad + \lambda\Big(1-\frac{\lambda}{\lambda+2\mu}\Big)^2\operatorname{tr}\varepsilon^{0P}(e,\ell_\Gamma)\operatorname{tr}\varepsilon^{0P}(\bar{e},\bar{\ell}_\Gamma) \\
&= \frac{2\mu\lambda}{\lambda+2\mu}\operatorname{tr}\varepsilon^{0P}(e,\ell_\Gamma)\operatorname{tr}\varepsilon^{0P}(\bar{e},\bar{\ell}_\Gamma).
\end{aligned}$$

Finally

$$a_0\big((e,\ell),(\bar{e},\bar{\ell})\big) = 2\mu\big[\varepsilon^{0P}(e,\ell_\Gamma)\cdot\cdot\,\varepsilon^{0P}(\bar{e},\bar{\ell}_\Gamma) + 2\varepsilon^0(e,\ell_\Gamma)n\cdot\varepsilon^0(\bar{e},\bar{\ell}_\Gamma)n\big] + \frac{2\mu\lambda}{\lambda+2\mu}\operatorname{tr}\varepsilon^{0P}(e,\ell_\Gamma)\operatorname{tr}\varepsilon^{0P}(\bar{e},\bar{\ell}_\Gamma),$$

and $\tilde{a}_2$ remains the same

$$\tilde{a}_2\big((e,\ell),(\bar{e},\bar{\ell})\big) = 2\mu\left[\varepsilon^{1P}(e,\ell_\Gamma)\cdot\cdot\,\varepsilon^{1P}(\bar{e},\bar{\ell}_\Gamma) + \frac{1}{2}\nabla_\Gamma\ell_n\cdot\nabla_\Gamma\bar{\ell}_n\right] + \lambda\operatorname{tr}\varepsilon^{1P}(e,\ell_\Gamma)\operatorname{tr}\varepsilon^{1P}(\bar{e},\bar{\ell}_\Gamma).$$

In case (4) we have

$$\ell_n = -\frac{\lambda}{\lambda+2\mu}\operatorname{div}_\Gamma(e) \quad\text{and}\quad \varepsilon^0(e,\ell_\Gamma)n = 0 \tag{8.11}$$

and a_0 further simplifies

$$a_0\big((e,\ell),(\bar{e},\bar{\ell})\big) = 2\mu\varepsilon^{0P}(e,\ell_\Gamma)\cdot\cdot\,\varepsilon^{0P}(\bar{e},\bar{\ell}_\Gamma) + \frac{2\mu\lambda}{\lambda+2\mu}\operatorname{tr}\varepsilon^{0P}(e,\ell_\Gamma)\operatorname{tr}\varepsilon^{0P}(\bar{e},\bar{\ell}_\Gamma).$$

In case (5) we have

$$\ell_n = -\frac{\lambda}{\lambda+2\mu}\operatorname{div}_\Gamma(e) \quad\text{and}$$
$$0 = \operatorname{tr}\varepsilon^1(e,\ell) = \operatorname{tr}\varepsilon^{1P}(e,\ell) = \operatorname{tr}\varepsilon^{1P}(e,\ell_\Gamma) + \ell_n\Delta b$$

and this introduces another constraint on the variable ℓ_Γ

$$\operatorname{div}_\Gamma(\ell_\Gamma) = D^2b\cdot\cdot\,D_\Gamma^P(e) + \frac{\lambda}{\lambda+2\mu}\Delta b\operatorname{div}_\Gamma(e).$$

The term a_0 is the same as in case (3)

$$a_0\big((e,\ell),(\bar{e},\bar{\ell})\big) = 2\mu\big[\varepsilon^{0P}(e,\ell_\Gamma)\cdot\cdot\,\varepsilon^{0P}(\bar{e},\bar{\ell}_\Gamma) + 2\varepsilon^0(e,\ell_\Gamma)n\cdot\varepsilon^0(\bar{e},\bar{\ell}_\Gamma)n\big] + \frac{2\mu\lambda}{\lambda+2\mu}\operatorname{tr}\varepsilon^{0P}(e,\ell_\Gamma)\operatorname{tr}\varepsilon^{0P}(\bar{e},\bar{\ell}_\Gamma).$$

The term a_2 simplifies since $\operatorname{tr}\varepsilon^1(e,\ell) = 0$

$$a_2\big((e,\ell),(\bar{e},\bar{\ell})\big) = 2\mu\big[\varepsilon^{1P}(e,\ell_\Gamma) + \ell_n D^2b\big]\cdot\cdot\,\big[\varepsilon^{1P}(\bar{e},\bar{\ell}_\Gamma) + \bar{\ell}_n D^2b\big] + 2\big(\frac{1}{2}\nabla_\Gamma\ell_n - D^2b\varepsilon^0(e,\ell_\Gamma)n\big)\cdot\left(\frac{1}{2}\nabla_\Gamma\bar{\ell}_n - D^2b\varepsilon^0(\bar{e},\bar{\ell}_\Gamma)n\right).$$

Moreover as previously discussed we can drop the term

$$2\frac{h^3}{3}\|\ell_n D^2b\|^2 \quad\text{and}\quad 2\frac{h^3}{3}|\ell_n\Delta b|^2,$$

as compared to $2h|\ell_n|^2$ and the term

$$2\frac{h^3}{3}\big|D^2b\varepsilon^0(e,\ell_\Gamma)n\big|^2,$$

as compared to $2h|\varepsilon^0(e,\ell_\Gamma)n|^2$ and use the following approximations

$$\tilde{a}_2\big((e,\ell),(\bar{e},\bar{\ell})\big) = 2\mu\left\{\varepsilon^{1P}(e,\ell_\Gamma)\cdot\cdot\,\varepsilon^{1P}(\bar{e},\bar{\ell}_\Gamma) + \frac{1}{2}\nabla_\Gamma\ell_n\cdot\nabla_\Gamma\bar{\ell}_n\right\}.$$

In case (6) we have

$$\ell_n = -\frac{\lambda}{\lambda+2\mu}\operatorname{div}_\Gamma(e) \quad\text{and}\quad \varepsilon^0(e,\ell_\Gamma)n = 0,$$
$$0 = \operatorname{tr}\varepsilon^1(e,\ell) \quad\text{and}\quad \nabla_\Gamma \ell_n = 0,$$

which means more constraints on (e,ℓ). First the vector function ℓ is specified by

$$\ell_n = -\frac{\lambda}{\lambda+2\mu}\operatorname{div}_\Gamma(e) \quad\text{and}\quad \ell_\Gamma = -\varepsilon_\Gamma(e)n,$$

and in addition we have the two constraints

$$\operatorname{div}_\Gamma(\ell_\Gamma) = D^2b\cdot\cdot D^P_\Gamma(e) + \frac{\lambda}{\lambda+2\mu}\Delta b \operatorname{div}_\Gamma(e) \quad\text{and}\quad \nabla_\Gamma(\operatorname{div}_\Gamma e) = 0.$$

The bilinear forms reduce to

$$a_0\big((e,\ell),(\bar e,\bar\ell)\big) = 2\mu\varepsilon^{0P}(e,\ell_\Gamma)\cdot\cdot\,\varepsilon^{0P}(\bar e,\bar\ell_\Gamma) + \frac{2\mu\lambda}{\lambda+2\mu}\operatorname{tr}\varepsilon^{0P}(e,\ell_\Gamma)\operatorname{tr}\varepsilon^{0P}(\bar e,\bar\ell_\Gamma),$$
$$\tilde a_2\big((e,\ell),(\bar e,\bar\ell)\big) = 2\mu\varepsilon^{1P}(e,\ell_\Gamma)\cdot\cdot\,\varepsilon^{1P}(\bar e,\bar\ell_\Gamma).$$

8.4. Comparison with the models of Naghdi and Koiter. We have seen in § 6 that Naghdi's model can be written in the form

$$\begin{aligned}(8.12)\quad 2h\int_\omega 2\mu\{\varepsilon^{0P}(e,\ell_\Gamma)\cdot\cdot\,\varepsilon^{0P}(\bar e,\bar\ell) + 2(\varepsilon^0(e,\ell)n)\cdot(\varepsilon^0(\bar e,\bar\ell)n)\}\\ + \frac{2\mu\lambda}{2\mu+\lambda}\operatorname{tr}\varepsilon^{0P}(e,\ell)\operatorname{tr}\varepsilon^{0P}(\bar e,\bar\ell)\,d\Gamma\\ + \frac{2h^3}{3}\int_\omega 2\mu\hat\varepsilon^{1P}(e,\ell)\cdot\cdot\,\hat\varepsilon^{1P}(\bar e,\bar\ell) + \frac{2\mu\lambda}{2\mu+\lambda}\operatorname{tr}\hat\varepsilon^{1P}(e,\ell)\operatorname{tr}\hat\varepsilon^{1P}(\bar e,\bar\ell)\,d\Gamma\\ = \int_\omega p\cdot\bar e\,d\Gamma + \int_\gamma N\cdot\bar e - M\cdot\bar\ell\,d\gamma,\end{aligned}$$

where ℓ and $\bar\ell$ are tangential, ε^0 is the same as before but

$$(8.13)\qquad \hat\varepsilon^1(v,\zeta) \overset{\text{def}}{=} \varepsilon_\Gamma(\zeta) + \frac{1}{2}[D^2bD_\Gamma v + {}^*D_\Gamma v D^2 b],$$

$$(8.14)\qquad \hat\varepsilon^{1P}(v,\zeta) \overset{\text{def}}{=} P\hat\varepsilon^1(v,\zeta)P.$$

The other difference is the absence of the normal component ℓ_n.

As was pointed out in § 7.2.4 of § 7, this model is the result of a *mathematical compromise*. It is assumed that the vector function ℓ is tangential ($\ell_n = 0$ and $\ell = \ell_\Gamma$), but that $\varepsilon(V)\nabla b\cdot\nabla b \neq 0$ in $S_h(\omega)$ and that its value is computed through the condition $\sigma(V)\nabla b\cdot\nabla b = 0$ in $S_h(\omega)$. This is equivalent to

$$\begin{aligned}0 = \sigma(V)\circ T_z n\cdot n = 2\mu\varepsilon(V)\circ T_z n\cdot n + \lambda\operatorname{tr}\varepsilon(V)\circ T_z\\ \Longrightarrow 0 = 2\mu\varepsilon(V)\circ T_z\nabla b\cdot\nabla b + \lambda\operatorname{tr}\varepsilon(V)\circ T_z\\ \Longrightarrow \varepsilon(V)\circ T_z\nabla b\cdot\nabla b = -\frac{\lambda}{2\mu+\lambda}\operatorname{tr}\varepsilon^P(V)\circ T_z.\end{aligned}$$

In local coordinates for our approximation

$$\varepsilon(V)\circ T_z \simeq \varepsilon^0(e,\ell_\Gamma) + z\varepsilon^1(e,\ell_\Gamma),$$

the above identities yields

$$\varepsilon^0(e,\ell_\Gamma)n\cdot n = -\frac{\lambda}{2\mu+\lambda}\operatorname{tr}\varepsilon^{0P}(e,\ell_\Gamma), \quad \varepsilon^1(e,\ell_\Gamma)n\cdot n = -\frac{\lambda}{2\mu+\lambda}\operatorname{tr}\varepsilon^{1P}(e,\ell_\Gamma),$$

In order to distinguish the real quantities $\varepsilon^0(e,\ell_\Gamma)n\cdot n$ and $\varepsilon^1(e,\ell_\Gamma)n\cdot n$ which are always zero we define the terms

$$\varepsilon^0_{nn}(e,\ell_\Gamma) \stackrel{\text{def}}{=} -\frac{\lambda}{2\mu+\lambda}\operatorname{tr}\varepsilon^{0P}(e,\ell_\Gamma), \quad \varepsilon^1_{nn}(e,\ell_\Gamma) \stackrel{\text{def}}{=} -\frac{\lambda}{2\mu+\lambda}\operatorname{tr}\varepsilon^{1P}(e,\ell_\Gamma).$$

The tangential part of the new strain tensors coincide with $\varepsilon^{0P}(e,\ell_\Gamma)$ and $\varepsilon^{1P}(e,\ell_\Gamma)$, the tangential part of $\varepsilon^0(e,\ell_\Gamma)n$ and $\varepsilon^1(e,\ell_\Gamma)n$ remain the same

$$P\varepsilon^0(e,\ell_\Gamma)n = \varepsilon^0(e,\ell_\Gamma)n, \quad P\varepsilon^1(e,\ell_\Gamma)n = \varepsilon^1(e,\ell_\Gamma)n,$$

since their image is in the tangent plane $T_x\Gamma$. So only the terms $\varepsilon^0(e,\ell_\Gamma)n\cdot n$ and $\varepsilon^1(e,\ell_\Gamma)n\cdot n$ have been modified.

We can now rewrite the bilinear forms a_0 and a_2 and the operator A for the modified strain tensors

$$\boxed{\tilde{\varepsilon}^0(e,\ell_\Gamma) \stackrel{\text{def}}{=} \varepsilon^0(e,\ell_\Gamma) + \varepsilon^0_{nn}(e,\ell_\Gamma)n^*n, \quad \tilde{\varepsilon}^1(e,\ell_\Gamma) \stackrel{\text{def}}{=} \varepsilon^1(e,\ell_\Gamma) + \varepsilon^1_{nn}(e,\ell_\Gamma)n^*n}$$

as follows

$$\begin{aligned}
a_0\big((e,\ell),(\bar e,\bar\ell)\big) &= 2\mu\big[\varepsilon^{0P}(e,\ell_\Gamma)\cdot\cdot\,\varepsilon^{0P}(\bar e,\bar\ell_\Gamma) + 2\varepsilon^0(e,\ell_\Gamma)n\cdot\varepsilon^0(\bar e,\bar\ell_\Gamma)n + \varepsilon^0_{nn}\bar\varepsilon^0_{nn}\big] \\
&\qquad + \lambda\big(\operatorname{tr}\varepsilon^{0P}(e,\ell_\Gamma) + \varepsilon^0_{nn}\big)\big(\operatorname{tr}\varepsilon^{0P}(\bar e,\bar\ell_\Gamma) + \bar\varepsilon^0_{nn}\big) \\
&= 2\mu\big[\varepsilon^{0P}(e,\ell_\Gamma)\cdot\cdot\,\varepsilon^{0P}(\bar e,\bar\ell_\Gamma) + 2\varepsilon^0(e,\ell_\Gamma)n\cdot\varepsilon^0(\bar e,\bar\ell_\Gamma)n\big] \\
&\qquad + \frac{2\mu\lambda}{2\mu+\lambda}\operatorname{tr}\varepsilon^{0P}(e,\ell_\Gamma)\operatorname{tr}\varepsilon^{0P}(\bar e,\bar\ell_\Gamma),
\end{aligned}$$

which is the same expression as the one we had in Naghdi's model. The bilinear form a_2 is similar to a_0

$$\begin{aligned}
a_2\big((e,\ell),(\bar e,\bar\ell)\big) &= 2\mu\big[\varepsilon^{1P}(e,\ell_\Gamma)\cdot\cdot\,\varepsilon^{1P}(\bar e,\bar\ell_\Gamma) + 2\varepsilon^1(e,\ell_\Gamma)n\cdot\varepsilon^1(\bar e,\bar\ell_\Gamma)n\big] \\
&\qquad + \frac{2\mu\lambda}{2\mu+\lambda}\operatorname{tr}\varepsilon^{1P}(e,\ell_\Gamma)\operatorname{tr}\varepsilon^{1P}(\bar e,\bar\ell_\Gamma),
\end{aligned}$$

where the term

$$\frac{2h^3}{3}\left|\varepsilon^1(e,\ell_\Gamma)n\right|^2 = \frac{2h^3}{3}\left|D^2b\varepsilon^0(e,\ell_\Gamma)n\right|^2,$$

in a_2 can be neglected when compared to the term $2h\left|\varepsilon^0(e,\ell_\Gamma)n\right|^2$ in a_0. As a result we can use the approximation

$$\tilde a_{2N}\big((e,\ell),(\bar e,\bar\ell)\big) = 2\mu\varepsilon^{1P}(e,\ell_\Gamma)\cdot\cdot\,\varepsilon^{1P}(\bar e,\bar\ell_\Gamma) + \frac{2\mu\lambda}{2\mu+\lambda}\operatorname{tr}\varepsilon^{1P}(e,\ell_\Gamma)\operatorname{tr}\varepsilon^{1P}(\bar e,\bar\ell_\Gamma).$$

Finally the associated linear operator is

$$
\begin{aligned}
&\langle A_N(e,\ell),(\bar{e},\bar{\ell})\rangle_{\mathcal{V}_N}\\
&\quad=\int_\omega 2h\Big\{2\mu\big[\varepsilon^{0P}(e,\ell_\Gamma)\cdot\cdot\,\varepsilon^{0P}(\bar{e},\bar{\ell}_\Gamma)+2\varepsilon^0(e,\ell_\Gamma)n\cdot\varepsilon^0(\bar{e},\bar{\ell}_\Gamma)n\big]\\
&\qquad\qquad\qquad+\frac{2\mu\lambda}{2\mu+\lambda}\operatorname{tr}\varepsilon^{0P}(e,\ell_\Gamma))\operatorname{tr}\varepsilon^{0P}(\bar{e},\bar{\ell}_\Gamma)\Big\}\\
&\quad+2\frac{h^3}{3}\Big\{2\mu\varepsilon^{1P}(e,\ell_\Gamma)\cdot\cdot\,\varepsilon^{1P}(\bar{e},\bar{\ell}_\Gamma)+\frac{2\mu\lambda}{2\mu+\lambda}\operatorname{tr}\varepsilon^{1P}(e,\ell_\Gamma)\operatorname{tr}\varepsilon^{1P}(\bar{e},\bar{\ell}_\Gamma)\Big\}\,d\Gamma.
\end{aligned}
$$

and up to the differences between ε^1 and $\hat{\varepsilon}^1$ *this looks very much like Naghdi's linear model.* All the theoretical results of § 7 applies to this model since

$$\|e\|_{L^2}^2+\|\ell_\Gamma\|_{L^2}^2+\big\|\varepsilon^0(e,\ell_\Gamma)\big\|_{L^2}^2+\big\|\varepsilon^1(e,\ell_\Gamma)\big\|_{L^2}^2,$$

is a norm equivalent to the $H^1\times H^1$-norm on

$$\mathcal{V}_N=\big\{(e,\ell)\in H^1(\omega)\times H^1(\omega)\colon \ell\cdot n=0 \text{ in } \omega\big\}.$$

Of course Koiter's model can be *mathematically interpreted* as a special case of Naghdi's model where $\varepsilon^0(e,\ell_\Gamma)n=0$. As was pointed out in § 6 Replacing $\hat{\varepsilon}^1$ by ε^1 introduces a negligible term which does not change the mathematical structure of the model and its mechanical interpretation. Recall that

$$
\begin{aligned}
\varepsilon^1(e,\ell)&\overset{\text{def}}{=}\varepsilon_\Gamma(\ell)-\frac{1}{2}\big[D_\Gamma(e)D^2b+D^2b^*D_\Gamma(e)\big],\\
\hat{\varepsilon}^1(e,\ell)&\overset{\text{def}}{=}\varepsilon_\Gamma(\ell)+\frac{1}{2}[D^2bD_\Gamma(e)+{}^*D_\Gamma(e)D^2b]\\
&\implies \hat{\varepsilon}^{1P}(e,\ell)-\varepsilon^{1P}(e,\ell)=\varepsilon_\Gamma^P(e)^*D^2b+D^2b^*\varepsilon_\Gamma^P(e).
\end{aligned}
$$

The squared norm of the right-hand side of the last expression is multiplied by h^3

$$2\frac{h^3}{3}\big\|\varepsilon_\Gamma^P(e)D^2b+D^2b\varepsilon_\Gamma^P(e)\big\|^2,$$

and is negligible compared to $2h\big\|\varepsilon_\Gamma^P(e)\big\|^2=2h\big\|\varepsilon^{0P}(e,\ell_\Gamma)\big\|^2$.

This last model is based on a *mathematical compromise.* Under the same assumptions the *uncompromised mathematical model* is model (5) which contains the normal variable ℓ_n and the two conditions

$$
\begin{aligned}
\ell_n&=-\frac{\lambda}{\lambda+2\mu}\operatorname{div}_\Gamma(e),\\
0&=\operatorname{tr}\varepsilon^1(e,\ell)=\operatorname{tr}\varepsilon^{1P}(e,\ell)=\operatorname{tr}\varepsilon^{1P}(e,\ell_\Gamma)+\ell_n\Delta b. \qquad (8.15)
\end{aligned}
$$

The first one is the explicit expression of ℓ_n which can now be eliminated from the equations. The second one introduces the following constraint on e and ℓ_Γ

$$\operatorname{div}_\Gamma(\ell_\Gamma)=D^2b\cdot\cdot\,D_\Gamma^P(e)+\frac{\lambda}{\lambda+2\mu}\Delta b\operatorname{div}_\Gamma(e),$$

where Δb is the mean curvature H of the mean surface. The bilinear form a_0 is the same but the other one

$$\tilde{a}_2\big((e,\ell),(\bar{e},\bar{\ell})\big)=2\mu\Big\{\varepsilon^{1P}(e,\ell_\Gamma)\cdot\cdot\,\varepsilon^{1P}(\bar{e},\bar{\ell}_\Gamma)+\frac{1}{2}\nabla_\Gamma\ell_n\cdot\nabla_\Gamma\bar{\ell}_n\Big\},$$

differs from a_{2N} by the term

$$2\mu\frac{1}{2}\nabla_\Gamma\ell_n\cdot\nabla_\Gamma\bar{\ell}_n,$$

which is to be compared with the term

$$\frac{2\mu\lambda}{2\mu+\lambda}\operatorname{tr}\varepsilon^{1P}(e,\ell_\Gamma)\operatorname{tr}\varepsilon^{1P}(\bar{e},\bar{\ell}_\Gamma).$$

From (8.15) $\operatorname{tr}\varepsilon^{1P}(e,\ell_\Gamma)=-\ell_n\Delta b$, but the term

$$\frac{2\mu\lambda}{2\mu+\lambda}\operatorname{tr}\varepsilon^{1P}(e,\ell_\Gamma)\operatorname{tr}\varepsilon^{1P}(\bar{e},\bar{\ell}_\Gamma)=\frac{2\mu\lambda}{2\mu+\lambda}(\Delta b)^2\ell_n\bar{\ell}_n,$$

still looks different from the term

$$2\mu\frac{1}{2}\nabla_\Gamma\ell_n\cdot\nabla_\Gamma\bar{\ell}_n.$$

So the model which corresponds to the mechanical assumptions of the Naghdi's model is characterized by the operator of case (5)

$$\boxed{\begin{aligned}&\langle A_5(e,\ell),(\bar{e},\bar{\ell})\rangle_{\mathcal{V}_5}\\ &\quad=\int_\omega 2h\Big\{2\mu\big[\varepsilon^{0P}(e,\ell_\Gamma)\cdot\cdot\,\varepsilon^{0P}(\bar{e},\bar{\ell}_\Gamma)+2\varepsilon^0(e,\ell_\Gamma)n\cdot\varepsilon^0(\bar{e},\bar{\ell}_\Gamma)n\big]\\ &\qquad\qquad+\frac{2\mu\lambda}{2\mu+\lambda}\operatorname{tr}\varepsilon^{0P}(e,\ell_\Gamma))\operatorname{tr}\varepsilon^{0P}(\bar{e},\bar{\ell}_\Gamma)\Big\}\\ &\qquad+2\frac{h^3}{3}2\mu\Big\{\varepsilon^{1P}(e,\ell_\Gamma)\cdot\cdot\,\varepsilon^{1P}(\bar{e},\bar{\ell}_\Gamma)\\ &\qquad\qquad+\frac{1}{2}\left(\frac{\lambda}{\lambda+2\mu}\right)^2\nabla_\Gamma\big(\operatorname{div}_\Gamma(e)\big)\cdot\nabla_\Gamma\big(\operatorname{div}_\Gamma(\bar{e})\big)\Big\}\,d\Gamma\end{aligned}}$$

with the constraint

$$\boxed{\operatorname{div}_\Gamma(\ell_\Gamma)=D^2b\cdot\cdot\,D_\Gamma^P(e)+\frac{\lambda}{\lambda+2\mu}\Delta b\operatorname{div}_\Gamma(e)}\tag{8.16}$$

and ℓ_n explicitly given by the expression

$$\boxed{\ell_n=-\frac{\lambda}{\lambda+2\mu}\operatorname{div}_\Gamma(e)}\tag{8.17}$$

This constraint can be removed by going to model (3) in which case the assumption $\sigma n\cdot n=0$ is only verified on the mean surface and the underlying operator is

$$\boxed{\begin{aligned}&\langle A_3(e,\ell),(\bar{e},\bar{\ell})\rangle_{\mathcal{V}_3}\\ &=\int_\omega 2h\Big\{2\mu\big[\varepsilon^{0P}(e,\ell_\Gamma)\cdot\cdot\,\varepsilon^{0P}(\bar{e},\bar{\ell}_\Gamma)+2\varepsilon^0(e,\ell_\Gamma)n\cdot\varepsilon^0(\bar{e},\bar{\ell}_\Gamma)n\big]\\ &\qquad+\frac{2\mu\lambda}{2\mu+\lambda}\operatorname{tr}\varepsilon^{0P}(e,\ell_\Gamma)\operatorname{tr}\varepsilon^{0P}(\bar{e},\bar{\ell}_\Gamma)\Big\}\\ &\quad+2\frac{h^3}{3}\Big\{2\mu\varepsilon^{1P}(e,\ell_\Gamma)\cdot\cdot\,\varepsilon^{1P}(\bar{e},\bar{\ell}_\Gamma)+\frac{1}{2}\left(\frac{\lambda}{\lambda+2\mu}\right)^2\nabla_\Gamma\big(\operatorname{div}_\Gamma(e)\big)\cdot\nabla_\Gamma\big(\operatorname{div}_\Gamma(\bar{e})\big)\\ &\qquad+\lambda\operatorname{tr}\varepsilon^{1P}(e,\ell_\Gamma)\operatorname{tr}\varepsilon^{1P}(\bar{e},\bar{\ell}_\Gamma)\Big\}\,d\Gamma\end{aligned}}$$

and ℓ_n explicitly is given by the expression

$$\boxed{\ell_n = -\frac{\lambda}{\lambda + 2\mu}\mathrm{div}_\Gamma(e)} \tag{8.18}$$

Of course additional assumptions can be used to reduce the complexity of the model. For instance we could assume that the mean surface has an almost constant tangential compressibility/extensibility, that is

$$\nabla_\Gamma\big(\mathrm{div}_\Gamma(e)\big) \simeq 0$$

and drop this second order term from the equation.

References

1. E. Acerbi, G. Buttazzo, and D. Percivale, *A variational definition for the strain energy of an elastic string*, J. Elasticity **25** (1991), 137–148.
2. S. M. Alessandrini, *Some two-dimensional plate models: Derivation, asymptotic properties, and numerical approximation*, Ph.D. thesis, Rutgers University, New Brunswick, New Jersey, 1991.
3. S. M. Alessandrini, D. N. Arnold, R. S. Falk, and A. Madureira, *Derivation and justification of plate models by variational methods*, Proc. Plates and shells: From theory to practice, CRM Proc. Lecture Note, Amer. Math. Soc., Providence, R.I. (to appear).
4. C. Amrouche, *Propriétés d'opérateurs de dérivation. Application au problème non homogène de Stokes*, C. R. Acad. Sci. Paris, Sér. I Math. **310** (1990), 367–370.
5. S. S. Antman, *Nonlinear problems of elasticity*, Springer-Verlag, New York 1991.
6. Th. Aubin, *Nonlinear analysis on manifolds. Monge-Ampère equations*, Springer-Verlag, Heidelberg, Berlin, 1982.
7. M. Bernadou, *Méthodes d'éléments finis pour les problèmes de coques minces*, Masson, Paris, Milan, Barcelone, 1994.
8. M. Bernadou and Ph. G. Ciarlet, *Sur l'ellipticité du modèle linéaire de coques de W. T. Koiter*, Computing Methods in Applied Sciences and engineering (R. Glowinski and J. L. Lions, eds.), Springer-Verlag, Berlin, Heidelberg, New York, 1976, pp. 89–136,
9. M. Bernadou, Ph. G. Ciarlet, and B. Miara, *Existence theorems for two-dimensional linear shell theories*, J. Elasticity **34** (1994), 111–138.
10. M. Bernadou and J. T. Oden, *An existence theorem for a class of nonlinear shallow shell problems*, J. Math. Pures Appl. **60** (1981), 285–308.
11. F. Bourquin, Ph. G. Ciarlet, G. Geymonat, and A. Raoult, *Γ-convergence et analyse asymptotique des plaques minces*, C. R. Acad. Sci. Paris Sér. I Math. **315** (1992), 1017–1024.
12. F. Brezzi and M. Fortin, *Mixed and hybrid finite element methods*, Springer-Verlag, New York, Berlin, Heidelberg, 1991.
13. C. Chen, *Asymptotic convergence rates for the Kirchhoff plate model*, Ph.D. Thesis, Penn. State, University Park, Pennsylvania, 1995.
14. Ph. G. Ciarlet, *Mathematical elasticity.* I. *Three-dimensional elasticity*, Stud. Math. Appl., vol. 20, North-Holland Publishing Co., Amsterdam-New York, 1988.
15. ______ *Plates and junctions in elastic multi-structures: An asymptotic analysis*, Masson, Paris, Milano, Barcelona, Mexico, and Springer-Verlag, Berlin, New York, 1990.
16. Ph. G. Ciarlet and V. Lods, *Ellipticité des équations membranaires d'une coque uniformément elliptique*, C. R. Acad. Sci. Paris Sér. I Math. **318** (1994), 195–200.
17. ______ *Analyse asymptotique des coques linéairement élastique.* I. *Coques membranaires*, C. R. Acad. Sci. Paris Sér. I Math. **318** (1994), 863–868.
18. ______ *Analyse asymptotique des coques linéairement élastique.* III. *Une justification du modèle de Koiter*, C. R. Acad. Sci. Paris Sér. I Math. **319** (1994), 299–304.
19. ______ *On the ellipticity of linear membrane shell equations*, Arch. Rational Mech. Anal., to appear.
20. ______ *Symptotic analysis of linearly elastic shells.* III. *Justification of Koiter's shell equations*, Arch. Rational Mech. Anal., to appear.
21. Ph. G. Ciarlet, V. Lods, and B. Miara, *Analyse asymptotique des coques linéairement élastique.* II. *Coques en flexion*, C. R. Acad. Sci. Paris Sér. I Math. **319** (1994), 95–100.

22. ——— *Symptotic analysis of linearly elastic shells. II. Justification of flexural ahell equations*, Arch. Rational Mech. Anal., to appear.
23. Ph. G. Ciarlet and E. Sanchez-Palencia, *Un théorème d'existence et d'unicité pour les équations des coques membranaires*, C. R. Acad. Sci. Paris Sér. I Math. **317** (1993), 801–805.
24. M. Dauge, *Complete asymptotics in thin elastic plates and optimal estimates for Kirchhoff-Love model*, Preprint no. 95-06, Institut de Recherche Mathématique de Rennes, Université de Rennes, Feb. 1995.
25. M. Dauge and I. Gruais, *Développement asymptotique d'ordre arbitraire pour une plaque élastique mince encastrée*, C. R. Acad. Sc. Paris Sér. I Math. **321**, (1995), no. 3, 375–380.
26. M. C. Delfour, *Méthodes intrinsèques pour le contrôle et l'optimisation des coques minces*, in "Élasticité, Viscoélasticité et Contrôle Optimal", Huitième Entretiens du Centre Jacques Cartier, ESAIM Proceedings, Volume I, 1996 (URL: http://www.emath.fr/proc/).
27. M. C. Delfour and J. Zhao, *Intrinsic nonlinear models of shells for Saint Venant-Kirchhoff materials*, CRM Report, Université de Montréal, April 1996.
28. M. C. Delfour and J.-P. Zolésio, *Shape sensitivity analysis via min max differentiability*, SIAM J. Control Optim. **26** (1988), 834–862.
29. ——— *Anatomy of the shape Hessian*, Ann. Mat. Pura Appl. (4) **CLVIII** (1991), 315–339.
30. ——— *Velocity method and Lagrangian Formulation for the computation of the shape Hessian*, SIAM J. Control Optim. **29** (1991), no. 6, 1414–1442.
31. ——— *Structure of shape derivatives for nonsmooth domains*, J. Funct. Anal. **104** (1992), 1–33.
32. ——— *Shape analysis via distance functions*, J. Funct. Anal. **123** (1994), 129–201.
33. ——— *Oriented distance function in shape analysis and optimization*, Control and Optimal Design of Distributed Parameter Systems (J. Lagnese, D. L. Russell, L. White, eds.), pp. 39–72, Springer-Verlag, Berlin, Heidelberg, New York, Tokyo 1994.
34. ——— *On a variational equation for thin shells*, Control and Optimal Design of Distributed Parameter Systems, J. Lagnese, D. L. Russell, L. White, eds.), Springer-Verlag, Berlin, Heidelberg, New York, Tokyo 1994, pp. 25–37.
35. ——— *A boundary differential equation for thin shells*, J. Differential Equations **119** (1995), 426–449.
36. ——— *Tangential differential equations for dynamical thin/shallow shells*, J. Differential Equations **128** (1996), 125–167.
37. ——— *Shape analysis via distance functions: local theory*, CRM Report 2299, March 1996, Université de Montréal.
38. ——— *On the design and control of systems governed by differential equations on submanifolds*, Control and Cybernet. **25** (1996), 497–514.
39. ——— *Hidden boundary smoothness for some classes of differential equations on submanifolds*, in "Proc. 1996 Joint Summer Research Conference on Optimization Methods in PDE's", S. Cox and I. Lasiecka, Contemporary Mathematics, AMS Publications, Providence, R.I., to appear.
40. ——— *Convergence to the asymptotic model for linear thin shells*, in 'Proc. 1996 Joint Summer Research Conference on Optimization Methods in PDE's", S. Cox and I. Lasiecka, Contemporary Mathematics, AMS Publications, Providence, R.I., to appear.
41. ——— *Intrinsic differential geometry and theory of thin shells*, Quaderni, Scuola Normale Superiore, Pisa (Italy), to appear; version 1.0, lecture notes, August 1996.
42. ——— *Convergence of the linear P(1,1) and P(2,1) thin shells to asymptotic shells*, in "Proc. Plates and Shells: from theory to practice", M. Fortin, ed., CRM Proc. Lect. Notes ser., AMS Publications, Providence, R.I. 1997, to appear.
43. ——— *Intrinsic differential geometry methods in the asymptotic analysis of linear thin shells*, in "Boundaries, interfaces and transitions", M. Delfour, ed., CRM Proc. Lect. Notes ser., AMS Publications, Providence, R.I. 1997, to appear.
44. Ph. Destuynder, *Sur la justification des modéles de plaques et de coques par les méthodes asymptotiques*, Doctoral dissertation, Université Pierre et Marie Curie, 1980.
45. ——— *Modélisation des coques minces élastiques*, Masson, Paris, Milan, Barcelone, 1990.
46. ——— *Une théorie asymptotique des plaques minces en elasticité linéaire*, Masson, Paris, Milan, Barcelone, 1986

47. L. C. Evans and R. F. Gariepy, *Measure theory and the properties of functions*, Stud. Adv. Math., CRC Press, Boca Raton, FL, 1992.
48. H. Federer, *Hausdorff measure and Lebesgue area*, Proc. Nat. Acad. Sci. U.S.A. **37** (1951), 90–94.
49. ______ *Curvature measures*, Trans. Amer. Math. Soc. **93** (1959), 418–419.
50. H. Federer and W. H. Fleming, *Normal and integral currents*, Ann. of Math. (2) **72** (1960), 458–520.
51. G. Fichera, *Linear elliptic differential systems and eigenvalue problems*, Lecture Notes in Math., vol. 8, Springer-Verlag, Berlin-New York, 1965.
52. I. M. N. Figueiredo, *Local existence and regularity of the solutions of the nonlinear thin shell model of Donnell-Mushtari-Vlasov*, Appl. Anal. **36** (1990), 221–234.
53. W. H. Fleming and R. Rishel, *An integral formula for total gradient variation*, Arch. Math. **11** (1960), 218–222.
54. D. D. Fox, A. Raoult, and J. C. Simo, *A justification of nonlinear properly invariant plate theories*, Arch. Rational Mech. Anal. **124** (1993), 157–199.
55. J. H. G. Fu, *Tubular neighborhoods in Euclidean spaces*, Duke Math. J. **52** (1985), 1025–1046.
56. P. Germain, *Mécanique*.I–II, Ellipses éditeur, Paris, 1986.
57. G. Geymonat, *Sui problemi ai limiti per i sistemi lieari ellitici*, Ann. Mat. Pura Appl. (4) **69** (1965), 207–284.
58. D. Gilbarg and N. S. Trudinger, *Elliptic partial differential equations of second order*, Springer-Verlag, Berlin, Heidelberg, New York, Tokyo, 1983.
59. A. L. Gol'denveizer, *Theory of elastic thin shells*, Pergamon Press, Oxford, 1961.
60. A. E. Green, P. M. Naghdi, and W. L. Wainwright, *A general theory of a Cosserat surface*, Arch. Rational Mech. Anal. **20** (1965), 287–308.
61. A. E. Green and W. Zerna, *Theoretical elasticity*, Oxford at the Clarendon Press, London, 1968.
62. O. V. Guseva, *On boundary problems for strongly elliptic systems*, Dokl. Akad. Nauk UzSSR **102** (1955), 1069–1072.
63. L. Hörmander, *The analysis of linear partial differential operators*. I, Springer-Verlag, Berlin, New York, Heidelberg, Tokyo, 1983.
64. ______ *The analysis of linear partial differential operators*. III, Springer-Verlag, Berlin, New York, Heidelberg, Tokyo, 1985.
65. G. Kirchhoff, *Vorlesungen über Mathematische Physik, Mechanik*, Leipzig, 1876.
66. W. T. Koiter, *On the foundations of the linear theory of thin elastic shells*. I, Proc. Konink. Nederl. Akad. Wetensch. **B73** (1970), 169–195.
67. ______ *On the nonlinear theory of thin elastic shells*, Proc. Konink. Nederl. Akad. Wetensch. **B69** (1966), 1–54.
68. J. E. Lagnese and J.-L. Lions, *Modeling, analysis and control of thin plates*, RMA, Vol. 6, Masson, Paris, Milan, Barcelone, Mexico 1988.
69. H. Le Dret and A. Raoult, *The nonlinear membrane model as variational limit of nonlinear three-dimensional elasticity*, J. Math. Pures Appl. (9) **74** (1995), 549–578.
70. ______ *The membrane shell model in nonlinear elasticity: A variational asymptotic derivation*, J. Nonlinear Sci. **6** (1996), 59–84.
71. J.-L. Lions, *Contrôle optimal de systèmes gouvernés par des équations aux dérivées partielles*, Dunod, Gauthier-Villars, Paris, 1968.
72. K. H. Lo, R. M. Christensen, and E. M. Wu, *A high-order theory of plate deformations*, J. Appl. Mech. **46** (1977), 663–676.
73. A. E. H. Love, *The mathematical theory of elasticity*, Cambridge University Press, 1934.
74. C. Mardare, *Modèles bi-dimensionnels de coques linéairement élastiques: Estimations de l'écart entre leurs solutions*, C. R. Acad. Sci. Paris Sér. I Math. **322** (1996), 793–796.
75. ______ *Estimation d'erreur dans l'analyse asymptotique des coques linéairement élastiques*, C. R. Acad. Sci. Paris Sér. I Math. **322** (1996), 895–898.
76. D. Morgenstern, *Herleitung der Plattentheorie aus der dreidimensionalen Elastizitätstheorie*, Arch. Rational Mech. Anal. **4** (1959), 145–152.
77. C. B. Morrey, Jr., *Multiple integrals in the calculus of variations*, Springer-Verlag, New York, Inc., New York, 1966.

78. P. M. Naghdi, *Foundations of elastic shell theory*, Progress in Solid Mechanics (I. N. Sueddon and R. Hill, eds.), vol. IV, North-Holland Publ. Co. Amsterdam, 1963, pp. 1–90.
79. ______ *The theory of shells and plates*, Handbuch der Physik, vol. VI a-2, Springer-Verlag, Berlin, 1972, pp. 425–640.
80. J. Nečas, *Les méthodes directes en théorie des équations elliptiques*, Masson, Paris and Academia, Prague, 1967.
81. V. V. Novozhilov, *Thin shell theory*, Wolters-Noordhoff Publishing, Groningen, 1970.
82. J.-C. Paumier and A. Raoult, *Asymptotic consistency of the polynomial approximation in the linearized plate theory. Application to the Reissner-Mindlin model*, Elasticity, Viscoelasticity and Optimal Control: Theoretical and Numerical Aspects, (Lyon, 1995), SMAI, France (to appear).
83. J. Sokolowski and J.-P. Zolésio, *Introduction to shape optimization: Shape sensitivity analysis*, Springer-Verlag, New York, Berlin, Heidelberg, 1992.
84. R. Temam, *Problèmes mathématiques en plasticité*, Méthodes Mathématiques de l'Informatique, vol. 12, Gauthier-Villars, Paris, 1983.
85. ______ *Monotone rearrangement of a function and the Grad-Mercier equation of plasmas physics*, Recent Methods in Non Linear Analysis and Application (E. De Giorgi, E. Magenes, and U. Mosco, eds.), Pitagora Editrice, Bologna, 1979, pp.83–98.
86. ______ *Navier-Stokes equations. Theory and numerical analysis*, North-Holland Publishing Co., Amsterdam-New York, 1979.
87. R. Valid, *Mechanics of continuous media and analysis of structures*, North-Holland Ser. Appl. Math. Mech. **26**, North-Holland Publishing Co., Amsterdam–New York, 1981.
88. ______ *The nonlinear theory of shells through variational principles*, Wiley, Chichester, 1995.
89. H. Whitney,*Geometric integration theory*, Princeton Math. Ser., vol. 21, Princeton Univ. Press, Princeton, 1957.
90. L. C. Young, *Partial area.* I, Riv. Mat. Univ. Parma **10** (1959), 103–113.
91. J. Zhao, *Natural theory of nonlinear shells*, Ph.D. Thesis, McGill University, April 1996.
92. J.-P. Zolésio, *Identification de domaines par déformation*, thèse de doctorat d'état, Université de Nice, France, 1979.
93. ______ *Solution variationnelle d'un problème de valeur propre non linéaire et frontière libre issu de la physique des plasmas*, C. R. Acad. Sci. Paris Sér. A–B **288** (1979), no. 19, A911–A914.
94. ______ *The material derivative (or speed) method for shape optimization*, Optimization of Distributed Parameter Structures (Iowa City, Iowa, 1980) (E. J. Haug and J. Céa, eds.), vol. II, NATO Adv. Study Inst. Ser. E: Appl. Sci., vol. 50, Nijhoff, The Hague, 1981, pp. 1089–1151.
95. ______ *Domain variational formulation for free boundary problems*, Optimization of Distributed Parameter Structures (Iowa City, Iowa, 1980) (E. J. Haug and J. Céa, eds.), vol. II, NATO Adv. Study Inst. Ser. E: Appl. Sci., vol. 50, Nijhoff, The Hague, 1981, pp. 1152–1194.

Centre de recherches mathématiques et Département de mathématiques et de statistique, Université de Montréal, C.P.6128, succ. Centre-ville, Montréal QC, Canada H3C 3J7

E-mail address: delfour@crm.UMontreal.ca

CNRS Institut Non Linéaire de Nice, 1361 route des lucioles, 06904 Sophia Antipolis Cedex, France, and Centre de Mathématiques Appliquées, Ecole des Mines de Paris (INRIA, CMA/MEIJE), 2004 route des Lucioles, BP 93, 06902 Sophia Antipolis Cedex, France

E-mail address: Jean-Paul.Zolesio@sophia.inria.fr

Centre de Recherches Mathématiques
CRM Proceedings and Lecture Notes
Volume **11**, 1997

From K-Theory to E-Theory

Peter A. Fillmore

We survey briefly the evolution of K-theory from its origins to some recent developments in operator algebra K-theory and E-theory.

1. Origins

The basic notion underlying K-theory is the integer-valued dimension function on the subspaces of a (finite-dimensional) vector space. Perhaps the first significant development occurs in the famous 1936 paper of F. J. Murray and J. von Neumann, "On rings of operators" [**20**]. They considered the more primitive notion of "equidimensionality" of subspaces, relative to certain subalgebras A of the algebra $\mathcal{B}(\mathcal{H})$ of bounded linear operators on a Hilbert space $\mathcal{H}$. This algebra has an involution $x \to x^*$ and Murray and von Neumann required in particular that A be closed under this operation. Only those subspaces of $\mathcal{H}$ for which the corresponding orthogonal projection operator is in A are considered. Equidimensionality relative to A is then the equivalence relation $\sim$ defined on the set $P(A)$ of projections in A by

$$p \sim q \text{ if } u^*u = p \text{ and } uu^* = q \text{ for some } u \in A.$$

It is called **Murray-von Neumann equivalence**. In terms of the range spaces $\mathcal{M}$ and $\mathcal{N}$ of p and q, this is equivalent to saying that A contains an element that takes $\mathcal{M}$ isometrically onto $\mathcal{N}$ and is zero on the orthogonal complement of $\mathcal{M}$. The set

$$D(A) = P(A)/\sim$$

of equivalence classes is the receptacle for the dimension function $d : P(A) \to D(A)$ that takes each projection to its equivalence class. It has the structure of an ordered abelian partial semigroup, a sum $[p] + [q] = [p' + q']$ being defined precisely when there exist $p' \sim p$ and $q' \sim q$ with p' and q' orthogonal (so that $p' + q'$ is a projection). The order comes from the inclusion relation on subspaces.

Restricting further to subalgebras A that are unital, closed in the weak operator topology and have trivial centre, Murray and von Neumann showed that $D(A)$ is isomorphic to a closed partial subsemigroup of $[0, +\infty]$, and hence that (up to a scale factor) there are only five possibilities for $D(A)$:

type $\mathbf{I}_n$: $\{0, 1, 2, \dots, n\}$
type $\mathbf{I}_\infty$: $\{0, 1, 2, \dots, +\infty\}$

1991 *Mathematics Subject Classification.* Primary: 46L80; Secondary: 19-02.
This is the final form of the paper.

type $\mathbf{II}_1$**:** $[0, 1]$
type $\mathbf{II}_\infty$**:** $[0, +\infty]$
type III: $\{0, +\infty\}$

Remarkably they were able to show that all of these possibilities actually occur. For example in the type II "continuous-dimensional" case there are no minimal subspaces, and in fact any nonzero subspace can be split into two subspaces each having dimension half that of the original.

2. Topological K-Theory

Grothendieck [**4**] defined a contravariant functor

$$K\colon \boldsymbol{V} \to \boldsymbol{Ab}$$

from the category of algebraic varieties to the category of abelian groups, as a natural setting for generalizations of the Riemann-Roch Theorem. If V is an algebraic variety then $K(V)$ is the abelian group generated by the set of coherent sheaves F on V with relations $F = F_1 + F_2$ whenever there is an exact sequence

$$0 \to F_1 \to F \to F_2 \to 0.$$

Atiyah and Hirzebruch [**2**] constructed the analogous functor

$$K\colon \boldsymbol{T} \to \boldsymbol{Ab}$$

on the category $\boldsymbol{T}$ of compact Hausdorff spaces, replacing sheaves by vector bundles. In fact, as they showed, much more is possible in this setting. By making use of the topological notion of suspension they were able to construct a generalized cohomology theory, based on their K-functor. In general this is a sequence of functors satisfying axioms designed to facilitate computations. In this case they showed that, because of Bott's fundamental result on the stable homotopy of the classical groups [**5**], the sequence of functors has period two.

It is not difficult to make the connection between the Atiyah-Hirzebruch and Murray-von Neumann constructions in the case $A = C(X), X$ compact and Hausdorff. In fact one can associate to any vector bundle E over X a projection in a matrix algebra $M_n\big(C(X)\big)$ over $C(X)$. The key fact here is that any E has a "complement" F; i.e. a bundle F such that $E \oplus F$ is a trivial bundle $X \times \mathbb{C}^n$. This means that $E_x \oplus F_x = \mathbb{C}^n$ for every $x \in X$, where E_x and F_x are the fibres of E and F over the point x. Then denoting by $p(x)$ the orthogonal projection of $\mathbb{C}^n$ on E_x we get a projection p in $C(X, M_n) = M_n\big(C(X)\big)$.

Strictly speaking the algebra in question here is not $C(X)$ but

$$M_\infty\big(C(X)\big) = \bigcup_{n=1}^{\infty} M_n\big(C(X)\big),$$

where M_n is regarded as a subalgebra of M_{n+1} by adding a row and column of zeros. This suggests defining, for any *-algebra A,

$$V(A) = D\big(M_\infty(A)\big) = P\big(M_\infty(A)\big)/\sim.$$

In $M_\infty(A)$ it is possible to find orthogonal representatives of any two projections; hence addition is always defined and $V(A)$ is a semigroup.

Let us go one step further here and consider the enveloping group, $K(A)$, of $V(A)$. This can be constructed from the cartesian product $V(A) \times V(A)$ in the same way that $\mathbb{Z}$ is constructed from $\mathbb{N}$; it has the universal property that any homomorphism from $V(A)$ to an abelian group factors uniquely through the canonical homomorphism $V(A) \to K(A)$.

We also note that this construction is functorial: to any homomorphism $\varphi : A \to B$ of *-algebras corresponds a homomorphism $M_\infty(A) \to M_\infty(B)$, obtained by applying φ entrywise, hence a homomorphism $V(A) \to V(B)$ of semigroups and finally a homomorphism $K(A) \to K(B)$ of groups.

EXAMPLES. (1) If $A = \mathcal{B}(\mathcal{H})$ then $D(A) = V(A) = \{0, 1, 2, \cdots, +\infty\}$ and so $K(A) = 0$, a not very promising beginning! In this connection note that in general $V(A)$ need not have cancellation, and that an equation $p + q = p$ in $V(A)$ forces q to become 0 in the group $K(A)$.

(2) If $A = \mathbb{C}$ then $D(A) = \{0, 1\}, V(A) = \mathbb{N} \cup \{0\}$ and $K(A) = \mathbb{Z}$. If $A = M_n(\mathbb{C})$ then $D(A) = \{0, 1, \ldots, n\}, V(A) = \mathbb{N} \cup \{0\}$ and $K(A) = \mathbb{Z}$. The same reasoning gives $K(A) = K\big(M_n(A)\big)$ for all n.

(3) Continuing as above, it is not difficult to see that $V\big(C(X)\big)$ is isomorphic to the semigroup of isomorphism classes of vector bundles over X, and hence that $K\big(C(X)\big) \cong K(X)$ as defined by Atiyah and Hirzebruch.

3. Operator K-Theory

It was realized early on, by Atiyah and others, that the category of Banach algebras has a good notion of suspension and hence that K-theory, as a generalized cohomology theory, extends to this category. (Note that K is contravariant on spaces but covariant on algebras.) However operator algebraists were slow to recognize the potential of this new tool and it was more than a decade later that the first significant applications were made.

A set of Université de Montréal lecture notes by Karoubi that included the Banach algebra case appeared in 1971, and Karoubi later published a book on the subject [**16**], but these were relatively inacessible to functional analysts. In 1975 J. L. Taylor published a careful account [**22**] intended for functional analysts; it focussed on the case of commutative Banach algebras A and included the result that

$$K(A) \cong K\big(\Delta(A)\big),$$

where $\Delta(A)$ is the maximal ideal space of A. With the publication 1973-77 of the results of Brown, Douglas and Fillmore [**6, 7**] and of Kasparov [**17, 18**] on C^*-algebra extensions and K-homology, and Elliott's classification in 1975 of AF C^*-algebras by K-theoretic methods [**11**], the attention of the operator algebra community was drawn to these new methods. For a good relatively recent account of operator K-theory, see Blackadar's book [**3**].

4. Properties of the Operator K-Functor

Recall that a C^*- algebra is a Banach algebra A with an involution $a \to a^*$ such that $\|a^*a\| = \|a\|^2$ for all $a \in A$. Denote by $\boldsymbol{SC^*}$ the category of separable C^*-algebras and $*$-homomorphisms, and by $\boldsymbol{CAb}$ the category of countable abelian groups and homomorphisms. Then

$$K\colon \boldsymbol{SC^*} \to \boldsymbol{CAb}$$

is a covariant functor with the following properties:

4.1. Homotopy invariance. For any C^*-algebras A and B and any $*$-homomorphism $A \to C([0,1],B)$, the compositions

$$A \to C\bigl([0,1],B\bigr) \underset{e_1}{\overset{e_0}{\rightrightarrows}} B$$

induce the same homomorphism $K(A) \to K(B)$. Here $C([0,1],B)$ is the algebra of continuous functions from $[0,1]$ to B; e_0 and e_1 are the evaluations at 0 and 1. In the commutative case, say $A = C(X)$ and $B = C(Y)$, $*$-homomorphisms from A to $C\bigl([0,1],B\bigr)$ correspond to continuous maps $[0,1] \times Y \to X$ and so one has the usual notion of homotopy from topology.

4.2. Continuity. Consider a direct sequence

$$A_1 \to A_2 \to A_3 \to \cdots$$

of C^*-algebras and $*$-homomorphisms with direct limit $A = \lim_{\to} A_n$, the completion of the algebraic direct limit. Then the canonical homomorphism

$$\varinjlim K(A_n) \to K(A),$$

induced by the maps $A_n \to A$ associated with the direct sequence, is an isomorphism. An important special case occurs when we have an increasing sequence

$$A_1 \subset A_2 \subset A_3 \subset \ldots$$

of C^*-subalgebras of a C^*-algebra A. We may as well assume that $\cup A_n$ is dense in A, and then we have an isomorphism $K(A) \cong \lim_{\to} K(A_n)$. For example, let $A_n = M_n(\mathbb{C})$. Then the direct limit is the completion of $M_\infty(\mathbb{C})$, and by viewing $M_n(\mathbb{C})$ as acting on the first n coordinates of the Hilbert space ℓ^2 this is seen to be precisely the closure of the finite rank operators, *i.e.* the algebra $\mathcal{K}$ of compact operators. Hence

$$K(\mathcal{K}) = K\bigl(\varinjlim M_n(\mathbb{C})\bigr) \cong \varinjlim K\bigl(M_n(\mathbb{C})\bigr) = \varinjlim \mathbb{Z} = \mathbb{Z}.$$

More generally, for any C^*-algebra A,

$$K(A \otimes \mathcal{K}) = K\bigl(\varinjlim A \otimes M_n \mathbb{C})\bigr) \cong \varinjlim K\bigl(M_n(A)\bigr) = K(A).$$

This special case of continuity is called **stability**. It was already noted in the examples that $K\bigl(A \otimes M_n(\mathbb{C})\bigr) = K\bigl(M_n(A)\bigr) = K(A)$.

4.3. Half-exactness. If $0 \to A \to B \to C \to 0$ is an exact sequence of C^*-algebras and $*$-homomorphisms then the induced sequence

$$K(A) \to K(B) \to K(C)$$

is exact. The "half" refers to the fact that we cannot in general put 0's at the ends of this sequence. In other words, the first map need not be injective and the second need not be surjective. A simple example is given by the inclusion of $\mathcal{K}$ as an ideal of $\mathcal{B}(\mathcal{H})$; since $K(\mathcal{K}) = \mathbb{Z}$ and $K\bigl(\mathcal{B}(\mathcal{H})\bigr) = 0$ the induced map is not injective.

4.4. Periodicity. The suspension functor of algebraic topology extends to the category of C^*-algebras, namely by defining

$$SA = C_0(\mathbb{R}) \otimes A \cong C_0(\mathbb{R}, A)$$

for any C^*-algebra A, where C_0 denotes the continuous functions vanishing at infinity. Iterated suspension then produces a sequence $K_n = K\big(S^n(\cdot)\big)$ of K functors $(n = 0, 1, \cdots)$. However this sequence is periodic: K_0 and K_2 are naturally isomorphic, so that all the even functors are "the same", as are all the odd ones. It is important to know that K_1 can be defined directly (without using the suspension). Let $GL_\infty(A)$ be the limit of the direct sequence

$$GL_1(A) \to GL_2(A) \to GL_3(A) \to \cdots$$

of topological groups, where $GL_n(A)$ is the group of invertible elements of $M_n(A)$ and the maps are of the form $a \to a \oplus 1$. Then $GL_\infty(A)$ is a topological group in the inductive limit topology and $K_1(A)$ is the group $GL_\infty(A)/GL_\infty(A)^0$ of connected components of $GL_\infty(A)$. For example, using the polar decomposition and the fact that any unitary operator can be written as an exponential, it is not difficult to see that $K_1\big(\mathcal{B}(\mathcal{H})\big) = 0$.

4.5. Long exact sequence. Each of the K_n's is half-exact and, given an exact sequence $0 \to A \to B \to C \to 0$, the induced sequences in (3) fit together via suitable "boundary maps" to yield a single infinite exact sequence. However periodicity means that this reduces to a 6-term exact sequence:

$$\begin{array}{ccccc} K_0(A) & \longrightarrow & K_0(B) & \longrightarrow & K_0(C) \\ {\scriptstyle\partial}\uparrow & & & & \downarrow{\scriptstyle\partial} \\ K_1(C) & \longleftarrow & K_1(B) & \longleftarrow & K_1(A) \end{array}$$

This is a basic tool for computations. Applying it to the exact sequence

$$0 \to \mathcal{K} \to \mathcal{B}(\mathcal{H}) \to \mathcal{C} \to 0,$$

where $\mathcal{C}$ is the Calkin algebra $\mathcal{B}(\mathcal{H})/\mathcal{K}$, yields isomorphisms

$$K_1(\mathcal{C}) \xrightarrow{\cong} K_0(\mathcal{K}) \cong \mathbb{Z}, \ K_0(\mathcal{C}) \xrightarrow{\cong} K_1(\mathcal{K}) = \{0\}$$

since both K-groups of $\mathcal{B}(\mathcal{H})$ vanish. The first of these essentially takes a Fredholm operator to its index, using the fact (Atkinson's Theorem) that the quotient map $\mathcal{B}(H) \to \mathcal{C}$ takes the Fredholm operators exactly to the invertible elements of $\mathcal{C}$. The groups in the second are trivial since any two nonzero projections in $\mathcal{C}$ are equivalent.

5. Axiomatic K-Theory

There is good evidence that any functor that is homotopy invariant, stable and half-exact on a large enough subcategory of $\boldsymbol{SC}^*$ must be closely related to the K-functor. Cuntz showed, for example, that any such functor is necessarily periodic [**10**]. This is in striking contrast with the situation for topological spaces (i.e. on the subcategory of commutative C^* algebras). Examples, other than K_0 and K_1, of functors with these properties are given by $F(A) = K(A \otimes B)$ for fixed B. If it is assumed in addition that $F(\mathbb{C}) = \mathbb{Z}$, Cuntz proved that there is a natural transformation from K to F; moreover if also $F(S\mathbb{C}) = 0$ (in other words

$F(C_0(\mathbb{R})) = 0$) then this is an isomorphism on a certain subcategory that includes most C^*-algebras of interest [**10**].

6. K-Homology

This term signifies a theory, covariant on spaces and contravariant on algebras, that is dual in an appropriate sense to K-theory. It exists as a result of abstract considerations, but its usefulness depends on the ability to construct it concretely. In 1970 Atiyah proposed a construction for $K^0(X)$ modelled on elliptic operators on a manifold [**1**]. Namely he considered triples (σ_1, σ_2, T) with T a Fredholm operator on a Hilbert space $\mathcal{H}$ and σ_1, σ_2 representations of $C(X)$ on $\mathcal{H}$ such that $T\sigma_1(f) - \sigma_2(f)T$ is compact for all $f \in C(X)$. Addition is defined for such objects by direct sum and $K^0(X)$ should result by imposing a suitable equivalence relation. Shortly afterward Brown, Douglas and Fillmore began an independent investigation of extensions of C^*-algebras which led to a concrete realization of $K^1(X)$ [**6, 7**].

7. Extensions of C^*-Algebras

This topic can be introduced by the following question. Let S be the forward shift on $\ell^2(\mathbb{N})$, so that $S^*S = I$ and $SS^* = I - P_0$, where P_0 is the projection of rank 1 on the first coordinate subspace. Thus S is very nearly unitary (i.e. we nearly have $S^*S = SS^* = I$) and we ask: is there a finite rank (or even compact) perturbation $S + F$ of S which is unitary?

To answer this, consider the C^*-subalgebra $\mathcal{T}$ of $\mathcal{B}(\ell^2(\mathbb{N}))$ generated by S. This algebra (the Toeplitz algebra) contains $\mathcal{K}$ as an ideal with quotient isomorphic to $C(S^1)$, so we have an extension (the Toeplitz extension)

$$0 \to \mathcal{K} \to \mathcal{T} \to C(S^1) \to 0$$

of $C(S^1)$ by $\mathcal{K}$. Any such extension is determined (in a suitable sense) by an integer, namely the Fredholm index of any $T \in \mathcal{T}$ that maps to the generator z of $C(S^1)$. The Toeplitz extension corresponds to -1, the Fredholm index of S, and hence the stability of the index under compact perturbations implies that $S + F$ can never be unitary.

The question and its answer can be extended to spaces more general than the circle. For any compact metrizable space X one can construct from extensions

$$0 \to \mathcal{K} \to \mathcal{E} \to C(X) \to 0$$

an abelian group $\operatorname{Ext} X$. The duality with K-theory takes the form of a bilinear pairing

$$K^1(X) \times \operatorname{Ext} X \to \mathbb{Z}$$

obtained from the K-theory long exact sequence (5) of the extension:

$$\begin{array}{ccccc} \mathbb{Z} = K_0(\mathcal{K}) & \longrightarrow & K_0(\mathcal{E}) & \longrightarrow & K_0(C(X)) \\ \partial\uparrow & & & & \downarrow \\ K^1(X) = K_1(C(X)) & \longleftarrow & K_1(\mathcal{E}) & \longleftarrow & K_1(\mathcal{K}) \end{array}$$

Thus given an element of $K^1(X)$ and an element of $\operatorname{Ext} X$ (an extension) one gets an integer from the boundary map.

The presence of $\mathcal{K}$ as the kernel plays a special role in the addition of extensions. Given extensions

$$0 \to \mathcal{K} \to \mathcal{E}_i \xrightarrow{p_i} C(X) \to 0 \qquad (i = 1, 2),$$

$$\mathcal{E} = \left\{ \begin{pmatrix} T_1 & K \\ L & T_2 \end{pmatrix} : T_1 \in E_1, T_2 \in E_2, K, L \in \mathcal{K} \text{ and } p_1(T_1) = p_2(T_2) \right\}$$

has the structure of a C^*-algebra that fits in an exact sequence

$$0 \to M_2(\mathcal{K}) \to \mathcal{E} \to C(X) \to 0$$

in an obvious way. If $\mathcal{K} = \mathcal{K}(\mathcal{H})$, the compact operators on $\mathcal{H}$, then

$$M_2(\mathcal{K}) \cong \mathcal{K}(\mathcal{H} \oplus \mathcal{H}) \cong \mathcal{K}(\mathcal{H}) = \mathcal{K}$$

and so we have an extension of the specified form.

To this point everything goes through for an arbitrary separable C^*-algebra A in place of $C(X)$, except that in general $\operatorname{Ext} A$ is only an abelian semigroup. However it is a group for the very large and useful class of **nuclear** C^*-algebras.

8. *KK*-Theory

The full power and utility of operator K-theory became available only with the development of the KK-functor by Kasparov in the period 1975–80 [**17, 18**]. This is a bifunctor KK from $\boldsymbol{SC^*}$ to $\boldsymbol{CAb}$, contravariant in the first variable and covariant in the second, that combines both K-theory and K-homology. In fact $KK(SA, \mathbb{C}) \cong \operatorname{Ext} A$ (for A nuclear) and $KK(\mathbb{C}, B) \cong K(B)$. It is homotopy invariant, stable and periodic, i.e.

$$KK(S^2A, B) = KK(A, S^2B) = KK(SA, SB),$$

in both variables.

Kasparov's definition was motivated by the ideas of Atiyah described above. Namely $\mathcal{K}$ is replaced by $B \otimes \mathcal{K}$ and its multiplier algebra $\mathcal{B}(\mathcal{H})$ by the multiplier algebra $\mathcal{M}(B \otimes \mathcal{K})$ of $B \otimes \mathcal{K}$, a C^*-algebra that contains $B \otimes \mathcal{K}$ as an ideal. Thus the elements of $KK(A, B)$ are determined by triples (σ_1, σ_2, T) with $T \in \mathcal{M}(B \otimes \mathcal{K})$ invertible modulo $B \otimes \mathcal{K}$ and $\sigma_1, \sigma_2 \colon A \to \mathcal{M}(B \otimes \mathcal{K})$ $*$-homomorphisms that are intertwined by T modulo $B \otimes K$. In particular (composing with the map $b \to b \otimes e$ from B to $B \otimes \mathcal{K}$, where e is a rank one projection in $\mathcal{K}$) any $*$-homomorphism $\sigma \colon A \to B$ determines an element (σ, σ, I) of $KK(A, B)$. Cuntz later showed that all elements of $KK(A, B)$ can be constructed as suitable pairs of $*$-homomorphisms. Alternatively, as Kasparov showed, when A is nuclear elements of $KK(SA, B)$ can be viewed as coming from extensions of the form

$$0 \to B \otimes \mathcal{K} \to E \to A \to 0.$$

The most important feature of KK-theory is the existence of an associative bilinear pairing, the **Kasparov product**,

$$KK(A, B) \times KK(B, C) \to KK(A, C)$$

which extends composition of $*$-homomorphisms, the cap and cup products of algebraic topology and also the pairing

$$K_1(A) \times \operatorname{Ext} A = KK(\mathbb{C}, SA) \times KK(SA, \mathbb{C}) \to \mathbb{Z}$$

described earlier. With it KK-theory becomes a powerful and flexible tool, used (for example) by Cuntz in his work on axiomatic K-theory and by Kasparov and others in reformulating (and in some cases establishing) Novikov's higher signature conjecture as a problem in operator K-theory.

The KK-functor is, however, not without its defects. The proofs are highly technical and it is in general not half-exact (so that there is no long exact sequence). The second point is connected with the fact, mentioned earlier, that $\operatorname{Ext} A$ is not in general a group. Both difficulties evaporate for nuclear algebras. In general it is true that KK is "split exact", in the sense that it behaves well (in either variable) on split exact sequences.

9. E-Theory

Recall that a category is **additive** if its hom sets are abelian groups in such a way that composition is bilinear. An example is the category $\boldsymbol{KK}$ having the same objects as $\boldsymbol{SC}^*$ but with the groups $KK(A,B)$ as the maps and the Kasparov product as composition. There is a functor $\boldsymbol{SC}^* \to \boldsymbol{KK}$, the identity on objects and taking $*$-homomorphisms to the corresponding KK-elements, and in 1987 Higson showed that this setup is characterized by a universal property: any homotopy-invariant, stable and split-exact functor from $\boldsymbol{SC}^*$ to an additive category factors uniquely through this one [**14**].

It is natural to ask if the KK-functor might be modified (for non-nuclear algebras) so as to become half-exact, while retaining homotopy invariance and stability. In 1990 Higson showed that such a bifunctor E from $\boldsymbol{SC}^*$ to $\boldsymbol{CAb}$ exists, and moreover that it is unique in the sense that there is a functor $\boldsymbol{SC}^* \to \boldsymbol{E}$ as above such that any homotopy invariant, stable and half-exact functor from $\boldsymbol{SC}^*$ to an additive category factors uniquely through $\boldsymbol{E}$ [**15**].

The basic idea here is familiar from homotopy theory: roughly speaking, for an ideal J of a C^*-algebra A such that A/J is contractible, one wants the inclusion $J \to A$ to induce an isomorphism. This is accomplished by making use of the notion of a category of fractions – a universal way of enlarging a given category so that a given set of morphisms become isomorphisms.

10. Concrete E-Theory

The usefulness of $\boldsymbol{E}$, defined in this way, is limited by its abstraction. One needs concrete cycles and this is precisely what Connes and Higson provided shortly thereafter [**8, 9**]. They showed that the elements of $E(A,B)$ arise from **asymptotic morphisms** $(\varphi_t)\colon A \to B$. Here $t \in [1,+\infty)$ and each φ_t is a map from A to B such that as $t \to +\infty$

$$\begin{aligned} \varphi_t(a+\lambda a') - \varphi_t(a) - \lambda\varphi_t(a') &\to 0 \\ \varphi_t(aa') - \varphi_t(a)\varphi_t(a') &\to 0 \\ \varphi_t(a^*) - \varphi_t(a)^* &\to 0 \end{aligned}$$

for all $a, a' \in A$ and $\lambda \in \mathbb{C}$, the only joint requirement being the continuity of $t \to \varphi_t(a)$ for all $a \in A$. Nothing else is assumed about the individual φ_t's, not even continuity or boundedness.

Homotopy for asymptotic morphisms $A \to B$ is defined in terms of asymptotic morphisms $A \to C\big([0,1],B\big)$ as in (1) above. If $\big[[A,B]\big]$ denotes the set of homotopy classes then there is an associative operation

$$\big[[A,B]\big] \times \big[[B,C]\big] \to \big[[A,C]\big]$$

that extends composition of $*$-homomorphisms. This operation is in fact just composition $(\varphi_t \circ \psi_t)$, but it is necessary to choose appropriate representatives of the homotopy classes in order to ensure that this defines an asymptotic morphism.

By forming direct sums

$$\begin{pmatrix} \varphi_t & 0 \\ 0 & \psi_t \end{pmatrix} : A \to M_2(B \otimes \mathcal{K}) \cong B \otimes \mathcal{K}$$

we can give $\big[[A, B \otimes K]\big]$ the structure of an abelian semigroup. Here we have used the facts $M_2(D) \cong M_2(\mathbb{C}) \otimes D$ for any D and $M_2(\mathcal{K}) \cong \mathcal{K}$. For the existence of inverses it is in general necessary to introduce suspensions; in particular

$$E(A, B) = \big[[SA \otimes \mathcal{K}, SB \otimes \mathcal{K}]\big]$$

is always an abelian group. This functor has all the properties of the functor E of the previous section, so by the universal property stated there they are naturally isomorphic. In particular $KK = E$ on the category of separable nuclear C^*-algebras and the Kasparov product is the composition described above.

11. Applications

Like Kasparov, Connes and Higson were interested in the Novikov conjecture and, more particularly, in a far-reaching generalization known as the **Baum-Connes conjecture** [**9**]. This conjecture asserts, very roughly speaking, that for suitable locally compact groups G the elements of the K-groups of the group C^*-algebra $C_r^*(G)$ arise in an essentially unique way as the index of elliptic operators on a certain G-space associated naturally with G. The expectation is that the use of E-theory will allow further progress to be made on these conjectures.

Connes and Higson also pointed out a connection between asymptotic morphisms and **deformations** of C^*-algebras [**8, 9**]. One way to look at deformations is in terms of bundles or continuous fields $\big(\{A_t\}_{t\in X}, \Gamma\big)$ of C^*-algebras over a base space X. Here Γ is a set of sections γ subject to certain requirements, notably that $t \to \|\gamma(t)\|$ be continuous and that the values fill up the fibres. We are interested in the case $X = [0, 1]$ and $A_t \equiv B$ for $t > 0$. Given such a field there is an asymptotic morphism $(\varphi_t)\colon A \to B$ (where $A = A_0$), constructed as follows: for each $a \in A$ choose $\gamma_a \in \Gamma$ with $\gamma_a(0) = a$ and set $\varphi_t(a) = \gamma_a(1/t)$. An example is the **Heisenberg deformation**, which arises from the group C^*-algebra $C^*(H)$ of the real Heisenberg group H, and for which $X = \mathbb{R}$, $A_t = \mathcal{K}$ for $t \neq 0$ and $A_0 = C_0(\mathbb{R}^2)$. Thus we have an asymptotic morphism $C_0(\mathbb{R}^2) \to \mathcal{K}$, or in other words an element of $E\big(C_0(\mathbb{R}^2), \mathbb{C}\big)$ (c.f. [**13**]). In fact this is the "Bott element": we can use the obvious map (tensor with id_A)

$$E(B, D) \to E(A \otimes B, A \otimes D)$$

to get an element $\beta \in E(S^2A, A)$ such that composition with β is the periodicity isomorphism

$$E(A, B) \to E(S^2A, B).$$

Finally, we mention that E. Kirchberg [**19**] and N.C. Phillips [**21**] recently used ideas from E-theory to complete a major step in Elliott's classification programme [**12**]. The objective of this programme is to use K-theoretical and related invariants to classify large families of (and perhaps all) separable nuclear C^*-algebras.

References

1. M. Atiyah, *Global theory of elliptic operators*, Proc. Internat. Conf. on Functional Analysis and Related Topics (Tokyo, 1969), Univ. of Tokyo Press, Tokyo, 1970, 21–30.
2. M. Atiyah and F. Hirzebruch, *Vector bundles and homogeneous spaces*, Proc. Sympos. Pure Math., vol. III, Amer. Math. Soc., Providence, R.I., 1961, 7–38.
3. B. Blackadar, *K-theory for operator algebras*, Math. Sci. Res. Inst. Publ., vol. 5, Springer-Verlag, New York-Berlin, 1986.
4. A. Borel and J. P. Serre,*Le théorème de Riemann-Roch*, Bull. Soc. Math. France **86** (1958), 97–136. (French)
5. R. Bott, *The stable homotopy of the classical groups*, Proc. Nat. Acad. Sci. U.S.A. **43** (1957), 933-935.
6. L. Brown, R. Douglas, and P. Fillmore, *Unitary equivalence modulo the compact operators and extensions of C^*-algebras*, Proc. of a Conference on Operator Theory (Dalhousie Univ., Halifax, N.S., 1973). Lecture Notes in Math., vol. 345, Springer, Berlin, 1973, 58–128.
7. ———, *Extensions of C^*-algebras and K-homology*, Ann. of Math. (2) **105** (1977), no. 2, 265–324.
8. A. Connes and N. Higson, *Déformations, morphismes asymptotiques et K-théorie bivariante*, C. R. Acad. Sci. Paris Sér. I Math. **311** (1990), no. 2, 101–106. (French)
9. A. Connes, *Noncommutative geometry*, Academic Press, Inc., San Diego, CA, 1994.
10. J. Cuntz, *K-theory and C^*-algebras*, Proc. of a Conference on K-theory (Bielefeld 1982), Lecture Notes in Math. **1046** (1983), Springer, New York, 55–79.
11. G. Elliott, *On the classification of inductive limits of sequences of semisimple finite-dimensional algebras*, J. Algebra **38** (1976), no. 1, 29–44.
12. ———, *The classification problem for amenable C^*-algebras*, Proceedings of ICM94, Birkhäuser Verlag, Basel, 1995.
13. ———, T. Natsume and R. Nest, *The Heisenberg group and K-theory*, K-Theory **7** (1993), no. 5. 409–428.
14. N. Higson, *A characterization of KK-theory*, Pacific J. Math **126** (1987), no. 2, 253–276.
15. ———, *Categories of fractions and excision in K-theory*, J. Pure Appl. Algebra **65** (1990), no. 2, 119–138.
16. M. Karoubi, *K-theory: An introduction*, Grundlehren der Mathematischen Wissenschaften, Band 226, Springer-Verlag, Berlin-New York, 1978.
17. G. Kasparov, *Topological invariants of elliptic operators* I: *K-homology*, Izv. Akad. Nauk SSR, Ser. Mat. **39** (1975), no. 4, 796–838.
18. ———, *The operator K-functor and extensions of C^*-algebras*, Izv. Akad. Nauk SSR, Ser. Mat. 44 (1980), no. 3, 571–636 (Russian); English transl., Math USSR Izv. **16** (1981) 513–572.
19. E. Kirchberg, *The classification of purely infinite C^*-algebras using Kasparov's theory*, preprint.
20. F. Murray and J. von Neumann, *On rings of operators*, Ann. Math. (2), **37** (1936) 116–229.
21. N. C. Phillips, *A classification theorem for nuclear purely infinite simple C^*-algebras*, preprint.
22. J. Taylor, *Banach algebras and topology*, Algebras in analysis (J. H. Williamson, ed.) (Proc. Instructional Conf. and NATO Advanced Study Inst., Birmingham, 1973), Academic Press, London, 1975, 118–186.

Department of Mathematics, Dalhousie University, Halifax, NS (Canada) B3H 3J5
E-mail address: fillmore@cs.dal.ca

Centre de Recherches Mathématiques
CRM Proceedings and Lecture Notes
Volume **11**, 1997

Partial Differential Equations with Differential Constraints

Alfred M. Grundland, Luigi Martina, and Guy Rideau

ABSTRACT. A new version of the nonclassical symmetry reduction method for the case of multiple differential constraints is proposed. The necessary and sufficient conditions for the existence of conditional symmetries for nonlinear systems of PDEs in many independent and dependent variables are discussed in detail. An algorithm for constructing conditionally invariant solutions is developed based on the adaptation of Lie's theorem on fundamental sets of solutions to the case of PDEs. A close connection between Bäcklund transformations and conditional symmetries is demonstrated. Finally, the case of multiple differential constraints containing higher order derivatives is investigated. The theoretical considerations are illustrated by several integrable and nonintegrable dynamical systems.

1. Introduction

The objective of this paper is to study the conditional symmetries in connection with integrability properties of nonlinear systems of PDEs with many independent and dependent variables. The term "conditional symmetry" constitutes an extension of the notion of "nonclassical symmetry" which originated from the work of G. W. Bluman and J. D. Cole [**2**]. It refers to the symmetries of an overdetermined system obtained by appending the initial system of PDEs with certain differential constraints (DCs) for which the symmetry criterion is identically satisfied. The crux of the matter is that, though these "new" symmetries are not symmetries of the original system, they can be used to construct new solutions of this sytem, not obtainable by the classical approach. This basic idea was developed and implemented recently by several authors (e.g. P. Olver and Ph. Rosenau [**22**], D. Levi and P. Winternitz [**16**], E. Pucci and G. Saccomandi [**24**]) who produced several variants of "nonclassical" symmetry reduction method. For a comprehensive review of this subject see the chapter by P. Olver and E. M. Vorobev in [**23**] (vol. 3 chapter 11).

The approach proposed in this work (initiated in [**11**] and [**9**]) is distinguished by the fact that we introduce multiple DCs compatible with the initial system of PDEs, the number of which is equal to the product of the numbers p and q of independent and dependent variables in this system. We postulate that these DCs

---

1991 *Mathematics Subject Classification.* Primary: 35Q58.
This is the final form of the paper.

take a specific form, for which all derivatives of unknown functions are expressible in terms of some functions of independent and dependent variables only. It follows from the above that for every set of $(p-1)$ fixed independent variables the DCs constitute a parametrized family of 1st order systems of ODEs. Furthermore, we demand that there exists a nonlinear superposition formula (SF) for each subsystem of this family of ODEs, i.e. that the assumptions of Lie's theorem on fundamental sets of solutions are satisfied. We prove that the zero gauge curvature conditions ensure the existence of solutions to the DCs of the postulated form. Some theorems useful for the construction of solutions based on conditional symmetries are given in Sections 2 and 3.

An interesting consequence of the adopted approach is that it leads to a direct connection between conditional symmetries and Bäcklund transformations (BTs) for the initial system of PDEs. In Section 4 we describe in detail the construction of solutions based on this connection and discuss some geometric aspects of our method.

Section 5 contains examples of applications of this method to the equations describing the three-wave resonant interaction, constant mean curvature surface equations and generalized Liouville equation. In Section 6 we propose a generalization of the conditional symmetry method (CSM) to the case of nonlinear PDEs admitting constraints which depend on higher order derivatives.

2. Conditional Symmetries for PDEs

Let us consider a system of 1st order PDEs, say Δ, in p independent variables $x = (x^1, \dots, x^p)$, which form local coordinates on the Euclidean space X and q dependent variables $u = (u^1, \dots, u^q)$ which are fiber coordinates on the space U. The derivatives of u^α are denoted by $u^\alpha_J = \partial^{\Sigma J} u^\alpha / \partial x^J$, where $J = (j_1, \dots, j_p)$ is a multi-index of order $k = \Sigma J$. Let U_k, with coordinates u^α_J, be the space representing all these k-th order derivatives. Let $U^{(k)} = U \times U_1 \times \cdots \times U_k$ be the space representing all partial derivatives of functions $f\colon X \mapsto U$ of order $\leq k$. According to the above notation a 1st order system of PDEs can be expressed as follows

$$\Delta^s(x, u^{(1)}) \equiv \Delta^s(x^i, u^\alpha, u^\alpha_i) = 0, \quad s = 1, \dots, m, \tag{2.1}$$

where $u^{(1)}$ are coordinates on 1st jet space $J^1 = J^1(X \times U)$ over X. All of our considerations are of a local character.

In this section we discuss in detail the necessary and sufficient conditions for existence of conditional symmetries for the system (2.1). In other words, we show how to construct a consistent set of 1st order DCs compatible with initial system (2.1).

Let us assume that the system (2.1) is subjected to $p \cdot q$ 1st order DCs

$$u^\alpha_i = \phi^\alpha_i(x, u), \quad i = 1, \dots, p, \; \alpha = 1, \dots, q, \tag{2.2}$$

for some differentiable functions ϕ^α_i defined on open subset of $X \times U$. The conditions (2.2) mean that all derivatives of u's with respect to x^i are expressible in terms of x and u only. The compatibility conditions for the overdetermined system composed of (2.1) and (2.2)

$$\Delta^s(x^i, u^\alpha, u^\alpha_i) = 0, \quad u^\alpha_i - \phi^\alpha_i(x, u) = 0, \tag{2.3}$$

give us an overdetermined system of $m + 1/2p(p-1)q$ equations for $p \cdot q$ unknown functions ϕ_i^α in $p+q$ independent variables x and u, i.e.

$$\Delta^s\big(x^i, u^\alpha, \phi_i^\alpha(x,u)\big) = 0, \quad s = 1, \dots, m, \tag{2.4}$$

$$\phi^\alpha_{[j,k]} + \phi^\beta_{[k}\phi^\alpha_{j],u^\beta} = 0, \quad \alpha, \beta = 1, \dots, q, \quad j,k = 1, \dots, p. \tag{2.5}$$

Here we adopt the summation convention over repeated indices unless otherwise stated. The bracket $[j,k]$ denotes the antisymmetrization with respect to the indices j and k. So, we have

$$\phi^\alpha_{[j,k]} := 2(\phi^\alpha_{j,k} - \phi^\alpha_{k,j}), \quad \phi^\beta_{[k}\phi^\alpha_{j],u^\beta} := 2(\phi^\beta_k\phi^\alpha_{j,u^\beta} - \phi^\beta_j\phi^\alpha_{k,u^\beta}). \tag{2.6}$$

If conditions (2.5) are satisfied then a solution of the system (2.2) depends on q integration constants. The equations (2.4) constitute a functional restriction on the solution space of the system (2.1).

Let us assume that there exists a set of p vector fields defined on $J^o = X \times U$

$$Z_k = \partial_{x^k} + \phi_k^\alpha(x,u)\partial_{u^\alpha}, \quad k = 1, \dots, p, \tag{2.7}$$

(where $\partial_{x^k} = \partial/\partial x^k$ and $\partial_{u^\alpha} = \partial/\partial u^\alpha$) such that the functions ϕ_k^α satisfy the conditions (2.4) and (2.5). We consider here the case when the number of vector fields Z_k is equal to the number of independent variables appearing in the initial system (2.1). From conditions (2.5) it follows that the commutators of Z_k, Z_j are equal to zero, i.e.

$$[Z_k, Z_j] = 0 \text{ for } 1 \leq k, j \leq p. \tag{2.8}$$

So, the vector fields $Z_1, \dots, Z_p$ span a p-dimensional Abelian Lie algebra L. Note that the requirement that the characteristics of the vector fields Z_k are equal to zero

$$Q_k^\alpha(x, u^{(1)}) := u_k^\alpha - \phi_k^\alpha(x,u) = 0, \tag{2.9}$$

is equivalent to the assumed form of DCs (2.2).

We now proceed to define basic notions of the conditional symmetry method.

Let us associate the systems of equations $\Delta^s \colon J^1 \to \mathbb{R}^m$ and $Q_k^\alpha \colon J^1 \to \mathbb{R}^{p\times q}$ with the subvarieties of the solution spaces

$$\begin{aligned} S_\Delta &= \big\{(x, u^{(1)}) \colon \Delta^s(x, u^{(1)}) = 0, \ s = 1, \dots, m\big\} \text{ and} \\ S_Q &= \big\{(x, u^{(1)}) \colon Q_k^\alpha(x, u^{(1)}) = 0; \ k = 1, \dots, p; \ \alpha = 1, \dots, q\big\}, \end{aligned} \tag{2.10}$$

respectively.

DEFINITION 2.1 (Equivalent to P. Olver's definition [**23**, page 311]). A vector field Z_k is called *a conditional symmetry of the system* $\Delta^s = 0$ if Z_k is tangent to $S = S_\Delta \cap S_Q$, i.e.

$$\mathrm{pr}^{(1)} Z_k\big|_S \in T_{(x,u^{(1)})}S, \tag{2.11}$$

where $\mathrm{pr}^{(1)} Z_k$ is the 1st prolongation of Z_k defined on J^2, i.e.

$$\mathrm{pr}^{(1)} Z_k = Z_k + (u^\alpha_{kj} - D_j Q_k^\alpha)\partial_{u_j^\alpha}, \tag{2.12}$$

D_j denote the total derivative operators in X restricted to J^1

$$D_j = \partial_{x^j} + u_j^\alpha \partial_{u^\alpha}, \quad j = 1, \dots, p \tag{2.13}$$

and $T_{(x,u^{(1)})}S$ is the tangent space to S at some point $(x, u^{(1)})$.

DEFINITION 2.2. An Abelian Lie algebra L spanned by the vector fields $Z_1, \dots, Z_p$ is called *a conditional symmetry algebra of the system* $\Delta^s = 0$ if the following condition

$$\operatorname{pr}^{(1)} Z_k(\Delta^s)\big|_S = 0, \quad k = 1, \dots, p, \tag{2.14}$$

is satisfied.

Note, that for an Abelian distribution of vector fields $Z_1, \dots, Z_p$ the invariance criterion for DCs (2.9) is identically satisfied by virtue of the conditions (2.5), i.e.

$$\operatorname{pr}^{(1)} Z_k(Q_j^\alpha) = \phi^\alpha_{[j,k]} + \phi^\beta_{[k}\phi^\alpha_{j],u^\beta} \equiv 0. \tag{2.15}$$

Every solution of (2.3) can be represented by its graph $\{bigl(x, u(x))\}$, which is a section of J^o. The conditional symmetry algebra L of $\Delta^s = 0$ defines locally the action of the corresponding Lie group G on J^o. Then the symmetry group G transforms solutions of (2.3) into solutions of (2.3). If the graph is preserved by G, then the related solution is G-invariant.

DEFINITION 2.3. Assume that L, spanned by vector fields $Z_1, \dots, Z_p$, is a conditional symmetry algebra of the system $\Delta^s = 0$. A solution $u = f(x)$ is said to be *a conditionally invariant solution of the system* $\Delta^s = 0$ if the graph $\{(x, f(x))\}$ is invariant under the vector fields $Z_1, \dots, Z_p$.

We have the following.

THEOREM 2.4. *A nondegenerate* 1st *order system of PDEs* (2.1) *admits a p-dimensional conditional symmetry algebra L iff p linearly independent vector fields* $Z_1, \dots, Z_p$ *satisfy the conditions* (2.4) *and* (2.5) *on some neighborhood of* (x_o, u_o) *of S.*

PROOF. The generators Z_k constitute the conditional symmetries for the system (2.1) if they are Lie point symmetries of the overdetermined system (2.3). Applying the invariance criterion (2.14) to (2.1) we obtain

$$\operatorname{pr}^{(1)} Z_k(\Delta^s) = \partial_{x^k}\Delta^s + \phi^\alpha_k \partial_{u^\alpha}\Delta^s + (\phi^\alpha_{k,j} + \phi^\beta_j \phi^\alpha_{k,u^\beta})\partial_{u^\alpha_j}\Delta^s. \tag{2.16}$$

On the other hand, carrying out the differentiations in (2.4) gives us

$$D_j\Delta^s = \partial_{x^j}\Delta^s + \phi^\alpha_{k,j}\partial_{u^\alpha_k}\Delta^s + \phi^\alpha_j(\partial_{u^\alpha}\Delta^s + \phi^\beta_{k,u^\alpha}\partial_{u^\beta_k}\Delta^s) = 0, \tag{2.17}$$

where $D_j = \partial_{x^j} + \phi^\alpha_j\partial_{u^\alpha}$ are the modified total derivative operators (2.13) whenever (2.9) holds. After substituting (2.17) into (2.16) and taking into account the assumption (2.5) we obtain that the equations (2.14) are identically satisfied.

Conversely, the assumption that the system (2.1) is nondegenerate means that it is locally solvable and takes a maximal rank at every point $(x_o, u_o^{(1)}) \in S_\Delta$. Therefore [**20**] the infinitesimal symmetry criterion is a necessary and sufficient condition for the existence of symmetry group G of the overdetermined system (2.3). Since the vector fields Z_k form an Abelian distribution, it follows that the condition (2.5) holds. Requiring that the characteristics of the vector fields Z_k are equal to zero and substituting them into (2.1) we obtain the condition (2.4). That ends the proof, since the solutions of the original system (2.1) are invariant under the algebra L generated by p vector fields $Z_1, \dots, Z_p$. □

Note that conditionally invariant solutions of (2.1) can be obtained through new reductions (i.e. new with respect to the results of the "classical" method).

Namely, if the set of solutions of the determining equations obtained by applying the symmetry criterion to the overdetermined system (2.3) is different than the set of solutions of the determining equations obtained for the system (2.1), then the system (2.3) admits other symmetries than the original system (2.1). So, new reductions for the system (2.1) can be constructed, using the same basic algorithm as in the classical method, since each solution of (2.3) is a solution of (2.1).

3. An Adaptation of Lie's Theorem to the Case of Nonlinear PDEs

The approach to the construction of conditionally invariant solutions proposed in this paper is based on Lie's theorem [**18**] on fundamental sets of solutions of 1st order system of q nonlinear ODEs

$$\frac{dy^\alpha}{dt}(t) = f^\alpha(t,y), \quad \alpha = 1,\dots,q. \tag{3.1}$$

We now show that Lie's theorem can be used to facilitate the construction of solutions of PDEs of the form (2.2)

$$\frac{\partial u^\alpha}{\partial x^i}(x) = \phi_i^\alpha(x,u); \quad i = 1,\dots,p; \ \alpha = 1,\dots,q. \tag{3.2}$$

We assume that the functions ϕ_i^α are analytic in u on some open domain $\mathcal{D} \subset X \times U$.

Note that for every set of $(p-1)$ fixed variables $x_o^1,\dots,\hat{x}^i,\dots,x_o^p$ the system (3.2) constitutes a parametrized family of ODEs with $(p-1)$ parameters $x_o^1,\dots,\hat{x}^i,\dots,x_o^p$ and independent variable x^i, i.e.

$$\forall i = 1,\dots,p, \quad \frac{du^\alpha}{dx^i}(x_o^1,\dots,\hat{x}^i,\dots,x_o^p) = \phi_i^\alpha(x_o^1,\dots,\hat{x}_i,\dots,x_o^p,u^1,\dots,u^q). \tag{3.3}$$

THEOREM 3.1. *The system of PDEs (3.2) possesses a nontrivial solution $u(x^1,\dots,x^p)$ such that for every $(p-1)$ fixed variables $x_o^1,\dots,\hat{x}^i,\dots,x_o^p$, this solution $u(x_o^1,\dots,\hat{x}^i,\dots,x_o^p)$ becomes a general solution of the system of ODEs (3.3) given by a superposition formula (i.e. this solution is expressed in terms of a finite number of particular solutions of (3.3) and a finite number of arbitrary constants) if the following conditions are satisfied*

1) *There exist two sets of C^1 functions $A^\ell\colon \mathcal{P} \to \mathbb{R}^p$ and $b_\ell\colon \mathcal{B} \to \mathbb{R}^q$ (where $\mathcal{P}$ and $\mathcal{B}$ are open subsets of X and U, respectively) such that all derivatives of u's with respect to x^i are decomposable in a following way*

$$\forall_{\substack{i=1,\dots,p,\\ \alpha=1,\dots,q}} \frac{\partial u^\alpha}{\partial x^i}(x) = \sum_{\ell=1}^{r} A_i^\ell(x) b_\ell^\alpha(u), \tag{3.4}$$

on some open domain $\mathcal{D} \subset X \times U$.

2) *There exists a set of r linearly independent vector fields on U*

$$\hat{b}_\ell = b_\ell^\alpha(u)\partial_{u^\alpha}, \quad \ell = 1,\dots,r, \tag{3.5}$$

generating a finite-dimensional Lie algebra $\mathfrak{g}$

$$[\hat{b}_k, \hat{b}_\ell] = C_{k\ell}^j \hat{b}_j, \quad 1 \le j,k,\ell \le r. \tag{3.6}$$

3) *There exists a vector valued differential 1-form on X*

$$\lambda = \sum_{i=1}^{p}\sum_{\ell=1}^{r} A_i^\ell(x) dx^i \hat{b}_\ell, \tag{3.7}$$

which takes its value in the Lie algebra $\mathfrak{g}$ and has a zero gauge curvature, i.e.

$$d\lambda + \frac{1}{2}[\lambda, \wedge\lambda]_{\mathfrak{g}} = 0, \tag{3.8}$$

where by the bracket $[\alpha, \wedge\beta]_{\mathfrak{g}}$ we denote the commutator of $\mathfrak{g}$-valued 1-forms α and β, i.e.

$$[\alpha, \wedge\beta]_{\mathfrak{g}}(u, v) = \big[\alpha(u), \beta(v)\big]_{\mathfrak{g}} - \big[\alpha(v), \beta(u)\big]_{\mathfrak{g}} \text{ for } u, v \in TX. \tag{3.9}$$

PROOF. To show that (3.2) possesses nontrivial solutions under the assumptions 1)-3) it is sufficient to demonstrate that the compatibility conditions for (3.4) are identically satisfied.

The system (3.4) can be written as a set of differential 1-forms

$$du = \sum_{\ell=1}^{r} \hat{b}_\ell \otimes \lambda^\ell, \text{ where } \lambda^\ell = A_i^\ell(x) dx^i, \tag{3.10}$$

at each point $x \in \mathcal{P}$. The compatibility conditions for (3.10) impose certain restrictions on the form of the vector functions b_ℓ and A^ℓ. Namely, closing (3.10) (by exterior differentiation) we obtain the equations

$$\sum_{\ell=1}^{r} (d\hat{b}_\ell \wedge \lambda^\ell + \hat{b}_\ell \otimes d\lambda^\ell) = 0, \tag{3.11}$$

which have to be satisfied whenever (3.10) holds.

Making use of (3.10) and denoting

$$\hat{b}_{\ell, \hat{b}_k} := \hat{b}_{\ell, u^\alpha} \hat{b}_k^\alpha, \tag{3.12}$$

we get

$$d\hat{b}_\ell = \sum_{k=1}^{r} \hat{b}_{\ell, \hat{b}_k} \otimes \lambda^k, \tag{3.13}$$

and

$$d\lambda^\ell = \sum_{j=1}^{p} A^\ell_{i, x^j} dx^j \wedge dx^i. \tag{3.14}$$

Substituting the above relations into the prolonged system (3.11) we obtain

$$\sum_{\ell \neq k=1}^{r} \big([\hat{b}_\ell, \hat{b}_k] \otimes \lambda^\ell \wedge \lambda^k + \hat{b}_\ell \otimes d\lambda^\ell\big) = 0. \tag{3.15}$$

Taking into account assumption 2), that is, the fact that the vector fields $\hat{b}_\ell$ generate a finite-dimensional Lie algebra $\mathfrak{g}$, we get

$$\sum_{\substack{s,\ell,k=1,\\ \ell \neq k}}^{r} \hat{b}_s \otimes \big(C^s_{\ell k} \lambda^\ell \wedge \lambda^k + d\lambda^s\big) = 0. \tag{3.16}$$

Next, from the assumed linear independency of the vector fields $\hat{b}_1, \ldots, \hat{b}_r$ on U space, we can construct the coframe ϕ of all 1-forms w^a on U satisfying

$$\langle w^a, \hat{b}_s \rangle = \delta^a_s. \tag{3.17}$$

By contracting (3.16) with $w^a, a = 1, \dots, r$ we get

$$\sum_{\substack{s,\ell,k=1 \\ \ell \neq k}}^{r} \langle w^a, \hat{b}_s \rangle \left(C^s_{\ell k} \lambda^\ell \wedge \lambda^k + d\lambda^s \right) = 0. \tag{3.18}$$

From (3.10) and (3.18) we obtain

$$\sum_{\ell,k=1}^{r} \frac{1}{2} C^a_{\ell k} A^\ell_i A^k_j + A^a_{[i,j]} = 0, \quad a = 1, \dots, r;\ \ell < k. \tag{3.19}$$

The above equations are equivalent to the zero gauge curvature conditions (3.8). Hence, by virtue of the assumption 3), the compatibility conditions (3.11) are identically satisfied.

The system of ODEs (3.3) has a superposition formula if the assumptions of the original Lie's theorem [**18**] are satisfied. Note, that these assumptions coincide with assumptions 1) and 2) of our theorem for the case of system (3.2) with $(p-1)$ fixed variables $x^1_o, \dots, \hat{x}^i, \dots, x^p_o$. In this instance the equations (3.4) takes the form

$$\forall i = 1, \dots, p, \quad \frac{du^\alpha}{dx^i}(x) = \sum_{\ell=1}^{r} A^\ell_i(x^1_o, \dots, \hat{x}^i, \dots x^p_o) b^\alpha_\ell(u^1, \dots, u^q), \tag{3.20}$$

and its general solution can be generated by SF. □

Now, we can derive the conditions on A^ℓ and b_ℓ (appearing in (3.4)) sufficient for the existence of the particular type of conditionally invariant solutions of the system (2.1).

THEOREM 3.2. *Suppose that there exist nonzero C^1 functions $A^\ell\colon X \to \mathbb{R}^p$ and $b_\ell\colon U \to \mathbb{R}^q, \ell = 1, \dots, r$ which satisfy:*

(i) *the assumptions* 1)-3) *of the Theorem* 3.1 *for the system*

$$\frac{\partial u^\alpha}{\partial x^i}(x) - \phi^\alpha_i(x, u) = 0$$

on some open subset $D \subset X \times U$.

(ii) *the conditions*

$$\Delta^s \left(x^i, u^\alpha, \sum_{\ell=1}^{r} A^\ell_i(x) b^\alpha_\ell(u) \right) = 0, \quad s = 1, \dots, m, \tag{3.21}$$

possessing nontrivial solutions $A^1, \dots, A^r$ for a given realization of a finite-dimensional Lie algebra $\mathfrak{g}$ in terms of vector fields (3.5).

Then the system (2.1) *possesses locally a conditionally invariant solution $u(x)$, such that its graph $\Gamma_u\colon x \mapsto \{(x, u(x))\}$ is invariant under a p-dimensional Abelian distribution of vector fields*

$$Z_i = \partial_{x^i} + \sum_{\ell=1}^{r} A^\ell_i(x) b^\alpha_\ell(u) \partial_{u^\alpha}, \quad i = 1, \dots, p. \tag{3.22}$$

PROOF. Indeed, from assumptions 1) and 2) of this theorem it follows that condition (2.4) holds. Under assumption 1) we obtain that the commutator of vector fields Z_i, Z_j vanishes, i.e.

$$
\begin{aligned}
[Z_i, Z_j] &= \left[\partial_{x^i} + \sum_{\ell=1}^{r} A_i^\ell(x) b_\ell^\alpha(u)\partial_{u^\alpha}, \partial_{x^j} + \sum_{k=1}^{r} A_j^k(x) b_k^\beta(u)\partial_{u^\beta}\right] \\
&= \left(A_{[i,j]}^a + \frac{1}{2} C_{\ell k}^a A_i^\ell A_j^k\right)\hat{b}_a = 0.
\end{aligned} \tag{3.23}
$$

This implies that condition (2.5) is also satisfied. So, by virtue of Theorem 2.4, Z_i constitute the conditional symmetry algebra L of the system (2.1). That ends the proof, since the solutions of the original system (2.1), which are invariant under the Abelian algebra L, are precisely the solutions of the overdetermined system (2.3) and they constitute conditionally invariant solutions of (2.1) in the sense of Definition 2.3. □

Note that the transformations

$$
A_i^\ell(x) \to c_{(\ell)} A_i^\ell \quad \text{and} \quad b_\ell^\alpha(u) \to \frac{1}{c_{(\ell)}} b_\ell^\alpha, \quad \ell = 1, \dots, r, \tag{3.24}
$$

where c_ℓ are arbitrary real (complex) constants, do not change the factorized form (3.4) and the equations (3.21). So, decomposition (3.4) proposed in Theorem 3.1 is given up to a choice of a set of constants.

If assumptions 1)-3) of Theorem 3.1 are satisfied then the system (3.4) can be solved by successive integrations, since (3.10) becomes an exact differential. The second assumption of Theorem 3.2 guarantees the compatibility of the DCs (3.4) with original system (2.1). This means that having the general solution of (3.4) we can then proceed to the construction of the solution of the system (2.1).

4. Bäcklund Transformations in the Context of Conditional Symmetries

We focus in this section on a close connection between the conditional symmetries and Bäcklund transformations associated with nonlinear systems of PDEs. This connection is mainly due to the fact that the set of DCs (3.2) admits a SF, as is also the case for BTs [**6**]. We make use of this link in the proposed procedure for constructing conditionally invariant solutions. It consists of the following steps:

1) We first look for a class of solutions of PDEs (2.1) which are invariant under a set of p linearly independent vector fields $Z_1, \dots, Z_p$ of the form (3.22) for which the set of the associated vector fields $\hat{b}_1, \dots, \hat{b}_r$ generates a finite-dimensional Lie algebra $\mathfrak{g}$, i.e.

$$
Z_i = \partial_{x^i} + \sum_{\ell=1}^{r} A_i^\ell(x)\hat{b}_\ell, \quad i = 1, \dots, p, \tag{4.1}
$$

where

$$
\hat{b}_\ell = b_\ell^\alpha(u)\partial_{u^\alpha}, \quad [\hat{b}_\ell, \hat{b}_k] = C_{\ell k}^j \hat{b}_j, \quad \ell = 1, \dots, r. \tag{4.2}
$$

We select a finite-dimensional Lie algebra $\mathfrak{g}$ in such a way that for a given representation of $\mathfrak{g}$ in u variables there exist nontrivial solutions $A^1, \dots, A^r$ of the system (3.21).

2) Having solved the system (3.21) we check if r vector fields $A^1, \dots, A^r$ on X space satisfy the zero curvature conditions (3.8). This leads to two different situations.

3) In the first case, the set of solutions $A^1, \dots, A^r$ of the overdetermined system composed of (3.8) and (3.21) depends on some functions v^α of x, their derivatives and some constants $c_1, \dots, c_n$.

Hence, the conditions (3.4) take the form

$$\frac{\partial u^\alpha}{\partial x^i}(x) = \sum_{\ell=1}^{r} A_i^\ell(v^{(1)}, c_1, \dots, c_n) b_\ell^\alpha(u). \tag{4.3}$$

If the functions v^α satisfy certain PDEs

$$\Delta_2^s(x, v^{(1)}) = 0, \quad s = 1, \dots, q, \tag{4.4}$$

then for any particular solution of (4.4), the system (4.3) defines a specific transformation between the systems (4.4) and (2.1). This means that from each solution of (4.4) we can derive a solution of (2.1) and vice versa. In other words, there exists a BT, determined by (4.3), between the sets of solutions of the system (2.1) and the system (4.4). Such a BT, defined by 1st order differential equations, will be called a *1st order BT.*

In this case the original system (2.1) admits infinitely many compatible constraints (4.3) because they are parametrized by functions v^α satisfying the PDEs (4.4).

Note that, for a chosen solution v^α of (4.4), the symmetry group G of the overdetermined system (2.3) with p-dimensional orbits has a complete set of q functionally independent invariants. So, the solutions u of the initial system (2.1) can be expressed in terms of these invariants. In other words, they are invariant under the finite Abelian Lie algebra L (3.23) parametrized by the functions v^α.

Note also that the system (4.4) is factorizable in the same sense as (3.21). Namely, the equations (4.3) can be solved for $\partial v^\alpha / \partial x^i$ by means of the Implicit Function Theorem. Substitution of those $\partial v^\alpha / \partial x^i$ into the system (4.4) yields the required decomposed form of this system.

4) In the second case, the set of solutions $A^1, \dots, A^r$ of the overdetermined system composed of (3.8) and (3.21) is uniquely defined (up to a choice of some constants). Then the functions ϕ_i^α are also uniquely defined by (3.4), up to a choice of some constants. Hence, a BT does not exist and only particular solutions to (2.1) can be constructed by applying the symmetry reduction technique. In this case we proceed to the integration of the equations (3.4) for given functions $\phi_i^\alpha = \sum_{\ell=1}^r A_i^\ell b_\ell^\alpha(u)$. The resulting solutions are invariant under the vector fields Z_i and constitute particular solutions to the initial system (2.1).

5. Examples of Applications

Now we present several examples of construction of conditionally invariant solutions. Our study concentrates on a class of completely integrable equations and we construct BTs for them. Due to the limited space we present only the main results, without going into details.

EXAMPLE 5.1. The three-wave resonant interaction (3WRI) in 1+1 dimensions can be described by the system of three PDEs of the form

$$v_{\alpha t} + c_{(\alpha)} v_{\alpha x} = i v_\beta^* v_\gamma^*, \tag{5.1}$$

where the indices $\alpha, \beta, \gamma = 1, 2, 3$ are cyclic, $v_\alpha(x,t)$ are the complex amplitudes of the wave packets and c_α denote their group velocities for which we have $c_1 < c_2 < c_3$.

In order to find the auto-BTs for (5.1) it is convenient, from the computational point of view, to perform an embedding transformation of (v_1, v_2, v_3) into two-dimensional space $(\wedge_1, \wedge_2)$ given by a specific Darboux transformation of the form **[12, 15, 7]**

$$v_\alpha = \rho_\alpha + u_\alpha, \quad \alpha = 1, 2, 3, \tag{5.2}$$

where $u_\alpha(x,t)$ satisfy (5.1) and the functions $\rho_\alpha(x,t)$ are such that

$$\begin{aligned} \rho_1 &= -i \wedge_2 \Delta^{-1}, \quad \text{where } \Delta := \lambda |\wedge_1|^2 - |\wedge_2|^2 - \mu \neq 0, \\ \rho_2 &= -i \wedge_1^* \Delta^{-1}, \qquad \lambda = \frac{c_3 - c_2}{c_3 - c_1}, \\ \rho_3 &= i \wedge_2^* \wedge_1 \Delta^{-1}, \qquad \mu = \frac{c_1 - c_2}{c_3 - c_1}. \end{aligned} \tag{5.3}$$

Substituting (5.2) and (5.3) into (5.1) we obtain

$$\begin{aligned} &\wedge_{2t} + c_1 \wedge_{2x} - \wedge_2 \Delta^{-1}(\Delta_t + c_1 \Delta_x) \\ &\qquad = - \wedge_2 \Delta^{-1} |\wedge_1|^2 - iu_3^* \wedge_1 + iu_2^* \wedge_1^* \wedge_2, \\ &\wedge_{1t}^* + c_2 \wedge_{1x}^* - \wedge_1^* \Delta^{-1}(\Delta_t + c_2 \Delta_x) \\ &\qquad = - \wedge_1^* \Delta^{-1} |\wedge_2|^2 + iu_1^* \wedge_1^* \wedge_2 - iu_3^* \wedge_2^*, \\ &\wedge_1(\wedge_{2t}^* + c_3 \wedge_{2x}^*) + \wedge_2^*(\wedge_{1t} + c_3 \wedge_{1x}) - \wedge_1 \wedge_2^* \Delta^{-1}(\Delta_t + c_3 \Delta_x) \\ &\qquad = \wedge_2 |\wedge_1|^2 \Delta^{-1} + iu_1^* \wedge_1 + iu_2^* \wedge_2^*, \end{aligned} \tag{5.4}$$

where

$$\begin{aligned} \Delta_t + c_\alpha \Delta_x = \lambda \big[\wedge_1^*(\wedge_{1t} + c_\alpha \wedge_{1x}) + \wedge_1(\wedge_{1t}^* + c_\alpha \wedge_{1x}^*) \big] \\ - \big[\wedge_2^*(\wedge_{2t} + c_\alpha \wedge_{2x}) + \wedge_2(\wedge_{2t}^* + c_\alpha \wedge_{2x}^*) \big]. \end{aligned}$$

We are going to construct the 2-dimensional Abelian Lie algebra L of conditional symmetries of 3WRI equations, making use of Theorem 3.2. So, we are looking for the solutions of (5.4) which are invariant under the vector fields

$$Z_1 = \partial_x + \sum_{\ell=1}^{r} (A_1^\ell \hat{b}_\ell + A_1^{\ell *} \hat{b}_\ell^*), \quad Z_2 = \partial_t + \sum_{\ell=1}^{r} (A_2^\ell \hat{b}_\ell + A_2^{\ell *} \hat{b}_\ell^*), \tag{5.5}$$

where the vector fields

$$\hat{b}_\ell = \sum_{i=1}^{2} b_{\ell i} \partial_{\wedge_i}, \quad \ell = 1, \dots, r, \tag{5.6}$$

generate a finite-dimensional Lie algebra $\mathfrak{g}$. The functions A_i^ℓ and $b_{\ell i}$ depend on (x,t) and $(\wedge_1, \wedge_1^*, \wedge_2, \wedge_2^*)$, respectively. The invariance requirement means that the equations (5.4) have to be complemented by the conditions

$$\begin{aligned} \wedge_{ix} &= \sum_{\ell=1}^{r} A_1^\ell b_{\ell i}, \quad \wedge_{ix}^* = \sum_{\ell=1}^{r} A_1^{\ell *} b_{\ell i}^*, \quad i = 1, 2, \\ \wedge_{it} &= \sum_{\ell=1}^{r} A_2^\ell b_{\ell i}, \quad \wedge_{it}^* = \sum_{\ell=1}^{r} A_2^{\ell *} b_{\ell i}^*. \end{aligned} \tag{5.7}$$

These conditions imply that $A_i^{\ell *}$ and $b_{\ell i}^*$ are the complex conjugates of A_i^ℓ and $b_{\ell i}$, respectively. Now, according to the procedure described in Section 4, we choose a realization of the Lie algebra $\mathfrak{g}$ through $\mathrm{sl}(3,\mathbb{C})$. It can be represented in terms of vector fields (5.6) of the form [**5**]

$$\begin{aligned} \hat{b}_1 &= \wedge_1^2\partial_{\wedge_1} + \wedge_1\wedge_2\,\partial_{\wedge_2}, & \hat{b}_5 &= \wedge_2\partial_{\wedge_1}, \\ \hat{b}_2 &= \partial_{\wedge_1}, & \hat{b}_6 &= \wedge_1\partial_{\wedge_2}, \\ \hat{b}_3 &= \partial_{\wedge_2}, & \hat{b}_7 &= \wedge_1\partial_{\wedge_1}, \\ \hat{b}_4 &= \wedge_1\wedge_2\,\partial_{\wedge_1} + \wedge_2^2\partial_{\wedge_2}, & \hat{b}_8 &= \wedge_2\partial_{\wedge_2}. \end{aligned} \tag{5.8}$$

The nonzero commutation relations between these operators are

$$\begin{aligned} &[\hat{b}_1,\hat{b}_2] = -2\hat{b}_7 - \hat{b}_8, [\hat{b}_1,\hat{b}_3] = -\hat{b}_6, [\hat{b}_1,\hat{b}_5] = -\hat{b}_4, [\hat{b}_1,\hat{b}_7] = -\hat{b}_1, \\ &[\hat{b}_2,\hat{b}_4] = \hat{b}_5, [\hat{b}_2,\hat{b}_6] = \hat{b}_3, [\hat{b}_2,\hat{b}_7] = \hat{b}_2, [\hat{b}_3,\hat{b}_4] = \hat{b}_7 + 2\hat{b}_8, \\ &[\hat{b}_3,\hat{b}_5] = \hat{b}_2, [\hat{b}_3,\hat{b}_8] = \hat{b}_3, [\hat{b}_4,\hat{b}_6] = -\hat{b}_1, [\hat{b}_4,\hat{b}_8] = -\hat{b}_4, \\ &[\hat{b}_5,\hat{b}_6] = \hat{b}_8 - \hat{b}_7, [\hat{b}_5,\hat{b}_7] = \hat{b}_5, [\hat{b}_5,\hat{b}_8] = -\hat{b}_5, [\hat{b}_6,\hat{b}_7] = -\hat{b}_6, \\ &[\hat{b}_6,\hat{b}_8] = \hat{b}_6. \end{aligned} \tag{5.9}$$

Equations (5.7) can be written as the matrix Riccati equations (MREs), associated with the vector fields (5.8)

(5.10)
$$W_x = WD_1W + WB_1 + C_1W + P_1, \quad W_t = WD_2W + WB_2 + C_2W + P_2,$$

where

$$W := (\wedge_1,\wedge_2)^T, \quad P_i = (A_i^2, A_i^3)^T, \quad B_i = \frac{1}{3}(A_i^7 + A_i^8),$$

$$C_i = \begin{vmatrix} \frac{1}{3}(2A_i^7 - A_i^8) & A_i^5 \\ A_i^6 & \frac{1}{3}(2A_i^8 - A_i^7) \end{vmatrix}, \quad D_i = (A_i^1, A_i^4), \quad i = 1,2.$$

The overdetermined system, composed by adding the zero gauge curvature conditions (3.8) to the functional system obtained by substituting (5.10) into the equations (5.4), possesses nontrivial solutions

$$\begin{aligned} A_1^1 &= i(c_1-c_2)^{-1}u_2, & A_2^1 &= -i(c_1-c_2)^{-1}c_2u_2, \\ A_1^2 &= i(c_1-c_2)^{-1}\frac{\mu}{\lambda}u_2^*, & A_2^2 &= -i(c_1-c_2)^{-1}c_2\frac{\mu}{\lambda}u_2^*, \\ A_1^3 &= -i(c_1-c_2)^{-1}\mu u_1, & A_2^3 &= i(c_1-c_2)^{-1}c_1\mu u_1, \\ A_1^4 &= i(c_1-c_2)^{-1}u_1^*, & A_2^4 &= -i(c_1-c_2)^{-1}c_1u_1^*, \\ A_1^5 &= -i(c_1-c_2)^{-1}\frac{\mu}{\lambda}u_3, & A_2^5 &= i(c_1-c_2)^{-1}\frac{\mu}{\lambda}c_3u_3, \\ A_1^6 &= i(c_1-c_2)^{-1}\mu u_3^*, & A_2^6 &= -i(c_1-c_2)^{-1}\mu c_3u_3^*, \\ A_1^7 &= -k(c_1-c_2)^{-1}\lambda^{-1}, & A_2^7 &= k(c_1-c_2)^{-1}\lambda^{-1}c_2, \\ A_1^8 &= -k(c_1-c_2)^{-1}, & A_2^8 &= k(c_1-c_2)^{-1}c_1, \end{aligned} \tag{5.11}$$

where $u_\alpha(x,t)$ satisfy the 3WRI equations (5.1) and k is an arbitrary complex constant—the Bäcklund parameter.

Next, substituting (5.11) into the system (5.10), we obtain the auto-BT in the form of projective MRE, which is a straight reconstruction of the known result [**12, 15**].

Finally, let us note that there exists a connection between the conditional symmetries for 3WRI (5.1) and the prolongation structure approach of Estabrook and Walquist [**27**]. The auxiliary functions $\wedge_i$ and $\wedge_i^*$ are the pseudopotentials appearing in the prolongation theory.

EXAMPLE 5.2. The 1st order system of PDEs describing constant mean curvature H of surfaces (CMCS), obtained by a generalization of the Weierstrass-Enneper formula, has the form [**13**]

$$u_{1y} - iu_{1x} = 2H\big(|u_1|^2 + |u_2|^2\big)u_2, \quad u_{2y} + iu_{2x} = -2H\big(|u_1|^2 + |u_2|^2\big)u_1. \tag{5.12}$$

Without loss of generality one can assume that $H = 1/2$. For computational purposes it is also useful to introduce the function

$$\rho = \frac{u_1}{u_2^*}. \tag{5.13}$$

The advantage here is that the differential equations for ρ turn out to be a system of coupled Riccati equations. In fact, after the transformation (5.13) the system (5.12) becomes the 2nd order PDE

$$\Delta\rho - \frac{2\rho^*}{1+|\rho|^2}(\nabla\rho)^2 = 0, \tag{5.14}$$

where

$$u_1 = \varepsilon\rho\frac{(\rho_y^* + i\rho_x^*)^{1/2}}{1+|\rho|^2} \quad \text{and} \quad u_2 = \varepsilon\frac{(\rho_y - i\rho_x)^{1/2}}{1+|\rho|^2}, \qquad \varepsilon = \pm 1. \tag{5.15}$$

Making use of stereographic projection

$$S = \begin{vmatrix} \frac{1-|\rho|^2}{1+|\rho|^2} & \frac{2\rho^*}{1+|\rho|^2} \\ \frac{2\rho}{1+|\rho|^2} & -\frac{1-|\rho|^2}{1+|\rho|^2} \end{vmatrix},$$

we can rewrite equation (5.14) in the equivalent form of a static σ-model in Euclidean space

$$[S, \partial\bar{\partial}S] = 0 \text{ where } \partial = \partial_x + i\partial_y \quad \text{and} \quad \bar{\partial} = \partial_x - i\partial_y. \tag{5.16}$$

A Bäcklund transformation can be established, via $\mathrm{sl}(2,\mathbb{C})$ realization, since it is the only finite-dimensional Lie algebra which can be constructed in terms of vector fields (3.5) in one complex variable ρ. The solution of the system composed of (3.8) and (3.21) in this case yields a BT in the Riccati form [**8**]

$$\rho_x = \frac{\lambda}{v_o - v}\eta_y\rho^2 + \frac{v_x}{v_o - v}\rho, \quad \rho_y = \frac{-\lambda}{v_o - v}\eta_x\rho^2 + \frac{v_y}{v_o - v}\rho, \tag{5.17}$$

where the functions η and v satisfy

$$\Delta\eta + \frac{2}{v_o - v}(\nabla\eta \mid \nabla v) = 0 \quad \text{and} \quad \Delta v = 0, \tag{5.18}$$

respectively. The bracket $(\cdot \mid \cdot)$ denotes the scalar product. The quantities λ and v_o are arbitrary complex parameters. Hence, from equations (5.15) we find

$$\begin{aligned} u_1 &= \frac{\varepsilon\rho}{(1+|\rho|^2)(v_o - v)}\left[-\lambda(\eta_x + i\eta_y)\rho + v_y - iv_x\right], \\ u_2 &= \frac{\varepsilon|\rho|^2}{(1+|\rho|^2)(v_o^* - v^*)}\left[-\lambda^*(\eta_x^* - i\eta_y^*)\rho^* + v_y^* + iv_x^*\right]. \end{aligned} \tag{5.19}$$

In particular, when $\rho = \alpha z^n$ $\alpha \in \mathbb{R}$, the solutions of (5.17) and, consequently, of (5.12) have the form of an algebraic soliton (bump) and a singular soliton

$$\begin{aligned} u_1 &= \frac{\varepsilon e^{i(\pi/4+m\theta)}(2m+1)^{1/2}\alpha^{1/2}|z|^{3m}}{1+\alpha^2|z|^{2(2m+1)}}, \qquad m \in \mathbb{Z}, \\ u_2 &= \frac{\varepsilon e^{i(\pi/4+m\theta)}(2m+1)^{1/2}\alpha^{1/2}|z|^{m}}{1+\alpha^2|z|^{2(2m+1)}}, \end{aligned} \tag{5.20}$$

and

$$\begin{aligned} u_1 &= \frac{\varepsilon e^{i/2(\pi/2+(2m-1)\theta)}(2m\alpha)^{1/2}|z|^{3(m-1/2)}}{1+\alpha^2|z|^{4m}}, \\ u_2 &= \frac{\varepsilon e^{i/2(\pi/2+(2m-1)\theta)}(2m\alpha)^{1/2}|z|^{m-1/2}}{1+\alpha^2|z|^{4m}}, \end{aligned} \tag{5.21}$$

respectively.

EXAMPLE 5.3. Let us consider now the generalized Liouville equation (GLE) in $(n+1)$ dimensions. It was formulated by P. Santini and has the form [**25**]

$$\frac{\partial}{\partial x_o}\left(\sum_{j=1}^{n}\left[u_{x_j x_j} - \frac{1}{2}(u_{x_j})^2\right]\right) = 0. \tag{5.22}$$

The analysis of this equation from the point of view of CSM was performed by P. Santini and one of the authors [**10**] and we present below its main results. According to Theorem 3.2 we look for the solutions of (5.22) invariant under an Abelian Lie algebra L spanned by $(n+1)$ vector fields

$$Z_\mu = \partial_{x_\mu} + \sum_{\ell=0}^{r} A_\mu^\ell(x)\hat{b}_\ell, \quad \mu = 0, 1, \ldots, n, \quad x \in \mathbb{R}^{n+1}, \tag{5.23}$$

where

$$\hat{b}_\ell = b_\ell(u)\partial_u, \quad \ell = 0, 1, \ldots, r.$$

The assumed realization of the finite-dimensional Lie algebra $\mathfrak{g}$ which can be constructed in terms of vector fields (3.5) in one real variable u is sl$(2,\mathbb{R})$ (see S. Lie [**17**]) and is given by the operators

$$\hat{b}_o = \frac{d}{du}, \quad \hat{b}_1 = u\frac{d}{du}, \quad \hat{b}_2 = u^2\frac{d}{du}. \tag{5.24}$$

So, (5.23) means that equation (5.22) is supplemented by the requirement that the characteristics of the vector fields Z_μ are equal to zero, i.e.

$$\frac{\partial u}{\partial x_\mu}(x) = A_\mu^o(x) + A_\mu^1(x)u + A_\mu^2(x)u^2. \tag{5.25}$$

Equations (5.25) become a matrix Riccati system which can be expected to possess the Painlevé property [**19**].

The functions A^ℓ_μ have to satisfy the overdetermined system obtained by adding the zero gauge curvature conditions (3.8), i.e.

$$A^\ell_{\mu,\nu} - A^\ell_{\nu,\mu} = C^\ell_{ab}(A^a_\mu A^b_\nu - A^a_\nu A^b_\mu), \text{ for } \ell, a < b = 0,1,2; \quad \mu,\nu = 0,1,\dots,n, \tag{5.26}$$

(where C^ℓ_{ab} are structure constants of $\mathrm{sl}(2,\mathbb{R})$), to the equation (3.21). Substituting (5.25) into (5.22) and then requiring that the coefficients of the successive powers of u in the obtained equations vanish, we obtain a system of 3 equations for the unknown functions A^ℓ_μ. These equations are subjected to conditions (5.26) and have nontrivial solutions. It appears that the functions A^ℓ_μ are parametrized by some functions v and w. The function w solves some linear differential equation, while v solves the same GLE (5.22). After the transformation $Y = \exp 1/2(u-v)$, the DCs (5.25) become a 1st order auto-BT for GLE

$$Y_{x_\mu} = \frac{1}{2} w_{x_\mu} Y^2, \tag{5.27}$$

where the function w satisfies

$$\sum_{j=1}^{n} (w_{x_j x_j} - v_{x_j} w_{x_j}) = 0. \tag{5.28}$$

An explicit solution of (5.27) in $(2+1)$ dimensions can be easily constructed from the constant solution v, i.e.

$$u = 2\ln\left| f(x_1 + ix_2, x_o) + g(x_1 - ix_2, x_o)\right| + h(x_o), \tag{5.29}$$

where f and g are arbitrary differentiable functions of their arguments and h is an arbitrary differentiable function of x_o. In the case when the function w has the form

$$w = w_o(x_o) + \sum_{j=1}^{n} w_j(x_j),$$

the following multi-soliton solution can be constructed in $(n+1)$ dimensions by successive applications of the auto-BT (5.27)

$$u = \ln\left| \frac{\dot F_o(x_o)\prod_{j=1}^n \dot F_j(x_j)}{\left(c_o G(x_o) + \sum_{j=1}^n c_j F_j(x_j)\right)^2}\right|, \tag{5.30}$$

where F_o, F_j and G are arbitrary functions of their arguments and a dot denotes the derivatives, i.e. $\dot F = dF/ds$.

Finally, let us consider the GL equation (5.22) on a $p = n+1$-dimensional ($n \geq 2$) pseudo-Riemannian manifold $\big(X, (\cdot \mid \cdot)\big)$, where the metric $(\cdot \mid \cdot)$ is nondegenerate but not necessarily positive definite. Then, for the function u (of the class C^3), defined on some neighborhood $\mathcal{D} \subset X$, the equation (5.22) takes the form

$$\frac{\partial}{\partial x_o}\left(\nabla^2 u - \frac{1}{2}(\nabla u \mid \nabla u)\right) = 0, \tag{5.31}$$

where ∇u denotes the gradient of the function u and ∇^2 is the Laplace-Beltrami operator with respect to the metric $(\cdot \mid \cdot)$. The 1st order auto-BT (5.27) for the equation (5.31) has the same form as in the case of a positively defined metric (i.e.

when X is an Euclidean space). Just as before, the function w has to satisfy the linear differential equation of the form

$$\nabla^2 w - \big(\nabla v \mid \nabla w\big) = 0. \tag{5.32}$$

Suppose now that the manifold X is a $(2+1)$-dimensional Minkowski space with Cartesian coordinates (x_o, x_1, x_2) such that $|x|^2 = x_o^2 - x_1^2 - x_2^2$. Then an explicit solution of (5.31) written in polar coordinates has the form

$$u = -2\ln\big|A(x_0)r^{-1/2}F(x_o - r, x_o)\big|, \tag{5.33}$$

where $x_1 = r\cos\varphi$, $x_2 = r\sin\varphi$, F is a real function of two variables, and A is an arbitrary function of one variable.

6. The Higher Order Differential Constraints

We now consider the possibility of constructing conditionally invariant solutions based on DCs containing derivatives of order higher than one. Here we adapt the results of Sections 2 and 3 to the case of a single kth order PDE admitting a certain class of compatible 2nd order DCs. Consider a scalar kth order PDE in $(p+1)$ independent variables $x = (\bar{x}, x_o) \in X \subset \mathbb{R}^{p+1}$, of the form

$$\Delta(x, u^{(k)}) = 0. \tag{6.1}$$

We postulate that the system (6.1) can be (possibly after a point transformation) developed into a finite power series in the variables u^α and in all their derivatives up to the kth order, with coefficients depending on x. We assume also that (6.1) admits a $(p+1)$-dimensional Lie module $\mathcal{A}$ of conditional symmetries with generators (defined on $J^o = X \times U^2$)

$$\begin{aligned} X_a &= \partial_{x_a} + \gamma_a \partial_u + \lambda_a \partial_v, \quad a = 1, \dots, p \\ X_o &= \partial_{x_o} + v\partial_u + \phi\partial_v, \end{aligned} \tag{6.2}$$

such that

$$[X_\mu, X_\nu] = \tau^\sigma_{\mu\nu} X_\sigma, \quad 0 \le \mu, \nu, \sigma \le p,$$

where we introduce a new real valued function $v\colon D \to U(D \subset X)$. We assume that the functions $\phi, \gamma_a, \lambda_a$ and $\tau^\sigma_{\mu\nu}$ depend on x, u and v only. We limit our considerations to the case when the initial PDE (6.1) is subjected to DCs, obtained by demanding that the characteristics of the vector fields X_μ are equal to zero

$$\begin{aligned} Q_o(x, u^{(1)}, v^{(1)}) &\equiv (u_o - v, v_o - \phi) = (0,0), \\ Q_a(x, u^{(1)}, v^{(1)}) &\equiv (u_a - \gamma_a, v_a - \lambda_a) = (0,0). \end{aligned} \tag{6.3}$$

Let us denote the overdetermined system composed of (6.1) and (6.3) by R. The elimination of the auxiliary variable v from the DCs (6.3) reduces the system R to the overdetermined system of PDEs for function u only, with 1st and 2nd order DCs, i.e.

$$\Delta(x, u^{(k)}) = 0, \tag{6.4a}$$

$$\begin{aligned} u_{oo} - \phi(x, u, u_o) = 0, \quad u_{oa} - \lambda_a(x, u, u_o) = 0, \\ u_a - \gamma_a(x, u, u_o) = 0. \end{aligned} \tag{6.4b}$$

The compatibility conditions for (6.4) give us

$$\begin{aligned} \gamma_{a,o} + u_o\gamma_{a,u} + \phi\gamma_{a,u_o} - \lambda_a &= 0, \\ \lambda_{a,o} + u_o\lambda_{a,u} + \phi\lambda_{a,u_o} - (\phi,_a + \gamma_a\phi,_u + \lambda_a\phi,_{u_o}) &= 0, \\ \gamma_{[b,a]} + \gamma_{[a}\gamma_{b],u} + \lambda_{[a}\gamma_{b],u_o} &= 0, \\ \lambda_{[b,a]} + \gamma_{[a}\lambda_{b],u} + \lambda_{[a}\lambda_{b],u_o} &= 0, \quad a,b = 1,\dots,p \end{aligned} \tag{6.5}$$

and

$$\Delta(x, u, u_o, \phi^{(k-2)}, \lambda_a^{(k-2)}, \gamma_a^{(k-1)}) = 0, \tag{6.6}$$

where we use the modified total derivative operators

$$\begin{aligned} X_o &= \partial_{x_o} + u_o\partial_u + \phi(x,u,u_o)\partial_{u_o}, \\ X_a &= \partial_{x_a} + \gamma_a(x,u,u_o)\partial_u + \lambda_a(x,u,u_o)\partial_{u_o}, \end{aligned} \tag{6.7}$$

instead of (6.2)

If the conditions (6.5) are satisfied, then the commutators of X_μ, X_ν are equal to zero. This means that the vector fields $\{X_o, X_1, \dots, X_p\}$ form a $(p+1)$-dimensional Abelian distribution defined on $X \times U^{(1)}$. The conditions (6.5) and (6.6) ensure that the obtained solutions of (6.4) constitute the solutions of the original PDE (6.1)

The theorem on the existence of conditional symmetries for the case of higher order DCs takes the following form.

THEOREM 6.1. *A nondegenerate kth order PDE* (6.1) *admits a p-dimensional conditional symmetry algebra L iff the* $(p+1)$ *linearly independent vector fields (defined on* $\mathcal{B} \subset X \times U^{(1)}$*)*

$$X_o = \partial_{x_o} + u_o\partial_u + \phi(x,u,u_o)\partial_{u_o}, \quad X_a = \partial_{x_a} + \gamma_a(x,u,u_o)\partial_u + \lambda_a(x,u,u_o), \partial_{u_o}$$

satisfy conditions (6.5) *and* (6.6) *on some neighborhood of* $(x_o, u_o^{(1)}) \in \mathcal{B}$.

The proof of this theorem proceeds in a way similar to the proof of Theorem 2.4. The generators X_μ constitute conditional symmetries for (6.1) if they are Lie point symmetries of the overdetermined system R. It is easy to check that under the conditions (6.5) the infinitesimal symmetry criterion for the system R is identically satisfied, i.e.

$$\mathrm{pr}^{(k)} X_\mu(\Delta) \equiv 0 \quad \text{and} \quad \mathrm{pr}^{(1)} X_\mu(Q_\nu) \equiv 0, \tag{6.8}$$

whenever $\Delta = 0$ and $Q_\nu = 0, u, \nu = 0, 1, \dots, p$.

We end this section with a simple example to illustrate the feasibility of the presented approach. Further theoretical elaboration on this subject can be found in [**1**].

EXAMPLE 6.2. We consider now the nonlinear parabolic evolution equation which arises in the study of heat propagation, i.e. the Galaktionov model for combustion [**4**]

$$u_t = u_{xx} + u_x^2 + u^2. \tag{6.9}$$

We look for a class of solutions of (6.9) which are invariat under the vector fields

$$Z_1 = \partial_x + v\partial_u, \quad Z_2 = \partial_t + \gamma\partial_u + \lambda\partial_v, \tag{6.10}$$

where we assume that γ and λ are some functions of x, t, u and v. We require that the characteristics of these vector fields are equal to zero

$$Q_1 = (v - u_x) = 0, \quad Q_2 = (\gamma - u_t, \lambda - v_t) = (0,0). \tag{6.11}$$

By adding the conditions (6.11) to the initial equation (6.9) we obtain an overdetermined system of the form

$$u_{xx} = \gamma - (v^2 + u^2), \quad u_x = v, \quad u_t = \gamma(x,t,u,v), \quad v_t = \lambda(x,t,u,v), \tag{6.12}$$

where the functions γ and λ can be determined from the compatibility conditions. The differential operators (6.10) form an Abelian distribution if

$$\begin{cases} \gamma_x + v\gamma_u - \lambda = 0, \\ \lambda_x + v\lambda_u = 0. \end{cases} \tag{6.13}$$

We eliminate λ from (6.13) and get a differential equation for γ, i.e.

$$\gamma_{xx} + v(2\gamma_{xu} + v\gamma_{uu}) = 0. \tag{6.14}$$

We assume that γ is analytic in the variables u and v and it can be represented in the form of a finite power series in u and v with nonconstant coefficients. We substitute this polynomial form of γ into (6.14) and then require that the coefficients of the successive powers of u and v in the obtained equation vanish. This leads to several cases. Let us consider one of them, when $\gamma_x = \gamma_v = 0$. Then the general solution of (6.13) is given by

$$\gamma = \alpha(t)u + \beta(t), \quad \lambda = \alpha(t)v, \tag{6.15}$$

where α and β are arbitrary functions of t. Substituting (6.15) into (6.12) and eliminating v from the obtained equations we get

$$u_{xx} = \alpha(t)u + \beta(t) - (u_x^2 + u^2), \tag{6.16a}$$

$$u_t = \alpha(t)u + \beta(t), \tag{6.16b}$$

$$u_{xt} = \alpha(t)u_x. \tag{6.16c}$$

Note that equation (6.16c) is a differential consequence of (6.16b). Equations (6.16a) and (6.16b) have the Painlevé property and can be solved by successive integrations, since the compatibility conditions for (6.16) are identically satisfied. The general solution of the system (6.16) has the form

$$u(x,t) = \alpha(t)[A\cos x + B\sin x + C] + \beta(t), \tag{6.17}$$

where the functions α and β satisfy the following ODEs

$$\begin{aligned} \frac{d\alpha}{dt} &= 2\alpha\beta - \alpha^2 + 2C\alpha, \quad A, B, C \in \mathbb{R}, \\ \frac{d\beta}{dt} &= \beta^2 + (A^2 + B^2 + C^2)\alpha^2 + 2C\alpha\beta, \end{aligned} \tag{6.18}$$

respectively. The solution (6.17) constitutes a conditionally invariant solution of (6.9). It is interesting to note that DCs (6.16) are equivalent to the "separation Ansatz"

$$u(x,t) = \alpha(t)\theta(x) + \beta(t), \tag{6.19}$$

proposed by P. Olver [**21**] for the equation (6.9). The 3rd order DC

$$u_x u_{xxt} - u_{xx} u_{xt} = 0, \tag{6.20}$$

corresponding to this Ansatz can be directly obtained by eliminating the functions α and β from (6.16). Indeed, after differentiating (6.16b) with respect to x and making use of (6.16a) and (6.16c) we obtain (6.20).

7. Final Remarks

The conditional symmetry method described in this paper is, obviously, less general than the nonclassical method (NM) requiring only one DC compatible with the initial system (2.1). However, introduction of multiple constraints satisfying several specific conditions (i.e. (3.4), (3.6), (3.8) and (3.21)) enables us to overcome the main problem of NM, which severely hinders its applications, namely the difficulties of solving the nonlinear determining equations (2.14) for the initial system (2.1). The proposed approach simplifies this task by providing us with an almost entirely algorithmic procedure. It is, of course, especially valuable in the case of multidimensional systems with an arbitrary number of independent and dependent variables. The most important advantage of this method is that it gives us effective tools for constructing certain classes of BTs in a systematic way. It is much simpler and faster than the traditional methods and its effectiveness has been demonstrated by the results, both reconstructed and new, obtained in Sections 5 and 6, as well as in [**9**].

From the point of view of possible applications it is important, especially given the quite restrictive form of conditions (3.8) and (3.21), to find a criterion for identifying the classes of initial systems which can be treated by CSM. We look for a class of equations (2.1) for which equations (3.8) and (3.21) form an involutive system for a given realization of a finite-dimensional Lie algebra $\mathfrak{g}$. This means that we fix linearly independent vector fields $\hat{b}_\ell, \ell = 1, \dots, r$ and so, the functions b_ℓ^α become given functions of u. Next, we choose the functions $A_i^\ell(x)$ in such a way that there exist nontrivial solutions of this involutive system. The Cartan-Kuranishi Theorem [**3, 14**] ensures that, under certain regularity requirements, the integrability conditions for (3.8) and (3.21) can be found after a finite number of steps. They give us specific differential conditions on the class of initial equations (2.1), thus providing us with the selection criteria for the systems (3.4) admitting SF. From the Cartan-Kuranishi Theorem it follows that if the system composed of (3.8) and (3.21) is in involution, then the overdetermined system (2.3) can be reduced to the sequence of Cauchy-Kovalevska equations which, after successive integrations, lead us to the solution of the problem with a proper degree of freedom.

Finally, it is worth noticing that the CSM can be used "in reverse" to address certain physical problems. Namely, we sometimes know the superposition rules for some functions in the physical systems for which analytical models are not fully developed. Using our approach we can select appropriate finite Lie algebras (3.6) corresponding to the given SF and characterize the class of equations describing the physical phenomena in question. This was attempted successfully for systems of ODEs [**26**] but not, to our knowledge, for PDEs.

Acknowledgements

The authors thank P. Olver and P. Winternitz for helpfull and interesting discussions on the topic of this paper.

The research reported in this paper was partly supported by research grants from NSERC of Canada, the "Fonds FCAR du Gouvernement du Québec" and the INFN of Italy.

References

1. I. Ayari and A. M. Grundland, *Conditionally invariant solutions*, (in preparation).
2. G. W. Bluman and J. D. Cole, *The general similarity solution of the heat equation*, J. Math. Mech. **18** (1969), 1025–1042.
3. E. Cartan, *Sur la structure des groupes infinis de transformations.* Chapitre I: *Les systèmes différentiels en involution*, Gauthier-Villars, Paris, 1953.
4. V. A. Galaktionov, *On new exact blow-up solutions for nonlinear heat conduction equations with source and applications*, Differential Integral Equations **3** (1990), no. 5, 863–874.
5. A. González-López, N. Kamran, and P. J. Olver, *Real Lie algebras of differential operators and quasi-exactly solvable potentials*, Philos. Trans. Roy. Soc. London Ser. A **354** (1996), no. 1710, 1165–1193.
6. E. Goursat, *Sur quelques transformations des équations aux dérivées partielles du second ordre*, Annales de la faculté des sciences de Toulouse, Série 2 (1902), Tome IV, 299–340.
7. A. M. Grundland and L. Martina, *The conditional symmetries for the three-wave resonant interaction*, (in preparation).
8. A. M. Grundland and L. Martina, *Constant mean curvature and completely integrable σ-model*, Lett. Math. Phys. (1996), (submitted for publication).
9. A. M. Grundland and G. Rideau, *Conditional symmetries for* 1st *order systems of* PDEs *in the context of the Clairin method*, Modern Group Theoretical Methods in Physics (J. Bertrand et al., eds.), Kluwer Academic Publishers, Netherlands (1995), 167–178.
10. A. M. Grundland and P. Santini, *The conditional symmetries for the generalized Liouville equation*, (in preparation).
11. A. M. Grundland and J. Tafel, *On the existence of nonclassical symmetries of partial differential equations*, J. Math. Phys. **36** (1995), no. 3, 1426–1434.
12. D. J. Kaup, *The lump solutions and the Bäcklund transformation for the three-dimensional three-wave resonant interaction*, J. Math. Phys. **22** (1981), no. 6, 1176–1181.
13. B. G. Konopelchenko and I. A. Taimanov, *Constant mean curvature surfaces via integrable dynamical system*, J. Phys. A **29** (1996), no. 6, 1261–1265.
14. M. Kuranischi, *Lectures on involutive systems of partial differential equations*, Publ. Brasil. Math. Soc., Sao Paulo, 1967.
15. M. Leo, R. A. Leo, G. Soliani, and L. Martina, *Prolongation theory of the three-wave resonant interaction*, Nuovo Cimento **88** (1985), no. 2, 81–101.
16. D. Levi and P. Winternitz, *Nonclassical symmetry reduction: Example of the Boussinesq equation*, J. Phys. A **22** (1989), no. 15, 2915–2924.
17. S. Lie and F. Engel, *Theorie der transformationgruppen*, B. G. Teubner, Leipzig, vol. 1 (1888); vol. 2 (1890), vol. 3 (1893); reprinted by Chelsea Publ. Comp., New York, 1967.
18. S. Lie and G. Scheffers, *Vorlesungen über continuierliche Gruppen mit Geometrischen und anderen Anwendungen*, B. G. Teubner, Leipzig, 1893; reprinted by Chelsea Publishing Co., New York, 1971.
19. M. Musette and R. Conte, *Algorithmic method for deriving Lax pairs from the invariant Painlevé analysis of nonlinear differential equations*, J. Math. Phys. **32** (1991), no. 6, 1450–1457.
20. P. J. Olver, *Applications of Lie groups to differential equations*, Graduate Texts in Math. **107**, Springer-Verlag, New York, 1986.
21. ______, *Direct reduction and differential constraints*, Proc. Roy. Soc. London Ser. A **444** (1994), no. 1922, 509–523.
22. P. J. Olver and Ph. Rosenau, *Group-invariant solutions of differential equations*, SIAM J. Appl. Math. **47** (1987), no. 2, 263–278.
23. P. J. Olver and E. M. Vorobev, *Nonclassical and conditional symmetries*, CRC Handbook of Lie Group Analysis (N. H. Ibragimov, ed.), CRC press, vol. 3, chapt. XI, London, 1995.
24. E. Pucci and G. Saccomandi, *On the weak symmetry groups of partial differential equations*, J. Math. Anal. Appl. **163** (1992), no. 2, 588–598.

25. P. Santini, *Linear theories, hidden variables and integrable nonlinear equations*, Inverse Problems (1996), (submitted for publication).
26. S. Shnider and P. Winternitz, *Classification of systems of nonlinear ordinary differential equations with superposition principles*, J. Math. Phys. **25** (1984), no. 11, 3155–3165.
27. H. D. Wahlquist and F. B. Estabrook, *Prolongation structures of nonlinear evolution equations*, J. Math. Phys. **16** (1975), no. 1, 1–7.

CENTRE DE RECHERCHES MATHÉMATIQUES, UNIVERSITÉ DE MONTRÉAL, C.P. 6128, SUCC. CENTRE-VILLE, MONTRÉAL (QUÉBEC) H3C 3J7, CANADA
E-mail address: grundlan@crm.umontreal.ca

DIPARTIMENTO DI FISICA DELL'UNIVERSITA AND INFN-SEZIONE DI LECCE, 73100 LECCE, ITALY
E-mail address: martina@le.infn.it

LABORATOIRE DE PHYSIQUE THÉORIQUE ET MATHÉMATIQUE, UNIVERSITÉ PARIS 7, DENIS DIDEROT, 2 PLACE JUSSIEU, F-75251, PARIS, CEDEX 05, FRANCE
E-mail address: rideau@ccr.jussieu.fr

Centre de Recherches Mathématiques
CRM Proceedings and Lecture Notes
Volume **11**, 1997

Bispectral Operators of Rank 1 and Dual Isomonodromic Deformations

J. Harnad

ABSTRACT. A comparison is made between bispectral operator pairs and dual pairs of isomonodromic deformation equations. Through examples, it is shown how operators belonging to rank one bispectral algebras may be viewed equivalently as defining 1-parameter families of rational first order differential operators with matricial coefficients on the Riemann sphere, whose monodromy is trivial. By interchanging the rôles of the two variables entering in the bispectral pair, a second 1-parameter family of operators with trivial monodromy is obtained, which may be viewed as the dual isomonodromic deformation system.

1. Bispectral Operators

1.1. Bispectral pairs. Consider pairs $L(x, \partial_x)$, $\Lambda(z, \partial_z)$ of differential operators in two variables x and z, for which there exists a function $\psi(x, z)$ that is simultaneously a parametric family of eigenfunctions of L and Λ.

$$L\psi(x,z) = f(z)\psi(x,z), \tag{1.1a}$$

$$\Lambda\psi(x,z) = \phi(x)\psi(x,z), \tag{1.1b}$$

where the eigenvalues $f(z)$ and $\phi(x)$ are nonconstant functions of the variables z and x. Such pairs of operators were named *bispectral* and studied by Duistermaat and Grünbaum in [**DG**]. Choosing a normalization in which both L and Λ have unit leading coefficients and constant next to leading coefficients (which is always possible, up to reparametrization), they were able to draw a number of conclusions about these operators. In particular, the coefficients of L and Λ are rational in the variables x and z, respectively, and the functions $f(x)$ and $\phi(z)$ are polynomials. Furthermore, for the case of second order operators:

$$L = \frac{d^2}{dx^2} + u(x) \tag{1.2}$$

1991 *Mathematics Subject Classification.* Primary: 32G34 ; Secondary: 58F07.

Research supported in part by the Natural Sciences and Engineering Research Council of Canada and the Fonds FCAR du Québec.

This is the final form of the paper.

[**DG**] were able to determine all possible $u(x)$; namely (up to translations in x or addition of a constant),

$$u(x) = x \quad \text{(Airy)} \quad \text{or} \quad u(x) = \frac{c}{x^2} \quad \text{(Bessel)} \tag{1.3}$$

or anything that can be obtained from the two cases

$$u(x) = 0, \quad u(x) = -\frac{1}{4x^2} \tag{1.4}$$

through the application of rational Darboux transformations.

1.2. Bispectral algebras of rank 1. Wilson [**W1,W2**] considered the case when L is embedded in a commutative algebra $\mathcal{A}$ of differential operators, sharing the same family of eigenfunctions $\psi(x,z)$, such that the orders of the elements of $\mathcal{A}$ are relatively prime (a rank 1 algebra). Such algebras can be characterized by the associated spectral data, consisting of an algebraic curve, the spectral curve of $\mathcal{A}$, denoted $\mathrm{spec}(\mathcal{A})$ and, in general, a line bundle whose fibres are the joint eigenvectors. This data may be viewed as determining a point $W \in \mathrm{Gr}$ in the Hilbert space Grassmannian of Sato [**SS**] and Segal-Wilson [**SW**], such that the bispectral wave function $\psi(x,z)$ is the corresponding Baker-Akhiezer function $\psi_W(\mathbf{t},z)$ at the point $(t_1 = x, t_i = 0, i \geq 2)$. We recall that the Baker-Akhiezer function is the unique function of the form

$$\psi_W(\mathbf{t},z) = \gamma(\mathbf{t})\left(1 + \sum_{i=1}^{\infty} \frac{b_i(\mathbf{t})}{z^i}\right), \tag{1.10}$$

where $\gamma(\mathbf{t}) := e^{\sum_{j=1}^{\infty} t_j z^j}$, taking values in the subspace W of the Hilbert space $\mathcal{H} := L^2(S^1, \mathbf{C})$ of square integrable functions on the unit circle in the complex z-plane. Here $\mathbf{t} = (t_1, t_2, \dots)$ is an infinite component vector and $\gamma(\mathbf{t})$ may be viewed as an element of the infinite abelian group Γ_+ consisting of elements of $L^2(S^1, \mathbf{C})$ admitting a holomorphic continuation to the interior of S^1 in the complex z-plane and taking value 1 at the origin. The Baker-Akhiezer function may, in turn, be expressed in terms of the Sato tau function associated to W, by the formula [**DJKM**]

$$\psi_W(\mathbf{t},z) = \gamma(\mathbf{t}) \frac{\tau_W(\mathbf{t} - [\mathbf{z}])}{\tau_W(\mathbf{t})}, \tag{1.11}$$

where the components of $[\mathbf{z}] = (z_1, z_2, \dots)$ are

$$z_k := \frac{1}{kz^k}. \tag{1.12}$$

Geometrically, τ_W is understood as the determinant, defined up to normalization, of the projection $P_+ : \gamma W \to \mathcal{H}_+$, where γW is the image under $\gamma \in \Gamma_+$ of the subspace $W \subset \mathcal{H}$ and $\mathcal{H}_+ \subset \mathcal{H}$ is the subspace consisting of elements admitting a holomorphic extension to the exterior of S^1.

Wilson [**W1**] gave the following equivalent characterizations of such W's corresponding to bispectral algebras of rank 1.

THEOREM (Wilson [**W1**]). *The algebra $\mathcal{A}$ is bispectral and of rank* 1 *if and only if the following equivalent conditions hold*:
(i) spec($\mathcal{A}$) is a rational algebraic curve with only cusp singularities.
(ii) Within the proper normalization, the Baker-Akhiezer function

$$\psi(x,z) = \psi_W(x = t_1, t_j = 0, j \geq 2, z) = e^{xz}\left(1 + \sum_{i=1}^{\infty} \frac{b_i(x)}{z^i}\right) \tag{1.13}$$

belongs to a plane W belonging to a subvariety $\mathrm{Gr}^{\mathrm{ad}}$ of the rational Segal-Wilson Grassmannian $\mathrm{Gr}^{\mathrm{rat}}$, called the adelic Grassmannian (defined below).

The adelic Grassmannian $\mathrm{Gr}^{\mathrm{ad}}$ is shown in [**W1**] to consist of the W-planes in the Hilbert space $\mathcal{H}$ of the form

$$W = \frac{1}{q(z)} W_+^{\mathbf{c}} \tag{1.14}$$

where $q(z)$ is a polynomial of suitable degree, rendering the projection $P_+ : W \rightarrow \mathcal{H}_+$ to the subspace $\mathcal{H}_+ \subset \mathcal{H}$ a Fredholm operator of index 0, and the subspace $W_+^{\mathbf{c}} \subset \mathcal{H}_+$ is the intersection of the kernels of a finite number of linear forms $\{c_i^\mu \in \mathcal{H}_+^*,\}$, each of which has support at one point λ_i in the complex z-plane. That is, each c_i^μ determines a *condition* of the form

$$c_i^\mu(g) = \sum_{a=1}^{m_i^\mu} c_{ia}^\mu g^{(a)}(\lambda_i) = 0, \tag{1.15}$$

where $g^{(a)}$ denotes the a-th derivative of $g \in \mathcal{H}_+$ and $\{c_{ia}^\mu\}$ is a finite set of coefficients. The λ_i's are just the roots of the polynomial $q(z)$, with multiplicities coinciding with the number of conditions localized at that root.

Wilson also showed that the property of bispectrality could be viewed as a consequence of the existence of a "bispectral involution"

$$b: \mathrm{Gr}^{\mathrm{ad}} \rightarrow \mathrm{Gr}^{\mathrm{ad}}, \quad b: W \mapsto W' \tag{1.16}$$

such that

$$\psi_{W'}(x,z) = \psi_W(z,x). \tag{1.17}$$

The larger, *rational* Grassmannian $\mathrm{Gr}^{\mathrm{rat}}$ contains all rational and soliton solutions to the KP hierarchy. It was shown by Krichever [**Kr**], following earlier work of Airault, Mckean and Moser [**AMM**] on rational solutions to the KdV equation, that rational solutions of the KP equation which tend to zero as $x \rightarrow \infty$ can be expressed as

$$u(x,y,t) = -\sum_{i=1}^{n} \frac{2}{(x - x_i(y,t))^2}, \tag{1.18}$$

where the location of the poles $x_i(y,t)$ is determined by the fact that they satisfy the equations of the first two commuting flows of the Calogero-Moser system [**M**]. It follows that these are associated to bispectral tau functions τ_W. In [**W2**], Wilson showed that, by suitably completing the complexified Calogero-Moser phase space so as to allow for particle collisions, this extends to a complete parametrization of the adelic Grassmannian $\mathrm{Gr}^{\mathrm{ad}}$, and hence the set of rank 1 bispectral algebras. More precisely, let C_n denote the completed n-particle rational Calogero-Moser phase space, defined as the space of pairs (X, Z) of complex $n \times n$ matrices whose commutator $[Z, X]$ is a rank 1 perturbation of the identity matrix $\mathbf{I}$, quotiented by the natural action of the general linear group $\mathfrak{Gl}(n, \mathbf{C})$ on such pairs;

$$g\colon (X, Z) \mapsto (gXg^{-1}, gZg^{-1}), \quad g \in \mathfrak{Gl}(n, \mathbf{C}). \tag{1.19}$$

Define a map from the union of the spaces C_n to the adelic Grassmannian

$$\mathcal{W}\colon \cup_n C_n \to \mathrm{Gr}^{\mathrm{ad}}, \quad \mathcal{W}\colon \big[(X, Z)\big] \mapsto W(X, Z), \tag{1.20}$$

where $W(X, Z) \in \mathrm{Gr}^{\mathrm{ad}}$ is determined by the following formula for the corresponding Baker function at $\mathbf{t} = (x, 0, \dots)$.

$$\psi_{W(X,Z)}(x, z) = e^{xz} \det\big(\mathbf{I} - (X + x\mathbf{I})^{-1}(Z + z\mathbf{I})^{-1}\big). \tag{1.21}$$

(Note that $\psi_W(x, z)$ and its derivatives at $x = 0$ span W.) Equivalently, the corresponding tau function may be expressed

$$\tau_{W(X,Z)} = \det\left(X + x\mathbf{I} + \sum_{j=2}^{\infty} j t_j (-Z)^{j-1}\right). \tag{1.22}$$

Wilson showed in [**W2**] that this identification gives an isomorphism of affine, nonsingular, irreducible algebraic varieties. It is clear from (1.16) that under this identification, the bispectral involution is equivalent to the symplectic involution

$$\begin{aligned} \tilde{b}\colon C_n &\to C_n, \\ \tilde{b}\colon (X, Z) &\mapsto (Z^t, X^t) \end{aligned} \tag{1.23}$$

on the completed Calogero-Moser phase space. (This was noted for the case of the original real Calogero-Moser phase by Kasman [**K**], with $\tilde{b}$ viewed as Moser's linearizing involution [**M**].)

This therefore provides a complete determination of all wave functions ψ_W corresponding to rank 1 bispectral algebras. The following are two of the simplest examples of these. (In each case, the algebra $\mathcal{A}$ is identified with the space of polynomials which, under multiplication, leave $W(X, Z)$ invariant. Only the simplest representative elements L and Λ, are given.)

EXAMPLE.

(i)

$$n = 1, \quad X = (\alpha), \quad Z = (0), \tag{1.24a}$$

$$\psi_{W(X,Z)}(x, z) = e^{xz}\left(1 - \frac{1}{(x + \alpha)z}\right), \tag{1.24b}$$

$$L = \frac{d^2}{dx^2} - \frac{2}{(x + \alpha)^2}, \quad f(z) = z^2, \tag{1.24c}$$

$$\Lambda = \frac{d^2}{dz^2} + 2\alpha\frac{d}{dz} - \frac{2}{z^2}, \quad \phi(x) = x^2 + 2\alpha x, \tag{1.24d}$$

(ii)

$$n = 2, \quad X = \begin{pmatrix} 0 & 0 \\ 1 & 0 \end{pmatrix}, \quad Z = \begin{pmatrix} 0 & -1 \\ 0 & 0 \end{pmatrix}, \tag{1.25a}$$

$$\psi_{W(X,Z)}(x,z) = e^{xz}\left(1 - \frac{2}{xz} + \frac{2}{x^2z^2}\right), \tag{1.25b}$$

$$L = \frac{d^3}{dx^3} - \frac{6}{x^2}\frac{d}{dx} + \frac{12}{x^3}, \quad f(z) = z^3, \tag{1.25c}$$

$$\Lambda = \frac{d^3}{dz^3} - \frac{6}{z^2}\frac{d}{dz} + \frac{12}{z^3}, \quad \phi(z) = x^3. \tag{1.25d}$$

2. Isomonodromic Deformations

2.1. Rational covariant derivatives operators. Consider a parametric family of matrix first order differential operators

$$\mathcal{D}_\lambda(\tau) := \frac{\partial}{\partial\lambda} - B(\lambda,\tau) - \sum_{i=1}^{n}\sum_{a=1}^{m_i} \frac{N_{ia}(\tau)}{(\lambda - \alpha_i(\tau))^a}, \tag{2.1}$$

where $B(\lambda,\tau)$ is an $r \times r$ matrix valued polynomial in λ that may depend also on a deformation parameter τ, $\{N_{ia}\}_{1\leq i\leq n, 1\leq a\leq m_i}$ is a set of $r \times r$ matrices, also dependent on τ, and $\{\alpha_i(\tau) \in \mathbf{C}\}_{1\leq i\leq n}$ are the location of the finite poles, also possibly τ dependent. This may be viewed as a τ-parametric family of rational covariant derivative operators defined on a rank r vector bundle over the Riemann sphere, punctured at the poles (including possibly ∞). The generalized monodromy data associated to the operator $\mathcal{D}_\lambda$ consists of the local monodromy matrices about each of the poles, together with the Stokes matrices and connection matrices [**JMU**]. Assuming local differentiability in the deformation parameter τ, it is a basic fact that this monodromy data is invariant under variations in τ provided there exists a second differential operator

$$\mathcal{D}_\tau := \frac{\partial}{\partial\tau} - E(\lambda,\tau), \tag{2.2}$$

where $E(\lambda,\tau)$ is also a τ-parametric family of $r \times r$ matrices, rational in λ, such the commutativity condition

$$[\mathcal{D}_\tau, \mathcal{D}_\lambda] = 0 \tag{2.3}$$

is satisfied. This means that the data $\{B(\lambda,\tau), \alpha_i(\tau), N_{ia}(\tau)\}$ determining $\mathcal{D}_\lambda(\tau)$ satisfy a set of first order ODE's in the deformation parameter τ. It also implies that, locally at least, an invertible matrix valued function $\Psi(\lambda,\tau)$ exists, uniquely determined by any given set of initial values, simultaneously satisfying

$$\mathcal{D}_\lambda\Psi = 0, \tag{2.4a}$$

$$\mathcal{D}_\tau\Psi = 0. \tag{2.4b}$$

The simplest case, studied since the beginning of this century, is when the operator $\mathcal{D}_\lambda$ is Fuchsian; i.e., it has only regular singular points. This means that all the finite poles are simple ($m_i = 1$, $\forall i$) and $B(\lambda, \tau) \equiv 0$, so $\mathcal{D}_\lambda$ is of the form

$$\mathcal{D}_\lambda = \frac{\partial}{\partial \lambda} - \sum_{i=1}^{n} \frac{N_i}{\lambda - \alpha_i}. \tag{2.5}$$

For this case, it has long been known that the only way the residue matrices $\{N_i\}_{i=1,\dots n}$ can vary is through their dependence on the pole locations $\{\alpha_i\}_{i=1,\dots n}$, and they are constrained to satisfy an overdetermined system of PDE's. Choosing the α_i's as the deformation parameters, the associated infinitesimal isomonodromic deformation operators are

$$\mathcal{D}_{\alpha_i} := \frac{\partial}{\partial \alpha_i} + \frac{N_i}{\lambda - \alpha_i}. \tag{2.6}$$

The corresponding commutativity conditions

$$[\mathcal{D}_\lambda, \mathcal{D}_{\alpha_i}] = 0, \quad [\mathcal{D}_{\alpha_i}, \mathcal{D}_{\alpha_j}] = 0, \quad 1 \leq i,\ j \leq n \tag{2.7}$$

are mutually compatible (i.e. integrable in the Frobenius sense), and equivalent to the following system of equations, known as the Schlesinger equations

$$\frac{\partial N_i}{\partial \alpha_j} = \frac{[N_i, N_j]}{\alpha_i - \alpha_j}. \quad i \neq j, \tag{2.8a}$$

$$\frac{\partial N_i}{\partial \alpha_i} = -\sum_{\substack{j=1\\ j\neq i}}^{n} \frac{[N_i, N_j]}{\alpha_i - \alpha_j}. \tag{2.8b}$$

(In particular, the simplest Fuchsian case involving nontrivial isomonodromic deformations is when $n = 3$ and $r = 2$, for which these reduce to the equations of the sixth Painlevé transcendent P_{VI}.)

The corresponding determination of all equations of the type (2.3), generating isomonodromic deformations of operators of the general form (2.1) was made in [**JMU**]. For the simplest nonfuchsian case, where the finite poles remain first order, but an irregular singularity of Poincaré index 1 is added at infinity by allowing B to be a nonzero diagonal matrix $B(\tau) = \mathrm{diag}\big(\beta_1(\tau), \dots \beta_r(\tau)\big)$ with distinct eigenvalues that are independent of λ, we have

$$\mathcal{D}_\lambda(\tau) = \frac{\partial}{\partial \lambda} - \mathcal{N}(\lambda, \tau) \tag{2.9}$$

where

$$\mathcal{N}(\lambda, \tau) := B(\tau) + \sum_{i=1}^{n} \frac{N_i(\tau)}{\lambda - \alpha_i}. \tag{2.10}$$

In this case, as shown in [**JMMS**], besides the original α_i's, the possible deformations parameters include the eigenvalues $\{\beta_a\}_{1\leq a\leq r}$ of the matrix B. The corresponding infinitesimal isomonodromic deformation operators are

$$\mathcal{D}_{\alpha_i} := \frac{\partial}{\partial\alpha_i} + \frac{N_i}{\lambda-\alpha_i}, \quad i=1,\dots,n \tag{2.11a}$$

$$\mathcal{D}_{\beta_a} := \frac{\partial}{\partial\beta_a} - \lambda E_a - \sum_{\substack{b=1\\ b\neq a}}^{r} \frac{E_a N_\infty E_b + E_b N_\infty E_a}{\beta_a-\beta_b}, \quad a=1,\dots r, \tag{2.11b}$$

where E_a is the elementary matrix with 1 in the aa position and 0's elsewhere, and

$$N_\infty := \sum_{i=1}^{n} N_i. \tag{2.12}$$

Again the mutual commutativity of the operators $\{\mathcal{D}_{\alpha_i}, \mathcal{D}_{\beta_a}, \mathcal{D}_\lambda\}_{i=1,\dots n, a=1,\dots r}$

$$[\mathcal{D}_{\alpha_i},\mathcal{D}_{\alpha_j}] = [\mathcal{D}_{\alpha_i},\mathcal{D}_{\beta_a}] = [\mathcal{D}_{\beta_a},\mathcal{D}_{\beta_b}] = [\mathcal{D}_\lambda,\mathcal{D}_{\alpha_i}] = [\mathcal{D}_\lambda,\mathcal{D}_{\beta_a}] = 0, \tag{2.13}$$

gives a Frobenius integrable system for the residue matrices N_i, locally determining their dependence on the parameters $\{\alpha_i, \beta_a\}$.

2.2. ***R*–matrix theory and duality.** The above isomonodromic deformation equations may be viewed as nonautonomous Hamiltonian equations ([**JM**], [**H**]) on the space $(\mathfrak{gl}(r,\mathbf{C}))^n$ of n-tuples of $r\times r$ matrices $\{N_1,\dots N_n\}$, with respect to the Lie Poisson bracket structure

$$\{(N_i)_{ab},(N_j)_{cd}\} = \delta_{ij}\left[\delta_{ab}N_{cb} - \delta_{bc}N_{da}\right]. \tag{2.14}$$

Equivalently, the rational matrix valued functions $\mathcal{N}(\lambda)$, $\mathcal{N}(\mu)$ for different values of the complex parameters λ and μ satisfy

$$\{\mathcal{N}(\lambda) \otimes \mathcal{N}(\mu)\} = \left[r(\lambda-\mu), \mathcal{N}(\lambda)\otimes \mathbf{I} + \mathbf{I}\otimes\mathcal{N}(\mu)\right] \tag{2.15}$$

where $r(\lambda-\mu)$ is the rational R-matrix, viewed as an element of $\mathrm{End}(\mathbf{C}^n\otimes\mathbf{C}^n)$, defined by

$$r(\lambda) := \frac{P_{12}}{\lambda}, \tag{2.16}$$

where

$$P_{12}(\mathbf{u}\otimes\mathbf{v}) = \mathbf{v}\otimes\mathbf{u} \tag{2.17}$$

is the transposition endomorphism. This defines a Poisson bracket structure on the space $\mathfrak{gl}_{\mathrm{rat}}(r,\mathbf{C})$ of rational $r\times r$ matrix valued functions of λ. The α_i's and β_a's may be viewed as multi-time parameters associated to the $n+r$ nonautonomous Hamiltonians $\{H_i, K_a\}_{\substack{i=1,\dots n\\ a=1,\dots r}}$ defined by

$$H_i := \frac{1}{4\pi i}\oint_{\lambda=\alpha_i} \mathrm{tr}(\mathcal{N}^2(\lambda))\,d\lambda = \mathrm{tr}(BN_i) + \sum_{\substack{j=1\\ j\neq i}}^{n} \frac{\mathrm{tr}(N_iN_j)}{\alpha_i-\alpha_j} \tag{2.18a}$$

$$\text{(2.18b)}\quad K_a := \frac{1}{4\pi i}\oint_{\lambda=\infty} d\lambda\lambda \oint_{z=\beta_a} dz\big[\operatorname{tr}\big((B - z\mathbf{I})^{-1}\mathcal{N}(\lambda)\big)^2 - z\operatorname{tr}\big((B - z\mathbf{I})^{-1}\mathcal{N}(\lambda)\big)\big]$$
$$= \sum_{j=1}^{n} \alpha_j (N_j)_{aa} + \sum_{\substack{n=1\\ b\neq a}}^{r} \frac{(N_\infty)_{ab}(N_\infty)_{ba}}{\beta_a - \beta_b}.$$

Defining a 1-form θ on the parameter space by

$$\text{(2.19)}\qquad \theta := \sum_{i=1}^{n} H_i d\alpha_i + \sum_{a=1}^{r} K_a d\beta_a,$$

the following results, which may be verified directly [**JMU**], are a consequence of the general R-matrix theory, adapted to the nonautonomous setting [**H**].

THEOREM. *The isomonodromic deformation equations* (2.13) *are Hamilton's equations for the set of Hamiltonians* $\{H_i, K_a\}$ *contained in the* 1*-form* θ

$$\text{(2.20)}\qquad dN_i := \sum_{j=1}^{n} \frac{\partial N_i}{\partial \alpha_j} d\alpha_j + \sum_{a=1}^{r} d\beta_a \frac{\partial N_i}{\partial \beta_a} = \{N_i, \theta\}.$$

Furthermore, the Hamiltonians $\{H_i, K_a\}$ *all Poisson commute amongst themselves, and hence* θ *is a closed form on the parameter space*

$$\text{(2.21)}\qquad d\theta = 0,$$

and there exists, at least locally, a function τ (the tau function) such that

$$\text{(2.22)}\qquad \theta = d(\ln \tau).$$

The Hamiltonian equations determining these isomonodromic deformations may in fact be lifted to an associated symplectic vector space, and then projected to *another* space of meromorphic differential operators, giving rise to a second, *dual* representation of these equations as isomonodromic deformations [**H**]. It is easy to see that all $\mathcal{N}(\lambda)$'s of the form (2.10) may be expressed as

$$\text{(2.23)}\qquad \mathcal{N}(\lambda) = B + G^T(A - \lambda\mathbf{I})^{-1}F,$$

where A is a diagonal $N \times N$ matrix,

$$\text{(2.24)}\qquad A := \operatorname{diag}(\alpha_1, \dots, \alpha_n),$$

$N = \sum_{i=1}^{n} k_i$, where $k_i = \operatorname{rk}(N_i)$, with eigenvalues α_i of multiplicity k_i at the poles of $\mathcal{N}(\lambda)$, and (F, Γ) are a pair of rectangular $N \times r$ matrices. If the space $\mathcal{M}^{N\times r} := \{(F, G)\}$ of such pairs is given the canonical symplectic structure

$$\text{(2.25)}\qquad \omega = \operatorname{tr}(dF^T \wedge dG),$$

the map to the space $\mathfrak{gl}_{\text{rat}}(r, \mathbf{C})$ defined, for fixed A and B, by

$$\text{(2.26)}\qquad J_B^A\colon (F, G) \mapsto \mathcal{N}(\lambda),$$

is a Poisson map. In fact, it may be viewed as defining a Poisson quotient of $\mathcal{M}^{N\times r}$ by the canonical Hamiltonian action of the stability subgroup $\mathfrak{G}_A \subset \mathfrak{Gl}(N,\mathbf{C})$ of A, given by

$$\begin{aligned} \mathfrak{G}_A \times \mathcal{M}^{N\times r} &\longrightarrow \mathcal{M}^{N\times r}, \\ \big(g,(F,G)\big) &\longmapsto \big(gF,(g^T)^{-1}G\big). \end{aligned} \tag{2.27}$$

(The $k_i \times r$ blocks (F_i, G_i) within the matrices F and G corresponding to the eigenvalues α_i determine the local monodromy of the operator $\mathcal{D}_\lambda$ about these points.) Denoting the pull-back of the 1-form θ under J_B^A by $\widetilde{\theta}$, the isomonodromic deformation equations (2.13) may be viewed as the projection to $\mathfrak{gl}_{\mathrm{rat}}(r,\mathbf{C})$ of the following nonautonomous Hamiltonian equations on the auxiliary space $\mathcal{M}^{N\times r}$

$$dF = \{F,\widetilde{\theta}\}, \quad dG = \{G,\widetilde{\theta}\}, \tag{2.28}$$

where

$$d := \sum_{j=1}^{n} d\alpha_j \frac{\partial}{\partial\alpha_j} + \sum_{a=1}^{r} d\beta_a \frac{\partial}{\partial\beta_a}. \tag{2.29}$$

To obtain the "dual" set of isomonodromic deformation equations, we define a new Hamiltonian quotient of $\mathcal{M}^{N\times r}$, this time by the stabilizer $\mathfrak{G}_B \subset \mathfrak{Gl}(r,\mathbf{C})$ of the element $B \in \mathfrak{Gl}(r)$ through the dual Poisson map $J_A^B\colon (F,G) \to \mathcal{M}(z)$, where

$$\mathcal{M}(z) := A + F(B - z\mathbf{I}_r)^{-1}G^T = A + \sum_{a=1}^{r} \frac{M_a}{z-\beta_a}. \tag{2.30}$$

Defining the associated set of differential operators

$$\widetilde{\mathcal{D}}_z := \frac{\partial}{\partial z} - \mathcal{M}(z), \tag{2.31a}$$

$$\widetilde{\mathcal{D}}_{\alpha_i} := \frac{\partial}{\partial\alpha_i} - zE_i - \sum_{\substack{j=1\\ j\neq i}}^{n} \frac{E_i M_\infty E_j + E_j M_\infty E_i}{\alpha_i - \alpha_j}, \tag{2.31b}$$

$$\widetilde{\mathcal{D}}_{\beta_a} := \frac{\partial}{\partial\beta_a} + \frac{M_a}{z-\beta_a}, \tag{2.31c}$$

Hamilton's equations (2.28) are projectible under the map J_A^B and imply the commutativity conditions

$$[\widetilde{\mathcal{D}}_{\alpha_i},\widetilde{\mathcal{D}}_{\alpha_j}] = [\widetilde{\mathcal{D}}_{\alpha_i},\widetilde{\mathcal{D}}_{\beta_a}] = [\widetilde{\mathcal{D}}_{\beta_a},\widetilde{\mathcal{D}}_{\beta_b}] = [\widetilde{\mathcal{D}}_z,\widetilde{\mathcal{D}}_{\alpha_i}] = [\widetilde{\mathcal{D}}_z,\widetilde{\mathcal{D}}_{\beta_a}] = 0, \tag{2.32}$$

which, in turn, imply the invariance of the monodromy data of the operator $\mathcal{D}_z$ under changes in parameter values. These equations may also be interpreted as Hamiltonian equations on the space of rational $N\times N$ matrix valued functions of $\mathcal{M}(z)$, generated by the set of Poisson commuting, nonautonomous Hamiltonians obtained by projection:

$$\widetilde{K}_a := \frac{1}{4\pi i}\oint_{z=\beta_a} \operatorname{tr}\big(\mathcal{M}^2(z)\big)\,dz = \operatorname{tr}(AB_a) + \sum_{\substack{b=1\\ b\neq a}}^{r} \frac{\operatorname{tr}(M_a M_b)}{\beta_a - \beta_b}, \tag{2.33a}$$

$$(2.33b)\quad \widetilde{H}_i := 14\pi i \oint_{z=\infty} dz\, z \oint_{\lambda=\alpha_i} d\lambda\big[\mathrm{tr}\big((A-\lambda\mathbf{I}_N)^{-1}\mathcal{M}(z)\big)^2 - \lambda\,\mathrm{tr}\big((A-\lambda\mathbf{I}_N)^{-1}\mathcal{M}(z)\big)\big]$$
$$= \sum_{b=1}^{r}\beta_b(M_b)_{ii} + \sum_{\substack{j=1\\ j\neq i}}^{n}\frac{(M_\infty)_{ij}(M_\infty)_{ji}}{\alpha_i-\alpha_j}.$$

The duality map

$$(2.34)\qquad (F, G, A, B, \lambda, z) \to (G^T, F^T, B, A, z, \lambda)$$

may be viewed, after projection, as transforming the system of isomonodromic deformations equations (2.13) into the dual system (2.32).

Clearly, this duality map has a close resemblance to the bispectral involution (1.23) expressed in terms of the matrices (X, Z) determining the Calogero-Moser system. A full explanation of the relationship between these two maps is not yet developed, but one certainly exists. What will be provided in the next section will be a reformulation of the two examples of bispectral pairs given in section 1 in terms of an equivalent, dual pair of isomonodromic deformation equations–in the very particular case where the monodromy happens to be trivial. These examples in fact illustrate a correspondence that can be made between *all* the bispectral pairs of the type studied by Wilson; i.e., all rank 1 bispectral algebras, and dual pairs of parametric families of matrix differential operators of first order, depending rationally on both parameters, and having trivial monodromy.

3. Relation Between Bispectrality and Dual Isomonodromy

The general structure of bispectral equations and infinitesimal isomonodromic deformation equations, both involving a wave function depending on two variables that simultaneously satisfies a linear differential equation in each of them, suggests that the two may be related. The similarity between the duality map (2.34) and the bispectral involution (1.23) further suggests a relation between bispectrality and duality. The deeper relation between bispectral pairs and dual isomonodromic deformations involves the theory of dressing transformations and the tau function, and will be not developed here. However, to demonstrate that such a relation does exist, we consider the two examples of bispectral rank 1 operators introduced in Section 1.2, and show how an equivalent infinitesimal isomonodromic deformation system can be derived for each.

3.1. Examples. (i) For the case of the Baker type bispectral wave function

$$(3.1)\qquad \psi(x,z) = e^{xz}\left(1 - \frac{1}{(x+\alpha)z}\right)$$

there exists a *second* bispectral wave function for the same pair of operators L and Λ given in (1.24c,d), namely

$$(3.2)\qquad \psi_1 := \psi(-x-2\alpha, z) = e^{-2\alpha z}\psi(x, -z).$$

Forming the Wronskian matrix

$$(3.3)\qquad \Psi := \begin{pmatrix} \psi & \psi_1 \\ \psi_x & \psi_{1,x} \end{pmatrix},$$

we may associate the following pair of rational 2×2 covariant derivative operators

$$\mathcal{D}_x := \frac{\partial}{\partial x} - \begin{pmatrix} 0 & 1 \\ z^2 + \frac{2}{(\alpha+x)^2} & 0 \end{pmatrix}, \tag{3.4a}$$

$$\mathcal{D}_z := \frac{\partial}{\partial z} - \begin{pmatrix} -\alpha & \frac{\alpha+x}{z} \\ (\alpha+x)z + \frac{2}{(\alpha+x)z} & \frac{1}{z} - \alpha \end{pmatrix}, \tag{3.4b}$$

which simultaneously annihilate Ψ

$$\mathcal{D}_x \Psi = 0, \quad \mathcal{D}_z \Psi = 0, \tag{3.5}$$

This is equivalent to the bispectral conditions (1.1a,b) for the operators L, Λ defined in (1.24c,d) and implies the commutativity condition

$$[\mathcal{D}_x, \mathcal{D}_z] = 0. \tag{3.6}$$

It follows that the monodromy of either of these operators, say $\mathcal{D}_z$, is invariant under the deformations determined by varying the other parameter, x. Note however the fact that there is not just *one* single-valued bispectral wave function, but a *basis* of single valued bispectral wave functions, which implies in this case that the operators $\mathcal{D}_x$ and $\mathcal{D}_z$ actually have *trivial* monodromy for all parameter values.

To obtain the *dual* system of (trivial) isomonodromic deformation equations, we just define new operators similarly, using the Wronskian matrix formed by differentiating with respect to the variable z:

$$\widetilde{\Psi} := \begin{pmatrix} \psi & \psi_1 \\ \psi_z & \psi_{1,z} \end{pmatrix}. \tag{3.7}$$

This is simultaneously annihilated by the operators

$$\widetilde{\mathcal{D}}_x := \frac{\partial}{\partial x} - \begin{pmatrix} \frac{\alpha z}{\alpha+x} & \frac{z}{\alpha+x}, \\ \frac{(x^2+2\alpha x)z}{\alpha+x} + \frac{2+\alpha z}{(\alpha+x)z} & \frac{1-\alpha z}{\alpha+x} \end{pmatrix}, \tag{3.8a}$$

$$\widetilde{\mathcal{D}}_z := \frac{\partial}{\partial z} - \begin{pmatrix} 0 & 1 \\ \frac{2}{z^2} + x^2 + 2\alpha x & -2\alpha \end{pmatrix}, \tag{3.8b}$$

implying the commutativity condition

$$[\widetilde{\mathcal{D}}_x, \widetilde{\mathcal{D}}_z] = 0, \tag{3.9}$$

from which again follows that the monodromy of either of these operators is invariant under the deformations determined by varying the other parameter.

Note however that the singularity structure of the operators $\mathcal{D}_x, \mathcal{D}_z, \widetilde{\mathcal{D}}_x$ and $\widetilde{\mathcal{D}}_z$ is such that the Hamiltonian R-matrix formulation discussed in Section 2.2 does not suffice. In order to include these, the R-matrix approach must be extended (cf. [**HTW**], [**HR**], [**HW1**], [**HW2**]) to allow for singularities of higher order at ∞.

(ii) For the case of the bispectral wave function

$$\psi_{W(X,Z)}(x,z) = e^{xz}\left(1 - \frac{2}{xz} + \frac{2}{x^2 z^2}\right), \tag{3.10}$$

because of the invariance of the eigenvector equations for the operators L and Λ in (1.25c,d) under the cyclic group generated by

$$z \mapsto \omega z, \quad \omega := e^{2\pi i/3}, \tag{3.11}$$

we may define a basis $\{\psi_0, \psi_1, \psi_2\}$ of bispectral wave functions by

$$\psi_j(x,z) := \psi(x, \omega^j z), \quad j = 0, 1, 2. \tag{3.12}$$

The Wronskian matrix

$$\Psi := \begin{pmatrix} \psi_0 & \psi_1 & \psi_2 \\ \psi_{0,x} & \psi_{1,x} & \psi_{2,x} \\ \psi_{0,xx} & \psi_{1,xx} & \psi_{2,xx} \end{pmatrix}, \tag{3.13}$$

is then simultaneously annihilated by the operators

$$\mathcal{D}_x := \frac{\partial}{\partial x} - \begin{pmatrix} 0 & 1 & 0 \\ 0 & 0 & 1 \\ z^3 - \frac{12}{x^3} & \frac{6}{x^2} & 0 \end{pmatrix}, \tag{3.14a}$$

$$\mathcal{D}_z := \frac{\partial}{\partial z} - \begin{pmatrix} 0 & \frac{x}{z} & 0 \\ 0 & \frac{1}{z} & \frac{x}{z} \\ xz^2 - \frac{12}{x^2 z} & \frac{6}{xz} & \frac{2}{z} \end{pmatrix}. \tag{3.14b}$$

This is equivalent to the eigenvector equations for the operators L and Λ of eqs. (1.25c,d), and implies the commutativity of the operators $\mathcal{D}_x, \mathcal{D}_z$. Hence the monodromy of each of these operators (which again is trivial because of the single valuedness of the matrix Ψ), remains invariant under changes in the parameter values. Because of the symmetry of the bispectral wave functions under the interchange $x \leftrightarrow z$, the corresponding "dual" isomonodromic system is obtained by simply interchanging the two variables in the definitions of the operators $\mathcal{D}_x, \mathcal{D}_z$. As in the previous example, the proper treatment of these isomonodromic systems within the R-matrix framework requires the inclusion of higher order singularities at ∞.

Proceeding similarly, we may associate to every rank 1 bispectral pair two mutually dual 1-parameter families of differential operators having trivial monodromy for all parameter values, since in each case we may form a basis of single-valued bispectral wave functions. This procedure may also be applied to the case of higher rank bispectral operators, except that the monodromy of the resulting dual families of operators is no longer necessarily trivial, and we obtain instead 1-parameter families of operators with constant monodromy under variation of the deformation parameters. These more general cases, as well as their relations to the evaluation of certain limits of Fredholm determinants, will be dealt with elsewhere.

Acknowledgements

The author would like to thank A. Grünbaum, A. Kasman and A. Orlov for helpful discussions.

References

[AMM] H. Airault, H. P. McKean, and J. Moser, *Rational and elliptic solutions of the Korteweg-de Vries equation and a related many-body problem*, Comm. Pure Appl. Math. **30** (1977), 95-148.

[DG] J. J. Duistermaat and A. Grünbaum, *Differential equations in the spectral parameter*, Comm. Math. Phys. **107** ((1986)), 177–240.

[DJKM] E. Date, M. Jimbo, M. Kashiwara, and T. Miwa, *Transformations groups for soliton equations.* I, Proc. Japan Acad. Ser. A Math. Sci. **57** (1981), 342–392; II, 387–392; III, J. Phys. Soc. Japan **50** (1981), 3806–3812; IV, Physica D **4** (1982), 343–365; Publ. Res. Inst. Math. Sci. **18**, 1111–1119; IV, J. Phys. Soc. Japan **50**, 3813–3818; Publ. Res. Inst. Math. Sci. **18** (1982), 1077–1110.

[H] J. Harnad, *Dual isomonodromic deformations and moment maps to loop algebras*, Comm. Math. Phys. **166** (1994), 337–365.

[HR] J. Harnad and M. Routhier, *R-matrix construction of electromagnetic models for the Painlevé transcendents*, J. Math. Phys. **36** (1995), 4863–4881.

[HTW] J. Harnad, C. A. Tracy, and H. Widom, *Hamiltonian structure of equations appearing in random matrices*, Low Dimensional Topology and Quantum Field Theory (H. Osborn, ed.), Plenum, New York, 1993, pp. 231–245.

[HW1] J. Harnad and M.-A. Wisse, *Loop algebra moment maps and Hamiltonian models for the Painlevé transcendents*, Mechanics Day Workshop Proceedings (P. S. Krishnaprasad, T. Ratiu, and W. F. Shadwick, eds.), vol. 7, Fields Inst. Comm., 1996, pp. 155–169.

[HW2] ———, *Moment maps to loop algebras, classical R-matrix and integrable systems*, Quantum Groups, Integrable Models and Statistical Systems (Proc. of the 1992 CAP Summer Workshop) (L. Vinet and J. Letourneux, eds.), World Scientific, Singapore, 1993.

[JMMS] M. Jimbo, T. Miwa, Y. Môri, and M. Sato, *Density matrix of an impenetrable Bose gas and the fifth Painlevé transcendent*, Physica D **1** (1980), 80–158.

[JMU] M. Jimbo, T. Miwa, K. Ueno, *Monodromy preserving deformation of linear ordinary differential equations with rational coefficients.* I, Physica D **2** (1981), 306–352.

[JM] M. Jimbo and T. Miwa, *Monodromy preserving deformation of linear ordinary differential equations with rational coefficients.* II, Physica D **2** (1981), 407–448; III, Physica D **4** (1981), 26–46.

[K] A. Kasman, *Bispectral KP solutions and linearization of Calogero–Moser particle systems*, Comm. Math. Phys. **172** (1995), 427–448.

[Kr] I. M. Krichever, *Rational solutions of the Kadomtsev-Petviashvili equation and integrable systems of N particles on a line*, Functional Anal. Appl **12** (1978), 59–61.

[M] J. Moser, *Three integrable Hamiltonian systems connected with isospectral deformations*, Adv. Math. **16** (1975), 197–220.

[SS] M. Sato and Y. Sato, *Soliton equations as dynamical systems on infinite dimensional Grassmann manifold*, Lecture Notes Numer. Appl. Anal. **5** (1982), 259–271.

[SW] G. Segal and G. Wilson, *Loop groups and equations of KdV type*, Inst. Hautes Études Sci. Publ. Math **61** (1985), 5–65.

[W1] G. Wilson, *Bispectral commutative ordinary differential operators*, J. Reine Angew. Math. **442** (1993), 179–204.

[W2] ———, *Collisions of Calogero-Moser particles and an adelic Grassmannian* (1996) (preprint).

Department of Mathematics and Statistics, Concordia University 7141 Sherbrooke W., Montréal, Canada H4B 1R6, and Centre de recherches mathématiques, Université de Montréal C. P. 6128, Succ. centre-ville, Montréal, Canada H3C 3J7

E-mail address: harnad@alcor.concordia.ca or harnad@crm.umontreal.ca

Centre de Recherches Mathématiques
CRM Proceedings and Lecture Notes
Volume **11**, 1997

Metaphor and Models in Science and Art

Philip Holmes

Without poetry our science will appear incomplete ... "
Matthew Arnold, The Study of Poetry.

ABSTRACT. I describe some similarities and differences between the worlds of applied mathematics and poetry. Scientific examples are drawn from celestial mechanics and dynamical systems theory, and literary examples from the verse of Gerard Manley Hopkins.

1. Introduction

Some thirty-six years ago, in his Rede Lectures [**Sno59**], C. P. Snow bemoaned the bifurcation of the arts and sciences into two cultures. Pointing out that his humanist colleagues were generally ignorant of such fundamental ideas as the second law of thermodynamics, he drew attention to a shameful state of affairs in which the citizens of democratic nations and many of their leaders operate in virtual ignorance of the scientific principles on which modern technology, and hence much of society, depends. Not only were many intelligent, cultured, and otherwise well-educated people ignorant of practically all science since the Renaissance, but some even seemed proud of their deficiency. Little has changed since then. Perhaps the split has widened, with scientists and engineers being cast as the architects of the Cold War and destroyers of the environment.

The dangers of such wilful ignorance do not need emphasizing to the readers of this volume. It should be clear that thinkers and leaders educated in the humanities can benefit from a scientific and mathematical perspective. Instead, in this essay I want to draw attention to some ways in which the two cultures are closer than we may realise. While the superficial languages of the arts and the sciences differ substantially, at a deeper level—and especially in the creative process itself—they have much in common. Both afford routes to understanding—or at least, coming to terms with—our world; both rely extensively on metaphor and are concerned with the building of "secondary worlds" to create and communicate information about the primary one; and both sometimes judge their creations as much by elegance as accuracy. Above all, both are human creations.

1991 *Mathematics Subject Classification.* Primary: 58FXX, Secondary: 34xx, 70F15.
This is the final form of the paper.

The affinities of (pure) mathematics and music, both generally regarded as abstract activities, are well-known and commonly discussed (e.g. [**Rot95**]). Here I shall address (slightly) more practical arts; I shall reflect on how and why one might enjoy doing and reading both applied mathematics and poetry. In revealing some common currents in these endeavors, we may learn something from humanists about the nature of mathematical models and their analysis. I start by discussing a poem, move on to a differential equation, and close with some comparative discussions. Although this essay focuses on poetry and differential equations, I believe that much of what I have to say applies generally to the arts and the sciences.

2. A Poem

The poet Gerard Manley Hopkins was born at Stratford, Essex, in 1844. The eldest of eight children in an artistic and cultured family, he began writing verse early and won the poetry prize at Highgate School in 1860. He went on to read Classics at Balliol College, Oxford, where his life-long friendship with Robert Bridges began. (Bridges was to become Poet Laureate and the first champion of Hopkins's own work.

Hopkins was an intensely religious man. At Oxford he abandoned his parents' Anglo-Catholicism for full membership in the church of Rome and, in 1868, entered the nine year training period for the Jesuit priesthood. On taking this decision, he destroyed his early poetry and vowed not to write again, as unfitting for a selfless servant of the Lord. However, in 1875, encouraged by his superiors, he did resume writing and continued to do so through the remainder of his short life, which he spent as a teacher, "missioner", and priest in various towns in England, Scotland and Wales. His final appointment was to the Chair in Classics at University College, Dublin, Ireland, where he died in 1889. His poetry remained unpublished and almost completely unknown until Bridges' edition of 1918 [**Hop18**].

Here is Hopkins's poem "The Sea and the Skylark", written at Rhyl, on the North Welsh coast, in May, 1877:

On ear and ear two noises too old to end
  Trench—right, the tide that ramps against the shore;
  With a flood or a fall, low lull-off or roar,
Frequenting there while moon shall wear and wend.
Left hand, off land, I hear the lark ascend,
  His rash-fresh, re-winded new-skeinèd score
  In crisps of curl off wild winch whirl, and pour
And pelt music, till none's to spill nor spend.
How these two shame this shallow and frail town!
  How ring right out our sordid turbid time,
Being pure! We, life's pride and cared-for crown,
  Have lost that cheer and charm of earth's past prime:
Our make and making break, are breaking, down
  To man's last dust, drain fast towards man's first slime.

The 'story' of the poem is simple and clear enough. Two natural things, the inanimate sea with its tides and the supremely animate skylark with his song, put to shame man's vain creations. In the context of English nature poetry and the Romantic movement, coming twenty-seven years after Wordsworth's death, these sentiments are scarcely original. Yet how much more is shown and hinted at in these fourteen lines, than this bare summary affords!

The first thing that must strike us about this or, indeed, any poem, even before reading, is its shape and appearance on the page: its *form*. This is far from prose. On reading, especially reading aloud (and most poetry should be read so), rhythm, rhyme, internal assonance and consonance further insist that it is not "normal language." All these effects are heightened in Hopkins, who, here as in all his work, brings them to bear on an exact description of the human and natural world. His imagery and diction may not be easy to grasp at first reading (just as a unfamiliar proof or piece of analysis may initially elude us), but I hope to show that they are precise and illuminating; that his effects are essential to the complex response he intends to evoke. To do this I shall first appeal to Hopkins's correspondence.

No matter how much the poet may wish to break the bonds of language, he must still make sense if he is to communicate. I hope you will see that the words and constructions of the poem that seem awkward, fall into place naturally on repeated acquaintance. Hopkins wanted his poetry to be "... the current language heightened ... and unlike itself, but not ... an obsolete one." Perhaps the best advice is his own: " ... read it with the ears, as I always wish to be read, and my verse becomes all right."

In a letter to a friend he elucidates the second quatrain, comparing the notes of the lark's song, as they tumble down towards earth, to ribs on a skein of silk unwound from a tight card, a fishing line unwound from a reel (winch), or pearls strung on a horsehair. These may seem forced similes, but the language brings them alive: in the central couplet we simultaneously *hear* the lark's bubbling song—the rhythm tripping on itself—as we *see* the notes flung out on the sky. The images of the first stanza: tides ascending the ramp of the beach driven by the gradually slowing moon, are equally intellectually and aurally precise.

Certainly there are puzzles here: for example, the word 'trench' in the fourth line is evidently the main (intransitive) verb of the sentence forming the first stanza. I take it to mean something like: "the noises entrench themselves in the mind." Trench is not commonly used as a verb, although the Shorter OED assigns it a dozen such uses, the one of major relevance here being: "to cut a trench or trenches (in the ground)." An even less common group of meanings includes "to come in thought, speech or action close upon (something); hence, to have a bearing upon or reference to." This may well have been Hopkins's primary sense. However, it is almost irresistible to read 'trench' as a noun, especially in its isolated, capitalized form at the beginning of the line. This unusual usage, typical of Hopkins, makes the action particularly concrete and vivid.

Although he has been seen as a precursor of modernism—the break with nineteenth century tradition associated with Pound and Eliot in poetry and Joyce in prose—Hopkins begins firmly within the conventions of his day. Indeed, precursors of his imagery and "way of using the body and movement of the language" are found in Shakespeare, as F. R. Leavis has pointed out [**Lea32**]. Almost all Hopkins's shorter poems are sonnets, the opening octet setting the scene and the closing sextet reflecting on it. Here too, the conventional form of the Petrarchan (or Miltonic) sonnet, with its strict *abba abba cdc dcd* rhyme scheme and iambic pentameter or "standard" rhythm, provides the framework on which Hopkins's verbal effects dance. (The perhaps better known Shakespearean sonnet—*abab cdcd efef gg*—employs three quatrains and a closing couplet.) Within this form, the regular rhythm is repeatedly broken, yielding to what Hopkins called *sprung rhythm*, and characterized as " ... the most natural of things ... the rhythm of common speech ... found in nursery rhymes, weather saws, and so on ... ." As Hopkins breaks our rhythmic expectations, so he surprises us with internal rhymes and half-rhymes within lines (rash / fresh; curl / whirl; sordid / turbid; make / break; last / fast). The line

How ríng ríght oút our sórdid túrbid time,

uses the three short, stressed syllables in the first half to aurally banish the dull plod of mercantile life in the second.

In addition to the Romantic tradition, Hopkins also draws on the much older traditions of oral Welsh and Anglo-Saxon verse: the poetry of Aneirin and the epic Beowulf. This is most evident in his use of internal consonance, rarely as regular as the traditional four stressed words beginning with the same letter in a line broken at its center, but often occurring in pairs and giving the line a special weight and force (flood / fall; land / lark; crisp / curl ... wild / winch / whirl; shame / shallow; turbid / time; cheer / charm; dust / drain ... fast / first). More subtly, Hopkins often chooses the word with Anglo-Saxon root over its French equivalent, giving his work a gritty and tactile quality. The number of monosyllables in this poem is particularly high, although they are far from monotonous: we hear them beating on the shore in the first stanza, but they ripple down as bird-song in the second. Such rhythmic and sonic effects permeate the poem, as breaking surf and bird-song throng the scene described.

Thus far I have focussed on the form of the poem; the technique. Now I turn to some thoughts about content and the poet's approach to it.

While he started from and used the conventions of earlier poetry, Hopkins aimed to produce poems "beautiful to individuation", capable of bringing out the unique individual qualities of the scene or event he wished to portray and comment on. In his journals he developed an elaborate theory of *instress* and *inscape*, which we may interpret roughly as the individual essence of a thing: its "thisness"[1]; and the unifying force or stress pattern that determines and maintains the thing. Along with these theoretical musings, he cultivated a skill and care in description of nature comparable to the greatest among his contemporary Victorian scientists. Here is a passage from his notebook of August 10th, 1872, five years before the poem above was composed:

> I was looking at high waves. The breakers always are parallel to the coast and shape themselves to it except where the curve is sharp however the wind blows. They are rolled out by the shallowing shore just as a piece of putty between the palms whatever its shape runs into a long roll. The slant ruck or crease one sees in them shows the way of the wind. The regularity of the barrels surprised and charmed the eye; the edge behind the comb or crest was as smooth and bright as glass. It may be noticed to be green behind and silver white in front: the silver marks where the air begins, the pure white is foam, the green/ solid water. Then looked at to the right or left they are scrolled over like mould-boards or feathers or jibsails seen by the edge. ...

[1]Hopkins was particularly attracted by the philosopher Duns Scotus (d. 1308), who stressed "haecceitas"; see Hopkins's poem "Duns Scotus' Oxford."

And earlier, in June 1871:

> The Horned Violet is a pretty thing, gracefully lashed. Even in withering the flower ran through beautiful inscapes by the screwing up of the petals into straight little barrels or tubes. It is not that inscape does not govern the behavior of things in slack and decay as one can see even in the pining of the skin in the old and even in a skeleton but that horror prepossesses the mind, but in this case there was nothing in itself to shew even whether the flower was shutting or opening.

Throughout the period when he forbore from writing verse, Hopkins continued to make such observations and gather material, always searching for the right words. Many of these sketches and part-formed images found their way into his later poems.

The poem I have chosen to discuss is by no means his greatest: in others, like "Spelt from Sybil's Leaves", his internal debates on the nature of faith and his piercing sincerity find even richer and more complex expression. As I have remarked, Hopkins was as concerned with the human condition as with nature itself. Spending much of his early working life in the great industrial towns, including London, Liverpool and Glasgow, he saw the miserable conditions in which the poor lived and worked at first hand. In poems like "Felix Randal" and "Harry Ploughman" he celebrated the skill and power of common people. In the poem quoted above, and others such as "Binsey Poplars" and "God's Grandeur", he criticizes the rapid progress of industrialism and destruction of the natural and social environment of rural Britain. At the same time he meditates on loss of faith in himself and others, particularly in his "black sonnets" such as "I wake and feel the fell of dark ..." and "No worst, there is none ...", written in Ireland, when he felt "... at a third/ Remove ... ." I regret that I do not have space to quote one of these terrible and moving pieces; I hope the reader will seek them out.

The main point I want to convey in this discussion of "The Sea and the Skylark" is the poet's careful study of nature (including human nature) and his search for precise, new and striking metaphors and images to capture and interpret it in language and so allow readers to reflect on and come to terms with it. Innovation in the context of tradition is essential in successful poetry. It is not dissimilar to the task the applied mathematician sets himself.

3. Differential Equations

Andrew Wiles [**Wil95**] has likened the mathematician to a person stumbling around in a darkened room, searching for the light switch. As he barks his shins on the furniture, the positions, functions and relationships of the various pieces gradually become clear. He envisages the result and a route to its proof. Finally he finds the switch, the theorem is proven, and he goes on to the next room.

Adapting this metaphor, the applied mathematician carries into the room an imperfect memory of the world outside. Feeling his way about, he is looking not for the light switch, but rather the cord to open the curtains. He must understand not only the nature and uses of the furniture, but also their relations to the "real" world outside. The making of new furnishings, which Wiles curiously omits, is also an important part of this: mathematics is not simply *discovered*, it is a supremely human *creation*. To resume the metaphor: the applied mathematician's creations are guided by his memories and, once the curtains are open, fresh observations of the world.

To illustrate the relationships among physical phenomena, their (mathematical) models, and the analysis of those models, I shall draw on a classical example from my own research area: dynamical systems theory, or, as it is more popularly and unfortunately misnamed: chaos theory (for chaos is only part of the story). I will remain mostly in the tradition of classical applied mathematics: mathematics applied to the analysis and understanding of the physical (inanimate) universe. This should not be taken to imply that mathematics cannot be effectively used to model aspects of biology, or the social, management, economic, and other human sciences; merely that I am not qualified to describe such applications (although I shall touch on neurobiology).

Dynamical systems theory has its origins in the work of Henri Poincaré, who developed the modern geometric and qualitative viewpoint in his attempts to solve the three-body problem of celestial mechanics. He built on the language of differential equations, already 200 years old, to which many of the greatest mathematicians of the seventeenth and eighteenth centuries had likewise contributed.

The question Poincaré addressed, a special case of the n-body problem, is a mathematical formulation of a problem that had engaged humanity for thousands of years: to predict and understand the motion of the heavenly bodies. Newton idealized the sun and planets of the solar system as point masses (in his *Principia* he proves that the gravitational field of a uniform solid sphere, outside the sphere, coincides with that of a point mass equal to the sphere, located at its center). In Proposition II of Book III: *The System of the World*, he formulated the principle [**New86**]:

> That the forces by which the primary planets are continually drawn off from rectilinear motions, and retained in their proper orbits, tend to the sun; and are inversely as the squares of the distances of the places of those planets from the sun's centre.

He turned this verbal model into a mathematical one by inventing the differential and integral calculus and appealing to his second law:

> The change in motion [linear momentum] is proportional to the motive force impressed; and is made in the direction of the right line in which that force is impressed.

The differential equations expressing Newton's inverse square gravitational law[2] are easy to state:

$$\ddot{\boldsymbol{q}}_i = \sum_{i=1}^{n} \frac{m_i m_j (\boldsymbol{q}_i - \boldsymbol{q}_j)}{|\boldsymbol{q}_i - \boldsymbol{q}_j|^3}; \quad i = 1, \ldots, n. \tag{1}$$

Here $m_1, m_2, \ldots, m_n$, denote the masses of n points and $\boldsymbol{q}_1, \boldsymbol{q}_2, \ldots, \boldsymbol{q}_n$, are 3-dimensional vector functions of time, t, describing their positions in space. This equation is the expression of Newton's second law in non-dimensional form; the second derivative with respect to time, $\ddot{\boldsymbol{q}}_i$, representing the rate of change of momentum, and the "force impressed" determined by mutual gravitational attraction. (The 'dot' notation is Newton's; Gottfried Wilhelm Leibniz, the co-inventor of calculus, preferred d/dt, which is now the standard in mathematics.) Classically, to

[2]It is customary to refer to the *inverse square law*, but it is perhaps more appropriate to say *inverse square model*. Laws can be made and enforced by humans, but nature does not obey laws in this sense, rather we are fortunate enough to observe natural behaviors that admit (simple) *models*.

solve the n-body problem means: given initial positions $\boldsymbol{q}_i(0)$ and velocities $\dot{\boldsymbol{q}}_i(0)$, find the functions $\boldsymbol{q}_i(t)$ and $\dot{\boldsymbol{q}}_i(t)$ for all future and past time.

Note that we have come some distance from the original physical problem. The motion of n bodies in physical space has become the *flow* of a point in $6n$-dimensional phase space, with coordinates $\boldsymbol{q}_i, \dot{\boldsymbol{q}}_i$. 'Flow' recalls Newton's term *fluent* for a dynamically varying quantity, although his *fluxion*, for the derivative, has not survived.

The solution of the two-body case due to Newton (and in complete generality in 1710 by Johann Bernouilli [**DH96**]) revealed Kepler's laws of planetary motion as direct consequences of gravitation, rather than unrelated empirical facts. Almost two centuries of attempts to solve the problem for $n \geq 3$ followed, until Poincaré [**Poi90**] proved that the restricted planar three body problem does not admit any analytic solution in the form of a (second) independent integral. By extension, this implies that classical solutions in the sense of Hamilton-Jacobi theory and integration by quadratures are generally precluded for all but the case $n = 2$. Series solutions can be found, albeit with very slow convergence properties, and, as the NASA space program attests, one can compute numerical approximations of great accuracy, although these do not confer much understanding of the nature of the problem. (The issue of solutions is discussed in some depth in [**Dia95**].) But in his great paper, Poincaré also revealed why the n-body problem (and many others) is in a deeper sense insoluble. He discovered the first example of chaos in a deterministic differential equation.

The fascinating story of Poincaré's paper, the first (and incorrect) version of which was awarded the prize endowed by King Oscar II of Sweden and Norway, has been told by Ivars Peterson in [**Pet93**]. More details can be found in a recent book by Florin Diacu and the present author [**DH96**].

Here I digress to describe a simple mathematical machine which generates chaos: iteration of a real-valued function (it will be linked to differential equations and the n-body problem later).

Take a number x_0 between 0 and 1, double it and subtract the integer part to produce x_1. Repeat *ad infinitum*. Calling the operation '$x \mapsto 2x \bmod 1$', '$x \mapsto f(x)$', we iterate the deterministic rule:

$$x_{n+1} = f(x_n), \tag{2}$$

given the initial datum x_0. Already the form is similar to equation (1). The beginning of an orbit of (2) might look like

$$0.3653 \mapsto 0.7306 \mapsto 1.4612 = 0.4612 \mapsto 0.9224 \mapsto 1.8448 = 0.8448 \mapsto \dots .$$

To understand the solutions of this dynamical system, we represent the numbers in $[0, 1]$ in binary form:

$$x_0 = \frac{a_1}{2} + \frac{a_2}{2^2} + \frac{a_3}{2^3} + \dots + \frac{a_k}{2^k} + \dots , \tag{3}$$

where each coefficient a_j takes the value 0 or 1. It is then easy to see that

$$2\left(\sum_{j=1}^{\infty} \frac{a_j}{2^j}\right) = a_1 + \frac{a_2}{2} + \frac{a_2}{2^2} + \frac{a_3}{2^3} + \dots .$$

Removing the integer part, regardless of whether $a_1 = 0$ or $a_1 = 1$, we have

$$x_1 = f(x_0) = \sum_{j=2}^{\infty} \frac{a_j}{2^{j-1}},$$

and in general the *orbit* $\{x_0, x_1, \ldots, x_k, \ldots\}$ of the point x_0 is given by:

$$x_k = f^k(x_0) = \sum_{j=k+1}^{\infty} \frac{a_j}{2^{j-k}}. \tag{4}$$

Thus, applying f is equivalent to shifting the binary point and dropping the leading coefficient:

$$(a_1 a_2 a_3 a_4 \ldots) \mapsto (a_2 a_3 a_4 \ldots), \tag{5}$$

just as multiplication by 10 shifts the decimal point. In this operation the leading symbol is removed and *information is lost.* If we knew x_0 to infinite accuracy (we had all the a_k's for k between 1 and infinity), this would not matter, for at each step we would still know the current state x_k exactly. But if we can store only, say, the first N binary places $(a_1, a_2, \ldots a_N)$, then, after N iterations, we cannot even say whether the state x_{N+1} lies above or below $1/2$! Moreover, even if two initial conditions differ only at the Nth binary place, so that they lie within distance $1/2^{N-1}$ of one another, after N iterations they fall on opposite sides of $1/2$ and thereafter behave essentially independently. This is referred to as *sensitive dependence on initial conditions*: the expansive dynamics amplifies small errors at the same time as the *modulo* 1 operation prevents orbits escaping from $[0, 1]$.

The symbolic binary description allows much more. To every infinite sequence of 0's and 1's there corresponds a point in $[0, 1]$, and *vice versa.* (The correspondence is not unique, but we can remedy this by identifying pairs of sequences whose heads coincide and whose tails are $1000\ldots0\ldots$ and $0111\ldots1\ldots$ respectively.) This implies that, given any random sequence (generated by tossing a coin and assigning Heads = 0, Tails = 1, for instance), there is an initial state x_0 such that the orbit $\{f^k(x_0)\}$ realizes that sequence. Hence our system has infinitely many such essentially random orbits. There are also infinitely many periodic orbits, corresponding to sequences such as $001001001\ldots$ and the set of all such periodic points is dense in $[0, 1]$, but since the irrational numbers form a set of full measure, the typical behavior is chaotic rather than periodic.

One can also use the binary representation to show that a dense orbit exists. Consider the sequence $\boldsymbol{a}^*$ formed by concatenating all possible sequences of lengths $1, 2, 3, \ldots$ end to end:

$$\boldsymbol{a}^* = 0\ 1\ 00\ 01\ 10\ 11\ 000\ 001 \ldots .$$

As one iterates and drops leading symbols, every possible subsequence of any given length appears at the head. This implies that the orbit of the point x^* corresponding to $\boldsymbol{a}^*$ contains points which approximate, to any desired accuracy, *every* point in the interval $[0, 1]$. Therefore $f^k(x^*)$ is a dense orbit.

The technical definition of chaos given in books such as [**Dev86, GH83**] requires just these properties: sensitive dependence, density of period orbits and a dense orbit (not all of them are independent).

How is this abstract example related to the n-body problem of equation (1)? For the planar restricted three body case of [**Poi90**], after reduction of dimension by use of conservation of momenta and transformation to a frame rotating with the (circular) orbits of the two primaries, the motion of the third (infinitesimal) mass is specified by two position and two velocity coordinates. The resulting two degree of freedom system still conserves a real-valued first integral (the Jacobi function), and so, for any given initial data, its solutions move on a three dimensional manifold. Poincaré considered special unstable periodic solutions and in their neighborhoods defined what we now call a *Poincaré map*. This is a function F defined on a two dimensional *cross section* Σ to the flow of the differential equation, which associates to each point $p \in \Sigma$ the point $q \in \Sigma$ in which the solution based at p next intersects Σ. Thus he turned a differential equation into an iterated mapping. In particular, the periodic orbit becomes a fixed point of the mapping: $s = F(s)$. He then demonstrated the existence of "solutions doublement asymptotiques," or as he later called them *homoclinic orbits*, to this periodic solution.

If these homoclinic orbits satisfy certain conditions (transversality), then it is possible to find, in any neighborhood of a homoclinic point, two regions in phase space which may be assigned the labels 0 and 1, such that orbits of the iterated mapping can be found which visit them in the preassigned order specified by any arbitrary infinite string of 0's and 1's. The unstable periodic orbit plays the same rôle as the point $x = 1/2$, which separates the domain of the map f into two regions. In fact there is a strict equivalence (a homeomorphism) between a subset of solutions of the original ordinary differential equation and the orbits of the doubling map f introduced above. Here I have summarized a story which only began with Poincaré and required the contributions of George Birkhoff [**Bir27**] and Stephen Smale [**Sma65, Sma67**] (cf. [**GH83, DH96**]) for its completion.

In this example the symbolic statement of the differential equation (1) is an exact and compact translation of Newton's laws into mathematical notation. But it is far more than that, for, with the tools of differential and integral calculus and all the theory of differential equations and dynamical systems that has been developed since Newton, it may also be (partially) solved to yield specific predictions and some understanding of both itself, and the process it models. In its analysis, the example of the doubling map is a metaphor for the 'true' behavior of solutions of the equation. Much easier to understand, the map admits us to the chaotic world of the original equation. Thus, while the bare statement of the equation is in a sense an equivalent to the poem, it implicitly carries within itself all the theory and tools of dynamical systems. It is not easy (or fruitful) to isolate it from the rest of mathematical literature, just as Hopkins's poem should not be read in isolation.

Thus far I have used the words metaphor and model somewhat interchangeably. It is time to draw a distinction. In an introductory article on turbulence [**LL86**], Leibovich and Lumley do so, as follows:

> We distinguish here between a metaphor and a model. In this context, a model is a description that comprises a mathematical statement of the physical laws governing the system, and embodies approximations of known accuracy that permit the prediction of system behavior. By contrast, a metaphor is a mathematical description that may get some important qualitative features right, but either is not derived from the physical laws controlling the system, or employs approximations of them that are not appropriate.

They go on to remark that metaphors are only needed if the models are too difficult to solve, and the metaphor much simpler.

In slight disagreement, I would propose that what may seem metaphors in the above sense can surprisingly often be rigorously shown to reside in models, much as the doubling map f resides in the Poincaré map F for the three body problem. They not only illuminate the model; they literally *are* an aspect of the model.

A recent instance of a fruitful metaphor is the use of simple chains of coupled oscillators, each described by a single phase variable θ_i, to represent central pattern generators (CPGs) in neurobiology [**CRG88**]. Here the analogue of Newton's Laws are the Hodgkin-Huxley equations [**HH52**], but these are far too complex to afford much understanding, particularly since as many sets of equations are required as there are neurons in the system. In the chain metaphor, one lumps together a group of neurons which collectively constitute an oscillator, and specifies its state at time t by the function $\theta_i(t)$. The state of each oscillator is then supposed to evolve according to a differential equation of the form

$$\dot{\theta}_i = \omega_i(\theta_i) + f(\theta_1, \ldots, \theta_n); \quad i = 1, \ldots, n. \tag{6}$$

To complete the metaphor, 'laws' are proposed for the intra- and inter-oscillator (coupling) functions ω_i and f. Here one is simplifying an interconnected network of thousands of neurons (even in a modest CPG, such as that for swimming in lamprey [**CHR82**]) to a few tens of differential equations. Nonetheless, under certain conditions, including weak or localized coupling, equations of the form (6) can be shown to emerge from almost *any* set of multidimensional coupled equations, provided only that each unit in isolation supports a non-degenerate periodic orbit (a limit cycle) [**EK84**].

In a more general sense, a number of metaphors have proven themselves useful in the construction and analysis of dynamical system models. Foremost among them is the notion of phase space and the global flow of solutions of differential equations, which grew out of Poincaré's work. His geometric viewpoint promotes a frame of mind which I find particularly helpful: using invariant manifolds and appealing to reduction theorems and (finite) classifications of generic bifurcations and normal forms [**GH83**] provides tools not only for the analysis of given differential equations, but also aids in the creation of new ones in our modeling endeavors. The CPG models of the form (6) above were constructed from this viewpoint [**CHR82**]. More recently, the introduction of knot theoretic methods into the study of three dimensional flows (e.g. [**GH93**]) suggests another class of topological metaphors which may help us approach differential equations.

4. Metaphors, Models and a Moral

In our example of a piece of applied mathematics, we find several parallels to the 'action' of Hopkins's poem. Perhaps the most obvious is that poem and differential equation alike use and adapt a ready-made language to describe and interpret something in the real world. In doing so, they construct, in the poet W.H. Auden's phrase, a "secondary world." As such, their creations are subject to all sorts of internal rules and conventions. Mathematics is no more a neutral language than English, and the modeler must clearly abide by its usages if sense is to be made. Much of the excitement and challenge in applied mathematics derives from the tension implicit in the use of "pure" mathematical objects and ideas as models for physical phenomena. One must bend and stretch the inhabitants of the secondary world to bring them closer to the primary one, but without breaking rules, and moreover one must coax the internal forms and structures of mathematics to illuminate the physics.

To pursue the linguistic metaphor in the context of differential equations: one's vocabulary and grammar surely influence the ability to speak and express oneself. After the elegant and successful development of linear algebra and analysis and before more recent progress in nonlinear analysis and dynamical systems theory, the temptation to construct *linear* models, capable of "complete" solution by known techniques, was often overwhelming. Now, with what we might call the phase space metaphor, the knowledge that surprisingly simple systems (like the doubling map) can exhibit rich behavior, and with new analytical tools at our disposal, we are emboldened to build nonlinear models. In a similar fashion, Hopkins's liberation from the standard (largely iambic) rhythms of English poetry helped prepare for the free verse of the twentieth century. (Actually, had his poems been published sooner, the history of English poetry might have been very different, as F. R. Leavis has speculated. Leavis's essay in [**Lea32**] is still one of the best introductions to Hopkins's work.)

A somewhat deeper parallel emerges from reflecting on the notions of *reading* the poem and *solving* the equation. Just as the poem only begins to reveal itself as a reader brings his or her own experience to bear on it, so the formulation of the model (here, equation (1)) is only a beginning. Clearly there is the practical desire to use the model to make predictions, but beyond this there are deeper issues of *understanding* the model itself and the illumination of the world that the model affords. Questioning some of his colleagues' preoccupation with uncovering fundamental laws to the exclusion of studying particular cases, the physicist Freeman Dyson has remarked that "... it often happens that the understanding of the mathematical nature of an equation is impossible without a detailed understanding of its solutions" [**Dys95**]. Only as particular solutions are found and related to the physical reality that the model supposedly reflects, do the capabilities and limitations of the model become clear. More subtly, the internal structure and implications of a model are seldom all transparent to its creator: drawing on the greater world of mathematics, the model generally contains more than has been intentionally put into it. In this sense, analyses or solutions of the model are analogous to critical readings and interpretations of the poem.

Implicit in this is the fact that, once stated (and published) a mathematical model becomes the common property of the global scientific community. Anyone may attempt to solve or analyze it. (How many authors have written on Laplace's equation, or on the n-body problem?) The model's creator, if he has chosen his problem well and done his work carefully, is setting a task for generations to come. This too is similar to the poem, which continues to resonate through time, acquiring new meanings and interpretations, including ones not envisaged by its author. Both poem and model are gifts to an unknown future.

There is however a notable difference. Applied mathematicians are apt to continually deconstruct and reconstruct their own models. Poets may be acute critics of the work of others, but once published, they generally leave their own poetry to fend for itself. (From this viewpoint, some of the less impressive applied mathematics—routine studies of "canonical" model equations, for example—closely resembles second-rate literary criticism.)

Here it is necessary to comment on the popular idea that science is indistinguishable from progress: new theories continually supplant old ones, and there is therefore little point in studying them, other than historical interest. In mathematics, this is of course utterly wrong. A theorem, once proven, remains true for ever. Shorter and more elegant proofs may be found, and the theorem may be generalized to include many more cases, but it remains a piece of the mathematical literature, just as a poem of Donne, Wordsworth or Hopkins will remain forever part of the canon of poetry in English. It may undergo revaluation, but it can never be supplanted. Applied mathematics is more problematic. Models certainly *can* be improved. Einstein's general relativity generalizes and improves Newtonian mechanics. However, the latter retains its value in describing and predicting events on 'reasonable' space-time scales, and relativity theory, which was essentially built on classical mechanics, cannot be understood or appreciated without first understanding the Newtonian world. This is not dissimilar from the way in which a study of Chaucer illuminates Shakespeare; or Donne, Hopkins. Poets do not replace their forebears; they build on them. In both cases, knowledge of the tradition illuminates practice and appreciation of the present. And, like poems that are forgotten and revived, mathematical models can reappear in new guises.

A third and related parallel concerns the interpretation of mathematical models and poems. I have already remarked on the question of what is meant by a *solution* of an equation such as (1). One can approach this particular question in at least four ways:

- Seek a closed form analytical solution in terms of elementary or special functions: the classical sense of integrating the equations of motion;
- Seek asymptotic or convergent series solutions;
- Seek qualitative information, such as the number and stability types of periodic orbits, the presence or absence of chaos, etc.;
- Seek numerical approximations to particular solutions.

For $n > 2$, (1) is generally insoluble in the first sense, but it is soluble in the remaining three. I once had a vehement argument with someone who insisted that *every* n degree of freedom Hamiltonian system is integrable. It turned out that he meant that local existence and uniqueness of solutions was guaranteed, while I was thinking of the (non-) existence of n independent integrals of motion. The point of this story is that one brings one's prejudices and preconceptions to the task of analyzing a differential equation just as much as to that of reading a poem.

Finally, and perhaps most significantly, in poem and model alike, the technique and tools force one to concentrate and to be clear. (A particularly vague and poor lecture was once reputedly criticized by the physicist Wolfgang Pauli for being "so bad that it wasn't even wrong.") A poem or a scientific model or theory are successful because they ignore and exclude far more than they embrace. Starting from the particular, both idealize and then focus on typical instances which stand for many such particular cases. (Newton's gravitational theory, quoted above, was couched in terms of the solar system, but is equally used to analyze and predict galactic motions.) Both share a strong aesthetic element; both prize compression: the single line of equation (1) stands for a model of any (Newtonian) universe we can imagine; the first stanza of "The Sea and the Skylark" resonates with all the waves and tides that have ever broken on the shores. And in both, the language and conventional forms pull one in particular directions. Indeed, the process thrives on this creative tension.

5. Afterword

The reader will note that I gave Hopkins's biography in brief, but not those of Poincaré or Smale. Biography is generally acknowledged to be useful in the interpretation of a poet's work, although not all scholars and critics agree (F.R. Leavis's "new criticism" [**Lea32**] stressed the work itself, with little or no reference to the author's life). In contrast, scientific creations, and particularly mathematics, are generally thought to be done quite independently of their authors' human sides. This is of course far from true; chance meetings and personal fortune or misfortune play a crucial rôle in scientists' decisions to work on particular projects, their successes or failures therein, and even on their decisions to become scientists at all. In a forthcoming book [**DH96**] Florin Diacu and I discuss some of the human factors behind the careers of Poincaré, Smale, Kolmogorov and other pioneers of dynamical systems.

Acknowledgments

This paper was developed from the 'Conférence publique' of the Aisenstadt Lecture Series at the Centre de Recherches Mathématiques, Montréal, on April 28, 1986, which was published as CRM-1387 Rapport. The author thanks the CRM for the opportunity to deliver that lecture. He also thanks Florin Diacu for his comments and corrections.

References

[Bir27] G. D. Birkhoff, *Dynamical systems*, Amer. Math. Soc., Providence, RI, 1927.

[CHR82] A. H. Cohen, P. Holmes, and R. H. Rand, *The nature of the coupling between segmental oscillators of the lamprey spinal generator for locomotion: A mathematical model*, J. Math. Biol. **13** (1982), 345–369.

[CRG88] A. H. Cohen, S. Rossignol, and S. Grillner (eds.), *Neural control of rhythmic movements in vertebrates*, John Wiley and Sons, New York, 1988.

[Dev86] R. L. Devaney, *An introduction to chotic dynamical systems*, Benjamin/Cummings, Menlo Park, CA, 1986.

[DH96] F. Diacu and P. Holmes, *Celestial encounters: The origins of chaos and stability*, Princeton University Press, Princeton, N.J., 1996.

[Dia95] F. Diacu, *The solution of the n-body problem*, Preprint DMS–704–IR, Dept. of Mathematics and Statistics, University of Victoria, B.C. V8W 3P4, Canada, 1995.

[Dys95] F. Dyson, *The scientist as rebel*, New York Review of Books **XLII (9)** (May 25, 1995), 31–33.

[EK84] G. B. Ermentrout and N. Kopell, *Frequency plateaus in a chain of weakly coupled oscillators*, I, SIAM J. Math. Anal. **15** (1984), 215–237.

[GH83] J. Guckenheimer and P. J. Holmes, *Nonlinear oscillations, dynamical systems and bifurcations of vector fields*, Springer-Verlag, New York, 1983.

[GH93] R. Ghrist and P. Holmes, *Knots and orbit genealogies in three dimensional flows*, Bifurcations and Periodic Orbits of Vector Fields (D. Schlomiuk, ed.), 1993, Proc. of the NATO Advanced Study Institute and Séminaire de mathématiques supérieures, Montréal, Canada, July 13-24, 1992, pp. 185–239.

[HH52] A.L. Hodgkin and A.F. Huxley, *A quantitative description of membrane current and its applications to conduction and excitation in nerve*, J. Physiol. **117** (1952), 500–544.

[Hop18] Robert Bridges (ed.), *Poems of Gerard Manley Hopkins*, Oxford University Press, Oxford, U.K., 1918; Fifth impression, (W. H. Gardner, ed.), 1956.

[Lea32] F. R. Leavis, *New bearings in English poetry*, Chatto and Windus, London, U.K., 1932; Reprinted with 'Retrospect 1950'; Republished by Penguin Books, 1963.

[LL86] S. Leibovich and J. L. Lumley, *Complex fluid motion: Models and metaphors*, Engineering: Cornell Quarterly **20** (1986), no. 3, 27–35.

[New86] Isaac Newton, *Philosophiae naturalis principia mathematica*, S. Pepys, Royal Society Press, London, 1686; Transl. Andrew Motte, 1729; revised, with commentary, Florian Cajori, University of California Press, Berkeley, CA, 1934.

[Pet93] I. Peterson, *Newton's clock: Chaos in the solar system*, W. H. Freeman and Company, New York, 1993.

[Poi90] H. Poincaré, *Sur le problème des trois corps et les équations de la dynamique*, Acta Mathematica **13** (1890), 1 270.

[Rot95] E. Rothstein, *Emblems of mind: The inner life of music*, Times Books/Random House, New York, 1995.

[Sma65] S. Smale, *Diffeomorphisms with many periodic points*, Differential and Combinatorial Topology: A Symposium in Honor of Marston Morse (Princeton, NJ) (S.S. Cairns, ed.), Princeton University Press, 1965, pp. 63–70.

[Sma67] ——— *Differentiable dynamical systems*, Bull. Amer. Math. Soc. **73** (1967), 747–817.

[Sno59] C.P. Snow, *The two csultures and the scientific revolution*, Cambridge University Press, Cambridge, U.K., 1959.

[Wil95] A. Wiles, *Fermat's last theorem: A personal account*, Lecture delivered at Princeton University on May 15th, 1995.

Program in Applied and Computational Mathematics and Department of Mechanical and Aerospace Engineering, Princeton University, Princeton, NJ 08544, U.S.A.

E-mail address: pholmes@rimbaud.princeton.edu

Centre de Recherches Mathématiques
CRM Proceedings and Lecture Notes
Volume **11**, 1997

Some Algebraic Geometry of Integrable Systems

J. C. Hurtubise

Algebraic geometry, through the presence of Abelian integrals, has been part of the theory of integrable Hamiltonian systems since the nineteenth century. One starts with an integrable system, for which the compact joint level sets of the Hamiltonian are tori. In good cases ("algebraically integrable systems" [**AvM2, vM**]), they remain tori after complexification and a suitable compactification, and are Abelian varieties. The flows are then linear on these varieties. In classical cases, these varieties are most often Jacobians of curves, and the linearisation is expressed in terms of Abelian integrals on the curves.

While this is the origin, it is by no means the end of the story, and there are many interesting links between symplectic and algebraic geometry, leading through the theory of curves to surfaces, and higher dimensional varieties. What follows is a fairly biased sample of some of these questions.

The CRM has been the focal point of a lot of research in integrable systems over the last fifteen years, both through the work of its members and its visitors, and through the hosting of some exciting conferences. I would like to express my gratitude to several members of the CRM, in particular John Harnad, for introducing me to this rich and diverse subject.

1. A Very Classical Integrable System

We begin with a classical integrable system, the Neumann oscillator, which, it is hoped, will provide some concrete illustration of some of the points we wish to discuss. This oscillator appeared in the work of C. Neumann [**N**] in 1859. It treats the motion of a particle on the $(n-1)$-dimensional sphere under the influence of a quadratic potential. Let

$$\begin{aligned} A &= \operatorname{diag}(\alpha_i), \quad i = 1, \dots, n \\ a(\lambda) &= \det(\lambda \mathbf{1} - A) = \prod_{i=1}^{n} (\lambda - \alpha_i) \end{aligned} \tag{1.1}$$

where the α_i are real and distinct. If $\boldsymbol{x} = (x_1, \dots, x_n)$ are standard coordinates on $\mathbb{R}^n$, and $\boldsymbol{y} = (y_1, \dots, y_n)$ the associated momenta, we can describe the cotangent

1991 *Mathematics Subject Classification.* Primary: 14K20; Secondary: 70H05.
The author of this article would like to thank NSERC and FCAR for their support.
This is the final form of the paper.

bundle $T^*(S^{n-1})$ as being cut out in $\mathbb{R}^{2n}$ by the constraints

$$\boldsymbol{x}^T\boldsymbol{x} = 1, \quad \boldsymbol{y}^T\boldsymbol{x} = 0. \tag{1.2}$$

The Hamiltonian on $T^*(S^{n-1})$ is

$$H(\boldsymbol{x},\boldsymbol{y}) = \frac{1}{2}[\boldsymbol{y}^T\boldsymbol{y} + \boldsymbol{x}^TA\boldsymbol{x}]. \tag{1.3}$$

Using the theory of constrained systems, one can instead consider the Hamiltonian flow on $\mathbb{R}^{2n}$ corresponding to

$$H_0(\boldsymbol{x},\boldsymbol{y}) = \frac{1}{2}\left[(\boldsymbol{x}^T\boldsymbol{x})(\boldsymbol{y}^T\boldsymbol{y}) + \boldsymbol{x}^TA\boldsymbol{x} - (\boldsymbol{x}^T\boldsymbol{y})^2\right], \tag{1.4}$$

which preserves the first of the constraints in (1.2); one then projects to satisfy the second constraint. This essentially amounts to a symplectic reduction. The Hamiltonian H_0 Poisson commutes with (and indeed is a combination of) the functions

$$I_i := \sum_{j=1,j\neq i}^{n} \frac{(x_iy_j - x_jy_i)^2}{\alpha_i - \alpha_j} + x_i^2. \tag{1.5}$$

These all commute among themselves, and so the system is completely integrable. We can define ellipsoidal coordinates $\lambda_\mu, \mu = 1,\dots,n-1$ by

$$\sum_{i=1}^{n} \frac{x_i^2}{\lambda - \alpha_i} = \frac{\prod_{\mu=1}^{n-1}(\lambda - \lambda_\mu)}{a(\lambda)}. \tag{1.6}$$

Let

$$\mathcal{P}(\lambda) = -\frac{a(\lambda)}{4}\sum_{i=1}^{n}\frac{I_i}{\lambda - \alpha_i} = P_{n-1}\lambda^{n-1} + P_{n-2}\lambda^{n-2} + \dots + P_0. \tag{1.7}$$

The coefficients of the polynomial $\mathcal{P}$ are functions of the commuting Hamiltonians I_i, and indeed, on the reduced manifold, up to an additive constant, $H_0 = 2P_{n-2}$. The conjugate momenta to λ_μ are

$$\zeta_\mu = \frac{1}{2}\sum_{i=1}^{n}\frac{x_iy_i}{\lambda_\mu - \alpha_i} = \sqrt{\frac{\mathcal{P}(\lambda_\mu)}{a(\lambda_\mu)}}. \tag{1.8}$$

Defining the Liouville generating function (see, e.g.,[**Ar**]),

$$S(\lambda_\mu, P_i) = \sum_{\mu=1}^{n-1}\int_0^{\lambda_\mu}\sqrt{\frac{\mathcal{P}(\lambda)}{a(\lambda)}}d\lambda. \tag{1.9}$$

We have that $\partial S/\partial\lambda_\mu = \zeta_\mu$; the canonically conjugate coordinates to the P_i, undergoing linear flow, are then:

$$Q_j := \frac{\partial S}{\partial P_j} = \frac{1}{2}\sum_{\mu=1}^{n-1}\int_0^{\lambda_\mu}\frac{\lambda^j d\lambda}{\sqrt{a(\lambda)\mathcal{P}(\lambda)}} = b_jt, \quad j = 0,\dots,n-2, \tag{1.10}$$

where, for our Hamiltonian $H_0 = 2P_{n-2}$,

$$b_{n-2} = 2, \quad b_j = 0, \quad j < n-2. \tag{1.11}$$

The integrands in (1.10) form a basis for the holomorphic differentials on a hyperelliptic curve $\mathcal{S}$ cut out by the equation

$$z^2 - a(\lambda)\mathcal{P}(\lambda) = 0. \tag{1.12}$$

One has a divisor $\sum p_\mu$ on the curve, where the p_μ have coordinates $(\lambda_\mu, z_\mu = a(\lambda_\mu)\zeta_\mu)$, and (1.10) tells us that the image of the divisor under the Abel map evolves linearly in the Jacobian of the curve.

2. Loop Algebras and Spectral Curves

This linear flow can be realised directly as a flow of line bundles. There have been several discussions of the algebraic geometry of this system; we refer to [**M, F, Ra, Sch**], and give the development of [**AHP**]. Consider the matricial function:

$$\mathcal{N}(\lambda) = \lambda^{-1}\begin{pmatrix} 0 & -1/2 \\ 0 & 0 \end{pmatrix} + \frac{\lambda^{-1}}{2}\begin{pmatrix} -\sum_{i=1}^n x_i y_i/(\lambda-\alpha_i) & -\sum_{i=1}^n y_i^2/(\lambda-\alpha_i) \\ \sum_{i=1}^n x_i^2/(\lambda-\alpha_i) & \sum_{i=1}^n x_i y_i/(\lambda-\alpha_i) \end{pmatrix}. \tag{2.1}$$

The equations of motion for our Hamiltonian system are equivalent to

$$\frac{d\mathcal{N}}{dt} = [\mathcal{B}, \mathcal{N}], \tag{2.2}$$

where

$$\mathcal{B} = \begin{pmatrix} \boldsymbol{x}^T\boldsymbol{y} & \lambda + \boldsymbol{y}^T\boldsymbol{y} \\ -\boldsymbol{x}^T\boldsymbol{x} & -\boldsymbol{x}^T\boldsymbol{y} \end{pmatrix}. \tag{2.3}$$

We remark that the equation is in Lax form, and so the infinitesimal evolution is by conjugation, and the spectrum of $\mathcal{N}$ is an invariant of the flow. This invariant, the *spectral curve*, is cut out in the plane by $\det(\mathcal{N}(\lambda) - \zeta I) = 0$, which by the change of variable $z = a(\lambda)\zeta$, is exactly the equation of the curve in (1.12).

While the spectrum, that is the set of eigenvalues, of $\mathcal{N}$ is invariant, the corresponding eigenlines defined for each point of the spectral curve gives us (generically) a line bundle over the spectral curve which is not invariant, and indeed flows linearly in a component of the Picard variety, or equivalently the Jacobian of the spectral curve.

We will try to see this in a more general context. For a finite dimensional complex Lie algebra $\mathfrak{g}$ corresponding to a group G, we consider the loop algebra of semi-infinite formal Laurent series

$$\tilde{\mathfrak{g}} = \left\{ \sum_{i=-\infty}^{k} g_i\lambda^i \mid k \in \mathbb{Z}, g_i \in \mathfrak{g} \right\}. \tag{2.4}$$

This can then be written as a sum of two subalgebras

$$\tilde{\mathfrak{g}}_+ = \left\{ \sum_{i=0}^{k} g_i\lambda^i \mid k \geq 0, g_i \in \mathfrak{g} \right\}, \quad \tilde{\mathfrak{g}}_- = \left\{ \sum_{i=-\infty}^{-1} g_i\lambda^i \mid g_i \in \mathfrak{g} \right\}. \tag{2.5}$$

There is a pairing

$$\begin{aligned} \tilde{\mathfrak{g}} \times \tilde{\mathfrak{g}} &\longrightarrow \mathbb{C} \\ (a,b) &\longmapsto \operatorname{res}\operatorname{tr}(ab), \end{aligned} \tag{2.6}$$

where tr denotes an invariant form on $\mathfrak{g}$, for example the Killing form, and res is the residue at $\lambda = \infty$. If tr is non degenerate, formula (2.6) identifies $\tilde{\mathfrak{g}}_-$ with a (formal) dual of $\tilde{\mathfrak{g}}_+$ (in what follows, it is in fact irrelevant to identify a precise functional-analytic dual; this formal dual suffices). Under this identification, the coadjoint action of $a \in \tilde{\mathfrak{g}}_+$ on $\tilde{\mathfrak{g}}_-$ is:

$$b \longmapsto [a, b]_- \tag{2.7}$$

i.e., the adjoint action in $\tilde{\mathfrak{g}}$, followed by projection to $\tilde{\mathfrak{g}}_-$.

As for any Lie algebra, the coadjoint orbits in $\tilde{\mathfrak{g}}_- = (\tilde{\mathfrak{g}}_+)^*$ have a natural (Kostant-Kirillov) symplectic structure. We are interested in finite-dimensional orbits in $(\tilde{\mathfrak{g}}_+)^*$; a theorem [**Hu1**] tells us that for $\mathfrak{g}$ semisimple, they are all of elements of $\tilde{\mathfrak{g}}_-$ of the form

$$\mathcal{N}(\lambda) = \frac{\mathcal{L}(\lambda)}{a(\lambda)} \tag{2.8}$$

where $\mathcal{L}(\lambda)$ lies in $\tilde{\mathfrak{g}}_+$, $a(\lambda)$ is a scalar polynomial, and degree$(\mathcal{L}) <$ degree(a). The scalar $a(\lambda)$ is an invariant of the orbit.

There are natural commuting flows on these orbits, given to us by the Adler Kostant Symes theorem [**A, K, S**]. This says that the invariant functions on $(\tilde{\mathfrak{g}})^*$, restricted to $(\tilde{\mathfrak{g}}_+)^*$, Poisson commute. For $\mathfrak{g} = \mathrm{sl}(n)$, for example, these functions are combinations of functions of the form

$$f_{i,k}\big(\mathcal{N}(\lambda)\big) = \operatorname{res}\operatorname{tr}\big(\lambda^i \mathcal{N}^k(\lambda)\big), \quad i, k \in \mathbb{Z}, k > 0. \tag{2.9}$$

They are essentially the coefficients of the equation of the *spectral curve* $\mathcal{S}$.

$$\det(\mathcal{L}(\lambda) - z\mathbf{1}) = 0. \tag{2.10}$$

The Hamiltonians (2.9) give flows in $\mathcal{N}$, or, equivalently, in $\mathcal{L}$:

$$\frac{d\mathcal{N}}{dt} = \big[(k\lambda^i \mathcal{N}^{k-1})_+, \mathcal{N}\big], \quad \frac{d\mathcal{L}}{dt} = \big[\big(k\lambda^i (a(\lambda)^{-1}\mathcal{L})^{k-1}\big)_+, \mathcal{L}\big] \tag{2.11}$$

with the spectral curve as an invariant of the flow. The system (2.2) is of this form.

These systems have a nice algebro-geometric integration [**AHH1, B1**]. If $a(\lambda)$ is of degree $(m+1)$, let $\mathcal{T}$ be the total space of m-th power $\mathcal{O}(m)$ of the hyperplane line bundle over $\mathbb{P}^1$. It is covered by two coordinate patches U_0 and U_1, with coordinates (z, λ) and $(z', \lambda') = (z\lambda^{-m}, \lambda^{-1})$ respectively. The equation (2.10) defines a curve in the patch U_0, which extends to a compact curve $\mathcal{S}$ in $\mathcal{T}$. Under the projection of $\mathcal{T}$ to $\mathbb{P}^1$, $\mathcal{S}$ is an r-fold branched cover of $\mathbb{P}^1$; the r points $\big(\lambda, z_i(\lambda)\big), i = 1, \ldots, r$ of $\mathcal{S}$ over $\lambda \in \mathbb{P}^1$ correspond to the r eigenvalues $z_i(\lambda)$ of $\mathcal{L}(\lambda)$. If $\mathcal{O}(j)$ denotes also the lift of $\mathcal{O}(j)$ to $\mathcal{T}$, then one can define a sheaf

$$0 \longrightarrow \mathcal{O}(-m)^{\oplus r} \xrightarrow{\big(z\mathbf{1} - L(\lambda)\big)} \mathcal{O}^{\oplus r} \xrightarrow{\pi} E \longrightarrow 0. \tag{2.12}$$

If $\mathcal{S}$ is smooth and reduced, as will be the case generically, then E is a line bundle over $\mathcal{S}$. We note that this line bundle is not the line bundle of eigenspaces, as it is built from the cokernel of $\big(z\mathbf{1} - L(\lambda)\big)$ rather than the kernel. It nevertheless encodes the same information and is more convenient from the point of view of algebraic geometry. We have:

LEMMA 2.1 ([**AHH1**]). a) *The degree of E is $g+r-1$ where $g = \mathrm{genus}(\mathcal{S})$.*
b) $H^0\big(\mathcal{S}, E(-1)\big) = H^1\big(\mathcal{S}, E(-1)\big) = 0$.

The line bundles satisfying (b) are dense in the degree $g + r - 1$ component of the Picard variety of line bundles over $\mathcal{S}$. If $D_{\lambda_0} = \mathcal{S} \cap \{\lambda = \lambda_0\}$ (so that D_{λ_0} is, generically, r points), lemma 2.1 and the long exact sequence in cohomology of

$$0 \to E(-1) \to E \to E|_{D_{\lambda_0}} \to 0 \tag{2.13}$$

imply that the restriction map

$$H^0(\mathcal{S}, E) \to H^0(D_{\lambda_0}, E) \simeq \mathbb{C}^r \tag{2.14}$$

is an isomorphism. Also, sections of E over $\mathcal{S}$ are all in the image of global sections of $\mathcal{O}^{\oplus r}$ under the map π of (2.12).

Up to conjugation by a constant matrix, one can reconstruct $\mathcal{L}(\lambda_0)$ for all λ_0 from an E satisfying a) and b) of lemma 2.1, by the commutative diagram

$$\begin{array}{ccc} H^0(\mathcal{S}, E) & \longrightarrow & H^0(D_{\lambda_0}, E) \\ {\scriptstyle \mathcal{L}(\lambda_0)}\downarrow & & \downarrow{\scriptstyle \times Z} \\ H^0(\mathcal{S}, E) & \longrightarrow & H^0(D_{\lambda_0}, E) \end{array} \tag{2.15}$$

Here $\times Z$ is the multiplication map by the fiber coordinate z on D_{λ_0}. Essentially, evaluation at D_{λ_0} diagonalises $\mathcal{L}(\lambda_0)$.

Explicitly, if $s_j, j = 1, \dots, r$ are a basis of $H^0(\mathcal{S}, E)$ and $z_i(\lambda), i = 1, \dots, r$ are the z-coordinates of the points of $\mathcal{S}$ above λ, and if ev is evaluation with respect to a fixed trivialisation of E over U_0, we define the matrix

$$\psi_{ij}(\lambda) = \operatorname{ev}(s_j)\big(\lambda, z_i(\lambda)\big) \tag{2.16}$$

and obtain, up to conjugation by a constant matrix (corresponding to different choices of basis for $H^0(\mathcal{S}, E)$) [**AHH1**]:

$$\mathcal{L}(\lambda) = \psi(\lambda)^{-1} \cdot \operatorname{diag}\big(z_i(\lambda)\big) \cdot \psi(\lambda). \tag{2.17}$$

Flows of $\mathcal{L}(\lambda)$ then correspond to flows of line bundles over the curve. Let us fix an initial line bundle E_0, with transition function $F(\lambda, z)$ from the open set U_0 to U_1. One can show [**AHH1**] that if one takes the flow of line bundles E_t corresponding to the transition functions $F(\lambda, z)\exp\big(t(k\lambda^i \tilde{a}(\lambda) z^{k-1})\big)$ (where $\tilde{a}(\lambda)$ is the Laurent expansion of $a(\lambda)^{-1}$ around infinity truncated at a suitable high order), and chooses the bases of $H^0(\mathcal{S}, E_t)$ in the proper fashion, then one solves equation (2.11). One can compute explicit theta-function formulae.

This construction of a pair (curve, line bundle on the curve) is invariant under conjugation of $\mathcal{N}$ by a constant matrix. Symplectically, this means that we are working on a reduction of the coadjoint orbit by this conjugation. This reduction amounts to fixing the highest order term $\mathcal{N}_{-1}$ of $\mathcal{N}$, and quotienting by its stabiliser. The resulting reduced orbit has dimension $2g$. The Hamiltonians (the coefficients of the defining equation of the spectral curves) descend to the reduction, and define there an integrable system whose g-dimensional Lagrangian leaves are open sets of the Picard varieties of the spectral curves. The reduced orbit is then a space of pairs (curves $\mathcal{S}$, sheaves E over $\mathcal{S}$).

3. Darboux Coordinates on the Coadjoint Orbits

Let us summarise what we have so far for the Neumann system.

1. An explicit solution in terms of Abelian integrals, obtained from a Liouville generating function, in the context of the classical symplectic geometry of $T^*(S^{n-1}) \subset \mathbb{R}^{2n}$, or, after complexification, in $\mathbb{C}^{2n}$.

2. A reformulation of the system on a (reduced) coadjoint orbit. Here one has an integrable system with an algebro-geometric integration scheme whose symplectic content is, however, not at all clear.

We would like to reconcile the two methods. To do this, we will associate some coordinates to our line bundles. We first note that a divisor of a section of a line bundle determines the bundle. Our bundles E come equipped with a family of sections, which are all projections of sections of $\mathcal{O}^{\oplus r}$ in (2.12). We choose one such section v corresponding to an eigenvector V of $\mathcal{N}_{-1}$, which is then left invariant up to scale by the symplectic reduction. The divisor of v as a section of E then consists of $g-1+r$ points on $\mathcal{S}$, $r-1$ of which are the points of the spectral curve over $\lambda = \infty$ which are not the eigenvalue of v. We write the other g points as a sum $\sum_\mu (\lambda_\mu, \zeta_\mu)$ of points in the plane, where $\zeta = a(\lambda)^{-1}z$.

The coordinates (λ_μ, ζ_μ) for a point on the curve are of course redundant: the equation of the curve determines ζ_μ as an algebraic function of λ_μ. We would like to think of the pairs $\sum_\mu (\lambda_\mu, \zeta_\mu)$ as not only determining the points on the curve, but indeed which curve in the family corresponding to the coadjoint orbit we are on: the curve must be such that it passes through the (λ_μ, ζ_μ). One proves that the g pairs $\sum_\mu (\lambda_\mu, \zeta_\mu)$ do indeed determine $(\mathcal{S}, E)$, over an open set, and moreover:

THEOREM 3.1 ([**AHH2**]). *The λ_μ, ζ_μ define a Darboux coordinate system on the reduced coadjoint orbit.*

There are also versions of this theorem for the unreduced orbit [**AHH2**]; also, if one replaces the standard Lie Poisson bracket on the loop algebra by the "quadratic" bracket of Sklyanin [**Sk**], one has similar coordinates on the symplectic leaves [**Sc**].

The divisor coordinates can be computed explicitly. Indeed, if we represent the section v by a constant section V of $\mathcal{O}^{\oplus r}$, v vanishes precisely when V lies in the image of $\big(z\mathbf{1} - L(\lambda)\big)$. This in turn, for simple eigenvalues, occurs when

$$[(z\mathbf{1} - L(\lambda))]_{\mathrm{adj}} V = 0, \tag{3.1}$$

where adj denotes the classical adjoint (matrix of cofactors). If we do this for our matrix (2.1) for the Neumann problem, with $V = (1,0)^T$, we find:

$$\sum_{i=1}^n \frac{x_i^2}{\lambda - \alpha_i} = \frac{\prod_{\mu=1}^{n-1}(\lambda - \lambda_\mu)}{a(\lambda)}, \qquad \zeta_\mu = \frac{1}{2}\sum_{i=1}^n \frac{x_i y_i}{\lambda_\mu - \alpha_i}, \tag{3.2}$$

precisely the coordinates of (1.6), (1.8).

Given the coordinates (λ_μ, ζ_μ), there is a simple, entirely classical integration on the reduced orbit using a Liouville generating function, which generalises the integration of section 1. If P_i are the Hamiltonians (=coefficients of the defining equation for the spectral curve), one has an algebraic function $\zeta(\lambda, P_1, \ldots, P_g)$ defined by the spectral curve. The generating function G is defined by

$$G(\lambda_\mu, P_j) = \sum_{\mu=1}^g \int_{\lambda_0}^{\lambda_\mu} \zeta(\lambda, P_1, \ldots, P_g)\, d\lambda. \tag{3.3}$$

One has $(\partial G)/(\partial \lambda_\mu) = \zeta_\mu$, and so, if P_j is the coefficient of $\lambda^\alpha z^\beta$ in $P(\lambda, z)$, setting

$$Q_j = \frac{\partial G}{\partial P_j} = \sum_{\mu=1}^g \int_{\lambda_0}^{\lambda_\mu} \frac{\lambda^\alpha z^\beta}{(\partial P_{L'}/\partial z)}\, d\lambda \tag{3.4}$$

gives us the coordinates in which the flow linearizes. One can show, using the Poincaré residue formula, that the integrands in (3.5) as P_i vary give a basis of the holomorphic differentials on the curve. One then has, from the symplectic point of view, that the flow of the divisor $\sum_\mu(\lambda_\mu, \zeta_\mu)$ maps to a linear flow in the Jacobian of the spectral curve under the Abel map, so that the line bundle E evolves linearly. Again, there is a version of this integration for the unreduced case [**AHH2**].

There is a more invariant version of the Darboux coordinate theorem, linking the algebraic geometry of the spectral curves and the symplectic geometry of the orbit. Indeed, the infinitesimal variations of the curve in $\mathcal{T}$ correspond to sections of the normal bundle N of the curve in $\mathcal{T}$, and one can show that the constraints on the curves associated to a coadjoint orbit restrict one to a subspace of $H^0(\mathcal{S}, N)$ which can be naturally identified with $H^0(\mathcal{S}, K)$, the sections of the canonical bundle. On the other hand, variations of the line bundle E correspond to the space $H^1(\mathcal{S}, \mathcal{O})$. One can then write down a decomposition of the tangent space to the orbit as a sum

$$H^0(\mathcal{S}, K) \oplus H^1(\mathcal{S}, \mathcal{O}). \tag{3.5}$$

The two spaces above are Serre duals of each other. There is a natural skew form one can define on the sum of a vector space and its dual, and one can prove that the Darboux coordinate theorem implies that this skew form is the Kostant-Kirillov form [**AHH3, AHH4**]. There is also a version of this for the $\tilde{\mathfrak{g}}_+$ associated to arbitrary semi-simple algebras $\mathfrak{g}$ [**Hu3**].

4. Symmetric Products of Surfaces

The loop algebra framework seems able to accommodate a large number of the classical known algebraically integrable systems such as geodesics on the ellipsoid, the various integrable tops, the Toda lattice, as well as the finite gap cases of integrable p.d.e.s. such as the KdV, NLS, CNLS, and Boussinesq equations. (We refer the reader to [**AvM, FRS, M, RS, AHP, HH**]). One chooses the correct classes of coadjoint orbits, and the desired equations appear as the equations for certain matrix coefficients. One could ask oneself if the loop algebra systems are not some sort of universal example, or more properly, if there is some sort of invariant which could allow one to determine when an integrable system fits into this framework.

The preceding section indicates a way to proceed. Indeed, the Darboux coordinate theorem told us that there was a symplectic equivalence between open sets of the coadjoint orbit and the g-fold symmetric product $SP^g(Q)$ of a symplectic surface Q with form $d\lambda \wedge d\zeta$. This symmetric product is not smooth, and it turns put that the correct symplectic desingularisation is given by the Hilbert scheme $\widetilde{SP}^g(Q)$ of length g 0-dimensional subschemes of Q. The integrable system expresses the orbit as an open set of a Lagrangian fibration $\mathbb{J}$ over a base U parametrising the spectral curves, with fibers the Jacobians of the spectral curves $\mathcal{S}$. These Jacobians are (generically bijective) images of the g-fold products $SP^g(\mathcal{S})$. If $\mathbb{S}$ is the $(g+1)$-dimensional "universal curve" over U, then let $SP^g_U(\mathbb{S})$ denote the ($2g$-dimensional) g-fold fibre-wise symmetric product over U. One has a diagram of maps between

$2g$-dimensional varieties:

$$\begin{array}{ccccc} & & SP^g_U(\mathbb{S}) & & \\ & \overset{\Pi}{\swarrow} & & \overset{\tilde{I}}{\searrow} & \\ \widetilde{SP}^g(Q) & & & & \mathbb{J} \end{array} \tag{4.1}$$

and this provides a birational symplectic equivalence between $SP^g_U(\mathbb{S})$ and $\widetilde{SP}^g(Q)$.

The universal curve $\mathbb{S}$ maps to $\mathbb{J}$ via the Abel map, choosing appropriate base points (i.e., a section $U \to \mathbb{J}$). There is also a natural map p to Q, so that one has:

$$\begin{array}{ccc} \mathbb{S} & \xrightarrow{A} & \mathbb{J} \\ \downarrow{\scriptstyle p} & & \\ Q & & \end{array} \tag{4.2}$$

If one chooses basepoints on $\mathbb{S}$ by lifting from Q, one has as a consequence of theorem 3.1 that the lift $A^*\Omega$ of the symplectic form on $\mathbb{J}$ is equal to the lift $p^*\omega$ of the form on Q. In turn, this implies that $\mathbb{S}$ is a coisotropic submanifold of $\mathbb{J}$, and that the quotient of $\mathbb{S}$ by the null foliation is Q.

This remark turns out to be the key to understanding a general Lagrangian foliation $\mathbb{J} \to U$ whose fibres are Jacobians of curves. Here U will some small ball in $\mathbb{C}^g$, so that we are only considering the local properties. Let Ω be the symplectic form on $\mathbb{J}$. One has, corresponding to $\mathbb{J}$, a universal curve $\mathbb{S}$, which as above is $(g+1)$-dimensional, and again, choosing base points, an Abel map $A : \mathbb{S} \to \mathbb{J}$. Let $S_h \subset \mathbb{S}$ be the curve corresponding to $h \in U$. We can show:

THEOREM 4.1 ([**Hu2**]). *Let $g \geq 3$.*

(i) *Let $A^*\Omega \wedge A^*\Omega = 0$. Under the embedding A, the variety $\mathbb{S}$ is coisotropic. Quotienting by the null foliation, one obtains, restricting U if necessary, a surface Q to which the form $A^*\Omega$ projects, defining a symplectic form ω on Q. The curves S_h all embed in Q.*

(ii) *If A, $\tilde{A}$ are two Abel maps with $A^*\Omega \wedge A^*\Omega = 0$, $\tilde{A}^*\Omega \wedge \tilde{A}^*\Omega = 0$, then $A^*\Omega = \tilde{A}^*\Omega$, and so Q depends only on $\mathbb{S}$ and not on the particular Abel map chosen.*

(iii) *There is a symplectic isomorphism*

$$\Phi : \widetilde{SP}^g(Q, \omega) \to \mathbb{J}, \tag{4.3}$$

defined over a Zariski open set, between $\mathbb{J}$ and a desingularisation $\widetilde{SP}^g(Q,\omega)$ of the g-fold symmetric product $SP^g(Q,\omega)$ of Q. The symmetric product $SP^g(S_h)$ of the curves is Lagrangian in $\widetilde{SP}^g(Q)$, and the restriction of Φ to $SP^g(S_h)$ is the extension A^g of the Abel map A to $SP^g(S_h)$

$$A^g : SP^g(S_h) \to J_h. \tag{4.4}$$

$\widetilde{SP}^g(Q,\omega)$ is the Hilbert scheme of length g 0-dimensional subschemes of Q.

There is some interesting projective geometry associated to Q. For this, we will distinguish two cases:

(a) none of the curves S_h are hyperelliptic,
(b) all of the curves S_h are hyperelliptic,

leaving the "mixed" case aside. We note that the fact that the surface Q is symplectic tells us that the normal bundle N of a curve S in the surface is the canonical bundle K_S of the curve. This then tells us that two "infinitesimally close" curves in our family intersect in a canonical divisor. We now suppose that this property holds macroscopically, that is that the intersection $S_h \cap S_{h'}$ of any two curves $S_h, S_{h'}$ in our family is a canonical divisor on either curve. Under this hypothesis, we have:

THEOREM 4.2. *Let $g \geq 3$.*

(i) *On Q, the line bundles $\mathcal{O}(S_h)$ are equivalent for all h, and restrict to the canonical bundle on any $S_{h'} \subset Q$. In case* (a) *the linear system $|\mathcal{O}(S_h)|$ embeds Q into $\mathbb{P}^g$, and the curves S_h are then given by hyperplane sections. The restriction of this embedding to any S_h is the canonical embedding. In case* (b)*, the hyperelliptic involutions on the S_h descend to $i : Q \to Q$, and the linear system $|\mathcal{O}(S_h)|$ embeds $Q/i \cong R$ in $\mathbb{P}^g$, in such a way that the hyperplane sections of R are the lines $S_h/i = \mathbb{P}^1$.*

(ii) *In case* (a)*, Q is embedded as an open subset of a complete algebraic surface $\overline{Q}$ of degree $2g-2$, with Kodaira dimension $-\infty$ or 0. In case* (b)*, R is embedded as an open subset of a complete rational algebraic surface $\overline{R}$ of degree $g-1$. The branching locus B of the involution also completes to an algebraic curve, and so one can define an algebraic branched cover $\overline{Q}$ completing Q. In both cases, the symplectic form ω is the restriction of a meromorphic form $\bar{\omega}$ on $\overline{Q}$.*

(iii) *Any surface Q of degree $(2g-2)$ in $\mathbb{P}^g$, along with a regular and nowhere zero section of its canonical bundle on a neighbourhood of a hyperplane section, defines an integrable system $\mathbb{J}$ of type* (a)*. Also, any surface R of degree $(g-1)$ in $\mathbb{P}^g$, along with a branching curve B of degree $2g+2$, defines a $\mathbb{J}$ of type* (b) *if the double cover Q of R branched along B admits a regular and nowhere zero section of its canonical bundle on a neighbourhood of the inverse image of a hyperplane section.*

There are then two sorts of invariants one can associate to our Lagrangian fibration. The first is local, and is given by the vanishing of $A^*\Omega \wedge A^*\Omega$; this must vanish for the system to be a symmetric product of surfaces. When this does vanish, one has a finer invariant, namely a surface with a family of curves, and in good situations we get a projective embedding of the surface such that the curves are cut out by hyperplanes. This projective class is then also an invariant of the system.

There are several examples of these surface systems (details are in [**Hu2**]):

1) **Coadjoint orbits:** The surface Q, or rather its closure $\overline{Q}$ in $\mathbb{P}^g$, is a rational surface, obtained by blowing up a compactification of the surface $\mathcal{T}$ at the points where the spectral curves $\mathbb{S}$ intersect $\lambda = \alpha_i$ or $\lambda = \infty$ (these points of intersection are constant along the reduced coadjoint orbit), and mapping the result to $\mathbb{P}^g$ via the appropriate linear system.

2) **Quadratic bracket:** There is more than one interesting Poisson bracket that one can define on the loop algebra $\tilde{\mathfrak{g}}_-$, when $\mathfrak{g} = \mathfrak{sl}(r, \mathbb{C})$. The standard Lie-Poisson bracket has the coadjoint orbits as symplectic leaves, with the Kostant-Kirillov form. There is also the quadratic bracket [**Sk**], defined on the matrix coefficient functions by:

$$\{N_{ij}(\mu), N_{kl}(\lambda)\} = \frac{N_{il}(\mu)N_{kj}(\lambda) - N_{il}(\lambda)N_{kj}(\mu)}{(\mu - \lambda)}. \tag{4.5}$$

For the symplectic leaves of this system, one again has the Darboux coordinates corresponding to divisors and line bundles, and again there is a surface Q obtained by blowing up $\mathcal{T}$, not at points along $\lambda = a_i, \infty$ but at points on $z = 0$ and along $\lambda = \infty$.

3) **Hitchin's stable pairs system:** In [**Hi1**], [**Hi2**], Hitchin considered the moduli spaces $\mathcal{M} = \mathcal{M}_{k,r}$ of stable pairs (E, ϕ) where E is a degree k rank r vector bundle over a Riemann surface Σ and ϕ a holomorphic section of $\mathrm{End}(E) \otimes K_\Sigma$ on Σ. A Zariski open subset $\mathcal{M}_\mathcal{N}$ of $\mathcal{M}$ lies over the moduli space $\mathcal{N}$ of stable E's, and one has that $M_N \simeq T^*\mathcal{N}$.

   $\mathcal{M}$ has a natural symplectic structure, which coincides with the cotangent structure over M_N [**Hi1**]. Hitchin defines an integrable system on $\mathcal{M}$, in a spirit which is very similar to the one on our coadjoint orbit: for each pair (E, φ) one has a spectral curve Σ lying in the total space $\mathcal{K}$ of the canonical line bundle K_Σ, defined by:

$$\det(\varphi - z\mathbf{1}) = 0.$$

   The coefficients of the spectral curve again define an integrable system, and one can again define a sheaf over the curve $\mathbb{S}$ in a way entirely analoguous to (2.12), so that the space $\mathcal{M}$ can be written as a space of pairs (spectral surve $\mathcal{S}$, sheaf E). There is a Darboux coordinate theorem similar to that given above, and one obtains a surface Q which is simply the surface $\mathcal{K}$. In particular these systems are not even locally equivalent to the loop algebra systems.

4) ***K*3 surfaces and Abelian surfaces:** We know that the canonical bundle of Q is trivial when Q is a $K3$ surface or an Abelian surface; and so the Hilbert schemes of points on these surfaces are also symplectic varieties, indeed closed symplectic varieties, which was not the case for the other examples considered. Our theorem gives a birational isomorphism between these and an integrable system of line bundles over a suitable family of curves in Q.

In [**B2**], Beauville considered these Hilbert schemes, and showed that while the $K3$ surfaces admit 19-parameter families of algebraic deformations, the Hilbert schemes have 20-parameter families, thereby exhibiting varieties which aren't symmetric products of surfaces.

5. Generalisations

The symplectic varieties associated to $K3$ surfaces or to Abelian varieties are part of a larger family considered by Mukai [**Mu**], who showed that the different components of the moduli space of simple sheaves on these surfaces all admitted natural symplectic forms. Our case corresponds to that of line bundles supported over curves; but in general he shows that if one represents deformations of the sheaf E by the group $\mathrm{Ext}^1_{\mathcal{O}_X}(E, E)$, then one has a natural pairing

$$\mathrm{Ext}^1_{\mathcal{O}_X}(E,E) \otimes \mathrm{Ext}^1_{\mathcal{O}_X}(E,E) \xrightarrow{\text{Yoneda}} \mathrm{Ext}^2_{\mathcal{O}_X}(E,E) \xrightarrow{\text{Serre duality}} \mathbb{C}$$

which defines a skew-form on the tangent space to the moduli; this form is non-degenerate and closed.

Another direction of generalisation has been explored by Markman [**Ma**]. If we consider our family of curves embedded in a symplectic surface, they define for purely dimensional reasons a family of Lagrangian subvarieties in our surface. One can think of our family of Jacobians as a family of Picard varieties parametrising line bundles over the curves, or in other words, a class of sheaves over the surface, supported on the curves. Markman has developed a beautiful generalisation of this. He considers for any projective symplectic variety X the Hilbert scheme $\overline{B}$ of Lagrangian subvarieties of X, which contains a smooth open quasi-projective subvariety B parametrising smooth deformations of a fixed smooth Lagrangian subvariety Z_0. Consider the relative Picard variety $\mathcal{M}^p$ of line bundles supported on elements of B with a fixed Hilbert polynomial p. If we assume given a line bundle over Z_0, whose Chern class deforms as a $(1,1)$ class over all of B, then the component $\mathcal{M}$ of $\mathcal{M}^p$ containing L_0 dominates B by the support map $h : \mathcal{M} \to B$, and one has

THEOREM 5.1 ([**Ma**]). *There exists a canonical symplectic structure $\sigma_{\mathcal{M}}$ on the relative Picard bundle $\mathcal{M} \to B$ over the open subset B of the Hilbert scheme of smooth projective Lagrangian subvarieties of X. The support map $h : \mathcal{M} \to B$ is a Lagrangian fibration.*

We refer, for this and other very interesting questions relating algebraic and symplectic geometry, to the CIME lecture notes of Donagi and Markman [**DM**].

References

[A] M. Adler, *On a trace functional for formal pseudo-differential operators and the symplectic structure of the Korteweg-de Vries equation*, Invent. Math. **50** (1978/79), no. 3, 219–248.

[Ar] V. I. Arnold, *Mathematical Methods of Classical Mechanics*, Translated from the Russian by K. Vogtmann and A. Weinstein. Grad. Texts in Math. **60**, Springer-Verlag, New York-Heidelberg, 1978.

[AHH1] M. R. Adams, J. Harnad, and J. Hurtubise, *Isospectral Hamiltonian flows in finite and infinite dimensions* II. *Integration of flows*, Comm. Math. Phys. **134** (1990), no. 3, 555–585.

[AHH2] ———, *Darboux coordinates and Liouville-Arnold integration in loop algebras*, Comm. Math. Phys. **155** (1993), no. 2, 385–413.

[AHH3] ———, *Darboux coordinates and Serre duality*, CRM preprint, 1994.

[AHH4] ———, *Coadjoint Orbits, Spectral Curves and Darboux Coordinates*, The Geometry of Hamiltonian Systems (Berkeley, CA, 1989) (T. Ratiu, ed.) , Math. Sci. Res. Inst. Publ., vol. 22, Springer, New York, 1991.

[AHP] M. R. Adams, J Harnad, and E. Previato, *Isospectral Hamiltonian flows in finite and infinite dimensions* I. *Generalised Moser systems and moment maps into loop algebras*, Comm. Math. Phys. **117** (1988), no. 3, 451–500.

[AvM] M. Adler and P. van Moerbeke, *Completely integrable systems, Euclidean Lie algebras, and curves*, Adv. in Math. **38** (1980), no. 3, 267-317; *Linearization of Hamiltonian systems, Jacobi varieties and representation theory*, ibid. **38** (1980), no. 3, 318–379.

[AvM2] ———, *The complex geometry of the Kowalewski-Painlevé analysis*, Invent. Math. **97** (1989), no. 1, 1–46.

[B1] A. Beauville, *Jacobiennes des courbes spectrales et systèmes Hamiltoniens complètement intégrables*, Acta Math. **164** (1990), no. 3–4, 211–235.

[B2] ———, *Variétés kähleriennes dont la première classe de Chern est nulle*, J. Differential Geom. **18** (1983), no. 4, 755–782. (French)

[DM] R. Donagi and E. Markman, *Spectral covers, algebraically completely integrable Hamiltonian systems, and moduli of bundles*, Integrable Systems and Quantum Groups (Montecatini Terme, 1993), 1–119, Lecture Notes in Math., 1620, Springer, Berlin, 1996.

[F] H. Flaschka, *Toward an algebro-geometrical interpretation of the Neumann system*, Tôhoku Math. J. (2) **36** (1984), no. 3, 407–426.

[FRS] I. B. Frenkel, A. G. Reiman, and M. A. Semenov-Tian-Shansky, *Graded Lie algebras and completely integrable Hamiltonian systems*, Sov. Math. Dokl. **20** (1979), no. 4, 811–814. (English)

[Hu1] J. Hurtubise, *Finite dimensional coadjoint orbits in a loop algebra*, Lett. Math. Phys. **30** (1994), no. 2, 99–104.

[Hu2] ———, *Integrable systems and algebraic surfaces*, Duke Math. J. **83** (1996), no. 1, 19–50.

[Hu3] ———, *The algebraic geometry of the Kostant-Kirillov form*, CRM preprint, Proc. London. Math. Soc. (to appear), 18p.

[HH] J. Harnad and J. Hurtubise, *Generalised tops and moment maps into loop algebras*, J. Math. Phys. **37** (1991), no. 7, 1780–1787.

[Hi1] N. J. Hitchin, *The self-duality equations on a Riemann surface*, Proc. London Math. Soc. (3) **55** (1987), no. 1, 59–126.

[Hi2] ———, *Stable bundles and integrable systems*, Duke Math. J. **54** (1987), no. 1, 91–114.

[K] B. Kostant, *The solution to a generalized Toda lattice and representation theory*, Adv. Math. **34** (1979), no. 3, 195–338.

[M] J. Moser, *Geometry of quadrics and spectral theory*, The Chern Symposium 1979 (Proc. Internat. Sympos., Berkeley, Calif., 1979), Springer, New York-Berlin, 1980, 147–188.

[Ma] E. Markman, (in preparation).

[Mu] S. Mukai, *Symplectic structure on the moduli space of sheaves on an abelian or $K-3$ surface*, Invent. Math. **77** (1984), no. 1, 101–116.

[N] C. Neumann, *De problemate quodam mechanico, quod ad primam integralium ultraellipticorum classem revocatur*, J. Reine Angew. Math. **56** (1859), 46–63.

[Ra] T. Ratiu, *The C. Neumann problem as a completely integrable system on an adjoint orbit*, Trans. Amer. Math. Soc. **439** (1981), 321–329.

[RS] A. G. Reiman and M. A. Semenov-Tian-Shansky, *Reduction of Hamiltonian systems, affine Lie algebras and lax equations* I, II, Invent. Math. **54** (1979), no. 1, 81–100; ibid. **63** (1981), no. 3, 423–432.

[S] W. Symes, *Systems of Toda type, inverse spectral problems and representation theory*, Invent. Math. **59** (1980), 13–51.

[Sc] D. R. D. Scott, *Classical functional Bethe ansatz for $SL(N)$: separation of variables for the magnetic chain*, J. Math. Phys. **35** (1994), no. 11, 5831–5843.

[Sch] R. Schilling, *Generalizations of the Neumann system. A curve theoretic approach.* I, Comm. Pure Appl. Math. **40** (1987), no. 4, 455–522.

[Sk] E. K. Sklyanin, *Separation of variables in the classical integrable* SL(3) *magnetic chain*, Comm. Math. Phys. **150** (1992), no. 1, 181–191.

[vM] P. van Moerbeke, *Introduction to algebraic integrable systems and their Painlevé analysis*, Theta functions—Bowdoin 1987, Part 1 (Brunswick, ME, 1987), 107–131, Proc. Sympos. Pure Math. **49**, Part 1, Amer. Math. Soc., Providence, RI, 1989.

Centre de recherches mathématiques, Université de Montréal, C.P. 6128, succ. Centre-ville, Montréal, Québec, Canada H3C 3J7

E-mail address: hurtubis@gauss.math.mcgill.ca

Centre de Recherches Mathématiques
CRM Proceedings and Lecture Notes
Volume **11**, 1997

My Encounters—as a Physicist—with Mathematics

R. Jackiw

It is a pleasure for me—as a physicist—to speak at the twenty–fifth anniversary of your mathematical institute, when also a new building is being dedicated. Such an occasion calls for accolades of appreciation, both to the profession and to its practitioners, and over the years many bouquets have been offered to mathematics.

For example, Gauss was quite pleased with his line of work, stating that

> "Mathematics is the queen of the sciences."
>
> (Gauss)

Even an outsider, the psychologist Havelock Ellis, evaluated mathematicians as having

> "...reached the highest rung on the ladder of human thought"
>
> (Ellis)

To be sure, there are also dissenters, for example Plato:

> "I have hardly ever known a mathematician who was capable of reasoning."
>
> (Plato)

Physicists' opinion lies between these extremes. Well known is Eugene Wigner's appreciation of the

> "unreasonable effectiveness of mathematics"
>
> (Wigner)

A forceful position in favor of mathematics in physics was stated by Paul Dirac:

> "The most powerful method of advance [in physics] ...is to employ all the resources of pure mathematics in attempts to perfect and generalize the mathematical formalism that forms the existing basis of theoretical physics, and ...to try to interpret the new mathematical features in terms of physical entities."
>
> (Dirac)

1991 *Mathematics Subject Classification.* Primary: 01A65; Secondary: 81T70.
This is the final form of the paper.

The list of apt quotations can be extended to great length, and I shall not attempt adding my own words. Let me merely repeat that mathematics is indeed good for us physicists, but also we are good for mathematics by providing new ideas for mathematical research and by finding fresh applications of old ideas.

These days there is intense cross fertilization between mathematics and physics, specifically between geometry and field theory. The contact, first established within Albert Einstein's general relativity, surged again about two decades ago. Some of my own research took place at this new beginning, so I thought I would present here a reminiscence, thereby providing a case history of a physics-mathematics encounter.

By early 1970's, quantum field theory was very much in favor with theoretical physicists, but the quantized equations resisted solution. It then occurred to many people that it would be worthwhile to ignore the quantal nature of the fields, and to solve the equations as if they describe non-linear, classical dynamical systems. Interesting, localized and non-dissipative solutions were found very quickly. These were the kinks in one dimension—relevant to physics on a line, vortices in two dimensions—in planar physics, Skyrmions and magnetic monopoles in the three dimensions of our physical world; collectively such solutions were called "solitons," the name being taken from applied mathematics. Another class of solutions comprised the "instantons" in four-dimensional space-time.

With colleagues at M.I.T., I addressed the question of how to extract from these *classical* results information on the *quantum* theory—i.e. we wanted to determine the quantum meaning of classical fields. Progress was made on this problem, and at a certain stage Claudio Rebbi and I decided that we needed to study both the linear small fluctuations about the non-linear soliton and instanton field profiles, and also the coupling of other linear systems, like fermions, to solitons and instantons. Thus we were led to linear eigenvalue equations and we realized that the "zero-modes", corresponding to vanishing eigenvalues, contain especially important information about the quantum physics. The zero-modes in the small fluctuation equations measure allowed deformations of the soliton or instanton, while the number of these modes gives the dimension of the moduli space for the solution of the non-linear equation. In the fermionic Dirac equation, the eigenvalues measure energy, and positive-energy modes describe quantum particles, negative-energy modes correspond to anti-particles, while zero-modes—when they exist—signal a degeneracy that gives rise to unexpected quantum numbers, e.g. fractional fermion number.

Rebbi and I were delighted to find precisely such zero-modes, and to establish their physical consequences. But we were surprised that the existence of these special solutions did not depend on the details of the localized profiles in the background solitons and instantons; rather only their large-distance behavior mattered. The long-range features of course characterize the topological properties of solitons and instantons, so we began to suspect that the occurrence of zero-modes was not an accident of our analysis, but a consequence of having non-trivial topological backgrounds.

We wanted to find out what mathematicians knew about this. At M.I.T. all buildings are connected and the math department is in the same structure as my work space. However, locked doors as well as the chemistry department intervene, so communication is obstructed. Nevertheless, we walked the corridors of the mathematics offices, but could not immediately find anyone who wanted to spend time understanding our questions, and answering them in a way comprehensible to non-specialists, to us physicists. Shortly later we met Barry Simon, who did not have specific information on our problem, but suggested that work of Michael Atiyah and Isadore Singer might be relevant.

Singer had temporarily moved from M.I.T. To Berkeley, but as it happened my colleague and collaborator Jeffrey Goldstone knew Atiyah from their student days in Cambridge, England, and had information that he was coming to visit his mathematics friends in Cambridge, Massachusetts.

So we arranged a meeting in my office. We invited physicists who were working on soliton-instanton questions, and we listened to Atiyah explain how his index theorem with Singer counts instanton zero-modes, and how their spectral flow theorem with V. Patodi is relevant to fractional charge. Learning that the four-dimensional index is given by an integral over the curvature-form F, specifically by $\int F \wedge F$, was especially thrilling to us since the integrand, $F \wedge F$, had also arisen in the physics literature as the anomalous divergence of the chiral fermion current, thereby controlling neutral pion decay. Evidently the chiral anomaly and the index theorem are related; they had been elaborated in the late 1960's at different ends of the same M.I.T. corridor, by people working in ignorance of each other!

We appreciated very much Atiyah's efforts to make his presentation understandable to us; still exchanging information was not easy. One young member of the audience impressed Atiyah, who encouraged the fellow to speak because he seemed to understand, better than anyone else, what Atiyah was saying. That person was Edward Witten; as all of you know, he has continued to impress Atiyah and other mathematicians.

Soon thereafter, I was asked to review these exciting new results about quantum field theory at a meeting of the American Physical Society. Since Singer was present, I yielded some of my time to him, with the suggestion that he describe the mathematical connection. But a detailed presentation could not be fit in, so he merely eulogized collaboration between mathematics and physics with the following ode.

> "In this day and age
> The physicist sage
> Writes page after page
> On the current rage
> The gauge
> Mathematicians so blind
> Follow slowly behind
> With their clever minds
> A theorem they'll find
> Duly written and signed
> But gauges have flaws
> God hems and haws
> As the curtain He draws
> O'er His physical laws
> It may be a lost cause"
>
> (Singer)

Index theory also received a contribution from physics. The Atiyah-Singer theorem applies to even-dimensional spaces on which a connection is defined. However, physicists are also interested in odd-dimensional spaces—where one-dimensional kinks or three-dimensional Skyrmions and monopoles reside. These configurations can lead to zero modes, even in the absence of a gauge connection. So we asked the mathematicians about odd dimensional indices; apparently nothing was known. At that time I had a mathematically-minded student, Costas Callias, and I asked him to prove an odd-dimensional index theorem. He succeeded and this further prompted Paul Bott and Robert Seeley to publish a mathematical exegesis of the result, immediately following Callias' paper in *Communications in Mathematical Physics*. Since then I have been happy to see the "Callias index theorem" used and cited.

The two approaches to solving problems—the explicit, goal-oriented methods of the physicists and the general theorems of the mathematicians—are well illustrated by the determination of the dimensionality for instanton moduli space: the n-instanton $SU(2)$ solution depends on $8n-3$ parameters. This result appears in the same issue of *Physics Letters*, once by Albert Schwarz, who used the Atiyah-Singer theorem, and a few pages later by Rebbi and me, who solved differential equations to find explicitly $8n-3$ zero-modes.

Gauge theories in general and instantons in particular continued to interest mathematicians. They established the topological properties of the instanton moduli space and produced a construction—but not an explicit formula—for the general solution. The most general *explicit* expression, which does not exhaust all the parameters, was given by Craig Nohl, Rebbi and me.

Further developments on four-dimensional gauge fields led mathematicians to Donaldson theory. In three dimensions, the Chern-Simons term—another gauge structure first emphasized in the physics literature by Jonathan Schonfeld, Steven Templeton and me—has been related by mathematicians to knot theory on manifolds with various topologies, while physicists applied this term to experimental phenomena on the plane, like the Hall effect.

These parallel investigations by physicists and mathematicians also highlight our differences: physicists use mathematics as a language for recording observations about physical systems, and this limits our interest in the full range of mathematical possibility, which fascinates mathematicians. For example, the general instanton solution does not appear to be physically relevant; only the original one-instanton and the explicit, but limited, multi-instanton solutions have illuminated physical theory. Even the physicists' language need not always be mathematical. The fractional charge phenomenon, which can be inferred from zero-modes or from spectral flow, was also independently established by Wu-Pei Su, J. Robert Schrieffer and Alan Heeger, who found a physical realization in linear polyacetylene chains. One of their derivations uses the pictorial language of chemical bonds, where the only mathematics consists of counting!

An analogy comes here to mind: the English language contains over 200,000 words and all of them interest the lexicographer; for William Shakespeare 20,000 words sufficed to express his ideas in plays and sonnets; while Winston Churchill used less than 2000 words in his historically decisive speeches. Physicists, like Churchill, achieve their goals by effective use of a limited vocabulary.

The process of gaining knowledge goes through the same steps in physics and mathematics: first there is the intuition/guess, then follows the proposal/conjecture and finally comes the verification. For the mathematician verification consists of constructing a proof, establishing a theorem according to rules whose legitimacy evolves slowly under the direction of the entire mathematics community. But the physicist verifies his ideas by finding a physical correlative: neutral pion decay validates chiral anomalies, properties of solitons in polyacetylene establish fractional charge. The rules for giving a proof are constantly and rapidly changing—presuppositions can become modified, experimental facts can evolve.

Because "proof" and "theorem" carry intellectual prestige and pleasure, occasionally there are attempts to employ them in physics. To my mind this is mostly futile and sterile. For example, physicists wanted very much to combine internal and space-time symmetries in a non-trivial fashion and were not daunted by a proven "impossibility theorem". Rather the "theorem" was circumvented by the simple device of replacing commutators with anti-commutators, by grading the algebra, and supersymmetry was born, which now is also influencing mathematics. Similarly, when field-theory "constructivists" proved the existence of quantum $\lambda\phi^4$ theory in $(1+1)$ dimensions, they were correct, but missed the entire quantum soliton phenomenon, which is the only physically interesting feature of that model.

A statement by Frank Yang accurately describes physicists' historical use of mathematics.

> "...physics is not mathematics, just as mathematics is not physics. Somehow nature chooses only a subset of the very beautiful and complex and intricate mathematics that mathematicians develop, and that precise subset is what the theoretical physicist is trying to look for."
>
> (Yang)

This conservative view of mathematics differs from Dirac's radical advice, cited earlier, that physicists should

> "...try to interpret ...mathematical features in terms of physical entities."
>
> (Dirac)

However, today faced with absence of new experimental data about fundamental phenomena, particle physics theory, as realized in the string program, is driven by mathematics in the manner advocated by Dirac. This was not the way things worked in the past, not even for Dirac: when first confronting his negative energy solutions, he identified them with the proton—the only then-known positively charged particle—as a physicist he did not at first trust his mathematics enough to postulate the existence of the positron!

I am immensely curious about the ultimate fate of the new physics, built entirely on mathematics, indeed on new mathematics that it helps to create. I trust that in the next quarter century the Centre de Recherches Mathématiques will play a role in settling this question.

Centre for Theoretical Physics, M.I.T 6–320, 77 Massachusetts Ave., Cambridge, MA02139-4307 (USA)

E-mail address: jackiw@mitlns.mit.edu

Centre de Recherches Mathématiques
CRM Proceedings and Lecture Notes
Volume **11**, 1997

Quelques points d'histoire des séries de Fourier

Jean-Pierre Kahane

Cet exposé ne contient rien d'original. C'est juste une présentation de quelques questions qui sont traitées dans mon livre avec Pierre-Gilles Lemarié-Rieusset, *Fourier Series And Wavelets*, à paraître chez Gordon and Breach. J'évoquerai la préhistoire des séries de Fourier, puis l'apport conceptuel majeur de Fourier, c'est-à-dire le va et vient entre l'analyse ($c_n = \dots$) et la synthèse ($f = \dots$), le point de vue de Riemann (partir de $f = \dots$), et le lien, dès Fourier, avec les équations de la physique mathématique. Le seul point un peu nouveau est la réhabilitation des premiers calculs de Fourier, pour obtenir les coefficients, que tout le monde à ma connaissance jugeait indéfendables (y compris moi dans *Fourier series and wavelets*). Ensuite, je me limiterai à trois grandes questions : la convergence des séries, l'intégration des fonctions, et les premières fractales de l'analyse mathématique.

1. L'essence de l'analyse harmonique consiste à décomposer un objet en objets plus simples, et à reconstituer l'objet à partir de sa décomposition. Le paradigme en est le programme de Platon réalisé dans l'astronomie de Ptolémée : décomposer des mouvements compliqués, ceux des planètes, en mouvements simples, circulaires, dont la superposition restitue le mouvement étudié. Aujourd'hui, la décomposition s'appelle l'analyse harmonique, et la recomposition se nomme la synthèse harmonique.

Le terme d'harmonie provient de la musique, et il apparaît aussi dans la cosmogonie de Platon : l'harmonie des sphères. Dans la vision de Platon, les mathématiques se composent de cinq parties : les nombres, la géométrie plane, la géométrie dans l'espace, l'astronomie et la musique. L'astronomie et la musique sont comme deux sœurs (République, VII, 530 d). Les mouvements harmonieux sont saisis par les yeux et les oreilles. Sons purs et cercles parfaits se font écho.

Les fonctions trigonométriques telles que nous les connaissons apparaissent beaucoup plus tard, avec l'imprimerie et l'édition des tables mathématiques utilisées en astronomie et en navigation. Les fonctions tabulées à l'époque (début du 16e siècle) ont été, et restent, les "fonctions élémentaires" de l'analyse.

La musique refait son apparition au 18^{e}siècle, avec Brook Taylor (1715) qui reconnaît le rôle des sinusoïdes dans la production des sons purs, Euler, dont la première publication (1727) traite de la propagation du son, d'Alembert, qui donne l'équation des cordes vibrantes (1747), Daniel Bernoulli, qui en donne une solution

1991 *Mathematics Subject Classification.* 01A55, 01A60.
This is the final form of the paper.

comme “mélange de plusieurs espèces de vibrations simples isochrones” (1753), et la célèbre controverse sur cette solution qui oppose à Daniel Bernoulli, en même temps d'entre eux, Euler, renforcé par Lagrange, et d'Alembert. L'exposé des différents points de vue est admirablement présenté par d'Alembert dans les mémoires 1 et 25 de ses *Opuscules mathématiques.*

Là, dirons-nous, s'arrête la préhistoire. Le dernier acteur en est Lagrange, qui a justifié la solution d'Euler en étudiant d'abord les vibrations d'un système de masses ponctuelles disposées sur un fil tendu. Là s'introduit une transformation de Fourier discrète, susceptible d'un traitement algébrique. L'impossibilité pour Lagrange d'étendre ce traitement au cas continu le conduit à condamner, avec Euler, la tentative de D. Bernoulli, et, plus tard, à une grande réticence à l'égard de Fourier.

2. L'œuvre de Fourier sur la théorie analytique de la chaleur, écrite et communiquée en 1807, publiée en 1822, s'est toujours heurtée en France à une certaine hostilité, ou incompréhension. Ce sont des Allemands, Dirichlet et Riemann, qui lui ont pleinement rendu justice. Pour Fourier, les séries trigonométriques sont à la fois un outil pour le traitement de l'équation de la chaleur, et un objet d'étude fascinant. Selon Riemann, Fourier a été le premier à comprendre la nature des séries trigonométriques de manière exacte et complète.

De quoi s'agit-il ? En notations modernes, de deux formules :

$$f(x) = \sum c_n e^{inx} \tag{1}$$

$$c_n = \int f(x) e^{-inx} \frac{dx}{2\pi}. \tag{2}$$

On part d'une fonction $f(x)$ définie sur un intervalle de longueur 2π, on applique la formule (2) et on considère la série dans le second membre de (1). Converge-t-elle vers $f(x)$? Fourier ne le démontre que dans un cas particulier, que nous allons voir dans un moment, mais l'énonce en général, pour une fonction “entièrement arbitraire”. Cet énoncé est à la fois faux et génial. Faux, d'abord, parce que l'intégrale dans (2) n'a pas de sens pour une fonction “entièrement arbitraire” ; il faut supposer, au minimum, que cette intégrale a un sens. Faux ensuite parce que, même si les formules (2) ont un sens, la série dans (1) n'est pas nécessairement convergente. En fait, ces questions, d'intégrabilité et de convergence, sont très délicates, et elles ont donné du travail aux mathématiciens pendant un siècle et demi, comme nous allons le voir.

Et pourtant, Fourier a raison. Les formules (1) et (2), ou plutôt leur couplage, sont plus importantes que les notions a priori sur l'intégrale des fonctions ou la convergence des séries. Dire qu'elles ont valeur universelle est mieux qu'une assertion (sans cesse démentie), c'est un programme (jamais atteint). Étant donné un phénomène périodique, de période 2π, comment donner un sens aux formules (2) ? Cela fait, comment donner un sens à la formule (1) ? On reconnaît là l'une des sources des extensions successives de la notion d'intégrale, de la sommabilité des séries, des espaces fonctionnels, et de la théorie des distributions de L. Schwartz. L'énoncé de Fourier sur les données “entièrement arbitraires”, faux dans la courte vue, s'est avéré génial au long terme.

Naturellement, ce commentaire extrapole à l'époque contemporaine le jugement de Riemann. Pour Riemann, l'apport conceptuel majeur de Fourier à la théorie des séries trigonométriques est le couplage des formules (1) et (2), de l'analyse et de la synthèse.

3. L'apport de Riemann va bien au delà. Il est contenu dans sa thèse sur les séries trigonométriques, présentée pour l'habilitation à Göttingen en 1854, et publiée seulement après sa mort, en 1867. Cette thèse comprend trois parties : une étude historique, la définition de l'intégrale (pour donner un sens aux formules (2)), et l'examen en toute généralité des fonctions données par une formule (1), c'est-à-dire sommes de séries trigonométriques partout convergentes (comme il s'agit pour Riemann de séries de sinus et de cosinus, il faut considérer comme sommes partielles de (1) les sommes symétriques, $\sum_{-N}^{N}$).

Cet examen aboutit à une caractérisation des fonctions en question. Elle n'est pas difficile à énoncer, et elle a servi plus tard, en particulier à Denjoy. Cependant je l'omettrai, pour insister sur les méthodes, et sur certains prolongements.

- Première méthode : intégrer deux fois la série (1). Riemann observe que, si la série (1) converge partout, les coefficients tendent vers zéro (la chose n'est pas évidente, et elle ne sera démontrée qu'en 1870, par Cantor). En intégrant deux fois, on obtient une fonction continue, $F(x)$. Riemann prouve que $F(x)$ admet une dérivée seconde symétrique généralisée, et qu'elle est égale à $f(x)$. La dérivée seconde symétrique généralisée est

$$\lim_{h\to 0}\Big(h^{-2}\big(F(x+h)+F(x-h)-2F(x)\big)\Big)$$

et l'égalité à $f(x)$ résulte du théorème de sommmabilité de Riemann : si la série $\sum A_n$ converge ($n \geq 0$), alors

$$\lim_{\alpha\to 0}\sum A_n \frac{\sin^2 n\alpha}{n^2\alpha^2} = \sum A_n.$$

- Deuxième méthode : supposant seulement que les coefficients tendent vers zéro, multiplier par une fonction test assez régulière portée par un intervalle donné, et obtenir une série du même type. Riemann prend la précaution d'intégrer d'abord deux fois, mais on peut aussi faire une multiplication formelle. Le théorème exprime que si une distribution de Schwartz a ses coefficients de Fourier tendant vers zéro (le livre de Kahane et Salem nomme pseudofonction une telle distribution), son produit par une fonction test a le même propriété. Ainsi, pour une distribution, le fait d'être une pseudofonction est une propriété locale. En passant, Riemann démontre que la fonction F, primitive double de la pseudofonction définie par la série (1), vérifie

$$\lim_{h\to 0}\Big(h^{-1}\big(F(x+h)+F(x-h)-2F(x)\big)\Big) = 0;$$

c'est donc une fonction lisse (smooth) au sens de Zygmund.

Fonctions lisses et pseudofonctions sont directement issues de la thèse de Riemann. Les deux méthodes fondamentales d'étude des distributions (intégrer autant de fois que nécessaire, multiplier par une fonction test à petit support) se trouvent aussi dans la thèse, ainsi que bien d'autres perles sous la forme d'exemples et de contre exemples.

4. Voici quelques prolongements directs de la thèse de Riemann.

4.1. *L'unicité selon Cantor.* Si deux séries trigonométriques partout convergentes ont partout la même somme, sont-elles égales ? La réponse est positive (Cantor, 1870). Que dire si on remplace "partout" par "sauf peut-être sur un ensemble E donné" ? La réponse est la même quand E est dénombrable (Cantor, Young), mais pas nécessairement quand E a une mesure de Lebesgue nulle (Menchoff). On dit que E est un ensemble d'unicité (U) quand la réponse est positive, et un ensemble de multiplicité (M) dans le cas contraire. Lorsque E est fermé, il est de type M ou U selon qu'il porte ou non une pseudofonction non nulle. A part cette caractérisation, il est hors d'espoir de décrire complètement les ensembles de type U. Jusqu'en 1987, c'était un fait d'expérience : de très beaux théorèmes avaient été démontrés, avec des rapports imprévus à la théorie des nombres (Salem–Zygmund), des exemples surprenants avaient été construits (Körner, Kaufman), et, malgré tout le travail fait, la description complète apparaissait hors de portée. Depuis 1987, c'est un théorème : étant donné un borélien quelconque dans $K(\mathbb{T})$, l'espace des parties compactes du cercle $\mathbb{T}$, il existe un compact de type U qui, non seulement n'appartient pas à ce borélien, mais n'appartient pas au σ-idéal qu'il engendre (Debs Saint-Raymond). Un σ-idéal, par définition, est stable par inclusion et par réunions dénombrables ; dans $K(\mathbb{T})$, les compacts de type U constituent un σ-idéal, et c'est une simple formulation d'un théorème de Nina Bari de 1923. Le théorème de Debs et Saint-Raymond fait partie d'un ensemble remarquable de travaux menés par des logiciens (Solovay, Kechris, Louveau et ses élèves) et par des analystes (R. Kaufman, R. Lyons), dans lesquels les classes U et M, et d'autres qui leur sont liées, sont explorées par les moyens modernes de la théorie descriptive des ensembles. C'est un retour remarquable car, historiquement, c'est le problème de l'unicité qui a amené Cantor à la création de la théorie des ensembles.

4.2. *La seconde totalisation de Denjoy.* Les totalisations de Denjoy généralisent l'intégrale de Lebesgue, en vue de résoudre complètement des problèmes spécifiques. La première totalisation (1912) permet le calcul de la primitive de la fonction dérivée la plus générale. La seconde totalisation (1921) permet le calcul des coefficients d'une série trigonométrique convergente quelconque dont la somme est donnée. Autrement dit, elle donne le sens à la formule (2) chaque fois que (1) a un sens usuel.

Les ingrédients de la première totalisation sont l'intégrale de Lebesgue, la théorie de Baire, et la théorie des ordinaux de Cantor. Ce sont aussi, mais de façon plus compliquée, les ingrédients essentiels de la seconde totalisation ; il faut y ajouter, naturellement, la théorie de Riemann des séries trigonométriques, et en particulier les propriétés de la dérivée seconde symétrique généralisée. L'exposé détaillé de l'ensemble se trouve dans un ouvrage monumental de Denjoy, en quatre volumes, intitulé *Leçons sur le calcul des coefficients d'une série trigonométrique* (1941–49).

Comme intermédiaire entre la première et la seconde totalisation, on peut exposer assez simplement une totalisation qui répond aux problèmes suivants, qui sont essentiellement équivalents :

a) calculer une fonction connaissant sa dérivée dyadique, définie comme

$$f'_d(x) = \lim_{n\to\infty} \left(2^n \left(f\left(\frac{p+1}{2^n}\right) - f\left(\frac{p}{2^n}\right)\right)\right), \quad \left(\frac{p}{2^n} \le x \le \frac{p+1}{2^n}\right).$$

b) Calculer les coefficients d'une série de Walsh convergente connaissant sa somme.

c) Calculer les termes d'une martingale dyadique définie sur $\{0,1\}^{\mathbb{N}}$ et partout convergente, connaissant sa limite.

J'ai fait cet exposé lors de l'une des réunions organisées par le CRM à Montréal (1989), et il est publié dans le n° 34 de l'*Enseignement mathématique.*

4.3. *Les séries trigonométriques qui convergent presque partout.* L'initiateur du programme est D. Menchoff (Mensŏv). En 1913, en relation avec le problème de l'unicité, il montre l'existence de séries trigonométriques non nulles qui convergent presque partout vers zéro. Que dire des fonctions qui sont sommes de séries trigonométriques convergeant presque partout ? Ce sont nécessairement des fonctions mesurables au sens de Lebesgue. En 1940 Menchoff établit la réciproque : toute fonction à valeurs réelles, mesurable au sens de Lebesgue, est représentable p.p. comme somme d'une série trigonométrique. Que dire alors des fonctions à valeurs dans $[-\infty,\infty]$? que dire, plus généralement, des limites inférieure et supérieure en chaque point x, $\underline{S}(x)$ et $\bar{S}(x)$, des sommes partielles d'une série trigonométrique générale ? Le problème est posé par Menchoff en 1954, avec les premiers éléments de solution. La solution finale n'a été donnée qu'en 1988, par S. V. Konyagin (Konjagin) et elle lui a valu le prix Salem : les couples $(\underline{S},\bar{S})$ sont caractérisés par les propriétés suivantes : $\underline{S}$ et $\bar{S}$ sont des fonctions mesurables à valeurs dans $[-\infty,\infty]$, $\underline{S}\le\bar{S}$, et $\underline{S}=-\infty \iff \bar{S}=+\infty$ p.p. . En particulier, les fonctions f telles que $f=\underline{S}=\bar{S}$, c'est-à-dire les sommes, finies ou non, de séries trigonométriques convergeant p.p. sont les fonctions mesurables finies p.p. .

5. Les exposés de J. Bourgain et de P. L. Lions au Congrès International de Zürich (1994) ont montré le lien étroit entre les développements modernes de l'analyse harmonique d'une part, et des équations aux dérivées partielles d'autre part. C'est là un retour à la tradition. La première équation des ondes est celle de l'Alembert pour les cordes vibrantes :

$$\frac{\partial^2 u}{\partial t^2}-\omega^2\frac{\partial^2 u}{\partial x^2}=0,$$

et c'est le prototype des équations hyperboliques. Les équations de Fourier pour la propagation et pour l'équilibre de la chaleur sont de la forme

$$\frac{\partial u}{\partial t}=K\Delta u,\qquad \Delta u=0$$

et ce sont les prototypes des équations paraboliques et elliptiques. C'est précisément pour la solution de ces différentes équations, complétées par des conditions aux limites, que D. Bernoulli d'abord et Fourier ensuite introduisent les séries trigonométriques. Il me semble intéressant d'indiquer le point de départ de Fourier, parce qu'il témoigne d'un aspect essentiel de son œuvre : l'inspiration puisée dans un problème de la physique ("l'étude approfondie de la nature est la source la plus féconde des découvertes mathématiques", dit-il dans son *Discours préliminaire*), et le traitement analytique le plus général et le plus puissant, le plus efficace dans les applications numériques.

Fourier examine d'abord comment se répartissent les températures, à l'équilibre, dans un corps homogène limité par une paroi horizontale portée à la température de 100 ° et deux parois verticales parallèles portées à la température de 0 ° . Il s'agit donc de résoudre l'équation

$$\frac{\partial^2 u}{\partial x^2} + \frac{\partial^2 u}{\partial y^2} = 0$$

sous les conditions aux limites $u(x,0) = 1$ pour $-\pi/2 < x < \pi/2, u(-\pi/2,y) = u(\pi/2,y) = 0$ pour $y > 0$ (les valeurs 1 et $\pi/2$ étant choisies par commodité), et la condition supplémentaire que $u(x,y)$ est borné. C'est ce qu'on appellera plus tard un problème de Dirichlet : déterminer une fonction harmonique, connaissant ses valeurs à la frontière. Fourier, inspiré sans doute par l'équation des cordes vibrantes, cherche d'abord les solutions de $\Delta u = 0$ qui sont de la forme $f(x)g(y)$. Celles qui sont compatibles avec les données verticales sont les $e^{-ny}\cos nx$, n impair. On obtient la solution au problème posé, sous la forme

$$(*) \qquad u(x,y) = \sum a_n e^{-ny} \cos nx \quad (n\colon \text{impair}), \qquad \left(-\frac{\pi}{2} < x < \frac{\pi}{2}, y > 0\right)$$

si l'on peut développer la constante 1, sur $(-\pi/2, \pi/2)$, comme

$$(**) \qquad 1 = \sum a_n \cos nx \quad (n\colon \text{impair}).$$

Fourier a conscience de l'étrangeté de cette dernière formule, car la série du second membre, si elle converge vers 1 sur $(-\pi/2, \pi/2)$ doit converger vers -1 sur $(\pi/2, 3\pi/2)$. Néanmoins, bravement, il calcule les coefficients a_n, par un procédé qu'aucun commentateur à ma connaissance n'a considéré comme justifiable, et qui pourtant l'est tout à fait. Il dérive la formule $2N-2$ fois, tronque en conservant N termes, fixe $x = 0$, résout, passe à la limite quand $N \to \infty$. Les dérivations formelles de séries divergentes étaient bien dans l'esprit du 18^{e}siècle, et les mathématiciens de l'époque avaient du flair pour savoir quoi en faire. Ce que fait Fourier revient à chercher les polynômes trigonométriques de la forme $1 - \sum \alpha_n \cos nx$ ($n$impair) de degré donné les plus plats possibles en 0, et à montrer que leurs coefficients tendent vers des limites, les a_n cherchés. Il trouve ainsi $a_n = (-1)^{n+1} 4/\pi n$ (n: impair). La méthode est correcte et même bonne, parce que les polynômes trigonométriques dérivés s'écrivent $K_N \sin x(1 - \cos 2x)^N$, et tendent au sens des distributions vers $2(\delta_{-\pi/2} - \delta_{\pi/2})$; il n'en faut pas plus pour justifier la formule (*), de la façon la plus rapide et naturelle. Fourier, cependant, tient à démontrer la formule (**), donc à établir la convergence de la série. C'est la seule démonstration de convergence dans tout l'ouvrage, mais elle est parfaitement correcte, et préfigure celle que Dirichlet donnera dans un cadre plus général. Il ajoute que la convergence n'est pas assez rapide pour fournir une bonne approximation, mais qu'elle suffit à établir la validité de l'équation. Puis il s'attache à la série (*) et constate qu'à cause du facteur exponentiel elle est *extrêmement convergente.*

A cet endroit Fourier n'a pas encore introduit ses formules pour le calcul des coefficients. Visiblement il attache du prix à sa méthode de calcul par dérivation et troncature, et j'espère avoir montré que cette méthode est loin d'être absurde. Il utilise librement des séries divergentes, à la manière du 18^{e}siècle, mais connaît aussi, et pratique, la notion de série convergente, et il démontre complètement la formule (**). Cependant les séries trigonométriques sont d'abord pour lui un outil de calcul, un algorithme efficace pour la résolution de problèmes intéressant la propagation ou l'équilibre de la chaleur. L'idée, très forte, qu'il y a des séries peu convergentes et des séries "extrêmement convergentes", a été trop longtemps négligée dans l'enseignement. Elle reprend vigueur avec l'usage des ordinateurs.

6. Depuis lors, la convergence des séries de Fourier est un sujet mathématique dont le statut est curieux : c'est une mine d'erreurs, d'exemples, de contre-exemples, et de théorèmes profonds.

Les premières erreurs, dues à Cauchy, ont inspiré Dirichlet et Riemann. L'article de Dirichlet (1829) relève d'abord les erreurs de Cauchy, qui portent sur deux questions où la maîtrise de Cauchy paraissait incontestable : la convergence des séries numériques, et les fonctions analytiques. Puis il établit, par étapes et de façon rigoureuse, son fameux théorème : si f est une fonction monotone par morceaux, sa série de Fourier converge en tout point x vers $1/2\big(f(x+0)+f(x-0)\big)$. Par sa précision, sa rigueur et son style cet article annonce une nouvelle vision de l'analyse, et sa portée a été considérable. A la fin de l'article Dirichlet, après quelques commentaires sur les fonctions intégrables, déclare qu'il reviendra plus tard au cas d'une fonction continue par morceaux. Il n'y est jamais revenu, et pour cause.

Riemann rend justice à Fourier et à Dirichlet dans la partie historique de sa thèse de 1854, et il ne se contente pas de relever les erreurs de Cauchy, il en dégage des distinctions importantes. S'agissant des séries, il distingue les séries absolument convergentes, où tout est permis, et les autres séries convergentes, où il faut se méfier de tout. L'essentiel de sa thèse, nous l'avons vu, traite des séries trigonométriques partout convergentes. Il donne et étudie de nombreux exemples. L'exemple le plus simple lui a cependant échappé ; c'est

$$\sum \frac{\sin nx}{\log n};$$

il est dû à Fatou et publié dans le livre de Lebesgue de 1906. Nous verrons son rôle dans les théories de l'intégration.

Il paraissait naturel que toute fonction continue soit somme de sa série de Fourier. En 1873, du Bois–Reymond a donné un contre-exemple. La meilleure généralisation du théorème de Dirichlet a été celle de Jordan en 1881, remplaçant les fonctions monotones par morceaux par des fonctions à variation bornée. Après 1873, et jusqu'en 1966, les séries de Fourier des fonctions continues ont constitué un mystère. Heureusement, en 1900, Fejér a montré qu'elles étaient sommables vers la fonction par le procédé de la moyenne arithmétique (procédé de Cesaro d'ordre 1). En 1965 j'ai montré avec Katznelson qu'elles pouvaient diverger sur un ensemble donné de mesure nulle. En 1966 Carleson, par un théorème profond et fameux, concluait l'affaire : elles convergent presque partout.

La convergence presque partout a lieu également pour les séries de Fourier des fonctions appartenant à L^2 (Carleson), et à la réunion des L^p ($1 < p \leq \infty$; R. Hunt 1967). Elle n'a pas lieu pour L^1 ; dès 1926, Kolmogorov avait produit une fonction intégrable au sens de Lebesgue dont la série de Fourier diverge partout. Le comportement presque partout des sommes partielles S_n de la série de Fourier d'une fonction intégrable est encore mal connu. On sait que p.p. $S_n(x) = o(\log n)$ et $S_n(x) = \Omega(\alpha_n \log\log n)$ si $\alpha_n \downarrow 0$ $(n \to 0)$. Le meilleur énoncé (publié sans démonstration par Rodin en 1992) est que p.p. $S_n(x)a$, comme fonction de n, une oscillation moyenne bornée, c'est-à-dire

$$\sup_{n,m} \frac{1}{m} \sum_{k=1}^{m} \left| S_{n+k} - \frac{1}{m} \sum_{j=0}^{m-1} S_{n+j} \right| < \infty.$$

Jusqu'ici nous n'avons parlé que des fonctions d'une variable, définies sur un intervalle de longueur 2π, ou 2π-périodiques. La scène change avec les fonctions de plusieurs variables, parce qu'il n'y a plus d'ordre évident sur les fréquences, qui sont des points de $\mathbb{Z}^\nu$. L'ordre le plus naturel est peut-être l'ordre sphérique : on prend la somme des termes pour lesquels la fréquence a une norme euclidienne inférieure à R, et on fait tendre R vers l'infini. Voici un phénomène curieux, découvert par M. A. Pinsky en 1993 : en dimension $\nu \geq 3$, la série de Fourier de la fonction indicatrice d'une boule diverge sphériquement au centre de la boule.

Pour les fonctions de carré intégrable, la bonne notion de convergence est la convergence dans L^2. C'est alors une convergence indépendante de l'ordre des termes. En particulier, la convergence numériquement la plus naturelle, qui consiste à sommer la série selon la taille des termes, a lieu dans L^2. On sait par contre qu'elle n'a pas lieu dans les L^p, $p \neq 2$ (Kahane 1966 pour $p = 1$; D. Rider 1969 pour $1 < p < 2$; Bachelis et Gilbert 1979 et Oberlin 1982 pour $2 < p < \infty$; Kahane et Katznelson 1978 pour $p = \infty$). L'une des raisons d'être des ondelettes d'Yves Meyer est que la convergence selon la taille des termes a lieu pour les séries en ondelettes dans la plupart des espaces fonctionnels usuels, et en particulier dans les L^p $(1 < p < \infty)$.

7. J'ai déjà évoqué l'intégrale de Riemann, l'intégrale de Lebesgue, l'intégrale de Denjoy, et aussi l'intégration des distributions. La plupart des notions d'intégrales ont un rapport intime avec les séries de Fourier. On pourrait poursuivre avec l'intégrale de Wiener $\int f dW$, dont la définition la plus rapide est $\sum \xi_n \hat{f}_n$, où les $\hat{f}_n$ sont les coefficients de Fourier de f (supposée ici définie sur l'intervalle $(0,1)$), et les ξ_n des variables gaussiennes complexes normalisées indépendantes. La série de Fourier-Wiener,

$$W(t) = \xi_0(t) + \sum_{n \in \mathbb{Z} \setminus \{0\}} \xi_n \frac{e^{2\pi int} - 1}{2\pi in}$$

correspond à $f = 1_{[0,t]}$.

Je veux insister sur un autre aspect, épistémologique. Les mathématiciens se flattent de *découvrir* et non d'*inventer*, parce que la découverte apparaît comme une nécessité, et l'invention comme un luxe. Ils ont ainsi, au début du 19[e]siècle, “découvert” ce que veut dire la convergence d'une série, et, du coup, ils ont laissé de côté pendant une longue période la véritable richesse de la notion de série, qui est le couplage entre une suite, donnée formellement, et un ou plusieurs procédés de sommation. Pour les fonctions intégrables, l'aventure est différente, et encore plus instructive.

Qu'est-ce qu'une fonction intégrable ? Cauchy s'en préoccupe, Dirichlet aussi, dans son mémoire de 1829. Pour qu'une fonction soit intégrable, dit-il en substance, il faut qu'elle soit continue sur un ouvert dense, et c'est, outre le fait d'être bornée, la seule restriction que voit Dirichlet (il pense sans doute à ajouter les intégrales sur les intervalles de continuité, ce qui serait une mauvaise définition de l'intégrale). C'est à cette occasion qu'il considère une fonction $\varphi(x)$ “égale à une constante déterminée c lorsque la variable x obtient une valeur rationnelle, et égale à une autre constante d, lorsque cette variable est irrationnelle”. C'est là typiquement pour Dirichlet une fonction non intégrable, donc non susceptible d'une analyse de Fourier.

C'est Riemann qui donne la première définition formelle, dans la seconde partie de sa thèse de 1854. Il distingue les intégrales propres et les intégrales généralisées, définies comme limites d'intégrales propres. Les intégrales propres, exigeant que la fonction soit bornée, sont définies en quelques lignes, et les fonctions intégrables au sens propre sont caractérisées (non pas, évidemment, par le théorème de Lebesgue sur la mesure de l'ensemble des points de discontinuité, mais par un énoncé équivalent). Puis il donne des exemples. Le premier est extrêmement instructif ; c'est

$$\sum_{n=1}^{\infty} \frac{(nx)}{n^2},$$

où (x) désigne la fonction périodique de période 1, égale à x sur $(-1/2, 1/2)$ et à 0 aux points $-1/2$ et $1/2$. Tout rationnel est point de discontinuité, tous les sauts sont positifs, et la somme des sauts est infinie. C'est ce que Paul Lévy, dans les années 1940, appellera une somme compensée de sauts, et les processus stochastiques à accroissements indépendants en donnent de nombreux exemples. Un second exemple, de fonction intégrable au sens généralisé cette fois, est

$$\frac{d}{dx}\left(x^{\nu} \cos \frac{1}{x}\right) \quad (\nu > 0).$$

C'est ce qu'il y a quelques années Yves Meyer a appelé un "chirp", une fonction qui vibre très fort au voisinage de 0. Riemann établit que les coefficients de Fourier des fonctions intégrables au sens propre tendent vers zéro, et il montre, par une habile estimation d'intégrale oscillante, qu'il n'en est pas nécessairement de même pour un "chirp" (à savoir, quand $\nu < 1/2$, les coefficients de Fourier tendent vers l'infini).

J'ai cité la note de Jordan de 1881. A la fin de cette note il construit une fonction à variation bornée dont l'ensemble des points de discontinuité est dense sur l'intervalle de définition. Cela n'est pas difficile, et Jordan aurait pu simplement, s'il l'avait connu, adapter pour son propos l'exemple de Riemann, par exemple sous la forme

$$\sum_{1}^{\infty} \frac{(nx)}{n^3}.$$

Jordan ajoute que cela contredit l'affirmation de Dirichlet, selon laquelle une fonction intégrable soit être continue sur au moins un intervalle, et, dit-il, il est surpris d'une telle erreur de la part d'un tel mathématicien. Aujourd'hui, nous sommes plutôt surpris par Jordan : le concept d'intégrabilité dépend de la définition de l'intégrale, et la définition (implicite) de Dirichlet n'était pas celle de Riemann, qui avait conquis le monde mathématique à l'époque. Pour longtemps d'ailleurs, "intégrable" allait signifier "intégrable au sens de Riemann". L'idée sous-jacente à l'affirmation de Jordan est que l'intégrabilité est un caractère intrinsèque des fonctions, comme, disons, la convergence pour les séries ; une fois mise en forme, la notion est intangible.

Quand Lebesgue donne sa définition de l'intégrale, en 1902, il évite d'ailleurs de parler de fonctions intégrables ; il parle de "fonctions sommables". Pour lui aussi, d'abord, les fonctions sommables doivent être bornées ; c'est ainsi qu'il expose la théorie dans sa thèse en 1902 et dans son livre sur l'intégration en 1904. Un des premiers succès de la théorie est une réponse partielle au programme laissé ouvert par les travaux de Riemann et de Cantor : une fonction somme de série trigonométrique partout convergente étant donnée, comment calculer les coefficients ? Réponse partielle : si elle est sommable, au moyen des formules de Fourier en utilisant l'intégrale de Lebesgue. C'est dans les *Leçons sur les séries trigonométriques* de 1906 que se trouve la définition générale des fonctions sommables (non nécessairement bornées) avec la démonstration de leurs principales propriétés. La fonction que donnait Dirichlet comme exemple de fonction non intégrale n'est intégrable en effet ni au sens de Dirichlet ni au sens de Riemann, mais elle l'est au sens de Lebesgue. Malgré l'importance fondamentale de l'intégrale de Lebesgue, il ne viendrait à l'esprit de personne aujourd'hui de dire que Dirichlet s'était trompé. Il y a plusieurs notions d'intégrale, c'est tout.

La fonction de Fatou, $\sum \sin nx \log n$, n'est intégrable ni au sens de Riemann ni au sens de Lebesgue. Elle l'est au sens de Denjoy, et c'est même la fonction test pour sa seconde totalisation. Je répète que l'exposé complet de la totalisation se trouve dans les *Leçons sur le calcul des coefficients d'une série trigonométrique.* Le rapport intime entre les séries trigonométriques et les différentes notions d'intégration, évident à priori quand on regarde les formules (2), correspond, à travers Dirichlet, Riemann, Lebesgue, Denjoy, à un mouvement historique.

8. Je terminerai avec l'évocation du rôle des séries trigonométriques dans la production d'objets fractals. On pense tout de suite aux fonctions de Weierstrass (1872)

$$\sum a^n \cos b^n x, \qquad 0 < a < 1, \quad ab > 1$$

(dont la dimension de Hausdorff du graphe est toujours inconnue !), ou de Hardy (1916)

$$\sum a^n \cos b^n x, \qquad 0 < a < 1, \quad ab = 1,$$

exemples classiques de fonctions continues nulle part dérivables. Les fonctions de Weierstrass ont partout la même irrégularité ; elles satisfont en tous les points la même condition de Hölder optimale, d'exposant $\alpha = -\log a/\log b$. Les fonctions de Hardy, comme l'a montré Geza Freud (1962), n'ont pas le même comportement en tous les points : presque partout, les accroissements sont en

$$0\left(h \log \frac{1}{h} \log\log\log \frac{1}{h}\right)$$

et rien de mieux ; en certains points (les points "lents"), les accroissements sont en $0(h)$, en d'autres (les points "rapides"), ils sont $0(h \log 1/h)$ et rien de mieux. Cette analyse locale d'une fonction, selon la rapidité de son oscillation au voisinage de chaque point, anticipe ce qui sera appelé dans les années 1980 l'"analyse multifractale".

L'idée de construire des fonctions nulle part dérivables au moyen de séries trigonométriques lacunaires a été inspirée à Weierstrass par des élèves de Riemann. Selon la tradition, Riemann aurait proposé la somme de la série

$$\sum_1^\infty \frac{\sin \pi n^2 x}{n^2}$$

comme exemple d'une fonction continue nulle part dérivable. Hardy (1916) montra en effet que la dérivée n'existe en aucun point x, si ce n'est peut-être les points rationnels de la forme $(2p+1)/(2q+1)$. Les choses en restèrent là jusqu'à ce que Joseph Gerver, alors étudiant à l'Université Columbia, montre qu'en ces points la fonction est dérivable (1970). C'est l'étude graphique de cette fonction au voisinage des points de dérivabilité qui a, de nouveau, attiré l'attention sur les "chirps" de Riemann (Jaffard et Meyer 1993). Par contre, les ondelettes de Meyer se sont révélées un moyen très utile pour l'analyse multifractale de la fonction, c'est-à-dire la détermination des dimension de Hausdorff des ensembles en lesquels la fonction satisfait une condition de Hölder optimale donnée. Voici les résultats de S. Jaffard, à paraître dans la *Revista Iberoamericana.*

Hors des points de dérivabilité $(2p+1)/(2q+1)$, la fonction de Riemann vérifie en chaque point x une condition de Hölder optimale d'exposant $\alpha(x)$, compris entre $1/2$ et $3/4$. On a

$$\alpha(x) = \frac{1}{2} + \frac{1}{2\tau(x)},$$

où $\tau(x)$ est le supremum des τ tels que, pour une infinité de rationnels p_m/q_m, donnés par le développement de x en fraction continue dont les numérateur et dénominateur ne sont pas impairs tous les deux, on ait $|x - p_m/q_m| \leq 1/q_m^\tau$. La dimension de Hausdorff de l'ensemble des x où $\alpha(x) = \alpha$ donné $(1/2 \leq \alpha \leq 3/4)$ est $d(\alpha) = 4\alpha - 2$.

Références

1. J. le Rond d'Alembert, *Recherches sur la courbe que forme une courbe tendue mise en vibration*, Mém. Acad. Berlin **49** (1747), 214.
2. ______, *Suite des recherches..*, Mém. Acad. Berlin **49** (1747), 220.
3. ______, *Suite des recherches..*, Mém. Acad. Berlin **52** (1750), 355.
4. ______, *Recherches sur différents points importants du système du monde*, Paris, 1754, voir vl. 2, p. 66.
5. ______, *Recherche sur les vibrations des cordes sonores*, Opuscules math. **1** (1761), no. 1, 1–73.
6. ______, *Nouvelles réflexions sur les vibrations des cordes sonores*, Opuscules math. **4** (1768), no. 25, 128–224.
7. G. Bachelis et J. Gilbert, *Banach algebras with Rider subalgebras*, Bull. Inst. Math. Acad. Sinica **7** (1979), no. 3, 333–347.
8. N. K. Bari, *Sur l'unicité du développement trigonométrique*, C. R. Acad. Sci. Paris **177** (1923), 1195–1197.
9. D. Bernoulli, *Réflexions et éclaircissements sur les nouvelles vibrations des cordes*, Mém. Acad. Berlin **55** (1753).
10. ______, *Sur le mélange de plusieurs espèces de vibrations simples isochrones, qui peuvent exister dans un même système de corps*, Mém. Acad. Berlin **55** (1753).
11. J. Bourgain, *Harmonic analysis and nonlinear partial differential equations*, (1995), 31–44.
12. G. Cantor, *Über die Ausdehnung eines Satzes aus der Theorie der trigonometrischen Reihen*, Math. Annalen **5** (1872), 123–132.
13. L. Carleson, *On convergence and growth of partial sums of Fourier series*, Acta Math. **116** (1966), 135–157.
14. G. Debs et J. Saint-Raymond, *Ensembles boréliens d'unicité et d'unicité au sens large*, Ann. Inst. Fourier (Grenoble) **37** (1987), no. 3, 217–239.
15. A. Denjoy, *Calcul de la primitive de la fonction dérivée la plus générale*, C. R. Acad. Sci. Paris **154** (1912), 1075.
16. ______, *Calcul des coefficients d'une série trigonométrique convergente quelconque dont la somme est donnée*, C. R. Acad. Sci. Paris **172** (1921), 1218.
17. ______, *Leçons sur le calcul des coefficients d'une série trigonométrique*. I. *La différentiation seconde mixte et son application aux séries trigonométriques*, Gauthier-Villars, Paris, 1941 ; II. *Métrique et topologie d'ensembles parfaits et de fonctions*, Gauthier-Villars, Paris, 1941 ; III. *Détermination d'une fonction par ses nombres dérivés seconds extrêmes finis*, Gauthier-Villars, Paris, 1941 ; IV. *Les totalisations. solution du problème de Fourier. 1. les totalisations ; 2. appendices et tables générales*, Gauthier-Villars, Paris, 1949.

18. P. G. Lejeune Dirichlet, *Sur la convergence des séries trigonométriques qui servent à représenter une fonction arbitraire entre des limites données*, J. Reine Angew. Math. **4** (1829), 157–169.

19. L. Euler, *Dissertatio physica de sono*, Opera omnia **1** (1727), 183–196.

20. L. Fejér, *Sur les fonctions bornées et intégrables*, C. R. Acad. Sci. Paris **131** (1900), 984–987.

21. J. Fourier, *Théorie analytique de la chaleur*, 1822.

22. G Freud, *Über trigonometrische Approximation und Fouriersche Reihen*, Math. Z. **78** (1962), 252–262.

23. J. Gerver, *The differentiability of the Riemann function at certain rational multiples of* π, Amer. J. Math. **92** (1970), 35–55.

24. Hardy, *Weierstrass's non-differentiable function*, Trans. Amer. Math. Soc. **17** (1916), 301–325.

25. R. A. Hunt, *On the convergence of Fourier series*, Orthogonal Expansions and their Continuous Analogues (Edwardsville, 1967), Southern Illinois Univ. Press, Carbondale, IL, 1968.

26. S. Jaffard, *The spectrum of singularities of Riemann's function*, Rev. Mat. Iberoamericana, **12** (1996), no. 2, 441–460.

27. S. Jaffard et Y. Meyer, *Wavelet methods for pointwise regularity and local oscillations of functions*, Mem. Amer. Math. Soc. (1995), à paraître.

28. C. Jordan, *Sur la série de Fourier*, C. R. Acad. Sci. Paris **92** (1881), 228–230.

29. J.-P. Kahane, *Idempotents and closed subalgebras of* $A(\mathbb{Z})$, Function Algebras (Tulane Univ., 1965) (F. Birtel, ed.), Scott-Foresman, Chicago, 1966, pp. 198–207.

30. ———, *Géza Freud and lacunary Fourier series*, J. Approx. Theory **46** (1986), no. 1, 51–57.

31. ———, *Une théorie de Denjoy des martingales dyadiques*, Enseign. Math. **34** (1988), no. 3-4, 255–268.

32. J.-P. Kahane et Y. Katznelson, *Sur les ensembles de divergence des séries trigonométriques*, Studia Math. **26** (1966), 305–306.

33. ———, *Mauvaise approximation par polynômes trigonométriques dans certaines algèbres de convolution*, Israel J. Math. **31** (1978), no. 3-4, 217–223.

34. J.-P. Kahane et P.-G. Lemarié-Rieusset, *Fourier series and wavelets*, Gordon & Breach, Londres, 1995.

35. J.-P. Kahane et R. Salem, *Ensembles parfaits et séries trigonométriques*, Actualités Sci. Indust., vol. 1301, Hermann, Paris, 1963.

36. R. Kaufman, *Absolutely convergent Fourier series and some classes of sets*, Bull. Sci. Math. (2) **109** (1985), no. 4, 363–372.

37. ———, *Bernoulli convolutions—an application of set theory in analysis*, Colloq. Math. **60/61** (1990), no. 1, 93–98.

38. A. S. Kechris et A. Louveau, *Descriptive set theory and the structure of sets of uniqueness*, London Math. Soc. Lecture Note Ser., vol. 128, Cambridge Univ. Press, 1987.

39. A. Kolmogorov, *Une série de Fourier-Lebesgue divergente partout*, C. R. Acad. Sci. Paris **183** (1926), 1327–1328.

40. J.-L. Lagrange, *Recherches sur la nature et la propagation du son*, Miscellanea Taurinensia **1** (1759).

41. ———, *Nouvelles recherches sur la nature et la propagation du son*, Miscellanea Taurinensia **2** (1760).

42. H. Lebesgue, *Intégrale, longueur, aire*, Thèse, 1902.

43. ———, *Leçons sur l'intégration*, Gauthier-Villars, Paris, 1904.

44. ———, *Leçons sur les séries trigonométriques*, Gauthier-Villars, Paris, 1906.

45. P. Lévy, *Processus stochastiques et mouvement brownien*, Gauthier-Villars, Paris, 1949.

46. P.-L. Lions, *On some recent methods for nonlinear partial differential equations*, Proceedings of the International Congress of Mathematicians, Vol. 1, 2 (Zürich, 1994), Birkäuser, Bâle, 1995, pp. 140–155.

47. D. Men′šov, *Sur l'unicité du développement trigonométrique*, C. R. Acad. Sci. Paris **163** (1916), 433–436.

48. ———, *Sur la représentation des fonctions mesurables par des séries trigonométriques*, C. R. (Doklady) Acad. Sci. URSS (N.S.) **26** (1940), 214–216.

49. ———, *On limits of indeterminacy in measure of partial sums of trigonometric series*, Mat. Sbornik (N.S.) **34** (1954), 557–574 (Russian).

50. D. Oberlin, *An approximation problem in* $L^p([0,2\pi])$, $2 < p < \infty$, Studia Math. **70** (1981), no. 3, 221–224.

51. M. A. Pinsky, *Fourier inversion for piecewise smooth functions in several variables*, Proc. Amer. Math. Soc. **118** (1993), no. 3, 903–910.

52. Platon, *République*, vol. VII, 530 d.

53. B. Riemann, *Über die Darstellbarkeit einer Funktion durch eine trigonometrische Reihe*, Habilitationsschrift, 1854.

54. V. A. Rodin, *The BMO-property of the partial sums of a Fourier series*, Dokl. Akad Nauk. SSR **319** (1991), no. 5, 1079–1081 (Russian) ; Trad. anglaise Soviet Math. Dokl. **44** (1992), no. 1, 294–296.

55. L. Schwartz, *Théorie des distributions I*, Publ. Inst. Math. Univ. Strasbourg, vol. 9, Hermann & Cie, Paris, 1950.

56. ———, *Théorie des distributions II*, Publ. Inst. Math. Univ. Strasbourg, vol. 10, Hermann & Cie, Paris, 1951.

57. B. Taylor, *De methodo incrementorum*, Londres, 1715.

58. K. Weierstrass, *Über continuirliche Functionen eines reellen Arguments, die für keinen Werth des letzeren einen bestimmten Differentialquotienten besitzen*, Math. Werke **2** (1872), 71–74.

59. W. H. Young, *A note on trigonometrical series*, Messenger Math. **38** (1909), 44–48.

60. A. Zygmund, *Smooth functions*, Duke Math. J. **12** (1945), 47–76.

Département de Mathématiques, Université de Paris XI, Orsay, France
E-mail address: jean-pierre.kahane@math.u-psud.fr

Centre de Recherches Mathématiques
CRM Proceedings and Lecture Notes
Volume **11**, 1997

Exterior Differential Systems and Normal Forms of Parabolic Systems

R. B. Gardner and N. Kamran

ABSTRACT. We present a review of a classification scheme and normal form results for fully non-linear C^∞ parabolic systems of two first order p.d.e.s for two functions of two variables. The approach is based on E. Cartan's theory of exterior differential systems. An existence theorem is also given for the initial value problem for C^∞ parabolic systems with completely integrable characteristics and non-focal Cauchy data.

1. Introduction

In this paper, we present a self-contained review of our recent classification and normal form results under the pseudo-group of contact transformations for fully non-linear determined systems of two first-order partial differential equations (p.d.e.s) in the plane [**GaK2**]. We shall limit ourselves to parabolic systems in the C^∞ category, for which the most interesting results are available, and refer the reader to [**GaK2**] for the hyperbolic and flat cases.

Our main tool will be the theory of exterior differential systems [**Ca3**], [**BC3G**], which gives a geometric and coordinate-free formulation of partial differential equations. Exterior differential systems are thus particularly well-suited to the study classification and normal form problems.

This class of p.d.e.s has been considered in the analytic (C^ω) category in Chapter VII of [**BC3G**], within the framework of Cartan-Kähler theory and the characteristic variety for exterior differential systems. Our approach will be different. We shall use normal form results for Pfaffian systems which are valid in the C^∞ case. We will work with the characteristic and focal Pfaffian systems associated to the given parabolic p.d.e. system. The characteristic system generalizes to the fully non-linear case the system defined by Friedrichs and Courant [**Fr**] in the quasilinear parabolic case . Focal systems were originally defined in [**Ga4**] as a differential geometrical generalization to the case of Pfaffian systems of the classical construction of envelopes of characteristic curves, for first-order p.d.e.s, known as focal curves.

1991 *Mathematics Subject Classification*. Primary: 58G37, 53C10; Secondary:35K45.

The work of the first author was supported by NSF Grants #DMS-9409037 and DMS-9204942.

The work of the second author was supported by NSERC Grant #OGP0105490.

This is the final form of the paper.

We shall demonstrate that geometrical assumptions about the parabolic system and its characteristic or focal systems are in many instances sufficient to characterize invariantly a single equivalence class of p.d.e.s or a stratum of equivalence classes with simple normal forms, whose properties can be readily analyzed.

In Section 2, we recall some of the basic results concerning Pfaffian systems that we shall need for our classification. These are mostly elementary generalizations of classical results such as the Frobenius theorem and the Goursat normal form [**BC3G**]. This material is included so as to make our paper essentially self-contained and therefore accessible to the mathematicians who are not familiar with the theory of exterior differential systems. In Section 3, we define the notion of parabolicity in a contact-invariant way and establish the fundamental structure equations for the Pfaffian systems associated to parabolic equations. We then present our classification, which is based on the analysis of the possible configurations for the characteristic system and its focal system. We establish normal form results in practically all of the non-generic branches. The generic branches require the implementation of E. Cartan's method of equivalence [**Ca1**], [**Ga5**]. Their treatment would not be particularly enlightening since the corresponding normal forms, even if they could be determined explicitly, would be of limited interest unless they happened to have additional properties, say of symmetry. This is why we have chosen to focus our attention instead on the non-generic branches, where the normal forms exhibit the richest array of features. We also show that the characteristic and focal systems play an important role in the analysis of the initial value problem by proving a smooth existence theorem for non-focal Cauchy data and completely integrable characteristics. This is closely related to the existence theorem outlined by E. Cartan [**Ca2**] and established in [**Ga2**] for scalar parabolic second-order p.d.e.s in the plane.

2. Some Normal Form Results for Pfaffian Systems

Our goal in this section is to recall some of the basic definitions and C^∞ normal form results from the theory of Pfaffian systems. The proofs are found in [**BC3G**], unless otherwise indicated.

We let M be a smooth (C^∞) n-dimensional manifold, $\Omega^p(M)$ denote the $C^\infty(M;\mathbb{R})$-module of smooth sections of the p-th exterior power bundle $\Lambda^p(T^*M)$ and $\Omega^*(M)$ be the direct sum $\bigoplus_{p=0}^n \Omega^p(M)$.

A *Pfaffian system* I is a submodule of $\Omega^1(M)$. Locally, I is thus generated by $s \leq n$ 1-forms

$$\omega^a = \sum_{i=1}^{n} A_i^a(x^1, \ldots, x^n)dx^i, \quad 1 \leq a \leq s. \tag{2.1}$$

The 1-forms ω^a, $1 \leq a \leq s$, will be assumed to be linearly independent. We will also assume that s is constant. The integer s will then be called the *dimension* of I. The Pfaffian system generated by $\omega^1, \ldots, \omega^s$ will be denoted by $I = \{\omega^1, \ldots, \omega^s\}$.

We define an *integral manifold* of I to be an immersion $f\colon N \to M$ such that

$$f^*\omega^a = 0, \quad 1 \leq a \leq s. \tag{2.2}$$

If the immersion f is given in a domain U of $\mathbb{R}^p$ with coordinates u^α, $1 \leq \alpha \leq p$, by an n-tuple of functions $\big(f^1(u^1, \ldots, u^p), \ldots, f^n(u^1, \ldots, u^p)\big)$, such that rank $\partial f^i/\partial u^\alpha = p$ in U, then its components are the solutions of the systems of partial differential equations given by

$$\sum_{i=1}^{n}\sum_{\alpha=1}^{p} A_i^a(f^1(u^1,\dots,u^p),\dots,f^n(u^1,\dots,u^p))\frac{\partial f^i}{\partial u^\alpha}=0, \quad 1\le a\le s. \tag{2.3}$$

The Frobenius Theorem is the basic existence theorem for integral manifolds:

THEOREM 2.1 (Frobenius). *Let I be a Pfaffian system generated by linearly independent 1-forms ω^a, $1\le a\le s$, such that*

$$d\omega^a\wedge\omega^1\wedge\cdots\wedge\omega^s=0, \quad 1\le a\le s. \tag{2.4}$$

There exist local coordinates $(u^1,\dots,u^n)$ such that I is generated by $du^1,\dots,du^s$.

The functions $u^1,\dots,u^s$ are called *first integrals* of the completely integrable system I. A Frobenius system admits thus, at least locally, $(n-s)$-dimensional integral manifolds of maximal dimension given by $u^1=c^1,\dots,u^s=c^s$, where $c^1,\dots,c^s$ are arbitrary real constants. A Pfaffian system which satisfies the Frobenius conditions (2.4) is said to be *completely integrable.* The Cauchy characteristics give a minimal set of coordinates for the generators of a Pfaffian system. The *Cauchy characteristic* vector fields system Char (I) is defined by

$$\mathrm{Char}(I)=\{X\mid X\in I^\perp, X\rfloor dI\subset I\}. \tag{2.5}$$

The Pfaffian system $C(I)$ defined by

$$C(I)=\mathrm{Char}(I)^\perp. \tag{2.6}$$

is called the *Cartan system* of I. The Cartan system $C(I)$ of any Pfaffian system I is completely integrable. The first integrals of $C(I)$ are the minimal set of coordinates The *class* of I is the dimension of its Cartan system. The *codimension* of a Pfaffian system is defined to be its class minus its dimension.

THEOREM 2.2 (E. Cartan). *Let I be a Pfaffian system of class r. Denote by $\{w^1,\dots,w^r\}$ a set of first integrals of the Cartan system $C(I)$. There exists a neighborhood U with local coordinates $(w^1,\dots,w^r;y^{r+1},\dots,y^n)$ such that I is generated by 1-forms in $w^1,\dots,w^r$ and their differentials.*

The *derived flag* of a Pfaffian system I, is defined as follows. Let $\{I\}$ denote the algebraic ideal generated by I in $\Omega^*(M)$. Let δ denote the $C^\infty(M;\mathbb{R})$-linear map $\delta\colon I\to\Omega^2(M)/\{I\}$ induced by the exterior derivative $d\colon I\to\Omega^2(M)$. The first derived system $I^{(1)}$ of I is the kernel of δ, which we assume to have constant dimension. We have therefore a short exact sequence of $C^\infty(M;\mathbb{R})$-modules

$$0\to I^{(1)}\to I\to dI/\{I\}\to 0. \tag{2.7}$$

It follows from the definition that $I^{(1)}=I$ if and only if I is completely integrable. The quotient $I/I^{(1)}$ is therefore a measure of the degree to which I fails to be completely integrable. The iteration of this construction gives rise to a flag of Pfaffian systems

$$\cdots\subset I^{(k)}\subset\cdots\subset I^{(1)}\subset I, \tag{2.8}$$

where

$$I^{(k)}=(I^{(k-1)})^{(1)}.$$

This defines the *derived flag* of I, which stabilizes at the maximal completely integrable subsystem of I. It follows from the commutativity of d and pull-backs that if $f\colon M \to M$ is a diffeomorphism and J is the Pfaffian system defined by $J = f^*I$, then $f^*I^{(i)} = J^{(i)}$, for all $i \geq 1$. The derived flag is thus invariantly associated to I.

For the sake of simplicity, the same symbol will be used to denote a Pfaffian system I and the ideal $\{I\}$ it generates in $\Omega^*(M)$.

There are normal form results similar to the Frobenius theorem for Pfaffian systems which are not completely integrable, but whose structure equations are of a particular form. An example of this is given by the *Pfaff problem*. Let I be a Pfaffian system generated by a single 1-form ω. The integer r defined by

$$(d\omega)^r \wedge \omega \neq 0, \quad (d\omega)^{r+1} \wedge \omega = 0. \tag{2.9}$$

is called the *rank* of $\{\omega\}$. In what follows, we shall assume r to be constant.

THEOREM 2.3 (The Pfaff problem). *If rank $\{\omega\} = r$, then there exist local coordinates $(z, p_1, \ldots, p_r, x^1, \ldots, x^r, u^{2r+2}, \ldots, u^n)$, such that*

$$\{\omega\} = \left\{dz - \sum_{i=1}^{r} p_i dx^i\right\}. \tag{2.10}$$

The integral manifolds of $I = \{\omega\}$ are thus parametrized by one arbitrary function of r variables and given by $z = f(x^1, \ldots, x^n)$, $p_i = \partial f/\partial x^i$, $1 \leq i \leq r$.

There is a result which is somewhat stronger than Theorem 2.3 and which concerns a relative version of the Pfaff problem.

THEOREM 2.4 (The Relative Pfaff problem). *Let I be the $(s+1)$-dimensional Pfaffian system given*

$$I = \{\theta^1, \ldots, \theta^s, \omega\}, \tag{2.11}$$

where $J = \{\theta^1, \ldots, \theta^s\}$ is completely integrable and the rank r of $I = \{\omega\}$ relative to J, defined by

$$(d\omega)^r \wedge \omega \not\equiv 0 \mod J, \quad (d\omega)^{r+1} \wedge \omega \equiv 0 \mod J, \tag{2.12}$$

is constant. There exist local coordinates $(z,\ p_1, \ldots, p_r,\ x^1, \ldots, x^r,\ w^1, \ldots, w^s,\ u^{2r+s+1}, \ldots, u^n)$, such that

$$I = \left\{dz - \sum_{i=1}^{r} p_i dx^i,\ dw^1, \ldots, dw^s\right\}. \tag{2.13}$$

It is straightforward to verify that the proof of Theorem 2.3 as given in [**BC3G**] can be suitably modified to give a proof of Theorem 2.4.

The integral manifolds of I are thus parametrized by one arbitrary function of r variables and s arbitrary constants. They are given by $z = f(x^1, \ldots, x^n)$, $p_i = \partial f/\partial x^i$, $w^i = c^i$, $1 \leq i \leq r$, $1 \leq \alpha \leq s$.

We state the relative version of the *Goursat normal form*[**GS**].

THEOREM 2.5 (Relative Goursat normal form). *Let I be an $(r+s)$-dimensional Pfaffian system of codimension two, of the form*

$$I = \{\omega^1, \ldots, \omega^r, \theta^1, \ldots, \theta^s\} \tag{2.14}$$

where $J = \{\theta^1, \ldots, \theta^s\}$ is completely integrable. Suppose that there exist 1-forms α and π such that $\alpha \not\equiv 0,$, $\pi \not\equiv 0 \mod I$ and such that the following structure *equations are valid*

$$
\begin{aligned}
&d\omega^1 \equiv \omega^2 \wedge \pi \mod \{\omega^1, \theta^1, \dots, \theta^s\}\\
&\vdots\\
&d\omega^i \equiv \omega^{i+1} \wedge \pi \mod \{\omega^1, \dots, \omega^i, \theta^1, \dots, \theta^s\}, \quad 1 \le i \le r-1\\
&\vdots\\
&d\omega^r \equiv \alpha \wedge \pi \mod \{\omega^1, \dots, \omega^r, \theta^1, \dots, \theta^s\} = I.
\end{aligned}
\tag{2.15}
$$

There exist local coordinates $(x, y, y', \dots, y^{(r)}, w^1, \dots, w^s, u^{r+s+3}, \dots, u^n)$, *such that*

$$I = \{dy - y'dx, \dots, dy^{(r-1)} - y^{(r)}dx, dw^1, \dots, dw^s\}. \tag{2.16}$$

Under the hypotheses of Theorem 2.5, the derived flag of I is given by

$$
\begin{aligned}
I^{(i)} &= \{\omega^1, \dots, \omega^{r-i}, \theta^1, \dots, \theta^s\},\\
&\vdots\\
I^{(r)} &= I^{(r+1)} = J,
\end{aligned}
\tag{2.17}
$$

and thus terminates at the completely integrable system $I^{(r)}$. The local integral manifolds of I are parametrized by one arbitrary function of one variable and r arbitrary constants. They are given by $y = f(x)$, $y' = f'(x), \dots, y^{(s)} = f^{(s)}(x)$, $u^1 = c^1, \dots, u^r = c^r$.

We shall need the following result of E. Cartan and Von Weber. A proof of it is given in [**Ga1**].

THEOREM 2.6 (E. Cartan, Von Weber). *Let I be an s-dimensional Pfaffian system satisfying*

$$\dim(I/I^{(1)}) = 1. \tag{2.18}$$

Let $\varphi^1, \dots, \varphi^{s-1}$ be generators for $I^{(1)}$ and α be a generator for a complement of $I^{(1)}$ in I. If $\varphi^1, \dots, \varphi^{s-1}, \alpha$ are such that

$$(d\alpha)^2 \wedge \alpha \wedge \varphi^1 \wedge \dots \wedge \varphi^{s-1} \neq 0, \tag{2.19}$$

then $I^{(1)}$ is completely integrable.

Let now $\pi \colon B \to M$ be a fibered manifold and let $V(B) = \ker \pi_*$ denote the vertical subbundle of TB, consisting of the π-vertical tangent vectors. For applications to partial differential equations, there is no loss of generality in considering semi-basic Pfaffian systems. These are Pfaffian systems I on B such that

$$V(B) \subset I^{\perp}. \tag{2.20}$$

If $J \subset I$ is a Pfaffian subsystem of I, then the *focal system* of J is the Pfaffian system $F(J)$ on B defined by [**Ga4**]

$$F(J) = \{Z \rfloor dJ, Z \in V(B)\}. \tag{2.21}$$

The system of *J-characteristic vector fields* of I, denoted by Char (I, dJ), is defined by

$$\mathrm{Char}(I, dJ) = \{X \mid X \in I^{\perp}, X \rfloor dJ \subset I\}, \tag{2.22}$$

and the J-*characteristic system* of I, $C(I, dJ)$, is then defined as the annihilator of $\mathrm{Char}(I, dJ)$, namely

$$C(I, dJ) = \mathrm{Char}(I, dJ)^{\perp}. \tag{2.23}$$

Let us illustrate these definitions on a simple example. Consider, on the fibered manifold $\pi\colon \mathbb{R}^6 \to \mathbb{R}^4\colon (x, y, z, w, p, q) \mapsto (x, y, z, w)$, the semi-basic Pfaffian system I generated by

$$\omega^1 = dz - p dx, \quad \omega^2 = dw - q dy.$$

We have

$$V(\mathbb{R}^6) = \left\{\frac{\partial}{\partial p}, \frac{\partial}{\partial q}\right\}, \quad \mathrm{Char}(I, dJ) = \left\{\frac{\partial}{\partial y} + q\frac{\partial}{\partial w}, \frac{\partial}{\partial q}\right\}$$

so that

$$C(I, dJ) = \{dw - q dy, dp, dx, dz\}, \quad F(J) = \{dx, dz\}. \tag{2.24}$$

The focal system of a subsystem J is thus the Pfaffian system obtained by Lie differentiation of J along the vector fields tangent to the fibers of π. The J-focal curves are by definition the integral curves of $F(J)$. They can be thought of as envelopes integral curves of $C(I, dJ)$.

It is straightforward to show that the focal system of J is a subsystem of the J-characteristic system,

$$F(J) \subset C(I, dJ). \tag{2.25}$$

This result also holds at the level of derived flags,

$$F(J)^{(i)} \subset C(I, dJ)^{(i)}, \quad \forall i \geq 1. \tag{2.26}$$

The geometry of the J-characteristic systems for scalar second-order hyperbolic equations in the plane has been studied in detail in [**GaK1**].

We conclude by recalling the following lemma of exterior algebra due to E. Cartan, a proof of which can be found in [**Ga2**].

LEMMA (The Quadratic Lemma). *Let V denote a finite-dimensional real vector space, let $\omega^1, \dots, \omega^s$, π be linearly independent elements of V^* and let Ω be an element of $\Lambda^2 V^*$. If, for some integer $p \geq 1$,*

$$\Omega^p \wedge \omega^1 \wedge \cdots \wedge \omega^s \wedge \pi = 0, \tag{2.27}$$

then

$$\Omega^{p+1} \wedge \omega^1 \wedge \cdots \wedge \omega^s = 0. \tag{2.28}$$

3. Some Normal Form Results for Parabolic Systems

The systems considered will be of the form

$$F\left(x, y, z, w, \frac{\partial z}{\partial x}, \frac{\partial z}{\partial y}, \frac{\partial w}{\partial x}, \frac{\partial w}{\partial y}\right) = 0, \quad G\left(x, y, z, w, \frac{\partial z}{\partial x}, \frac{\partial z}{\partial y}, \frac{\partial w}{\partial x}, \frac{\partial w}{\partial y}\right) = 0. \tag{3.1}$$

We shall be working in the vein of E. Cartan's method of equivalence [**Ca1**], [**Ga5**], by choosing adapted moving coframes and using the normal form results on Pfaffian systems reviewed in Section 2.

To express the system (3.1) as a Pfaffian system, we consider the bundle $J^1(\mathbb{R}^2, \mathbb{R}^2)$ of 1-jets of maps from $\mathbb{R}^2$ to $\mathbb{R}^2$ and local coordinates (x, y, z, w, p, q, m, n) in which the Pfaffian system $\Omega^1_{\text{cont}}(\mathbb{R}^2, \mathbb{R}^2)$ of contact 1-forms is generated by

$$\theta^1 = dz - pdx - qdy, \quad \theta^2 = dw - mdx - ndy. \tag{3.2}$$

This means that if $f\colon \mathbb{R}^2 \to J^0(\mathbb{R}^2, \mathbb{R})\colon (x, y) \mapsto \big(x, y, z(x, y), w(x, y)\big)$ is a section, then

$$p(j^1 f) = \frac{\partial z}{\partial x}, \quad q(j^1 f) = \frac{\partial z}{\partial y}, \quad m(j^1 f) = \frac{\partial w}{\partial x}, \quad n(j^1 f) = \frac{\partial w}{\partial y}.$$

The equations

$$F(x, y, z, w, p, q, m, n) = 0, \quad G(x, y, z, w, p, q, m, n) = 0, \tag{3.3}$$

corresponding to the p.d.e.s (3.1) will be assumed to define a 6-dimensional submanifold $i\colon \Sigma_6 \to J^1(\mathbb{R}^2, \mathbb{R}^2)$ which fibers over the bundle $J^0(\mathbb{R}^2, \mathbb{R}^2)$ of 0-jets, with local coordinates (x, y, z, w).

$$\begin{array}{ccc} \Sigma_6 & \xrightarrow{\ i\ } & J^1(\mathbb{R}^2, \mathbb{R}^2) \\ & \searrow^{\pi} & \downarrow \pi^1_0 \\ & & J^0(\mathbb{R}^2, \mathbb{R}^2) \end{array}$$

Therefore, not all of the jacobian determinants

$$\frac{\partial(F, G)}{\partial(p, q)}, \quad \frac{\partial(F, G)}{\partial(p, m)}, \quad \frac{\partial(F, G)}{\partial(p, n)}, \quad \frac{\partial(F, G)}{\partial(q, m)}, \quad \frac{\partial(F, G)}{\partial(q, n)}, \quad \frac{\partial(F, G)}{\partial(m, n)}$$

are zero on the locus $F = 0$, $G = 0$.

The Pfaffian system obtained by pulling back the contact system $\Omega^1_{\text{cont}}(\mathbb{R}^2, \mathbb{R}^2)$ to Σ_6 will be denoted by I. It is easy to see that I is semi-basic for $\pi\colon \Sigma_6 \to J^0(\mathbb{R}^2, \mathbb{R}^2)$.

The solutions of (3.1) are characterized geometrically [**Ga2**] as the integral manifolds $\sigma\colon \mathbb{R}^2 \to \Sigma_6$ of I, which are transversal to $V(\Sigma_6)$.

An *internal contact transformation* between two systems of p.d.e.s with Pfaffian systems I and $\bar{I}$ is defined as a local diffeomorphism $\Phi\colon \Sigma_6 \to \bar{\Sigma}_6$ such that $\Phi^* \bar{I} = I$. This notion is generally distinct from that of an *external contact transformation*, meaning a local diffeomorphism $\Psi\colon J^1(\mathbb{R}^2, \mathbb{R}^2) \to J^1(\mathbb{R}^2, \mathbb{R}^2)$ of the ambient jet bundles which satisfies $\Phi(i(\Sigma_6)) = \bar{i}(\bar{\Sigma}_6)$ and preserves the contact systems $\Omega^1_{\text{cont}}(\mathbb{R}^2, \mathbb{R}^2)$. Every external contact transformation gives rise to an internal one by restriction, but the converse is not always true [**AKO**].

The generators of I will be denoted by

$$\omega^1 = i^* \theta^1, \quad \omega^2 = i^* \theta^2, \tag{3.4}$$

The Pfaffian system defined by (3.4) is of codimension at most four since $\dim C(I) = 5$ or 6 and $\dim I = 2$. Therefore, [**Ga1**],there exists an intrinsic conformal symmetric $C^\infty(\Sigma_6; \mathbb{R})$-bilinear form $\langle \cdot, \cdot \rangle$ on I, defined by

$$\langle \varphi, \psi \rangle \mathrm{Vol}_{\Sigma_6} = d\varphi \wedge d\psi \wedge \omega^1 \wedge \omega^2, \tag{3.5}$$

where Vol_{Σ_6} denotes any volume form on Σ_6.

The matrix of $\langle\cdot,\cdot\rangle$ is given in terms of F and G by:

$$\begin{pmatrix} \langle\omega^1,\omega^1\rangle & \langle\omega^1,\omega^2\rangle \\ \langle\omega^1,\omega^2\rangle & \langle\omega^2,\omega^2\rangle \end{pmatrix} = \begin{pmatrix} \frac{\partial(F,G)}{\partial(m,n)} & \frac{1}{2}\left(\frac{\partial(F,G)}{\partial(q,m)} - \frac{\partial(F,G)}{\partial(p,n)}\right) \\ \frac{1}{2}\left(\frac{\partial(F,G)}{\partial(q,m)} - \frac{\partial(F,G)}{\partial(p,n)}\right) & \frac{\partial(F,G)}{\partial(p,q)} \end{pmatrix} \tag{3.6}$$

The only invariants of $\langle\cdot,\cdot\rangle$ being its rank and index, we have a first pointwise classification of (3.1) into the following four types

Rank	Index	Type
2	0	Elliptic
2	1	Hyperbolic
1	0	Parabolic
0	0	Flat

This classification is invariant under contact transformations since $\langle\cdot,\cdot\rangle$ is intrinsic. The determinant of the matrix (3.6) can be written as

$$\Delta = \frac{\partial(F,G)}{\partial(p,m)}\frac{\partial(F,G)}{\partial(q,n)} - \frac{1}{4}\left(\frac{\partial(F,G)}{\partial(p,n)} + \frac{\partial(F,G)}{\partial(q,m)}\right)^2 \tag{3.7}$$

so that the classification of the type of $\langle\cdot,\cdot\rangle$ corresponds in the quasi-linear case to that of Friedrichs and Courant [**Fr**].

We shall assume from now on that the symmetric bilinear form $\langle\cdot,\cdot\rangle$ is of parabolic type. We can therefore choose, at least locally, generators π^1 and π^2 for I in which the matrix of $\langle\cdot,\cdot\rangle$ is in the Witt normal forms

$$\begin{pmatrix} 0 & 0 \\ 0 & 1 \end{pmatrix}. \tag{3.8}$$

The conjugate subspace of I with respect to $\langle\cdot,\cdot\rangle$ defines a subsystem N of I which is invariantly associated to I,

$$N = \{\pi^1\}.$$

The N-characteristic system $C(I, dN)$ generalizes the Friedrichs characteristic system from quasilinear parabolic systems [**Fr**] to the fully non-linear case. In honour of the original classification, we shall call this the *Friedrichs characteristic system.*

For example, the system

$$p = w, \quad m = f(x,y,z,w,n,q),$$

is parabolic whenever

$$f_q \neq 0,$$

in which case the subsystem N is given by

$$\{\pi^1\} = \{ndx - dq\}.$$

We obtain structure equations for parabolic systems by choosing generators for I in which the matrix of $\langle\cdot,\cdot\rangle$ is in the Witt normal form (3.8). From (3.5) we obtain:

THEOREM 3.1. *If I is a two-dimensional Pfaffian system on a six-dimensional manifold, whose associated conformal symmetric bilinear form $\langle\cdot,\cdot\rangle$ defined by* (3.5) *is of constant parabolic type on* Σ_6*, then there exist, locally, generators* π^1, π^2 *for* I *and* 1*-forms* $\omega^3, \omega^4, \omega^5, \omega^6$ *on* Σ_6 *such that:*

$$
\begin{aligned} d\pi^1 &\equiv 0, \\ d\pi^2 &\equiv \omega^3 \wedge \omega^5 + \omega^4 \wedge \omega^6, \end{aligned} \quad \mod \{\pi^1, \pi^2\} \tag{3.9}
$$

or

$$
\begin{aligned} d\pi^1 &\equiv \omega^3 \wedge \omega^4\ , \\ d\pi^2 &\equiv \omega^3 \wedge \omega^5 + \omega^4 \wedge \omega^6\ , \end{aligned} \quad \mod \{\pi^1, \pi^2\} \tag{3.10}
$$

where

$$
\pi^1 \wedge \pi^2 \wedge \omega^3 \wedge \omega^4 \wedge \omega^5 \wedge \omega^6 \neq 0.
$$

The proof of Lemma 3.5 of [**GaK1**] can be adapted to prove that for parabolic systems, the image of every two dimensional-integral manifold $\sigma\colon \mathbb{R}^2 \to \Sigma_6$ admits a vector field in Char (I, dN) as a tangent vector field.

The Pfaffian system $I = i^*\Omega^1_{\text{cont}}(\mathbb{R}^2, \mathbb{R}^2)$ corresponding to the p.d.e. system (3.1) will of course *not* be the generic Pfaffian system of two equations in six variables. The Pfaffian systems of two equations in six variables have many more branches in their classification than the p.d.e. systems (3.1). A number of these additional branches can be ruled out by showing that these Pfaffian systems admit normal forms which are manifestly incompatible with a p.d.e. system of the form (3.1). The subspace N and its focal system $F(N)$ will be essential tools in proving these results.

We now proceed to develop our classification scheme and to establish some normal form results for parabolic systems.

It is easy to rule out the case when I is completely integrable since $\langle\cdot,\cdot\rangle$ is rank 1. Therefore the first case to be considered is when the first derived system is 1-dimensional, that is

$$
\dim(I/I^{(1)}) = 1, \tag{3.11}
$$

From the structure equations (3.9) and from (3.11), we conclude that $I^{(1)} = N = \{\pi^1\}$. The assumption of parabolicity together with the Witt normal form (3.8) gives

$$
(d\pi^2)^2 \wedge \pi^1 \wedge \pi^2 \neq 0. \tag{3.12}
$$

We have the following unique normal form result for the the parabolic systems satisfying (3.11).

THEOREM 3.2. *Every parabolic system* (3.1) *such that* $I^{(1)}$ *is* 1*-dimensional can be locally transformed by a contact transformation to*

$$
\frac{\partial z}{\partial x} = 0, \quad \frac{\partial z}{\partial y} = 0. \tag{3.13}
$$

PROOF. The conditions (3.11) and (3.12) allow us to apply the Cartan-Von Weber theorem (Theorem 2.5) to conclude that $I^{(1)} = \{\pi^1\} = N$ must be completely integrable. We can therefore apply the solution to the relative Pfaff problem (Theorem 2.4) to argue that there exist local coordinates (x, y, z, w, m, n) such that

$$
\{\pi^1, \pi^2\} = \{dw - m dx - n dy, dz\}. \tag{3.14}
$$

In this case, the focal system $F(N)$ is a one-dimensional completely integrable system

$$F(N) = \{dz\}\ .$$

The generic parabolic systems are such that I has a trivial first derived system

$$I^{(1)} = 0.$$

These systems are divided into two subclasses depending on whether the class of the subsystem $N = \{\pi^1\}$ is equal to three or five. The first subclass corresponds to a single equivalence class under contact transformations. □

THEOREM 3.3. *Every parabolic p.d.e. system* (3.1) *such that I and the subsystem N satisfy the conditions*

$$I^{(1)} = 0, \quad \operatorname{class} N = 3,$$

is locally equivalent under a contact transformation to

$$\frac{\partial z}{\partial x} = 0, \quad \frac{\partial w}{\partial x} = \frac{\partial z}{\partial y}.$$

We refer the reader to [**GaK2**] for the proof.

The hypotheses of Theorem 3.3 can also be formulated in terms of the focal system $F(N)$ of the isotropic subsystem N. Let us choose the 1-form ω^3 in the structure equations (3.16) so that

$$\omega^3 \equiv 0 \text{ mod base},$$

where "mod base" means " modulo the ideal of $\Omega^*(\Sigma_6)$ generated by the of the 1-forms on $J^0(\mathbb{R}^2, \mathbb{R}^2)$ pulled-back to Σ_6. (In local coordinates, this ideal is generated by dx, dy, dz, dw.) it follows from (3.10) that

$$\omega^3 \wedge \omega^4 \equiv 0 \text{ mod base},$$

since $\pi^1 \equiv 0$ mod base. Therefore, we have

$$a\omega^3 + b\omega^4 \equiv 0 \text{ mod base},$$

for some functions a and b. If $a \neq 0$, we let

$$\overline{\pi}^1 = a\pi^1, \quad \overline{\omega}^3 = a\omega^3 + b\omega^4,$$

which implies

$$d\overline{\pi}^1 \equiv \overline{\omega}^3 \wedge \omega^4 \quad \text{mod } \{\pi^1, \pi^2\}$$

as claimed. The structure equation for $d\pi^2$ is preserved if we let

$$\overline{\omega}^5 = \frac{\omega^5}{a}, \quad \overline{\omega}^6 = \omega^6 - \frac{b}{a}\omega^5.$$

A similar argument applies if $a = 0$.

The first of the structure equations (3.10) implies that

$$d\pi^1 \equiv \omega^3 \wedge \omega^4 + \alpha \wedge \pi^2 \quad \text{mod } \{\pi^1\}\ , \tag{3.15}$$

so that class $\{\pi^1\} = 3$, or equivalently

$$(d\pi^1)^2 \wedge \pi^1 = 0,$$

if and only if

$$\pi^1 \wedge \pi^2 \wedge \omega^3 \wedge \omega^4 \wedge \alpha = 0. \tag{3.16}$$

It follows from (3.15) and (3.16) that we can achieve the congruence

$$d\pi^1 \equiv \omega^3 \wedge \omega^4 \mod \{\pi^1\},$$

by adding multiples of π^2 to ω^3 and ω^4. We now compute the focal system $F(N)$. Let Z be π-vertical. We have

$$Z \rfloor d\pi^1 \equiv -\langle Z, \omega^4\rangle \omega^3 \mod \{\pi^1\} .$$

This implies that if $F(N)$ is 2-dimensional, then

$$F(N) = \{\pi^1, \omega^3\}.$$

Furthermore, if $F(N)$ is completely integrable, then the Frobenius condition

$$d\pi^1 \wedge \pi^1 \wedge \omega^3 = 0$$

combined with the Quadratic Lemma of Section 2 gives

$$(d\pi^1)^2 \wedge \pi^1 = 0,$$

or, in other words, class $\{\pi^1\} = 3$. Thus we have proved

COROLLARY 1. *Every parabolic p.d.e. system* (3.1) *such that the Pfaffian system I and the subsystem N satisfy*

$$I^{(1)} = 0, \quad \dim F(N) = 2, \quad F(N)^{(1)} = F(N),$$

$$\frac{\partial z}{\partial x} = 0, \quad \frac{\partial w}{\partial x} = \frac{\partial z}{\partial y}.$$

Another characterization of the parabolic systems satisfying the hypotheses of Theorem 3.3 can be given in terms of the Friedrichs-characteristic system $C(I, dN)$, defined by (2.21) and (2.22).

COROLLARY 2. *Every p.d.e. system* (3.1) *of parabolic type whose associated Pfaffian system I and Friedrichs characteristic system N satisfy*

$$I^{(1)} = 0, \quad C(I, dN)^{(1)} = C(I, dN),$$

is locally equivalent under a contact transformation to

$$\frac{\partial z}{\partial x} = 0, \quad \frac{\partial w}{\partial x} = \frac{\partial z}{\partial y}.$$

We refer the reader to [**GaK2**] for a proof of Corollary 3.2.

In [**Ca2**], E. Cartan outlined an existence theorem for solutions to the Cauchy problem for scalar parabolic second-order p.d.e.s in the plane whose characteristics are completely integrable. This result was proved by one of us in [**Ga4**] by applying his general method of solution of the Cauchy problem for C^∞ Pfaffian systems by means of k-stable vector fields [**Ga3**], [**GaK1**]. We now prove a similar theorem for C^∞ parabolic first-order systems (3.1) with completely Friedrichs integrable characteristics $C(I, dN)$.

THEOREM 3.4. *Consider a system* (3.1) *of parabolic type whose associated Pfaffian system I and Friedrichs characteristic system N satisfy*

$$I^{(1)} = 0, \quad C(I, dN)^{(1)} = C(I, dN).$$

If $\theta \colon (a, b) \subset \mathbb{R} \to \Sigma_6$ is a one-to-one smooth map which defines non-focal Cauchy data for I,

$$\theta^* I = 0, \quad \theta^* F(N) \neq 0,$$

then there exists locally a solution of (3.1) *extending* θ, *in other words a one-to-one smooth map* $\sigma\colon (a,b)\times(c,d)\subseteq \mathbb{R}\times\mathbb{R}\to\Sigma_6$ *satisfying*

$$\sigma^* I = 0, \quad \sigma(u,0)=\theta(u).$$

PROOF. We assume without loss of generality that the Cauchy data are extended by the flow of a vector field X defined on Σ_6, in other words that

$$\sigma(u,t)=\exp_{\theta(u)}(tX),$$

where the right-hand side denotes the point at time t on the integral curve of X starting from $\theta(u)$ at time zero. We now recall from [**Ga3**] and [**GaK1**] that the necessary and sufficient condition on X for σ to satisfy $\sigma^*I=0$ is that X be k-stable and compatible with the Cauchy data. The k-stability condition means that $X\in I^{\perp}$, X is transversal to the Cauchy data, and that the Pfaffian systems $D^p_X(I)$, $p\geq 0$, defined by

$$D^0_X(I) := I$$
$$D^p_X(I)=\{D^{p-1}_X(I), \mathcal{L}_X D^{p-1}_X(I)\}$$

stabilize for $p=k$, meaning that

$$D^k_X(I)=D^{k+1}_X(I).$$

The compatibility with the Cauchy data amounts to the condition

$$\theta^* D^k_X(I)=0.$$

We now show that under our assumptions, there exists a 1-stable vector field X. The parabolic structure equations (3.16) are thus valid and the N-characteristic system $C(I,dN)=\{\pi^1,\pi^2,\omega^3,\omega^4\}$ is assumed to be completely integrable. We choose a non-zero $X\in \mathrm{Char}\,(I,dN)$, which we normalize by letting

$$X\rfloor\omega^5=\lambda, \quad X\rfloor\omega^6=1.$$

This gives

$$D^1_X(I)=\{\pi^1,\pi^2,\lambda\omega^3+\omega^4\}.$$

Since $C(I,dN)$ is completely integrable, there exist functions A,B,C,D such that

$$X\rfloor d\omega^3\equiv A\omega^3+B\omega^4 \mod I, \quad X\rfloor d\omega^4\equiv C\omega^3+D\omega^4 \mod I,$$

Therefore, the 1-stability condition $D^1_X(I)=D^2_X(I)$ is equivalent to

$$X\rfloor d(\lambda\omega^3+\omega^4)\wedge(\lambda\omega^3+\omega^4)\equiv 0 \mod I,$$

which can be rewritten as

$$X(\lambda)+\lambda(A-D)-\lambda^2B+C=0.$$

This is an inhomogeneous quasi-linear first order p.d.e. for λ.

The assumption that the initial data are not N-focal is equivalent to

$$\theta^*\omega^4\neq 0.$$

We can choose λ to be the solution of the above p.d.e., satisfying the initial condition

$$\theta^* D^1_X(I)=0.$$

Indeed, since $\theta^*I=0$, we have

$$\theta^* D^1_X(I)=\theta^*(\omega^3+\lambda\omega^4)=\theta^*\omega^3+(\lambda\circ\theta)\theta^*\omega^4.$$

This concludes the proof of the theorem. □

Finally, we have the most general class of parabolic systems, characterized by the conditions

$$I^{(1)} = 0, \quad \text{class } N = 5.$$

In this case, one can always normalize the structure equations (3.10) so as to have

$$d\pi^1 \equiv \omega^3 \wedge \omega^4 + (a_5\omega^5 + a_6\omega^6) \wedge \pi^2 \mod \{\pi^1\}, \tag{3.17}$$

where

$$\omega^3 \equiv 0, \quad \omega^6 \equiv 0 \text{ mod base}. \tag{3.18}$$

This can be established as follows. We can choose ω^3 to be congruent to zero mod base in (3.10), by the same argument as the one used in the proof of Corollary 3.1. Since $\pi^2 \equiv 0$ mod base, we then obtain from (3.10) that

$$\omega^4 \wedge \omega^6 \equiv 0 \text{ mod base}.$$

This implies that

$$k\omega^4 + l\omega^6 \equiv 0 \text{ mod base},$$

for some functions k and l which do not both vanish. If $l = 0$, then the congruences

$$\pi^1 \equiv 0, \quad \pi^2 \equiv 0, \quad \omega^3 \equiv 0, \quad \omega^4 \equiv 0 \text{ mod base},$$

and the fact that $\dim J^0(\mathbb{R}^2, \mathbb{R}^2) = 4$ would imply that $\{\pi^1, \pi^2, \omega^3, \omega^4\}$ is completely integrable. From Corollary 3.2, we know that this would force the class of N to be equal to three, which is a contradiction. Therefore l must be non-zero.

We have

$$d\pi^1 \equiv \omega^3 \wedge \omega^4 + (a_5\omega^5 + a_6\omega^6) \wedge \pi^2 \mod \{\pi^1\},$$

after suitable translation of ω^3 and ω^4 in the direction of π^2. For some functions a_5 and a_6, so that

$$d(l\pi^1) \equiv l\left(\omega^3 - \frac{k}{l}a_6\pi^2\right) \wedge \omega^4 + \left(la_5\omega^5 + a_6(k\omega^4 + l\omega^6)\right) \wedge \pi^2 \mod \{\pi^1\}.$$

The claim is established by defining

$$\overline{\pi}^1 = l\pi^1, \quad \overline{\omega}^3 = \omega^3 - \frac{k}{l}a_6\pi^2, \quad \overline{\omega}^4 = l\omega^4, \quad \overline{\omega}^6 = k\omega^4 + l\omega^6, \quad \overline{a}_5 = la_5,$$

and dropping overbars.

We are now in a position to compute the focal system $F(N)$. From (3.18), we see that the annihilator of the vertical bundle of Σ_6 is given by

$$V(\Sigma_6)^\perp = \{\pi^1, \pi^2, \omega^3, \omega^6\}.$$

Let Z_1 and Z_2 be linearly independent vertical vector fields. We have

$$Z_i \rfloor d\pi^1 \equiv -\langle Z_i, \omega^4\rangle\omega^3 + a_5\langle Z_i, \omega^5\rangle\pi^2 \mod \{\pi^1\}, \quad 1 \le i \le 2.$$

Let

$$\Lambda = \begin{pmatrix} -\langle Z_1, \omega^4\rangle & a_5\langle Z_1, \omega^5\rangle \\ -\langle Z_2, \omega^4\rangle & a_5\langle Z_2, \omega^5\rangle \end{pmatrix}.$$

If $\operatorname{rank} \Lambda = 0$, then we have $\omega^4 \in V(\Sigma_6)^\perp$, which is a contradiction with the assumption that $\pi^1 \wedge \pi^2 \wedge \omega^3 \wedge \omega^4 \wedge \omega^6 \neq 0$. Indeed, this would force the fibers of π to be of dimension strictly less than two.

If rank $\Lambda = 1$, then

$$0 = a_5 \det \begin{pmatrix} -\langle Z_1, \omega^4 \rangle & \langle Z_1, \omega^5 \rangle \\ -\langle Z_2, \omega^4 \rangle & \langle Z_2, \omega^5 \rangle \end{pmatrix}.$$

This can only happen if $a_5 = 0$. It is easy to see that this can only happen if $a_5 = 0$.Therefore the focal system $F(N)$ is two-dimensional

$$F(N) = \{\pi^1, \omega^3\},$$

and (3.17) becomes

$$d\pi^1 \equiv \omega^3 \wedge \omega^4 + a_6 \omega^6 \wedge \pi^2 \mod \{\pi^1\}.$$

We now show that the class of first derived system $F(N)^{(1)}$ is either 1 or 3. First of all, $F(N)$ cannot be completely integrable. Indeed, if that were the case, then the Frobenius condition $d\pi^1 \wedge \pi^1 \wedge \omega^3 = 0$, which is equivalent by (3) to $a_6 = 0$, would imply using the Quadratic Lemma of Section 2 that $(d\pi^1)^2 \wedge \pi^1 = 0$. But then, the class of $\{\pi^1\}$ = class N would be strictly less than five, which is a contradiction. We therefore have $a_6 \neq 0$ and the only possibilities are $F(N)^{(1)} = \{\overline{\omega}^3\}$, where $\overline{\omega}^3$ is obtained from ω^3 by a suitable translation in the π^1 direction, or $F(N)^{(1)} = 0$. (We will drop this overbar from now on.) In the first case, it is easy to see using the the Quadratic Lemma that the class of $F(N)^{(1)} = \{\omega^3\}$ can be at most equal to three. We therefore have either class $F(N)^{(1)} = 1$ or class $F(N)^{(1)} = 3$. The first case corresponds to a whole stratum of orbits with simple normal forms.

THEOREM 3.5. *Every parabolic p.d.e. system* (3.1) *such that the Pfaffian system I and the subsystem N satisfy*

$$I^{(1)} = 0, \quad \text{class } N = 5, \quad \dim F(N) = 2, \quad \text{class } F(N)^{(1)} = 1,$$

is locally equivalent under a contact transformation to a system of the form

$$\frac{\partial z}{\partial x} = w, \quad \frac{\partial z}{\partial y} = f\left(x, y, z, w, \frac{\partial w}{\partial x}\right),$$

where f depends explicitly on $\partial w/\partial x$.

PROOF. By letting $\overline{\pi}^2 = -a_6\pi^2$ and $\overline{\omega}^4 = -a_6\omega^4$ (recall that $a_6 \neq 0$), the structure equations (3.10) and (3) take the form

$$d\pi^1 \equiv \pi^2 \wedge \omega^6 \mod \{\pi^1, \omega^3\}, \quad d\pi^2 \equiv \omega^4 \wedge \omega^6 \mod \{\pi^1, \pi^2, \omega^3\}$$

where $\{\omega^3\} = F(N)^{(1)}$ is completely integrable by assumption. We know from Theorem 2.5 (the relative Goursat normal form) that there exist functions u^1, u^2, u^3, u^4, v^1, v^2, v^3 such that $\{\omega^3\} = \{dv^1\}$ and

$$\{\pi^1, \pi^2\} = \{du^1 - u^2du^3 + v^2dv^1, du^2 - u^4du^3 + v^3dv^1\}.$$

From the fact that class N = class $\{\pi^1\} = 5$, we deduce that $du^1 \wedge du^2 \wedge du^3 \wedge dv^1 \wedge dv^2 \neq 0$. Moreover, since the system is parabolic, we have $d\pi^2 \wedge d\pi^2 \wedge \pi^1 \wedge \pi^2 \neq 0$, which implies that $du^1 \wedge du^2 \wedge du^3 \wedge du^4 \wedge dv^1 \wedge dv^3 \neq 0$.

Let $x = u^3$, $y = v^1$, $z = u^1$, $w = u^2$, $q = f(x, y, z, w, m, n) = -v^2$, $m = u^4$, $n = v^3$. The condition $d\pi^1 \wedge d\pi^2 \wedge \pi^1 \wedge \pi^2 = 0$ which follows from parabolicity gives $f_n = 0$. This concludes the proof. □

It is not hard to prove that the case class $F(N)^{(1)} = 3$ cannot occur. Indeed if this were the case, then $\{\omega^3\} = \{dv_2 - v_3 dv_1\}$ and $d\omega^3 \wedge \omega^3 = dv_1 \wedge dv_2 \wedge dv_3$ up to multiple. But $d\omega^3 \wedge \omega^3 \wedge \pi^1 = 0$, so

$$\pi^1 \wedge dv_1 \wedge dv_2 \wedge dv_3 = 0$$

and

$$d\pi^1 \wedge dv_1 \wedge dv_2 \wedge dv_3 = 0.$$

But then $d\pi^1 \equiv a_6\omega^6 \wedge \pi^2 \mod \pi^1, \omega^3$ gives

$$a_6\omega^6 \wedge \pi^2 \wedge dv_1 \wedge dv_2 \wedge dv_3 = 0.$$

This implies, since $a_6 \neq 0$, that all of $\pi^1, \pi^2, \omega^3, \omega^6$ are expressible as linear combinations of dv_1, dv_2, dv_3, giving a contradiction.

The remaining case, namely class $F(N)^{(1)} = 0$, does not admit a normal form which we can identify.

The case when rank $\Lambda = 2$ does not admit any normal forms which we can identify either.

References

[AKO] I. Anderson, N. Kamran, and P. J. Olver, *Internal, external and generalized symmetries*, Adv. Math. **100** (1993), no. 1, 53–100.

[BC3G] R. Bryant, S. S. Chern, R. B. Gardner, H. Goldschmidt, and P.A. Griffiths, *Exterior differential systems*, Math. Sci. Res. Inst. Publ., vol. 18. Springer-Verlag, New York, 1991.

[BGH] R. Bryant, P. A. Griffiths, and L. Hsu, *Hyperbolic exterior differential systems and their conservation laws*, Inst. for Adv. Stud., Princeton, preprint.

[Ca1] E. Cartan, *Les sous-groupes des groupes continus de transformations*, Ann. École Norm. **25** (1908), 57–194.

[Ca2] ______, *Les systèmes de Pfaff à cinq variables et les équations aux dérivées partielles du second ordre*, Ann. École Norm. **27** (1910), 109–192.

[Ca3] ______, *Les systèmes différentiels extérieurs et leurs applications géométriques*, Hermann, Paris, 1947.

[Fr] K. O. Friedrichs and R. Courant, *Supersonic flow and shock waves*, Interscience Publishers, New York, 1948.

[Ga1] R. B. Gardner, *Invariants of Pfaffian systems*, Trans. Amer. Math. Soc. **126** (1967), 514–533.

[Ga2] ______, *Some applications of the retraction theorem in exterior algebra*, J. Differential Geom. **2** (1968), 25–31.

[Ga3] ______, *The Cauchy problem for Pfaffian systems*, Comm. Pure Appl. Math. **22** (1969), 587–596.

[Ga4] ______, *A differential geometrical generalization of characteristics*, Comm. Pure Appl. Math. **22** (1969), 597–626.

[Ga5] ______, *The method of equivalence and its applications*, CBMS-NSF Regional Conf. Ser. in Appl. Math., vol. 58, SIAM, Philadelphia, 1989.

[GaK1] R. B. Gardner and N. Kamran, *Characteristics and the geometry of hyperbolic equations in the plane*, J. Differential Equations **104** (1993), 60–116.

[GaK2] ______, *Normal forms and focal systems for determined systems of two first-order partial differential equations in the plane*, Indiana Univ. Math. J. **44** (1996), 1127–1162.

[GS] R. B. Gardner and W. F. Shadwick, *The GS algorithm for exact linearization to the Brunovsky normal form*, IEEE Trans. Automat. Control **37** (1992), 224–230.

[O] P. J. Olver, *Applications of Lie groups to differential equations*, 2nd ed., Grad. Texts in Math. **107**, Springer-Verlag, New York, 1993.

Department of Mathematics, University of North Carolina, Chapel Hill, NC 27599, USA.

E-mail address: rbg@math.unc.edu

Department of Mathematics and Statistics, 805 Sherbrooke W., Pavillon Burnside Hall, McGill University, Montréal, Québec H3A 2K6, Canada.

E-mail address: nkamran@Math.Mcgill.CA

Centre de Recherches Mathématiques
CRM Proceedings and Lecture Notes
Volume **11**, 1997

Aspects combinatoires des équations de Bethe

Robert Langlands et Yvan Saint-Aubin

1. Introduction

Il y a plus de soixante ans Bethe [**B**] introduisait sa stratégie pour résoudre un problème de valeurs propres issu de la théorie des métaux. Cette méthode et tous ses raffinements se sont avérés par la suite d'une grande importance. Pourtant on n'a jamais cherché de lui donner une base rigoureuse. C'est le but du papier [**BL**] dont nous avons donné une esquisse dans [**AG**]. En même temps, à notre insu, le problème était considéré par V. Tarasov et A. Varchenko [**TV**]. Ils ont traité la méthode dans un cadre bien plus large que nous et, ce qui est plus pertinent ici, leurs arguments sont nettement plus directs et plus faciles que les nôtres. La question se pose alors s'il vaut la peine de poursuivre le problème avec nos méthodes.

Avant d'y répondre nous décrivons brièvement les équations de Bethe et la méthode. Quoique nous n'ayons jamais essayé de traiter les cas généraux de l'*Ansatz* de Bethe, il nous faut malgré tout traiter des cas plus généraux que celui du papier de Bethe lui-même car il s'agit ici de questions de géométrie algébrique et il s'est avéré nécessaire de passer aux équations génériques plutôt que de rester avec les équations originales. Ces nouvelles équations sont aussi attachées aux problèmes spectraux que l'on peut déduire des travaux de l'école de Leningrad [TΦ].

Soit R une fonction rationnelle

$$R(z) = \kappa \frac{\prod_{m=1}^N (z-\alpha_m)}{\prod_{m=1}^N (z-\beta_m)}, \tag{1.1}$$

où $\beta_m = 2\Delta - 1/\alpha_m$. Parfois on admet des cas dégénérés

$$R(z) = \kappa \frac{\prod_{m=1}^N (\gamma_m z-\alpha_m)}{\prod_{m=1}^N (\delta_m z-\beta_m)},$$

1991 *Mathematics Subject Classification.* Primary: 14E05; Secondary: 82B23.

Une partie importante du travail du premier auteur sur ce papier a été accomplie pendant un séjour de quelques mois au Max-Planck-Institut à Bonn et pendant des visites répétées au Centre de recherches mathématiques à Montréal. Il est reconnaissant aux directeurs de ces deux institutions, Friedrich Hirzebruch et Luc Vinet, pour un accueil d'une obligeance exceptionnelle.

This is the final form of the paper.

où $\beta_m\alpha_m = 2\Delta\alpha_m\delta_m - \gamma_m\delta_m$. En particulier le cas

$$R(z) = z^N,$$

qui est celui de Bethe, est admis. Dans cet exposé, nous ne considérons toutefois que (1.1). Soient $F(z,z') = zz' - 2\Delta z + 1$ et $Q(z,z') = -F(z,z')/F(z',z)$. Nous appelons équations de Bethe le système

$$R(z_i) = \prod_{j\neq i} Q(z_i, z_j), \quad i = 1, \ldots, r, \tag{1.2}$$

pour les r inconnues z_i. Le cas générique de ces équations est le cas pour lequel Δ, κ, et $\alpha_1, \ldots, \alpha_N$ sont tous génériques. Pour le moment toutes les variables z_i sont censées être finies. Observons que la dépendance des fonctions R et Q en Δ et en les autres paramètres n'est qu'implicite dans la notation !

Une solution est dite *admissible* si les z_i sont tous différents et ne sont ni des zéros ni des pôles de R ; sinon elle est *inadmissible.* A chaque solution de ces équations on attache un vecteur propre du problème spectral que nous ne décrivons pas explicitement. Ce vecteur est zéro si la solution est inadmissible, et les deux vecteurs issus à deux solutions admissibles dont l'une est une permutation de l'autre sont les mêmes à une constante près. Il s'agit de vérifier le théorème suivant.

THÉORÈME. *Pour des valeurs génériques des paramètres, il y a* $N(N-1)\cdots(N-r+1)$ *solutions admissibles et les vecteurs obtenus de celles-ci engendrent un espace de dimension* $N!/(N-r)!r!$ *et sont par conséquent tous les vecteurs propres du problème.*

Il s'avère que la partie difficile de cet énoncé est de montrer qu'il y a exactement $N(N-1)\cdots(N-r+1)$ solutions admissibles des équations de Bethe.

La stratégie de [**BL**] est d'observer que les solutions des équations de Bethe sont les points fixes d'une correspondance algébrique et de les compter en utilisant la formule de Lefschetz. Un obstacle majeur à surmonter est qu'il n'est pas du tout évident que tous les points fixes sont de multiplicité un, et sans ce fait, la formule de Lefschetz est peu utile. Notre stratégie était de vérifier que la multiplicité est toujours un en examinant la correspondance autour de valeurs obtenues en faisant dégénérer les paramètres d'une façon convenable, ce qui est en soi assez raisonnable, mais notre choix de ces paramètres, $\Delta = 0$ mais κ et les α_k génériques, était à maints égards malencontreux. Il soulève plusieurs difficultés d'un ordre strictement mathématique qui n'apportent, en autant que nous sachions, rien à la compréhension des problèmes spectraux de départ. Par contre le choix qu'ont fait Tarasov et Varchenko, bien plus astucieux que le nôtre, permet, comme nous expliquons dans le second appendice, non seulement une démonstration facile que la multiplicité est un, mais une démonstration directe et élémentaire de l'énoncé lui-même sans aucun appareil technique.

En revanche l'étude de la correspondance autour de $\Delta = 0$ n'est pas sans intérêt d'un point de vue combinatoire ou géométrique. La trace qui apparaît en utilisant la formule de Lefschetz s'exprime à partir d'objets combinatoires, en particulier des graphes pondérés, et une partie importante de cette structure combinatoire est encore présente dans l'ensemble des points fixes de la correspondance autour de

$\Delta = 0$. Le but de cet exposé — en grande partie plutôt informel — est de la révéler, même si les questions soulevées ne seront pas toutes complètement résolues.

La définition technique des correspondances est reléguée au premier appendice. Le lecteur est invité à consulter cet appendice pour les définitions et notations qu'il ne trouve pas dans le texte, mais, sauf s'il est spécialiste en géométrie algébrique, à ne pas trop s'en préoccuper. Notons toutefois que, pour compter non seulement les points fixes mais aussi les points fixes inadmissibles, il faut introduire, pour diverses raisons, des espaces $X^{A,B}$ où les coordonnées sont soumises à des conditions supplémentaires : certaines des coordonnées sont supposées soit égales entre elles soit égales à des zéros ou à des pôles de R. La correspondance est alors définies par deux applications $\varphi^{A,B}$ et $\psi^{A,B}$ d'une variété $C_\Delta^{A,B}$ dans $X^{A,B}$. Ceci est la notation du premier appendice, mais dans le corps du papier A et B sont fixés et par conséquent nous les supprimerons de la notation.

Il y a néanmoins une convention de notation qu'il faut remarquer explicitement. Les contraintes imposées par A et B exigent que quelques coordonnées soient égales. Nous pouvons en tenir compte soit en les comptant séparément soit en attachant à chaque coordonnée un poids $\big(|A_k|, |\mathbb{B}'_k|, |\mathbb{B}''_k|\big)$. Dans la seconde partie nous employons la première de ces conventions. Mais ensuite dans les parties suivantes, lorsque il s'agit de résoudre les équations pertinentes, il est préférable d'utiliser les poids.

Notre première tâche est d'expliciter les définitions de l'appendice pour la fibre $\Delta = 0$. Ensuite nous trouvons les points fixes de la correspondance dégénérée sur cette fibre et calculons leur multiplicité. Il faudra ensuite, en poursuivant l'argument esquissé dans [**AG**], montrer qu'il y a au moins le nombre requis de déformations en série de Puiseux autour de ces points fixes. Les arguments algébriques et combinatoires menant à cette conclusion ne sont guère évidents et dans ce papier nous nous contentons d'exemples, assez nombreux nous croyons pour convaincre le lecteur que les arguments sont bien fondés. Puisque c'est à notre avis la richesse combinatoire de ces arguments qui justifie à présent l'emploi de la dégénérescence de la correspondance que nous avons choisie plutôt que celle implicite dans [**TV**], une introduction directe et concrète à ces arguments sera peut-être plus attrayante que la présentation formelle que de toute façon nous ne sommes pas encore en état de donner. En effet avant d'approfondir ces idées nous voulons réfléchir encore à leur utilité.

2. Étude d'une fibre spéciale ($\Delta = 0$)

Quoique la fibre $\Delta = 0$ ne soit pas générique puisque le paramètre Δ y est soumis à une condition explicite, nous exigeons que les autres paramètres demeurent génériques. Explicitons d'emblée quelques conditions de cette généricité. On suppose qu'aucun α_k ou β_k n'est égal à $\pm\sqrt{-1}$ au point $\Delta = 0$.

Pour tout z nous posons $\tilde{z} = -1/z$. On a partout

$$R(z)R(\tilde{z}) = \kappa^2 \frac{\prod_1^N (z - \alpha_m) \prod_1^N (\tilde{z} - \alpha_m)}{\prod_1^N (z - \beta_m) \prod_1^N (\tilde{z} - \beta_m)}$$

mais à $\Delta = 0$, $\beta_m = \tilde{\alpha}_m$ de sorte que le quotient à droite est indépendant de z et on a

$$R(z)R(\tilde{z}) = \kappa^2 \prod_1^N (-\alpha_m^2) = \rho. \tag{2.1}$$

Le nombre ρ est supposé générique. Puisque $\tilde{z} = z$ si $z = \pm\sqrt{-1}$, les deux valeurs $R(\pm\sqrt{-1})$ sont $\sqrt{\rho}$ et par conséquent génériques. La constante ρ est aussi le produit de $\rho/\kappa = R(0)$ et de $\kappa = R(\infty)$. Puisque κ et ρ sont génériques et indépendants, nous pouvons supposer que $\kappa^m \rho^n$ avec m et n dans $\mathbb{Z}$ et $0 \leq m + n$, $n \leq 2N$ n'est pas 1 sauf si $m = n = 0$.

Ce n'est qu'en $\Delta = 0$ que les coordonnées $p'_{i,k}$, $q'_{i,k}$, $p''_{i,k}$, $q''_{i,k}$ ne sont pas superflues. Nous commençons en réduisant leur nombre. Nous rappelons d'abord que ces coordonnées sont nécessaires seulement autour de $z_i = \alpha_k$ ou $z_i = \beta_k$. Par généricité $\alpha_k \neq \beta_k$. Autour de $z_i = \alpha_k$ les deux coordonnées $z_i - \beta_k$ et $z_i - \gamma_k$ ne sont pas nulles de sorte que les coordonnées $p'_{i,k}$, $q'_{i,k}$ et $\xi'_{i,k}$ sont superflues aussi bien que $p''_{i,l}$ $q''_{i,l}$, $\xi''_{i,l}$, $l \neq k$. Les trois coordonnées $p''_{i,k}$, $q''_{i,k}$ et $\xi''_{i,k}$ qui restent seront dénotées p_i, q_i et ξ_i. Donc

$$z_i - \alpha_k = \xi_i p_i, \quad \Delta = \xi_i q_i. \tag{2.2}$$

Si par contre nous étudions un voisinage de $z_i = \beta_k$, nous n'avons besoin que de $p'_{i,k}$, $q'_{i,k}$ et $\xi'_{i,k}$ qui pourraient être remplacées par p_i, q_i et ξ_i telles que

$$z_i - \beta_k = \xi_i p_i, \quad \Delta = \xi_i q_i.$$

Mais il s'avère plus utile de choisir autour de $z_i = \beta_k$ la coordonnée $\tilde{z}_i = -1/z_i$. On a

$$\tilde{z}_i - \tilde{\beta}_k = \frac{z_i - \beta_k}{z_i \beta_k}$$

de sorte que nous préférons introduire plutôt p_i, q_i et ξ_i comme suit

$$\tilde{z}_i - \tilde{\beta}_k = \xi_i p_i, \quad \Delta = \xi_i q_i. \tag{2.3}$$

Nous utiliserons désormais cette définition.

Nous avons convenu de supprimer les indices A et B. Puisqu'il s'agit toujours de la fibre $\Delta = 0$ nous supprimons parfois l'indice Δ aussi, de sorte que la fibre de la correspondance $\mathcal{C}^{A,B}$ à $\Delta = 0$, dénotée C_0 dans la premier appendice, devient C.

Disons qu'un point p sur C est *spécial* si $z_i = \pm\sqrt{-1}$, 0 ou ∞ pour une des coordonnées de $(z, w) = \pi_1(p)$. Dans le lemme suivant la fonction R est toujours supposée générique.

LEMME 2.1. *Il n'existe pas de point spécial p sur C tel que*

$$\varphi(p) = \psi(p). \tag{2.4}$$

Rappelons pour la dernière fois que l'équation est en effet $\varphi^{A,B}(p) = \psi^{A,B}(p)$. Soit $\pi_1(p) = (z, w)$ et supposons que l'équation (2.4) soit satisfaite. Fixons $\varepsilon = \pm\sqrt{-1}$. Soit E l'ensemble des indices i tels que $z_i = \varepsilon$. Le Lemme 2.1 affirme en particulier que E est vide.

Observons qu'au point $\Delta = 0$, $F(z, z') = F(z', z) = zz' + 1$. Par conséquent si $z = \pm\sqrt{-1}$ et $z' \neq z$, le quotient $Q(z, z')$ est égal à -1 et cela, même si $z' = \infty$.

Donc pour $i \in E$ et $j \notin E$ le quotient $Q(z_i, z_j) = -1$. Soit $(z', w') = \varphi(p)$ et $(z'', w'') = \psi(p)$. Les équations $w'_{i,j} = w''_{i,j}$ impliquent alors que $w_{i,j} = -1$ si $i \in E$ et $j \notin E$. De plus, puisque $z''_i = z'_i$, $i \in E$, est égal à $R(\varepsilon)$ et partant fini, le point p est dans la partie de l'éclatement définie par un élément de la décomposition polyédrique sur lequel on a $f_i \geq 0$, $i \in E$ et $e_{i,j} = 0$, $i \in E$, $j \notin E$. Sur cette partie le produit

$$\prod_{i \in E} \prod_{j \neq i} w_{i,j} \tag{2.5}$$

est défini, car il correspond au caractère

$$\sum_{i \in E} f_i. \tag{2.6}$$

Il est égal à $\prod_{i \in E} z''_i$.

Puisque $e_{i,j} = -e_{j,i}$ la somme (2.6) est

$$\sum_{i \in E} \sum_{j \notin E} e_{i,j}.$$

Donc (2.5) n'est que

$$\prod_{i \in E} \prod_{j \notin E} w_{i,j} = \pm 1.$$

Des équations

$$z'_i = z''_i$$

on déduit que

$$R(\varepsilon)^{|E|} = \pm 1. \tag{2.7}$$

Nous avons exclu cette possibilité par généricité.

Soit $(z, w) = \big(z(p), w(p)\big)$ l'image d'un point p et tel qu'aucun z_i n'est égal à $\pm\sqrt{-1}$. (Lorsqu'il n'est question que des coordonnées du point p nous préférons employer simplement la notation (z, w) mais lorsque nous voulons traiter les coordonnées dans un voisinage de p nous écrivons $\big(z(p), w(p)\big)$ pour que (z, w) puissent être des coordonnées variables.) Nous introduisons la partition $D_1, \dots, D_m$ de $\{1, \dots, r\}$ définie par l'équivalence $\sim$: $i \sim j$ si et seulement si $z_i = z_j$ ou $z_i = -1/z_j = \tilde{z}_j$. Cette équivalence est plus grossière que celle définie par les A_k, les B'_l, et les B''_l qui définissent (A, B). Donc chacun de ces ensembles est contenu dans une classe d'équivalence. Nous répartissons chaque D_k en deux ensembles E_k et F_k de façon que si $i \in E_k$ et $j \in F_k$ alors $z_i = \tilde{z}_j$, une équation dans laquelle les valeurs 0 et ∞ sont permises. Puisqu'aucun z_i n'est égal à $\pm\sqrt{-1}$, il n'arrive jamais que $E_k = F_k$, quoique la possibilité que E_k ou F_k soit vide n'est pas exclue.

Supposons que $i \in E_k$ mais que $j \notin F_k$ ou que $j \in F_k$ mais $i \notin E_k$. Alors pour $\Delta = 0$ (qui est la seule valeur de Δ qui nous intéresse à présent) $F(z_i, z_j) = F(z_j, z_i) = z_i z_j + 1 \neq 0$ de sorte que les coordonnées $u_{i,j}$ et $u_{j,i}$ sont superflues. Donc la structure de la variété $E = E_0^{A,B}$ est donnée par les coordonnées $(u_{i,j}, u_{j,i})$, $i \in E_k$, $j \in F_k$, et en sus toutes les coordonnées z_i et quelques-unes des p_i, q_i, ξ_i. De plus, sauf pour ces paires, la valeur de $w_{i,j}$ à un point fixe est -1.

Supposons que la valeur commune a de z_i, $i \in E_k$, n'est ni une racine ni un pôle de R_0. L'équation (2.4) implique alors une équation semblable à (2.7),

$$R(a)^{e_k} R\Big(-\frac{1}{a}\Big)^{f_k} = \pm 1, \tag{2.8}$$

si $e_k = |E_k|$ et $f_k = |F_k|$. Il résulte de cette équation et de l'hypothèse de généricité de R que a n'est ni 0 ni ∞. Donc l'équation (2.4) n'est satisfaite pour aucun point spécial.

A chacune de ces paires $\{E, F\}$ sont attachées les valeurs communes z_E et $z_F = \tilde{z}_E$ des coordonnées z_i, $i \in E$ et des z_j, $j \in F$. Nous dirons qu'une paire $\{E, F\}$ est *libre* si z_E et z_F ne sont pas des zéros ou des pôles de R. Par contre si z_E et z_F sont des zéros ou pôles de R la paire sera *liée*. Dans ce cas nous supposerons d'habitude que c'est z_E qui est le zéro de R.

Afin d'examiner de plus proche la correspondance autour de $\Delta = 0$ et en particulier autour des points fixes, nous remplaçons $(u_{i,j}, u_{j,i})$ par des coordonnées plus convenables $t_{i,j}$ et $v_{i,j}$. Puisque le point p n'est pas spécial, pour les paires (i, j), $i \in E_k$, $j \in F_k$, on a $z_i\tilde{z}_j + 1 = z_i^2 + 1 \neq 0$. Fixons k et ne considérons que les $i, i', \dots$ dans E_k et les $j, j', \dots$ dans F_k. Soit $t_{i,j} = u_{i,j} - u_{j,i}$ et $t_{j,i} = -t_{i,j}$. On a

$$\begin{aligned} F(z_i, z_j)t_{i,j} &= \big(F(z_i, z_j) - F(z_j, z_i)\big)u_{i,j} \\ &= 2\Delta(z_j - z_i)u_{i,j}. \end{aligned} \tag{2.9}$$

Si par exemple on utilise la coordonnée $\tilde{z}_j$ plutôt que z_j, cette relation devient

$$F(z_i, \tilde{z}_j)t_{i,j} = -2\Delta(z_i\tilde{z}_j + 1)u_{i,j}. \tag{2.10}$$

Il y a une troisième possibilité et même une quatrième si on admet que z_i soit remplacé par $\tilde{z}_i$. Nous nous restreignons à l'équation (2.10) en laissant au lecteur le soin de s'assurer que les arguments sont aussi valables pour les trois autres cas. En effet nous nous occuperons surtout des points ou $z_i \neq 0$ et $z_j \neq 0$ de sorte que le choix entre z_i et $\tilde{z}_i$ importe peu.

La relation

$$F(z_i, \tilde{z}_j) = z_i - \tilde{z}_j + 2\Delta z_i\tilde{z}_j$$

s'écrit d'une façon moins formelle comme

$$z_i - \tilde{z}_j = F(z_i, \tilde{z}_j) - 2\Delta z_i\tilde{z}_j = F(z_i, \tilde{z}_j)\left(1 - \frac{z_i\tilde{z}_j}{z_i\tilde{z}_j + 1}\frac{t_{i,j}}{u_{i,j}}\right),$$

de sorte que si nous remplaçons la paire $t_{i,j}$ et $u_{i,j}$ par la paire $t_{i,j}$ et

$$v_{i,j} = -2(z_i\tilde{z}_j + 1)\left(u_{i,j} - \frac{z_i\tilde{z}_j}{z_i\tilde{z}_j + 1}t_{i,j}\right)$$

on a

$$\Delta v_{i,j} = (z_i - \tilde{z}_j)t_{i,j}. \tag{2.11}$$

Observons que

$$\begin{aligned} v_{i,j} &= -2(z_i\tilde{z}_j+1)\left(u_{i,j}-u_{j,i}+u_{j,i}+\frac{z_i\tilde{z}_j}{z_i\tilde{z}_j+1}t_{j,i}\right) \\ &= -z_i\tilde{z}_j\{2(z_j\tilde{z}_i+1)\left(u_{j,i}-\frac{1}{z_i\tilde{z}_j+1}t_{j,i}\right)\} \\ &= z_i\tilde{z}_j v_{j,i}. \end{aligned}$$

La coordonnée supplémentaire $\lambda_{i,j}$ définie par les équations

$$F(z_i,\tilde{z}_j)=\lambda_{i,j}u_{i,j}, \quad F(\tilde{z}_j,z_i)=\lambda_{i,j}u_{j,i}$$

est remplacée par une coordonnée $\mu_{i,j}$ telle que

$$z_i-\tilde{z}_j=\mu_{i,j}v_{i,j}, \quad \Delta=\mu_{i,j}t_{i,j}. \tag{2.12}$$

Par conséquent

$$\mu_{i,j}=-\frac{\lambda_{i,j}}{2(z_i\tilde{z}_j+1)}.$$

Observons que

$$\begin{aligned} 2(z_i\tilde{z}_j+1)u_{i,j} &= v_{i,j}+2z_i\tilde{z}_jt_{i,j} \\ 2(z_i\tilde{z}_j+1)u_{j,i} &= v_{i,j}-2t_{i,j}. \end{aligned}$$

Par conséquent

$$w''_{i,j}=-\frac{u_{i,j}}{u_{j,i}}=-\frac{v_{i,j}+2z_i\tilde{z}_jt_{i,j}}{v_{i,j}-2t_{i,j}}. \tag{2.13}$$

Autour d'un point p où $z_i(p)=\alpha=\alpha_k$ est un zéro de R et $z_j(p)=\beta=\tilde{\alpha}$ un pôle, il faudra en plus savoir comment exprimer $w''_{i,j}$ en fonction des coordonnées (p_i,q_i) et (p_j,q_j) de (2.3) et (2.4). Puisqu'il s'agit des coordonnées projectives on peut supposer qu'une des coordonnées p_i et q_i est 1. Puisqu'en plus nous nous intéressons aux points par lesquels passent des courbes qui s'étendent en dehors de $\Delta=0$ où la coordonnée q_i n'est pas 0, nous supposons que $q_i=1$ et de la même façon que $q_j=1$. En revanche il faut accepter que p_i ou p_j puissent tendre vers ∞ lorsque $\Delta\to 0$. Il s'avérera que ceci n'arrive pas !

Les deux fonctions $F(z_i,\tilde{z}_j)$ et $F(\tilde{z}_j,z_i)$ s'expriment à partir des coordonnées p_i et ξ_i. Observons que $\tilde{\beta}=\alpha$ à $\Delta=0$ mais en général $\tilde{\beta}=\alpha/(1-2\Delta\alpha)$ de sorte que $-\tilde{\beta}+\alpha=-2\Delta\alpha\tilde{\beta}$. Ainsi

$$\begin{aligned} F(z_i,\tilde{z}_j) &= z_i-\alpha+\alpha+2\Delta z_i\tilde{z}_j-\tilde{z}_j+\tilde{\beta}-\tilde{\beta} \\ &= \xi_ip_i-\xi_jp_j+2\Delta z_i\tilde{z}_j-2\Delta\alpha\tilde{\beta} \\ &= \xi_ip_i-\xi_jp_j+2\Delta(\xi_ip_i\tilde{\beta}+\xi_jp_j\alpha+\xi_i\xi_jp_ip_j). \end{aligned}$$

De la même façon

$$\begin{aligned} F(\tilde{z}_j,z_i) &= z_i-2\Delta-\tilde{z}_j \\ &= \xi_ip_i-\xi_jp_j-2\Delta(1+\alpha\tilde{\beta}). \end{aligned}$$

Puisque $q_i = q_j = 1$ on a $\xi_i = \xi_j = \Delta$ et il en résulte que

$$w''_{i,j} = Q(z_i, \tilde{z}_j) = -\frac{p_i - p_j + 2\Delta(p_i\tilde{\beta} + p_j\alpha + \Delta p_i p_j)}{p_i - p_j - 2(1+\alpha\tilde{\beta})}. \tag{2.14}$$

La définition des coordonnées $u_{i,j}$ ainsi que la condition que (A.11) résulte de (A.10) impliquent que l'on peut remplacer toute relation du genre

$$\sum_{\{(i_1,j_1),(j_1,i_1)\}} \cdots \sum_{\{(i_M,j_M),(j_M,i_M)\}} g_{i'_1,j'_1,\ldots,i'_M,j'_M} H_{i'_1,j'_1}, \ldots, H_{i'_M,j'_M} = 0,$$

avec $H_{i'_l,j'_l}$ égal à $z_i - \tilde{z}_j$ ou à Δ, par une relation semblable où ces deux expressions possibles pour $H_{i'_l,j'_l}$ sont remplacées, l'une par $v_{i',j'}$, l'autre par $t_{i',j'}$. On se souvient que chaque somme porte sur deux éléments (i_l, j_l) et (j_l, i_l). Elle est homogène en toutes les coordonnées $v_{i,j}, t_{i,j}$. Par exemple, il résulte de la relation triviale

$$(z_i - \tilde{z}_j)\Delta^3 + (z_{i'} - \tilde{z}_{j'})\Delta^3 = (z_i - \tilde{z}_{j'})\Delta^3 + (z_{i'} - \tilde{z}_j)\Delta^3$$

que

$$v_{i,j}t_{i',j'}t_{i',j}t_{i,j'} + v_{i',j'}t_{i,j}t_{i',j}t_{i,j'} = v_{i,j'}t_{i,j}t_{i',j}t_{i',j'} + v_{i',j}t_{i,j}t_{i,j'}t_{i',j'}. \tag{2.15}$$

Cette relation s'avère importante.

A un point p donné nous avons réparti l'ensemble des indices en des sous-ensembles $D_k = E_k \cup F_k$. Nous introduisons en plus une partition de chaque $D = D_k = E \cup F$. L'ensemble des indices (i,j) tels que $i \in E_k$ et $j \in F_k$ se répartit en deux sous-ensembles : l'ensemble où $t_{i,j} = t_{i,j}(p) \neq 0$ et l'ensemble où $t_{i,j} = t_{i,j}(p) = 0$. On observe que

$$F\big(z_i(p), \tilde{z}_j(p)\big) = F\big(\tilde{z}_j(p), z_i(p)\big) = 0 \tag{2.16}$$

si $\Delta = 0$ et (i,j) est une telle paire. Nous écrivons $i \leftrightarrow j$ si $t_{i,j} \neq 0$.

LEMME 2.2. *Si $i \leftrightarrow j$, $i \leftrightarrow j'$ et $i' \leftrightarrow j$ alors $i' \leftrightarrow j'$.*

Ce lemme nous permet de définir une relation d'équivalence sur D_k en posant $i \leftrightarrow i'$ s'il existe un j tel que $i \leftrightarrow j$ et $i' \leftrightarrow j$. On définit $j \leftrightarrow j'$ de la même façon. Cette relation d'équivalence sera dite *fine*.

Supposons que $t_{i',j'} = 0$. La relation (2.15) devient

$$v_{i',j'}t_{i,j}t_{i',j}t_{i,j'} = 0,$$

d'où il résulte que $v_{i',j'} = 0$. Puisque $v_{i',j'}$ et $t_{i',j'}$ ne peuvent pas être tous les deux nuls, le lemme est vérifié.

Si la valeur commune z_E de z_i, $i \in E$ est un zéro α et par conséquent z_F est le β correspondant, il y a en plus les coordonnées p_i, q_i et ξ_i et p_j, q_j et ξ_j. La relation homogène

$$(z_i - \tilde{z}_j + g\Delta)\Delta^2 = (z_i - \alpha)\Delta^2 - (\tilde{z}_j - \tilde{\beta})\Delta^2$$

avec $g = 2\alpha^2/(1 - 2\Delta\alpha)$ découle de

$$(z_i - \tilde{z}_j)\Delta^2 = z_i\Delta^2 - \tilde{z}_j\Delta^2$$

et

$$\alpha - \tilde{\beta} = \alpha - \frac{\alpha}{1-2\Delta\alpha} = -\frac{2\Delta\alpha^2}{1-2\Delta\alpha}.$$

Elle implique que

$$(v_{i,j} + g t_{i,j}) q_i q_j = p_i q_j t_{i,j} + p_j q_i t_{i,j}. \tag{2.17}$$

Il s'ensuit que si deux des trois coordonnées q_i, q_j et $t_{i,j}$ ne sont pas égales à zéro, alors la troisième ne l'est pas non plus. Toutefois cette relation est bien moins importante que (2.15).

Le but de ce chapitre est de trouver des formes convenables des équations de Bethe ou des points fixes de la correspondance autour d'un point fixe p dans C donné. Nous avons fixé $D = D_k$ car, au moins à $\Delta = 0$, nous pouvons isoler les coordonnées z_i, $\tilde{z}_j$, $w_{i,j}$, $i \in E$ et $j \in F$ des autres. Une paire (i,j) sera dite *pertinente* si $i \in E$ et $j \in F$. Si (i,j) n'est pas pertinente, alors $w''_{i,j} = -1$ et, en p même, la relation (2.11) fixe les variables $t_{i,j}$ et $v_{i,j}$ à 0 et 1. Pour ces paires qui ne sont pas pertinentes, nous supposons donc que, sur la fibre $\Delta = 0$ la coordonnée $t_{i,j} = 0$ et $v_{i,j} = 1$ dans un voisinage de p.

A ce point nous voulons introduire le premier argument informel en affirmant que le découplage évident à $\Delta = 0$ a pour conséquence qu'il suffit d'étudier le cas $D = \{1, \ldots, r\}$. Pour les paires qui ne sont pas pertinentes l'équation (2.11) nous permet de trouver, dans un voisinage de p, $t_{i,j}$ en fonction de z_i et $\tilde{z}_j$, de sorte que la coordonnée $t_{i,j}$ devient superflue. Il en résulte que les facteurs $w''_{i,j}$ attachés à ces indices, qui s'expriment selon (2.13) en fonction de $t_{i,j}$, peuvent être intégrés au côté gauche des équations de Bethe. Sans nous préoccuper davantage de ce découplage nous l'acceptons. Ainsi, sauf pour les arguments combinatoires, nous faisons semblant que $D = \{1, \ldots, r\}$.

Fixons D et ne considérons que des indices dans D. Pour $i \leftrightarrow j$ nous supposons que $t_{i,j} = 1$ de sorte que $v_{i,j}/t_{i,j} = v_{i,j}$. Selon les équations (2.12) on n'a besoin dans une classe d'équivalence fine que d'un seul z_i ou $\tilde{z}_j$. Les autres s'expriment à partir de cette coordonnée et de Δ et des $v_{i,j}$. En particulier $\mu_{i,j} = \Delta$. Si i, i', j, j' sont tous dans la même classe d'équivalence fine l'équation (2.15) devient

$$v_{i,j} + v_{i',j'} = v_{i,j'} + v_{i',j}. \tag{2.18}$$

Cette relation implique l'existence de nombres r_i et s_j tels que

$$v_{i,j} = r_i - s_j.$$

Ces nombres sont uniques à une constante additive près.

Pour $i \in E$ et $j \in F$ mais $i \not\leftrightarrow j$ on met $v_{i,j} = 1$ et on garde les coordonnées $t_{i,j}$ aussi bien que $z_i - \tilde{z}_j$. Au point p lui-même $t_{i,j} = 0$. Les relations entre les $t_{i,j}$ ne sont plus linéaires ; elles sont plutôt quadratiques. Supposons que $i \leftrightarrow j$ et $i' \leftrightarrow j$ mais que $i \not\leftrightarrow j'$. Il résulte de (2.15) que

$$v_{i,j} t_{i',j'} t_{i,j'} + t_{i,j'} = t_{i',j'} + v_{i',j} t_{i,j'} t_{i',j'}. \tag{2.19}$$

Si par contre $i \leftrightarrow j$ et $i \leftrightarrow j'$ mais $i' \not\leftrightarrow j$, alors

$$v_{i,j} t_{i',j'} t_{i',j} + t_{i',j} = v_{i,j'} t_{i',j} t_{i',j'} + t_{i',j'}. \tag{2.20}$$

Si enfin $i \leftrightarrow j$ et $i' \leftrightarrow j'$ mais $i \not\leftrightarrow j'$, alors

$$v_{i,j}t_{i',j}t_{i,j'} + v_{i',j'}t_{i',j}t_{i,j'} = t_{i',j} + t_{i,j'}. \tag{2.21}$$

On obtient deux autres relations semblables en prenant d'abord $i \leftrightarrow j$ mais $i \not\leftrightarrow j'$ et $i' \not\leftrightarrow j$ et $i' \not\leftrightarrow j'$ et ensuite en supposant que les quatre indices sont tous indépendants. La première, la plus importante, est

$$v_{i,j}t_{i',j'}t_{i',j}t_{i,j'} + t_{i',j}t_{i,j'} = t_{i',j}t_{i',j'} + t_{i,j'}t_{i',j'} \tag{2.22}$$

alors que la seconde est cubique

$$t_{i',j'}t_{i',j}t_{i,j'} + t_{i,j}t_{i',j}t_{i,j'} = t_{i,j}t_{i',j}t_{i',j'} + t_{i,j}t_{i,j'}t_{i',j'}. \tag{2.23}$$

Si la paire (z_E, z_F) est liée, c'est-à-dire si $z_E = \alpha$ et $z_F = \beta$ où α est un zéro de R et $\beta = \tilde{\alpha}$ un pôle, il y a des coordonnées supplémentaires. Si $i \leftrightarrow j$ il y a deux possibilités : soit $q_i \neq 0$, $q_j \neq 0$ de sorte qu'on peut mettre $q_i = q_j = 1$ et la relation (2.17) devient

$$v_{i,j} + g = p_i + p_j; \tag{2.24}$$

soit $q_i = q_j = 0$ au point p de sorte qu'on peut mettre $p_i = p_j = 1$ et la relation (2.17) devient

$$(v_{i,j} + g)q_iq_j = q_i + q_j, \tag{2.25}$$

une relation quadratique. Rappelons enfin que les cas $q_i = p_j = 0$ ou $q_j = p_i = 0$ sont exclus si $i \leftrightarrow j$ par l'équation (2.17).

Si par contre $i \not\leftrightarrow j$ alors on met $v_{i,j} = 1$. Si $q_i = p_j = 1$ la relation (2.17) devient

$$(1 + gt_{i,j})q_j = p_iq_jt_{i,j} + t_{i,j}, \tag{2.26}$$

une relation qui lie q_j et $t_{i,j}$. Si $q_j = p_i = 1$ il y a une relation pareille

$$(1 + gt_{i,j})q_i = t_{i,j} + p_jq_it_{i,j}. \tag{2.27}$$

Si enfin $p_i = p_j = 1$ alors

$$(1 + gt_{i,j})q_iq_j = (q_i + q_j)t_{i,j}, \tag{2.28}$$

relation pour laquelle, contrairement aux deux relations précédentes, les termes de plus bas degré sont quadratiques.

Après ces préparatifs nous venons à l'équation $\varphi(q) = \psi(q)$ autour d'un point fixe $p \in C$ donné. Les coordonnées z_i et $w_{i,j}$ sont celles de q. Supposons aux fins de simplicité que $D = \{1, \ldots, r\}$.

Soit $e = |E|$ le nombre d'éléments de E, $f = |F|$ celui de F. Les équations $\varphi(p) = \psi(p)$ qui s'écrivent

$$\begin{aligned} R(z_i) &= \prod_{m \neq i} w''_{i,m}, \quad i \in E, \\ R(z_j)^{-1} &= \prod_{m \neq j} w''_{m,j} \quad j \in F, \end{aligned} \tag{2.29}$$

deviennent

$$(2.30)\qquad \begin{aligned} R(z_i) &= (-1)^{r-f-1}\prod_{j\in F} w''_{i,j}, \quad i\in E,\\ R(z_j)^{-1} &= (-1)^{r-e-1}\prod_{i\in E} w''_{i,j}. \quad j\in F. \end{aligned}$$

Si F est vide ces équations deviennent simplement

$$(2.31)\qquad R(z_i) = (-1)^{r-1}.$$

Dans le cas générique cette équation a N racines distinctes. Si E est vide on obtient une équation semblable pour $R(z_j)$. Donc si un des deux ensembles E et F est vide, nous pourrons traiter facilement les équations pour les coordonnées dans ces ensembles. Supposons, par exemple, que tous les F_k sont vides et que chaque E_k ne contient qu'un seul élément ; nous obtenons $N(N-1)\dots(N-r+1)$ solutions qui se déforment facilement et qui donnent toutes les solutions admissibles, au moins si le théorème est valable. Même sans le théorème il est évident que génériquement les déformations de ces solutions donnent le nombre voulu de solutions des équations de Bethe et que les vecteurs attachés par Bethe à ces solutions sont linéairement indépendants. Donc par déformation, qui sur l'ordinateur se fait facilement, on résout vite par la méthode de Bethe le problème de départ, c'est-à-dire celui de trouver les vecteurs et les valeurs propres de l'application linéaire de [**B**].

Nous savons exprimer les équations (2.30) en termes des coordonnées de C. Les deux cas pour (E,F), le libre et le lié, sont foncièrement différents.

Supposons d'abord que (E,F) soit libre. La relation (2.1) donne

$$(2.32)\qquad R(z_j)^{-1} = \frac{R(\tilde z_j)}{\rho},$$

et la seconde des équations (2.30) deviennent

$$(2.33)\qquad \frac{R(\tilde z_j)}{\rho} = (-1)^{r-e-1}\prod_{i\in E} w''_{i,j}.$$

En particulier, au point p, où tous les z_i sont égaux à z_E ainsi que tous les $\tilde z_j$, on obtient en prenant des produits de (2.30) et (2.33) que

$$(2.34)\qquad R(z_E)^f(-1)^{e(r-f-1)} = R(z_E)^e(-1)^{f(r-e-1)}\rho^f.$$

Puisqu'on peut supposer ρ générique, e et f sont distincts et la valeur de $R(z_E)$ est générique. Les cas $e=0$ et $f=0$ ayant été écartés, il résulte de l'inégalité $f\neq e$ qu'un de ces deux entiers est plus grand que 1. Par conséquent le point p n'est pas un point fixe admis! On verra qu'il n'admet pas de déformation en solution admissible.

Il faudra examiner les équations (2.30) et leurs solutions en $\Delta=0$ et autour de $\Delta=0$. L'analyse est quelque peu différente dans les deux cas. Commençons par $\Delta=0$. Il résulte de la relation (2.13) que $w''_{i,j}=-1$ si $i\not\leftrightarrow j$ puisqu'alors $t_{i,j}(p)=0$ et $v_{i,j}=1$. Les équations (2.30) deviennent par conséquent un ensemble d'équations, une pour chaque classe ω d'équivalence fine dans D. Soit e_ω le nombre d'éléments dans l'intersection de ω et E, f_ω celui dans l'intersection de ω et F. En écrivant $v_{i,j} = r_i + s_j$ lorsque $i \leftrightarrow j$ et en posant

$$a = 2z_E^2 \neq -2, \quad b = -2, \quad c = (-1)^{r-1} R(z_E), \quad d = (-1)^{r-1} \frac{R(z_E)}{\rho},$$

on obtient

$$\begin{aligned} c &= \prod_{j \in \omega \cap F} \frac{r_i - s_j + a}{r_i - s_j + b}, \quad i \in \omega \cap E, \\ d &= \prod_{i \in \omega \cap E} \frac{r_i - s_j + a}{r_i - s_j + b}, \quad j \in \omega \cap F. \end{aligned} \tag{2.35}$$

Nous observons d'emblée que la généricité de ρ et par conséquent de z_E, de c, et de d ainsi que la relation

$$c^{e_\omega} = d^{f_\omega} \tag{2.36}$$

impliquent l'existence d'entiers m_ω, m, e et f tels que $e_\omega / m_\omega = e/m$ et $f_\omega / m_\omega = f/m$. Observons que $\sum m_\omega = m$. De l'équation (2.36) résulte l'équation

$$(-1)^{e_\omega (r-1)} R(z_E)^{f_\omega} = (-1)^{f_\omega (r-1)} R(z_E)^{e_\omega} \rho^{f_\omega}. \tag{2.37}$$

Nous revenons encore dans le chapitre suivant aux équations que satisfont c et d, n'observant ici que l'on peut trouver N valeurs z_E distinctes telles que $c = (-1)^{r-1} R(z_E)$ parce que c est générique.

Tournons-nous maintenant vers le cas où (E, F) est une paire liée. Ce sont alors les équations (2.14) qui sont pertinentes. On suppose que z_E est un zéro α de R et z_F le pôle β correspondant. Alors autour de $\Delta = 0$ on a $z_i - \alpha = \Delta p_i$ et

$$\begin{aligned} R(z_i) &= A \Delta p_i + O(\Delta^2), \quad A = R'(z_E), \\ R(z_j)^{-1} &= B \Delta p_j + O(\Delta^2), \quad B = \frac{A}{\rho}. \end{aligned} \tag{2.38}$$

car $R(z)^{-1} = R(\tilde{z})/\rho$. Nous soulignons que le quotient A/B est générique.

Dans cet exposé que nous avouons être informel nous n'étudions pas les équations précises qui résultent de (2.14) et de (2.37). Nous les simplifions. Soit

$$S(p, q) = \frac{p - q}{p - q - a}, \quad a = 2(1 + \alpha^2) \neq 0. \tag{2.39}$$

Les équations qui seront examinées sont

$$\begin{aligned} A \Delta p_i &= (-1)^{r-1} \prod_{j \in F} S(p_i, p_j), \quad i \in E, \\ B \Delta p_j &= (-1)^{r-1} \prod_{i \in E} S(p_i, p_j), \quad j \in F. \end{aligned} \tag{2.40}$$

Observons que $p_i = p_{i'}$ si par exemple i et i' appartiennent au même A_k.

Notre étude de ces équations se divise en celle des équations (2.35) qui définissent une variété dans la fibre de C à $\Delta = 0$, ensuite celle des déformations des points sur ces variétés, autour de $\Delta = 0$, et celle des équations (2.40). Observons déjà qu'en $\Delta = 0$ les équations (2.40) ont des solutions triviales qui ne nous intéressent point, celles pour lesquelles à chaque i est attaché un j tel que $p_i + p_j = 0$. En effet l'ensemble de ces points fixes n'est plus de dimension 0 à $\Delta = 0$ quoique l'ensemble de points qui se déforment à $\Delta \neq 0$ l'est.

3. Des variétés spéciales

Ce chapitre est consacré à une étude préliminaire des paires libres (E, F). La fibre C de la correspondance à $\Delta = 0$ n'est plus irréductible. Pour les paires libres, z_E et z_F ne sont ni des zéros ni des pôles de R, et les composantes irréductibles pertinentes sont celles définies à partir des équations (2.11). Pour chaque paire (i, j), on a soit $z_i = \tilde{z}_j$ soit $t_{i,j} = 0$. Donc sur chaque composante il y a des égalités entre les coordonnées z_i ou $\tilde{z}_j$ qui sont forcément satisfaites. Aux intersections des composantes, les deux facteurs qui interviennent dans une des équations (2.11), donc $z_i - \tilde{z}_j$ et $t_{i,j}$, sont pour certaines paires (i, j) tous les deux nuls. Pour qu'un point p appartienne à une composante donnée il faut évidemment que $z_i(p) = \tilde{z}_j(p)$ si cette relation est satisfaite partout sur la composante, mais il pourrait bien avoir des paires (i, j) telles que $z_i(p) = \tilde{z}_j(p)$ sans que $z_i = \tilde{z}_j$ partout sur la composante. Dans ce cas, $t_{i,j}$ est 0 partout sur la composante et en particulier au point p. Par conséquent p peut appartenir à plusieurs composantes irréductibles, celles où $z_i = z_j$ ou $z_i = \tilde{z}_j$ partout si i et j sont équivalents dans le sens fin par rapport à p.

Nous fixons désormais implicitement les ensembles de A et de B, donc les ensembles A_i, $1 \leq i \leq s$ et B_j, $1 \leq j \leq t$; et nous désignons par r le nombre total de ces ensembles, $r = s + t$. (Ce nombre est donc plus petit ou égal au nombre r introduit dans l'introduction. Il faut l'en distinguer.) Le poids de chacun de ces ensembles est le nombre de coordonnées qu'il contient. On peut attacher à chaque composante irréductible une décomposition de $\{1, \ldots, r\}$ en paires (E_k, F_k). L'ensemble E_k, ainsi que l'ensemble F_k, est un ensemble maximal dans lequel toutes les coordonnées z_i, $i \in E_k$ sont égales entre elles. Si $i \in E_k$ et $j \in F_k$ alors $z_j = \tilde{z}_i$. Dans ce chapitre nous étudions les équations des points fixes sur une composante donnée mais avec l'hypothèse supplémentaire que $E \cup F = D = \{1, \ldots, r\}$. Il s'agit surtout de trouver une borne supérieure de la multiplicité d'un tel point, qui s'avérera par la suite être la valeur exacte de cette multiplicité.

Soit $e = |E|$ et $f = |F|$ et $a \neq b$, $c \neq 0$ et $d \neq 0$ quatre nombres complexes génériques. Soient μ_i, $1 \leq i \leq e$ et $\mu_{\underline{j}}$, $1 \leq \underline{j} \leq f$ des entiers positifs. Nous considérons les équations

$$c = \prod_{\underline{j}=1}^{f} \left(\frac{r_i - s_{\underline{j}} + a}{r_i - s_{\underline{j}} + b} \right)^{\mu_{\underline{j}}}, \quad d = \prod_{i=1}^{e} \left(\frac{r_i - s_{\underline{j}} + a}{r_i - s_j + b} \right)^{\mu_i}, \tag{3.1}$$

dont les inconnues sont les r_i, $1 \leq i \leq e$ et les $s_{\underline{j}}$, $1 \leq \underline{j} \leq f$. Ces équations découlent des équations (2.35). Cependant, comme mentionné dans l'introduction nous tenons explicitement compte des conditions imposées par la partition A. (Puisqu'il est supposé que $D = \{1, \ldots, r\}$ l'ensemble B est vide.) Si i et i' appartiennent au même A_k et j et j' au même A_l alors $v_{i,j} = v_{i',j'}$ de sorte que $r_i = r_{i'}$ et $s_j = s_{j'}$. Les entiers μ_i sont les entiers $|A_i|$, $A_i \subset E$, et les entiers $\mu_{\underline{j}}$ les entiers $|A_{\underline{j}}|$, $A_{\underline{j}} \subset F$. Nous avons trouvé utile de distinguer entre les éléments de la partition contenus dans E et ceux contenus dans F en soulignant l'indice de ces derniers.

Soient $\mu(E) = \sum_i \mu_i$ et $\mu(F) = \sum_{\underline{j}} \mu_{\underline{j}}$. Observons qu'avec les conventions du chapitre précédent $\mu(E)$ serait e. L'entier e a acquis une signification nouvelle qu'il gardera par la suite. Pour que les équations (3.1) aient des solutions il faut que $c^{\mu(E)} = d^{\mu(F)}$. Une des $e+f$ équations est alors superflue. Quoique nous les avons déduites à partir des équations comme des équations de points fixes aux points de la composante irréductible où la coordonnée $t_{i,\underline{j}}/v_{i,\underline{j}}$ n'est nulle que si elle est identiquement nulle sur la composante. On pourrait donner les équations des points fixes pour lesquels cette condition n'est pas satisfaite, mais nous préférons dans ce chapitre introduire à partir des équations (3.1) des variétés projectives abstraites. Nous revenons aux correspondances dans le prochain chapitre.

Rappelons qu'il est entendu que les coordonnées r_i et $s_{\underline{j}}$ ne sont définies qu'à une constante additive près,

$$r_i \longrightarrow r_i + t, \quad s_{\underline{j}} \longrightarrow s_{\underline{j}} + t.$$

Si on permet aux coordonnées r_i et $s_{\underline{j}}$ de tendre vers l'infini, ces équations définissent une variété algébrique projective V. Plus précisément il s'agit d'un schéma dans le sens de Grothendieck. Sinon la notion de multiplicité d'un point dans une variété de dimension zéro n'aurait pas de sens. Nous préférons utiliser une terminologie moins exacte mais qui, en revanche, décourage moins le lecteur inexpérimenté.

Soit

$$X = \prod_{i=1}^{e} \mathbb{P}^1 \times \prod_{\underline{j}=1}^{f} \mathbb{P}^1$$

avec des coordonnées x_i et $x_{\underline{j}}$ assujetties à l'équation

$$\prod_{i=1}^{e} x_i^{\mu_i/\eta} = \prod_{\underline{j}=1}^{f} x_{\underline{j}}^{\mu_{\underline{j}}/\eta}$$

où η est le plus grand diviseur commun de tous les μ_i et $\mu_{\underline{j}}$. Posons

$$x_i = \prod_{\underline{j}=1}^{f} \left(\frac{r_i - s_{\underline{j}} + a}{r_i - s_{\underline{j}} + b} \right)^{\mu_{\underline{j}}}, \quad x_{\underline{j}} = \prod_{i=1}^{e} \left(\frac{r_i - s_{\underline{j}} + a}{r_i - s_{\underline{j}} + b} \right)^{\mu_i}. \tag{3.2}$$

Il n'est guère surprenant qu'il soit parfois convenable et même souvent nécessaire de remplacer (en revenant sur nos pas) les coordonnées r_i et $s_{\underline{j}}$ par des coordonnées $v_{i,\underline{j}}$, $1 \leq i \leq e$, $1 \leq \underline{j} \leq f$ en posant $r_i - s_{\underline{j}} = v_{i,\underline{j}}$. Les variables $v_{i,\underline{j}}$ sont assujetties aux relations

$$v_{i,\underline{j}} + v_{i',\underline{j}'} = v_{i,\underline{j}'} + v_{i',\underline{j}}. \tag{3.3}$$

Les équations (3.2) deviennent

$$x_i = \prod_{\underline{j}=1}^{f} \left(\frac{v_{i,\underline{j}} + a}{v_{i,\underline{j}} + b} \right)^{\mu_{\underline{j}}}, \quad x_{\underline{j}} = \prod_{i=1}^{e} \left(\frac{v_{i,\underline{j}} + a}{v_{i,\underline{j}} + b} \right)^{\mu_i}. \tag{3.4}$$

On écrit alors les équations (3.1) d'une façon simple

$$x_i = c, \quad x_{\underline{j}} = d. \tag{3.5}$$

En écrivant les équations de cette façon on obtient immédiatement la variété projective V. On remplace $v_{i,\underline{j}}$ par $v_{i,\underline{j}}/t_{i,\underline{j}}$ où $(t_{i,\underline{j}}, v_{i,\underline{j}})$ sont des coordonnées projectives. Lorsque quelqu'unes des coordonnées $t_{i,\underline{j}}$ deviennent 0 les équations se découplent de sorte qu'en fin de compte on revient aux équations (3.1). Donc au début nous ne considérons que des solutions telles que tous les $t_{i,\underline{j}}$ sont 1 et par conséquent tous les r_i et $s_{\underline{j}}$ finis.

On pourrait essayer de résoudre les équations (3.5) en admettant des valeurs $v_{i,j} = -a$ et $v_{i,j} = -b$ et en donnant des valeurs arbitraires aux produits (3.4) qui contiennent au moins un facteur 0 et au moins un facteur ∞. Cela est impossible. Appelons en effet une telle solution une solution *spectrale*. Le lemme suivant est tout à fait semblable au lemme A.6.

LEMME 3.1. *Il n'existe pas de solution spectrale des équations* (3.4) *et* (3.5).

Pour une solution spectrale il existe une paire (i, j) telle que $v_{i,\underline{j}} = -a$. Puisque c n'est ni 0 ni ∞ il existe en plus un indice j' tel que $v_{i,\underline{j}'} = -b$. Puisque d non plus n'est ni 0 ni ∞ il existe un i' tel que $v_{i',\underline{j}} = -b$. Alors

$$v_{i',\underline{j}} - v_{i,\underline{j}} = a - b$$

et

$$v_{i,\underline{j}'} - v_{i,\underline{j}} = a - b.$$

De l'équation

$$v_{i,\underline{j}} + v_{i',\underline{j}'} = v_{i,\underline{j}'} + v_{i',j}$$

et du fait que $a \neq b$, il résulte que $v_{i',\underline{j}'} \neq -a$. Mais un indice $\underline{j}''$ tel que $v_{i',\underline{j}''} = -a$ doit exister parce que $c \neq 0$.

En effet nous pouvons construire une suite

$$(3.6) \qquad (i_0, \underline{j}_0), (i_0, \underline{j}_1), (i_1, \underline{j}_1), (i_1, \underline{j}_2), \ldots, (i_{n-1}, \underline{j}_n), (i_n, \underline{j}_n)$$

telle que

$$v_{i_m,\underline{j}_m} = -a, \quad 1 \leq m \leq n, \qquad v_{i_m,\underline{j}_{m+1}} = -b, \quad 1 \leq m < n$$

et telle que $(i_0, \underline{j}_0) = (i_n, \underline{j}_n)$. Il est convenable de définir $(i_m, \underline{j}_m)$ pour tout entier m en posant $(i_{m+sn}, \underline{j}_{m+sn}) = (i_m, \underline{j}_m)$, $s \in \mathbb{Z}$. La relation

$$(3.7) \qquad v_{i_m,\underline{j}_{m'}} = (m' - m)a + (m - m')b - a$$

est évidente pour $m' = m$ et $m' = m + 1$. Elle résulte en général par récurrence sur $|m' - m|$ de la relation

$$v_{i_{m-1},\underline{j}_{m'+1}} + v_{i_m,j_{m'}} = v_{i_{m-1},\underline{j}_{m'}} + v_{i_m,\underline{j}_{m'+1}},$$

qui n'est qu'un cas particulier de (3.3). Mais puisque la suite des $(i_m, \underline{j}_m)$ est périodique et $a \neq b$, la relation (3.7) est manifestement impossible.

Les équations admettent une modification simple mais utile. On a

$$\frac{(v_{i,j} + \varepsilon)/\lambda + a}{(v_{i,j} + \varepsilon)/\lambda + b} = \frac{v_{i,j} + a'}{v_{i,j} + b'}$$

avec $a' = \lambda a + \varepsilon$ et $b' = \lambda b + \varepsilon$. Donc en utilisant de nouvelles variables on peut supposer que $a = -b = \sqrt{-1}$, ce que nous ferons souvent. Alors des valeurs réelles pour des $v_{i,j}$ mènent à des quotients dont la valeur absolue est 1 et il sera utile de supposer que $|c| = |d| = 1$. Puisque c'est le cas générique qui nous importe, nous admettrons cette hypothèse lorsqu'elle est convenable mais seulement après avoir vérifié deux lemmes qui sont valables sous la seule hypothèse que $c \neq 0$ et $d \neq 0$.

Le lemme suivant est une forme plus forte du lemme précédent.

LEMME 3.2. *Supposons que $b = -a = \bar{a}$. Il existe une constante ε telle que, pour tout point fixe de la correspondance (φ, ψ), on ait*

$$|v_{i,\underline{j}} \pm a| \geq \varepsilon$$

pour tout i et j.

Choisissons δ tel que

$$\delta < \min\{|c|, |d|\} \leq \max\{|c|, |d|\} < \frac{1}{\delta}$$

et soit $M = \max\{e, f\}$. Choisissons ε en sorte que $|v_{i,\underline{j}} + a| < \varepsilon$ implique

$$\left| \frac{v_{i,\underline{j}} + a}{v_{i,\underline{j}} - a} \right| < \delta^{M^{2M}}$$

et $|v_{i,\underline{j}} - a| < \varepsilon$ implique

$$\left| \frac{v_{i,\underline{j}} + a}{v_{i,\underline{j}} - a} \right| > \delta^{-M^{2M}}.$$

Supposons qu'il existe une solution des équations (3.4) et (3.5) et un i et un $\underline{j}$ tel que

$$\left| \frac{v_{i,\underline{j}} + a}{v_{i,\underline{j}} - a} \right| < \delta^{M^m}, \quad m \geq 1.$$

Alors il existe un $\underline{j}'$ tel que

$$\left| \frac{v_{i,\underline{j}'} + a}{v_{i,\underline{j}'} - a} \right| > \delta^{-M^{m-1}}$$

car sinon le produit

$$\prod_{\underline{j}=1}^{f} \left| \frac{v_{i,\underline{j}} + a}{v_{i,\underline{j}} - a} \right| \leq \delta^{M^{m-1}} \leq \delta.$$

Il existe aussi un i' tel que la même inégalité soit satisfaite lorsque on substitue $(i', \underline{j})$ à $(i, \underline{j}')$. De la même façon, s'il existe i et $\underline{j}$ tel que

$$\left| \frac{v_{i,\underline{j}} + a}{v_{i,\underline{j}} - a} \right| > \delta^{-M^m}$$

alors il existe un $\underline{j}'$ tel que

$$\left| \frac{v_{i,\underline{j}'} + a}{v_{i,\underline{j}'} - a} \right| < \delta^{M^{m-1}}.$$

Ces remarques nous permettent de construire par récurrence à partir d'une paire $(i_1, \underline{j}_1)$ qui satisfait à une des inégalités

$$|v_{i,\underline{j}} \pm a| < \varepsilon$$

une suite

$$(i_0, \underline{j}_0), (i_0, \underline{j}_1), \ldots, (i_{n-1}, \underline{j}_n), (i_n, \underline{j}_n), \qquad n \leq 2M,$$

telle que $(i_0, \underline{j}_0) = (i_n, \underline{j}_n)$ et

$$\left|\frac{v_{i_m,\underline{j}_m} + a}{v_{i_m,\underline{j}_m} - a}\right| < \delta, \quad \left|\frac{v_{i_m,\underline{j}_{m+1}} - a}{v_{i_m,\underline{j}_{m+1}} + a}\right| < \delta$$

pour $0 \leq m \leq n$. Puisqu'il nous est permis de choisir δ très petit, nous pouvons supposer que les premières de ces inégalités impliquent que

$$|v_{i_m,\underline{j}_m} + a| < \left|\frac{a}{K}\right|, \tag{3.8}$$

et les deuxièmes que

$$|v_{i_m,\underline{j}_{m+1}} - a| < \left|\frac{a}{K}\right| \tag{3.9}$$

où K est un entier demeurant à notre disposition.

En imitant la démonstration de l'égalité (3.7) nous définissons la suite $(i_m, \underline{j}_m)$ pour tout m en posant $(i_{m+rn}, \underline{j}_{m+rn}) = (i_m, \underline{j}_m)$ et vérifions que

$$\left|v_{i_m,\underline{j}_{m'}} + (1 - 2m' + 2m)a\right| < \left|\frac{3^m a}{K}\right|.$$

En prenant $K = 2 \cdot 3^M$ et d'abord $m = m' = 0$ mais ensuite $m = 0$ et $m' = n$ nous obtenons de

$$2na = (v_{i_0,\underline{j}_0} + a) - \left(v_{i_0,\underline{j}_n} + (1 - 2n)a\right)$$

que

$$|2na| \leq \left|\frac{a}{K}\right| + \left|\frac{3^n a}{K}\right| < |a|,$$

dont l'impossibilité conclut la preuve.

COROLLAIRE 3.3. *La variété des solutions de* (3.4) *et* (3.5) *est un ensemble fini.*

La variété des solutions de (3.4) et (3.5) est une variété algébrique complexe affine. Sur une telle variété une fonction rationnelle qui omet un ensemble ouvert est constante sur chaque composante irréductible. Les fonctions $v_{i,\underline{j}}/t_{i,\underline{j}}$ sont des fonctions rationnelles sur cette variété et, par le lemme, l'ensemble de leurs valeurs a une intersection vide avec des voisinages ouverts de $\pm a$. Donc chacune de ces fonctions est constante sur chacune des composantes irréductibles qui sont en nombre fini. Il résulte alors facilement du découplage des équations lorsque quelques-unes des coordonnées $t_{i,\underline{j}}$ sont zéro que la variété projective V est aussi finie.

Nous supposons ensuite que $|c| = |d| = 1$.

LEMME 3.4. *Si $b = \bar{a} = -a$ et $|c| = |d| = 1$ alors toutes les solutions des équations* (3.4) *et* (3.5) *sont réelles.*

Le lemme se vérifie comme les deux lemmes précédents. Supposons que, parmi les $v_{i,j}$, quelques-uns ne soient pas réels. Soit ε le signe de $a\sqrt{-1}$. On a

$$\left|\frac{v+a}{v-a}\right| < 1 \Longleftrightarrow \varepsilon\Im(v) > 0, \qquad \left|\frac{v+a}{v-a}\right| > 1 \Longleftrightarrow \varepsilon\Im(v) < 0.$$

Puisque $|c| = |d| = 1$ il résulte de l'existence d'une paire $(i, \underline{j})$ telle que $\Im(v_{i,\underline{j}}) > 0$ qu'il existe un $i' \neq i$ et un $\underline{j}' \neq j$ tels que

$$\Im(v_{i',\underline{j}}) < 0, \quad \Im(v_{i,\underline{j}'}) < 0.$$

Par conséquent nous pouvons à partir de cette paire construire encore une fois une suite (3.6) mais telle que

$$\Im(v_{i_m,\underline{j}_m}) > 0, \quad \Im(v_{i_m,\underline{j}_{m+1}}) < 0.$$

Soit

$$\delta = \min\{\Im(v_{i_m,\underline{j}_m}), -\Im(v_{i_m,\underline{j}_{m+1}})\}.$$

Puisque

$$v_{i_m,\underline{j}_{m'}} = v_{i_m,\underline{j}_{m'-1}} + v_{i_{m'-1},\underline{j}_{m'}} - v_{i_{m'-1},\underline{j}_{m'-1}}$$

on peut démontrer par récurrence que si $l > 0$ alors

$$\Im(v_{i_m,\underline{j}_{m+l}}) < -(2l-1)\delta.$$

De la même façon

$$v_{i_m,\underline{j}_{m'}} = v_{i_{m'},\underline{j}_{m'}} + v_{i_m,\underline{j}_{m'+1}} - v_{i_{m'},\underline{j}_{m'+1}}.$$

Il résulte de cette équation que si $l > 0$ alors

$$v_{i_{m+l},\underline{j}_m} > (2l-1)\delta.$$

Ces deux inégalités sont impossibles et nous avons déduit une contradiction de l'hypothèse qu'il existe un $v_{i,\underline{j}}$ avec $\Im(v_{i,\underline{j}}) > 0$. De la même façon il n'existe pas de $v_{i,\underline{j}}$ avec $\Im(v_{i,\underline{j}}) < 0$. Le lemme s'ensuit.

Soit ϕ l'application définie par (3.4). L'espace défini par les variables $v_{i,\underline{j}}$ est un espace affine de dimension $e+f-1$ dans lequel les $v_{1,\underline{j}}$, $1 \leq \underline{j} \leq f$ et $v_{i,\underline{1}}$, $1 < i \leq e$ sont des coordonnées possibles. Il est plus simple d'utiliser toutes les variables $v_{i,\underline{j}}$ aussi bien que les différentielles $dv_{i,\underline{j}}$. La dimension de X est aussi $e+f-1$.

LEMME 3.5. *Supposons que $b = \bar{a} = -a$ et que $|c| = |d| = 1$. En toute solution finie $(v_{i,\underline{j}})$ des équations* (3.4) *et* (3.5)*, l'application tangentielle $d\phi$ est un isomorphisme.*

Soulignons que l'application ϕ est aussi définie sur la variété projective attachée aux coordonnées $(t_{i,\underline{j}}, v_{i,\underline{j}})$. Nous n'affirmons pas que le lemme est vrai aux points où certains des $t_{i,\underline{j}}$ sont zéro.

Il faut montrer que les équations

$$0=\sum_{\underline{j}=1}^{f}\frac{\partial x_i}{\partial v_{i,\underline{j}}}dv_{i,\underline{j}},\quad 0=\sum_{i=1}^{e}\frac{\partial x_{\underline{j}}}{\partial v_{i,\underline{j}}}dv_{i,\underline{j}}$$

n'ont que la solution triviale. Dans ces équations on peut remplacer les coefficients

$$\frac{\partial x_i}{\partial v_{i,\underline{j}}},\quad \frac{\partial x_{\underline{j}}}{\partial v_{i,\underline{j}}}$$

par

$$\frac{1}{x_i}\frac{\partial x_i}{\partial v_{i,\underline{j}}},\quad \frac{1}{x_{\underline{j}}}\frac{\partial x_{\underline{j}}}{\partial v_{i,\underline{j}}}.$$

Puisque

$$\frac{v-a}{v+a}\frac{d}{dv}\left(\frac{v+a}{v-a}\right)=-2a\frac{1}{v^2-a^2}$$

les équations deviennent

$$0=\sum_{\underline{j}=1}^{f}A_{i,\underline{j}}dv_{i,\underline{j}},\quad 0=\sum_{i=1}^{e}A_{i,\underline{j}}dv_{i,\underline{j}}. \tag{3.10}$$

On a mis

$$A_{i,\underline{j}}=\frac{\mu_i\mu_{\underline{j}}}{v_{i,\underline{j}}^2-a^2}.$$

Les nombres $v_{i,\underline{j}}^2$ sont positifs et $-a^2=|a|^2$. Par conséquent tous les éléments $A_{i,\underline{j}}$ sont positifs. Puisque

$$dv_{i,\underline{j}}+dv_{i',\underline{j}'}=dv_{i,\underline{j}'}+dv_{i',\underline{j}},$$

il existe des nombres dr_i et $ds_{\underline{j}}$ tels que $dv_{i,\underline{j}}=dr_i-ds_{\underline{j}}$. Il suffit de montrer que $dr_i=ds_{\underline{j}}$ pour tout i et $\underline{j}$.

Soient

$$\sum_{\underline{j}}A_{i,\underline{j}}=U_i,\quad \sum_i A_{i,\underline{j}}=V_{\underline{j}}.$$

Alors les équations (3.10) deviennent

$$U_i dr_i-\sum_j A_{i,j}ds_{\underline{j}}=0,\quad \sum_i A_{i,j}dr_i-V_{\underline{j}}ds_{\underline{j}}=0. \tag{3.11}$$

Nous introduisons deux matrices,

$$B=\left(\frac{A_{i,\underline{j}}}{V_{\underline{j}}}\right),\quad C=\left(\frac{A_{i,\underline{j}}}{U_i}\right).$$

et deux vecteurs colonnes $dr'=(dr'_i)$, $dr'_i=U_i dr_i$, et $ds'=(ds'_{\underline{j}})$, $ds'_{\underline{j}}=V_{\underline{j}}ds_{\underline{j}}$. On a $dr'=Bds'$ et $ds'=C^t dr'$ de sorte que

$$dr'=BC^t dr'.$$

Mais

$$BC^t = \left(\sum_{\underline{j}} \frac{A_{i,\underline{j}}}{V_{\underline{j}}} \frac{A_{i',\underline{j}}}{U_{i'}}\right) = (D_{ii'}) = D$$

est une matrice à éléments positifs.

On vérifie que

$$\sum_{i'} D_{ii'} U_{i'} = U_i.$$

Selon le théorème de Perron-Frobenius la valeur propre 1 de D est de multiplicité 1. Il en résulte qu'il existe une constante γ telle que $dr'_i = \gamma U_i$ pour tout i. Par conséquent $dr_i = \gamma$ pour tout i et $ds_{\underline{j}} = \gamma$ pour tout $\underline{j}$.

Le lemme 3.5 implique le lemme suivant qui est plus faible.

LEMME 3.6. *Supposons que c et d soient génériques. Alors pour toute solution* $v_{i,\underline{j}}$ *des équations* (3.4) *et* (3.5) *l'application tangentielle* $d\phi$ *est un isomorphisme.*

Les lemmes 3.5 et 3.6 affirment que les points de V où toutes les coordonnées $t_{i,\underline{j}}$ sont non nulles sont de multiplicité un. En général la relation (3.3) nous permet en imitant la définition du chapitre précédent d'attacher à chaque point v de V une relation d'équivalence dans la réunion de $\{1, \dots, e\}$ et $\{\underline{1}, \dots, \underline{f}\}$. Nous posons $i \leftrightarrow \underline{j}$ si $t_{i,\underline{j}} \neq 0$ au point v et $i \leftrightarrow i'$ s'il existe un $\underline{j}$ tel que $i \leftrightarrow \underline{j}$ et $i' \leftrightarrow \underline{j}$. Nous définissons $\underline{j} \leftrightarrow \underline{j}'$ de la même façon.

LEMME 3.7. *Soit M le nombre de classes d'équivalence pour la relation d'équivalence attachée au point v. Alors, pour c et d génériques, la multiplicité du point v sur la variété V est au plus* $m(v) = (M-1)!$

Par la suite nous saurons déduire de ce lemme un corollaire plus fort.

COROLLAIRE 3.8. *Soit M le nombre de classes d'équivalence pour la relation d'équivalence attachée au point v. Alors, pour c et d génériques, la multiplicité du point v sur la variété V est précisément* $m(v) = (M-1)!$

Pour démontrer le lemme il s'avère utile de supposer que $a = -b = -\sqrt{-1}$ et ensuite de définir des variétés plus générales dont V est une spécialisation. Ces nouvelles variétés (dénotés aussi $V = V(\{\varepsilon_{i,\underline{j}}\}))$ sont toujours des sous-variétés d'un produit sur $\{(i,\underline{j})\}$ de droites projectives. Soient $(t_{i,\underline{j}}, v_{i,\underline{j}})$ les coordonnées sur une de ces droites. Soient $\varepsilon_{i,\underline{j}}$ des nombres positifs. Les équations auxquelles sont soumises les coordonnées d'un point sur la sous-variété sont les équations (2.15)

$$v_{i,\underline{j}} t_{i',\underline{j}'} t_{i',\underline{j}} t_{i,\underline{j}'} + v_{i',\underline{j}'} t_{i,\underline{j}} t_{i',\underline{j}} t_{i,\underline{j}'} = v_{i,\underline{j}'} t_{i,\underline{j}} t_{i',\underline{j}} t_{i',\underline{j}'} + v_{i',\underline{j}} t_{i,\underline{j}} t_{i,\underline{j}'} t_{i',\underline{j}'} \tag{3.12}$$

et les équations (2.19-23) qui en découlent. Les équations de la variété sont alors

$$\begin{aligned} c &= \prod_{\underline{j}} \left(\frac{v_{i,\underline{j}} - \varepsilon_{i,\underline{j}} t_{i,\underline{j}} \sqrt{-1}}{v_{i,\underline{j}} + \varepsilon_{i,\underline{j}} t_{i,\underline{j}} \sqrt{-1}} \right)^{\mu_{\underline{j}}}, \quad i \in E, \\ d &= \prod_{i} \left(\frac{v_{i,\underline{j}} - \varepsilon_{i,\underline{j}} t_{i,\underline{j}} \sqrt{-1}}{v_{i,\underline{j}} + \varepsilon_{i,\underline{j}} t_{i,\underline{j}} \sqrt{-1}} \right)^{\mu_i}, \quad \underline{j} \in F \end{aligned} \tag{3.13}$$

où c et d sont des constantes génériques.

Comme ci-dessus il est possible d'introduire la notion d'une solution spectrale de ces équations.

LEMME 3.9. *Il n'existe pas de solution spectrale sur les variétés* $V(\{\varepsilon_{i,\underline{j}}\})$.

Ce lemme se démontre comme le lemme 3.1.

Ces variétés sont aussi de dimension zéro mais avant d'énoncer le lemme pertinent nous observons qu'une équivalence fine $i \leftrightarrow \underline{j}$ (ou $\underline{i} \leftrightarrow i'$ et $\underline{j} \leftrightarrow \underline{j}'$) est attachée à chaque point de ces nouvelles variétés. Si $t_{i,\underline{j}} = 0$ alors

$$\frac{v_{i,\underline{j}} - \varepsilon_{i,\underline{j}} t_{i,\underline{j}} \sqrt{-1}}{v_{i,\underline{j}} + \varepsilon_{i,\underline{j}} t_{i,\underline{j}} \sqrt{-1}} = 1$$

de sorte que les équations découplent selon les classes d'équivalence. Ce découplage n'a toutefois qu'une validité limitée.

LEMME 3.10. *Les variétés* $V(\{\varepsilon_{i,\underline{j}}\})$ *sont de dimension* 0.

Pour nous amuser et en même temps pour nous préparer aux arguments à suivre nous vérifions ce lemme en considérant les déformations autour d'un point donné v.

En utilisant l'équivalence définie par v, en prenant les variations logarithmiques et en enlevant un facteur $-2\sqrt{-1}$ inutile, on obtient pour les équations de déformation

$$\begin{aligned} 0 &= \sum_{\underline{j} | \underline{j} \leftrightarrow i} \frac{\mu_i \mu_{\underline{j}} \varepsilon_{i,\underline{j}} dv_{i,\underline{j}}}{v_{i,\underline{j}}^2 + \varepsilon_{i,\underline{j}}^2} - \sum_{\underline{j} | \underline{j} \not\leftrightarrow i} \frac{\mu_i \mu_{\underline{j}} \varepsilon_{i,\underline{j}} dt_{i,\underline{j}}}{1 + \varepsilon_{i,\underline{j}}^2}, \\ 0 &= \sum_{i | i \leftrightarrow \underline{j}} \frac{\mu_i \mu_{\underline{j}} \varepsilon_{i,\underline{j}} dv_{i,\underline{j}}}{v_{i,\underline{j}}^2 + \varepsilon_{i,\underline{j}}^2} - \sum_{i | i \not\leftrightarrow \underline{j}} \frac{\mu_i \mu_{\underline{j}} \varepsilon_{i,\underline{j}} dt_{i,\underline{j}}}{1 + \varepsilon_{i,\underline{j}}^2}. \end{aligned} \tag{3.14}$$

Il est entendu que $t_{i,\underline{j}} = 1$ si $i \leftrightarrow \underline{j}$ mais que $v_{i,\underline{j}} = 1$ sinon.

Les coordonnées locales au point v sont les $\delta v_{i,\underline{j}}$, $i \leftrightarrow \underline{j}$ et les $\delta t_{i,\underline{j}}$, $i \not\leftrightarrow \underline{j}$. Puisque les $t_{i,\underline{j}}$ pertinents sont tous 0 en v, on a $\delta t_{i,\underline{j}} = t_{i,\underline{j}}$. Soient $\mathfrak{a}$ l'anneau local complet défini par ces paramètres et soit $\mathfrak{m}$ l'anneau maximal en v. La variété V est de dimension 0 en v si et seulement si $\mathfrak{a}/\mathfrak{m}$ est de dimension finie. La dimension est la multiplicité $m(v)$. Puisque nous ne cherchons qu'une borne supérieure pour $m(v)$ il est permis de le remplacer par un idéal plus petit $\mathfrak{m}'$, celui engendré par les termes de plus bas degré dans les équations qui définissent $\mathfrak{m}$. (Observons que si quelques équations sont passées inaperçues rien n'est perdu car on ne cherche à présent qu'une borne supérieure.) Les équations obtenues de (3.14) en remplaçant $dv_{i,\underline{j}}$ et $dt_{i,\underline{j}}$ par $\delta v_{i,\underline{j}}$ et $\delta t_{i,\underline{j}}$ s'obtiennent aussi de (3.13) en supprimant tout terme sauf ceux du plus bas degré. On obtient d'autres équations de (2.19), (2.20), (2.21) de la même façon, à savoir

$$t_{i',\underline{j}'} = t_{i,\underline{j}}, \quad i' \leftrightarrow i, \quad \underline{j} \leftrightarrow \underline{j}', \qquad t_{i',\underline{j}} = -t_{i,\underline{j}'}, \quad i' \leftrightarrow \underline{j}, \quad i \leftrightarrow \underline{j}'. \tag{3.15}$$

Il résulte de ces équations que $t_{i,\underline{j}} = t_{\omega,\eta}$ ne dépend que des classes d'équivalence ω et η qui contiennent i et $\underline{j}$ et que $t_{\eta,\omega} = -t_{\omega,\eta}$. Pour vérifier que $\mathfrak{a}/\mathfrak{m}'$ est de dimension finie il suffit de montrer que la seule solution des équations (3.14) et (3.15) est la solution triviale $\delta v_{i,\underline{j}} = 0$ et $t_{i,\underline{j}} = 0$.

Fixons une classe d'équivalence ω et faisons la somme des équations attachées aux i de cette classe ainsi que la somme de celles attachées aux $\underline{j}$ de la même classe. Nous nommerons les équations ainsi obtenues les équations horizontales et verticales.

Prenons ensuite la différence de l'équation horizontale et de la verticale. Il ne reste que

$$0 = \sum_{\eta|\eta\neq\omega} \varepsilon_{\omega,\eta} t_{\omega,\eta} \tag{3.16}$$

si

$$\varepsilon_{\omega,\eta} = \sum_{i\in\omega,\underline{j}\in\eta} \mu_i\mu_{\underline{j}} \frac{\varepsilon_{i,\underline{j}}}{1+\varepsilon_{i,\underline{j}}^2} + \sum_{\underline{j}\in\omega,i\in\eta} \mu_i\mu_{\underline{j}} \frac{\varepsilon_{i,\underline{j}}}{1+\varepsilon_{i,\underline{j}}^2} > 0. \tag{3.17}$$

Observons que $\varepsilon_{\omega,\eta} = \varepsilon_{\eta,\omega}$. En remplaçant l'équation (2.22) par ses termes de degré deux nous obtenons

$$t_{\omega,\eta}t_{\eta,\xi} = t_{\omega,\eta}t_{\omega,\xi} + t_{\eta,\xi}t_{\omega,\xi}. \tag{3.18}$$

Il est entendu que ω, η et ξ sont tous différents.

Nous établissons d'abord qu'il résulte de ces équations que tous les $t_{\omega,\eta}$ sont zéro. On en déduit tout de suite que si $t_{\omega,\eta} \neq 0$ et $t_{\eta,\xi} \neq 0$ alors $t_{\omega,\xi} \neq 0$. Cela nous permet de répartir les classes ω en classes d'équivalence. Les équations (3.16) restent évidemment valables si on restreint la somme à une seule classe. Pour ω et η dans la même classe on pose

$$\lambda_{\omega,\eta} = \frac{1}{t_{\omega,\eta}}.$$

La relation (3.18) devient

$$\lambda_{\omega,\xi} = \lambda_{\eta,\xi} + \lambda_{\omega,\eta}.$$

Par conséquent il existe des nombres μ_ω tels que

$$\lambda_{\omega,\eta} = \mu_\omega - \mu_\eta.$$

La relation (3.16) dans laquelle les $\varepsilon_{\omega,\eta}$ sont des nombres positifs implique que

$$\sum_{\eta\neq\omega} \varepsilon_{\omega,\eta}\Re(t_{\omega,\eta}) = 0, \quad \sum_{\eta\neq\omega} \varepsilon_{\omega,\eta}\Im(t_{\omega,\eta}) = 0. \tag{3.19}$$

Il résulte de (3.19) que s'il existe un η tel que $\Re(t_{\omega,\eta}) > 0$ il existe aussi un ξ tel que $\Re(t_{\omega,\xi}) < 0$.

Nous introduisons une relation d'ordre en posant $\omega \succeq \eta$ si et seulement si $\Re(t_{\omega,\eta}) \geq 0$, donc si et seulement si $\Re(\lambda_{\omega,\eta}) \geq 0$, ou même si et seulement si $\Re(\mu_\omega) \geq \Re(\mu_\eta)$. Il résulte de (3.19) que s'il existe un η tel que $\omega \succeq \eta$ alors il existe un ξ tel que $\xi \succeq \omega$. Ceci n'est possible que si la valeur de $\Re(\mu_\omega)$ est constante dans cette classe, car sinon on choisit ω tel que $\Re(\mu_\omega)$ est maximal pour en déduire une contradiction. Puisque le signe de $\Im(\lambda_{\omega,\eta})$ est opposé à celui de $\Im(t_{\omega,\eta})$ un argument pareil montre que $\Im(\mu_\omega)$ est constant sur la classe d'équivalence. Il en résulte que $\lambda_{\omega,\eta} = 0$ si $\omega \neq \eta$. Puisque ceci est impossible toute classe d'équivalence se réduit à un seul élément. Autrement dit $t_{\omega,\eta} = 0$ si $\omega \neq \eta$. Les équations (3.16) seront appelées les équations *antisymétriques*.

Il reste à traiter les équations

$$0=\sum_{\underline{j}\leftrightarrow i}\frac{\mu_i\mu_{\underline{j}}\varepsilon_{i,\underline{j}}\delta v_{i,\underline{j}}}{v_{i,\underline{j}}^2+\varepsilon_{i,\underline{j}}^2},\qquad 0=\sum_{i\leftrightarrow \underline{j}}\frac{\mu_i\mu_{\underline{j}}\varepsilon_{i,\underline{j}}\delta v_{i,\underline{j}}}{v_{i,\underline{j}}^2+\varepsilon_{i,\underline{j}}^2}.$$

Nous les appelons les équations *symétriques*. Pour montrer qu'il n'y a pas de solution non triviale de ces équations et par conséquent terminer la démonstration du lemme 3.10, il suffit de répéter la démonstration du lemme 3.5.

L'argument que nous venons d'utiliser a des conséquences. Supposons que nous déformions les équations en variant les $\varepsilon_{i,\underline{j}}$. Dans ce cas on obtient les équations de déformation en ajoutant aux quantités à gauche $\sum_{\underline{j}\leftrightarrow i} v_{i,\underline{j}}\delta\varepsilon_{i,\underline{j}}$ et $\sum_{i\leftrightarrow\underline{j}} v_{i,\underline{j}}\delta\varepsilon_{i,\underline{j}}$. Ces termes s'annulent lorsqu'on forme les équations antisymétriques. On cherche à trouver des déformations des solutions telles que les $t_{i,\underline{j}}$ restent constants, donc égaux à leur valeur en v qui est soit 0 soit 1. Il résulte immédiatement de la démonstration du lemme 3.5 et du théorème des fonctions implicites qu'une telle déformation existe et qu'elle est unique.

Puisqu'il n'y a qu'une seule déformation du point $v\in V$ et le type du point, donc la relation d'équivalence, reste constant le long de la déformation on peut en vérifiant le lemme 3.7 remplacer V par $V\big(\{\varepsilon_{i,\underline{j}}\}\big)$ et v par le point correspondant $v\big(\{\varepsilon_{i,\underline{j}}\}\big)$, car la multiplicité est conservée. L'avantage des paramètres est qu'on peut les spécialiser, ce qui, au pire, ne fait qu'accroître la multiplicité.

Il y a une transformation triviale des équations. On remplace chaque $t_{i,\underline{j}}$ pertinent par $\lambda_{i,\underline{j}}t_{i,\underline{j}}$ et chaque $v_{i,\underline{j}}$ par $v_{i,\underline{j}}/\lambda_{i,\underline{j}}$; donc on ne change qu'une des coordonnées projectives $v_{i,\underline{j}}$ et $t_{i,\underline{j}}$, l'autre reste 1. L'effet est d'abord de remplacer $\varepsilon_{i,\underline{j}}$ par $\lambda_{i,\underline{j}}\varepsilon_{i,\underline{j}}$, et donc, en effet, après un simple changement de notation, de ne rien changer du tout dans les équations (3.13). Par contre les autres équations, celles déduites de (3.12) subissent des modifications importantes.

On se souvient que nous essayons de trouver une borne supérieure pour la dimension de $\mathfrak{a}/\mathfrak{m}$. Nous avons remarqué qu'il est possible de remplacer $\mathfrak{m}$ par $\mathfrak{m}'$. Ces deux idéaux dépendent maintenant des paramètres $\mathfrak{m}=\mathfrak{m}\big(\{\varepsilon_{i,\underline{j}},\lambda_{i,\underline{j}}\}\big)$ et $\mathfrak{m}'=\mathfrak{m}'\big(\{\varepsilon_{i,\underline{j}},\lambda_{i,\underline{j}}\}\big)$. Nous trouverons une borne supérieure pour la dimension de $\mathfrak{a}/\mathfrak{m}'$ en faisant dégénérer les paramètres convenablement, car une telle dégénérescence ne fait qu'accroître la dimension.

Puisque les coordonnées $\varepsilon_{i,\underline{j}}$ et $\lambda_{i,\underline{j}}$ seront choisies de sorte qu'elles ne dépendent que des classes auxquelles appartiennent i et $\underline{j}$, nous pourrons remplacer tout de suite les variables $t_{i,\underline{j}}$ par les $t_{\omega,\eta}$ et ensuite passer aux équations symétriques et antisymétriques, parce que ces équations resteront valables. Les équations symétriques montrent que les variables $\delta v_{i,\underline{j}}$, $i\leftrightarrow\underline{j}$ sont dans $\mathfrak{m}'$. Tout revient alors aux variables $t_{\omega,\eta}$. Il y a $M=M(v)$ classes ω et l'argument, qui ne donne qu'une borne supérieure, est par récurrence sur M.

Pour $M=1$ il n'y a plus de coordonnées de façon que la multiplicité est évidemment $1=0!$ Fixons ω. Nous posons $\varepsilon_{i,\underline{j}}=\lambda_{\eta,\xi}=1$ sauf si $i\in\omega$ ou $\underline{j}\in\omega$; dans ce cas nous posons $\varepsilon_{i,\underline{j}}=\varepsilon$ et $\lambda_{i,\underline{j}}=\lambda$, les deux paramètres ε et λ étant libres. On enverra λ à ∞ et $\varepsilon\lambda$ à 0.

Soit désormais $\varepsilon_{\eta,\xi}$ en général le facteur obtenu en supposant tous les coefficients $\varepsilon_{i,\underline{j}}$ égaux à 1. C'est un nombre positif. Les équations qui n'impliquent pas les $t_{\omega,\eta}$ ne sont guère touchées. Les $\varepsilon_{\eta,\xi}$ de (3.17) sont ceux qu'on vient de définir.

L'équation antisymétrique qui change devient, à part un facteur $2\sqrt{-1}\varepsilon\lambda$ que l'on supprime,

$$0 = \sum_{\eta \neq \omega} \varepsilon_{\omega,\eta} t_{\omega,\eta}. \tag{3.20}$$

La somme porte sur η, mais ω reste fixe.

Il y a toutefois aussi une équation antisymétrique pour chaque $\eta \neq \omega$ L'équation pour un η donné est

$$0 = \sum_{\xi | \xi \neq \omega,\eta} \{\varepsilon_{\eta,\xi} t_{\eta,\xi} + \varepsilon\lambda\varepsilon_{\eta,\omega} t_{\eta,\omega}\}.$$

Dans la limite $\varepsilon = 0$ cette équation devient

$$0 = \sum_{\xi \neq \omega,\eta} \varepsilon_{\eta,\xi} t_{\eta,\xi}. \tag{3.21}$$

Considérons ensuite les équations (2.19) à (2.23). Il est convenu que tout terme dans ces équations sauf ceux de degré minimal sera supprimé. Nous avons déjà anticipé que les équations (2.19), (2.20) et (2.21) deviendraient alors des équations linéaires dont il résulte que les $t_{i,\underline{j}}$ ne dépendent que des classes auxquelles appartiennent i et $\underline{j}$ et que $t_{\omega,\eta} = -t_{\eta,\omega}$. Les équations (2.22) deviennent des équations de degré deux. Si aucun des indices n'appartient à ω on a simplement

$$t_{i',\underline{j}} t_{i,\underline{j}'} = t_{i',\underline{j}} t_{i',\underline{j}'} + t_{i,\underline{j}'} t_{i',\underline{j}'}.$$

Si par contre $i' \in \omega$ on obtient

$$\lambda t_{i',\underline{j}} t_{i,\underline{j}'} = \lambda^2 t_{i',\underline{j}} t_{i',\underline{j}'} + \lambda t_{i,\underline{j}'} t_{i',\underline{j}'}.$$

Lorsque $\lambda \to \infty$ on obtient

$$0 = t_{i',\underline{j}} t_{i',\underline{j}'}.$$

Donc si ω, η, et ξ sont tous différents

$$0 = t_{\omega,\eta} t_{\omega,\xi}. \tag{3.22}$$

De ces relations et de la relation (3.20) il découle que (3.22) est aussi valable lorsque $\eta = \xi$.

Il résulte que nous avons les équations pour l'étape antérieure, dans lesquelles ω n'intervient pas de sorte que M devient $M - 1$; ces équations sont supplées par les équations en les variables $t_{\omega,\eta}$ qui sont alors

$$t^2_{\omega,\eta} = 0, \quad \sum_{\eta} t_{\omega,\eta} = 0. \tag{3.23}$$

Donc la variété définit par $\mathfrak{m}'$ devient une variété produit. Nous pouvons calculer la dimension de $\mathfrak{a}/\mathfrak{m}'$ comme le produit de la dimension à l'étape antérieure qui, par l'hypothèse de récurrence, est $(M-2)!$, et de la multiplicité de la variété locale définie par (3.23), qui est évidemment $M - 1$. Le lemme 3.7 s'ensuit.

Soit η le plus grand diviseur commun des entiers μ_i et $\mu_{\underline{j}}$, $i \in E$, $\underline{j} \in F$.

LEMME 3.11. *Le nombre de points sur la variété V comptés avec multiplicité est $\eta\mu(E)^{f-1}\mu(F)^{e-1}$.*

Nous avons défini l'application ϕ d'image X. Nous choisissons un ensemble de $e+f-1$ vecteurs

$$(\lambda_1, \dots, \lambda_e, \lambda_{\underline{1}}, \dots, \lambda_{\underline{f}})$$

qui avec

$$\left(\frac{\lambda_1}{\eta}, \dots, \frac{\lambda_e}{\eta}, \frac{\lambda_{\underline{1}}}{\eta}, \dots, \frac{\lambda_{\underline{f}}}{\eta}\right)$$

forment une base du réseau de vecteurs entiers. Les coordonnées sur X seront

$$\prod_i x_i^{\lambda_i} \prod_{\underline{j}} x_{\underline{j}}^{\lambda_{\underline{j}}}$$

de sorte que X est un produit de $e+f-1$ exemplaires de $\mathbb{P}^1$.

Avant de démontrer le lemme 3.11 nous établirons un lemme simple.

LEMME 3.12. *Le nombre de points dans X tels que $x_i = \rho x_{\underline{j}}$ pour tout i et $\underline{j}$ est $\left|\mu(E) - \mu(F)\right|/\eta$.*

La seule condition sur $x = x_i = \rho x_{\underline{j}}$ est que

$$x^{(\mu(E)-\mu(F))/\eta} = \gamma$$

où γ est un nombre bien déterminé mais générique. Nous rappelons que $\mu(E) \neq \mu(F)$.

Pour établir le lemme 3.11 il suffit de montrer que le degré de l'application ϕ est $\eta\mu(E)^{f-1}\mu(F)^{e-1}$. Observons que nous n'avons jamais défini cette application précisément. En effet pour la définir là où les produits à gauche de (3.4) contiennent des facteurs 0 et des facteurs ∞ il faut des éclatements. A cause du lemme 3.1 leur forme précise n'est guère importante. D'ailleurs le degré est un invariant birationnel.

Puisque le degré est invariant dans une famille d'applications, nous calculons celui de ϕ en le plongeant dans une famille dont le degré du membre générique est facile à calculer. On se souvient que ϕ est donné par des produits

$$x_i = \prod_{uj}\left(\frac{f(r_i, s_{\underline{j}})}{g(r_i, s_{\underline{j}})}\right)^{\mu_{\underline{j}}}, \quad x_{\underline{j}} = \prod_i\left(\frac{f(r_i, s_{\underline{j}})}{g(r_i, s_{\underline{j}})}\right)^{\mu_i}. \tag{3.24}$$

Nous allons prendre

$$f_{i,\underline{j}} = c'(r_i s_{\underline{j}} + d'r_i - d's_{\underline{j}} + a'), \quad g_{i,\underline{j}} = (r_i s_{\underline{j}} + d'r_i - d's_{\underline{j}} + b'), \tag{3.25}$$

de sorte que ϕ s'obtient en prenant $c' = 1$ et en faisant $a', b', d' \to \infty$ de façon que $a'/d' \to a$ et $b'/d' \to b$.

Observons que les nouvelles fonctions sont des combinaisons linéaires de la fonction constante et de

$$(r_i - u)(s_{\underline{j}} + u) \tag{3.26}$$

où $u = d'$. Ces fonctions sont invariantes par rapport à l'application

$$r_i' - u = \lambda(r_i - u), \quad s_{\underline{j}}' + u = \frac{(s_{\underline{j}} + u)}{\lambda} \tag{3.27}$$

qui dans la limite $u \to \infty$, $(1-\lambda)u \to t$ deviennent

$$r_i' = r_i + t, \quad s_{\underline{j}}' = s_{\underline{j}} + t.$$

Donc même avec ces nouvelles fonctions dans les applications (3.4) il y a une des variables r_i ou $s_{\underline{j}}$ qui est superflue.

Nous faisons dégénérer la variété et l'application (ce qui ne change pas le degré) en prenant

$$c' = \infty, \quad d' = a' = 0, \quad \frac{c'}{b'} = 1.$$

L'application (3.24) devient

$$x_i = \prod_{\underline{j}} (r_i s_{\underline{j}})^{\mu_{\underline{j}}}, \quad x_{\underline{j}} = \prod_i (r_i s_{\underline{j}})^{\mu_i}. \tag{3.28}$$

Pour calculer le degré il suffit de trouver un point $(x_i, x_{\underline{j}})$ dont l'image inverse est finie et de calculer le nombre de points qu'elle contient. L'avantage de (3.28) est qu'il s'agit d'une application de tores algébriques et pour des tores le degré se calcule à partir d'une matrice. Puisque $r_i s_{\underline{j}}$ ne change pas si on remplace r_i par λr_i et $s_{\underline{j}}$ par $s_{\underline{j}}/\lambda$ il y a toujours une variable superflue que l'on met égale à 1.

La matrice d'exposants attachée à cette application de tores algébriques est

$$M = \begin{pmatrix} \mu(F) & 0 & \dots & 0 & \mu_{\underline{1}} & \mu_{\underline{2}} & \dots & \mu_{\underline{f}} \\ 0 & \mu(F) & \dots & 0 & \mu_{\underline{1}} & \mu_{\underline{2}} & \dots & \mu_{\underline{f}} \\ \vdots & \vdots & \ddots & \vdots & \vdots & \vdots & & \vdots \\ 0 & 0 & \dots & \mu(F) & \mu_{\underline{1}} & \mu_{\underline{2}} & \dots & \mu_{\underline{f}} \\ \mu_1 & \mu_2 & \dots & \mu_e & \mu(E) & 0 & \dots & 0 \\ \mu_1 & \mu_2 & \dots & \mu_e & 0 & \mu(E) & \dots & 0 \\ \vdots & \vdots & & \vdots & \vdots & \vdots & \ddots & \vdots \\ \mu_1 & \mu_2 & \dots & \mu_e & 0 & 0 & \dots & \mu(E) \end{pmatrix}.$$

Cette matrice à $e+f$ lignes et colonnes est de rang $e+f-1$ car son déterminant est égal à 0. (Notons que le vecteur $(1,\dots,1,-1,\dots,-1)$ avec e composantes 1 et f composantes -1 engendre le noyau de M.) Puisque nous avons mis une des variables r_i ou $s_{\underline{j}}$ égale à 1 nous supprimons une colonne pour obtenir une matrice non singulière. Posons par exemple $s_{\underline{f}} = 1$. En ce qui concerne les lignes nous pouvons prendre des combinaisons linéaires à coefficients entiers pour construire des vecteurs de la forme

$$(0,\dots,0,l,\dots,0,*).$$

Le dernier coefficient est sans importance car nous posons $s_{\underline{f}} = 1$ mais l est un entier positif et tous les autres coefficients sont 0. Par conséquent on peut calculer tous les r_i et $s_{\underline{j}}$ — ou plutôt une puissance de ces nombres — par des produits de la forme

$$\prod_i x_i^{a_i} \prod_{\underline{j}} x_{\underline{j}}^{a_{\underline{j}}}$$

où les exposants sont des entiers positifs ou négatifs. Si tous les x_i et $x_{\underline{j}}$ sont finis et non nuls, alors tous les r_i et $s_{\underline{j}}$ sont aussi finis et non nuls.

Calculons maintenant le degré. Pour cela prenons tous les vecteurs entiers dont une composante est 0 et calculons l'indice de son image (par rapport à l'application linéaire donnée par M) dans l'ensemble des vecteurs entiers $(a_i, b_{\underline{j}})$ tels que

$$\sum_i \mu_i a_i = \sum_{\underline{j}} \mu_{\underline{j}} b_{\underline{j}}. \tag{3.29}$$

Puisque le noyau de M n'est que l'ensemble de vecteurs $(a, \ldots, a, -a, \ldots, -a)$, avec e coefficients égaux à a et f à $-a$ l'espace de vecteurs entiers est la somme directe du noyau et des vecteurs dont une coordonnée donnée est 0. Donc nous pouvons appliquer M à tout l'espace pour la traiter comme une application d'un premier espace $\mathfrak{X}$, obtenu en divisant l'espace des vecteurs entiers par le noyau de M, dans un second espace $\mathfrak{Y}$ défini par (3.29). On a donc un plongement

$$M\colon \mathfrak{X} \hookrightarrow \mathfrak{Y},$$

dont nous désirons calculer l'indice $[\mathfrak{Y}\colon \mathfrak{X}]$ qui est égal au degré recherché. Nous allons maintenant définir deux espaces $\mathfrak{X}'$ et $\mathfrak{Y}'$ tels que les flèches du diagramme suivant soient des plongements :

$$\begin{array}{ccc} \mathfrak{X} & \longrightarrow & \mathfrak{Y} \\ \uparrow & & \uparrow \\ \mathfrak{X}' & \longrightarrow & \mathfrak{Y}' \end{array}$$

Il en résultera que

$$[\mathfrak{Y}\colon \mathfrak{X}] = \frac{[\mathfrak{Y}\colon \mathfrak{Y}'][\mathfrak{Y}'\colon \mathfrak{X}']}{[\mathfrak{X}\colon \mathfrak{X}']}.$$

Soit $\mathfrak{Y}' = \mathfrak{Y}_1 + \mathfrak{Y}_2$ la somme directe de $\mathfrak{Y}_1$, l'ensemble de vecteurs tels que

$$\sum \mu_i a_i = \sum \mu_{\underline{j}} b_{\underline{j}} = 0$$

et de $\mathfrak{Y}_2$, l'ensemble de ceux tels que $a_i = a\mu(F)$ et $b_{\underline{j}} = a\mu(E)$ avec $a \in \mathbb{Z}$. A cause de l'équation (3.29), pour les vecteurs dans $\mathfrak{Y}$ la somme $\sum \mu_i x_i$ est divisible par le plus grand diviseur commun μ de tous les μ_i aussi bien que par le plus grand diviseur commun ν de tous les $\mu_{\underline{j}}$ et donc par $\mu\nu/\eta$. Il en résulte que

$$[\mathfrak{Y}\colon \mathfrak{Y}'] = \frac{\mu(E)\mu(F)}{(\mu\nu/\eta)} = \frac{\eta\mu(E)\mu(F)}{\mu\nu}.$$

Soit $\mathfrak{X}' = \mathfrak{X}_1 + \mathfrak{X}_2$ la somme directe de $\mathfrak{X}_1$, l'ensemble de vecteurs $(a_i, b_{\underline{j}})$ tels que

$$\sum \mu_i a_i = \sum \mu_{\underline{j}} b_{\underline{j}} = 0,$$

et de $\mathfrak{X}_2$, l'ensemble de ceux tels que a_i sont tous égaux au même entier a et tous les $b_{\underline{j}}$ au même entier b. L'espace $\mathfrak{X}_1$ est le noyau dans $\mathfrak{X}$ de l'application

$$(a_i, b_{\underline{j}}) \to \Bigl(\sum \mu_i a_i, \sum \mu_{\underline{j}} b_{\underline{j}}\Bigr)$$

L'image de $\mathfrak{X}$ est l'ensemble des (a, b) tels que $\mu|a$ et $\nu|b$ et celle de $\mathfrak{X}'$ l'ensemble de $\bigl(\mu(E)a, \mu(F)b\bigr)$, $a, b \in \mathbb{Z}$. Par conséquent $[\mathfrak{X}\colon \mathfrak{X}'] = \mu(E)\mu(F)/\mu\nu$.

On a $\mathfrak{Y}_1 \supset \mathfrak{X}_1$ et $\mathfrak{Y}_2 \supset \mathfrak{X}_2$ de sorte que $[\mathfrak{Y}'\colon \mathfrak{X}'] = [\mathfrak{Y}_1\colon \mathfrak{X}_1][\mathfrak{Y}_2\colon \mathfrak{X}_2]$. Évidemment $[\mathfrak{Y}_1\colon \mathfrak{X}_1] = \mu(E)^{f-1}\mu(F)^{e-1}$. Le plongement de $\mathfrak{X}_2$ dans $\mathfrak{Y}_2$ est

$$(a, b) \to \bigl(\mu(F)(a+b), \mu(E)(a+b)\bigr)$$

de sorte que $[\mathfrak{Y}_2\colon \mathfrak{X}_2] = 1$. Il s'ensuit que

$$[\mathfrak{Y}\colon \mathfrak{X}] = \mu(E)^{f-1}\mu(F)^{e-1}\eta,$$

qui est la formule voulue.

4. Les paires libres

La variété algébrique V était attachée à une paire E et F dont les éléments étaient supposés être $1, \dots, e$ et $\underline{1}, \dots, \underline{f}$. Nous supposons maintenant que E et F sont tous les deux des sous-ensembles de $\{1, \dots, r\}$ et nous dénotons V par $V(E, F)$.

Supposons que $D_1, \dots, D_l$ sont des sous-ensembles disjoints de $\{1, \dots, r\}$ et que D_k, $1 \le k \le l$, est la réunion de deux sous-ensembles disjoints E_k et F_k. Pour le moment nous supposons pour simplifier la notation que $\{1, \dots, r\}$ est la réunion des ensembles D_k. Alors, pour chaque D_k et chaque choix des paramètres c_k et d_k (qui remplacent c et d) les équations (3.1) avec $i \in E_k$ et $\underline{j} \in F_k$ définissent une variété $V(E_k, F_k)$. Le produit des variétés $V(E_k, F_k)$ est presqu'une sous-variété de la fibre C de la correspondance en $\Delta = 0$. Pour la plonger dans la correspondance il faut trouver des coordonnées z_{E_k} et z_{F_k} telles que $c_k = R(z_{E_k})$ et $d_k = R(z_{F_k})$. Pour que cela soit possible il faut et il suffit que $c_k = d_k\rho$. Disons que le produit des variétés est une sous-variété potentielle de la correspondance à $\Delta = 0$. Observons qu'il y a plusieurs plongements du produit dans la correspondance selon le choix des coordonnées z_{E_k} et z_{F_k}.

Soit p un point fixe de la correspondance et supposons en profitant du découplage du chapitre 2 qu'il n'y ait qu'un seul ensemble $D = \{1, \dots, r\}$ dans la partition attachée à p. Pour que p appartienne à l'image d'un plongement du produit

$$\prod_k V(E_k, F_k)$$

dans la correspondance à $\Delta = 0$, il faut d'abord que la partition $\{D_k\}$ soit plus grossière que celle attachée à l'équivalence fine définie par p, et alors il suffit que

$$c_k = R(z_E), \quad d_k = \frac{R(z_E)}{\rho}.$$

Il en résulte immédiatement un lemme que nous signalons explicitement.

LEMME 4.1. *Si un point de l'image d'un plongement dans la variété C du produit $\prod_k V(E_k, F_k)$ est un point fixe de la correspondance, alors tous les points dans les images de tous les plongements sont aussi des points fixes.*

La multiplicité du point p dans le produit est le produit des multiplicités et nous pouvons la majorer à l'aide du lemme 3.7.

LEMME 4.2. *Soit m_k le nombre de classes d'équivalence fine par rapport à p dans D_k. Alors la multiplicité de p dans $\prod_k V(E_k, F_k)$ est au plus $\prod_k (m_k - 1)!$*

Le point $p = \prod_k v_k$ et $m(v_k)$ est évidemment égal à m_k.

Il est possible qu'il y ait plusieurs décompositions $\{D_1, \ldots, D_l\}$ de $\{1, \ldots, r\}$ telles que p appartiennent à $\prod_k V(E_k, F_k)$.

LEMME 4.3. *Soit $n(p) = n(p, D)$ le nombre de classes d'équivalence fines par rapport à p dans l'ensemble D. Pour chaque décomposition $\{D_1, \ldots, D_l\}$ de $D = \{1, \ldots, r\}$ et chaque point p qui appartient au produit $\prod_k V(E_k, F_k)$, soit*

$$m(p; D_1, \ldots, D_l)$$

la multiplicité de p dans le produit. Alors pour p donné la somme sur toutes ces décompositions de cette multiplicité satisfait

$$\sum_{\{D_1, \ldots, D_l\}} m(p; D_1, \ldots, D_l) \leq n(p)!$$

Soit $n = n(p)$. D'après ce que l'on a déjà dit la somme de ces multiplicités est au plus la somme de $\prod_k (|C_k| - 1)!$ sur toutes les décompositions de n en sous-ensembles C_k disjoints. Le résultat de cette somme purement combinatoire est donnée par le lemme suivant.

LEMME 4.4. *Pour tout entier positif n, la somme sur toutes les partitions de $\{1, \ldots, n\}$ en des sous-ensembles disjoints $C_1, \ldots, C_l$ de $\prod(|C_k| - 1)!$ est $n!$*

Ce lemme est évident pour $n = 1, 2$. On le vérifie par récurrence. Comme d'habitude on construit les partitions de $\{1, \ldots, n\}$ en ajoutant n à un des sous-ensembles d'une partition $C_1, \ldots, C_k$ de $\{1, \ldots, n-1\}$ ou en ajoutant $\{n\}$ à une telle partition. Donc on multiplie $\prod(|C_k| - 1)!$ par $\sum |C_k|$ et par 1. Puisque la somme de ces deux nombres est n le lemme en résulte.

Avant de revenir aux structures combinatoires de [**BL**] nous vérifions un dernier lemme. Supposons que la paire (E, F) qui donne D soit libre. Nous cherchons des points fixes de la correspondance autour de p qui se développent en séries de Puiseux en Δ dont les exposants ne sont pas trop compliqués. En fait nous les supposerons demi-entiers de sorte que les séries seront des séries de Taylor en $\Delta^{1/2}$. Donc pour $i \not\leftrightarrow \underline{j}$

$$\begin{aligned} t_{i,\underline{j}} &= a(t_{i,\underline{j}})\Delta^{1/2} + b(t_{i,\underline{j}})\Delta + \ldots, \\ z_i &= z_i^0 + a(z_i)\Delta^{1/2} + b(z_i)\Delta + \ldots, \\ \tilde{z}_{\underline{j}} &= \tilde{z}_{\underline{j}}^0 + a(\tilde{z}_{\underline{j}})\Delta^{1/2} + b(\tilde{z}_{\underline{j}})\Delta + \ldots. \end{aligned} \tag{4.1.a}$$

Les $v_{i,\underline{j}}$, $i \leftrightarrow \underline{j}$, seront par contre des séries en Δ lui-même

$$v_{i,\underline{j}} = v_{i,\underline{j}}^0 + a(v_{i,\underline{j}})\Delta + \ldots \tag{4.1.b}$$

En général pour n'importe quelle coordonnée x qui se développe en série de Puiseux nous posons $x = x^0 + a(x)\Delta^{\alpha(x)} + \dots$, où $\alpha(a)$ est censé être la plus petite puissance positive qui intervient avec un coefficient $a(x) \neq 0$.

LEMME 4.5. *Il existe au moins* $n(p)!$ *solutions distinctes de la forme* (4.1) *des équations pour les points fixes.*

Ici comme par la suite les énoncés sur l'existence de solutions en séries de Puiseux de telle ou telle équation ne sont pas démontrés d'une fa¸con complète. Donc le caractère informel de cet exposé qui était annoncé au début se révèle enfin. Nous n'examinons que les premiers termes.

Nous supposons que $a(t_{i,\underline{j}}) \neq 0$. Il résulte des équations (2.19), (2.20) et (2.21) que $a(t_{i,\underline{j}})$ ne dépend que des classes d'équivalence fine auxquelles appartiennent i et $\underline{j}$ de sorte que l'on écrit $a(t_{i,\underline{j}}) = 1/\mu_{\omega,\eta}$ avec $\mu_{\omega,\eta} = -\mu_{\eta,\omega}$. (Ces facteurs $\mu_{\omega,\eta}$ sont étroitement liés aux facteurs $\mu_{i,\underline{j}}$ de l'équation (2.12).) Il résulte de ces mêmes équations que si i et $\underline{j}$ sont équivalents dans le sens fin alors $a(z_i) = a(\tilde{z}_{\underline{j}})$ de sorte que l'on peut poser $a(z_i) = a(\tilde{z}_{\underline{j}}) = \nu_\omega$, $i \in \omega$, $\underline{j} \in \omega$. Les équations (2.12) donnent enfin

$$\nu_\omega - \nu_\eta = \mu_{\omega,\eta}, \quad \text{pour } \omega \neq \eta. \tag{4.2}$$

Il y a d'autres équations de déformation qui proviennent des équations (2.30). Puisqu'aucun $R(z_i)$ n'est 0 en p on calcule plutôt les dérivées logarithmiques. Soient

$$A' = \frac{1}{R(z)}\frac{dR(z)}{dz}, \quad z = z_E, \quad \Delta = 0,$$

$$B' = -\frac{1}{R(z)}\frac{dR(z)}{d\tilde{z}} = \frac{1}{R(\tilde{z})}\frac{dR(\tilde{z})}{d\tilde{z}} = A', \quad z = z_F = \tilde{z}_E, \quad \Delta = 0,$$

$$C = \frac{1}{R}\frac{dR}{d\Delta} = +2\sum_{k=1}^{N}\frac{1}{z-\beta_k}, \quad z = z_E, \quad \Delta = 0,$$

$$C' = -\frac{1}{R}\frac{dR(z)}{d\Delta} = -2\sum_{k=1}^{N}\frac{1}{z-\beta_k}, \quad z = z_F = \tilde{z}_E, \quad \Delta = 0.$$

Nous supposons que les α_k sont des constantes de sorte que les $\beta_k = 2\Delta - 1/\alpha_k$ sont des fonctions linéaires de Δ.

Les équations de déformation sont pour $a = -b$

$$A'\delta z_i + C\Delta = -2a\sum_{\underline{j}\leftrightarrow i}\frac{\mu_i\mu_{\underline{j}}\delta v_{i,\underline{j}}}{v_{i,\underline{j}}^2 - a^2} + 2a\sum_{\underline{j}\not\leftrightarrow i}\frac{\mu_i\mu_{\underline{j}}\delta t_{i,\underline{j}}}{1-a^2},$$

$$A'\delta\tilde{z}_{\underline{j}} + C'\Delta = -2a\sum_{i\leftrightarrow\underline{j}}\frac{\mu_i\mu_{\underline{j}}\delta v_{i,\underline{j}}}{v_{i,\underline{j}}^2 - a^2} + 2a\sum_{i\not\leftrightarrow\underline{j}}\frac{\mu_i\mu_{\underline{j}}\delta t_{i,\underline{j}}}{1-a^2},$$

car $\delta\Delta = \Delta$. Dans ces équations nous remplaçons δz_i et $\delta\tilde{z}_{\underline{j}}$ par $\nu_\omega\Delta^{1/2}$ si ω est la classe de i ou de $\underline{j}$.

De ces équations nous déduisons d'abord les équations antisymétriques dont nous ne gardons que les termes en $\Delta^{1/2}$. Donc nous fixons ω et nous prenons la différence entre la somme des premières équations sur tout $i \in \omega$ et la somme des secondes équations sur tout $\underline{j} \in \omega$ ne gardant que les termes en $\Delta^{1/2}$ déduits de (4.1). Rappelons d'abord que les entiers e et f qui interviennent dans la formule (2.34) sont devenus $\mu(E)$ et $\mu(F)$. Les entiers e_ω et f_ω de (2.37) sont donnés par

$$e_\omega = \sum_{i\in\omega} \mu_i, \quad f_\omega = \sum_{\underline{j}\in\omega} \mu_{\underline{j}}.$$

On se souvient que $e_\omega = m_\omega \mu(E)/m$ et $f_\omega = m_\omega \mu(F)/m$. Puisque $\delta v_{i,\underline{j}}$ n'a pas de terme en Δ, le côté gauche de l'équation ainsi obtenu est

$$A'(e_\omega - f_\omega)\nu_\omega = A' m_\omega \frac{\mu(E) - \mu(F)}{m}.$$

Le côté droit est

$$\frac{4a}{1-a^2} \sum_{\eta\neq\omega} \frac{e_\omega f_\eta}{\nu_\omega - \nu_\eta} = \frac{4a}{1-a^2} \frac{m_\omega}{m^2} \sum_{\eta\neq\omega} \frac{m_\eta}{\nu_\omega - \nu_\eta}.$$

Donc les équations antisymétriques sont

$$A' m_\omega m\big(\mu(E) - \mu(F)\big)\nu_\omega = \frac{4a}{1-a^2} \mu(E)\mu(F) m_\omega \sum_{\eta\neq\omega} m_\eta \frac{1}{\nu_\omega - \nu_\eta}.$$

Rappelons que $\mu(E) \neq \mu(F)$. Un facteur $1/m^2$ a été supprimé aux deux côtés.

Nous nous contentons de la preuve que ces équations en les $n(p)$ variables ν_ω ont $n(p)!$ solutions. Si on remplace chaque ν_ω par $\alpha\nu_\omega$, le coefficient $A'\big(\mu(E) - \mu(F)\big)$ à gauche est multiplié par α^2 ; ainsi, pour étudier les solutions de ces équations, nous pouvons supprimer ce coefficient aussi bien que le coefficient $4a\mu(E)\mu(F)/(1-a^2)$ qui intervient à droite. L'équation qui reste est

$$m_\omega \nu_\omega = m_\omega \sum_{\eta\neq\omega} m_\eta \frac{1}{\nu_\omega - \nu_\eta}. \tag{4.3}$$

LEMME 4.6. *Si $n = n(p)$ est le nombre de classes, alors l'équation* (4.3) *possède au moins $n!$ solutions.*

Nous ordonnons les classes d'une façon arbitraire, $\omega_1, \dots, \omega_n$. Il suffit de montrer qu'il y a une solution dans la région

$$\nu_{\omega_1} > \nu_{\omega_2} > \cdots > \nu_{\omega_n}. \tag{4.4}$$

Considérons la fonction

$$e^{-2\sum_\rho m_\rho \nu_\rho^2} \prod_{\substack{\omega,\eta\\ \omega\neq\eta}} (\nu_\omega - \nu_\eta)^{2m_\omega m_\eta}$$

dans la fermeture de l'ensemble (4.4). Elle est 0 à la frontière. En même temps il est évident qu'elle est très petite en dehors d'une grande boule centrée à l'origine. Il en résulte qu'elle atteint une valeur maximale à un point où toutes les dérivées de son logarithme seront 0. Ces dérivées sont

$$4\sum_\eta \frac{m_\omega m_\eta}{\nu_\omega - \nu_\eta} - 4m_\omega\nu_\omega.$$

Chaque point de valeur maximale donne alors une solution des équations (4.3).

Observons en particulier que d'une solution ν_ω on obtient une autre en remplaçant tous les ν_ω par $-\nu_\omega$. S'il est le cas qu'il y a exactement $n!$ solutions il s'ensuit que chaque solution est telle que si les valeurs sont ordonnées comme dans (4.4) alors $\nu_{\omega_k} = -\nu_{\omega_l}$ si $k = n - l + 1$.

5. La structure combinatoire

Nous avons esquissé une description de la structure combinatoire des points fixes autour de $\Delta = 0$ dans [**AG**]. Les équations pertinentes de ce papier, corrigées dans l'appendice 3, sont (12), (14) et (15). La structure combinatoire que ces équations expriment est déjà présente dans les descriptions des points fixes à $\Delta = 0$ introduites au chapitre 2.

A chaque point fixe p est attachée une première décomposition $\{D_1, \dots, D_m\}$ de $\{1, \dots, r\}$ qui est définie à partir des valeurs des coordonnées $z_i(p)$, $D_k = E_k \cup F_k$ et E_k est par exemple l'ensemble de tous les indices i tels que $z_i(p)$ a une valeur donnée z_{E_k}. Rappelons que le lemme 2.2 permettait d'introduire, dans chaque ensemble D_k, une équivalence fine. Nous considérons des décompositions $\{D'_1, \dots, D'_n\}$ qui sont plus fines que $\{D_1, \dots, D_m\}$. Donc chaque D_k est la réunion des D'_l qu'il contient. De plus chaque D'_l est lui-même la réunion des classes d'équivalence fine qu'il contient. Évidemment $n \geq m$. Soit $D'_l = E'_l \cup F'_l$.

Nous savons attacher à chaque paire (E'_l, F'_l) une variété $V(E'_l, F'_l)$ définie à partir d'équations semblables à (3.1) mais avec $c = c'_l$ et $d = d'_l$. Pour qu'un point du produit $\prod_l V(E'_l, F'_l)$ appartienne à la variété de points fixes de la correspondance, il faut et il suffit que $c'_l = d'_l \rho$ pour tout l et alors chaque point du produit y appartient.

Selon les lemmes (3.11) et (3.12), la somme sur tous les c'_l et d'_l possibles du nombre de points sur le produit est

$$\prod_l \left|\mu(E'_l) - \mu(F'_l)\right| \mu(E'_l)^{f'_l - 1} \mu(F'_l)^{e'_l - 1}. \tag{5.1}$$

Mais pour définir un plongement du produit dans la correspondance il faut choisir les $z_{E'_l}$ et $z_{F'_l}$. Puisque $z_{F'_l} = \tilde{z}_{E'_l}$ et $R(z_{E'_l}) = c'_l$ il y a N^n choix supplémentaires si n est le nombre d'éléments D'_l dans la décomposition.

L'expression (5.1) multipliée par N^n est le produit des termes

$$N\left|\mu(E'_l) - \mu(F'_l)\right| \mu(E'_l)^{f'_l - 1} \mu(F'_l)^{e'_l - 1} \tag{5.2}$$

et correspond donc à la formule (12) de [**AG**].

Il y a cependant dans [**AG**] une imprécision provenant de l'incompréhension des auteurs de leur propre formule. Les deux décompositions $\{D_1, \dots, D_m\}$ et $\{D'_1, \dots, D'_n\}$ ne sont pas toujours les mêmes car il est bien possible que $z_{E'_k} = z_{E'_l}$ pour $k \neq l$. Un lecteur de [**AG**] pourrait penser que dans (12) il s'agit de la première décomposition et non de la seconde. En fait, il ne s'agit ni de l'une ni de l'autre. Il s'agit d'une décomposition intermédiaire, celle qui définit la variété produit du chapitre 4.

La même confusion est possible avec (14) et (15), mais avant de passer à ces deux formules, observons que les formules (5.1) et (5.2) proviennent des formules pour le nombre de points sur les variétés $V(E, F)$, chacun compté avec sa multiplicité. Sous divers plongements des produits de ces variétés plusieurs points peuvent avoir la même image dans la correspondance, mais si nous comptons le nombre de solutions en séries de Puiseux qui émergent de cette image nous trouvons selon les lemmes (4.3) et (4.5) au moins le nombre prédit par la formule (12). Nous obtiendrons le même résultat pour les formules (14) et (15). Il en résulte qu'en comptant toutes les solutions trouvées autour de $\Delta = 0$ sans tenir compte de leur multiplicité nous obtiendrons un nombre n_2 qui est *au moins* le nombre n_3 donné par (12), (14) et (15) en suivant la procédure combinatoire décrite dans [**AG**] ; mais ce dernier nombre n_3 est celui prédit par la formule de Lefschetz, qui compte les solutions avec multiplicité. Mais n_2 est certainement plus petit ou égal au nombre total n_1 de solutions car, en trouvant n_2, il est bien possible qu'il y ait des solutions restées inaperçues. Donc $n_1 \geq n_3$ mais n_3 compte les solutions avec multiplicité de sorte que n_3 est nécessairement plus grand ou égal à n_1. Il en résulte que $n_1 = n_3$ et que tous nos résultats antérieurs admettent des précisions. En particulier le corollaire 3.8 est valable.

Il serait utile en traitant les équations (14) et (15) pour les paires liées d'introduire quelque notation supplémentaire. Pour une paire $\{E, F\}$ sans sommet interdit dans le sens de [**AG**], que nous distinguons de la paire $\{F, E\}$ de sorte que le facteur 2 dans (14) s'enlève, nous introduisons

$$\chi(E,F) = \min\{\mu(E), \mu(F)\}\mu(E)^{f-1}\mu(F)^{e-1}. \tag{5.3}$$

Pour distinguer les deux ordres possibles nous convenons que la valeur commune z_E de z_i, $i \in E$ est un zéro de R. Observons tout de suite que (5.3) est 0 si un des deux ensembles E et F est vide.

Si une paire contient un sommet interdit nous la dénotons $\{E, F, B\}$ et convenons que les éléments de B' sont dans E et ceux de B'' dans F. Pour une telle paire nous introduisons

$$\chi(E,F,B) = \big(\mu(E)\omega'' + \mu(F)\omega' + \omega'\omega''\big)\big(\mu(E)+\omega'\big)^{f-1}\big(\mu(F)+\omega''\big)^{e-1}. \tag{5.4}$$

Dans la formule (15) de [**AG**] il y a une erreur que la formule (5.4) corrige. Supposons que D contienne un sommet interdit $B = \{B', B''\}$ aux poids $\omega' = |B'|$ et $\omega'' = |B''|$ et soit $\omega = \omega' + \omega''$. Soit $\{A_1, \dots, A_{d-1}\}$ l'ensemble des autres sommets aux poids $\mu_i = |A_i|$. Puisque D contient d éléments alors le poids attaché à D par le lemme de Fan Chung [**AG**] est $\omega\mu(D)^{d-2}$ avec $\mu(D) = \omega + \sum_i \mu_i$. Ce poids apparaît à gauche dans la formule (15) de [**AG**] et il nous faut changer la somme à droite qui porte sur toutes les partitions $\{E, F\}$ de $\{A_1, \dots, A_{d-1}\}$ car telle quelle la formule n'est malheureusement pas correcte. La formule correcte suit. Soient comme d'habitude $e = |E|$ et $f = |F|$.

LEMME 5.1.

$$\omega\mu(D)^{d-2} = \sum\big(\mu(E)\omega'' + \mu(F)\omega' + \omega'\omega''\big)\big(\mu(E)+\omega'\big)^{f-1}\big(\mu(F)+\omega''\big)^{e-1}. \tag{5.5}$$

Si $d = 1$ alors le côté gauche n'est que $\omega\omega^{-1} = 1$ et à droite il n'y a que la partition avec E et F vides. Donc $\mu(E) = \mu(F) = 0$ et le côté droit est aussi 1. Supposons donc que $d > 1$.

Il est plus convenable d'écrire le côté droit de l'équation comme

$$\sum\big\{(\mu(E)+\omega')^f\big(\mu(F)+\omega''\big)^e - \mu(E)\mu(F)\big(\mu(E)+\omega'\big)^{f-1}\big(\mu(F)+\omega''\big)^{e-1}\big\}.$$

Si $d = 2$ il n'y a que deux possibilités à droite : soit $e = 1$ et $f = 0$; soit $e = 0$ et $f = 1$. La seconde expression de chaque terme est 0 et l'égalité devient

$$\omega = \omega' + \omega''$$

qui s'avère exact. Supposons donc que $d > 2$ et procédons par récurrence.

Les deux côtés de l'égalité à vérifier sont des polynômes en $\mu_i = |A_i|$, ω' et ω'' de degré $d-1$. Puisque leur différence s'annule si $\omega' = \omega'' = 0$, il suffit de vérifier que le produit $\prod_1^{d-1} \mu_i$ la divise, donc qu'elle s'annule si $\mu_i = 0$ pour un i donné.

On peut appliquer l'identité à l'ensemble de variables μ_j, $j \neq i$, ω' et ω''. Puisque $\mu_i = 0$, le côté gauche de (5.5) devient $\omega(\omega + \sum_i \mu_j)^{d-3} = \omega\mu(D)^{d-3}$. En le multipliant par $\omega + \sum_j \mu_j$ on obtient le côté gauche de (5.5). A toute partition $\{E_1, F_1\}$ de $\{1, \ldots, i-1, i+1, \ldots, d-1\}$ correspond deux partitions $\{E, F\}$ de $\{1, \ldots, d-1\}$: on ajoute i soit à E_1 soit à F_1. On obtient de cette façon chaque partition $\{E, F\}$. La somme sur ces deux partitions de $\big(\mu(E) + \omega'\big)^f \big(\mu(F) + \omega''\big)^e$ n'est que

$$\mu(D)\big(\mu(E_1) + \omega'\big)^{f_1}\big(\mu(F_1) + \omega''\big)^{e_1}.$$

Le même argument s'applique au second terme

$$\mu(E)\mu(F)(\mu(E) + \omega')^{f-1}(\mu(F) + \omega'')^{e-1}$$

de sorte que les deux côtés de l'identité se multiplie par $\mu(D)$ en passant de l'ensemble $\{1, \ldots, i-1, i+1, \ldots, d-1\}$ à l'ensemble $\{1, \ldots, d-1\}$ pourvu que la variable μ_i soit 0. Le lemme en résulte.

La confusion déjà mentionnée provient du fait qu'il est parfois nécessaire, pendant le compte, de décomposer une paire (E, F) définie par les valeurs des z_i au point fixe en deux paires (E', F') et (E'', F''). Nous avons su tenir compte de cette possibilité pour des paires libres. Sa cause est même évidente pour des paires liées car alors le facteur N attaché à chaque paire provient de l'existence des N zéros de R. Mais il est bien possible de choisir le même zéro pour E' que pour E'' de sorte qu'ils se confondent dans le compte naïf. Heureusement les paires libres et les paires liées ne se confondent pas de sorte qu'il est permis en traitant les paires libres de les découpler des paires liées comme nous avons fait implicitement.

Puisque nous n'avons pas encore d'arguments généraux pour les paires liées, nous ne tenons compte que d'une façon explicite et maladroite de la possibilité des paires diagonales. A $\Delta = 0$ tous les indices i tels que z_i est un zéro donné α de R sont mis ensemble pour obtenir E, et alors F est l'ensemble de j tels que $z_j = \beta = \tilde{\alpha}$. S'il n'y pas de sommet interdit, alors il faut examiner toutes les décompositions possibles $\big\{(E_1, F_1), \ldots, (E_m, F_m)\big\}$ de (E, F) et vérifier qu'il y a

$$\sum \prod \chi(E_k, F_k) \tag{5.6}$$

solutions en séries de Puiseux qui émergent du point fixe à $\Delta = 0$ aux coordonnées $z_i = \alpha$, $i \in E$, et $z_j = \beta$, $j \in F$, au moins dans le cas que $E \cup F = \{1, \ldots, r\}$. Il suffit de traiter ce cas car les indices attachées aux zéros différents de α ou aux paires libres sont certainement découplés de ceux attachés à α et β. La somme porte sur toutes les décompositions possibles.

S'il y a un sommet interdit B nous définissons E et F de la même façon sauf que nous écartons les indices i qui appartiennent à B. Dans ce cas il y a une paire préférée (E_1, F_1) dans chaque décomposition, celle à laquelle sont attachés les éléments de B. Cependant il n'est pas exclu que E ou F soit vide ou même qu'ils soient tous les deux vides. Le compte pour la paire $\{E, F\}$ doit donner

$$\sum \chi(E_1, F_1, B) \prod_{i>1} \chi(E_i, F_i) \tag{5.7}$$

où le produit porte sur toutes les décompositions simultanées de E et de F.

Nous ne sommes pas en état de traiter les problèmes autour des formules (5.6) et (5.7) d'une façon générale. Il faut nous contenter d'exemples assez simples et même, pour ces exemples, de n'examiner que les premiers terme des séries de Puiseux qui y apparaissent. Nous ne vérifions pas que les séries elles-mêmes existent.

6. Les paires liées sans sommet interdit

Dans ce chapitre nous examinons les séries de Puiseux pour les paires liées qui n'ont pas de sommets interdits, donc des paires telles qu'aucune des coordonnées z_i, $i \in E \cup F$, n'est contrainte à demeurer un zéro ou un pôle de R même lors de la déformation. Cependant, à $\Delta = 0$, toutes les coordonnées sont égales soit à un zéro donné de R soit au pôle correspondant. La formule pertinente est (2.40) dans laquelle nous remplaçons A et B par $(-1)^{r-1}A$ et $(-1)^{r-1}B$ pour alléger la notation. Les éléments de F seront dénotés $\underline{j}$. Les équations deviennent

$$A\Delta p_i = \prod_{\underline{j} \in F} S(p_i, p_{\underline{j}})^{\mu_{\underline{j}}}, \quad B\Delta p_{\underline{j}} = \prod_{i \in E} S(p_i, p_{\underline{j}})^{\mu_i}. \tag{6.1}$$

Les poids ne sont qu'implicites dans (2.40). Nous ne traitons que quelques cas simples de ces équations, qui ne sont d'ailleurs que des équations simplifiées par rapport aux vraies équations de déformation. La valeur de a dans la formule (2.39) n'est guère importante car les degrés d'homogénéité des deux côtés sont différents, et nous supposons que $a = -1$. Nous soulignons enfin que l'argument global est qu'il suffit de trouver *assez* de solutions ! Il s'avérera alors à la toute fin de l'argument que nous les avons toutes trouvées. Pour faire le compte nous trouvons les parties dominantes possibles $x^0 + a(x)\Delta^\alpha$ des solutions en séries de Puiseux mais nous acceptons qu'il est possible de trouver par récurrence les termes suivants.

A : e=f=1. Soient $\mu = \mu_1$ et $\nu = \nu_1$. La somme (5.6) ne contient qu'un seul terme et si on utilise une transformation simple ($p \to -ap$, $q \to -aq$) pour éliminer $-a$ de (2.39), les équations simplifiées (2.40) deviennent

$$A\Delta p_1 = \left(\frac{p_1 + p_{\underline{1}}}{p_1 + p_{\underline{1}} + 1}\right)^\nu, \quad B\Delta p_{\underline{1}} = \left(\frac{p_1 + p_{\underline{1}}}{p_1 + p_{\underline{1}} + 1}\right)^\mu,$$

avec de nouvelles constantes A et B. Il y a deux types de solutions de ces équations en série de Puiseux en puissances de Δ. D'abord $p_1 \equiv p_{\underline{1}} \equiv 0$ (c'est-à-dire en fonction de Δ) donne une seule solution.

Soit autrement $\alpha = \alpha(p_1)$, $\beta = \alpha(p_{\underline{1}})$ et $\gamma = \alpha(p_1 + p_{\underline{1}})$ les exposants de p_1, $p_{\underline{1}}$ et $p_1 + p_{\underline{1}}$ définis selon nos conventions. Évidemment $\gamma > 0$ de sorte que $p_1^0 = -p_{\underline{1}}^0$. On a $1 + \alpha = \nu\gamma$ et $1 + \beta = \mu\gamma$. Supposons que $\nu \leq \mu$ de sorte que $\chi(E, F) = \nu$.

Si $\nu < \mu$ alors $\beta > \alpha$ de sorte que $\gamma = \alpha = 1/(\nu - 1)$. En comparant les coefficients des termes du plus bas degré dans la première des équations on obtient

$$Aa(p_1) = a(p_1)^\nu.$$

Puisque la solution $a(p_1) = 0$ est exclue par la définition même de $a(p_1)$, il y a $\nu - 1$ solutions admises. Elles sont différentes. L'équation pour $a(p_{\underline{1}})$ est alors

$$Ba(p_{\underline{1}}) = a(p_1)^\mu$$

qui détermine $a(p_{\underline{1}})$. Par conséquent nous obtenons $\nu - 1 + 1 = \chi(E, F)$ solutions.

Si $\mu = \nu$ alors $\alpha = \beta$. En divisant les deux équations on obtient une équation pour $p_1/p_{\underline{1}}$ dont -1 n'est pas une solution puisque A/B est générique. Par conséquent $\gamma = \alpha = \beta = 1/(\nu - 1)$. On obtient à nouveau ν solutions.

B : f=1. Soit $\nu = \nu_{\underline{1}}$. Puisque $\chi(E, \varnothing) = 0$ pour tout E, il n'y a toujours pas de décomposition à examiner et il suffit de trouver

$$\chi(E, F) = \min\left\{\sum \mu_i, \nu\right\}\nu^{e-1}$$

solutions.

Soit, lorsqu'ils sont définis, $\alpha_i = \alpha(p_i)$, $\beta = \alpha(p_{\underline{1}})$ et $\gamma_i = \alpha(p_i + p_{\underline{1}})$. Commençons par les solutions avec $p_{\underline{1}} \equiv 0$. Soit R l'ensemble des i tels que $p_i \not\equiv 0$. A cause de l'équation

$$B\Delta p_{\underline{1}} = \prod_i \left(\frac{p_i + p_{\underline{1}}}{p_i + p_{\underline{1}} + 1}\right)^{\mu_i}$$

l'ensemble R n'est pas tout l'ensemble E. Pour $i \in R$ on a $1 + \alpha = \gamma$ et l'équation pour $a(p_i)$ est

$$Aa(p_i) = a(p_i)^\nu,$$

qui possède $\nu - 1$ solutions à part la solution 0. On obtient donc au total

$$\sum_R (\nu - 1)^{|R|} = \nu^e - (\nu - 1)^e,$$

solutions car la somme porte sur tout sous-ensemble R de $\{1, \ldots, e\}$ sauf l'ensemble lui-même.

Supposons maintenant que $p_{\underline{1}} \not\equiv 0$ de sorte qu'aucun p_i n'est identiquement nul. Nous répartissons l'ensemble des indices en quatre :

$$\begin{aligned} R &= \{i \mid \alpha_i = \gamma_i < \beta\}, \\ S &= \{i \mid \alpha_i = \beta < \gamma_i\}, \\ T &= \{i \mid \beta = \gamma_i < \alpha_i\}, \\ U &= \{i \mid \alpha_i = \gamma_i = \beta\}. \end{aligned}$$

Des équations

$$A\Delta p_i = \left(\frac{p_i + p_{\underline{1}}}{p_i + p_{\underline{1}} + 1}\right)^\nu, \quad B\Delta p_{\underline{1}} = \prod_i \left(\frac{p_i + p_{\underline{1}}}{p_i + p_{\underline{1}} + 1}\right)^{\mu_i},$$

il résulte que

$$\begin{aligned}
\alpha_i &= \gamma_i = \frac{1}{\nu-1}, \quad (\nu-1)\beta > 1, && i \in R,\\
\gamma_i &= \frac{1+\beta}{\nu} > \beta, \quad 1 > (\nu-1)\beta, && i \in S,\\
\alpha_i &= \nu\gamma_i - 1 = \nu\beta - 1 > \beta, \quad (\nu-1)\beta > 1, && i \in T,\\
\nu &= \sum \mu_i, \quad \beta = \frac{1}{(\nu-1)}, && i \in U.
\end{aligned}$$

Par conséquent on a soit $S = E$, soit $U = E$, soit quelque partition de $\{1, \dots, e\}$ entre R et T, S et U étant vides. Le cas $\sum \mu_i = \nu$ étant exceptionnel, nous le remettons à la fin de l'argument. Dans tout autre cas on a

$$1 + \beta = \sum_{i \in T} \mu_i \beta + \sum_{i \in S} \frac{1+\beta}{\nu} \mu_i + \sum_{i \in R} \frac{\mu_i}{\nu - 1}$$

ou

$$\beta \Big(\sum_T \mu_i + \sum_S \frac{\mu_i}{\nu-1} \Big) = 1 - \sum_S \frac{\mu_i}{\nu} - \sum_R \frac{\mu_i}{\nu-1}.$$

Observons que R est vide si $\nu = 1$. De plus si R et T sont vides, on obtient de cette équation, sauf si $\sum \mu_i = \nu$, que la quantité non négative β est -1.

Si R est tout l'ensemble E, on obtient

$$\beta = \frac{\sum \mu_i - \nu + 1}{\nu - 1}$$

et l'équation $(\nu - 1)\beta > 1$ implique que $\sum \mu_i > \nu$ ou $\sum \mu_i \geq \nu$. L'équation pour $a(p_i)$ est

$$Aa(p_i) = a(p_i)^\nu$$

dont il y a $\nu - 1$ solutions pour chaque i. Ajoutant à celles-ci les solutions $p_{\underline{1}} \equiv 0$, nous avons trouvé

$$\nu^e - (\nu-1)^e + (\nu-1)^e = \nu^e$$

solutions qui, si $\sum \mu_i > \nu$, est le nombre voulu $\chi(E, F)$.

Si T est non vide, alors

$$\beta \Big(\sum_T \mu_i - 1 \Big) = 1 - \sum_R \frac{\mu_i}{\nu - 1}.$$

De l'équation $(\nu - 1)\beta > 1$, il résulte que

$$\sum_T \mu_i - 1 < \nu - 1 - \sum_R \mu_i \Leftrightarrow \sum \mu_i < \nu.$$

Si cette condition est satisfaite, alors l'ensemble T ne peut pas être vide, mais c'est la seule condition sur T. Les équations pour $a(p_i)$, $a(p_{\underline{1}})$ sont

$$\begin{aligned}
Aa(p_i) &= a(p_i)^\nu, \quad i \in R,\\
Aa(p_i) &= a(p_{\underline{1}})^\nu, \quad i \in T,\\
Ba(p_{\underline{1}}) &= \prod_T a(p_{\underline{1}})^{\mu_i} \prod_R a(p_i)^{\mu_i},
\end{aligned}$$

qui possèdent $(\nu-1)^{|R|}(\sum_T \mu_i - 1)$ solutions. Sommant sur toutes les partitions, on obtient le nombre de solutions :

$$\sum_{R\subsetneq E}(\nu-1)^{|R|}\Big(\sum_{i\in E\setminus R}\mu_i - 1\Big) = \sum_i \mu_i \sum_{i\notin R}(\nu-1)^{|R|} - \nu^e + (\nu-1)^e.$$

Si nous nous rappelons les solutions déjà trouvées, cela donne

$$\Big(\sum \mu_i\Big)\nu^{e-1} = \chi(E,F)$$

solutions. Observons que ces arguments restent valables même pour $\nu = 1$.

Il ne reste que le cas que $\sum \mu_i = \nu$. Dans ce cas nous n'avons trouvé que ν^e solutions à part celles données par le cas où chaque i est dans U. Alors en éliminant des puissances de Δ des équations nous obtenons d'abord

$$A'x_i = (x_i+1)^\nu, \quad x_i = \frac{a(p_i)}{a(p_{\underline{1}})}, \quad A' = Aa(p_{\underline{1}})^{1-\nu}, \tag{6.2}$$

ce qui nous permet de trouver toutes les valeurs possibles des x_i comme fonction de A'. En plus il y a l'équation

$$B' = \prod (x_i+1)^{\mu_i}, \quad B' = Ba(p_{\underline{1}})^{1-\nu}. \tag{6.3}$$

Ces équations exigent malheureusement une discussion un peu fastidieuse. Observons tout de suite que les solutions de l'équation en x

$$A'x = (x+1)^\nu \tag{6.4}$$

sont toutes distinctes sauf si pour une d'elles $A' = \nu(x+1)^{\nu-1}$, donc si

$$x = \left(\frac{A'}{\nu}\right)^{1/(\nu-1)}.$$

Il résulte de cette dernière équation que si les équations (6.2) et (6.3) sont satisfaites et si les solutions de (6.4) ne sont pas toutes distinctes alors A' et B' appartiennent à des ensembles finis donnés. Puisque $A'/B' = A/B$ est générique cela n'est pas possible.

Soit $a = a(p_{\underline{1}})$. Nous remplaçons dans (6.2) et (6.3) le facteur x_i+1 par $x_i+\delta/a$ où δ est un paramètre que nous mettons ensuite égal à 0. Si on met $y_i = ax_i + \delta$ cela donne des équations

$$Ay_i = y_i^\nu, \qquad Ba = \prod y_i^{\mu_i},$$

qui sont des équations de type multiplicatif et dont le nombre de solutions non nulles, chacune de multiplicité un, est (à signe près) le déterminant

$$\begin{vmatrix} \nu-1 & 0 & \dots & 0 & \nu-1 \\ 0 & \nu-1 & \dots & 0 & \nu-1 \\ 0 & 0 & \ddots & \vdots & \vdots \\ 0 & 0 & \dots & \nu-1 & \nu-1 \\ \mu_1 & \mu_2 & \dots & \mu_e & \nu-1 \end{vmatrix}.$$

Un calcul facile montre que ce déterminant est $-(\nu-1)^e$. En admettant aussi des solutions telles que quelques y_i sont nuls de façon que a est forcément nul, nous obtenons en tout ν^e solutions, toutes de multiplicité un. Puisque $\nu^e = \chi(E,F)$ est exactement le nombre voulu de solutions, il ne reste qu'à se convaincre que toutes ces solutions peuvent être déformées de $\delta = 0$ à $\delta/a = 1$. Hors des points où $a = 0$ on peut récupérer les x_i à partir des y_i.

Puisqu'elles toutes ces solutions sont de multiplicité un à $\delta = 0$ une petite déformation est certainement possible. En examinant les équations on vérifie que aucun y_i n'est nul près de $\delta = 0$ sauf à $\delta = 0$. Donc a n'est nul hors de $\delta = 0$. Ayant déformé les $\chi(E,F)$ solutions de $\delta = 0$ dans un petit voisinage où elles restent forcément distinctes et en particulier à un point particulier $\delta \neq 0$ de ce voisinage, nous remplaçons δ/a par δ pour continuer jusqu'à $\delta = 1$ en évitant tout point de ramification. Le seul accroc possible est que deux solutions $\big(x_i(\delta), a(\delta)\big)$ et $\big(x_i'(\delta), a'(\delta)\big)$ distinctes en $\delta = 0$ coïncident en $\delta = 1$. Il résulte de notre discussion de l'équation (6.4) que chaque $y_i(0)$ est alors égal à $y_i'(0)$, et que cette égalité demeure valable le long de tout chemin de déformation. Mais alors $\big(a(\delta)/a'(\delta)\big)^{1-\nu}$ reste constant sur le chemin le long lequel on déforme en faisant varier δ, car

$$Ba = \prod (y_i - \delta)^{\mu_i}.$$

La constante $a(\delta)/a'(\delta)$ est nécessairement égale à 1.

C : e=f=2 avec tous les poids égaux à 1. Dans ce cas il y a des répartitions possibles de la forme $\{E_1, F_1\} \mid \{E_2, F_2\}$ pour lesquelles chacun des quatre ensembles n'a qu'un seul élément. Il y a deux telles répartitions et $\chi(E_1,F_1) = \chi(E_2,F_2) = 1$. Puisque $\mu_1 + \mu_2 = \nu_1 + \nu_2 = 2$ on a $\chi(E,F) = 2\cdot 2\cdot 2 = 8$. Nous cherchons par conséquent $8+2 = 10$ solutions.

Il y a d'abord $p_1 = p_2 = p_{\underline{1}} = p_{\underline{2}} \equiv 0$. Puis ce sont les solutions pour lesquelles une seule des variables n'est pas identiquement zéro. Pour cette variable il y a quatre choix. Si, par exemple, elle est p_2 alors

$$A\Delta p_2 = \frac{p_2^2}{(p_2+1)^2}.$$

Il en résulte que $\alpha(p_2) = 1$ et que $a(p_2) = A$.

Ensuite ce sont celles avec un seul $p_i \equiv 0$ et un seul $p_{\underline{i}} \equiv 0$. Il y a deux possibilités pour i et deux pour $\underline{i}$. Si $i = 1$ et $\underline{i} = \underline{1}$ alors les équations pour p_2 et $p_{\underline{2}}$ deviennent

$$A\Delta p_2 = \frac{p_2 + p_{\underline{2}}}{p_2 + p_{\underline{2}} + 1}\frac{p_2}{p_2+1}, \quad B\Delta p_{\underline{2}} = \frac{p_2 + p_{\underline{2}}}{p_2 + p_{\underline{2}} + 1}\frac{p_{\underline{2}}}{p_{\underline{2}}+1}.$$

Ces équations impliquent de plus que $\alpha(p_2 + p_{\underline{2}}) = 1$. En divisant ces deux équations nous trouvons l'équation $A/B = (p_{\underline{2}}+1)/(p_2+1)$. Puisque A/B est générique il est exclu que $\alpha(p_2) > 0$ et $\alpha(p_{\underline{2}}) > 0$. Par conséquent $a(p_2) = -a(p_{\underline{2}})$ et

$$\frac{A}{B} = -\frac{a(p_2)+1}{a(p_2)-1}.$$

De cette équation on trouve la valeur de $a(p_2)$.

Nous avons trouvé, à ce point, neuf solutions. Il nous en faut une de plus. Supposons qu'aucune des quatre fonctions p_i et $p_{\underline{j}}$ n'est identiquement zéro. On obtient alors

$$(6.5)\qquad A\Delta p_i = \prod_{\underline{j}} \frac{p_i + p_{\underline{j}}}{p_i + p_{\underline{j}} + 1}, \quad B\Delta p_{\underline{j}} = \prod_i \frac{p_i + p_{\underline{j}}}{p_i + p_{\underline{j}} + 1}.$$

Nous cherchons des solutions telles que $\alpha(p_i) = \alpha(p_{\underline{j}}) = 1$ pour tout i et $\underline{j}$ et telles que $a(p_i) = a$ est indépendant de i et $a(p_{\underline{j}}) = b$ indépendant de $\underline{j}$. On obtient

$$Aa = (a+b)^2, \quad Bb = (a+b)^2,$$

et par conséquent $a = c/A$, $b = c/B$, $1 = c(1/A + 1/B)^2$. Puisque nous avons trouvé 10 solutions il ne faut pas en chercher plus ! Nous reviendrons à ce cas mais seulement après en avoir examiné un autre.

D : e=3, f=2 mais tout poids est 1. Les décompositions possibles sont de la forme $\{E_1, F_1\} \mid \{E_2, F_2\}$ et elles auront $f_1 = e_2 = f_2 = 1$ et $e_1 = 2$ et pour chacune on a $\chi(E_1, F_1) = 1 \cdot 2^0 \cdot 1^1 = 1$, tandis que $\chi(E, F) = 2 \cdot 3^1 \cdot 2^2 = 24$. Nous cherchons alors $24 + 6 \cdot 1 = 30$ solutions. Elles se distribueront de la façon suivante. Dans cette liste il est entendu que toutes les variables sauf celles données explicitement ne sont pas identiquement zéro.

(1) $p_1 = p_2 = p_3 = p_{\underline{1}} = p_{\underline{2}} \equiv 0$ — une solution : 1.
(2) $p_1 = p_{\underline{1}} \equiv 0$ — une solution et six choix d'indices : $1 \times 6 = 6$.
(3) $p_1 = p_{\underline{1}} = p_{\underline{2}} \equiv 0$ — une solution et trois choix d'indices : $1 \times 3 = 3$.
(4) $p_1 = p_2 = p_{\underline{1}} \equiv 0$ — deux solutions et six choix d'indices : $2 \times 6 = 12$.
(5) $p_1 = p_2 = p_{\underline{1}} = p_{\underline{2}} \equiv 0$ — une solution et trois choix d'indices : $1 \times 3 = 3$.
(6) $p_1 = p_2 = p_3 = p_{\underline{1}} \equiv 0$ — deux solutions et deux choix d'indices : $2 \times 2 = 4$.
(7) Aucune des variables n'est identiquement zéro — une solution : 1.

La somme de ces nombres est 30. Donc ayant trouvé ces solution, il n'est pas nécessaire d'en chercher plus. Ce qui frappe le plus (pour des raisons que nous discuterons plus tard) est qu'il n'est pas du tout évident qu'il n'y ait qu'une seule solution telle qu'aucune des variables n'est identiquement zéro. Si $e = f = 2$, il s'avère possible de manipuler les équations de façon à se convaincre qu'il n'y pas que la solution trouvée mais, pour $e = 2$ et $f = 3$, nous n'avons pas su le faire.

Examinons les divers cas. Le premier est évident. Le deuxième donne

$$A\Delta p_i = S(p_i)S(p_i + p_{\underline{2}}), \quad i = 2, 3,$$
$$B\Delta p_{\underline{2}} = S(p_{\underline{2}}) \prod_{i=2}^{3} S(p_i + p_{\underline{2}}).$$

De ces équations il résulte que

$$\frac{A^2\Delta}{B} = \frac{p_{\underline{2}} + 1}{(p_2 + 1)(p_3 + 1)}$$

et par conséquent que $p_{\underline{2}} = -1 + a(p_{\underline{2}})\Delta^{\alpha(p_{\underline{2}})} + \dots$, $\alpha(p_{\underline{2}}) = 1$. Nous concluons de plus que $p_i = 1 + a(p_i)\Delta + \dots$, $i = 2, 3$. On obtient enfin l'équation

$$A = \big(a(p_i) + a(p_{\underline{2}})\big), \qquad i = 2, 3,$$

d'où il résulte que $a(p_2) = a(p_3)$ et

$$B = \frac{(a(p_2) + a(p_{\underline{2}}))^2}{a(p_{\underline{2}})}.$$

Ces équations admettent une solution unique ! En particulier, $a(p_{\underline{2}}) = A^2/B$.

Dans le troisième cas on a

$$A\Delta p_i = S(p_i)^2, \quad i = 2, 3,$$

ou

$$A\Delta = \frac{p_i}{(p_i+1)^2}.$$

Ces équations admettent évidemment une seule solution en série de Puiseux et $p_2 = p_3$.

Dans le quatrième cas on a

$$A\Delta = \frac{p_3 + p_{\underline{2}}}{(p_3 + p_{\underline{2}} + 1)(p_3 + 1)}, \quad B\Delta = \frac{(p_3 + p_{\underline{2}})p_{\underline{2}}}{(p_3 + p_{\underline{2}} + 1)(p_{\underline{2}} + 1)^2}.$$

En divisant les deux équations nous trouvons

$$\frac{A}{B} = \frac{(p_{\underline{2}} + 1)^2}{p_{\underline{2}}(p_3 + 1)}.$$

Nous concluons de la première équation que $p_3 + p_{\underline{2}} = a(p_3 + p_{\underline{2}})\Delta^\alpha + \dots$, $\alpha = \alpha(p_3 + p_{\underline{2}}) = 1$. Puisque A/B est générique il résulte de la dernière que $p_3 = -a + a(p_3)\Delta^\alpha + \dots$ et que $p_{\underline{2}} = a + a(p_{\underline{2}})\Delta^\alpha + \dots$ où $a \neq 0$ est une des deux solutions de $A/B = -(a+1)^2/a(a-1)$. La valeur de $a(p_3 + p_{\underline{2}})$ se déduit alors de la première équation, mais pour trouver $a(p_3)$ et $a(p_{\underline{2}})$ il faut utiliser la dernière équation une deuxième fois en examinant les termes linéaires en Δ.

Dans le cinquième cas il n'y a que l'équation

$$A\Delta = \frac{p_3}{(p_3+1)^2}.$$

Il en résulte que $\alpha(p_3) = 1$ et que $a(p_3) = A$.

Dans le sixième cas l'équation est

$$B\Delta = \frac{p_{\underline{2}}^2}{(p_{\underline{2}} + 1)^3}$$

de sorte que $\alpha(p_{\underline{2}}) = 1/2$ et $B = a(p_{\underline{2}})^2$.

Il reste enfin le cas qu'aucune des séries n'est identiquement nulle. Puisque nous n'avons besoin que d'une seule solution, nous supposons que $p_1 = p_2 = p_3$ et que $p_{\underline{1}} = p_{\underline{2}}$. Les deux équations deviennent

$$A\Delta p_1 = \frac{(p_1 + p_{\underline{1}})^2}{(p_1 + p_{\underline{1}} + 1)^2}, \quad B\Delta p_{\underline{1}} = \frac{(p_1 + p_{\underline{1}})^3}{(p_1 + p_{\underline{1}} + 1)^3}.$$

En général les équations

$$A\Delta p_1 = \frac{(p_1 + p_{\underline{1}})^\nu}{(p_1 + p_{\underline{1}} + 1)^\nu}, \quad B\Delta p_{\underline{1}} = \frac{(p_1 + p_{\underline{1}})^\mu}{(p_1 + p_{\underline{1}} + 1)^\mu}.$$

avec $\mu > \nu > 1$ admettent des solutions $p_1 = a\Delta^\alpha + \dots$, $p_{\underline{1}} = b\Delta^\beta + \dots$ avec $\alpha = 1/(\nu - 1)$, $\beta = \alpha\mu - 1 = (\mu - \nu + 1)/(\nu - 1)$. Les valeurs de $a \neq 0$ et $b \neq 0$ sont déterminées par

$$Aa = a^\nu, \quad Bb = a^\mu.$$

Il y a par conséquent $\nu - 1$ possibilités. Si $\mu = \nu$ on arrive au même résultat mais les équations sont un peu plus compliquées.

Nous ne traitons pas d'autres cas de paires libres mais avant de passer aux paires liées nous examinons les équations

$$A\Delta p_i = \prod_{\underline{j}\in F} S(p_i + p_{\underline{j}}), \quad B\Delta p_{\underline{j}} = \prod_{i\in E} S(p_i + p_{\underline{j}}) \tag{6.6}$$

sous l'hypothèse qu'aucune des séries p_i et $p_{\underline{j}}$ n'est identiquement zéro. Ce sont les équations (6.5) généralisées. Nous montrons, pour $e = f = 2$, comment en déduire que toute telle solution satisfait aux conditions : $p_1 = p_2$ et $p_{\underline{1}} = p_{\underline{2}}$. Cependant, même pour $e = 3$, $f = 2$ nous ne savons pas établir ce fait directement.

En divisant les deux équations nous obtenons

$$\frac{A}{B}\frac{p_i}{p_{\underline{j}}} = \frac{\prod_{\underline{l}\neq\underline{j}} S(p_i + p_{\underline{l}})}{\prod_{k\neq i} S(p_k + p_{\underline{j}})}.$$

Nous supposons que $p_i^0 = 0$ pour tout i et que $p_{\underline{j}}^0 = 0$ pour tout $\underline{j}$. Cette équation s'écrit

$$Ap_i \prod_{k\neq i} S(p_k + p_{\underline{j}}) = Bp_{\underline{j}} \prod_{\underline{l}\neq\underline{j}} S(p_i + p_{\underline{l}}). \tag{6.7}$$

Les premières des équations (6.6) s'écrivent

$$\begin{aligned} Ap_i\Delta &= \frac{p_i}{p_i + p_{\underline{j}} + 1} \prod_{\underline{l}\neq\underline{j}} S(p_i + p_{\underline{l}}) + \frac{p_{\underline{j}}}{p_i + p_{\underline{j}} + 1} \prod_{\underline{l}\neq\underline{j}} S(p_i + p_{\underline{l}}) \\ &= \frac{p_i}{p_i + p_{\underline{j}} + 1} \prod_{\underline{l}\neq\underline{j}} S(p_i + p_{\underline{l}}) + \frac{A}{B}\frac{p_i}{p_i + p_{\underline{j}} + 1} \prod_{k\neq i} S(p_k + p_{\underline{j}}). \end{aligned}$$

Puisque $p_i \not\equiv 0$ il est permis de le supprimer.

Si $e = 2$ et $f = 2$, l'équation qui reste peut être considérée comme une équation linéaire pour p_1, p_2, $p_{\underline{1}}$, $p_{\underline{2}}$ dans laquelle les coefficients dépendent de ces mêmes variables. Si i et $\underline{j}$ sont donnés, supposons que $k \neq i$ et $\underline{l} \neq \underline{j}$

$$A\Delta = (p_i + p_{\underline{l}}) \prod_{\underline{n}} \frac{1}{p_i + p_{\underline{n}} + 1} + \frac{A}{B}(p_k + p_{\underline{j}}) \prod_m \frac{1}{p_m + p_{\underline{j}} + 1}. \tag{6.8}$$

Puisqu'il y a quatre choix de la paire $(i, \underline{j})$, il y a quatre équations pour les quatre variables. La matrice des coefficients, dans laquelle la première correspond à la paire $(1, \underline{1})$, la deuxième à $(2, \underline{1})$ et ainsi de suite, est

$$M = \begin{pmatrix} \rho_1 & \sigma_{\underline{1}} & \sigma_{\underline{1}} & \rho_1 \\ \sigma_{\underline{1}} & \rho_2 & \sigma_{\underline{1}} & \rho_2 \\ \rho_1 & \sigma_{\underline{2}} & \rho_1 & \sigma_{\underline{2}} \\ \sigma_{\underline{2}} & \rho_2 & \rho_2 & \sigma_{\underline{2}} \end{pmatrix}$$

où

$$\rho_i = \prod_{\underline{n}} \frac{1}{p_i + p_{\underline{n}} + 1}, \quad \sigma_{\underline{j}} = \frac{A}{B} \prod_m \frac{1}{p_m + p_{\underline{j}} + 1}.$$

Soulignons qu'il s'agit des équations linéaires en $p_1, p_2, p_{\underline{1}}, p_{\underline{2}}$ dont les coefficients dépendent de ces mêmes variables.

À $\Delta = 0$ on a

$$\rho_i = \rho = 1, \quad \sigma_{\underline{j}} = \sigma = \frac{A}{B}.$$

Si $\rho^2 \neq \sigma^2$ la matrice

$$\begin{pmatrix} \rho & \sigma & \sigma & \rho \\ \sigma & \rho & \sigma & \rho \\ \rho & \sigma & \rho & \sigma \\ \sigma & \rho & \rho & \sigma \end{pmatrix}$$

est de rang trois. Le seul vecteur qu'elle annule est le transposé de $(1, 1, -1, -1)$. Puisque A/B est générique, il en résulte que les équations (6.8) déterminent toutes les quantités $p_i + p_{\underline{j}}$ à partir des coefficients ρ_i et $\sigma_{\underline{j}}$.

En particulier

$$p_i + p_{\underline{j}} = \frac{\rho_k - \sigma_{\underline{l}}}{\rho_1 \rho_2 - \sigma_{\underline{1}} \sigma_{\underline{2}}} A\Delta$$

de sorte que $\alpha(p_i + p_{\underline{j}}) = 1$ pour tout i et $\underline{j}$. Puisque ρ_k et $\sigma_{\underline{l}}$ sont des fonctions des $p_m + p_{\underline{n}}$, nous pouvons utiliser ces équations pour calculer par récurrence le développement en série de Puiseux des fonctions $p_m + p_{\underline{n}}$. Il en résulte qu'elles ne dépendent pas des indices. Enfin en utilisant les équations (6.6) nous concluons que $p_1 = p_2$, que $p_{\underline{1}} = p_{\underline{2}}$ et que $\alpha(p_i) = \alpha(p_{\underline{j}}) = 1$.

7. Les paires liées à sommet interdit

La somme pertinente est (5.7). Les équations pertinentes sont des modifications de (6.1)

$$(7.1) \qquad A\Delta p_i = S(p_i)^{\omega''} \prod_{\underline{j} \in F} S(p_i, p_{\underline{j}})^{\mu_{\underline{j}}}, \quad B\Delta p_{\underline{j}} = S(p_{\underline{j}})^{\omega'} \prod_{i \in E} S(p_i, p_{\underline{j}})^{\mu_i},$$

où $S(p) = S(p, 0)$. Nous nous permettons d'utiliser deux notations déjà introduites bien qu'elles soient en conflit : la constante B de (2.40) et l'ensemble B de sommets interdits. Cela n'entraîne guère d'inconvénients. On se souvient aussi que selon les conventions de la section 5, $B = B' \cup B''$ et que B' est rattaché à E et B'' à F, mais que E et B' aussi bien que F et B'' sont disjoints. Les facteurs supplémentaires à droite dans les équations sont attachés au sommets interdits B'' et B', car pour les indices i et $\underline{j}$ dans $\mathbb{B}$ les coordonnées p_i et $p_{\underline{j}}$ restent 0. Ces équations sont valables même pour $i \in B'$ ou $\underline{j} \in B''$, mais elles ne donnent rien car alors p_i et $S(p_i)$ ou $p_{\underline{j}}$ et $S(p_{\underline{j}})$ sont nuls. Pour tout i, $p_i^0 = 0$ et pour tout $\underline{j}$, $p_{\underline{j}}^0 = 0$.

A : f=0. C'est le cas où F est vide. Alors la seule décomposition à considérer est $\{E, \varnothing, B\}$ et

$$\chi(E, \varnothing, B) = (\omega'')^e.$$

Les équations à résoudre sont

$$A\Delta p_i = \left(\frac{p_i}{p_i+1}\right)^{\omega''}.$$

Ces équations ne sont pas couplées. On peut choisir $p_i \equiv 0$, qui est la seule solution si $\omega''=1$. Si $\omega''>1$ toute autre solution est de la forme

$$p_i = a(p_i)\Delta^{\alpha(p_i)}+\dots.$$

Évidemment $\alpha(p_i)=1/(\omega''-1)$ et l'équation pour $a(p_i)$ est

$$Aa(p_i) = a(p_i)^{\omega''},$$

dont il y a $\omega''-1$ solutions distinctes et distinctes de zéro. Donc chaque équation a ω'' solutions. Puisqu'il y a e équations indépendantes on obtient le nombre voulu de solutions.

Nous ne traitons ensuite que le cas où F ne contient qu'un seul élément, donc où $f=1$. A part la partition principale qui correspond à un diagramme connexe, il y a d'autres partitions qui correspondent aux diagrammes non connexes et donc aux termes diagonaux. On a donc dans (5.7) des termes attachés aux décompositions

$$\{E,F,B\} \quad \{E,F\},\{B\} \quad \{E_1,B\},\{E_2,F\} \quad \{E,\varnothing\},\{F,B\}.$$

Puisque $\chi(E,\varnothing)=0$ la somme (5.7) est

$$\chi(E,F,B)+\chi(E,F)\chi(B)+\sum\chi(E_1,\varnothing,B)\chi(E_2,F). \tag{7.2}$$

Il est entendu que dans la somme ni E_1 ni E_2 n'est vide et que leur réunion est E. En plus de l'hypothèse $f=1$ nous supposons soit que $e=1$ soit que $\omega'=\omega''=1$.

B : f=1 et e=1. Soit μ le poids du seul élément de E et ν le poids du seul élément de F. D'abord

$$\chi(E,F,B)=\omega'\omega''+\omega'\nu+\omega''\mu$$

tandis que

$$\chi(E,F)\chi(B)=\min\{\mu,\nu\}.$$

Puisqu'il n'existe pas de décomposition $\{E_1,B\},\{E_2,F\}$ avec $E_1\neq\varnothing$ et $E_2\neq\varnothing$ nous cherchons

$$\chi(E,F,B)+\chi(E,F)\chi(B)=\omega'\omega''+\omega'\nu+\omega''\mu+\min\{\mu,\nu\} \tag{7.3}$$

solutions liées qui se développent en séries de Puiseux

$$p_i = p_i^0 + a(p_i)\Delta^{\alpha(p_i)}+\dots,$$
$$p_{\underline{j}} = p_{\underline{j}}^0 + a(p_{\underline{j}})\Delta^{\alpha(p_{\underline{j}})}+\dots.$$

Puisque $e=f=1$ il n'y a que deux équations, celles pour $i=1$ et pour $\underline{j}=\underline{1}$. Observons comme première conséquence des équations (7.1) que, si $p_1^0\neq 0$ ou $p_{\underline{1}}^0\neq 0$, alors $p_1^0+p_{\underline{1}}^0=0$. Cette dernière équation est donc toujours valable et nous écrivons

$$p_1+p_{\underline{1}} = a(p_1+p_{\underline{1}})\Delta^\gamma+\dots.$$

Pour pouvoir plus facilement analyser les équations pour les exposantes $\alpha = \alpha(p_1)$, $\beta = \alpha(p_{\underline{1}})$ et γ nous posons

$$a = \omega' - 1, \quad b = \mu, \quad c = \omega'' - 1, \quad d = \nu.$$

Comme nous le verrons par la suite, les conditions sur ces quantités qui s'avèrent pertinentes sont,

$$\text{(7.4)} \qquad \operatorname{sgn}(bc - ad) = \operatorname{sgn}(b - d) = \operatorname{sgn}(c - a), \quad |a - c| \geq |b - d|.$$

Supposons que $b \leq d$, donc que $\mu \leq \nu$. Si la condition (7.4) n'est pas satisfaite nous obtiendrons par la suite $a(c+d) + bc$. Par contre si elle est satisfaite alors $a - c \geq d - b$ de sorte que $a + b \geq c + d$ et nous obtiendrons

$$c(a + b) + ad - bc + bc = a(c + d) + bc.$$

Donc deux expressions à première vue différentes qui apparaîtront sont en fait les mêmes — une remarque facile mais utile !

Nous cherchons d'abord des solutions avec $p_1^0 = p_{\underline{1}}^0 = 0$. Si ni p_1 ni $p_{\underline{1}}$ n'est identiquement nul, il y a quatre possibilités

$$\text{(7.5.a)} \qquad \alpha = \gamma < \beta,$$
$$\text{(7.5.b)} \qquad \beta = \gamma < \alpha,$$
$$\text{(7.5.c)} \qquad \alpha = \beta < \gamma,$$
$$\text{(7.5.d)} \qquad \alpha = \beta = \gamma.$$

Pour trois des cas $\beta \geq \alpha$ et les inconnues convenables sont $p_{\underline{1}}$ et $p_1 + p_{\underline{1}}$. Les équations pour α et β sont dans les cas correspondants à (7.5.a) et (7.5.c) :

$$\text{(7.6.a)} \qquad 1 + \alpha = \omega''\alpha + \nu\alpha, \quad 1 + \beta = \omega'\beta + \mu\alpha,$$
$$\text{(7.6.c)} \qquad 1 + \alpha = \omega''\alpha + \nu\gamma, \quad 1 + \alpha = \omega'\alpha + \mu\gamma.$$

Dans le cas (7.5.d) qui n'arrive que si $\omega' + \mu = \omega'' + \nu$, tous les exposants sont égaux à $1/(\omega' + \mu - 1)$.

Considérons les contraintes qu'impose l'existence de solutions positives des équations (7.6.a) et (7.6.c) qui satisfont aux inégalités (7.5.a) et (7.5.c). Pour (7.6.a) on trouve

$$\alpha = \frac{1}{\omega'' + \nu - 1}, \quad \beta = \frac{1 - \mu\alpha}{\omega' - 1}.$$

Alors on a $\alpha < \beta$ (et par conséquent $\beta > 0$) si et seulement si $\omega' + \mu < \omega'' + \nu$. Il en résulte que les deux cas (7.5.a) et (7.6.b) s'excluent l'un l'autre.

La solution des équations (7.6.c) est

$$\text{(7.7)} \qquad \alpha = \frac{\mu - \nu}{\mu(\omega'' - 1) - \nu(\omega' - 1)}, \quad \gamma = \frac{\omega'' - \omega'}{\mu(\omega'' - 1) - \nu(\omega' - 1)}$$

sauf peut-être dans le cas exceptionnel $\mu(\omega'' - 1) - \nu(\omega' - 1) = 0$. A part ce cas, les équations (7.7) n'ont de solution positive que si les conditions

$$\operatorname{sgn}(\mu - \nu) = \operatorname{sgn}(\omega'' - \omega') = \operatorname{sgn}\bigl(\mu(\omega'' - 1) - \nu(\omega' - 1)\bigr)$$

sont satisfaites. Donc (7.5.c) est possible si et seulement si $\mu < \nu$ et $\omega' + \mu > \omega'' + \nu$ ou $\nu < \mu$ et $\omega'' + \nu > \omega' + \mu$. Dans le cas exceptionnel où le déterminant des équations (7.6.c) s'annule, il n'y pas de solution sauf si $\omega' = \omega''$ et $\mu = \nu$ et alors elle n'est pas bien définie. Nous reviendrons à ce cas. Les autres possibilités sont donc :

$$\omega'' + \nu > \omega' + \mu, \quad \mu \leq \nu, \tag{7.8.a}$$
$$\omega' + \mu > \omega'' + \nu, \quad \nu \leq \mu, \tag{7.8.b}$$
$$\omega'' + \nu > \omega' + \mu, \quad \nu < \mu, \tag{7.8.c}$$
$$\omega' + \mu > \omega'' + \nu, \quad \mu < \nu. \tag{7.8.d}$$

Dans chacun des cas nous comptons le nombre de solutions de nos équations ou plutôt le nombre de possibilités pour $a(p_1)$, $a(p_{\underline{1}})$, et $a(p_1 + p_{\underline{1}})$ sans nous occuper trop des termes d'ordre plus élevé. Par symétrie le cas (7.5.b) est compris dans le cas (7.5.a). Encore par symétrie, mais une autre, il suffit de ne considérer dans les quatre cas (7.5.a), (7.5.b) (7.5.c) et (7.5.d) que (7.8.a) et (7.8.d). On suppose alors par la suite qu'une de ces deux inégalités, qui sont celles de la remarque qui suit (7.4), est satisfaite ! Dans le cas (7.8.d) les conditions (7.4) sont satisfaites. Elles ne le sont pas dans le cas (7.8.a). Observons de plus que les quantités qui figurent dans les inégalités de (7.8) sont des données, tandis que celles de (7.5) sont libres, de sorte que chacune des quatre possibilités envisagées doit être considérée.

Avant de traiter ces cas, nous comptons les autres solutions, celles pour lesquelles $p_1 \equiv 0$ ou $p_{\underline{1}} \equiv 0$. Si $p_1 \equiv 0$, l'équation pour $p_{\underline{1}}$ devient

$$\Delta B p_{\underline{1}} = \left(\frac{p_{\underline{1}}}{p_{\underline{1}} + 1}\right)^{\omega' + \mu}$$

dont il y a $\omega' + \mu$ solutions. Si $p_{\underline{1}} \equiv 0$ on obtient une équation semblable, mais alors en comptant les solutions il faut écarter celle telle que $p_1 \equiv 0$. Donc on n'en obtient que $\omega'' + \nu - 1$. Au total on obtient $\omega' + \mu + \omega'' + \nu - 1$.

Commençons maintenant avec (7.5.a). On a déjà observé que les égalités (7.6.a) ne sont pas compatibles avec les inégalités de (7.8.d). Nous supposons par conséquent que les inégalités (7.8.a) sont satisfaites. Les équations pour $a(p_1)$ et $a(p_{\underline{1}})$ sont

$$Aa(p_1) = a(p_1)^{\omega'' + \nu}, \quad Ba(p_{\underline{1}}) = a(p_{\underline{1}})^{\omega'} a(p_1)^{\mu}.$$

Le nombre de solutions qui ne sont pas identiquement 0 de ces équations est évidemment $(\omega' - 1)(\omega'' + \nu - 1)$. Pour cette équation comme pour les équations suivantes il est évident que les solutions sont distinctes.

Considérons ensuite les solutions de type (7.5.c). Elles n'existent que dans le cas (7.8.d), mais alors il faut ajouter le nombre de solutions de type (7.5.b) qui est $(\omega'' - 1)(\omega' + \mu - 1)$. Dans le cas (7.5.c) les équations pour $a(p_1)$ et $a(p_1 + p_{\underline{1}})$ deviennent

$$Aa(p_1) = a(p_1)^{\omega''} a(p_1 + p_{\underline{1}})^{\nu}, \quad Ba(p_1) = a(p_1)^{\omega'} a(p_1 + p_{\underline{1}})^{\mu}$$

dont le nombre de solutions non-zéro est

$$\left|(\omega'' - 1)\mu - (\omega' - 1)\nu\right|.$$

Il résulte du commentaire suivant (7.4) que dans les deux cas on obtient au total

$$(\omega' - 1)(\omega'' - 1 + \nu) + (\omega' + \mu) + (\omega'' + \nu) - 1 = \omega'\omega'' + \omega'\nu + \mu$$

solutions. Par conséquent les conditions $p_1^0 = p_{\underline{1}}^0 = 0$ ne donnent pas toutes les solutions cherchées.

Nous supposons toujours que $\mu \leq \nu$ et considérons les solutions avec $p_1^0 = -1$, $p_{\underline{1}}^0 = 1$. Les équations pour les exposants sont

$$1 = -\omega''\alpha + \nu\gamma, \quad 1 = \mu\gamma.$$

Donc $\gamma = 1/\mu$ et

$$\alpha = \frac{\nu/\mu - 1}{\omega''} \geq 0.$$

Le cas $\mu = \nu$ est exceptionnel. Si $\nu > \mu$ les équations pour $a(p_1)$ et u sont

$$-Aa(p_1)^{\omega''} = (-1)^{\omega''} a(p_1 + p_{\underline{1}})^{\nu}, \quad B = \left(\frac{1}{2}\right)^{\omega'} a(p_1 + p_{\underline{1}})^{\mu}.$$

Il y a $\mu\omega''$ solutions de ces équations. On trouve d'abord $a(p_1 + p_{\underline{1}})$ et ensuite $a(p_1)$. Sauf peut-être dans les cas exceptionnels, nous trouvons au total

$$\omega'\omega'' + \omega'\nu + \omega''\mu + \min\{\mu, \nu\}$$

solutions, précisément le nombre (7.3) escompté.

Si $\mu = \nu$ nous cherchons des solutions avec $p_1^0 \neq 0, \pm 1$. Alors $\gamma = 1/\mu$ et il s'agit de trouver les valeurs permises de $x = p_1^0 = -p_{\underline{1}}^0 = -\underline{x}$. Soit $u = a(p_1 + p_{\underline{1}})$. Les équations pertinentes sont

$$Ax = \left(\frac{x}{x+1}\right)^{\omega''} u^{\nu}, \quad B\underline{x} = \left(\frac{\underline{x}}{\underline{x}+1}\right)^{\omega'} u^{\mu}.$$

En formant le quotient des deux équations on obtient une équation pour x car $x = -\underline{x}$. Supposons que $\omega'' > \omega'$. Alors

$$-\frac{A}{B} = x^{\omega'' - \omega'} \frac{(x-1)^{\omega'}}{(x+1)^{\omega''}}.$$

Selon le prochain lemme cette équation a ω'' solutions distinctes. Après avoir trouvé x on résout les premières équations pour trouver μ valeurs de u. On trouve donc les $\omega''\mu$ solutions supplémentaires cherchées et ce, même si $\omega' = \omega''$.

LEMME 7.1. *Supposons que a_i, $1 \leq i \leq m$ et b_j, $1 \leq j \leq n$ sont des paramètres donnés mais que C est générique. Si $a_i \neq b_j$ pour tout i et j, alors l'équation*

$$C \prod_{i=1}^{m} (x - a_i)^{\mu_i} = \prod_{j=1}^{n} (x - b_j)^{\nu_j} \tag{7.9}$$

n'a pas de solution multiple.

En effet chaque solution multiple satisfait à l'équation obtenue en prenant la dérivée logarithmique des deux côtés, donc à

$$\sum_{i=1}^{m} \frac{\mu_i}{x - a_i} = \sum_{j=1}^{n} \frac{\nu_j}{x - b_j}. \tag{7.10}$$

Cette équation a un nombre fini de solutions qui ne satisfont pas (7.9) si C et générique.

Il reste toujours le cas exceptionnel $\omega' + \mu = \omega'' + \nu$. Si $\mu < \nu$ et $p_1^0 = p_{\underline{1}}^0 = 0$, on a le cas (7.5.d). Pour $a(p_1)$ et $a(p_{\underline{1}})$ les équations sont

$$(7.11)\qquad Aa(p_1) = a(p_1)^{\omega''}\big(a(p_1) + a(p_{\underline{1}})\big)^{\nu}, \quad Ba(p_{\underline{1}}) = a(p_{\underline{1}})^{\omega'}\big(a(p_1) + a(p_{\underline{1}})\big)^{\mu}.$$

En divisant la première des équations par la seconde on obtient une équation pour $x = a(p_1)/a(p_{\underline{1}})$ de la forme $(1+x)^m x^n = C$ où $m = \nu - \mu$, $n = \omega'' - 1$ sont des entiers non négatifs et C est générique. Selon le lemme 7.1 cette équation a $\nu - \mu + \omega'' - 1 = \omega' - 1$ solutions distinctes de sorte que (7.11) a $(\omega'' + \nu - 1)(\omega' - 1)$ solutions distinctes et différentes de 0. C'est le nombre déjà trouvé pour les cas non exceptionnels. Puisque le nombre des solutions avec $p_1^0 = -1$, $p_{\underline{1}}^0 = 1$ ne change pas, il ne reste que le cas $\mu = \nu$ et $\omega' = \omega''$. Dans ce cas nous écartons arbitrairement la possibilité (7.6.c) et nous obtenons encore le nombre de solutions voulu.

C : $\omega' = \omega''$=1. Supposons enfin que $\omega' = \omega'' = 1$ mais que e soit arbitraire. La valeur de f est toujours 1. Il résulte immédiatement des premières équations (7.1) que si $p_i^0 + p_{\underline{1}}^0 \neq 0$ alors $p_i \equiv 0$.

Si $\nu = \mu(F)$ et les valeurs de $|A_i|$, $A_i \in E$, sont μ_i, on a

$$(7.12)\qquad \begin{aligned}\chi(E,F,B) &= \omega'(\nu+\omega'')^e + \omega''\Big(\sum \mu_i\Big)(\nu+\omega'')^{e-1}\\ &= (\nu+1)^e + \mu(E)(\nu+1)^{e-1}.\end{aligned}$$

On a de plus pour une décomposition $E = E_1 \cup E_2$,

$$(7.13)\qquad \begin{aligned}\chi(E_1,\varnothing,B) &= (\omega'')^{e_1} = 1,\\ \chi(E_2,F) &= \min\{\mu(E_2),\nu\}\nu^{e_2-1},\\ \chi(E_1,\varnothing,B)\chi(E_2,F) &= (\omega'')^{e_1}\min\{\mu(E_2),\nu\}\nu^{e_2-1}\\ &= \min\{\mu(E_2),\nu\}\nu^{e_2-1}.\end{aligned}$$

Observons encore que $\chi(E_2,F) = 0$ si $e_2 = 0$ et que

$$(7.14)\qquad \sum_{E=E_1\cup E_2} \mu(E_2)\nu^{e_1} = \sum_i \mu_i \sum_{i\notin E_1} \nu^{e_1} = \mu(E)(\nu+1)^{e-1}$$

est le second terme de (7.12), que nous remplaçons désormais par le côté gauche de (7.14).

Nous considérons d'abord les solutions avec $p_{\underline{1}} \equiv 0$. Puisque $\omega'' = 1$, l'équation pour p_i devient

$$A\Delta p_i = \left(\frac{p_i}{p_i+1}\right)^{\nu+1}.$$

Ces équations sont indépendantes et elles ont par conséquent $(\nu+1)^e$ solutions dont quelques-unes auront $p_i \equiv 0$. De toute façon on a ici le premier terme de (7.12) de sorte qu'il ne reste qu'à couvrir (7.13) et (7.14).

Nous examinons ensuite les solutions avec $p_{\underline{1}}^0 = 0$ mais $p_{\underline{1}} \not\equiv 0$, et par conséquent tous les $p_i^0 = 0$. Soit E_2 un sous-ensemble de E. Nous cherchons les solutions telles que $p_i \equiv 0$ si et seulement si $i \in E_2$. Si le complément de E_2 est E_1 posons

$$p_i + p_{\underline{1}} = u_i\Delta^{\gamma_i} + \dots, \qquad i \in E_1.$$

On a $\gamma_i = 1/\nu$ pour tout $i \in E_1$. Soit $\beta = \alpha(p_{\underline{1}})$. De l'équation pour $p_{\underline{1}}$ résulte

$$1 = \sum_{i \in 2} \mu_i \beta + \sum_{i \in E_1} \frac{\mu_i}{\nu},$$

et cette équation admet une solution positive en β si et seulement si

$$\sum_{i \in E_1} \mu_i < \nu. \tag{7.15}$$

Alors les équations pour les u_i et $\underline{a} = a(p_{\underline{1}})$ deviennent

$$A = u_i^{\nu}, \quad i \in E_1, \qquad B = \underline{a}^{\sum_{E_2} \mu_i} \prod_{E_1} u_i^{\mu_i}$$

dont le nombre de solutions est

$$\nu^{e_1} \sum_{E_2} \mu_i = \nu^{e_1} \mu(E_2). \tag{7.14$'$}$$

L'expression (7.14$'$) n'est qu'un terme de (7.14). On a par conséquent trouvé assez de solutions pour couvrir tout terme de (7.14) attaché à une décomposition qui satisfait à (7.15).

Nous supposons ensuite que $p_{\underline{1}}^0 = -1$ de sorte que $E = E_1 \cup E_2$ et $p_i \equiv 0$, $i \in E_2$, $p_i^0 = 1$, $i \in E_1$. Plutôt que p_i ce sont les variables $p_i + p_{\underline{1}}$, $i \in E_1$, qui sont primaires. On a forcément $\gamma_i = 1/\nu$, $i \in E_1$. L'équation qui détermine $\beta = \alpha(p_{\underline{1}})$ est

$$1 = -\sum_{E_2} \mu_i \beta + \sum_{E_1} \frac{\mu_i}{\nu},$$

qui admet une solution positive en β seulement si

$$\sum_{i \in E_1} \mu_i > \nu. \tag{7.15$'$}$$

Cette condition est à comparer avec celle de (7.15). Si elle est satisfaite l'équation pour $u_i = a(p_i + p_{\underline{1}})$ est

$$2A = u_i^{\nu}$$

et celle pour $\underline{a} = a(p_{\underline{1}})$ est alors

$$-B = \left(-\frac{1}{\underline{a}}\right)^{\mu(E_2)+1} \prod_{E_1} u_i^{\mu_i}.$$

Donc le nombre de solutions est $\nu^{e_1}\big(\mu(E_2)+1\big)$, ce qui est la somme de

$$\nu^{e_1} \mu(E_2) \tag{7.14$''$}$$

et de

$$\nu^{e_1}. \tag{7.13$'$}$$

En particulier il en résulte qu'il ne manque à (7.14) que les termes pour lesquels

$$\sum_{i \in E_1} \mu_i = \nu, \tag{7.15$''$}$$

car (7.14″) donne les termes voulus lorsque (7.15′) est satisfait et (7.14′) couvre les autres. Nous revenons au cas (7.15″) à la toute fin. Observons tout de suite que l'expression (7.13′) est égale à (7.13) lorsque (7.15′) est valable.

Nous cherchons enfin des solutions avec $p_{\underline{1}}^0 = 1$. Nous supposons encore que $p_i \equiv 0$, $i \in E_2$ et que les variables $p_i + p_{\underline{1}}$ sont primaires pour $i \in E_1$. Soient $\alpha_i = \alpha(P_i)$. Nous répartissons E_1 en quatre sous-ensembles

$$\begin{aligned} E_1^1&: \quad \gamma_i > \alpha_i = \beta, \\ E_1^2&: \quad \alpha_i > \gamma_i = \beta, \\ E_1^3&: \quad \beta > \alpha_i = \gamma_i, \\ E_1^4&: \quad \alpha_i = \gamma_i = \beta. \end{aligned}$$

Pour $i \in E_1^1$ on a

$$1 = -\beta + \nu\gamma_i, \quad \gamma_i = \frac{1+\beta}{\nu}.$$

Puisque γ_i est censé être plus grand que β, on exige que

$$\frac{1+\beta}{\nu} > \beta \iff 1 > (\nu-1)\beta.$$

Pour $i \in E_1^2$,

$$1 = -\alpha_i + \nu\beta, \qquad \alpha_i = \nu\beta - 1 \implies (\nu-1)\beta > 1.$$

Il en résulte qu'un des deux ensembles E_1^1 et E_1^2 est vide. Pour $i \in E_1^3$,

$$1 = -\alpha_i + \nu\alpha_i, \quad \alpha_i = \frac{1}{(\nu-1)} \implies (\nu-1)\beta > 1.$$

Si E_1^4 n'est pas vide alors $(\nu-1)\beta = 1$.

Supposons d'abord que E_1^1 n'est pas vide de sorte que les autres sous-ensembles le sont. L'équation pour β est

$$1 = \sum_{E_1} \mu_i \frac{1+\beta}{\nu}$$

qui a une solution positive seulement si

$$\nu > \sum_{E_1} \mu_i = \mu(E_1). \tag{7.15‴}$$

Pour que $1 > (\nu-1)\beta$ il faut que

$$1 > (\nu-1)\frac{\nu-\mu(E_1)}{\mu(E_1)}$$

ce qui n'est pas possible car $(\nu-1)/\mu(E_1) \geq 1$ et $\bigl(\nu-\mu(E_1)\bigr)$ est un entier positif.

Supposons ensuite que $E_1 = E_1^2 \cup E_1^3$. L'équation pour β est

$$1 = \sum_{E_1^2} \mu_i\beta + \sum_{E_1^3} \frac{\mu_i}{(\nu-1)}.$$

Pour que cette équation admette une solution positive, il faut d'abord que

$$1 > \frac{\mu(E_1^3)}{(\nu-1)}. \tag{7.16}$$

De plus pour que $(\nu-1)\beta > 1$, il faut que

$$\frac{\nu - 1 - \mu(E_1^3)}{\mu(E_1^2)} > 1 \qquad \Longrightarrow \nu - 1 > \mu(E_1).$$

Cette dernière inégalité implique (7.16).

Si elle est satisfaite, les équations pour les $a_i = a(p_i)$, $i \in E_1^2$ et $a_i = a(p_i)$, $i \in E_1^3$ sont

$$-A = -\frac{1}{a_i}\underline{a}^{\nu}, \quad i \in E_1^2, \qquad -A = -\frac{1}{a_i}a_i^{\nu}, \quad i \in E_1^3,$$
$$B = \left(\frac{1}{2}\right)^{1+\mu(E_2)} \underline{a}^{\mu(E_1^2)} \prod_{E_1^3} a_i^{\mu_i}.$$

Les deuxièmes équations admettent chacune $\nu - 1$ solutions distinctes en a_i. La dernière donne ensuite $\mu(E_1^2)$ valeurs pour $\underline{a}$. Les premières équations nous permettent enfin de trouver à tour de rôle chaque a_i, $i \in E_i^2$.

Donc si $\nu - 1 > \mu(E_1)$ on a

$$\sum_{E_1^2 \cup E_1^3 = E_1} \mu(E_1^2)(\nu-1)^{e_1^3} = \mu(E_1)\nu^{e_1-1} \tag{7.13''}$$

où l'égalité résulte de (7.14). Cette expression est (7.13) lorsque (7.15''') est satisfait.

Supposons enfin que $\nu - 1 = \mu(E_1)$. Alors $E_1 = E_1^4$ et les équations pour a_i et $\underline{a}$ sont

$$A = \frac{1}{a_i}(a_i + \underline{a})^{\nu}, \quad B = \frac{1}{2}\prod_{j \in E_1}(a_j + \underline{a})^{\mu_j}.$$

Il s'avère que ces équations admettent $\nu^{e_1-1}\mu(E_1)$ solutions.

Pour $e_1 = 1$ c'est évident car alors on trouve d'abord $a_1 + \underline{a}$ et ensuite a_1. En général nous divisons les premières équations par la dernière pour obtenir des équations pour $v_i = a_i/\underline{a}$

$$\frac{A}{2B}\prod_{j \in E_1}(v_j + 1)^{\mu_j} = \frac{(v_i+1)^{\nu}}{v_i}. \tag{7.17}$$

Une équation plus facile à traiter est

$$\frac{A}{2B}\prod_{j \in E_1}(v_j + 1)^{\mu_j} = (v_i+1)^{\nu}. \tag{7.18}$$

Pour la résoudre nous posons

$$v_i + 1 = \varepsilon_i w$$

où w ne dépend pas de i et $\varepsilon_i^\nu = 1$. Il faut donc que

$$\frac{A}{2B}\prod_{j\in E_1}\varepsilon_j^{\mu_j} = w,$$

une équation qui sert à fixer w. Il y a ν^{e_1} choix pour les ε_i. Mais si $\varepsilon^\nu = 1$ on peut remplacer chaque ε_i par $\varepsilon\varepsilon_i$ et w par $\varepsilon^{-1}w$. Cela nous donne ν^{e_1-1} possibilités, mais ensuite il faut trouver $\underline{a}$ à partir de l'équation

$$2B = \underline{a}^{\mu(E_1)}\prod_{j\in E_1}(v_i+1)^{\mu_j}. \tag{7.19}$$

Cette équation a $\mu(E_1)$ solutions.

Les deux systèmes d'équations (7.17) et (7.18) sont deux éléments d'une famille définie par

$$\frac{A}{2B}(\alpha+\beta v_i)\prod_{j\in E_1}(v_j+1)^{\mu_j} = (v_i+1)^\nu \tag{7.20}$$

auxquelles on ajoute (7.19). Puisque le déterminant de la matrice

$$\begin{pmatrix} \nu-\mu_1 & -\mu_2 & \dots & -\mu_r \\ -\mu_1 & \nu-\mu_2 & \dots & -\mu_r \\ \vdots & \vdots & \ddots & -\mu_r \\ -\mu_1 & -\mu_2 & \dots & \nu-\mu_r \end{pmatrix}$$

n'est pas nul, les ν^{e_1-1} solutions de (7.18) admettent des déformations autour de $\alpha = 1$, $\beta = 0$. Une solution v_i de l'équation

$$\lambda v_i = (v_i+1)^\nu \tag{7.21}$$

n'est une solution multiple que si $v_i + 1 = \nu v_i$, et alors λ et v_i sont des nombres rationnels. Le λ qui est pertinent pour (7.17) est

$$\lambda = \frac{A}{2B}\prod_{j\in E_1}(v_j+1)^{\mu_j}.$$

Si ce λ est rationnel et si le système (7.17) est satisfait alors tous les v_i sont algébriques et par conséquent $A/2B$ est algébrique et appartient même a un corps de nombres que l'on peut déterminer. Puisque $A/2B$ est supposé être générique ceci est exclu.

Supposons maintenant que nous commençons avec la collection de solutions de (7.18) et (7.19) que nous avons trouvées et que nous les déformons tout en évitant les valeurs de

$$\lambda = \frac{A}{2B}(\alpha+\beta v_i)\prod_{j\in E_1}(v_j+1)^{\mu_j}$$

où (7.21) a des solutions multiples. Cela est possible (par des principes généraux) sur une courbe qui va de $\alpha = 1$, $\beta = 0$ à $\alpha = 0$, $\beta = 1$ car elle n'a de solution multiple ni au point de départ ni au point final. Il en résulte que si deux solutions arrivent à des solutions avec les mêmes valeurs pour toutes les variables v_i alors elles commençaient avec les mêmes valeurs de v_i. Il résulte alors de (7.19) que le quotient des valeurs de $\underline{a}$ pour les deux solutions reste constant pendant la déformation. En particulier s'il est 1 à la fin, il est 1 au début et nous avons trouvé $\nu^{e_1-1}\mu(E_1)$ solutions distinctes de (7.17-7.19) en déformant celles de (7.18-7.19).

Il ne nous reste qu'à traiter le cas (7.15″). Si $\mu(E_1)=\nu$, nous cherchons des solutions telles que $p_{\underline{1}}^0$ n'est ni 0 ni ± 1. Les équations pour $x_i = p_i^0$ et $\underline{x} = p_{\underline{1}}^0$ et $u_i = p_i + p_{\underline{1}}$ sont

$$\begin{aligned} &x_i = -\underline{x}, \quad Ax_i = \frac{x_i}{x_i+1}\left(\frac{u_i}{u_i+1}\right)^{\nu}, \\ &B\underline{x} = \left(\frac{\underline{x}}{\underline{x}+1}\right)^{\mu(E_2)+1}\prod_{E_1}\left(\frac{u_i}{u_i+1}\right)^{\mu_i}. \end{aligned} \tag{7.22}$$

On résout les secondes équations pour trouver u_i en fonction de $\underline{x}$ pour obtenir enfin l'équation

$$B(\underline{x}+1) = \left(\frac{\underline{x}}{\underline{x}+1}\right)^{\mu(E_2)}\prod_{E_1}\bigl(A(1-\underline{x})\bigr)^{\mu_i/\nu},$$

ou

$$B(\underline{x}+1) = cA(1-\underline{x})\left(\frac{\underline{x}}{\underline{x}+1}\right)^{\mu(E_2)}. \tag{7.23}$$

La constante c dépend des racines choisies en résolvant les premières équations. Puisque A et B sont génériques, il y a $\mu(E_2)+1$ solutions distinctes de (7.23). Le nombre de solutions de (7.22) est donc $\nu^{e_1}\bigl(\mu(E_2)+1\bigr)$. Cette expression est la somme de ν^{e_1} qui est (13) et $\nu^{e_1}\mu(E_2)$ qui est la contribution pertinente à (14).

8. Derniers commentaires

Quoique nous ne soyons pas en état de traiter le cas général dans cet exposé, nous voulons tirer quelques conséquences des exemples. Le numérateur et le dénominateur de $S(p_i+p_{\underline{j}})$ ne sont 0 que si $p_i+p_{\underline{j}}$ est 0 ou -1. Nous répartissons par conséquent les indices en classes, deux indices i et i' (ou $\underline{j}$ et $\underline{j}'$) étant dans la même classe si la différence $p_i - p_{i'}$ est un entier. Par contre i et $\underline{j}$ sont dans la même classe si $p_i+p_{\underline{j}}$ est un entier. Les équations pour les exposants $\alpha(p_i)$ et $\alpha(p_{\underline{j}})$ découple car $\alpha(p_i+p_{\underline{j}})$ et $\alpha(p_i+p_{\underline{j}}+1)$ sont 0 si i et $\underline{j}$ n'appartiennent pas à la même classe. La classe attachée aux entiers, donc la classe à laquelle 0 appartient s'il se trouve parmi les p_i et $p_{\underline{j}}$, contient tous les i tels que $p_i \equiv 0$ et tous les $\underline{j}$ tels que $p_{\underline{j}} \equiv 0$.

Nous avons trouvé dans les exemples que $\alpha(p_i)$ ne dépend que de p_i^0 et que $\alpha(p_{\underline{j}})$ ne dépend que de $p_{\underline{j}}^0$. Nous ne savons pas si il s'agit ici d'un fait général, mais une telle hypothèse est certainement raisonnable.

Examinons les conséquences de cette hypothèse pour les exposants. Une classe Q correspond à une suite $q_k = q+k$, $K_1 \le k \le K_2$. Soit P_k l'ensemble d'indices i tels que $p_i = q-k$ mais $p_i \not\equiv 0$ et $\underline{P}_k$ l'ensemble d'indices $\underline{j}$ tels que $p_{\underline{j}} = -q+k$ mais $p_{\underline{j}} \not\equiv 0$. Soit R l'ensemble d'indices tels que $p_i \equiv 0$ et $\underline{R}$ l'ensemble tels que $p_{\underline{j}} \equiv 0$. Soit $r=|R|$ et $\underline{r}=|\underline{R}|$. En changeant q au besoin nous pouvons supposer que $K_1=0$. Posons alors $K=K_2$.

Supposons d'abord que q n'est pas un entier. Alors $p_i^0 \neq 0$ et $p_{\underline{j}}^0 \neq 0$ pour tout i et $\underline{j}$ dans Q. Soit $M_k = \sum_{i \in P_k} \mu_i$ et $N_k = \sum_{\underline{j} \in \underline{P}_k} \nu_{\underline{j}}$. Supposons que $\alpha(p_i + p_{\underline{j}})$ est égal à γ_k pour tout $i \in P_k$ et $\underline{j} \in \underline{P}_k$ tandis que $\alpha(p_i + p_{\underline{j}} + 1) = \delta_k$ pour tout $i \in P_k$ et tout $\underline{j} \in \underline{P}_{k-1}$. Les équations sont

$$
\begin{aligned}
1 &= N_k \gamma_k, && k = 0, \\
1 &= -N_{k-1}\delta_k + N_k \gamma_k, && 0 < k \leq K, \\
1 &= M_k \gamma_k - M_{k+1}\delta_{k+1}, && 0 \leq k < K, \\
1 &= M_k \gamma_k, && k = K.
\end{aligned}
\tag{8.1}
$$

Il y a $2K + 2$ équations et $2K + 1$ inconnues car δ_0 n'est pas défini.

La solution est

$$
\begin{aligned}
\gamma_0 &= \frac{1}{N_0}, \\
\delta_1 &= \frac{(M_0\gamma_0 - 1)}{M_1} = \frac{(M_0 - N_0)}{M_1 N_0}, \\
\gamma_1 &= \frac{(1 + N_0\delta_1)}{N_1} = \frac{(M_0 + M_1 - N_0)}{M_1 N_1}, \\
\delta_k &= \frac{(M_0 + \cdots + M_{k-1} - N_0 - \cdots - N_{k-1})}{M_k N_{k-1}}, \quad 0 < k \leq K, \\
\gamma_k &= \frac{(M_0 + \cdots + M_k - N_0 - \cdots - N_{k-1})}{M_k N_k}, \quad 0 \leq k \leq K, \\
\gamma_K &= \frac{1}{M_k}.
\end{aligned}
$$

En comparant les deux expressions pour γ_K nous concluons qu'il y a une solution si et seulement si $\sum_k M_k = \sum_k N_k$, ce qui correspond aux résultats des exemples.

Supposons maintenant que $q + k = 0$ pour $K_1 \leq K_2$. Sans perte de généralité nous pouvons supposer que $q = 0$. Les équations (8.1) se modifient légèrement. Supposons que $\alpha(p_i) = \alpha$ pour $i \in P_0$ et que $\alpha(p_{\underline{j}}) = \beta$ pour $\underline{j} \in \underline{P}_0$. Pour $k \neq 0$ on a toujours

$$
\begin{aligned}
1 &= N_k \gamma_k, && k = K_1, \\
1 &= -N_{k-1}\delta_k + N_k \gamma_k, && K_1 < k \leq K_2, \\
1 &= M_k \gamma_k - M_{k+1}\delta_{k+1}, && K_1 \leq k < K_2, \\
1 &= M_k \gamma_k, && k = K_2.
\end{aligned}
\tag{8.2}
$$

Mais pour $k = 0$ il faut tenir compte des éléments de R et $\underline{R}$,

$$
1 + \alpha = N\alpha - N_{-1}\delta_0 + N_0\gamma_0, \quad 1 + \beta = M\beta + M_0\gamma_0 - M_1\delta_1,
$$

où $N = \sum_{\underline{j} \in \underline{R}} \nu_{\underline{j}}$ et $M = \sum_{i \in R} \mu_i$. Si $K_1 = 0$ posons $N_{-1} = 0$ et si $K_2 = 0$ posons $M_1 = 0$. Il y a maintenant plus d'inconnues que d'équations mais il y a des conditions supplémentaires car une des conditions suivantes est satisfaite :

(1) $\alpha = \beta < \gamma_0$;
(2) $\alpha = \gamma_0 < \beta$;
(3) $\beta = \gamma_0 < \alpha$;
(4) $\alpha = \beta = \gamma_0$.

Avec ces conditions le nombre d'inconnues devient égal au nombre d'équations. Nous avons vu qu'il est même nécessaire de permettre des combinaisons de ces conditions en divisant P_0 et $\underline{P}_0$ en sous-ensembles.

Les arguments combinatoires qui permettent de traiter tous les cas d'une manière uniforme et de montrer que les conclusions des exemples sont toujours valables restent à faire ! Il s'agit certainement d'un problème séduisant pour un mathématicien mais dont la pertinence pour l'étude du problème de départ, le calcul des valeurs propres d'une matrice, n'est point évidente.

Appendice A

Les équations de Bethe peuvent être définies à partir de deux applications rationnelles $\varphi = \varphi_\Delta$ et $\psi = \psi_\Delta$ d'une variété X dans elle-même. Cette variété est un produit de droites projectives,

$$X = \prod_{i=1}^{r} \mathbb{P}^1 \times \prod_{1 \le i < j \le r} \mathbb{P}^1 = Z \times W.$$

Les coordonnées sont z_i, $1 \le i \le r$, et $w_{i,j}$, $1 \le i < j \le r$. Bien qu'elles soient superflues, nous utiliserons également les coordonnées $w_{i,j} = w_{j,i}^{-1}$ pour $i > j$.

L'application φ est définie par $\varphi\colon (z,w) \to (z',w')$ où $z'_i = R(z_i)$ et $w'_{i,j} = w_{i,j}$ alors que l'application ψ l'est par $\psi\colon (z,w) \to (z'',w'')$ où

$$z''_i = \prod_{j \ne i} w_{i,j} \tag{A.1}$$

et

$$w''_{i,j} = Q(z_i, z_j) = \frac{-F(z_i,z_j)}{F(z_j,z_i)}, \tag{A.2}$$

avec $F(z,z') = F_\Delta(z,z') = zz' - 2\Delta z + 1$. Les solutions de l'équation de Bethe sont les points (z,w) tels que

$$(z',w') = (z'',w'').$$

Cependant, pour appliquer la formule de Lefschetz, on a besoin d'une correspondance

$$(\varphi,\psi)\colon C \to X \times X$$

définie par des applications régulières partout. Il est claire que l'application ψ n'est pas bien définie aux points où $F(z_i,z_j) = F(z_j,z_i) = 0$.

Nous reviendrons aux éclatements nécessaires pour qu'elle le soit. Puisqu'il faut dénombrer non seulement l'ensemble de toutes les solutions des équations de Bethe mais également le sous-ensemble des solutions inadmissibles pour ensuite prendre la différence, nous introduisons d'abord des sous-variétés $X^{A,B}$ de X et des correspondances (encore mal définies) sur ces sous-variétés. Soit $\{1,\dots,r\}$ la réunion disjointe de sous-ensembles $A_1,\dots,A_s$, $B'_1,\dots,B'_t$, et $B''_1,\dots,B''_t$. Soit $B_l = B'_l \cup B''_l$. Nous exigeons que tous les A_l, B'_l et B''_l soient non vides. A chaque l est attaché un entier $k = k(l)$, $1 \le k \le N$. Nous posons $i \equiv j$ si i et j appartiennent au même A_l, ou si $i \in B'_l$ et $j \in B'_m$ avec $k(l) = k(m)$ ou enfin

$i \in B''_l$ et $j \in B''_m$ avec encore $k(l) = k(m)$. Parfois, pour être plus clairs, nous écrivons $i \equiv j(A, B)$, où

$$A = \{A_1, \dots, A_s\}, \quad B' = \{B'_1, \dots, B'_t\}, \quad B'' = \{B''_1, \dots, B''_t\}, \quad B = \{B', B''\}.$$

Soit $X^{A,B}$ la sous-variété de X définie par les conditions :
(1) $z_i = z_j$ et $w_{i,m} = w_{j,m}$ si $i \equiv j$;
(2) si $i \equiv j$ alors $w_{i,j} = -1$;
(3) si $i \in B'_l$ alors $z_i = 0$;
(4) si $i \in B''_l$ alors $z_i = \infty$;
(5) si $i \in B'_{l'}$ et $j \in B''_{l''}$ avec $k(l') = k(l'')$ alors $w_{i,j} = 0$ et $w_{j,i} = \infty$;
(6) supposons que $k' = k(l') \neq k(l'') = k''$ alors
(a) $w_{i,j} = Q(\alpha_{k'}, \alpha_{k''})$ pour $i \in B'_{l'}$ et $j \in B'_{l''}$,
(b) $w_{i,j} = Q(\alpha_{k'}, \beta_{k''})$ pour $i \in B'_{l'}$ et $j \in B''_{l''}$,
(c) $w_{i,j} = Q(\beta_{k'}, \beta_{k''})$ pour $i \in B''_{l'}$ et $j \in B''_{l''}$,
(d) $w_{i,j} = Q(\beta_{k'}, \alpha_{k''})$ pour $i \in B''_{l'}$ et $j \in B'_{l''}$.

On observe que pour obtenir X lui-même il suffit de prendre $s = r$ (de sorte que $A = \{\{1\}, \dots, \{r\}\}$) et $t = 0$. Nous utilisons plusieurs B'_l et B''_l avec le même $k = k(l)$ parce que cela simplifie les arguments combinatoires mais à maints égards il est plus utile de ne considérer que

$$\mathbb{B}'_k = \cup_{\{l|k(l)=k\}} B'_l, \quad \mathbb{B}''_k = \cup_{\{l|k(l)=k\}} B''_l.$$

Soit $\mathbb{B}_k = \mathbb{B}'_k \cup \mathbb{B}''_k$. Il n'est pas exclu que $\mathbb{B}'_k$ ou $\mathbb{B}''_k$ soit vide mais alors tous les deux le sont.

La variété $C^{A,B}$ est définie par des conditions quelque peu différentes :
(1) $z_i = z_j$ et $w_{i,m} = w_{j,m}$ si $i \equiv j$;
(2) si $i \equiv j$ alors $w_{i,j} = -1$;
(3) si $i \in B'_l$ et $k = k(l)$ alors $z_i = \alpha_k$;
(4) si $i \in B''_l$ et $k = k(l)$ alors $z_i = \beta_k$;
(5) si $i \in B'_{l'}$ et $j \in B''_{l''}$ avec $k(l') = k(l'')$ alors $w_{i,j} = 0$ et $w_{j,i} = \infty$.
(6) supposons que $k' = k(l') \neq k(l'') = k''$ alors
(a) $w_{i,j} = Q(\alpha_{k'}, \alpha_{k''})$ pour $i \in B'_{l'}$ et $j \in B'_{l''}$,
(b) $w_{i,j} = Q(\alpha_{k'}, \beta_{k''})$ pour $i \in B'_{l'}$ et $j \in B''_{l''}$,
(c) $w_{i,j} = Q(\beta_{k'}, \beta_{k''})$ pour $i \in B''_{l'}$ et $j \in B''_{l''}$,
(d) $w_{i,j} = Q(\beta_{k'}, \alpha_{k''})$ pour $i \in B''_{l'}$ et $j \in B'_{l''}$.

La condition (5) sur la variété $C^{A,B}$ nous permet de définir l'application $\psi^{A,B}$ de sorte qu'elle l'envoie dans une sous-variété de X qui satisfait aux conditions (2) et (3) de la définition de $X^{A,B}$. Pour que $C^{A,B}$ soit de la même dimension que $X^{A,B}$ il faut que $X^{A,B}$ satisfasse à la même condition. Les conditions (6) sont facultatives. Nous les avons introduites simplement pour ne pas être encombrés par des coordonnées qui pourraient sembler arbitraires mais dont les valeurs aux points fixes sont prescrites.

Les deux variétés $X^{A,B}$ et $C^{A,B}$ sont des produits de deux variétés, celle avec les coordonnées z_i et celle avec les coordonnées $w_{i,j}$. La deuxième variété est la même dans les deux cas et sera dénotée $W^{A,B}$. Par contre les conditions sur les premières coordonnées sont différentes et on dénote le premier facteur de $X^{A,B}$ par $Z^{A,B}$ et le premier facteur de $C^{A,B}$ par $E^{A,B}$

L'ensemble des paires (A, B) est (partiellement) ordonné. La paire $(\tilde{A}, \tilde{B})$ est plus profonde (ou plus grossière) que (A, B) si chaque A_k est contenu dans un $\tilde{A}_{l'}$, un $\widetilde{B}'_l$ ou un $\widetilde{B}''_l$, chaque B'_l dans un $\widetilde{B}'_{l'}$, et chaque B''_l dans un $\widetilde{B}''_{l'}$. Si $(\tilde{A}, \tilde{B})$ est plus profond que (A, B) alors $C^{\tilde{A},\tilde{B}}$ est une sous-variété de $C^{A,B}$. En particulier si $s = r$ et $t = 0$ alors $C = X$ et toute variété $C^{A,B}$ est une sous-variété de X.

L'application φ envoie évidemment $C^{A,B}$ dans $X^{A,B}$ et devient

$$\varphi = \varphi^{A,B} = \varphi_\Delta^{A,B}\colon (z, w) \to (z', w')$$

avec

$$z'_i = R(z_i) \mid i \in A_k, \quad z'_i = 0 \mid i \in B'_l, \quad z'_i = \infty \mid i \in B''_l,$$

et $w'_{i,j} = w_{i,j}$ pour tout i et j. Nous dénotons la restriction de φ à $C^{A,B}$ par $\varphi = \varphi^{A,B} = \varphi_\Delta^{A,B}$.

Pour définir l'application

$$\psi = \psi^{A,B} = \psi_\Delta^{A,B}\colon (z, w) \to (z'', w''),$$

nous posons d'abord $z''_i = 0$ s'il existe l tel que $i \in \mathbb{B}'_l$ et $z''_i = \infty$ s'il existe k tel que $i \in \mathbb{B}''_k$. Nous posons autrement

$$z''_i = \prod_{j \neq i} w_{i,j}.$$

De la même façon nous posons $w''_{i,j} = 0$ et $w''_{j,i} = \infty$ si $i \in \mathbb{B}'_k$ et $j \in \mathbb{B}''_k$ pour un certain k et $w''_{i,j} = -1$ si $i \equiv j$. Autrement nous définissons $w''_{i,j}$ par (A.2).

Dans le cas générique, qui exige en particulier que $\Delta \neq 0$, aucun α_i n'est un point fixe ni de l'application linéaire

$$\alpha \to A(\alpha) = 2\Delta - \frac{1}{\alpha} \tag{A.3}$$

ni de son carré. Par conséquent si $i \in \mathbb{B}'_k$ et $j \in \mathbb{B}''_k$ alors $F(z_i, z_j) = 0$ mais $F(z_j, z_i) \neq 0$ pour $(z, w) \in C^{A,B}$. Donc $Q(z_i, z_j) = 0$ de sorte que l'application $\psi^{A,B}$ est la restriction de ψ à $C^{A,B}$.

Nous observons en particulier que si δ_1 et δ_2 sont les deux racines de $z^2 - 2\Delta z + 1 = 0$ alors δ_1 et δ_2 sont fixés par l'application (A.3). Sauf à $\Delta = \pm 1$ ces deux racines sont différentes. Dans le cas générique on a

$$R(\delta_i) \neq 0, \quad i = 1, 2.$$

L'application linéaire (A.3) est définie par la matrice

$$\begin{pmatrix} 2\Delta & -1 \\ 1 & 0 \end{pmatrix}$$

dont les valeurs propres sont $\exp(\pm\eta\sqrt{-1})$ si $\cos\eta = \Delta$. Si $\Delta = \pm 1$ l'application linéaire est parabolique de sorte qu'elle n'a pas de point d'ordre fini sauf le point fixe $\delta_1 = \delta_2$. Pour les autres valeurs de Δ cette application n'a pas de point d'ordre deux sauf si $\Delta = 0$, et alors tout point sauf δ_1 et δ_2 est d'ordre deux.

Si $\mathbb{B}_k$ n'est pas vide alors les applications

$$z \to \frac{F(z, \alpha_k)}{F(\alpha_k, z)}, \quad z \to \frac{F(z, \beta_k)}{F(\beta_k, z)}$$

interviennent dans la définition de $\psi^{A,B}$. Elles sont des fonctions linéaires de z qui s'écrivent

$$\frac{\alpha_k - 2\Delta}{\alpha_k}\frac{z-\gamma_k}{z-\beta_k}, \quad \frac{\beta_k - 2\Delta}{\beta_k}\frac{z-\alpha_k}{z-\delta_k} \tag{A.4}$$

où $A(\gamma_k) = \alpha_k$ et $\delta_k = A(\beta_k)$. Donc pourvu que $\Delta \neq 0$ elles sont bien définies car $\gamma_k \neq \beta_k$ et $\alpha_k \neq \delta_k$. Pour les cas exceptionnels où $\alpha_k = 0$ ou $\beta_k = 0$ on remplace (A.4) par

$$-2\Delta(z-\beta_k), \quad -2\Delta(z-\alpha_k).$$

Par contre si $\alpha_k - 2\Delta = 0$ ou $\beta_k - 2\Delta = 0$ on a

$$\frac{1}{\alpha_k}\frac{1}{z-\beta_k}, \quad \frac{1}{\beta_k}\frac{1}{z-\delta_k}.$$

Ainsi, ce fonctions sont toujours de vraies fonctions linéaires et non pas simplement des constantes car le dénominateur et le numérateur ne peuvent pas s'annuler en même temps.

Puisque l'applications $\psi^{A,B}$ est mal définie il faut modifier la variété $C^{A,B} \times \mathbb{P}^1$ qui est une variété fibrée sur $\mathbb{P}^1$ en sorte que $\psi^{A,B} = \psi_\Delta^{A,B}$ soit bien défini sur les nouvelles fibres $C_\Delta^{A,B}$. Les problèmes proviennent principalement de deux sources. Parfois, pour un indice i qui n'est ni dans un $\mathbb{B}'_k$ ni dans un $\mathbb{B}''_k$, il y a des indices supplémentaires j et m tels que $w_{i,j} = 0$ et $w_{i,m} = \infty$. Alors la formule (A.1) n'a plus de sens. D'un autre côté il peut arriver que pour deux indices i et j on ait

$$F(z_i, z_j) = F(z_j, z_i) = 0 \tag{A.5}$$

de sorte que le quotient dans (A.2) n'est pas défini. Autour de $\Delta = 0$ un troisième obstacle se présente parce que quelques-uns des quotients (A.4) peuvent ne pas être définis.

Nous remarquons tout d'abord que les modifications nécessaires pour enlever les premières difficultés ne font intervenir que les coordonnées $w_{i,j}$ tandis que les modifications pour les secondes n'impliquent que les coordonnées z_i et Δ. Cela nous permet de les traiter séparément et donc d'éclater les deux facteurs

$$W^{A,B} \subseteq W = \prod_{i<j} \mathbb{P}^1$$

et

$$E^{A,B} \times \mathbb{P}^1 \subseteq Z \times \mathbb{P}^1,$$

de $C^{A,B} \times \mathbb{P}^1$ et de prendre ensuite le produit des variétés ainsi obtenues. On se souvient qu'il y a des contraintes sur les coordonnées de $E^{A,B}$ et de $W^{A,B}$. Quelques-unes parmi ces coordonnées sont nécessairement égales. Les deux variétés $E^{A,B}$ et $W^{A,B}$ sont néanmoins aussi des produits d'espaces projectifs de dimension un. Les produits portent sur les coefficients qui ont des valeurs variables et à chaque famille de coordonnées égales (donc pour $E^{A,B}$ à chaque A_k) ne correspond qu'un seul facteur.

Nous avertissons le lecteur que la fibre au-dessus de $\Delta = \infty$ ne nous intéresse pas quoique nous examinerons de près son voisinage dans un autre papier. Nous avons inclus cette fibre par force d'habitude, pour ne pas sortir du cadre des variétés projectives.

Les modifications qui impliquent les coordonnées z_i servent des fins différentes que celles qui impliquent les coordonnées $w_{i,j}$. Les variétés données par les premières exigeront une étude locale soigneuse pour comprendre les multiplicités des points fixes. Par contre les variétés obtenues par les modifications des coordonnées $w_{i,j}$ sont plutôt abstraites et leur but est surtout de permettre l'utilisation rigoureuse de la formule de Lefschetz. Soit $\mathcal{C}^{A,B}$ la variété obtenue en modifiant les coordonnées z_i et en suivant ces modifications par celles des coordonnées $w_{i,j}$. Donc $\mathcal{C}^{A,B}$ s'obtient en modifiant d'abord $E^{A,B} \times \mathbb{P}^1$ pour obtenir $\mathcal{E}^{A,B}$ et en modifiant ensuite $W^{A,B}$ pour obtenir $\mathcal{W}^{A,B}$ et en posant enfin

$$\mathcal{C}^{A,B} = \mathcal{E}^{A,B} \times \mathcal{W}^{A,B}. \tag{A.6}$$

Soit π_1 et π_2 les compositions de

$$\mathcal{C}^{A,B} \to C^{A,B} \times \mathbb{P}^1$$

avec les deux projections $C^{A,B} \times \mathbb{P}^1 \to C^{A,B}$ et $C^{A,B} \times \mathbb{P}^1 \to \mathbb{P}^1$. Nous dénotons la fibre de π_2 au-dessus du point Δ par $C_\Delta = C_\Delta^{A,B}$.

On aura

Lemme A.1. *L'application π_2 est lisse sur les fibres génériques.*

En particulier la variété $C_\Delta^{A,B}$ est lisse pour des valeurs génériques de Δ.

Soit $E_\Delta^{A,B}$ la fibre de $\mathcal{E}^{A,B}$ au-dessus de Δ. Les définitions seront telles que

$$\mathcal{E}^{\tilde{A},\widetilde{B}} \subseteq \mathcal{E}^{A,B}, \quad E_\Delta^{\tilde{A},\widetilde{B}} \subseteq E_\Delta^{A,B}$$

si $(\tilde{A},\widetilde{B})$ est plus profond que (A,B). La relation entre $\mathcal{C}^{\tilde{A},\widetilde{B}}$ et $\mathcal{C}^{A,B}$ sera par contre plus compliquée. Il existera une sous-variété $\mathcal{C}^{A,B}(\tilde{A},\widetilde{B})$ de $\mathcal{C}^{A,B}$ pourvue d'une application

$$\eta: \mathcal{C}^{A,B}(\tilde{A},\widetilde{B}) \to \mathcal{C}^{\tilde{A},\widetilde{B}} \tag{A.7}$$

mais cette sous-variété n'est pas toujours lisse et l'application η n'est pas toujours un isomorphisme. Elle est toutefois surjective. Soit

$$C_\Delta^{A,B}(\tilde{A},\widetilde{B})$$

la fibre de $\mathcal{C}^{A,B}(\tilde{A},\widetilde{B})$ au point Δ. La restriction de η à la fibre sera aussi dénotée η. Le diagramme

$$\begin{array}{ccc} C_\Delta^{A,B}(\tilde{A},\widetilde{B}) & \xrightarrow{\ \eta\ } & C_\Delta^{\tilde{A},\widetilde{B}} \\ {\scriptstyle \varphi_\Delta^{A,B}\times\psi_\Delta^{A,B}}\downarrow & & \downarrow{\scriptstyle \varphi_\Delta^{\tilde{A},\widetilde{B}}\times\psi_\Delta^{\tilde{A},\widetilde{B}}} \\ X^{A,B} \times X^{A,B} & \longleftarrow & X^{\tilde{A},\widetilde{B}} \times X^{\tilde{A},\widetilde{B}} \end{array} \tag{A.8}$$

sera commutatif.

Bien qu'il existe sans doute des références utiles pour la construction simple qui résolvent les difficultés qui proviennent des points où (A.5) est satisfait pour une paire (i,j), nous ne les avons pas trouvées. Notre discussion n'en sera pas moins aussi brève que possible. La nouvelle variété $\mathcal{E}^{A,B}$ sera définie en ajoutant des coordonnées supplémentaires à Δ et aux z_i. Pour chaque paire (i,j) (telle que $i < j$) nous ajoutons des coordonnées projectives $(u_{i,j}, u_{j,i})$ qu'on choisit habituellement tel qu'une des deux coordonnées $u_{i,j}$ et $u_{j,i}$ est égale à 1. On introduit en plus des relations

$$u_{i,j}F(z_j,z_i)=u_{j,i}F(z_i,z_j) \tag{A.9}$$

et des coordonnées $\lambda_{i,j}=\lambda_{j,i}$ supplémentaires telles que

$$F(z_i,z_j)=\lambda_{i,j}u_{i,j},\quad F(z_j,z_i)=\lambda_{j,i}u_{j,i}.$$

Par exemple, autour d'un point où $z_j=\infty$ il convient d'utiliser plutôt la coordonnée $\tilde z_j=-1/z_j$ de sorte que $F(z_i,z_j)$ est remplacée par $F(z_i,\tilde z_j)=z_i+2\Delta z_i\tilde z_j-\tilde z_j$ et $F(z_j,z_i)$ par $F(\tilde z_j,z_i)=z_i-2\Delta-\tilde z_j$. L'équation (A.9) devient

$$u_{i,j}F(\tilde z_j,z_i)=u_{j,i}F(z_i,\tilde z_j).$$

Soulignons qu'en un point où $F(z_i,z_j)\neq 0$ ou $F(z_j,z_i)\neq 0$ l'équation (A.9) nous permet d'exprimer le point projectif $(u_{i,j},u_{j,i})$ en termes des z. Donc pour cette paire (i,j), ces nouvelles coordonnées sont alors superflues. En particulier si $i\in\mathbb{B}'_k$ et $j\in\mathbb{B}''_k$ on a $(u_{i,j},u_{j,i})=(0,1)$.

Pour que les quotients (A.4) soient toujours définis il faut ajouter des coordonnées projectives $(p'_{i,k},q'_{i,k})$ et $(p''_{i,k},q''_{i,k})$ aussi bien que $\xi'_{i,k}$ et $\xi''_{i,k}$ telles que

$$z_i-\gamma_k=\xi'_{i,k}p'_{i,k},\quad z_i-\beta_k=\xi'_{i,k}q'_{i,k},\quad z_i-\alpha_k=\xi''_{i,k}p''_{i,k},\quad z_i-\delta_k=\xi''_{i,k}q''_{i,k}.$$

Ces coordonnées sont toutes superflues sauf à $\Delta=0$. Évidemment on suppose comme condition de généricité que ni α_k ni β_k n'est égal à 0 en $\Delta=0$.

Les relations (A.9) ne sont pas les seules auxquelles sont assujetties les nouvelles coordonnées. Soient $\{(i_1,j_1),\dots,(i_M,j_M)\}$ un ensemble de paires fixées. Supposons que localement on ait des fonctions régulières

$$g_{i'_1,j'_1,\dots,i'_M,j'_M}$$

définies pour (i'_m,j'_m) égal à (i_m,j_m) ou à (j_m,i_m) et une équation multihomogène

$$\sum_{\{(i_1,j_1),(j_1,i_1)\}}\cdots\sum_{\{(i_M,j_M),(j_M,i_M)\}} g_{i'_1,j'_1,\dots,i'_M,j'_M}\times F(z_{i'_1},z_{j'_1})\dots F(z_{i'_M},z_{j'_M})=0. \tag{A.10}$$

Alors

$$\sum_{\{(i_1,j_1),(j_1,i_1)\}}\cdots\sum_{\{(i_M,j_M),(j_M,i_M)\}} g_{i'_1,j'_1,\dots,i'_M,j'_M}u_{i'_1,j'_1}\dots u_{i'_M,j'_M}=0. \tag{A.11}$$

Il y a des conditions semblables dans lesquelles interviennent en plus les coordonnées $p'_{i,k}$, $q'_{i,k}$, $p''_{i,k}$, $q''_{i,k}$. En particulier, si $i\in\mathbb{B}'_k$, de la relation $z_i-\alpha_k=0$ résulte $p''_{i,k}=0$ et si $i\in\mathbb{B}''_k$, on a $q'_{i,k}=0$. De plus si i et j sont dans le même A_l, les égalités évidentes entre ces coordonnées supplémentaires doivent alors être satisfaites.

Les définitions sont évidemment telles que si $(\tilde A,\widetilde B)$ est plus profond que (A,B) alors $\mathcal{E}^{\tilde A,\widetilde B}$ est une sous-variété de $\mathcal{E}^{A,B}$ car les éclatements que nous venons de décrire sont définis d'une façon cohérente. Les nouvelles coordonnées sont les mêmes pour les deux variétés, mais pour la paire plus profonde elles sont assujetties à plus de contraintes. Par conséquent la nouvelle variété attachée à une paire $(\tilde A,\widetilde B)$ plus profonde que (A, B) est trivialement une sous-variété de celle attachée à (A, B). Par exemple, si $i \equiv i'$ alors

$$F(z_i, z_j)F(z_j, z_{i'}) = F(z_{i'}, z_j)F(z_j, z_i).$$

Il en résulte que $(u_{i,j}, u_{j,i})$ et $(u_{i',j}, u_{j,i'})$ définissent le même point projectif et donc qu'un des deux points est superflu.

On remarque aussi que dans le cas où z_i est égal à ∞, de sorte que la coordonnée convenable est $\tilde{z}_i$, on remplace $F(z_i, z_j)$ par $F(\tilde{z}_i, z_j) = z_j + 2\Delta - \tilde{z}_i$ ou même $F(\tilde{z}_i, \tilde{z}_j) = 1 - 2\Delta\tilde{z}_j + \tilde{z}_i\tilde{z}_j$ et $F(z_j, z_i)$ par $F(z_j, \tilde{z}_i) = z_j + 2\Delta\tilde{z}_i z_j - \tilde{z}_i$ ou $F(\tilde{z}_j, \tilde{z}_i) = 1 - 2\Delta\tilde{z}_i + \tilde{z}_i\tilde{z}_j$. Il s'agit de changements faciles dont on tiendra compte aux endroits appropriés. Il y a une ambiguïté malencontreuse dans cette notation dont le lecteur est prié de se rendre compte. Lorsque un tilde apparaît parmi les arguments de la fonction F, la fonction elle-même se modifie.

Puisque l'application π_2 du lemme A.1 est définie à partir d'une application de $\mathcal{E}^{A,B}$ dans $\mathbb{P}^1$ la démonstration du lemme remonte au lemme suivant, dont l'énoncé anticipe la définition des variétés $\mathcal{W}^{A,B}$. Nous la donnerons par la suite.

LEMME A.2. *Les fibres de l'application* $\mathcal{E}^{A,B} \to \mathbb{P}^1$ *sont lisses sauf en* $\Delta = 0, \pm 1, \infty$; *et les variétés* $\mathcal{W}^{A,B}$ *sont lisses.*

Supposons que Δ n'appartienne pas à $\{0, \pm 1, \infty\}$. On peut remplacer les coordonnées projectives $(u_{i,j}, u_{j,i})$, $i < j$, par $(x_{i,j}, y_{i,j})$ avec

$$x_{i,j} = \frac{u_{j,i} - u_{i,j}}{2\Delta}, \quad y_{i,j} = u_{i,j} + z_i x_{i,j}. \tag{A.12}$$

Les relations (A.9) sont remplacées par

$$x_{i,j}(z_i^2 - 2\Delta z_i + 1) = y_{i,j}(z_i - z_j). \tag{A.13}$$

En effet, puisque les relations (A.12) définissent une transformation linéaire bijective, la paire $(x_{i,j}, y_{i,j})$ sert aussi bien que $(u_{i,j}, u_{j,i})$ comme coordonnées projectives.

La fonction

$$z^2 - 2\Delta z + 1 = (z - \delta_1)(z - \delta_2)$$

n'est 0 qu'aux deux points δ_1 et δ_2 qui sont différents puisque $\Delta \neq \pm 1$. Autour de $z_i = \delta_1$ on peut remplacer $y_{i,j}$ par

$$y'_{i,j} = \frac{y_{i,j}}{z_i - \delta_2}$$

et $x_{i,j}$ par

$$x'_{i,j} = -x_{i,j} + y'_{i,j}.$$

Alors (A.13) est remplacé par

$$x'_{i,j}(z_i - \delta_1) = y'_{i,j}(z_j - \delta_1). \tag{A.14}$$

En particulier les coordonnées $x'_{i,j}$ et $y'_{i,j}$ sont superflues sauf si z_j est aussi près de δ_1. On traite les coordonnées qui sont près de δ_2 de la même façon.

Il en résulte qu'autour d'un point p donné il est utile de répartir les coordonnées z_i dans trois sous-ensembles : D_1 qui contient celles telles que $z_i(p) = \delta_1$; D_2 qui contient celles telles que $z_i(p) = \delta_2$; et D_0 qui contient celles telles que $z_i(p)$ n'est ni δ_1 ni δ_2. Les seules coordonnées $(x'_{i,j}, y'_{i,j})$ qu'il faut retenir autour de p sont celles telles que i et j sont tous les deux dans D_1 ou D_2. Pour elles on a soit (A.14) soit

$$x'_{i,j}(z_i - \delta_2) = y'_{i,j}(z_j - \delta_2). \tag{A.15}$$

Si $i < j$ et si on a retenu $y'_{i,j}$, soit $y'_{j,i} = x'_{i,j}$. Les relations du genre (A.10) sont remplacées par

$$\sum_{\{(i_1,j_1),(j_1,i_1)\}} \cdots \sum_{\{(i_M,j_M),(j_M,i_M)\}} g_{i_1,j_1,\ldots,i_M,j_M}(z_{i_1} - \delta_{\varepsilon(1)}) \ldots (z_{i_M} - \delta_{\varepsilon(M)}) = 0,$$

où $\varepsilon(m)$ est 1 ou 2. Dans chaque somme les indices $i_m \neq j_m$ parcourent l'ensemble $D_{\varepsilon(m)}$ Chacune de ces relations implique une relation semblable entre les $y'_{i,j}$

$$\sum_{\{(i_1,j_1),(j_1,i_1)\}} \cdots \sum_{\{(i_M,j_M),(j_M,i_M)\}} g_{i_1,j_1,\ldots,i_M,j_M} y'_{i_1,j_1} \ldots y'_{i_M,j_M} = 0.$$

Puisque

$$(z_i - \delta_\varepsilon)(z_j - \delta_\varepsilon)(z_k - \delta_\varepsilon) = (z_i - \delta_\varepsilon)(z_j - \delta_\varepsilon)(z_k - \delta_\varepsilon),$$

il en résulte que

$$y'_{i,j}y'_{j,k}y'_{k,i} = y'_{i,k}y'_{k,j}y'_{j,i}. \tag{A.16}$$

Il s'agit ici des coordonnées projectives. Posons, pour un point p donné, $j \succeq i$ si $y'_{j,i} \neq 0$. Alors on a soit $j \succeq i$ soit $i \succeq j$ et même, dans la plupart des cas, les deux simultanément. Supposons que $k \succeq j$ et $j \succeq i$. Si $y'_{k,i} = 0$ il résulte de (A.16) que $y'_{i,k} = 0$; mais ceci est exclu puisqu'il s'agit des coordonnées projectives. Par conséquent $k \succeq i$.

Pour chaque point il a des éléments maximaux dans D_1 et dans D_2 par rapport à cet ordre et un élément qui est maximal à un point donné l'est aussi dans un voisinage de ce point. Soient

$$D_1 = \{i_1, i_2, \ldots, i_{s_1}\}, \quad D_2 = \{j_1, j_2, \ldots, j_{s_2}\},$$

où les deux suites (i_k) et (j_k) sont décroissantes. Alors

$$z_{i_{k+1}} - \delta_1 = (z_{i_k} - \delta_1)\frac{y'_{i_{k+1},i_k}}{y'_{i_k,i_{k+1}}}.$$

et

$$z_{j_{k+1}} - \delta_2 = (z_{j_k} - \delta_2)\frac{y'_{j_{k+1},j_k}}{y'_{j_k,j_{k+1}}}.$$

Par conséquent, si on met tous les $y'_{i,j}$, $i \succeq j$, égaux à 1, ce qui est possible car il s'agit de coordonnées projectives, alors toutes les coordonnées sont des fonctions régulières de z_{i_1}, y'_{i_{k+1},i_k}, $1 \leq k < s_1$, z_{j_1}, y'_{j_{k+1},j_k}, $1 \leq k < s_2$, et de z_i, $i \in D_0$ et (pour ne pas l'oublier) de Δ. Il en résulte que la variété éclatée de cette façon est lisse et, qui est plus, que l'application π_2 de cette variété dans $\mathbb{P}^1$ déduite de la projection $E^{A,B} \times \mathbb{P}^1 \to \mathbb{P}^1$ est aussi lisse. Donc la première affirmation du lemme est valable.

Il faut ensuite nous occuper des difficultés qui proviennent des quotients qui définissent les fonctions z''_i. Seules les coordonnées $w_{i,j}$ sont pertinentes. Les paires

(i,j) pour lesquelles les coordonnées $w_{i,j}$ sont égales à 0 ou à l'infini sur $C^{A,B}$ par la définition même de cette variété, exigeront une attention particulière.

Nous commençons en introduisant la variété $V = V^{A,B}$ définie par les coordonnées $w_{i,j}$ en admettant $i > j$ mais en exigeant que $w_{i,j}w_{j,i} = 1$ et en excluant toute paire telle que $w_{i,j}$ est par définition égal à 0 ou à l'infini ou à -1 sur $W = W^{A,B}$. (Nous supprimerons parfois les indices A et B.) Si $i \equiv i'$ et $j \equiv j'$ alors $w_{i,j}$ ne diffère pas de $w_{i',j'}$. Les deux coordonnées sont les mêmes. Les valeurs de toutes les coordonnées sont nécessairement différentes de 0. Il s'agit d'un tore algébrique dans le sens de [**TE**].

Le groupe $U = U^{A,B}$ de caractères du tore pertinent est engendré par des éléments $e_{i,j}$ sauf que les paires (i,j) telles que $i \in \mathbb{B}'_k$ ou $j \in \mathbb{B}''_k$ ne sont pas admises. Les seules relations sont $e_{i,j} + e_{j,i} = 0$, $e_{i,j} = e_{i',j'}$ si $i \equiv i'$ et $j \equiv j'$ et $e_{i,j} = 0$ si $i \equiv j$. La première de ces relations n'est pertinente que si les deux éléments qui y interviennent sont définis. Soit $Y = Y^{A,B} = \mathrm{Hom}(U, \mathbb{R})$.

L'application évidente

$$V \to W = W^{A,B} = \prod \mathbb{P}^1$$

est un plongement normal équivariant dans le sens de [**TE**]. Il est associé dans le sens de [**TE**, §I.2, th. 6] à la décomposition polyédrique de Y définie par des hyperplans $e_{i,j}(y) = 0$. Chaque élément ouvert de la décomposition est donc défini par des inégalités

$$\varepsilon_{i,j} e_{i,j}(y) \geq 0, \tag{A.17}$$

où $\varepsilon_{i,j} = \pm 1$ et $\varepsilon_{i,j}\varepsilon_{j,i} = 1$. Pour (i,j) et (i',j') tels que $e_{i,j} = e_{i',j'}$, on exige en plus que $\varepsilon_{i,j} = \varepsilon_{i',j'}$. Si $i \equiv j(A,B)$ le facteur $\varepsilon_{i,j}$ n'est pas défini. Nous observons que les coordonnées supplémentaires qui apparaissent dans la condition (5) de la définition de $C^{A,B}$ et qui sont des constantes apparaissent parmi celles de $W^{A,B}$ mais ont été supprimées en définissant $V^{A,B}$.

Notre éclatement de W sera défini par une décomposition polyédrique plus fine que (A.17) (voir [**TE**, §I.2, th. 7]). Pour chaque i dans la réunion des ensembles A_k soit f_i la forme linéaire

$$f_i(y) = \sum_{j \neq i} e_{i,j}(y).$$

Nous raffinons la décomposition en ajoutant tous les hyperplans $f_i(y) = 0$, donc en ajoutant les inégalités

$$\eta_i f_i(y) \geq 0, \tag{A.18}$$

où $\eta_i = \pm 1$ et $\eta_i = \eta_j$ si i et j appartiennent au même A_k. La variété ainsi obtenue est telle que les $\exp(f_i)$ aussi bien que les fonctions $\exp(e_{i,j}) = w_{i,j}$ y sont bien définie quoique la valeur ∞ est admise. Nous pouvons même définir $\exp(f_i)$ pour $i \in B'_l$ en posant $\exp(f_i) \equiv 0$ et pour $i \in B''_l$ en posant $\exp(f_i) \equiv \infty$.

La variété associée à cette décomposition est complète (voir [**TE**, §I.2, th. 7]) mais en général singulière. Selon la définition même, à chaque point de cette nouvelle variété une des deux fonctions

$$z''_i = \exp\big(f_i(w)\big) = \prod_{i \neq j} w_{i,j}, \qquad \tilde{z}''_i = \exp\big(-f_i(w)\big) = \prod_{i \neq j} w_{j,i}$$

est régulière. (Si $i \in \mathbb{B}'_k$ et $j \in \mathbb{B}''_k$ nous posons $z''_i = \tilde{z}''_j = 0$ et $\tilde{z}''_i = z''_j = \infty$.) Par conséquent cet éclatement de W suffit pour que l'application $\psi^{A,B}$ soit définie. De plus à cette étape l'éclatement de $W^{\tilde{A},\widetilde{B}}$ est une sous-variété de $W^{A,B}$ si $(\tilde{A}, \widetilde{B})$ est plus profond que (A, B). Mais un éclatement de plus est nécessaire pour obtenir une variété lisse.

L'existence d'un tel éclatement $\mathcal{W}^{A,B}$ est vérifiée dans [**TE**, §I.2, lem. 1, 2, 3]. (Nous observons que la démonstration du th. 11 de [**TE**] est indépendante des th. 9 et 10. L'éclatement pertinent est obtenu en utilisant les trois lemmes et le th. 4.) Pour construire cet éclatement il suffit de raffiner la première décomposition en une décomposition simpliciale.

A certains égards la décomposition précise est sans importance. Mais une partie de la décomposition doit être choisie soigneusement. On observe que les constructions de [**TE**] sont telles que la modification commence avec les cellules (ou cônes) de la dimension la plus basse, donc de dimension 0, puis passe aux autres cellules en augmentant la dimension k à la dimension $k+1$; les cellules de dimension k ne sont plus modifiées après l'étape k. De plus, si une cellule de dimension k est déjà un cône simplicial, il n'est pas nécessaire de la modifier ou de modifier les cellules de sa frontière.

La décomposition donnée par (A.17) est une décomposition en cônes fermés simpliciaux. Par conséquent si quelqu'uns de ces cônes ou des cônes de leurs frontières sont telles que les inégalités (A.18) sont satisfaites avec des choix convenables des constantes η_i alors il est admissible de ne pas les modifier ; et nous ne les modifions pas.

Supposons que la réunion $\cup A_k$, donc le complément de la réunion $\cup \mathbb{B}_k$, soit la réunion de deux ensembles disjoints D' et D''. On attache à une telle paire une cellule *spéciale* en posant $\varepsilon_{i,j} = 1$ si $i \in D'$ et $\varepsilon_{i,j} = -1$ si $i \in D''$. Pour satisfaire à (A.18) on choisit $\eta_i = 1$, $i \in D'$ et $\eta_i = -1$, $i \in D''$. La cellule ainsi obtenue est une cellule (A.17) et ni elle ni les composantes de sa frontière ne seront modifiées. Il est utile d'étendre la notion d'une cellule spéciale en n'exigeant pas que la réunion de D' et D'' soit le complément de celle des $\mathbb{B}_k$. Alors il est entendu que $e_{i,j}(y) = 0$ et $f_i(y) = 0$ pour $i \notin D' \cap D''$. Dans un sens plus large, ces cellules spéciales sont de toute façon des cellules de la frontière des cellules spéciales dans un sens restreint.

A chaque cellule Σ qui est spéciale ou qui appartient à la frontière d'une cellule spéciale correspond une sous-variété W_Σ ouverte, dans le sens de Zariski, de l'éclatement et la restriction à W_Σ de l'application qui envoie l'éclatement dans W définit un isomorphisme de W_Σ avec son image. Nous dénotons cet image par le même symbole W_Σ.

La variété W_Σ attachée à une cellule spéciale peut être définie d'une façon plus concrète. Elle est définie par les conditions : (i) $w_{i,j}$ est fini pour i dans D' ; (ii) $w_{j,i}$ est fini si $i \in D''$. Les variétés attachées aux cellules de sa frontière satisfont à ces conditions et en plus à la condition que $w_{i,j}$ est fini et pas zéro pour certaines paires (i, j). Il résulte de [**TE**, th. 6] (légèrement corrigé — dans la dernière phrase les orbites devraient être remplacées par leurs fermetures) qu'un point de W qui est contenu dans un sous-ensemble W_Σ est l'image d'un point unique de $\mathcal{W}$.

A nos fins il faut nous assurer en plus que les décompositions de [**TE**] peuvent être choisies en sorte que $\mathcal{C}^{AB}(\tilde{A}, \widetilde{B})$ et l'application η de (A.7) soient définis. Puisque $\mathcal{E}^{\tilde{A},\widetilde{B}}$ est une sous-variété de $\mathcal{E}^{A,B}$, il suffira de définir une sous-variété

$\mathcal{W}^{A,B}(\tilde{A},\widetilde{B})$ de $\mathcal{W}^{A,B}$ et une application (notée encore η)

$$\eta\colon\quad \mathcal{W}^{A,B}(\tilde{A},\widetilde{B}) \to \mathcal{W}^{\tilde{A},\widetilde{B}}.$$

On aura alors

$$\mathcal{C}^{A,B}(\tilde{A},\widetilde{B}) = \mathcal{E}^{\tilde{A},\widetilde{B}} \times \mathcal{W}^{A,B}(\tilde{A},\widetilde{B}).$$

En particulier le lemme A.1 sera valable.

A chaque paire (A,B) nous avons attaché l'espace vectoriel $Y^{A,B}$. Si $(\tilde{A},\widetilde{B})$ est plus profond que (A,B) nous définissons un sous-espace $Y^{A,B}(\tilde{A},\widetilde{B})$ de $Y^{A,B}$ par la condition que $y \in Y^{A,B}$est dans $Y^{A,B}(\tilde{A},\widetilde{B})$ si et seulement si

(1) $e_{i,j}(y) = e_{i',j'}(y)$ chaque fois que $i \equiv i'(\tilde{A},\widetilde{B})$ et $j \equiv j'(\tilde{A},\widetilde{B})$;

(2) $e_{i,j}(y) = 0$ s'il existe $k \neq k'$ tels que $i \in \tilde{\mathbb{B}}_k$ et $j \in \tilde{\mathbb{B}}_{k'}$.

Ces conditions sont cependant pertinentes seulement si les éléments qui y interviennent sont définis. La première implique que $e_{i,i'}$ s'annule sur $Y^{A,B}(\tilde{A},\widetilde{B})$ si $i \equiv i'(\tilde{A},\widetilde{B})$. Nous nous intéressons surtout au sous-ensemble $Y_+^{A,B}(\tilde{A},\widetilde{B})$ de $Y^{A,B}(\tilde{A},\widetilde{B})$ défini par les conditions

(1) $e_{i,j}(y) \geq 0$ si $i \in \tilde{\mathbb{B}}'_k$, $j \in \tilde{\mathbb{B}}''_k$;

(2) $e_{i,j}(y) \leq 0$ si $i \in \tilde{\mathbb{B}}''_k$, $j \in \tilde{\mathbb{B}}'_k$;

(3) si $i \in \tilde{\mathbb{B}}'_k$ alors $f_i(y) \geq 0$ et si $j \in \tilde{\mathbb{B}}''_k$ alors $f_j(y) \leq 0$.

Nous observons pour la dernière fois que les première et seconde conditions ne sont pertinentes que si $e_{i,j}$ est défini, donc si $i \notin \mathbb{B}_k$.

Par l'intérieur d'un cône fermé Σ d'une décomposition polyédrique nous entendons son intérieur dans l'espace vectoriel qu'il engendre. Donc si $e_{i,j}(y) = 0$ pour un point intérieur du cône alors $e_{i,j}$ s'annule sur le cône entier.

LEMME A.3. *Si l'intersection de l'intérieur Σ° d'un cône fermé Σ avec l'espace $Y_+^{A,B}(\tilde{A},\widetilde{B})$ n'est pas vide alors cette intersection est égale à $\Sigma^\circ \cap Y^{A,B}(\tilde{A},\widetilde{B})$.*

Supposons que l'intersection $\Sigma^\circ \cap Y_+^{A,B}(\tilde{A},\widetilde{B})$ ne soit pas vide. Il faut vérifier que, si $i \in \tilde{\mathbb{B}}'_k$ et $j \in \tilde{\mathbb{B}}''_k$ et si $i \notin \tilde{\mathbb{B}}_k$ de sorte que $e_{i,j}$ est un élément bien défini de $U^{A,B}$, alors $e_{i,j}(y) \geq 0$ pour chaque élément y de l'intersection $\Sigma^\circ \cap Y^{A,B}(\tilde{A},\widetilde{B})$. Si $e_{i,j}(y) = 0$ alors $e_{i,j}$ est 0 partout sur cette intersection parce que y appartient à Σ°. Si $e_{i,j}(y) \neq 0$ et $i \in \mathbb{B}'_k$, $j \in \mathbb{B}''_k$ il faut que $e_{i,j}(y) > 0$ car le signe de $e_{i,j}$ est constant sur Σ° et $\Sigma^\circ \cap Y_+^{A,B}(\tilde{A},\widetilde{B})$ n'est pas vide. De la même façon si $f_i(y)$ n'est pas identiquement 0 sur Σ alors il est de signe constant et ce signe est celui exigé par la définition de $Y_+^{A,B}(\tilde{A},\widetilde{B})$.

Il existe une application linéaire λ bien définie de $Y^{A,B}(\tilde{A},\widetilde{B})$ sur $Y^{\tilde{A},\widetilde{B}}$ car si $\tilde{e}_{i,j}$ et $\tilde{e}_{i',j'}$ sont deux éléments égaux de $U^{\tilde{A},\widetilde{B}}$, donc deux éléments définis par des paires (i,j) et (i',j') avec $i \equiv i'(\tilde{A},\widetilde{B})$ et $j \equiv j'(\tilde{A},\widetilde{B})$, alors les deux éléments $e_{i,j}$ et $e_{i',j'}$ satisfont les mêmes restrictions dans $Y^{A,B}(\tilde{A},\widetilde{B})$. Nous voulons construire les décompositions polyédriques tel que toute intersection $\Sigma \cap Y_+^{A,B}(\tilde{A},\widetilde{B})$ est contenue dans l'image réciproque $\lambda^{-1}(\tilde{\Sigma})$ d'un cône de la décomposition de $Y^{\tilde{A},\widetilde{B}}$. Puisque les décompositions des cellules sont définies par récurrence sur leur dimension et la dimension de $Y^{A,B}$ est plus grande que celle de $Y^{\tilde{A},\widetilde{B}}$, cela s'avère possible.

Il faut cependant tenir compte de l'intersection des raffinements des cellules spéciales et de leurs frontières.

LEMME A.4. *Soit Σ une cellule spéciale de $Y^{A,B}$ ou une cellule de sa frontière. Alors*

$$\Sigma^\circ \cap Y_+^{A,B}(\tilde{A}, \widetilde{B}) \tag{A.19}$$

est contenu dans l'image réciproque d'une cellule spéciale de $Y^{\tilde{A},\widetilde{B}}$ ou d'une cellule de sa frontière.

Il suffit évidemment de montrer que cette intersection est contenue dans l'image réciproque d'une cellule spéciale dans le sens large. Supposons qu'une cellule spéciale qui contient Σ soit définie par des ensembles D' et D''. Soient $\hat{D}'$ l'ensemble de tous les i tel que $i \equiv i'(\tilde{A}, \widetilde{B})$ et $i' \in D'$. Définissons $\hat{D}''$ de la même façon. Puisque nous admettons les cellules spéciales dans le sens large nous pouvons remplacer D' et D'' par $\hat{D}'$ et $\hat{D}''$ car Σ est contenu dans la cellule plus petite qu'ils définissent. Soient $\tilde{D}'$ et $\tilde{D}''$ les intersections de $\hat{D}'$ et de $\hat{D}''$ avec $\cup \tilde{A}_k$ Il est clair que (A.19) est contenu dans l'image réciproque de la cellule spéciale attachée à $\tilde{D}'$ et $\tilde{D}''$.

PROPOSITION A.5. *Il est possible de construire simultanément pour toute paire (A, B) des raffinements en cônes simpliciaux des décompositions de $Y^{A,B}$ données par* (A.17) *et* (A.18) *tels que*

(1) *les cellules spéciales et les cônes de leur frontières ne soient pas modifiés* ;
(2) si la paire $(\tilde{A}, \widetilde{B})$ est plus profonde que la paire (A, B) alors l'intersection de chaque cône du raffinement de la décomposition sur $Y^{A,B}$ avec $Y_+^{A,B}(\tilde{A}, \widetilde{B})$ est contenue dans l'image réciproque d'un cône du raffinement de la décomposition de $Y^{\tilde{A},\widetilde{B}}$.

La construction se fait par récurrence sur la dimension de $Y^{A,B}$. Pour (A, B) donné elle se fait par récurrence sur la dimension du cône. Soit Σ un cône donné dans $Y^{A,B}$ et supposons que tout cône de la frontière de Σ a déjà été modifié. Si l'intersection de Σ° et $Y_+^{A,B}(\tilde{A}, \widetilde{B})$ est vide, alors on décompose Σ à partir de la décomposition des cônes de sa frontière. Si Σ est spécial ou appartient à la frontière d'une cellule spéciale, on n'y touche pas. Par contre s'il ne l'est pas, on choisit un point dans Σ° et on prend la décomposition simpliciale définie par ce point et la décomposition de la frontière.

Supposons alors que

$$\Sigma^\circ \cap Y_+^{A,B}(\tilde{A}, \widetilde{B})$$

n'est pas vide. Il y a deux possibilités :

(1) $\Sigma^\circ \subseteq Y_+^{A,B}(\tilde{A}, \widetilde{B})$;
(2) $\Sigma^\circ \not\subseteq Y_+^{A,B}(\tilde{A}, \widetilde{B})$.

Dans le premier des deux cas l'intersection de l'image réciproque du raffinement de la décomposition de $Y^{\tilde{A},\widetilde{B}}$ est une décomposition de Σ dont les cônes ne sont pas nécessairement simpliciaux sauf si Σ appartient à la frontière d'une cellule spéciale car, selon le lemme A.4, Σ est alors contenu dans l'image réciproque d'un cône qui n'est pas raffiné. Si les cônes ne sont pas simpliciaux nous pouvons raffiner cette décomposition de Σ en une décomposition en cônes simpliciaux.

Dans le second cas nous pouvons, encore à cause du lemme A.4, écarter le cas où Σ est spécial ou contenu dans la frontière d'une cellule spéciale. Ensuite en décomposant Σ d'une façon préliminaire (et presqu'arbitraire car les cas importants ont été écartés), nous pouvons supposer qu'il est devenu la réunion de cônes simpliciaux. Nous décomposons ces nouveaux cônes Θ par récurrence sur leur dimension. Seuls les Θ qui ne sont pas contenus dans une cellule spéciale et dont l'intersection

$$\Theta^\circ \cap Y_+^{A,B}(\tilde{A}, \widetilde{B})$$

n'est pas vide et ne contient pas Θ° posent des problèmes. Dans ce cas nous raffinons la décomposition de

$$\Theta \cap Y_+^{A,B}(\tilde{A}, \widetilde{B})$$

donnée par l'image réciproque de celle de $Y^{\tilde{A},\widetilde{B}}$. Ensuite nous utilisons les suspensions des cônes de cette décomposition avec ceux de la frontière de Θ pour obtenir la décomposition définitive de Θ.

En général les plongements toroïdaux sont tels qu'à chaque cône fermé de la décomposition et de n'importe quelle dimension est attachée une sous-variété localement fermée. Par exemple au point $\{0\}$ est attaché le tore lui-même et à un cône ouvert un seul point. Nous dénotons par $W_\Sigma^{A,B}$ la sous-variété de $\mathcal{W}^{A,B}$ attachée au cône Σ.

En particulier chaque cône fermé de la décomposition de $Y^{A,B}$ dont l'intérieur rencontre $Y_+^{A,B}(\tilde{A}, \widetilde{B})$ définit une sous-variété localement fermée de l'éclatement $\mathcal{W}^{A,B}$. Sur cette sous-variété $\exp(e_{i,j}-e_{i',j'})$ est défini si $i \equiv i'(\tilde{A}, \widetilde{B})$ et $j \equiv j'(\tilde{A}, \widetilde{B})$ et n'est ni 0 ni ∞. D'ailleurs $\exp e_{i,j}$ est fini sur cette sous-variété si $i \in \widetilde{\mathbb{B}}_l'$ et $j \in \widetilde{\mathbb{B}}_l''$. (Observons que nous avons ressuscité les coordonnées de $W^{A,B}$ qui n'apparaissent pas dans la définition de $V^{A,B}$.) Si $i \in \mathbb{B}_k$ et $j \in \mathbb{B}_{k'}$ avec $k \neq k'$ alors $\exp e_{i,j}$ est fini et différent de zéro sur cette sous-variété. De plus, $\exp f_i$, $i \in \mathbb{B}_k'$ et $\exp(-f_j)$, $j \in \mathbb{B}_k''$ sont finis.

Les équations $\exp(e_{i,j} - e_{i',j'}) = 1$, $i \equiv i'(\tilde{A}, \widetilde{B})$, $j \equiv j'(\tilde{A}, \widetilde{B})$, $\exp e_{i,j} = -1$, $i \in \mathbb{B}_k'$ et $j \in \mathbb{B}_k''$, $\exp f_i = 0$, $i \in \mathbb{B}_k'$, $\exp(-f_j) = 0$, $j \in \mathbb{B}_k''$, et enfin pour $k \neq k'$

$$\begin{aligned}
\exp e_{i,j} &= Q(\alpha_k, \alpha_{k'}), \quad & i \in \mathbb{B}_k', \quad & j \in \mathbb{B}_{k'}' \\
\exp e_{i,j} &= Q(\alpha_k, \beta_{k'}), \quad & i \in \mathbb{B}_k', \quad & j \in \mathbb{B}_{k'}'' \\
\exp e_{i,j} &= Q(\beta_k, \beta_{k'}), \quad & i \in \mathbb{B}_k'', \quad & j \in \mathbb{B}_{k'}'' \\
\exp e_{i,j} &= Q(\beta_k, \alpha_{k'}), \quad & i \in \mathbb{B}_k'', \quad & j \in \mathbb{B}_{k'}'
\end{aligned}$$

définissent alors une sous-variété fermée $\mathcal{W}^{A,B}(\tilde{A}, \widetilde{B})$ de $\mathcal{W}^{A,B}$. Elle est fermée parce qu'elle est contenue dans la réunion des W_Σ tels que l'intérieur de Σ rencontre $Y_+^{A,B}(\tilde{A}, \widetilde{B})$.

Il y a une application η de $\mathcal{W}^{A,B}(\tilde{A}, \widetilde{B})$ dans $\mathcal{W}^{\tilde{A},\widetilde{B}}$. En effet $\mathcal{W}^{A,B}(\tilde{A}, \widetilde{B})$ est la réunion de sous-variétés ouvertes, chacune attachée à un cône fermé Σ dont l'intérieur rencontre $Y_+^{A,B}(\tilde{A}, \widetilde{B})$. L'intersection

$$\Sigma^\circ \cap Y_+^{A,B}(\tilde{A}, \widetilde{B})$$

est contenue dans l'image réciproque d'un cône fermé $\tilde{\Sigma}$ de $Y^{\tilde{A},\widetilde{B}}$.

Nous envoyons la sous-variété de $\mathcal{W}^{A,B}(\tilde{A},\widetilde{B})$ attachée à Σ dans $W_{\tilde{\Sigma}}^{\tilde{A},\widetilde{B}}$ en posant

$$\exp\tilde{e}\big(\eta(p)\big) = \exp\big(e(p)\big) \tag{A.20}$$

pour chaque élément $\tilde{e}$ de $U^{\tilde{A},\widetilde{B}}$ tel que $\tilde{e}(y) \geq 0$ pour tout $y \in \tilde{\Sigma}$. Si

$$\tilde{e} = \sum m_{i,j} e_{i,j}$$

dans $U^{\tilde{A},\widetilde{B}}$ alors e est défini par la même somme dans $U^{A,B}$. Quoique e n'est pas unique les valeurs de $\exp e$ sur $W^{A,B}(\tilde{A},\widetilde{B})$ le sont.

Puisque n'importe quelle fonction régulière sur la sous-variété ouverte $W_{\tilde{\Sigma}}$ de $\mathcal{W}^{\tilde{A},\widetilde{B}}$ attachée à $\tilde{\Sigma}$ est une combinaison linéaire des fonctions $\exp\tilde{e}$ telles que $\tilde{e}(y) \geq 0$ pour tout $y \in \tilde{\Sigma}$, l'application η est bien définie en sorte que le diagramme

$$\begin{array}{ccc} W^{A,B}(\tilde{A},\widetilde{B}) & \xrightarrow{\eta} & W^{\tilde{A},\widetilde{B}} \\ \downarrow & & \downarrow \\ W^{A,B}\times E^{A,B} & \longleftarrow & W^{\tilde{A},\widetilde{B}}\times E^{\tilde{A},\widetilde{B}} \end{array}$$

est commutatif. Par conséquent le diagramme (A.8) est aussi commutatif.

Nous observons enfin que l'éclatement du deuxième facteur n'implique pas Δ. Par conséquent l'application π_2 du lemme A.1 reste lisse au-dessus des points où elle était lisse après le premier éclatement.

L'importance des cellules spéciales découle du prochain lemme.

LEMME A.6. *Supposons que Δ soit générique. Soit p un point de $C_{\Delta}^{A,B}$ et posons $(z',w') = \varphi(p)$ et $(z'',w'') = \psi(p)$. Supposons que $(z',w') = (z'',w'')$. Alors il existe une cellule spéciale Σ (dans le sens large) telle que W_{Σ} contient $w' = w''$. Si la cellule spéciale est définie par (D',D'') alors D' est contenu dans l'ensemble des $i \notin \cup\mathbb{B}_k$ tels que $z_i \in \{\alpha_1,\dots,\alpha_N\}$ et D'' dans l'ensemble des $j \notin \cup\mathbb{B}_k$ tels que $z_j \in \{\beta_1,\dots,\beta_N\}$.*

Soit $(z,w) = (z',w') = (z'',w'')$ et soient $Z_1, Z_2, \dots$ les valeurs différentes des coordonnées z_i. Nous écrivons $Z_i \to Z_j$ si

$$Z_j = A(Z_i) = 2\Delta - \frac{1}{Z_i}.$$

Puisqu'il s'agit du cas générique nous pouvons supposer que l'application (A.3) n'a pas de point cyclique d'ordre fini et plus petit ou égal à N, sauf pour les deux points fixes δ_1 et δ_2. Par conséquent l'ensemble $\{Z_k\}$ se décompose dans des sous-ensembles disjoints $\{Z_{i_1},\dots,Z_{i_n}\}$ tels que

$$Z_{i_1} \to Z_{i_2} \to \cdots \to Z_{i_n}. \tag{A.21}$$

De plus il n'existe aucun k tel que $Z_{i_n} \to Z_k$. L'indice n peut être différent pour des sous-ensembles différents.

Si Z_k et Z_l n'appartiennent pas au même sous-ensemble et $z_i = Z_k$ et $z_j = Z_l$ alors ni $F(z_i,z_j)$ ni $F(z_j,z_i)$ n'est égal à 0 et $w_{i,j}$ n'est ni 0 ni ∞. Cette affirmation reste vraie si Z_k et Z_l appartiennent au même sous-ensemble mais si $Z_k \not\to Z_l$.

Il n'y a donc que deux cas intéressants : ou il y a une suite maximale (A.21) avec $n > 1$; ou il y a un k tel que $Z_k = \delta_i$, $i = 1, 2$. Les suites maximales S de longueur deux donnent une paire (D'_S, D''_S) en prenant $D'_S = \{i \mid z_i = Z_{i_1}\}$, $D''_S = \{j \mid z_j = Z_{i_2}\}$. Dans ce cas $w_{i,j} = 0$ si $i \in D'_S$ et $j \in D''_S$. Par contre si $i \in D'_S$ et $j \notin D''_S$ alors $w_{i,j}$ n'est ni 0 ni ∞. Il résulte de l'égalité $z'_i = z''_i$ qu'il existe un indice k tel que $Z_{i_1} = \alpha_k$. Pour des raisons semblables il existe un indice k' tel que $Z_{i_2} = \beta_{k'}$. Évidemment on a $k = k'$. Nous posons

$$D' = \cup_S D'_S, \quad D''_S = \cup D''_S.$$

Supposons qu'il y a une suite maximale pour laquelle $n > 2$. Le même argument montre que $Z_{i_1} = \alpha_k$, et que $Z_{i_n} = \beta_{k'}$. Mais on peut évidemment exiger que dans le cas générique aucun $\beta_{k'}$ ne soit dans l'ensemble engendré par α_k sous des puissances de l'application (A.3). Il en résulte que $n = 2$ et que $k = k'$.

Supposons enfin qu'il y ait une suite maximale avec $n = 1$. Supposons que $Z = Z_{i_1} = \delta_1$. Le cas $Z_{i_1} = \delta_2$ est tout à fait semblable. Si $z \neq \delta_1$ alors

$$F(\delta_1, z) = \delta_1(z - 2\Delta + \frac{1}{\delta_1}), \quad F(z, \delta_1) = (\delta_1 - 2\Delta)\Big(z + \frac{1}{\delta_1 - 2\Delta}\Big).$$

Puisque

$$\frac{1}{\delta_1 - 2\Delta} = -\delta_1 = -2\Delta + \frac{1}{\delta_1},$$

le quotient $F(\delta_1, z)/F(z, \delta_1)$ est égal à $-\delta_1^2$. D'autre part

$$\frac{\delta_1 - \alpha_i}{\delta_1 - \beta_i} = \delta_1 \alpha_i$$

parce que

$$\delta_1 - \beta_i = \delta_1 - 2\Delta + \frac{1}{\alpha_i} = \frac{1}{\alpha_i} - \frac{1}{\delta_1} = \frac{\delta_1 - \alpha_i}{\alpha_i \delta_1}.$$

Nous multiplions toutes les équations $z'_i = z''_i$ pour $z_i = Z$ ensemble. D'abord

$$\prod z''_i = \prod_{z_i = Z, z_j \neq Z} \delta_1^2 = \delta_1^{2s(N-s)},$$

si s est le nombre d'indices i tels que $z_i = Z$, et ensuite

$$\prod z'_i = R(\delta_1)^s = \left(\frac{\delta_1^N \sqrt{(-1)^N \rho}}{\kappa}\right)^s,$$

où ρ est défini par (2.1). Dans le cas générique l'équation

$$\delta_1^{2s(N-s)} = \left(\frac{\delta_1^N \sqrt{(-1)^N \rho}}{\kappa}\right)^s,$$

$0 < s \leq N$, n'est pas satisfaite.

Appendice B

La fonction R qui définit la correspondance est

$$R(z) = \kappa \frac{\prod(z-\alpha_k)}{\prod(z-\beta_k)}.$$

Comme nous avons déjà remarqué, les auteurs de [**TV**] permettent la possibilité que la correspondance dégénère en faisant $\kappa \to 0$ tout en gardant Δ et tous les α_k et β_k génériques. Pour pouvoir exploiter cette dégénérescence de la correspondance, il faut le lemme suivant.

LEMME A.7. *Fixons les α_k, les β_k et Δ. Alors lorsque $\kappa \to 0$ tout point fixe de la correspondance reste fini et tend vers une solution des équations*

$$\prod_k (z_i - \beta_k) \prod_{j\neq i} (z_i z_j - 2\Delta z_i + 1) = 0. \tag{A.22}$$

Supposons que $z_i \to \infty$. Alors $F(z_i, z_j)/F(z_j, z_i)$ reste fini sauf si $z_j \to 0$ et il approche un nombre non nul sauf si $z_j \to 2\Delta$, donc sauf si $z_j \to A(z_i)$ où A est toujours l'application linéaire $z \to 2\Delta - 1/z$.

Si z_i reste éloigné de tout pôle de R, en particulier si $z_i \to 0$ ou si $z_i \to 2\Delta$ car les β_i et Δ sont génériques, alors $R(z_i) \to 0$. Par conséquent il y a un indice j tel que

$$z_j \to A(z_i).$$

Puisque Δ est générique de sorte que A est d'ordre infini, on arrive enfin à un indice l tel que z_l approche un pôle β_k. En particulier le point de départ n'était pas 0. Le lemme en résulte.

Les solutions de (A.22) ont une structure combinatoire simple qui ressemble beaucoup aux structures déjà rencontrées. Il résulte de la démonstration du lemme que pour une solution (z_i) de (A.22) l'ensemble des z_i tels que $z_i \in \{\beta_1, \dots, \beta_N\}$ n'est pas vide. Appelons ces z_i les racines. Toute autre coordonnée z_j s'obtient d'une racine en appliquant A^{-1} plusieurs fois $z_j = A^{-1}(A^{-1}(\dots(z)\dots))$ où z est une racine. Dans le cas générique ces racines déterminent des sous-ensembles différents, donc des composantes connexes mais dans un sens nouveau. A chaque z_j dans une composante connexe est attaché un entier, sa distance de la racine.

La multiplicité des solutions des équations (A.22) n'est pas nécessairement un. Il est d'abord évident que les équations découplent localement selon les composantes connexes. Donc, pour calculer la multiplicité, on peut supposer qu'il n'y a qu'une seule composante. Si la racine est égale à $\beta = \beta_l$ et si E_k est l'ensemble de z_i à la distance k de la racine, alors localement les équations (A.22) sont équivalentes à

$$\begin{aligned} 0 &= (z_i - \beta), && i \in E_0, \\ 0 &= \prod_{j \in E_{k-1}} z_j z_i - 2\Delta z_j + 1, && i \in E_k, \quad k \geq 1. \end{aligned}$$

La multiplicité de la solution $z_i = \beta$ pour $i \in E_0$, $z_j = A^{-1}(z_i)$ pour $i \in E_k$ et $j \in E_{k+1}$ est évidemment

$$\prod_{k=0}^{\infty} |E_k|^{E_{k+1}}.$$

Avec quelques moments de réflexion on se persuade que ce produit est exactement le nombre de réunions d'arbres connexes, enracinés aux racines données et dans lesquels la distance d'un sommet z_j de la racine est k si $j \in E_k$. Nous avons par conséquent retrouvé la structure de [**AG**].

Appendice C

Vers la fin de l'article [**AG**] (qui comme celui-ci a été écrit trop hâtivement pour ne pas trop dépasser l'échéance imposée) quelques coquilles et petites erreurs se sont glissées dans le texte et nous voulons profiter de cette occasion pour les corriger.

D'abord la formule (12) devient

$$N \sum_{E \cup F = D} \left|\mu(E) - \mu(F)\right| \mu(E)^{f-1} \mu(F)^{e-1} \tag{A.24}$$

et la formule (14)

$$2 \sum_{E \cup F = D} \min\{\mu(E), \mu(F)\} \mu(E)^{f-1} \mu(F)^{e-1}. \tag{A.25}$$

La formule (15) est fausse et devrait être remplacée par la formule suivante qui est vérifiée dans le corps de cet exposé

$$\omega\mu(D)^{d-2} = \sum \big(\mu(E)\omega'' + \mu(F)\omega' + \omega'\omega''\big)\big(\mu(E) + \omega'\big)^{f-1}\big(\mu(F) + \omega''\big)^{e-1}. \tag{A.26}$$

Il y a enfin la formule (17) qui n'intervient pas dans ce papier et que nous corrigeons en un autre lieu.

References

[B] H. Bethe, *Zur theorie der metalle*, Zeit. f. Physik **71** (1931), 205–226.

[TE] G. Kempf, Finn Faye Knudsen, D. Mumford, and B. Saint-Donat, *Toroidal embeddings. I*, Lecture Notes in Math., vol. 339, Springer-Verlag, Berlin, New York, 1973.

[AG] Robert P. Langlands et Yvan Saint-Aubin, *Algebro-geometric aspects of the Bethe equations*, Strings and Symmetries, Lecture Notes in Phys., vol. 447, Springer-Verlag, Berlin, New York, 1995.

[BL] ———, *Les sommes de Bethe et la formule de Lefschetz* (en préparation).

[TΦ] Л. А. Тахтаджян и Л. Д. Фаддеев, Квантовой метод обратной задачи у *XYZ* модель Гейзенберга, Успехи Мат. Наух **34** (1979), no. 5, 13–63.

[TV] V. Tarasov et A. Varchenko, *Bases of Bethe vectors and difference equations with regular singular points*, (1995) (prépublication).

School of Mathematics, Institute for Advanced Study, Olden Lane, Princeton, NJ 08540, USA.

E-mail address: rpl@Math.ias.edu

Centre de recherches mathématiques, Université de Montréal, C. P. 6128, succ. Centre-ville, Montréal, Québec H3C 3J7, Canada.

E-mail address: saint@CRM.UMontreal.CA

Centre de Recherches Mathématiques
CRM Proceedings and Lecture Notes
Volume **11**, 1997

Metric Dimension and Statistical Estimation

L. Le Cam

1. Introduction

The purpose of this paper is to emphasize a connection between a concept of metric dimension and the achievable risks in statistical estimation.

Metric dimension, a variation on Kolmogorov's metric entropy, is defined in Section 2. In Section 3 we present a particular method of estimation and give a bound on its risks. Section 4 assumes that the experiment at hand is generated by independent observations and gives a bound in terms of a metric defined from Hellinger distances.

Section 5 explains that, in spite of the fact that our bounds are obtained by crude methods, they are not far from the "best possible".

Much of the material presented here is already available in published form, in Le Cam (1975, 1986) or Birgé (1983a, b) but the emphasis is new.

2. Metric Dimension

Let Θ be a space with a metric or pseudometric w. Let τ be a number, $\tau \geq 0$.

The dimension of Θ at the level τ for w is a number $D(\tau)$ defined as the smallest number such that every subset of Θ of diameter $2x$, $x \geq \tau$, can be covered by at most $2^{D(\tau)}$ sets of diameter at most x.

This concept is closely related to the metric entropy of A. N. Kolmogorov. For a set S, this is a number $K(S, \tau)$ equal to the $\log_2$ of the minimal number of sets of diameter at most τ needed to cover S. A simple argument shows that if S has dimension $D(\tau)$ and S has diameter $2^m\tau$ then

$$K(S,\tau) \leq \sum_{j=0}^{m} D(2^j\tau).$$

For most nonparametric statistical purposes one can use either concept almost interchangeably. Note that both numbers increase as τ decreases. If $D(\tau)$ behaves approximately as $\tau^{-\alpha}$, $\alpha > 0$, as $\tau \to 0$ then the above formula shows that for $m > 1$ and for a set S of diameter $\operatorname{diam} S \in [2\tau, 2^m\tau]$ one will have

$$K(S,\tau) \leq CD(\tau)$$

1991 *Mathematics Subject Classification*. Primary: 62G05; Secondary: 62B15.
This is the final form of the paper.

for a constant C that depends on the exponent α. Furthermore, from the definition of $D(\tau)$, there will be sets S of diameter 2τ or more such that $D(\tau) \leq K(S, \tau)$.

Since Kolmogorov's entropies have been calculated for a variety of functional spaces, see Kolmogorov and Tichomirov (1961) and Vitushkin (1959), this remark yields easily accessible bounds for dimensions.

The "dimension" defined here is a metric concept, not a topological one. As such it changes if one changes the metric. However it behaves well under some transformations. Suppose for instance that one considers two metrics w_1 and w_2 on the set Θ. Suppose that w_1 and w_2 are related in that there exists a strictly increasing continuous map ξ varying regularly, in the sense of Karamata, at zero and infinity so that

$$aw_2 \leq \xi(w_1) \leq bw_2$$

for numbers $0 < a < b < \infty$.

Then the dimension for w_1 is bounded by a constant times the dimension for w_2 and conversely.

For supplementary information on these points, see P. Assouad (1976).

The number $D(\tau)$ is called "dimension" because $\mathbb{R}^m$ has $D(\tau) \equiv m$ for the sup norm metric. Kolmogorov's entropy was meant to be used for precompact spaces, but we shall let his numbers $K(S, \tau)$ or our $D(\tau)$ jump to infinity if that happens to be necessary. Note that if Θ has diameter d than $D(d) = 0$.

3. Estimation and Testing

A statistical experiment $\mathcal{E} = \{P_\theta; \theta \in \Theta\}$ is a family of probability measures on a σ-field $\mathcal{A}$ carried by a set $\mathcal{X}$. The set Θ is usually called the "parameter space". It can be entirely arbitrary. For our purposes here we shall assume that it carries a metric or pseudo-metric w.

An estimate T will be a function $x \rightsquigarrow T(x)$ from $\mathcal{X}$ to Θ, with the restriction that, for each $\theta \in \Theta$, the function $x \rightsquigarrow w[T(x), \theta]$ is $\mathcal{A}$-measurable. One could also consider "randomized" estimates at the cost of some notational complications. There are various criteria to express that estimates are "good". For instance one could ask that $\int g[w[T(x), \theta]] P_\theta(dx)$ be small uniformly in θ for some positive increasing function g. Another possibility is to require that for some $b \in (0, \infty)$ the probabilities $P_\theta\{w(T, \theta) \geq b\}$ do not exceed a small $\varepsilon > 0$, uniformly in θ. (One can contemplate various criteria that are not uniform in θ. We shall not do so, for simplicity).

A "test function" or simply a "test" is an $\mathcal{A}$-measurable function φ from $\mathcal{X}$ to $[0, 1]$. If testing between two subsets A and B of Θ, the value $\varphi(x)$ is the conditional probability of rejecting B given that x has been observed. Thus the probability of rejecting a value $s \in A$ will be $\int [1 - \varphi(x)] P_s(dx)$. That of rejecting at $t \in B$ is $\int \varphi(x) P_t(dx)$. We shall often work with the number

$$\pi(A, B, \varphi) = \sup_{s,t} \left\{ \int [1 - \varphi(x)] P_s(dx) + \int \varphi(x) P_t(dx); s \in A t \in B \right\}.$$

There are two simple facts relating tests and estimates. The first is:

FACT 1. A good estimate can be used to generate many good tests.

The second is more delicate:

FACT 2. If one can generate a sufficiently large family of very good tests, one can also generate good estimates.

For the first assertion suppose that A and B are two subsets of Θ whose distance $\inf\{w(s,t); s \in A, t \in B\}$ is at least $2b > 0$. Suppose that $P_\theta\{w(T,\theta) \geq b\} \leq \varepsilon$ for all $\theta \in \Theta$. Take a function φ so that $\varphi(x) = 1$ if $T(x)$ is closer to A than to B. Put it equal to zero otherwise. Then $\int [1 - \varphi(x)] P_s(dx) \leq \varepsilon$ for $s \in A$ and $\int \varphi(x) P_t(dx) \leq \varepsilon$ for $t \in B$. The many tests one can generate that way are organized in a complex manner. We do not have a simple description of their mutual relations.

For the existence of tests one does have a fairly simple criterion. To state it let $\mathcal{M}$ be the space of bounded $\mathcal{A}$-measurable functions. Consider the P_θ as elements of the dual L of $\mathcal{M}$ for the uniform norm and define a norm on L by $\|\mu\| = \sup\{|\int \psi d\mu|; \psi \in \mathcal{M}, |\psi| \leq 1\}$.

This makes L into a Banach lattice. In fact it becomes an L-space in the sense of Kakutani. If P and Q are two positive elements of L one can define their minimum $P \wedge Q$ by

$$\int f d(P \wedge Q) = \inf_{g,h} \left\{ \int g dP + \int h dQ; g \geq 0, h \geq 0, g + h = f \right\}.$$

If $\|P\| = \|Q\| = 1$ this gives $\|P \wedge Q\| = 1 - 1/2\|P - Q\|$.

Now for any subset $A \subset \Theta$ define A^* as the closure in L for the weak topology induced on L by $\mathcal{M}$ of the convex hull of $\{P_\theta; \theta \in A\}$.

Let $\pi(A, B, \varphi)$ be the number defined above and let $\pi(A, B) = \inf_\varphi \pi(A, B, \varphi)$ the infimum being taken over all tests functions φ.

This gives

THEOREM 1. *The number $\pi(A, B)$ is precisely equal to*

$$\pi(A, B) = \sup\{\|P \wedge Q\|; P \in A^*, Q \in B^*\}.$$

(For a proof see for instance Le Cam (1986) page 476).

The fact that A^* is the closure of a *convex hull* makes the theorem difficult to apply in practice. We shall go around the difficulty in Section 4 by using only very special convex sets.

To construct an estimate from tests proceed as follows.

Select a number $q \in (0, 1)$ and an integer m. Assume that the diameter of Θ for w does not exceed a number b_0. Let $b_v = b_0 q^v$ and, for some $r > 2$, let $a_v = b_v / r$.

For each $v = 0, 1, 2, \ldots, m$ let assume that there is a finite partition $\{A_{v,j}; j \in J_v\}$ of Θ by sets $A_{v,j}$ of diameter at most a_v. If $A_{v,j}$ and $A_{v,k}$ are two sets of the vth-partition call them v-distant if there are points $s \in A_{v,j}$, $t \in A_{v,k}$ such that $w(s,t) \geq b_v$. For such pairs, select test functions $\varphi_{v,j,k}$ so that

$$\varphi_{v,j,k} = \varphi_{v,j,k}^2 = 1 - \varphi_{v,k,j}.$$

Construct a sequence of sets S_n; $n = 0, 1, \ldots, m$ as follows. Take $S_0 = \Theta$. If $S_0, S_1, \ldots, S_{v-1}$ have been constructed, let J_v^- be the set of indices $j \in J_v$ such that $A_{v,j}$ intersects S_{v-1}. For all pairs $j, k \in J_v^-$ such that $A_{v,j}$ and $A_{v,k}$ are v-distant perform the test $\varphi_{v,j,k}$ and eliminate the rejected set. A set $A_{v,j}$ will remain only if $\psi_{v,j} = \inf_k \varphi_{v,j,k} = 1$. Take the union of the sets that remain and its intersection with S_{v-1}. This will be the set S_v.

Note that according to this procedure the diameter of S_v is at most b_v.

To get a point estimate select a point in the last S_v, $v \leq m$ that is not empty.

(In the procedure, if a set $A_{v,j}$ $j \in J_v^-$ is not part of a distant pair, it is retained). The performance of the procedure will, of course, depend on the selection of the partitions $\{A_{v,j}\}$ and of the tests $\varphi_{v,j,k}$.

To describe it let $\beta(v)$ be a number $\beta(v) > \sup \pi(A_{v,j}, A_{v,k})$ where the sup is taken over all v-distant pairs $(A_{v,j}, A_{v,k})$. Let $C(v)$ be the minimum number of sets of diameter $\leq a_v$ needed to cover a hull of center θ and radius $b_{v-1} + a_v$, whatever may be $\theta \in \Theta$.

Then one can prove the following

THEOREM 2. *There exists choices of partitions $\{A_{v,j}\}$ and test functions $\varphi_{v,j,k}$ such that, if the S_v are constructed as above, then for every $k \leq m$ the diameter of S_k does not exceed b_v and*

$$P_\theta[\theta \notin S_k] \leq 2 \sum_{v=1}^{k} C(v)\beta(v).$$

For a proof see Le Cam (1975) or Le Cam and Yang (1990) page 132.

In this formula the numbers $C(v)$ can be bounded in terms of the dimension numbers $D(\tau)$ of Section 2. Indeed the diameter of a ball of radius $b_{v-1} + a_v$ is at most $2(b_{v-1} + a_v)$ and $2(b_{v-1} + a_v)/a_v = 2(1 + r/q)$. Thus $\log_2 C(v) \leq k_0 D(a_v)$ where k_0 is the smallest integer at least as large as $1 + \log_2(1 + r/q)$.

The bound of Theorem 2 yields bounds on the risks of the estimates T. A simple computation yields the following inequality.

COROLLARY. *Let g be an increasing function from $[0, \infty)$ to $[0, \infty)$ with $g(0) = 0$. Then*

$$E_\theta g\big[w(T, \theta)\big] \leq g(b_m) + 2 \sum_{k=1}^{m} g(b_{m-k}) C(m+1-k)\beta(m+1-k).$$

These formulas call for some comments. Note first that in most cases the numbers $C(v)$ and $\beta(v)$, or the bounds available for them, will increase, sometimes rapidly, as v increases. Thus in the sum of Theorem 2 the last term will be dominant. Similarly, in the sum of the Corollary the first term will dominate. Thus the choice of the value q^m will be important.

Second, note that the procedure is very crude. It does not take into account any of the relations that may exist between the various tests $\varphi_{v,i,k}$. It seems difficult to obtain any information on this without further assumptions.

Third, one could hurry along the decrease of the sets S_v as follows. If for a pair $(A_{v,j}, A_{v,k})$ the set $A_{v,k}$ is rejected, one could remove its entire closed convex hull without affecting the probabilities. To take advantage of this one needs information on the diameter of the union of the sets that remain. This is not available under our assumptions.

Fourth, in order that the procedure have some chance of success, there must be some relation between the distances of two sets $A_{v,j}$, $A_{v,k}$ and the performance of the statistical tests. In Section 4 we shall use a distance for which a good relation can be established.

Fifth and finally the procedure as described assumes for its start that the w-diameter of Θ is finite. This is not too serious. There are several possible extensions to sets of infinite diameter. One such extension, but a complex one, is described in Le Cam (1986), page 501. There are many other, simpler possibilities. We shall not insist on this point to keep this paper within reasonable bounds.

4. The Case of Independent Observations

Consider a set of indices I and a family $(\mathcal{X}_i, \mathcal{A}_i)$ of measurable spaces. Let $(\mathcal{X}, \mathcal{A})$ be their product. Take an experiment $\mathcal{E} = \{P_\theta; \theta \in \Theta\}$ where for each θ the measure P_θ is the product of probability measures $p_{\theta,i}$, $i \in I$ with $p_{\theta,i}$ on $(\mathcal{X}_i, \mathcal{A}_i)$.

In such a situation L_1-norms are not easily tractable but some other distances are. For each $i \in I$ let L_i be the dual of the space $\mathcal{M}_i$ of bounded $\mathcal{A}_i$ measurable functions. Consider the $p_{\theta,i}$ as elements of that dual. The Hellinger distance h between two such measures is defined by

$$h_i^2(s,t) = h_i^2(p_{s,i}, p_{t,i}) = \frac{1}{2} \int \left(\sqrt{dp_{s,i}} - \sqrt{dp_{t,i}} \right)^2.$$

The Hellinger affinity is

$$\rho(p_{s,i}, p_{t,i}) = \int \sqrt{dp_{s,i} dp_{t,i}} = 1 - h_i^2(s,t).$$

For the corresponding product measures P_s and P_t one has

$$\rho(P_s, P_t) = \Pi_i \rho(p_{s,i}, p_{t,i}).$$

In other words

$$\log \rho(P_s, P_t) = \sum_i \log\left[1 - h^2(p_{s,i}, p_{t,i})\right].$$

Aiming at situations where the individual h_j^2 are small, we shall use the function

$$H^2(s,t) = H^2(P_s, P_t) = \sum_i h_i^2(p_{s,i}, p_{t,i}),$$

and assume that our measures are such that $H^2(P_s, P_t) < \infty$ so that H is indeed a distance. (One could admit infinite values, but this introduces several complications of a measure theoretic nature, such as the possibility that the P_s are mutually disjoint but that there is no function T such that $P_s(T = s) = 1$ for all s).

The function H is defined for all pairs of product measures on the $(\mathcal{X}_i, \mathcal{A}_i)$ even if they do not occur in the experiment $\mathcal{E}$.

In this case a theorem of L. Birgé (1983b) implies the following:

THEOREM 3. *Let P_s and P_t be two particular product measures s.t. $H(P_s, P_t) \geq b > 0$. Let A be the set of product measures whose H-distance to P_s does not exceed $b/15$. Let B be the set of product measures whose H-distance to P_t does not exceed $b/15$. Then there is a test function φ for which*

$$\int (1 - \varphi) dP \leq \exp\left\{-\frac{1}{2} b^2\right\}, P \in A, \quad \int \varphi dQ \leq \exp\left\{-\frac{1}{2} b^2\right\}, Q \in B.$$

For a proof and for a recipe for the construction of φ, see Birgé (1983b) or Le Cam (1986) pages 481-492.

If one replaces the arbitrary metric w of Section 3 by the distance H, Theorem 3 yields bounds on the risk $\sup_\theta E_\theta H^2(T, \theta)$ of an estimate T constructed according to the method of Section 3 or extensions of it.

Using the formula of Corollary 1, Section 3 with a suitable choice of q and m one obtain a result of the following type

THEOREM 4. *There are universal constants C_1 and C_2 such that, if the experiment $\mathcal{E}$ satisfies the conditions of* Theorem 3, *then for any τ such that $\tau^2 \geq C_1 D(\tau)$ there are estimates T such that*

$$\sup_\theta E_\theta H^2(P_T, P_\theta) \leq C_2 \tau^2.$$

The estimates can be constructed by an extension of the method of Section 3.

It can happen that there is a τ such that $t^2 \geq c_1 D(t)$ for all $t > \tau$ but that τ itself is a point of discontinuity of D so that $\tau^2 < C_1 D(\tau)$. If not, one can just take τ so that $\tau^2 = C_1 D(\tau)$ and the bound on the risk becomes

$$E_\theta H^2(P_T, P_\theta) \leq C_1 C_2 D(\tau).$$

The choice of τ was meant to insure that in the formula of Corollary 1, Section 3 the term $C(m)\beta(m)$ is not larger than one.

We have not made a definite attempt to get optimal values for C_1 and C_2. Rough computations suggest values $C_1 \leq 1/22.5$ and $C_2 \leq 5(15)^2$. However L. Birgé (1983a) uses a different method of construction that give a similar result with better constants. This is contraintuitive since Birgé covers the entire space Θ by balls of radius τ and uses tests between pairs of them. This is wasteful in terms of the number of tests used. Here for points (s,t) such that $H(P_s, P_t) = 15c$ one tests between balls of radius c centered at P_s and P_t respectively. If c is much larger than τ the number of such balls needed for a covering can be much smaller. Thus further investigation seems necessary.

Here we have applied the formula of Corollary 1, Section 3 to a function g: $w \rightsquigarrow g(w) = w^2$. The formula can also be applied to other functions, such as, for instance, $g(w) = w$. It would give

$$E_\theta H(P_T, P_\theta) \leq C_3 \tau \text{ for } \tau \text{ such that } \tau^2 \geq C_1 D(\tau).$$

One important feature of the construction is that the estimates possess a certain quality of "robustness". Suppose that P is a product measure that is *not* one of the P_θ, $\theta \in \Theta$ but is such that $\inf_\theta H(P_\theta, P) = d \leq \tau$. Then the estimate T would be such that

$$E\{H^2(P_T, P) \mid P\} \leq C_2(\tau + d)^2.$$

This is visible from the statement of Theorem 3.

In certain cases one can use instead of the distance H another distance K that would give the same bounds (but with better constants). This is the case when the experiment at hand is "infinitely divisible", that is when it can be represented for every integer n as a product with $I = \{1, 2, \ldots, n\}$ and the $p_{s,i}$ replicates of $p_{s,1}$.

In such a case $\big[-\log \rho(P_s, P_t)\big] = K^2(s,t)$ is the square of a distance, and the inequalities of Theorem 3 remain valid with K substituted to H.

If for instance the P_θ are Gaussian measures obtained from each other by shifts then $\big[-8 \log \rho(P_s, P_t)\big] = \|m_s - m_t\|^2$ where m_s and m_t are the respective centers of P_s and P_t.

In this Gaussian shift case one knows the minimax risk. If the centers vary in an entirely unrestricted fashion, the minimax risk would be the ordinary linear dimension of the linear space in which the centers vary.

Thus, in such a case, the minimax risk bound of Theorem 4 is too large, but only by a factor $8C_1C_2$.

This is so even though the method we used is crude and the computation of upper bounds very crude.

5. Relations to Rates of Convergence of Estimates

The reader will observe that in the previous sections we did not carry out any sort of asymptotic arguments. We gave actual bounds that are admittedly not sharp.

One of the favorite domains of study of Statisticians, especially in the past decade, has been the study of rates of convergence of estimates. This is done mostly in the independent identically distributed case (i.i.d.), that is in a product situation, as in Section 4, but with $I = \{1, 2, \ldots, n\}$ and with $(\mathcal{X}_i, \mathcal{A}_i, p_{s,i})$ a copy of $(\mathcal{X}_1, \mathcal{A}_1, p_{s,1})$.

One studies the rate of convergence of estimates as $n \to \infty$.

The inequalities of Section 4 are easily adaptable to such a study. For $I = \{1, 2, \ldots, n\}$ and the i.i.d. case our square distance H^2 becomes $H_n^2(s,t) = nh^2(s,t)$ where h is the Hellinger distance between $p_{s,1}$ and $p_{t,1}$. The inequality of Theorem 4 becomes

$$\sup_\theta nh^2(s,t) \le C_2 \tau^2$$

where for simplicity we shall assume $\tau^2 = C_1 D(\tau)$. However D is relative to the distance H_n.

If d is the dimension function of Θ for the distance h one has $D(\tau) = d(\tau/\sqrt{n})$. Thus τ would be determined by the equality

$$\tau^2 = C_1 d\left(\frac{\tau}{\sqrt{n}}\right).$$

One could call "parametric" those situations where $\sup_y d(y) < \infty$. In other situations, to be called "non parametric", $d(y)$ will tend to infinity as $y \to 0$. (It may jump to ∞ at some $y > 0$, but we shall argue as if this does not happen, so that bounded subsets of Θ are precompact).

Let M_n be the minimax risk computed for the square distance h^2. Birgé's inequalities and ours say that $M_n \le C_2 y_n^2$ with

$$y_n^2 = C_1 \frac{1}{n} d(y_n).$$

For instance if $d(y) = C_4\ y^{-\alpha}$, $\alpha > 0$, this would give

$$y_n = C_6 n^{-1/2+\alpha}.$$

Now most of the standard arguments on rates of convergence have been carried out for special cases where the $p_{s,i}$ have densities with respect to the Lebesgue measure on some space $\mathbb{R}^k$ and in particular on $\mathbb{R}$ itself or even on the interval $[0,1]$. This is in contrast with the arguments of Section 3 or Section 4 where the nature of the underlying spaces $\mathcal{X}_i$ did not play any role.

If, for instance, the underlying space is $[0,1]$ one can contemplate estimates of the kernel type, nearest neighborhood estimates, estimates based on wavelets, or any number of other procedures. This is more practical than the procedure of Section 3 that may require a very fast computer to obtain covers by partitions $A_{v,j}$ that are (nearly) minimal.

For densities, especially on $[0,1)$ or on $\mathbb{R}$, the assumptions made by various authors are "smoothness" assumptions. For instance one may assume that the densities $f_\theta(x)$ admit m derivatives in x, the mth one being subject to a Hölder assumption: $\left|f_\theta^{(m)}(x) - f_\theta^{(m)}(y)\right| \le K|x-y|^\alpha$. Some assume that the f_θ are in a Sobolev ball, or a Besov ball and so forth. The variations on the smoothness assumptions are numerous and so are the variations on the metrics corresponding to the w of Section 3.

The possibilities are endless. However, when the metric used can be related to the distance H of Section 4, the speeds of convergence obtained coincide, except for a constant factor, with the speed obtained here by our crude method. See for instance Birgé (1983a). Birgé (1983a) has emphasized that we did not have at that time good procedures to obtain *lower* bounds on minimax risks. The tools are mostly limited to the use of Fano's lemma or Assouad's lemma, for which see for instance Le Cam (1986), page 524. The situation has remained much the same since that time, except that fine *asymptotic* results have been obtained by Donoho and coauthors. See Donoho and Liu (1991) and especially by M. Nussbaum (1994).

Birgé also emphasizes that the method described here on instead of on his own cannot be expected to give adequate results in certain special cases. He gives in his (1983a) paper examples where the dimension $D(\tau)$ is actually infinite for small but the attainable speed of convergence is the familiar parametric $Cn^{-1/2}$. Simple examples where $D(\tau) \to \infty$ as $\tau \to 0$ but where this same speed is achievable can be obtained as follows: Take on $\mathbb{R}$ a function f such that $f(x) = 0$ for $x \leq 0$ and $f(x) > 0$ for $x > 0$, the behavior of $1/f(x)$ near zero being that of

$$x\ell_1\left(\frac{1}{x}\right)\ell_2\left(\frac{1}{x}\right)\dots\ell_{m-1}\left(\frac{1}{x}\right)\ell_m\left(\frac{1}{x}\right)^{\alpha+1}$$

with $\alpha > 0$ and for ℓ_k the function log iterated k-times. Take for densities f_θ a function $kf(x-\theta)$, with k such that $k\int_0^\infty f(x)dx = 1$.

However Birgé (1984) shows that problems in which the speed of convergence is not controlled by our dimension function D are "unstable" in the sense that small perturbations can totally change the speed of convergence.

All this was done with a view to estimate a θ that is either the density itself or a sequence of such densities in the cadre of Section 4. There are other problems one can consider. The most common one would be to estimate only a smooth real function, say γ of the $p_{\theta,i}$. Two such cases have recently been described.

The first case, in the i.i.d. situation, is where γ is linear in $p_{\theta,1}$. This was covered by Donoho and Liu (1991) but it is a special case in that the sets $\{p; \gamma(p) \leq c_1\}$ and $\{p; \gamma(p) \geq c_2\}$ are *convex* sets.

The second case, considered by Bickel and Ritov (1990) is a case where γ is quadratic in p, for instance $\gamma(p) = \int [f'(x)]^2 dx$ where f is the density of p.

The possible behavior of estimates in that second case is very different from that in the linear case.

Various other special possibilities can be studied. However we need to point out that we do not have at the present time any general theory. The results of Section 3 suggest that it may be necessary to study distances between the *convex* hulls of sets such as $\{p; \gamma(p) \leq c_1\}$ and $\{p; \gamma(p) \geq c_2\}$, or more precisely the distances between the *convex* hulls of the product measures $P_\theta = \Pi p_{\theta,i}$ such that $\gamma(p_{s,i}) = c_1$ or $\gamma(p_{t,i}) \geq c_2$, respectively.

Such convex hulls are not ordinarily very tractable in product situations. Thus some new techniques need to be developed.

References

1. P. Assouad, *Espaces métriques plongements, facteurs*, thèse de doctorat, Orsay, 1976.

2. P. Bickel and Y. Ritov, *Achieving information bounds in non and semiparametric models*, Ann. Statist. **18** (1990), no. 2, 925–938.
3. L. Birgé, *Approximation dans les espaces métriques et théorie de l'estimation*, Z. Wahrsch. Verw. Gebiete **65** (1983a), no. 2, 181–237.
4. ———, *Robust testing for independent nonidentically distributed variables and Markov chains*, Specifying Statistical Models, (Louvain-la-Neuve, 1981), Lecture Notes in Statist. **16**, Springer, New York-Berlin, 1983b, pp. 134–162,
5. ———, *Stabilité et instabilité du risque minimax pour des variables indépendantes équidistribuées*, Ann. Inst. H. Poincaré Probab. Statist. **20** (1984), no. 3, 201–223.
6. D. Donoho and R. Liu, *Geometrizing rates of convergence.* II & III, Ann. Statist. **19**, (1991), no. 2, 633–701.
7. D. Donoho, R. Liu, and B. MacGibbon, *Minimax risk for hyperrectangles*, Ann. Statist. **18** (1990), no. 3, 1416–1437.
8. A. N. Kolmogorov and V. N. Tichomirov, *ε-entropy and ε-capacity of sets in function spaces*, Amer. Math. Soc. Transl. Ser. 2 **17** (1961), 277–364.
9. L. Le Cam, *On local and global properties in the theory of asymptotic normality of experiments*, (M. Puri, ed.) (Proc. Summer Res. Inst. Statist. Inference for Stochastic Processes, Indiana Univ., Bloomington, Ind., 1974, Vol. 1; dedicated to Jerzy Neyman), Academic Press, New York, 1975, pp. 13–54.
10. ———, *Asymptotic methods in statistical decision theory*, 1986, Springer Verlag.
11. L. Le Cam and G. L. Yang, *Asymptotics in statistics*, Some Basic Concepts, Springer Ser. Statist., Springer-Verlag, New York, 1990.
12. M. Nussbaum, *Asymptotic equivalence of density estimation and white noise*, 1994, preprint.
13. A. G. Vitushkin, *Bounds on the complexity of tabulation problems*, 1959 (Russian); Contemp. Problems of Mathematics.

DEPARTMENT OF STATISTICS, UNIVERSITY OF CALIFORNIA, BERKELEY, 367 EVANS HALL #3860, BERKELEY, CA94720-3860, U.S.A.

E-mail address: `lecam@stat.Berkeley.edu`

Centre de Recherches Mathématiques
CRM Proceedings and Lecture Notes
Volume **11**, 1997

Mordell-Weil Problem for Cubic Surfaces

Yu. I. Manin

0. Introduction

Let V be a plane non-singular geometrically irreducible cubic curve over a finitely generated field k. The Mordell-Weil theorem for V can be restated in the following geometric form: there is a finite subset $B \subset V(k)$ such that the whole $V(k)$ can be obtained from B by drawing secants (and tangents) through pairs of previously constructed points and consecutively adding their new intersection points with V.

In this note I address the question of validity of this statement for cubic surfaces. After reminding some constructions from the book [M], I analyze a numerical example, and then prove a different version of the Mordell-Weil statement for split cubic surfaces. A shameless change of the composition law allows me to reduce this problem to the classical theorem on the structure of abstract projective planes. Unfortunately, the initial question, which is more natural to ask for minimal surfaces, remains unanswered. I would like to call attention to this problem and its calculational aspects.

I am grateful to Don Zagier and V. E. Govorov whose tables are quoted in § 2, and to M. Rovinsky and A. Skorobogatov, discussions with whom helped me to state and prove the main theorem.

1. A Summary of Known Results

1.1. Notation. Let V be a cubic hypersurface without multiple components over a field k in $\mathbb{P}^d$, $d \geq 2$. If $x, y, z \in V(k)$ are three points (possibly coinciding) lying on a line in $\mathbb{P}^d$ not belonging to V, we write $x = y \circ z$. Thus $\circ$ is a (partial and multivalued) composition law on $V(k)$. We will also consider its restriction on subsets of $V(k)$, e.g. that of smooth points.

If $x \in V(k)$ is smooth, and does not lie on a hyperplane component of V, the birational map $t_x \colon V \to V, y \mapsto x \circ y$, is well defined. Denote by $\operatorname{Bir} V$ the full group of birational automorphisms of V.

1991 *Mathematics Subject Classification.* Primary: 14G05; Secondary: 11G05.
This is the final form of the paper.

The following two results summarize the properties of $\{t_x\}$ for curves and surfaces respectively. The first one is classical, and the second one is proved in [M].

1.2. THEOREM. *Let V be a smooth cubic curve. Then:*

a) $\operatorname{Bir} V$ *is a semidirect product of a finite group and the subgroup consisting of products of an even number of reflections* $\{t_x \mid x \in V(k)\}$.

b) *We have identically*

$$t_x^2 = (t_x t_y t_z)^2 = 1 \tag{1.1}$$

for all $x, y, z \in V(k)$.

If in addition k is finitely generated over a prime field, then:

c) $\operatorname{Bir} V$ *is finitely generated.*

d) *All points of $V(k)$ can be obtained from a finite subset of them by drawing secants and tangents and adding the intersection points.*

1.3. THEOREM. *Let V be a minimal smooth cubic surface over a perfect non-closed field k. Then:*

a) $\operatorname{Bir} V$ *is a semi-direct product of the group of projective automorphisms and the subgroup generated by*

$$\{t_x \mid x \in V(k)\}$$

and

$$\{s_{u,v} \mid u, v \in V(K); [K : k] = 2; u, v \text{ are conjugate over } k\}$$

where

$$s_{u,v} := t_u t_{u \circ v} t_v,$$

and u, v do not lie on lines of V.

b) *We have identically*

$$t_x^2 = (t_x t_{x \circ y} t_y)^2 = (s_{u,v})^2 = 1, \quad s t_x s^{-1} = t_{s(x)}, \tag{1.2}$$

for all pairs u, v not lying on lines in V, and projective automorphisms s.

c) *The relations* (1.2) *form a presentation of* $\operatorname{Bir} V$.

We remind that V is called *minimal* if one cannot blow down some lines of V by a birational morphism defined over k. The opposite class consists of *split* surfaces upon which all lines are k-rational.

1.4. Discussion. Although the two theorems are strikingly parallel, there is an important difference between finiteness properties in one- and two-dimensional cases.

Basically, (1.1) means only that $x + y := e \circ (x \circ y)$ is an Abelian group law with identity e: see [M], Theorem I.2.1. The statements c) and d) of the Theorem 1.2 additionally assert that this group is finitely generated. Therefore, (1.1) generally is not a complete system of relations between $\{t_x\}$.

On the contrary (1.2) is complete.

Comparing Theorems 1.2 and 1.3 we see that Mordell-Weil problem for cubic surfaces has at least two different versions:

QUESTION 1. *Is $(V(k), \circ)$ finitely generated?*

QUESTION 2. *Is the group $\langle t_x \mid x \in V(k)\rangle$ finitely generated?*

One implication can be easily established:

1.4.1. PROPOSITION. *If $\langle t_x \mid x \in V(k)\rangle$ is finitely generated, then $\big(V(k), \circ\big)$ is as well.*

PROOF. Let $\{t_{x_1}, \dots, t_{x_n}\}$ generate $\langle t_x \mid x \in V(k)\rangle$. We will show how to represent a point $y \in V(k)$ as a monomial in x_i w.r.t. $\circ$. Represent t_y as $t_{x_{i_m}} \dots t_{x_{i_1}}$. Assume that there exists another point $z \in V(k)$ lying in the subset of $\big(V(k), \circ\big)$ generated by $\{x_i\}$ and such that

$$x_{i_1} \neq z, \quad t_{x_{i_1}}(z) \neq x_{i_2}, \quad t_{x_{i_2}} t_{x_{i_1}}(z) \neq x_{i_3}, \dots, t_{x_{i_m}} \dots t_{x_{i_1}}(z) \neq y.$$

Then we have $t_y(z) = t_{x_{i_m}} \dots t_{x_{i_1}}(z)$, both sides being well defined, and finally

$$y = \big[x_{i_m} \circ (x_{i_{m-1}} \circ \dots \circ (x_{i_1} \circ z) \dots)\big] \circ z.$$

It remains to find z. To this end it suffices to take $\{x_i\}$ large enough so that it generates a Zariski dense subset of $V(k)$.

I cannot deduce the inverse implication. Moreover, I am inclined to believe that if $V(k)$ is infinite, then $\langle t_x \mid x \in V(k)\rangle$ *is not finitely generated.* This is suggested (but not proved) by a direct group theoretical argument establishing a canonical form of a word in $\{t_x, s_{u,v}\}$ (cf. [M], sections 39.8.1 and 39.8.2). The set of all canonical forms generally involves infinitely many generators. This is explained by the following argument.

Consider an element $s \in \operatorname{Bir} V$ and apply it to the linear system of hyperplane sections of V. We will get a new linear system with multiple base points. In [M] I prove in particular that the point of maximal multiplicity of this system can lie anywhere in $V(k) \setminus \{$ lines$\}$ if one chooses s appropriately (cf. Example 39.8.4). But if this point x is defined over k, then the canonical representative of s starts with t_x.

If (as I expect) the Question 1 admits a positive answer, whereas the Question 2 has a negative one, this must reflect a deep difference between the relations among $\{t_x, s_{u,v}\}$ in $\operatorname{Bir} V$ and relations among $\{x\}$ in $\big(V(k), \circ\big)$. The latter are less well understood than the former. One reason is that the exceptional subvarieties of birational automorphisms are rationally parametrized curves in V which presumably should be treated as a whole in a reasonable finiteness statement. In fact, a typical example of such subset is a cubic curve $C(x)$ with double point $x \in V(k)$ obtained as intersection of V with the tangent plane at x. Now, the set $\big(C(x)(k) \setminus \{x\}, \circ\big)$ with a composition law $x + y = e \circ (x \circ y)$ is isomorphic to the group of k-points of a form of the multiplicative group. Such a group is not finitely generated even for $k = \mathbb{Q}$. On the other hand, in $\big(V(k), \circ\big)$ this whole set must be considered as the domain of multivalued expression $x \circ x$, because geometrically all its points can be obtained by drawing tangents in k-rational directions to x. Therefore finite generation is still conceivable.

This comment might also help the reader to accept the definition of a generalized operation $\circ_{(C,p)}$ in § 3, which is another way to deal with the same difficulty.

Notice in conclusion that D. Kanevskii established that $\operatorname{Bir} V$ has good algorithmic properties, in particular, both word problem and conjugacy problem is solvable, if k is solvable. He has also proved that the only elements of finite order are conjugate to the generators described above.

2. Minimal Cubic Surfaces: Some Numerical Data

2.1. The structure of data. Let V be a smooth cubic surface over a field k such that $V(k)$ is infinite. Let $h: V(k) \to \mathbb{R}_+$ be a counting function (i.e. for all $H > 0$, the set $V_H := \{x \in V(k) \mid h(x) < H\}$ is finite). In order to find a generating subset in $\big(V(k), \circ\big)$, one can proceed as follows.

(A) Choose a large H and compile the list of all elements of V_H. Let points x in it be ordered by increasing $h(x)$. We will write $x < y$ if x precedes y, and use the number of a point in this list as its name.

(B) For every x and every $y < x$, calculate points $x \circ y$ and choose among them those $z = x \circ y$ for which $z < x$. Rewrite every such relation as $x = y \circ z$, $y, z < x$, and register it at the same line as (coordinates and number of) x. Notice that if by chance $y = z$, the last relation means exactly that x lies in the tangent section of V with double point y.

If such a relation exists for x, we will call x *strongly decomposable.*

If all points x with sufficiently large $h(x)$ were strongly decomposable, then the ones which are not would form a finite generating set. This is the case for cubic curves with height as counting function. For cubic surfaces the tables strongly indicate that it is not the case.

Therefore we have to consider decompositions of length ≥ 3, $x = M(x_1, \dots, x_n)$, $x_i < x$, where M is a non-associative monomial w.r.t. $\circ$. We will call *weakly decomposable* points admitting such a decomposition. Again, if all points in our list with sufficiently large height are weakly decomposable, $V(k)$ is finitely generated.

A direct search of such decompositions can be very time-consuming (as well as a direct search of points). One problem is that intermediate results can have height much larger than H; another is that we have no a priori bound for the length of decomposition.

In the example discussed below we used simple search algorithms allowing to list those monomials $M(x_1, \dots, x_n) < H$ for which there is a computation scheme representing it as an iteration of double compositions with all intermediate results registered in V_H. For example, if we have two strong decompositions $x = y \circ z = u \circ v$ with, say, $y > z, u, v$, then we get a weak decomposition $y = z \circ (u \circ v)$.

2.2. An example. D. Zagier produced a table of all primitive integral solutions of $\sum_{i=1}^4 i x_i^3 = 0$ with $h(x) := \sum_{i=1}^4 |x_i| \leq 1100$. He found 379 such points and strong decompositions of 339 among them.

By the search described above we found weak decompositions of further 24 points. This left us with 16 generators for 379 points, probably too many to state a finiteness conjecture. However, there remained a possibility that this number would diminish if decompositions with larger intermediate results were taken into account.

Here are some numerical illustrations. The first three points $1 = (1, 0, 1, -1)$, $2 = (1, 1, -1, 0)$, $3 = (1, -1, -1, 1)$ are indecomposable. However, 1 and 2 both belong to $3 \circ 3$. The next 26 points are strongly decomposable, e.g.

$$24 = (1, 28, -19, -18) = 2 \circ 2 = 13 \circ 13 = 14 \circ 21 = 5 \circ 23.$$

Points 27, 28, and 29 are only weakly decomposable, and $30 = (15, -37, 5, 29)$ of height 86 stubbornly resisted decomposition for several years.

One of the longer weak decompositions found was

$$77 = 5 \circ \big(1 \circ \big[35 \circ (2 \circ [33 \circ ([2 \circ 11] \circ [12 \circ (21 \circ 70)])])\big]\big).$$

In the Spring 1995, V. E. Govorov extended Zagier's tables listing the first 623 points of height ≤ 1800 and respective strong decompositions. Using them, I found that the point number 30 does indeed admit a weak decomposition, but one of the intermediate points in the computation scheme has height 1719. The same happened to several other points of Zagier's list which previously remained undecomposed. Of course, Govorov's list also added a number of new undecomposed points.

Summarizing our numerical findings to this day (June 1995), we have:

2.2.1. CLAIM. All points of height ≤ 506 (162 of them) are generated by a single point (1,-1,-1,1).

2.2.2. QUESTION. *Does this point generate the whole of* $V(\mathbb{Q})$?

3. Birationally Trivial Cubic Surfaces: A Finiteness Theorem

3.1. Modified composition. Let V be a smooth cubic surface, and $x, y \in V(k)$. Let $C \subset V$ be a curve on V passing through x, y, and $p\colon C \to \mathbb{P}^2$ an embedding of C into a projective plane such that $p(C)$ is cubic, and $p(x) \circ p(y)$ is defined in $p(C)$. We assume that C and p are defined over k.

In this situation we will put

$$x \circ_{(C,p)} y := p^{-1}\big(p(x) \circ p(y)\big).$$

EXAMPLE 1. Choose $C =$ a plane section of V containing x, y. If p is the embedding of C into the secant plane, then $x \circ_{(C,p)} y = x \circ y$ in the standard notation. Notice that the result does not depend on C if $x \neq y$. If $x = y$, then the choice of C is equivalent to the choice of a tangent line to V at x so that the multivaluedness of $\circ$ is taken care of by the introduction of this new parameter.

EXAMPLE 2. Assume now that V admits a birational morphism $p: V \to \mathbb{P}^2$ defined over k (e.g., V is split). We will choose and fix p once for all. Then any plane section C of V not containing one of the blown down lines as a component is embedded by p into $\mathbb{P}^2$ as a cubic curve. Therefore we can apply to (C, p) the previous construction. Notice that this time $x \circ_{(C,p)} y$ depends on C even if $x \neq y$.

The following Theorem is the main result of this note:

3.2 THEOREM. *Assume that k is a finitely generated field. In the situation of* Example 2, *the complement* $U(k)$ *to the blown down lines in* $V(k)$ *is finitely generated with respect to operations* $\circ_{(C,p)}$.

PROOF. Let us start with the following auxiliary construction. Choose a k-rational line $l \subset \mathbb{P}^2$. Then $\Gamma := p^{-1}(l)$ is a twisted rational cubic in V. The family of all such cubics reflects properties of that of lines:

a) any two different points a, b of $U(k)$ belong to a unique $\Gamma(a, b)$;
b) any two different Γ's either have one common k-point, or intersect a common blown down line.

Define now a (partial) quaternary operation on $U(k)$:

$$*(a,b;c,d) := \Gamma(a,b) \cap \Gamma(c,d).$$

It is defined for a Zariski dense open subset in $U(k)^4$.

CLAIM 1. If $x = *(a,b;c,d)$ is well-defined, then there exists a plane section C of V such that

$$*(a,b;c,d) = a \circ_{(C,p)} b.$$

In fact, choose C containing a, b, and x. Then p maps $\Gamma(a,b)$ to a line intersecting $p(C)$ at a, b, x.

It suffices now to establish the following fact:

CLAIM 2. $U(k)$ is finitely generated with respect to $*$.

To prove this, it suffices to demonstrate that $\mathbb{P}^2(k)$ is finitely generated with respect to the similar quaternary operation

$$*(a,b;c,d) := l(a,b) \cap l(c,d)$$

where $l(a,b)$ is the line containing a, b.

In fact, start with four points in general position in $\mathbb{P}^2(k)$. Introduce projective coordinates using these four points as a basis. Generate all points starting with these four and adding intersections of lines passing through pairs of previously constructed points. Obviously, the resulting set will be an abstract projective plane satisfying the Desargues axiom. Hence it will coincide with $\mathbb{P}^2(k_0)$ where k_0 is the prime subfield. Represent k as $k_0(t_1, \ldots, t_n)$. Add to the initial four points the ones with coordinates $(1{:}\, t_i{:}\, 0)$ and generate a new abstract projective plane as earlier. It will contain $\mathbb{P}^1(k)$ and hence coincide with $\mathbb{P}^2(k)$, by a classical reasoning: cf. [H].

References

[H] R. Hartshorne, *Foundations of projective geometry*, Lecture Notes, Harvard University, 1966/67 W. A. Benjamin, Inc., New York, 1967.

[M] Yu. Manin, *Cubic forms: Algebra, geometry, arithmetic*, North Holland, 1974 and 1986.

MPI FUR MATHEMATIK, GOTTFRIED-CLAREN-STRASSE 26, D-53225 BONN, GERMANY
E-mail address: manin@mpim-bonn.mpg.de

Centre de Recherches Mathématiques
CRM Proceedings and Lecture Notes
Volume **11**, 1997

Non-Vanishing of L-Functions

M. Ram Murty

The L-function is a convenient mechanism through which one can understand the arithmetic and the algebraic universe. More precisely, analytic properties of the L-function convey to us arithmetic and algebraic information. This has been the dominant theme in number theory ever since Riemann introduced his famous ζ-function

$$\zeta(s) = \sum_{n=1}^{\infty} \frac{1}{n^s}$$

defined above for $\Re(s) > 1$. In his celebrated paper of 1860, Riemann showed how one can extend $(s-1)\zeta(s)$ to an entire function for all complex values of s. He further established the functional equation

$$\pi^{-s/2}\Gamma\left(\frac{s}{2}\right)\zeta(s) = \pi^{-(1-s)/2}\Gamma\left(\frac{1-s}{2}\right)\zeta(1-s).$$

In addition, he stressed that the unique factorization property of the natural numbers is equivalent to the equation

$$\zeta(s) = \prod_p \left(1 - \frac{1}{p^s}\right)^{-1}, \quad \Re(s) > 1,$$

where the product is over the prime numbers. Earlier, Euler had noticed this factorization as well as the functional equation but he treated $\zeta(s)$ as a function of a real variable. Because of this identity, many intricate questions about the distribution of prime numbers can be answered by studying the ζ-function and the location of its complex zeros. Indeed, in his paper, Riemann outlined how one can go about trying to prove an asymptotic formula for the distribution of prime numbers, commonly referred to as the prime number theorem. It took several decades for subsequent mathematicians to develop the necessary ideas of complex analysis to vindicate Riemann's approach. Hadamard and de la Vallée Poussin proved that $\zeta(1+it) \neq 0$ for all $t \in \mathbb{R}$ and thereby deduced the celebrated prime number theorem: if $\pi(x)$ denotes the number of primes up to x, then

$$\pi(x) \sim \frac{x}{\log x}$$

1991 *Mathematics Subject Classification*. Primary: 11F66; Secondary: 11F70.
Research partially supported by FCAR and NSERC grants.
This is the final form of the paper.

as $x \to \infty$.

Hadamard's proof that $\zeta(1+it) \neq 0$ is exceedingly simple and can be explained intuitively as follows. Write

$$\log \zeta(s) = \sum_{n=1}^{\infty} \frac{a_n}{n^s}$$

and note that $a_n \geq 0$. Since $\zeta(s)$ has a simple pole at $s = 1$,

$$\log \zeta(1+\varepsilon) = \log \frac{1}{\varepsilon} + \mathrm{O}(1)$$

as $\varepsilon \to 0^+$. If $\zeta(s)$ had a zero at $s = 1 + it$, then

$$\log \zeta(1+it+\varepsilon) = -m \log \frac{1}{\varepsilon} + \mathrm{O}(1)$$

for some positive integer m. Therefore, using $a_n \geq 0$, we deduce that $n^{it} \simeq -m$ for most n's. So $m = 1$. Also, $n^{2it} \simeq 1$ for most n's so that

$$\zeta(1+2it+\varepsilon) = \log \frac{1}{\varepsilon}$$

which is impossible since $1 + 2it$ is a regular point of $\zeta(s)$ for $t \neq 0$. This proves the non-vanishing.

Later, Norbert Wiener, as a consequence of his famous Tauberian theorem, showed that the Prime Number Theorem is equivalent to the non-vanishing of the ζ-function on $\Re(s) = 1$.

Another instance in which non-vanishing of L-functions leads to arithmetic information is the celebrated theorem of Dirichlet. Namely, Dirichlet proved that for any integer m, there are infinitely many primes in a given arithmetic progression mod m corresponding to a coprime residue class. More precisely, he introduced the L-functions

$$L(s,\chi) = \sum_{n=1}^{\infty} \frac{\chi(n)}{n^s}$$

for each character $\chi\colon (\mathbb{Z}/m\mathbb{Z})^* \to \mathbb{C}^*$. (The character is extended to all of $\mathbb{Z}$ by setting $\chi(n) = 0$ whenever n and m have a common prime factor.) Dirichlet proved that

$$L(1,\chi) \neq 0$$

for every character χ. This enabled him to show the infinitude of primes in a given arithmetic progression which is coprime to the modulus. Historically, this theorem preceded the prime number theorem described in the previous paragraph. The non-vanishing theorem of Dirichlet did not allow him to establish the asymptotic law for the number of such primes. For this, it was later shown that

$$L(s,\chi) \neq 0 \quad \text{if} \quad \Re(s) = 1.$$

This implies (either via a Tauberian argument of Wiener or by classical methods of contour integration in analytic number theory) that the number of primes $p \leq x$ such that $p \equiv a \pmod{m}$ with $(a,m) = 1$ satisfies

$$\sim \frac{\pi(x)}{\phi(m)}$$

as $x \to \infty$.

This theme has gained a larger prominence in recent times. In the 1950's, Birch and Swinnerton-Dyer made a fundamental observation that the non-vanishing of

L-functions attached to an elliptic curve implies arithmetic algebraic information about the curve. More precisely, consider the equation

$$\text{E:} \qquad y^2 = x^3 + ax + b, \quad a, b \in \mathbb{Z}.$$

The question they asked themselves was: how can one decide if there are infinitely many rational solutions for E? Their conjectural answer, based on extensive numerical calculations, can be described as follows. For each prime p, let N_p be the number of solutions of the congruence

$$y^2 = x^3 + ax + b \pmod{p}.$$

Define $a_p(E)$ by

$$a_p(E) = p + 1 - N_p.$$

Now define the L-function of E as

$$\tilde{L}_E(s) = \prod_{p \nmid 4a^3+27b^2} \left(1 - \frac{a_p(E)}{p^s} + \frac{1}{p^{2s-1}}\right)^{-1}.$$

One can show that the product converges absolutely for $\Re(s) > 3/2$. It is possible to define the 'missing Euler factors' corresponding to primes dividing $4a^3 + 27b^2$ so that we obtain what is called the L-function of E. Birch and Swinnerton-Dyer conjecture that $L_E(s)$ extends to an entire function for all s in the complex plane. They further conjecture that $L_E(1) \neq 0$ if and only if E does not have infinitely many rational solutions.

This conjecture is now known to be true in many cases. It is a deeper example of how the analytic property of non-vanishing of an L-function leads to arithmetic (or diophantine) information on rational points on a curve. In fact, Birch and Swinnerton-Dyer make more precise conjectures. To describe these, we need to acquaint the reader with a few basic facts about elliptic curves.

If K is a field and we consider all solutions (x, y) of the equation

$$\text{E:} \qquad y^2 = x^3 + ax + b, \quad a, b \in K$$

with $x, y \in K$, then the set of solutions (together with a formal identity) can be given the structure of an additive Abelian group which we will denote by $E(K)$. Over the complex numbers, the points on the curve are parametrized by $(\wp(z), \wp'(z))$ where $\wp$ denotes the Weierstrass $\wp$-function associated to a n appropriate lattice. The addition law on $E(\mathbb{C})$ then corresponds to the famous addition formula of the Weierstrass $\wp$-function. The celebrated Mordell-Weil theorem is that if K is an algebraic number field, then $E(K)$ is finitely generated. Thus, by the structure theorem for finitely generated Abelian groups,

$$E(K) \simeq \mathbb{Z}^r \oplus E(K)_{\text{tors}}$$

where r denotes the number of generators and $E(K)_{\text{tors}}$ is the torsion subgroup. If K is an algebraic number field, one can define for each prime ideal $\mathfrak{p}$, $a_{\mathfrak{p}}(E)$ as before and thus form the L-series of E over K as

$$\tilde{L}_{E/K}(s) = \prod_{\mathfrak{p} \nmid (4a^3+27b^2)} \left(1 - \frac{a_{\mathfrak{p}}(E)}{\boldsymbol{N}\mathfrak{p}^s} + \frac{1}{\boldsymbol{N}\mathfrak{p}^{2s-1}}\right)^{-1},$$

where the product is over the prime ideals $\mathfrak{p}$ of K and $\boldsymbol{N}\mathfrak{p}$ denotes the absolute norm of $\mathfrak{p}$. Birch and Swinnerton-Dyer conjecture that $\tilde{L}_{E/K}(s)$ extends to an entire function and has a zero at $s = 1$ of order r equal to the rank of $E(K)$.

If the Birch and Swinnerton-Dyer conjecture is true, then we would have the beginnings of an effective procedure of obtaining all the solutions of a diophantine equation by means of analysis. A conjectural algorithm for determining all the rational points on an elliptic curve was first described by Manin [**Ma**] and then refined by Lang [**La**]. An accurate description appears in a paper by Gebel and Zimmer [**GZ**].

The torsion subgroup is easily determined. Indeed, the classical theorem of Lutz and Nagell gives an effective procedure for determining $E(K)_{\mathrm{tors}}$. In case $K = \mathbb{Q}$, Mazur's theorem gives a complete description of what groups can arise as $E(\mathbb{Q})_{\mathrm{tors}}$. Mazur [**Ma2**] showed that $E(\mathbb{Q})_{\mathrm{tors}}$ is one of

$$\mathbb{Z}/n\mathbb{Z}, \quad 1 \le n \le 10, \text{ or } n = 12, \qquad \mathbb{Z}/2\mathbb{Z} \oplus \mathbb{Z}/n\mathbb{Z}, \quad 1 \le n \le 4.$$

Kamienney and later Kamienney and Mazur extended this result to other algebraic number fields of degree ≤ 8 over $\mathbb{Q}$. But a major breakthrough came recently when L. Merel [**Me**] proved the conjecture for all algebraic number fields K. A part of his result can be stated very precisely.

Let $[K\colon \mathbb{Q}] = d > 1$. Suppose p is a prime dividing the order of $E(K)_{\mathrm{tors}}$. Then, $p \le d^{3d^2}$. There is a subsequent refinement by Oesterlé of this estimate. He proved $p \le (3^{d/2}+1)^2$. Again Merel's theorem relied for its proof on a crucial non-vanishing theorem of L-functions attached to modular forms of prime level. (More precisely, it needs a higher dimensional generalisation of the result of Kolyvagin described below.)

Since the topic of modular elliptic curves has been in the forefront of mathematics recently, we give below a quick review of the topic.

Let

$$\Gamma_0(N) = \left\{ \begin{pmatrix} a & b \\ c & d \end{pmatrix} : a, b, c, d \in \mathbb{Z}, ad - bc = 1, N \mid c \right\}.$$

This is a subgroup of finite index in $SL_2(\mathbb{Z})$ which can also be thought of as $\Gamma_0(1)$.

Let $\mathcal{H}$ denote the upper half-plane

$$\{z = x + iy\colon y > 0\}.$$

The extended upper half-plane will be denoted by $\mathcal{H}^*$ and it is

$$\mathcal{H}^* = \mathcal{H} \cup \mathbb{Q} \cup \{i\infty\}.$$

A holomorphic function

$$f\colon \mathcal{H}^* \to \mathbb{C}$$

is called a **modular form** of **weight** k and **level** N if f satisfies the transformation property

$$f\left(\frac{az+b}{cz+d}\right) = (cz+d)^k f(z)$$

for all matrices

$$\begin{pmatrix} a & b \\ c & d \end{pmatrix} \in \Gamma_0(N).$$

Since

$$\begin{pmatrix} 1 & 1 \\ 0 & 1 \end{pmatrix} \in \Gamma_0(N),$$

any such f has a Fourier expansion

$$f(z) = \sum_{n=1}^{\infty} a_n(f) e^{2\pi i n z}.$$

For the sake of illustration, let us consider the following example. Let

$$g(z) = q\prod_{n=1}^{\infty}(1-q^n)^2(1-q^{11n})^2$$

with $q = e^{2\pi i z}$. Then g is a modular form of weight 2 and level 11.

One of the spectacular turning points in number theory was the formulation of the Taniyama-Shimura conjecture (originally formulated by Taniyama in the 1950's and made more precise by Shimura later on). This conjecture makes the following prediction. If E is an elliptic curve over $\mathbb{Q}$, then there is a modular form f_E of weight 2 and of some level N such that

$$a_p(E) = a_p(f_E).$$

If the Taniyama-Shimura conjecture is true for the elliptic curve E, we call E a **modular elliptic curve**. Thus all elliptic curves over $\mathbb{Q}$ are conjectured to be modular.

In June 1993, Andrew Wiles announced a proof of an important special case of this conjecture (the so-called semi-stable case). The interest in this special case is heightened by the fact that Ribet had shown in 1987 that the semi-stable case of the Taniyama-Shimura conjecture implies the celebrated Fermat's Last Theorem. Though the approach of Wiles was nothing short of brilliant, the original proof contained a serious gap which was later circumvented by another ingenious work of Taylor and Wiles [**TW**]. The complete work now appears in Wiles [**W**] and Taylor-Wiles [**TW**]. This work has recently been extended to include the famous Frey curves:

E: $$y^2 = x(x-A)(x+B),$$

by Diamond and Kramer [**DK**]. So we now know that for each such Frey curve, there is a modular form of weight 2 such that the Taniyama-Shimura conjecture holds. Since the opening has been made, it may be only a question of time that the full Taniyama-Shimura conjecture is finally resolved.

To illustrate, consider the curve

$$y^2 + y = x^3 - x^2 - 10x - 20.$$

One can show that the modular form corresponding to this curve is $g(z)$ defined above.

If E is a modular elliptic curve with corresponding modular form f_E, then the L-function attached to E coincides with the Dirichlet series

$$\sum_{n=1}^{\infty}\frac{a_n(f_E)}{n^s}.$$

By writing this series as a Mellin transform and using the fact that f_E satisfies the modular transformation property, one can derive an analytic continuation of the above Dirichlet series and a functional equation for it, very much in the spirit of Riemann's derivation of the functional equation of the ζ-function. Thus for modular elliptic curves, one has a functional equation of the form

$$\left(\frac{\sqrt{N}}{2\pi}\right)^s\Gamma(s)L_E(s) = w\left(\frac{\sqrt{N}}{2\pi}\right)^{2-s}\Gamma(2-s)L_E(2-s)$$

where $w = \pm 1$ and N is the **conductor** of E. (Here, $L_E(s)$ is essentially the same as $\tilde{L}_E(s)$ introduced before. We modify the latter by introducing Euler factors for primes $p \mid 4a^3 + 27b^2$.)

Hence, if all elliptic curves over $\mathbb{Q}$ are modular, we can establish the analytic continuation and functional equation for $L_E(s)$, thereby proving the first part of the Birch and Swinnerton-Dyer's conjectures. Indeed, their conjectures can be seen as an attempt to generalize local-global (or Hasse) principle of the classical theory of quadratic forms. Recall that the Hasse-Minkowski theorem states that a quadratic form over $\mathbb{Q}$ represents zero non-trivially if and only if it represents zero non-trivially over $\mathbb{R}$ and all the p-adic number fields $\mathbb{Q}_p$. A more general version of the same theorem states that two quadratic forms over $\mathbb{Z}$ are equivalent[1] if and only if they are equivalent over $\mathbb{R}$ and all the p-adic number fields $\mathbb{Q}_p$.

To what extent is the structure of an algebraic object determined once it is known over $\mathbb{R}$ and all the p-adic fields $\mathbb{Q}_p$? This attitude of considering arithmetic questions and relating them to p-adic information is often loosely referred to as the Hasse principle.

Thus motivated, we will say that two curves C and C' of genus one are **locally isomorphic** if they are isomorphic as varieties over $\mathbb{R}$ and all the p-adic number fields $\mathbb{Q}_p$.

Let us fix an elliptic curve E. It is known that the Hasse principle is false for E. Namely, there are examples of curves C over $\mathbb{Q}$ which are not determined by their behaviour over $\mathbb{R}$ and all the $\mathbb{Q}_p$. However, one conjectures that there are only finitely many curves C' of genus one over $\mathbb{Q}$ which are locally isomorphic to a given elliptic curve E. (For the cognoscenti, this conjecture is equivalent to the celebrated conjecture about the finiteness of the Shafarevich-Tate group (see Mazur [**Ma2**]).)

Recently, V. Kolyvagin proved the following important theorem:

THEOREM. *Let E be a modular elliptic curve such that*

$$\operatorname{ord}_{s=1} L_E(s) \leq 1.$$

Then,

$$\operatorname{rank} E(\mathbb{Q}) = \operatorname{ord}_{s=1} L_E(s);$$

moreover, there are only finitely many curves C' of genus one over $\mathbb{Q}$ which are locally isomorphic to E **provided** *there exists a squarefree $D \equiv 1 \pmod{4N}$ such that the Dirichlet series*

$$\sum_{n=1}^{\infty} \frac{a_n(f_E)}{n^s} \left(\frac{D}{n}\right)$$

has a zero of order $1 - \operatorname{ord}_{s=1} L_E(s)$ at $s = 1$.

Here (D/n) denotes the Kronecker symbol which extends the definition of the usual quadratic residue symbol of Legendre. Let us remark at the outset that this 'twisted' Dirichlet series above does admit an analytic continuation as an entire function and so one can discuss its behaviour at $s = 1$. In fact, if E is described by the equation

$$\text{E:} \qquad y^2 = x^3 + ax + b$$

[1]Two quadratic forms of n variables and with coefficients in an integral domain R are said to be **equivalent** if there is a $GL_n(R)$-linear transformation of variables taking one form into the other. That is, the transforming matrix should be invertible over R.

the 'twisted curve' is described by

$$E_D: \qquad Dy^2 = x^3 + ax + b.$$

Moreover, if the original curve is modular, so is the twisted curve and

$$a_p(E_D) = a_p(E)\left(\frac{D}{n}\right).$$

Here again, we see analysis intervening to establish an arithmetic algebraic result. Fortunately, the proviso has now been established in two separate approaches. One is due to Kumar and Ram Murty [**MM**] and another is due to D. Bump, S. Friedberg and J. Hoffstein [**BFH**]. (The case when $\mathrm{ord}_{s=1} L_E(s) = 1$ can also be handled by a representation-theoretic result of Waldspurger [**Wa**].) So the theorem is now unconditional. In fact, the theorem proved in [**MM**] is stronger than the condition required by Kolyvagin. The authors in [**MM**] prove that there is a non-zero constant c such that

$$\sum_{\substack{D \leq Y \\ D \equiv 1 \pmod{4N}}} \mu^2(D) L'_{E_D}(1) = cY \log Y + o(Y)$$

as $Y \to \infty$, where μ denotes the Möbius function. The congruence condition on D is one imposed by Kolyvagin in his work and it implies the twisted L-function has a zero at $s = 1$. From the above result, we deduce that there are infinitely many D such that the the derivative does not vanish at $s = 1$. Hence, there is a squarefree D satisfying the Kolyvagin hypothesis.

If in the definition of a modular form, we drop the holomorphy condition and replace it by

$$-y^2\left(\frac{\partial^2}{\partial x^2} + \frac{\partial^2}{\partial y^2}\right) f = \lambda f$$

we arrive at the notion of Maass forms (first studied by Maass in the 1940's). Again, as in the case of modular forms, we can attach an L-function to such a Maass form f and establish its analytic continuation for all $s \in \mathbb{C}$. Moreover, the L-function attached to f, let us denote it by $L(s, f)$, satisfies the functional equation

$$A^s \Gamma\left(\frac{s - ir}{2}\right) \Gamma\left(\frac{s + ir}{2}\right) L(s, f) = w A^{1-s} \Gamma\left(\frac{1 - s - ir}{2}\right) \Gamma\left(\frac{1 - s + ir}{2}\right) L(1 - s, \bar{f})$$

where A is a positive real number, w is a complex number of modulus 1 and

$$\lambda = \frac{1}{4} + r^2$$

is the eigenvalue corresponding to f.

The eigenvalue conjecture of Selberg predicts that for any congruence subgroup of $SL_2(\mathbb{Z})$, we must have

$$\lambda \geq \frac{1}{4}.$$

In his panoramic vision, Langlands [**L**] has reinterpreted this conjecture as the 'Ramanujan conjecture at infinity'. This interpretation has earlier origins in the work of Satake [**S**]. For clarification on this theme, we refer to the survey article [**M**].

There are many classical problems in number theory which are related to the Selberg eigenvalue conjecture. For example, it is unknown if there are infinitely many primes of the form $n^2 + 1$. A related question would be to consider the largest prime factor of n^2+1 and show that it is large. For example, the Selberg conjecture implies that the largest prime factor of n^2+1 is greater than $n\sqrt{3/2}$ for infinitely many n. (See Iwaniec [**DI**]).

In passing, let us mention that there are other applications of this conjecture. One is to the distribution of Kloosterman sums. Another is the distribution of prime closed geodesics.

Recently, W. Luo, Z. Rudnick, and P. Sarnak [**LRS**] improved previous lower bound estimates for the eigenvalue by showing

$$\lambda \geq \frac{171}{784}$$

for any congruence subgroup of $SL_2(\mathbb{Z})$.

Again, this is proved by a non-vanishing theorem of the kind needed in Kolyvagin's theorem above. We will briefly indicate the ideas.

As stated above, one knows that

$$A^s\Gamma\left(\frac{s-ir}{2}\right)\Gamma\left(\frac{s+ir}{2}\right)L(s,f)$$

is entire. Hence, $L(s,f)$ has a 'trivial' zero at $s=ir$. Therefore, $L(ir,f)=0$. Now we consider 'twists' by Dirichlet characters χ mod m satisfying $\chi(-1)=1$. (Such characters are called 'even'.) By the theory of automorphic forms, one knows that twisting the L-function

$$\sum_n \frac{a_n(f)}{|n|^s}$$

by χ to obtain

$$\sum_n \frac{a_n(f)\chi(n)}{|n|^s}$$

gives again an L-function $L(s,f\otimes\chi)$ which satisfies a functional equation of the form

$$A_\chi^s\Gamma\left(\frac{s-ir}{2}\right)\Gamma\left(\frac{s+ir}{2}\right)L(s,f\otimes\chi)$$
$$= w_\chi A_\chi^{1-s}\Gamma\left(\frac{1-s-ir}{2}\right)\Gamma\left(\frac{1-s+ir}{2}\right)L(1-s,\bar{f}\otimes\bar{\chi})$$

with w_χ of modulus 1 and A_χ a positive real number. The important point to note is that the Gamma functions have not changed and therefore, we must have a trivial zero of $L(s,f\otimes\chi)$ at $s=ir$. The secret now is to consider the average

$$\sum_{\chi \text{even}} L(ir,f\otimes\chi).$$

By the methods of [**MM**] one can now show that this sum is in fact large provided

$$\lambda < \frac{3}{16}.$$

This contradiction implies $\lambda \geq 3/16$. But this is an old result of Selberg (obtained here by a new method). Luo, Rudnick and Sarnak noticed that this method should be applied to $L(s,\mathrm{Sym}^2(f))$ instead of $L(s,f)$ to get the improvement

$$\lambda \geq \frac{171}{784}.$$

One can find an introduction to the L-function $L(s, \mathrm{Sym}^2(f))$ in [**M**]. Here is where the representation-theoretic interpretation proves expedient. Indeed, realizing f as an automorphic representation π, one can consider

$$\pi \otimes \pi = \mathrm{Sym}^2(\pi) \oplus \mathrm{Alt}^2(\pi)$$

as a direct sum of the symmetric tensors and the alternating tensors. A fundamental theorem of Gelbart and Jacquet [**GJ**] is that if f corresponds to an automorphic representation, then $\mathrm{Sym}^2(f)$ is also an automorphic representation. It is this fact which intervenes in a fundamental way in the above argument because then we can attach an L-function to it. One then considers the average

$$\sum_{\chi \text{even}} L(2ir, \mathrm{Sym}^2 f \otimes \chi)$$

and shows that this is large if $\lambda < 171/784$. Therefore, one of the summands must be non-zero. This non-vanishing result establishes the stated lower bound for λ.

References

[BFH] D. Bump, S. Friedberg, and J. Hoffstein, *Nonvanishing theorems for L-functions of modular forms and their derivatives*, Invent. Math. **102** (1990), no. 3, pp. 543–618.

[DI] J.-M. Deshouillers and H. Iwaniec, *On the greatest prime factor of n^2+1*, Ann. Inst. Fourier (Grenoble) **32** (1982), no. 4, pp. 1–11.

[DK] F. Diamond and K. Kramer, *The modularity of Frey curves*, Math. Res. Lett. **2** (1995), 299–304.

[GZ] J. Gebel and H. Zimmer, *Computing the Mordell-Weil group of an elliptic curve over $\mathbb{Q}$*, Elliptic Curves and Related Objects, (H. Kisilevsky and M. Ram Murty, eds.), CRM Proc. and Lecture Notes, vol. 4, Amer. Math. Soc., Providence, RI, (1994), pp. 61-84.

[GJ] S Gelbart and H. Jacquet, *A relation between automorphic representations of $GL(2)$ and $GL(3)$*, Ann. Sci. École Norm. Sup. (4) **11** (1978), no. 4, pp. 471–552.

[Ka] Jean-Pierre Kahane, *Jacques Hadamard*, Math. Intelligencer **13** (1991), no. 1, pp. 23–29.

[K] V. Kolyvagin, *Finiteness of $E(\mathbb{Q})$ and $\text{Ш}(E/\mathbb{Q})$ for a subclass of Weil curves*, Izv. Akad. Nauk. SSSR Ser. Math. **52** (1988), no. 3, pp. 522–540; English transl. in Math. USSR-Izv. **32** (1989), pp. 523–541.

[La] S. Lang, *Conjectured Diophantine estimates*, Arithmetic and Geometry, Conference in honour of I. R. Shafarevic, (M. Artin and J. Tate, eds.), vol. I, 155–171, Progr. Math. **35**, Birkhäuser, 1983, pp. 155–171.

[L] R. Langlands, *Problems in the theory of automorphic forms*, Lectures in Modern Analysis and Applications, Lecture Notes in Math., vol. 170, Springer, Berlin, 1970, pp. 18–86.

[LRS] W. Luo, Z. Rudnick, P. Sarnak, *On Selberg's eigenvalue conjecture*, preprint, 1994.

[Ma1] B. Mazur, *Modular curves and the Eisenstein ideal*, Inst. Hautes Études Sci. Publ. Math. **47** (1978), pp. 33–186.

[Ma2] ———, *On the passage from the local to the global in number theory*, Bull. Amer. Math. Soc. (N.S.) **29** (1993), no. 1, pp. 14–50.

[Me] L. Merel, *Bornes pour la torsion des courbes elliptiques sur les corps de nombres*, Invent. Math. (to appear).

[Ma] Y. Manin, *Cyclotomic fields and modular curves*, Uspehi Mat. Nauk **26** (1971) no. 6, (162), pp. 7–71; English transl. in Russian Math. Surveys **26** (1971), no. 6, pp. 7–78.

[MM] M. Ram Murty and V. Kumar Murty, *Mean values of derivatives of modular L-series*, Ann. of Math. (2) **133** (1991), no. 3, pp. 447–475.

[M] M. Ram Murty, *A motivated introduction to the Langlands program*, Advances in Number Theory (Kingston, ON, 1991), (F. Gouvea and N. Yui, eds.), Oxford Sci. Publ., Oxford Univ. Press, New York, 1993, pp. 37–66.

[S] I. Satake, *Spherical functions and Ramanujan conjecture*, Algebraic Groups and Discontinuous Subgroups (Proc. Symp. Pure Math. Boulder, Colo.), vol. 9, Amer. Math. Soc., Providence, R.I., 1966, pp. 258–264.

[TW] R. Taylor and A. Wiles, *Ring-theoretic properties of certain Hecke algebras*, Ann. of Math. (2), **141** (1995), no. 3, pp. 553–572

[W] A. Wiles, *Modular elliptic curves and Fermat's last theorem*, Ann. of Math. (2) **141** (1995), no. 3, pp. 443–551.

[Wa] J.-L. Waldspurger, *Correspondance de Shimura*, J. Math. Pures Appl. (9) **59** (1980), pp. 1–133. (French)

Department of Mathematics and Statistics, McGill University, Burnside Hall, 805 Sherbrooke Street West, Montreal, Quebec H3A 2K6, Canada

Current address: Department of Mathematics, Queen's University, Kingston, Ontario K7L 3N6, Canada

E-mail address: `murty@mast.queensu.ca`

Centre de Recherches Mathématiques
CRM Proceedings and Lecture Notes
Volume **11**, 1997

Decomposition of Tensor Products of the Fundamental Representations of E_8

S. Grimm and J. Patera

ABSTRACT. The tensor products of all pairs of the eight fundamental representations of the simple Lie algebra of type E_8 are decomposed into the direct sums of irreducible representations. Simultaneously the tensor products of pairs of Weyl group orbits of fundamental weights are decomposed into the direct sums of irreducible orbits. Both decompositions are presented in the same table.

1. Introduction

The interest in the simple Lie algebra E_8 and its finite dimensional representations can be traced to two sources besides the most obvious one which takes E_8 as a challenging testing ground for efficiency of various algorithms in Lie theory.

A number of classification problems in mathematics has a solution in the form of a list of cases symbolized by the Coxeter–Dynkin diagrams corresponding to the Weyl groups of the finite dimensional simple Lie algebras, equivalently to finite indecomposable crystallographic root systems [1, 2, 13]. The ubiquity of Coxeter–Dynkin diagrams, among which E_8 is the largest exceptional one, is well known and was pointed out before. It implies that E_8 dominates in problems related in no obvious way to Lie theory. In general it is difficult to avoid asking the intriguing question about the role of E_8 in Nature. At present we are hardly close to having a satisfactory answer to it.

The second source of interest is the unique position E_8 with respect to some physics phenomena and other mathematical objects. Let us single out heterotic string theory in particle physics [8, 12], the theory of quasicrystals [6, 23, 24], the finite subgroups of $SU(2)$ and the McKay correspondence [14–16, 29].

In many other cases properties of the simple Lie group/algebra were subject of a study of physicists and mathematicians. See for example [4, 7–12, 17–28]

1991 *Mathematics Subject Classification.* Primary: 17B20.

Key words and phrases. E_8, fundamental representation, tensor product decomposition, irreducible character, orbit sum, elements of finite order.

The work of the first author was partially supported by the Schweizerischer Nationalfonds.

The work of the second author was supported, in part, by the Natural Sciences and Engineering Research Council of Canada and by the FCAR of Quebec.

This is the final form of the paper.

The purpose of this contribution to the 25-th anniversary volume of the Centre de recherches mathématiques at the Université de Montréal, is to provide yet another piece of basic information about the representations of E_8 which is not available elsewhere. In the final table of the article solutions of two problems are presented simultaneously:

(i) decomposition of tensor products of all pairs of fundamental representations of E_8 into the direct sum of the irreducible ones,
(ii) decomposition of tensor products of the Weyl group orbits of fundamental weights of E_8 into direct sum of Weyl group orbits.

A decade ago such a undertaking would have been a major one. At present the task has become easier.

Our computation was done exploiting the general method of decomposition of class functions on compact simple Lie groups [22] with the improved efficiency brought to it in [9, 10]. The main idea of the method is to work with a suitably chosen set of elements of finite order in the Lie group rather than with the whole Lie group. Then using the character values of these elements on all pertinent irreducible representations, calculated following [22], the decomposition becomes feasible. The theory and computing of character values of elements of finite order on irreducible representations of simple Lie groups were developed in [21] based on [20]. The calculation of the decompositions in this article was done by evaluating characters (and orbit sums) on all conjugacy classes of elements of order 19 in the compact simple Lie group E_8 and solving the resulting system of linear equations. It turns out that the character values of these elements split, i.e. recognize as distinct, all the irreducible representations which appear in the decompositions of the tensor product of the fundamental ones.

The indicated method is incorporated into the program *simpLie* [11], though for performance reasons the actual calculations for the Table were done using a separate program running on a UNIX platform.

In Section 2 some properties of the eight fundamental representations of E_8 and their systems of weights are summarized. In Section 3 the decomposition of tensor products of representations and Weyl group orbits is discussed. In particular it is explained how to find by 'finger tip' methods the unique (multiplicity 1) highest and the lowest representations of the decomposition as well as the second highest ones. In Section 4 the method used in the calculations is explained in more detail and the resulting decompositions are found in the Table for both the tensor products of fundamental representations as well as those of fundamental Weyl group orbits. Note that the multiplicities in the case of representations are much lower than in the case of W-orbits in spite of the fact that the orbits are simpler objects than the weight systems of the representations.

The decompositions in this paper could have been done also using the software LiE [3]. Our main goal, motivated mainly by recent developments in particle physics, is to decompose as well various plethysms based on the fundamental representations of E_8 and other high rank semisimple Lie groups in addition to the tensor products of this paper. At present such a task cannot be solved efficiently by other methods.

2. Fundamental Representations

The numbering convention of the simple roots of E_8, of the fundamental weights, of the reflections generating the Weyl group of E_8, etc. is that of Dynkin. It is shown on the following diagram:

(1)

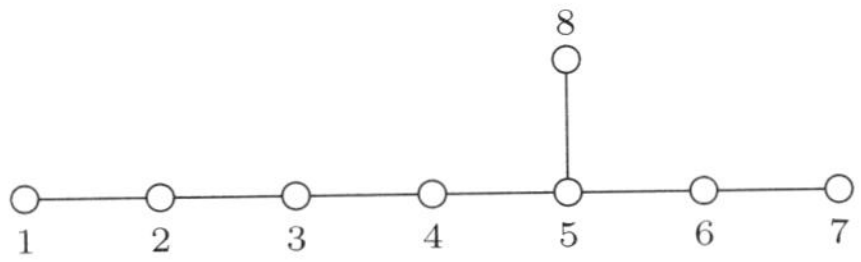

Nodes of the diagram stand for the simple roots $\{\alpha_1, \ldots, \alpha_8\}$, spanning an 8-dimensional real Euclidean space (the root space or the weight space). Two nodes which are not linked directly are mutually orthogonal, a direct line connecting two nodes indicates the angle 120° between the corresponding simple roots. It is standard to normalize the simple roots by requiring $\alpha_j \cdot \alpha_j = 2$ for all $j \in \{1, \ldots, 8\}$.

The ω-basis $\{\omega_1, \ldots, \omega_8\}$ is the basis of fundamental weights defined by the scalar products

$$\alpha_j \cdot \omega_k = \delta_{jk}, \qquad j, k \in \{1, \ldots, 8\}, \tag{2}$$

where δ_{jk} is the Kronecker symbol. The normalization of simple roots fixes the lengths of ω's.

An irreducible representation is given, up to the conjugacy under the Lie group of E_8, by its highest weight which is a set of eight non-negative integers (coordinates of the highest weight in the ω-basis) which can be understood as attached to the nodes of the diagram (1). Each such set indicates precisely one (finite dimensional) irreducible representation. The fundamental representations are the representations whose highest weights are the fundamental weights.

Using a standard algorithm, one calculates from a given highest weight $\lambda := (a_1, \ldots, a_8)$ the set of all weights, called the weight system $\Omega(\lambda)$, of the E_8-representation for which λ is the highest weight. Each weight of E_8 is an integer linear combination of simple roots. Nevertheless it is useful to use the ω-basis. Among all the weights of a weight system $\Omega(\lambda)$ one singles out the dominant weights by the property that their coordinates in the ω-basis are non-negative. In particular, any highest weight is dominant, so are the fundamental weights, as well as the single weight $(0, \ldots, 0)$ of the 1-dimensional representation.

The Weyl group W of E_8 is the finite group (of order 696,729,600) generated by the eight reflections, $W(E_8) = \langle r_1, \ldots, r_8 \rangle$, in planes orthogonal to simple roots and containing the origin of coordinates. A reflection r_k acts on any weight $\lambda := (a_1, \ldots, a_8)$ according to

$$r_k \lambda = \lambda - (\lambda \cdot \alpha_k)\alpha_k = \lambda - a_k \alpha_k. \tag{3}$$

In particular, for $\lambda = \omega_m$, we have $r_k \omega_m = \omega_m - \delta_{km}\alpha_k$.

Each Weyl group orbit $\mathcal{O}(\mu) := W\mu$ of a weight μ of E_8 contains one dominant weight. The practical value of the notion of dominant weight is in specifying its W-orbit. All the simple roots of E_8 are in the same W-orbit. The weights in that orbit are called the roots of E_8. The dominant weight of that orbit is called the highest root. In the case of E_8 the highest root happens to be the fundamental weight $\omega_1 = (1, 0, 0, 0, 0, 0, 0)$.

A weight system $\Omega(\lambda)$ is the union of Weyl group orbits of weights. Denoting the set of *distinct* dominant weights of $\Omega(\lambda)$ by $\Delta(\lambda)$, we have

$$\Omega(\lambda) = \bigcup_{\mu\in\Delta(\lambda)} m_\lambda^\mu \mathcal{O}(\mu) = \mathcal{O}(\lambda) \cup \left(\bigcup_{\mu\in\Delta(\lambda)\setminus\lambda} m_\lambda^\mu \mathcal{O}(\mu) \right), \tag{4}$$

where m_λ^μ is a positive integer called the multiplicity of the dominant weight μ in $\Omega(\lambda)$. The second equality in (4) underlines the fact that the multiplicity of the highest weight in its weight system is always one.

Properties of fundamental representations of E_8 are shown in Table 1. The fundamental representations are indicated by their highest weights ω_k. The entry in the column 'level' is the number of simple roots one needs to subtract from the highest weight ω_k in order to reach the lowest one $-\omega_k$. Note that the actual number of levels in the weight system is one higher (the entry gives the number of 'gaps' between the levels). The 'dimension' is the dimension of the irreducible representation with the highest weight ω_k. It is also the total number of weights in the weight system $\Omega(\omega_k)$, each weight counted with its proper multiplicity. The orbit size is the number of weights found in the Weyl group orbit of the fundamental weight ω_k. The column ω^2 contains the square-lengths of the fundamental weights. The columns I_2 and I_4 contain the second and fourth indices of representation. For their definition and properties the reader can consult [27]. Let us point out here that I_2 equals up to a known factor the eigenvalue of the second order Casimir operator on the representation. Finally note also that the dimensions of all fundamental representations of E_8 are divisible by 31. In general the dimension $\dim(\lambda)$ of an irreducible representations λ of E_8 has the property [27]

$$\dim(\lambda) = 0, \pm 1 \mod 31. \tag{5}$$

3. Decomposition of Tensor Products of Fundamental Representations of E_8

It is convenient to denote irreducible representations by their highest weights. Whenever explicit coordinates of the highest weight need to be given, they are written relative to the basis of fundamental weights.

TABLE 1. The fundamental representations of E_8 given by the highest weights ω_k in Dynkin numbering and some integers associated with them.

#	level	dimension	orbit size	ω^2	I_2	I_4
ω_1	58	248	240	2	60	960
ω_7	92	3875	2160	4	1500	41280
ω_2	114	30380	6720	6	14700	517440
ω_8	136	147250	17280	8	85500	3666240
ω_3	168	2450240	60480	12	1778400	96744960
ω_6	182	6696000	69120	14	5292000	314979840
ω_4	220	146325270	241920	20	141605100	10422135360
ω_5	270	6899079264	483840	30	8345660400	774477285120

From the complete reducibility of the tensor products of representations of simple Lie algebras, one has $\omega_j \otimes \omega_k = \omega_k \otimes \omega_j$. Our problem here is to determine the summands in the direct sum

$$\omega_j \otimes \omega_k = \bigoplus_{\lambda \in \mathcal{D}} m^{\lambda}_{\omega_j \omega_k} \lambda, \qquad \text{for all } j, k \in \{1, \dots, 8\}. \tag{6}$$

The summation extends over (the highest weights of) the *distinct* irreducible representations of the decomposition. Equivalently one wants to determine the set $\mathcal{D}$ of the highest weights λ of the representations present in the decomposition and the values of the multiplicities $m^{\lambda}_{\omega_j \omega_k}$ as functions of j and k. The elements of $\mathcal{D}$ are partially ordered by the level of the highest weight. The level of a weight $(a_1, \dots, a_8)$ is defined as the integer

$$L((a_1, \dots, a_8)) := 58a_1 + 114a_2 + 168a_3 + 220a_4 + 270a_5 + 182a_6 + 92a_7 + 136a_8. \tag{7}$$

If λ is the highest weight, then L takes all the values $L(\lambda), L(\lambda) - 2, L(\lambda) - 4, \dots, -L(\lambda)$ on the weights of the representation λ.

There are three terms of the direct sum of (6) that are obtained readily by hand computation. The multiplicity of each of the three for any product of two irreducible representations is one. The first is the highest term $\lambda_{\max} = \omega_j + \omega_k$. This term is the highest one meaning that $L(\lambda_{\max})$ is the largest value L takes on the weights of $\mathcal{D}$.

The second term in (6) which can be obtained readily is the lowest one. Its highest weight $\lambda_{\min}$ is found by starting from either of the weights

$$\omega_j - \omega_k, \qquad -\omega_j + \omega_k. \tag{8}$$

Using the known fact that both these weights are on the Weyl group orbit of $\lambda_{\min}$, one needs to transform one of the weights (8) into the dominant dominant weight $\lambda_{\min}$. Note that the weights (8) are dominant only if $j = k$. In that case one has $\lambda_{\min} = \{0, \dots, 0\}$; the lowest representation in the direct sum (6) is 1-dimensional. If $\omega_k \neq \omega_j$ one needs to apply a suitably chosen sequence $r_{i_1} \cdots r_{i_N}$ of the reflections (3) to one of the weights (8) so that the weight is transformed into $\lambda_{\min}$. The sequence of required reflections is shorter for $\omega_j - \omega_k$ if $L(\omega_j) > L(\omega_k)$. In order to choose among the many sequences of reflections which lead to the desired dominant weight, it is practical to build the sequence step-by-step as follows. A reflection r_m is applied to a weight μ if the k-th coordinate (relative to the ω-basis) of μ is negative. Thus for example the first reflection applied to $\omega_j - \omega_k$ should be r_k. When there are no negative coordinates in μ, the sequence is complete, i.e. we have found the dominant weight, $\mu = \lambda_{\min}$.

The third term in (6) which can be obtained readily is the second highest one, λ_2. From Theorem 3.1 of [5] one has

$$\lambda_2 = \omega_j + \omega_k - (\alpha_j + \cdots + \alpha_k). \tag{9}$$

Here $\alpha_j + \cdots + \alpha_k$ is a minimal chain of simple roots linking ω_j with ω_k along the diagram (1). The longest chain is between ω_1 and ω_7. It consists of all simple roots but α_8. When $j = k$ the chain consists of α_j, i.e. $\lambda_2 = 2\omega_j - \alpha_j$.

4. Calculation of the Decomposition of Tensor Products of Fundamental Representations and Weyl Group Orbits of E_8

To an irreducible representation λ of E_8 corresponds the character χ_λ, which is given as follows: Let $\rho_\lambda \colon E_8 \longrightarrow GL(V)$ then the character is given by

$$\chi_\lambda := \mathrm{tr}(\rho_\lambda). \tag{10}$$

The representation ring $R(E_8)$, i.e. the ring generated by the irreducible representations of E_8 with $\oplus$ and $\otimes$ as addition and multiplication,is generated by the eight fundamental representations/characters $\chi_{\omega_1}, \ldots, \chi_{\omega_8}$. Furthermore any character is a class function and the set of irreducible characters forms a basis of the space of class functions (for the compact Lie group of type E_8), which is isomorphic to $R(E_8)$. Thus from a decomposition (6) follows for the characters

$$\chi_{\omega_j}\chi_{\omega_k} = \sum_{\lambda\in\mathcal{D}} m^{\lambda}_{\omega_j\omega_k}\chi_\lambda. \tag{11}$$

The simplest W-invariant objects are the orbit sums

$$\phi_\lambda := \sum_{\mu\in\mathcal{O}(\lambda)} e^{\mu}. \tag{12}$$

They form another basis of $R(E_8)$, and as before the fundamental orbit sums $\phi_{\omega_1}, \ldots, \phi_{\omega_8}$ generate $R(E_8)$ (as a ring). The orbit sums ϕ_λ are related to the characters by the (inner) multiplicities (see also (4)):

$$\chi_\lambda = \sum m^{\mu}_{\lambda}\phi_\mu. \tag{13}$$

The m^{μ}_{λ} can be efficiently calculated (see [20]). An equation of class functions

$$f = \sum_{\lambda\in\mathcal{D}} a_\lambda\phi_\lambda, \quad \text{with } \mathcal{D} \text{ finite}, \tag{14}$$

can be solved by evaluating both sides on a suitable set of elements. We choose this set of elements to be $T_n := \{x \in E_8 \mid x^n = 1\}$, the elements of a fixed finite order n, and define a sesquilinear form on the space of class functions by

$$\langle f, g\rangle_{T_n} := \sum_{x\in T_n} f(x)\overline{g(x)}. \tag{15}$$

With respect to this form the ϕ_λ are orthogonal for a specific set of weights (see [9]), allowing us to use Fourier like methods to solve the above equation, i.e. applying $\langle\cdot, \phi_\mu\rangle_{T_n}$ to (14) gives

$$\langle f, \phi_\mu\rangle_{T_n} = \sum_{\lambda\in\mathcal{D}} a_\lambda \frac{\langle\phi_\lambda, \phi_\mu\rangle_{T_n}}{\langle\phi_\mu, \phi_\mu\rangle_{T_n}} = a_\mu, \tag{16}$$

provided the ϕ_λ, $\lambda \in \mathcal{D}$ are orthogonal with respect to (15).

In order to avoid having to do the calculations in the complex numbers, where rounding errors are difficult to avoid—especially by summing over orbits of big sizes like in E_8—we use the following fact: The values of the orbit sums on an element of order n (and thus also for characters) all lie in the cyclotomic field $\mathbb{Z}[\zeta]$, where ζ is a primitive n-th root of unity. Moreover the cyclotomic field $\mathbb{Z}[\zeta]$ can be mapped onto a finite field $\mathbb{F}_p$ where p is a prime, for which n divides $p-1$. On the integers the kernel of this map is $p\mathbb{Z}$, and all calculations can be done in integers rather than in $\mathbb{Z}[\zeta]$, given the resulting coefficients are known to be in $\mathbb{Z}^{>0}$ and are smaller than p, which is the case in the considered cases.

The tables have been calculated by using the 1976 conjugacy classes of elements of order 19 in E_8. For the prime number p we chose 1 000 000 021 and for the 19-th root of unity in $\mathbb{F}_{1000000021}$ we choose 27′062′880, i.e. we have

$$27\,062\,880^{19} \equiv 1 \pmod{1\,000\,000\,021}. \tag{17}$$

Following [20] the values of the 253 orbit sums corresponding to the weights of the irreducible representation of highest weight $2\omega_5$ on T_{19} have been calculated. From this and (13) the values of the fundamental characters are easily computed. Furthermore the 253 orbit sums are orthogonal by using the above linear form (15) for $n = 19$.

We now multiply the values of the characters χ_{ω_i} and χ_{ω_j} $(1 \leq i, j \leq 8)$ and apply $\langle \cdot, \phi_\lambda \rangle / \langle \phi_\lambda, \phi_\lambda \rangle_{T_{19}}$ for all 253 weights, giving the multiplicities $m^\lambda_{\omega_i \omega_j}$. These are listed in the tables in each column on the left side. The same procedure is done for the orbit sums ϕ_{ω_i} resulting in the coefficients of the decomposition of

$$\phi_{\omega_i}\phi_{\omega_j} = \sum_{\lambda \in \mathcal{D}} o^\lambda_{\omega_i\omega_j}\phi_\lambda. \tag{18}$$

The $o^\lambda_{\omega_i\omega_j}$ are listed in each column on the right side.

References

1. V. I. Arnold, *Singularities, bifurcations and catastrophes*, Russian Math. Surveys **141** (1983), 569–590.
2. N. Bernstein, I. M. Gelfand, V. A. Ponomarev, *Coxeter functors and Gabriel's theorem* (Russian); Uspehi Mat. Nauk **28** (1973), no. 2 (170), 19-33.
3. A. Cohen, B. Lisser, M. van Leeuwen, *Software* LiE, Computer Algebra Group, CWI, Amsterdam, 1994.
4. J. H. Conway and L. Queen, *Computing the character table of a Lie group group*, Finite groups—coming of age (Montreal, Que., 1982), Contemp. Math. **45**, Amer. Math. Soc., Providnece, R.I., (1985), 51–87.
5. E. B. Dynkin, *Maximal subgroups of the classical groups*, Amer. Math. Soc. Transl. Ser. 2 **6** (1965), 245–378.
6. V. Elser, N. J. A. Sloane, *A highly symmetric four-dimensional quasicrystal*, J. Phys. A **20** (1987), no. 18, 6161–6168.
7. I. Frenkel, J. Lepowsky, and A. Meurman, *An E_8 approach to F_1*, Finite Groups—Coming of age (Montreal, Que., 1982), Contemp. Math. **45**, Amer. Math. Soc., Providence, R.I., 1985, 99–120.
8. M. B. Green and J. H. Schwarz, *Anomaly cancellations in supersymmetric $D = 10$ gauge theory and superstrong theory*, Phys. Lett. B **149** (1984), no. 1-3, 117–122.
9. S. Grimm, *On orbit sum values of elements of finite order* (to appear).
10. ______, *Elements of finite order in representation theory* (to appear).
11. R. V. Moody, J. Patera, D. Rand, *simpLie, Macintosh software for representations of simple Lie algebras*, Les Publications CRM, Montréal, 1996.
12. D. J. Gross, J. A. Harvey, E. Martinec, R. Rhom, *Heterotic string*, Phys. Rev. Lett. **54** (1985), 502–505.
13. H. Hazewinkel, W. Hesselink, D. Siersma, and F. D. Veldkamp, *The ubiquity of Coxeter–Dynkin diagrams*, Nieuw Arch. Wisk. **XXV** (1977), 257–307.
14. B. Kostant, *On finite subgroups of $SU(2)$, simple Lie algebras, and the McKay correspondence*, Proc. Nat. Acad. Sci. U.S.A **81** (1984), no. 16, Phys. Sci., 5275–5277.
15. J. McKay, *Graphs, singularities, and finite groups*, Proc. Sympos. Pure Math., The Santa Cruz conference on finite groups (Univ. California, Santa Cruz, Calif., 1979) **37** (1980), 183–186.

16. ———, *Cartan matrices, finite groups of quaternions, and Kleinian singularities*, Proc. Amer. Math. Soc. **81** (1981), no. 1, 153–154

17. W. G. McKay, R. V. Moody, J. Patera, *Tables of E_8 characters and decomposition of plethysms*, Lie Algebras and Related Topics, (D. J. Britten, F. W. Lemire, R. V. Moody, eds.), Amer. Math. Soc., Providence R.I. (1985), 227–264.

18. ———, *Decomposition of tensor products of E_8 representations*, Algebras Groups Geom. **3** (1986), 286–328.

19. W. G. McKay, R. V. Moody, J. Patera, A. Pianzola, *The* 785 *conjugacy classes of rational elements of finite order in E_8*, Contemp. Math. **110** (1990), 79–123.

20. R. V. Moody, J. Patera, *Fast recursion formula for weight multiplicities*, Bull. Amer. Math. Soc. (N.S.) **7** (1982), 237–242.

21. ———, *Characters of elements of finite order in simple Lie groups*, SIAM J. Algebraic Discrete Methods **5** (1984), 359–383.

22. ———, *Computation of character decompositions of class functions on compact semisimple Lie groups*, Math. Comp. **48** (1987), 799–827.

23. ———, *Quasicrystals and icosians*, J. Phys. A **26** (1993), 2829–2853.

24. ———, *The E_8 family of quasicrystals*, Proc. NATO ASI Noncompact Lie groups and some of their apllications (San Antonio, TX, 1993) (E. A. Tanner and R. Wilson, eds.), NATO Adv. Sci. Inst. Ser. C Math. Phys. Sci. **429** (1993), 341–348.

25. W. G. McKay, J. Patera, D. Rand, *Tables of representations of simple Lie algebras* I: *Exceptional simple Lie algebras*, Les Publications CRM, Montréal, 1990.

26. W. G. McKay, J. Patera, R.T. Sharp, *Branching rules and Clebsch-Gordan series for E_8*, J. Math. Phys. **17** (1976), 1371–1375.

27. J. McKay, J. Patera, R. T. Sharp, *Second and fourth indices of plethysms*, J. Math. Phys. **22** (1981), 2770–2774.

28. J. Patera, M. Rodriguez, M. Zaoui, *On intersections of A_1 subgroups in the exceptional simple Lie groups E_6, E_7, E_8*, J. Phys. A **23** (1990), 5695–5705.

29. P. Slodowy, *Platonic solids, Kleinian singularities and Lie groups*, Algebraic geometry (Ann Arbor, Mich., 1981), Lecture Notes in Math., vol. 1008, Springer, Berlin-New York, 1983, pp. 102–138.

TABLE 2. Decomposition of the fundamental representations in E_8

	$\omega_1 \otimes \omega_1$		$\omega_1 \otimes \omega_7$		$\omega_1 \otimes \omega_2$		$\omega_7 \otimes \omega_7$		$\omega_1 \otimes \omega_8$		$\omega_2 \otimes \omega_7$		$\omega_1 \otimes \omega_3$	
(1,0,1,0,0,0,0,0)													1	1
(0,0,0,1,0,0,0,0)													1	4
(2,0,0,0,0,0,1,0)														
(0,1,0,0,0,0,1,0)											1	1	1	10
(1,0,0,0,0,0,0,1)									1	1	1	7	1	21
(0,0,0,0,0,0,2,0)							1	1						
(0,0,0,0,0,1,0,0)							1	2	1	7	1	21	1	35
(3,0,0,0,0,0,0,0)														
(1,1,0,0,0,0,0,0)					1	1					1	27	1	27
(0,0,1,0,0,0,0,0)					1	3	1	10	1	16	1	48	1	48
(1,0,0,0,0,0,1,0)			1	1	1	12	1	32	1	32	2	92	1	60
(0,0,0,0,0,0,0,1)			1	8	1	28	1	70	1	56	1	168	1	56
(2,0,0,0,0,0,0,0)	1	1			1	56	1	126			1	0		
(0,1,0,0,0,0,0,0)	1	2	1	27	1	54	1	144	1	72	2	270	1	27
(0,0,0,0,0,0,1,0)	1	14	1	64	1	84	1	280	1	64	1	448		
(1,0,0,0,0,0,0,0)	1	56	1	126	1	56	1	576			1	756		
(0,0,0,0,0,0,0,0)	1	240					1	2160						

	$\omega_2 \otimes \omega_2$		$\omega_7 \otimes \omega_8$		$\omega_1 \otimes \omega_6$		$\omega_2 \otimes \omega_8$		$\omega_3 \otimes \omega_7$		$\omega_8 \otimes \omega_8$		$\omega_6 \otimes \omega_7$	
(0,0,0,0,0,1,1,0)													1	1
(0,0,0,0,0,0,0,2)											1	1		
(0,0,0,0,1,0,0,0)											1	2	1	3
(3,0,0,0,0,0,1,0)														
(1,1,0,0,0,0,1,0)														
(0,0,1,0,0,0,1,0)									1	1	1	8	1	8
(2,0,0,0,0,0,0,1)														
(0,1,0,0,0,0,0,1)							1	1	1	6	1	20	1	15
(1,0,0,0,0,0,2,0)											1	32	1	32
(1,0,0,0,0,1,0,0)					1	1	1	6	1	15	1	40	1	30
(4,0,0,0,0,0,0,0)														
(2,1,0,0,0,0,0,0)														
(0,0,0,0,0,0,1,1)			1	1	1	7	1	21	1	35	2	70	2	56
(0,2,0,0,0,0,0,0)	1	1							1	27	1	72		
(1,0,1,0,0,0,0,0)	1	2					1	16	1	32	1	80	1	40
(0,0,0,1,0,0,0,0)	1	6	1	5	1	10	1	40	1	60	1	120	1	80
(2,0,0,0,0,0,1,0)	1	12					1	32	1	60	1	160	1	32
(0,1,0,0,0,0,1,0)	1	20	1	16	1	16	2	72	2	96	3	192	2	112
(1,0,0,0,0,0,0,1)	2	42	1	35	1	21	2	126	2	140	2	280	2	147
(0,0,0,0,0,0,2,0)	1	84	1	64	1	64	1	0	1	280	2	560	1	0
(0,0,0,0,0,1,0,0)	1	84	1	70	1	42	2	182	1	210	2	422	2	175
(3,0,0,0,0,0,0,0)	1	56									1	576		
(1,1,0,0,0,0,0,0)	2	54	1	72			2	216	2	216	2	432	1	216
(0,0,1,0,0,0,0,0)	2	140	1	120	1	40	2	288	2	270	2	560	1	240
(1,0,0,0,0,0,1,0)	3	216	2	192	1	32	3	432	2	352	3	800	2	272
(0,0,0,0,0,0,0,1)	2	336	1	280	1	28	2	616	1	420	1	1120	1	280

	$\omega_2 \otimes \omega_2$		$\omega_7 \otimes \omega_8$		$\omega_1 \otimes \omega_6$		$\omega_2 \otimes \omega_8$		$\omega_3 \otimes \omega_7$		$\omega_8 \otimes \omega_8$		$\omega_6 \otimes \omega_7$	
(2,0,0,0,0,0,0,0)	2	756	1	576			1	0	1	756	2	2016		
(0,1,0,0,0,0,0,0)	2	542	1	432			2	864	1	432	2	1584	1	216
(0,0,0,0,0,0,1,0)	2	840	1	560			1	1344	1	280	1	2240	1	64
(1,0,0,0,0,0,0,0)	1	1512	1	576			1	2016			1	4032		
(0,0,0,0,0,0,0,0)	1	6720									1	17280		

	$\omega_1 \otimes \omega_4$		$\omega_2 \otimes \omega_3$		$\omega_2 \otimes \omega_6$		$\omega_3 \otimes \omega_8$		$\omega_4 \otimes \omega_7$		$\omega_6 \otimes \omega_8$		$\omega_1 \otimes \omega_5$		$\omega_2 \otimes \omega_4$	
(0,1,0,1,0,0,0,0)															1	1
(1,0,0,0,0,1,1,0)																
(1,0,0,0,0,0,0,2)																
(1,0,0,0,1,0,0,0)													1	1	1	4
(4,0,0,0,0,0,1,0)																
(2,1,0,0,0,0,1,0)																
(0,0,0,0,0,0,2,1)																
(0,2,0,0,0,0,1,0)																
(1,0,1,0,0,0,1,0)															1	8
(0,0,0,0,0,1,0,1)											1	1	1	6	1	15
(0,0,0,1,0,0,1,0)									1	1	1	4	1	6	1	24
(3,0,0,0,0,0,0,1)																
(1,1,0,0,0,0,0,1)															1	15
(0,0,1,0,0,0,0,1)							1	1	1	5	1	10	1	10	2	40
(2,0,0,0,0,0,2,0)																
(0,1,0,0,0,0,2,0)											1	16			1	40
(2,0,0,0,0,1,0,0)															1	20
(0,1,0,0,0,1,0,0)					1	1	1	5	1	10	1	20	1	10	2	60
(5,0,0,0,0,0,0,0)																

	$\omega_1 \otimes \omega_4$		$\omega_2 \otimes \omega_3$		$\omega_2 \otimes \omega_6$		$\omega_3 \otimes \omega_8$		$\omega_4 \otimes \omega_7$		$\omega_6 \otimes \omega_8$		$\omega_1 \otimes \omega_5$		$\omega_2 \otimes \omega_4$	
(3,1,0,0,0,0,0,0)																
(1,0,0,0,0,0,1,1)					1	6	1	15	1	20	2	35	1	15	3	90
(1,2,0,0,0,0,0,0)																
(2,0,1,0,0,0,0,0)															1	16
(0,1,1,0,0,0,0,0)			1	1			1	16	1	16	1	40			2	80
(1,0,0,1,0,0,0,0)	1	1	1	3	1	10	1	30	1	30	1	60	1	10	3	115
(0,0,0,0,0,0,3,0)											1	64				
(0,0,0,0,0,1,1,0)					1	21	1	35	1	35	2	56	1	21	2	140
(0,0,0,0,0,0,0,2)					1	28	1	56	1	70	1	0	1	56	1	0
(0,0,0,0,1,0,0,0)	1	5	1	10	1	30	1	60	1	60	2	80	1	30	3	150
(3,0,0,0,0,0,1,0)																
(1,1,0,0,0,0,1,0)			1	10	1	16	2	56	1	40	2	96			3	160
(0,0,1,0,0,0,1,0)	1	8	1	24	2	48	2	96	2	80	3	128	1	24	4	216
(2,0,0,0,0,0,0,1)			1	21	1	21	1	105	1	35	1	140			2	210
(0,1,0,0,0,0,0,1)	1	15	2	45	2	72	3	146	2	100	3	181	1	20	5	261
(1,0,0,0,0,0,2,0)			1	60	2	32	2	160	1	160	3	240			2	240
(1,0,0,0,0,1,0,0)	1	20	2	80	3	102	3	210	2	135	4	252	1	15	5	330
(4,0,0,0,0,0,0,0)																
(2,1,0,0,0,0,0,0)			1	27			1	216			1	216			1	216
(0,0,0,0,0,0,1,1)	1	35	2	126	3	126	3	315	2	175	4	350	1	21	4	420
(0,2,0,0,0,0,0,0)			1	0	1	216	2	432	1	216	1	0			1	0
(1,0,1,0,0,0,0,0)	1	16	3	116	2	160	3	320	2	160	3	320			4	400
(0,0,0,1,0,0,0,0)	1	40	2	180	3	190	3	410	2	200	3	440	1	10	4	480
(2,0,0,0,0,0,1,0)			2	192	2	240	3	400	1	160	3	512			2	480
(0,1,0,0,0,0,1,0)	1	40	4	261	4	256	5	560	2	240	5	536			5	560
(1,0,0,0,0,0,0,1)	1	35	4	357	4	357	4	735	2	245	4	700			4	651

	$\omega_1 \otimes \omega_4$		$\omega_2 \otimes \omega_3$		$\omega_2 \otimes \omega_6$		$\omega_3 \otimes \omega_8$		$\omega_4 \otimes \omega_7$		$\omega_6 \otimes \omega_8$		$\omega_1 \otimes \omega_5$		$\omega_2 \otimes \omega_4$	
(0,0,0,0,0,0,2,0)			1	0	2	672	2	1344	1	560	2	0			1	0
(0,0,0,0,0,1,0,0)	1	35	3	462	3	434	3	952	1	280	3	840			3	665
(3,0,0,0,0,0,0,0)			1	756			1	0			1	2016				
(1,1,0,0,0,0,0,0)			3	486	2	432	4	936	1	216	3	1152			2	720
(0,0,1,0,0,0,0,0)	1	16	3	640	3	528	3	1200	1	240	3	1088			2	720
(1,0,0,0,0,0,1,0)			4	792	3	672	4	1632	1	160	4	1360			2	720
(0,0,0,0,0,0,0,1)			2	1008	2	728	2	1960	1	70	2	1688			1	560
(2,0,0,0,0,0,0,0)			1	0	1	2016	2	4032			1	0				
(0,1,0,0,0,0,0,0)			2	1260	1	864	2	2592			2	1872			1	216
(0,0,0,0,0,0,1,0)			1	1344	1	672	1	3360			1	2240				
(1,0,0,0,0,0,0,0)			1	756			1	4032			1	2016				
(0,0,0,0,0,0,0,0)																

	$\omega_3 \otimes \omega_3$		$\omega_3 \otimes \omega_6$		$\omega_4 \otimes \omega_8$		$\omega_5 \otimes \omega_7$		$\omega_6 \otimes \omega_6$		$\omega_2 \otimes \omega_5$		$\omega_3 \otimes \omega_4$	
(0,0,1,1,0,0,0,0)													1	1
(0,1,0,0,0,0,0,2)														
(2,0,0,0,1,0,0,0)														
(0,1,0,0,1,0,0,0)											1	1	1	3
(5,0,0,0,0,0,1,0)														
(3,1,0,0,0,0,1,0)														
(1,0,0,0,0,0,2,1)														
(1,2,0,0,0,0,1,0)														
(2,0,1,0,0,0,1,0)														
(1,0,0,0,0,1,0,1)											1	5	1	10
(0,1,1,0,0,0,1,0)													1	8
(1,0,0,1,0,0,1,0)											1	6	1	18

	$\omega_3 \otimes \omega_3$		$\omega_3 \otimes \omega_6$		$\omega_4 \otimes \omega_8$		$\omega_5 \otimes \omega_7$		$\omega_6 \otimes \omega_6$		$\omega_2 \otimes \omega_5$		$\omega_3 \otimes \omega_4$	
(0,0,0,0,0,0,4,0)														
(4,0,0,0,0,0,0,1)														
(0,0,0,0,0,1,2,0)														
(2,1,0,0,0,0,0,1)														
(0,0,0,0,0,0,1,2)											1	21	1	35
(0,2,0,0,0,0,0,1)													1	15
(0,0,0,0,0,2,0,0)									1	1	1	21	1	35
(0,0,0,0,1,0,1,0)							1	1	1	2	1	20	1	40
(1,0,1,0,0,0,0,1)											1	10	2	30
(3,0,0,0,0,0,2,0)														
(1,1,0,0,0,0,2,0)													1	40
(0,0,0,1,0,0,0,1)					1	1	1	4	1	6	2	30	2	60
(3,0,0,0,0,1,0,0)														
(1,1,0,0,0,1,0,0)											1	10	2	50
(0,0,1,0,0,0,2,0)									1	8	1	24	1	72
(0,0,1,0,0,1,0,0)			1	1	1	4	1	6	1	12	2	36	3	88
(6,0,0,0,0,0,0,0)														
(4,1,0,0,0,0,0,0)														
(2,0,0,0,0,0,1,1)											1	15	2	75
(2,2,0,0,0,0,0,0)														
(0,1,0,0,0,0,1,1)			1	5	1	10	1	10	2	20	3	50	5	125
(3,0,1,0,0,0,0,0)														
(0,3,0,0,0,0,0,0)														
(1,1,1,0,0,0,0,0)													2	80
(2,0,0,1,0,0,0,0)											1	10	2	105
(0,0,2,0,0,0,0,0)	1	1			1	16			1	40	1	80	2	240

	$\omega_3 \otimes \omega_3$		$\omega_3 \otimes \omega_6$		$\omega_4 \otimes \omega_8$		$\omega_5 \otimes \omega_7$		$\omega_6 \otimes \omega_6$		$\omega_2 \otimes \omega_5$		$\omega_3 \otimes \omega_4$	
(1,0,0,0,0,0,3,0)									1	32			1	160
(0,1,0,1,0,0,0,0)	1	2	1	10	1	20	1	10	1	40	2	60	4	180
(1,0,0,0,0,1,1,0)			1	15	1	20	1	15	2	30	3	66	4	180
(1,0,0,0,0,0,0,2)			1	21	1	35	1	35	1	0	2	21	3	210
(1,0,0,0,1,0,0,0)	1	6	1	24	1	36	1	24	2	48	4	72	6	224
(4,0,0,0,0,0,1,0)														
(2,1,0,0,0,0,1,0)													2	160
(0,0,0,0,0,0,2,1)			1	35	1	35	1	21	3	42	2	105	3	245
(0,2,0,0,0,0,1,0)	1	10	1	16	1	40			1	80	1	80	4	240
(1,0,1,0,0,0,1,0)	1	16	2	40	2	56	1	24	3	80	3	96	8	304
(0,0,0,0,0,1,0,1)	1	20	2	45	2	60	2	45	3	40	4	80	5	300
(0,0,0,1,0,0,1,0)	1	36	2	72	2	88	2	52	4	80	5	120	7	378
(3,0,0,0,0,0,0,1)													1	210
(1,1,0,0,0,0,0,1)	2	30	2	66	2	80	1	20	2	120	3	105	8	405
(0,0,1,0,0,0,0,1)	2	60	3	103	3	121	2	60	4	120	5	135	9	475
(2,0,0,0,0,0,2,0)	1	60	1	0	1	160			2	240	1	240	4	720
(0,1,0,0,0,0,2,0)	1	80	3	96	2	160	1	80	4	160	2	160	6	480
(2,0,0,0,0,1,0,0)	1	60	2	90	1	120	1	15	2	182	2	120	6	520
(0,1,0,0,0,1,0,0)	3	100	4	142	4	165	2	70	6	174	6	171	12	590
(5,0,0,0,0,0,0,0)														
(3,1,0,0,0,0,0,0)													1	216
(1,0,0,0,0,0,1,1)	4	150	6	195	5	230	3	86	8	240	7	195	14	765
(1,2,0,0,0,0,0,0)	1	0	1	216	1	216			1	0			3	720
(2,0,1,0,0,0,0,0)	2	100	1	160	1	160			2	240	1	80	5	640
(0,1,1,0,0,0,0,0)	3	160	3	240	3	240	1	80	3	160	2	160	8	800
(1,0,0,1,0,0,0,0)	4	210	5	275	4	290	2	90	6	300	5	210	12	875

	$\omega_3 \otimes \omega_3$		$\omega_3 \otimes \omega_6$		$\omega_4 \otimes \omega_8$		$\omega_5 \otimes \omega_7$		$\omega_6 \otimes \omega_6$		$\omega_2 \otimes \omega_5$		$\omega_3 \otimes \omega_4$	
(0,0,0,0,0,0,3,0)	1	280	1	0	1	560			2	672	1	672	2	0
(0,0,0,0,0,1,1,0)	2	210	4	245	3	350	2	105	6	420	4	315	8	875
(0,0,0,0,0,0,0,2)	2	420	2	0	2	560	1	0	3	560	3	420	5	1736
(0,0,0,0,1,0,0,0)	3	302	5	330	4	390	3	105	6	400	5	260	9	1080
(3,0,0,0,0,0,1,0)	1	192	1	240	1	160			1	480			3	960
(1,1,0,0,0,0,1,0)	6	320	7	376	5	360	1	80	8	352	3	240	15	1080
(0,0,1,0,0,0,1,0)	6	390	7	456	6	464	2	128	9	512	6	288	14	1296
(2,0,0,0,0,0,0,1)	4	420	4	525	3	455	1	35	5	490	2	210	9	1365
(0,1,0,0,0,0,0,1)	7	540	8	545	6	600	2	120	8	600	5	306	14	1520
(1,0,0,0,0,0,2,0)	4	480	6	720	4	640	1	240	7	640	2	192	9	1600
(1,0,0,0,0,1,0,0)	8	700	9	720	6	680	2	126	10	720	5	320	15	1755
(4,0,0,0,0,0,0,0)	1	756							1	2016			1	4032
(2,1,0,0,0,0,0,0)	3	432	3	720	2	720			3	864			5	1800
(0,0,0,0,0,0,1,1)	6	882	7	987	5	805	2	140	8	840	4	315	11	2121
(0,2,0,0,0,0,0,0)	4	1260	3	0	3	1440			4	1512	1	720	6	3240
(1,0,1,0,0,0,0,0)	7	880	8	880	5	880	1	80	8	880	2	320	12	2240
(0,0,0,1,0,0,0,0)	6	1080	6	1100	4	965	1	120	7	1086	3	300	10	2360
(2,0,0,0,0,0,1,0)	7	1080	6	1120	4	1200			7	1152	1	240	9	2560
(0,1,0,0,0,0,1,0)	9	1440	10	1320	6	1136	1	80	10	1280	2	320	13	2880
(1,0,0,0,0,0,0,1)	8	1680	8	1701	5	1330	1	35	8	1540	2	231	10	3360
(0,0,0,0,0,0,2,0)	4	3444	3	0	2	2240			4	3360	1	672	4	6720
(0,0,0,0,0,1,0,0)	5	2184	6	1974	3	1540	1	21	6	1960	1	210	6	3850
(3,0,0,0,0,0,0,0)	2	4032	2	0	1	4032			2	4032			2	0
(1,1,0,0,0,0,0,0)	6	2160	6	2016	3	1800			6	2160			7	3960
(0,0,1,0,0,0,0,0)	6	2710	5	2496	3	1640			6	2256	1	80	6	4320
(1,0,0,0,0,0,1,0)	6	3360	6	3120	3	1760			6	2880			6	5280

	$\omega_3 \otimes \omega_3$		$\omega_3 \otimes \omega_6$		$\omega_4 \otimes \omega_8$		$\omega_5 \otimes \omega_7$	$\omega_6 \otimes \omega_6$		$\omega_2 \otimes \omega_5$	$\omega_3 \otimes \omega_4$	
(0,0,0,0,0,0,0,1)	3	4200	3	3808	1	1680		3	3360		3	5740
(2,0,0,0,0,0,0,0)	3	7560	2	0	1	4032		3	10080		2	10080
(0,1,0,0,0,0,0,0)	3	5760	3	4752	1	1440		3	4464		2	6480
(0,0,0,0,0,0,1,0)	2	7560	1	6720	1	560		2	5600		1	6720
(1,0,0,0,0,0,0,0)	1	12096	1	10080				1	12096		1	4032
(0,0,0,0,0,0,0,0)	1	60480						1	69120			

	$\omega_4 \otimes \omega_6$	$\omega_5 \otimes \omega_8$	$\omega_3 \otimes \omega_5$	$\omega_4 \otimes \omega_4$	$\omega_5 \otimes \omega_6$	$\omega_4 \otimes \omega_5$	$\omega_5 \otimes \omega_5$	
(0,0,0,0,2,0,0,0)							1	1
(0,0,0,1,0,1,0,1)							1	2
(0,0,1,0,0,0,1,2)							1	6
(0,0,0,2,0,0,1,0)							1	6
(0,0,1,0,0,2,0,0)							1	6
(0,0,1,0,1,0,1,0)							1	8
(0,1,0,0,0,1,1,1)							2	12
(0,0,1,1,0,0,0,1)							2	12
(0,1,0,0,0,0,0,3)							1	20
(0,0,2,0,0,0,2,0)							1	24
(0,1,0,0,1,0,0,1)							2	18
(0,1,0,1,0,0,2,0)							1	24
(0,0,2,0,0,1,0,0)							1	24
(0,1,0,1,0,1,0,0)							3	26
(1,0,0,0,0,0,2,2)							1	0
(1,0,0,0,0,2,1,0)							1	0
(1,0,0,0,1,0,2,0)							2	24
(1,0,0,0,0,1,0,2)							3	20

	$\omega_4 \otimes \omega_6$		$\omega_5 \otimes \omega_8$		$\omega_3 \otimes \omega_5$		$\omega_4 \otimes \omega_4$		$\omega_5 \otimes \omega_6$		$\omega_4 \otimes \omega_5$		$\omega_5 \otimes \omega_5$	
(0,1,1,0,0,0,1,1)													4	36
(1,0,0,0,1,1,0,0)													3	24
(1,0,0,1,0,0,1,1)													6	36
(0,2,0,0,0,0,3,0)													1	80
(0,0,3,0,0,0,0,0)													1	80
(1,0,1,0,0,0,3,0)													1	64
(0,0,0,0,0,1,2,1)													2	0
(0,2,0,0,0,1,1,0)													2	60
(0,1,1,1,0,0,0,0)													2	60
(0,0,0,0,0,0,1,3)													2	0
(1,0,1,0,0,1,1,0)													6	56
(0,2,0,0,0,0,0,2)													2	90
(0,0,0,0,0,2,0,1)													3	0
(0,0,0,0,1,0,1,1)													7	20
(1,0,1,0,0,0,0,2)													4	60
(0,2,0,0,1,0,0,0)													2	74
(1,0,0,2,0,0,0,0)													2	0
(0,0,0,1,0,0,3,0)													2	48
(1,0,1,0,1,0,0,0)													7	70
(0,0,0,1,0,1,1,0)													8	36
(1,1,0,0,0,0,2,1)													6	80
(0,0,0,1,0,0,0,2)													6	36
(1,1,0,0,0,1,0,1)													9	96
(0,0,0,1,1,0,0,0)											1	1	7	12
(0,0,1,0,0,0,2,1)													10	60
(0,2,1,0,0,0,1,0)													3	96
(1,0,2,0,0,0,1,0)													4	96

	$\omega_4 \otimes \omega_6$		$\omega_5 \otimes \omega_8$		$\omega_3 \otimes \omega_5$		$\omega_4 \otimes \omega_4$		$\omega_5 \otimes \omega_6$		$\omega_4 \otimes \omega_5$		$\omega_5 \otimes \omega_5$	
(0,0,1,0,0,1,0,1)											1	3	16	74
(2,0,0,0,0,0,4,0)													1	240
(1,1,0,1,0,0,1,0)													9	108
(0,1,0,0,0,0,4,0)													1	160
(2,0,0,0,0,1,2,0)													3	120
(0,1,0,0,0,1,2,0)													9	100
(2,0,0,0,0,0,1,2)													6	180
(0,0,1,1,0,0,1,0)											1	6	14	72
(2,0,0,0,0,2,0,0)													4	200
(0,1,0,0,0,0,1,2)											1	10	15	120
(2,0,0,0,1,0,1,0)													8	144
(0,3,0,0,0,0,0,1)													1	180
(0,1,0,0,0,2,0,0)											1	10	9	120
(0,1,0,0,1,0,1,0)											1	12	25	110
(1,1,1,0,0,0,0,1)													8	140
(2,0,0,1,0,0,0,1)													8	204
(0,0,2,0,0,0,0,1)											1	10	9	60
(1,0,0,0,0,0,3,1)													8	150
(1,2,0,0,0,0,2,0)													4	0
(0,1,0,1,0,0,0,1)											2	18	22	152
(2,0,1,0,0,0,2,0)													8	176
(1,0,0,0,0,1,1,1)											2	20	28	180
(1,2,0,0,0,1,0,0)													6	180
(0,1,1,0,0,0,2,0)											1	24	16	64
(1,0,0,0,0,0,0,3)											1	35	6	0
(2,0,1,0,0,1,0,0)													9	216
(1,0,0,0,1,0,0,1)											2	30	26	192

	$\omega_4 \otimes \omega_6$		$\omega_5 \otimes \omega_8$		$\omega_3 \otimes \omega_5$		$\omega_4 \otimes \omega_4$		$\omega_5 \otimes \omega_6$		$\omega_4 \otimes \omega_5$		$\omega_5 \otimes \omega_5$	
(0,1,1,0,0,1,0,0)											2	28	25	144
(1,0,0,1,0,0,2,0)											1	36	24	180
(0,0,0,0,0,0,5,0)													1	672
(1,0,0,1,0,1,0,0)											3	40	34	216
(0,0,0,0,0,1,3,0)													4	210
(2,1,0,0,0,0,1,1)													12	240
(0,0,0,0,0,0,2,2)											1	0	13	420
(0,2,0,0,0,0,1,1)											2	40	24	140
(0,0,0,0,0,2,1,0)											1	0	10	420
(0,4,0,0,0,0,0,0)													1	720
(0,0,0,0,1,0,2,0)											2	40	19	220
(1,0,1,0,0,0,1,1)											5	55	54	244
(0,0,0,0,0,1,0,2)											3	35	17	180
(1,2,1,0,0,0,0,0)													3	320
(0,0,0,0,1,1,0,0)									1	1	3	40	27	380
(2,0,2,0,0,0,0,0)													4	520
(3,0,0,0,0,0,3,0)													2	640
(2,1,0,1,0,0,0,0)													7	340
(0,1,2,0,0,0,0,0)											1	80	7	80
(1,1,0,0,0,0,3,0)											1	80	14	320
(0,0,0,1,0,0,1,1)									1	3	6	60	48	300
(3,0,0,0,0,1,1,0)													6	360
(0,2,0,1,0,0,0,0)											1	60	15	260
(1,1,0,0,0,1,1,0)											4	80	48	260
(3,0,0,0,0,0,0,2)													3	0
(1,0,1,1,0,0,0,0)											3	90	28	320
(0,0,1,0,0,0,3,0)											1	96	17	248

	$\omega_4 \otimes \omega_6$		$\omega_5 \otimes \omega_8$		$\omega_3 \otimes \omega_5$		$\omega_4 \otimes \omega_4$		$\omega_5 \otimes \omega_6$		$\omega_4 \otimes \omega_5$		$\omega_5 \otimes \omega_5$	
(1,1,0,0,0,0,0,2)											3	90	26	120
(3,0,0,0,1,0,0,0)													7	366
(0,0,1,0,0,1,1,0)									1	6	7	88	54	384
(1,1,0,0,1,0,0,0)											5	101	46	318
(0,0,1,0,0,0,0,2)									1	10	5	91	31	300
(0,0,0,2,0,0,0,0)							1	1	1	10	2	0	18	600
(0,0,1,0,1,0,0,0)					1	1	1	2	1	12	8	114	49	456
(2,0,0,0,0,0,2,1)											3	110	31	300
(2,2,0,0,0,0,1,0)													6	400
(0,1,0,0,0,0,2,1)									1	10	8	125	59	380
(3,0,1,0,0,0,1,0)													7	480
(0,3,0,0,0,0,1,0)											1	80	8	480
(2,0,0,0,0,1,0,1)											5	130	42	280
(1,1,1,0,0,0,1,0)											5	128	46	416
(0,1,0,0,0,1,0,1)					1	4	1	6	2	18	13	148	82	452
(2,0,0,1,0,0,1,0)											5	162	49	420
(0,0,2,0,0,0,1,0)							1	8	1	24	5	144	36	768
(1,0,0,0,0,0,4,0)											1	240	7	480
(0,1,0,1,0,0,1,0)					1	6	1	12	2	28	14	170	87	536
(1,0,0,0,0,1,2,0)									1	15	6	180	44	420
(3,1,0,0,0,0,0,1)													5	540
(1,0,0,0,0,0,1,2)					1	15	1	20	2	20	11	195	58	480
(1,2,0,0,0,0,0,1)											3	195	26	480
(1,0,0,0,0,2,0,0)					1	15	1	20	1	0	7	200	40	360
(1,0,0,0,1,0,1,0)					1	16	1	24	3	32	18	224	105	560
(2,0,1,0,0,0,0,1)											5	200	42	440
(0,1,1,0,0,0,0,1)					1	10	2	20	2	40	13	215	70	720

	$\omega_4 \otimes \omega_6$		$\omega_5 \otimes \omega_8$		$\omega_3 \otimes \omega_5$		$\omega_4 \otimes \omega_4$		$\omega_5 \otimes \omega_6$		$\omega_4 \otimes \omega_5$		$\omega_5 \otimes \omega_5$	
(4,0,0,0,0,0,2,0)													3	1440
(2,1,0,0,0,0,2,0)											4	240	30	480
(1,0,0,1,0,0,0,1)					2	24	2	36	3	48	18	276	93	648
(4,0,0,0,0,1,0,0)													3	760
(0,0,0,0,0,0,3,1)									1	21	4	245	22	840
(0,2,0,0,0,0,2,0)							1	40	1	80	6	0	42	1200
(2,1,0,0,0,1,0,0)											5	250	44	600
(1,0,1,0,0,0,2,0)					1	24	1	48	2	64	15	288	84	688
(0,0,0,0,0,1,1,1)					2	30	2	40	4	15	15	300	71	540
(0,2,0,0,0,1,0,0)					1	10	1	40	1	60	11	271	56	722
(0,0,0,0,0,0,0,3)					1	56	1	70	1	0	3	0	13	1120
(1,0,1,0,0,1,0,0)					2	32	3	56	4	65	23	312	121	762
(0,0,0,0,1,0,0,1)			1	1	2	45	2	60	4	40	15	310	60	600
(0,0,0,1,0,0,2,0)					1	48	1	72	4	52	13	328	67	804
(3,0,0,0,0,0,1,1)											4	330	32	720
(0,0,0,1,0,1,0,0)	1	1	1	3	3	55	3	80	5	60	21	390	87	744
(3,2,0,0,0,0,0,0)													2	0
(1,1,0,0,0,0,1,1)					3	45	4	80	5	90	31	365	152	860
(4,0,1,0,0,0,0,0)													3	880
(1,3,0,0,0,0,0,0)											1	720	6	0
(2,1,1,0,0,0,0,0)											3	400	23	480
(0,0,1,0,0,0,1,1)	1	4	1	6	5	75	6	110	9	84	35	456	143	852
(0,2,1,0,0,0,0,0)							1	80	1	80	5	480	29	560
(3,0,0,1,0,0,0,0)											3	395	22	780
(1,0,2,0,0,0,0,0)					1	80	2	160	1	0	8	520	38	640
(2,0,0,0,0,0,3,0)							1	160	1	240	5	640	28	960
(1,1,0,1,0,0,0,0)					2	60	3	120	3	110	20	495	96	920

	$\omega_4 \otimes \omega_6$		$\omega_5 \otimes \omega_8$		$\omega_3 \otimes \omega_5$		$\omega_4 \otimes \omega_4$		$\omega_5 \otimes \omega_6$		$\omega_4 \otimes \omega_5$		$\omega_5 \otimes \omega_5$	
(0,1,0,0,0,0,3,0)					1	80	1	160	2	160	10	400	42	1440
(2,0,0,0,0,1,1,0)					2	60	2	120	3	150	19	480	90	1080
(0,1,0,0,0,1,1,0)	1	10	1	10	5	101	6	160	9	125	36	510	144	1020
(2,0,0,0,0,0,0,2)					1	0	2	210	2	210	10	0	46	1680
(0,0,1,1,0,0,0,0)	1	10	1	10	3	120	4	180	5	60	20	600	76	840
(0,1,0,0,0,0,0,2)	1	15	1	20	4	96	4	180	5	120	21	515	71	960
(2,0,0,0,1,0,0,0)					3	66	3	158	3	152	20	522	86	1032
(0,1,0,0,1,0,0,0)	1	18	1	18	6	128	8	202	10	138	36	624	141	1142
(5,0,0,0,0,0,1,0)													1	1920
(3,1,0,0,0,0,1,0)											3	480	22	800
(1,0,0,0,0,0,2,1)	1	20	1	15	5	140	6	220	9	195	32	630	118	1420
(1,2,0,0,0,0,1,0)					1	80	3	160	2	160	14	560	63	960
(2,0,1,0,0,0,1,0)					2	96	4	208	4	176	23	656	106	1280
(1,0,0,0,0,1,0,1)	2	30	2	30	9	150	10	260	13	200	46	690	160	1200
(0,1,1,0,0,0,1,0)	2	32	1	24	6	160	9	256	9	160	39	760	141	1072
(1,0,0,1,0,0,1,0)	2	48	2	40	10	198	12	324	15	220	55	811	192	1356
(0,0,0,0,0,0,4,0)							1	560	1	672	2	0	10	6720
(4,0,0,0,0,0,0,1)											1	630	10	1260
(0,0,0,0,0,1,2,0)	1	35	1	21	3	210	3	280	5	315	16	840	51	2520
(2,1,0,0,0,0,0,1)					2	105	3	300	3	240	18	735	76	1320
(0,0,0,0,0,0,1,2)	2	35	2	35	5	105	6	420	8	420	22	980	65	1260
(0,2,0,0,0,0,0,1)	1	60	1	20	4	210	7	390	6	200	24	800	82	1200
(0,0,0,0,0,2,0,0)	1	0	1	0	3	0	4	560	6	455	15	0	50	2730
(0,0,0,0,1,0,1,0)	3	60	3	50	10	240	10	400	14	300	41	875	120	1680
(1,0,1,0,0,0,0,1)	3	71	2	50	10	240	14	400	14	260	51	915	168	1440
(3,0,0,0,0,0,2,0)					1	240	2	480	1	0	8	1440	32	1920
(1,1,0,0,0,0,2,0)	2	80	1	80	6	320	10	480	10	176	38	1120	126	1440

	$\omega_4 \otimes \omega_6$		$\omega_5 \otimes \omega_8$		$\omega_3 \otimes \omega_5$		$\omega_4 \otimes \omega_4$		$\omega_5 \otimes \omega_6$		$\omega_4 \otimes \omega_5$		$\omega_5 \otimes \omega_5$	
(0,0,0,1,0,0,0,1)	3	88	3	72	10	264	12	512	14	420	38	1104	113	1788
(3,0,0,0,0,1,0,0)					1	120	2	400	2	240	12	940	50	1440
(1,1,0,0,0,1,0,0)	3	100	2	65	11	291	16	500	15	310	57	1070	184	1580
(0,0,1,0,0,0,2,0)	3	104	2	96	7	336	10	464	13	360	36	1200	112	1920
(0,0,1,0,0,1,0,0)	5	121	4	91	14	312	17	624	18	402	54	1159	153	1848
(6,0,0,0,0,0,0,0)													1	10080
(4,1,0,0,0,0,0,0)											1	720	5	2160
(2,0,0,0,0,0,1,1)	3	150	2	80	10	375	14	660	14	300	48	1280	148	1680
(2,2,0,0,0,0,0,0)							2	720	1	720	5	0	23	4320
(0,1,0,0,0,0,1,1)	8	165	6	121	20	430	26	700	27	460	78	1340	214	2060
(3,0,1,0,0,0,0,0)					1	80	2	560	1	320	8	1040	32	1600
(0,3,0,0,0,0,0,0)	1	216			1	720	2	1440	1	0	6	0	18	4320
(1,1,1,0,0,0,0,0)	2	160	1	80	5	400	10	800	8	360	31	1040	98	1760
(2,0,0,1,0,0,0,0)	2	200	1	90	7	425	11	790	10	420	34	1380	106	2000
(0,0,2,0,0,0,0,0)	2	0	1	0	4	0	8	1480	7	840	18	0	57	3640
(1,0,0,0,0,0,3,0)	2	240	1	240	4	240	6	640	6	640	18	2080	46	2880
(0,1,0,1,0,0,0,0)	6	215	4	140	14	460	19	900	17	560	53	1465	139	2080
(1,0,0,0,0,1,1,0)	7	230	5	180	17	480	22	880	23	525	63	1605	166	2100
(1,0,0,0,0,0,0,2)	4	210	3	210	10	735	13	910	12	420	32	2100	80	2520
(1,0,0,0,1,0,0,0)	8	284	6	188	18	564	24	1014	24	606	65	1660	167	2400
(4,0,0,0,0,0,1,0)							1	960	1	240	4	1920	18	1920
(2,1,0,0,0,0,1,0)	3	280	1	80	6	480	13	960	9	480	35	1520	102	2080
(0,0,0,0,0,0,2,1)	5	350	4	245	10	525	13	1260	14	560	32	2065	76	2100
(0,2,0,0,0,0,1,0)	5	336	2	160	8	480	17	1200	13	736	37	1760	101	2160
(1,0,1,0,0,0,1,0)	11	344	6	208	22	632	33	1232	28	688	81	1768	202	2560
(0,0,0,0,0,1,0,1)	8	380	6	255	16	770	19	1180	19	600	46	2070	104	2680
(0,0,0,1,0,0,1,0)	10	452	7	288	19	744	25	1430	24	768	58	2208	134	2784

	$\omega_4 \otimes \omega_6$		$\omega_5 \otimes \omega_8$		$\omega_3 \otimes \omega_5$		$\omega_4 \otimes \omega_4$		$\omega_5 \otimes \omega_6$		$\omega_4 \otimes \omega_5$		$\omega_5 \otimes \omega_5$	
(3,0,0,0,0,0,0,1)	1	420	1	35	3	630	6	1260	5	581	17	2030	48	2310
(1,1,0,0,0,0,0,1)	10	455	5	255	18	735	29	1420	23	840	64	2220	158	2760
(0,0,1,0,0,0,0,1)	11	570	7	310	19	885	26	1640	23	840	56	2320	124	3000
(2,0,0,0,0,0,2,0)	5	0	2	0	7	0	15	2720	12	1680	30	0	76	6240
(0,1,0,0,0,0,2,0)	10	600	5	336	15	720	22	2080	18	800	47	2160	99	2960
(2,0,0,0,0,1,0,0)	8	540	4	246	14	840	22	1720	17	910	48	2255	111	3060
(0,1,0,0,0,1,0,0)	14	660	8	370	23	1000	35	1840	29	985	69	2530	153	3440
(5,0,0,0,0,0,0,0)							1	4032			1	10080	3	12096
(3,1,0,0,0,0,0,0)	1	720			1	720	4	1440	2	1080	8	3240	24	4320
(1,0,0,0,0,0,1,1)	18	770	10	405	27	1020	40	2260	33	1080	76	2760	158	3600
(1,2,0,0,0,0,0,0)	4	720	1	720	4	1080	11	2160	6	1080	19	4320	45	4320
(2,0,1,0,0,0,0,0)	6	720	2	320	7	1120	17	1840	12	1120	32	2880	79	3520
(0,1,1,0,0,0,0,0)	10	840	4	560	13	1320	22	2080	15	1120	41	3440	86	3840
(1,0,0,1,0,0,0,0)	14	965	7	485	20	1260	32	2460	24	1210	57	3120	118	3900
(0,0,0,0,0,0,3,0)	3	2240	1	0	3	2240	6	6720	5	0	9	0	20	8960
(0,0,0,0,0,1,1,0)	10	1155	6	385	14	1071	20	3010	17	1190	36	3115	68	3780
(0,0,0,0,0,0,0,2)	5	0	3	0	7	0	11	5040	9	2240	17	0	35	8400
(0,0,0,0,1,0,0,0)	11	1085	6	480	17	1300	22	2922	17	1320	38	3393	70	4720
(3,0,0,0,0,0,1,0)	4	880	1	240	4	1440	11	2240	6	2112	18	3600	40	6240
(1,1,0,0,0,0,1,0)	18	1120	6	560	19	1520	38	2880	25	1400	61	3760	124	4640
(0,0,1,0,0,0,1,0)	16	1328	7	576	19	1464	33	3200	24	1536	50	3888	99	4656
(2,0,0,0,0,0,0,1)	10	1330	4	651	11	1785	21	3220	14	1785	33	4515	66	5040
(0,1,0,0,0,0,0,1)	16	1520	7	615	18	1605	30	3800	21	1680	46	4130	84	5280
(1,0,0,0,0,0,2,0)	11	1920	4	880	11	1920	20	4320	13	1600	28	5440	49	6720
(1,0,0,0,0,1,0,0)	17	1766	7	690	17	1920	31	4150	21	1800	44	4770	80	5680
(4,0,0,0,0,0,0,0)	1	0					3	10080	1	10080	3	0	8	30240
(2,1,0,0,0,0,0,0)	7	1800	1	720	4	2160	13	4320	6	3240	17	5040	33	8640

	$\omega_4 \otimes \omega_6$		$\omega_5 \otimes \omega_8$		$\omega_3 \otimes \omega_5$		$\omega_4 \otimes \omega_4$		$\omega_5 \otimes \omega_6$		$\omega_4 \otimes \omega_5$		$\omega_5 \otimes \omega_5$	
(0,0,0,0,0,0,1,1)	12	2135	5	840	12	2121	20	4900	14	2100	27	5775	46	5720
(0,2,0,0,0,0,0,0)	6	0	1	0	4	0	13	7200	7	4536	14	0	31	15120
(1,0,1,0,0,0,0,0)	14	2040	4	720	12	1936	25	4880	14	2240	33	5280	58	6400
(0,0,0,1,0,0,0,0)	10	2400	4	780	9	2160	17	5680	11	2170	22	5844	38	6786
(2,0,0,0,0,0,1,0)	10	3040	2	640	6	2400	18	6400	10	2320	20	6080	38	7680
(0,1,0,0,0,0,1,0)	15	2760	4	736	11	2320	23	6400	13	2480	28	6200	46	7840
(1,0,0,0,0,0,0,1)	11	3325	3	735	8	2415	18	7210	10	2520	20	7245	34	7980
(0,0,0,0,0,0,2,0)	4	0	1	0	2	0	7	14000	4	6720	6	0	12	23520
(0,0,0,0,0,1,0,0)	7	3801	2	560	5	2310	10	8400	6	2800	11	7595	17	9240
(3,0,0,0,0,0,0,0)	3	10080			1	10080	5	20160	1	0	4	0	7	40320
(1,1,0,0,0,0,0,0)	8	4536	1	720	3	3240	12	10080	5	2376	12	7200	20	12960
(0,0,1,0,0,0,0,0)	6	4400	1	480	3	2416	10	9440	5	2640	9	8640	17	10400
(1,0,0,0,0,0,1,0)	7	4960	1	240	3	2160	9	11840	4	2880	9	9600	14	13440
(0,0,0,0,0,0,0,1)	3	6160	1	56	1	1680	4	13510	2	2240	4	10920	6	13440
(2,0,0,0,0,0,0,0)	2	0					4	30240	1	10080	2	0	5	60480
(0,1,0,0,0,0,0,0)	3	6840			1	720	4	17280	1	2160	3	10800	5	18720
(0,0,0,0,0,0,1,0)	1	8960					2	22400	1	672	1	13440	3	23520
(1,0,0,0,0,0,0,0)	1	10080					1	40320			1	10080	1	60480
(0,0,0,0,0,0,0,0)							1	241920					1	483840

Mathematisches Institut, Universität Basel, Rheinsprung 21, 4051 Basel, Schweiz
E-mail address: grimm@math.unibas.ch

Département de mathématique et de statistique and Centre de recherches mathématiques, Université de Montréal, C.P. centre-ville, Montréal, Québec, H3C 3J7, Canada.
E-mail address: patera@crm.umontreal.ca

Centre de Recherches Mathématiques
CRM Proceedings and Lecture Notes
Volume **11**, 1997

The One-loop Graviton Scattering Amplitude in String Theory

D. H. Phong

ABSTRACT. The one-loop graviton scattering amplitude in string theory is the simplest non-vanishing amplitude incorporating quantum effects in a consistent theory of gravity. We discuss its analytic continuation to physical values of the external momenta, and the issue of analyticity in string theory vs. in field theories. A key ingredient of our analysis is a uniform description of the singularities of the Green's function near the boundary of the moduli space of Riemann surfaces.

1. Graviton Scattering in String Theory

1.1. Introduction. The problem of quantizing the force of gravity is one of the major challenges of modern theoretical physics. At large scales, the gravitational field is described by a metric $g_{\mu\nu}$ and the Einstein-Hilbert action. However, all attempts at quantizing this theory within the otherwise very successful framework of quantum field theory have failed so far to produce a renormalizable and unitary theory. By giving up the concept of point particles, superstring theories [**17, 19, 31, 32, 33**] offer at present the only known way out of this impasse. The basic ingredient in these theories is a string, and their spectrum an infinite tower of states corresponding to the vibrational modes of the string. At the lowest level are massless multiplets, which incorporate in particular a spin 2 particle identifiable as the graviton, and spin 1 particles potentially identifiable as the gauge bosons of the known interactions. In addition to these remarkable features which make them the only candidates for a unified theory of all interactions, supertring theories have been shown to be unitary in perturbation theory [**3**], and are expected to be finite.

A fundamental reason for the expected tame ultraviolet behavior of string theories can be easily understood geometrically. Strings interact by joining and splitting (Fig. 1). However, the interaction point depends on the Lorentz frame, so in effect the interactions have been smeared out and can only be detected by the topology of the world-sheet. This is unlike the case of point particles, where interactions occur by introducing vertices at a well-defined point, which can lead to uncontrollable

---

1991 *Mathematics Subject Classification*. Primary: 30Fxx, 31Axx, 81T30, 81T40, 83E30, 81T60; Secondary: 32G15, 33Cxx, 53C80, 83C45, 83C47.

Research partially supported by NSF grand DMS-95-05399.

This is the final form of the paper.

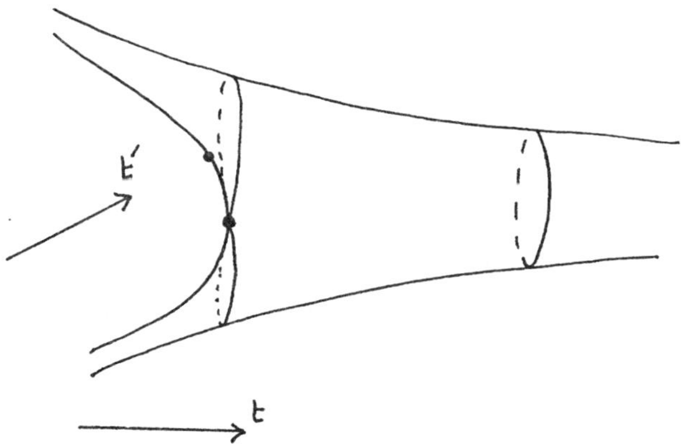

FIGURE 1. Interaction of closed strings. The interaction point depends on the choice of Lorentz frame.

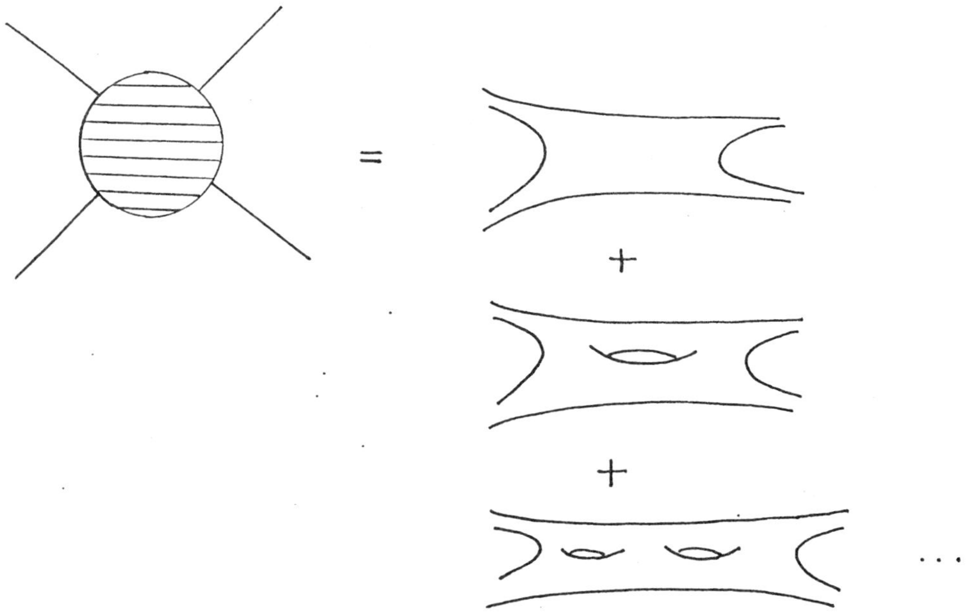

FIGURE 2. The loop expansion in string perturbation theory.

ultraviolet divergences. In spite of this intuitive explanation, the program of formulating rules for evaluating string scattering amplitudes in perturbation theory and establishing their finiteness is still incomplete at the present time. In fact, even the finiteness of the most basic scattering amplitudes incorporating quantum effects has not been elucidated until very recently [**9, 10, 11**].

In perturbation theory, string scattering amplitudes are given by a sum over the genus g (i.e. number of handles) of the world-sheet spanned by the strings (Fig. 2), with each handle corresponding to the creation and annihilation of a virtual pair

$$\mathcal{A}(k_1, \ldots, k_N) = \sum_{g=0}^{\infty} \lambda^{2-2g} \mathcal{A}_g(k_1, \ldots, k_N). \tag{1.1}$$

Here $k_1, \ldots, k_N$ are the external momenta of the incoming and outgoing strings, λ is the string coupling constant, and $\mathcal{A}_g(k_1, \ldots, k_N)$ is the g-loop contribution to the full amplitude. Quantum effects take place only at $g \geq 1$. By supersymmetry, the $g = 1$ amplitudes vanish identically for $N \leq 3$ external particles. Thus the simplest non-trivial amplitude incorporating quantum gravity effects in a causal and unitary theory in Physics is the string scattering amplitude to one loop of 4 massless on-shell states in the graviton multiplet, whose explicit form (c.f. equation (1.8) below) was first obtained by Green and Schwarz in one of the early triumphs of superstring theory [**16**]. Indeed the amplitude did not exhibit in its integrand the severe divergences which had plagued earlier attempts at quantizing gravity. However, it still diverged for physical values of the momenta, and was formally real, seemingly in contradiction with the optical theorem, which would relate its imaginary part to the non-vanishing values of tree-level ($g = 0$) amplitudes.

These apparent contradictions are symptoms of the same illness: the amplitudes require an analytic continuation in the external momenta. In quantum field theory, the same basic problem is cured by Feynman's $i\varepsilon$ prescription for the propagators. No such prescription was available in string theory, where amplitudes are calculated through a first quantized formalism, and string field theory is still in its infancy. In particular, the issue of whether an analytic continuation does exist, and whether its singularities would be physically acceptable, is a crucial issue which has to be addressed. This last question about the singularities of the physical amplitude gives some measure of the complexity of the problem: in string perturbation theory, the amplitude at each loop order is given by a single diagram, which must then by itself account for all the physical poles and cuts of both the box graph and the one-particle reducible graphs (Fig. 3) in an infinite superposition of quantum field theories!

In this talk, we wish to report on the recent progress in these directions [**9, 10, 11**], and in particular on a method of analytic continuation which resolves satisfactorily all the above problems. In addition, it sheds additional light on the stringy effects which separate superstring theory from a mere superposition of field theories. We also indicate a few directions for further investigation. On the mathematical side, this circle of problems is of particular interest, since it leads to the study of new generalized functions, which can be viewed as suitable extensions of the well-known distribution $(x - i\varepsilon)^{-1}$ of field theory. Due to the fundamentally geometric nature of string interactions however, these problems force us to probe deeply into sophisticated mathematical structures such as the geometry of moduli spaces, and should prove to be a fertile ground for new interactions between Analysis and Geometry.

1.2. Explicit formulae for the graviton scattering amplitudes. At each loop level, the amplitude $\mathcal{A}_g(k_1, \ldots, k_N)$ is of the form

$$\begin{aligned} \mathcal{A}_g(k_1, \ldots, k_N) &\\ = (2\pi)^{10} \delta(k_1 + \cdots + k_N) &\prod_{i=1}^{N} \varepsilon_i^{\mu_i \bar{\mu}_i} K_{\mu_1 \ldots \mu_N} \overline{K}_{\bar{\mu}_1 \ldots \bar{\mu}_N} A_g(k_1, \ldots, k_N) \end{aligned} \tag{1.2}$$

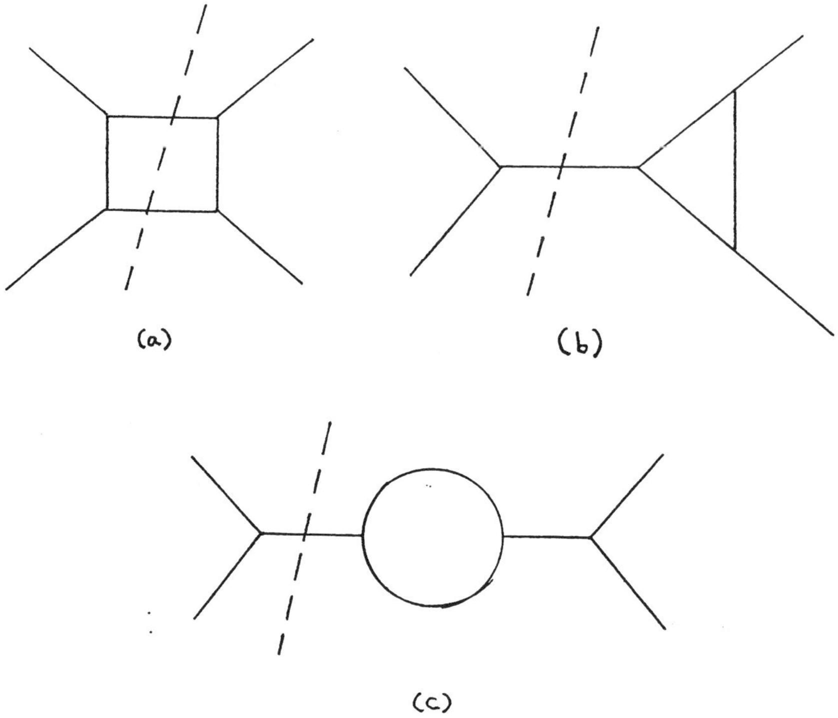

FIGURE 3. The box graph and 1-particle reducible graph. (a) branch cuts; (b) branch cuts on top of simple poles; (c) branch cuts on top of double poles.

where $\varepsilon_i^{\mu_i\bar{\mu}_i}$ are polarization tensors, $K_{\mu_1\ldots\mu_N}\overline{K}_{\bar{\mu}_1\ldots\bar{\mu}_N}$ kinematic factors which depend polynomially in the momenta, and $A_g(k_1,\ldots,k_N)$ the key Lorentz invariant factors incorporating the physical singularities of the amplitude. The factor $(2\pi)^{10}$ is due to the fact that superstrings can propagate consistenly only in 10-dimensional space-time. To discuss analyticity in the external momenta, we introduce the Mandelstam variables $s_{ij} = -(k_i + k_j)^2$, and restrict ourselves to the case of interest, namely the scattering of 4 massless states in the graviton multiplet. Since momentum is conserved, $\sum_{i=1}^4 k_i = 0$, and the external states are massless, $k_i^2 = 0$, the number of Mandestam variables reduces to just three, $s_{12} \equiv s$, $s_{23} \equiv t$, $s_{13} \equiv u$, constrained by the relation

$$s + t + u = 0. \tag{1.3}$$

We can give now an explicit formula for the tree level amplitude

$$A_0(s,t,u) = -\frac{2}{s^2}\int_{\mathbb{C}} d^2z\,|z|^{-t-2}\,|1-z|^{-s}. \tag{1.4}$$

To derive (1.4) from first principles would take us too far afield (c.f. [**12**, Section III.L]). Nevertheless, some of its features could have been foreseen. In Feynman-Dirac quantization, the amplitude is obtained by summing over all possible paths. For strings, this means summing over all surfaces (with a fixed topology corresponding to the order of perturbation theory). A fundamental requirement of string theory is *conformal invariance*, which means that the contribution of each surface depends only on the conformal class of its metric, and not on the metric itself. Unlike the space of metrics which is infinite-dimensional, the space of conformal classes can be described by a finite number of parameters, referred to as the *moduli* of the surface. Thus first-quantized string amplitudes are given by integrals over moduli space of correlation functions of two-dimensional conformal field theories, whose propagators are just the Green's functions on the surface.

Now the world-sheet at tree level for the scattering of 4 external states is the sphere with 4 punctures $z_1, \dots, z_4$. It is well-known that all metrics on the sphere are conformally equivalent. Furthermore, by $SL(2,\mathbb{C})$ invariance, we may set $z_1 = 0$, $z_2 = 1$, $z_3 = z$, and $z_4 = \infty$. We can now see that (1.4) incorporates all the features we expect, since the d^2z integral is just an integral over the moduli parameter z, and the integrand can be recognized as $\prod_{i<j} \exp\big(1/2 s_{ij} G(z_i, z_j)\big)$, where $G(z,w) = -\ln|z-w|^2/(1+|z|^2)(1+|w|^2)$ is the Green's function on the sphere.

Even at tree-level, the amplitude converges only for

$$\text{(1.5)} \qquad \operatorname{Re}(s) < 0, \quad \operatorname{Re}(t) < 2, \quad \operatorname{Re}(s+t) > 0$$

and otherwise requires an analytic continuation. In this case, the solution is provided by the explicit formula

$$\text{(1.6)} \qquad A_0(s,t,u) = \frac{\Gamma(-s/2)\Gamma(-t/2)\Gamma(-u/2)}{\Gamma(1+s/2)\Gamma(1+t/2)\Gamma(1+u/2)}$$

which exhibits then the infinite number of poles in s, t, and $u = -(s+t)$ at even integers, corresponding to massive intermediate states going on mass shell.

For one-loop amplitudes, the world-sheet is a torus. Each torus is conformally equivalent to a quotient space $M_\tau = \mathbb{C}/\{m + n\tau\}$, where τ is a complex number with positive imaginary part, and $\{m + n\tau; m, n \in \mathbb{Z}\}$ is the lattice spanned by the two vectors 1 and τ. Since $-1/\tau$ and $1+\tau$ lead to the same lattice, the moduli space of the torus is $F = \{\tau; \operatorname{Im}\tau > 0\}/\operatorname{PSL}(2,\mathbb{Z})$. The emission and absorption points for the external states correspond to four points $z_1, \dots, z_4$ in M_τ. On each torus M_τ, we define the Green's function $G(z,w)$ by

$$\text{(1.7)} \qquad -\partial_z \partial_{\bar z} G(z,w) = 2\pi\delta(z,w) - \frac{2\pi g_{z\bar z}}{\int d^2z g_{z\bar z}}, \qquad \int d^2z g_{z\bar z} G(z,w) = 0.$$

The one-loop amplitude for the scattering of 4 external states in the graviton multiplet in the Type IIB superstring theory is then given by [**16**] and [**12**, Section III.M]

$$\text{(1.8)} \qquad A_1(s,t,u) = \int_F \frac{d^2\tau}{\tau_2^2} \int_{M_\tau} \prod_{i=1}^4 \frac{d^2 z_i}{\tau_2} \prod_{i<j} \exp\Big(\frac{1}{2} s_{ij} G(z_i, z_j)\Big).$$

By translation invariance, we may set $z_4 = 0$. It is also convenient to select the following fundamental domain

$$\text{(1.9)} \qquad F = \big\{\tau = \tau_1 + i\tau_2 \in \mathbb{C}; \tau_2 > 0, |\tau_1| \le 1/2, |\tau| \ge 1\big\}$$

for the moduli space $\{\tau_2 > 0\}/\operatorname{PSL}(2,\mathbb{Z})$. We note that its area with respect to the measure $d^2\tau/\tau_2^2$ is finite and equal to $\pi/2$.

To determine the values of s, t, u for which (1.8) converges, we split the region of integration in z_1, z_2, z_3 into 6 regions, corresponding to the different orderings of $\mathrm{Im} z_1$, $\mathrm{Im} z_2$, $\mathrm{Im} z_3$. The contributions of these regions are equal pairwise, so that the one-loop amplitude $A_1(s,t,u)$ can be reexpressed as **[18, 1, 27]**

$$A_1(s,t,u) = 2A_1(s,t) + 2A_1(t,u) + 2A_1(u,s) \tag{1.10}$$

with $A_1(s,t)$ being given by (1.8), but with the region of integration restricted to

$$0 \leq \mathrm{Im}(z_1) \leq \mathrm{Im}(z_2) \leq \mathrm{Im}(z_3) \leq \tau_2 \tag{1.11}$$

and the value of u set to $u = -(s+t)$. Next we need a precise description of the singularities of the Green's function $G(z,w)$ in the various regions where z may tend to w, or τ may tend to ∞ separately or simultaneously. First, in terms of the complex coordinates z, $G(z,w)$ can be expressed in terms of θ-functions

$$G(z,w) = -\ln\left|\frac{\vartheta_1(z-w \mid \tau)}{\vartheta_1'(0 \mid \tau)}\right|^2 + \frac{2\pi}{\tau_2}\big(\mathrm{Im}(z-w)\big)^2. \tag{1.12}$$

To compare the behavior of $G(z,w)$ as τ varies, we introduce the real coordinates $(x,y) \in [0,1]^2$ by

$$z = x + y\tau \tag{1.13}$$

so that all tori M_τ are parametrized by the unit square. If we set $q = e^{2\pi i\tau}$, (2.6) can be rewritten as

$$\exp\big(-G(z,0)\big) = \frac{1}{4\pi^2}|q|^{-y(1-y)}\,|1-q^y e^{2\pi i x}|^2\,|1-q^{1-y}e^{-2\pi i x}|^2 R(z) \tag{1.14}$$

with

$$R(z) = \prod_{n=1}^{\infty} |1-q^{n+y}e^{2\pi i x}|^2\,|1-q^{n+1-y}e^{-2\pi i x}|^2\,|1-q^n|^{-4}. \tag{1.15}$$

In the fundamental region (1.9), we have $|q| < e^{-\pi\sqrt{3}}$, and $R(z)$ is bounded from above and below uniformly in all variables. Thus (1.14)–(1.15) provides the precise description of the singularities of $G(z,w)$ which we need. The resulting expression for the amplitude (1.8) is most succinct in terms of the following variables

$$\begin{aligned} u_1 &= y_1, & \alpha_1 &= 2\pi(x_1 + u_1\tau_1) \\ u_2 &= y_2 - y_1, & \alpha_2 &= 2\pi(x_2 - x_1 + u_2\tau_1) \\ u_3 &= y_3 - y_2, & \alpha_3 &= 2\pi(x_3 - x_2 + u_3\tau_1) \\ u_4 &= 1 - y_3, & \alpha_4 &= 2\pi\tau_1 - \alpha_1 - \alpha_2 - \alpha_3. \end{aligned} \tag{1.16}$$

If we also set $s_i = s$ for i even and $s_i = t$ for i odd, the amplitude $A_1(s,t)$ can be recast as

$$\begin{aligned} A_1(s,t) = &\int_F \frac{d^2\tau}{\tau_2^2}\int_0^{2\pi}\prod_{i=1}^4 \frac{d\alpha_i}{2\pi}\,\delta\Big(2\pi\tau_1 - \sum_{i=1}^4 \alpha_i\Big) \\ &\times |q|^{-(su_1u_3+tu_2u_4)}\prod_{i=1}^4 \big|1-e^{i\alpha_i}|q|^{u_i}\big|^{-s_i}\Big|1-\prod_{j\neq i} e^{i\alpha_j}|q|^{u_j}\Big|^{-s_j} \\ &\times \prod_{i=2k}\prod_{j=2m+1}\big|1-e^{i(\alpha_i+\alpha_j)}|q|^{u_i+u_j}\big|^{s_i+s_j} \\ &\times \left(\frac{R(z_2-z_1)R(z_3)}{R(z_3-z_1)R(z_2)}\right)^{-s}\left(\frac{R(z_3-z_2)R(z_1)}{R(z_3-z_1)R(z_2)}\right)^{-t}. \end{aligned} \tag{1.17}$$

It is now readily seen that the integral expression for the partial amplitude $A_1(s,t)$ converges only in the following infinite strip

$$\text{Re}(s) \leq 0, \quad \text{Re}(t) \leq 0, \quad \text{Re}(s+t) > -2. \tag{1.18}$$

Since the full amplitude $A_1(s,t,u)$ requires all three amplitudes $A_1(s,t)$, $A_1(t,u)$, $A_1(u,s)$ with the additional constraint $s+t+u=0$, it follows that $A_1(s,t,u)$ converges only when

$$\text{Re}(s) = \text{Re}(t) = \text{Re}(u) = 0. \tag{1.19}$$

Conversely, at these values, the integrand in (1.8) has modulus 1, and the integral converges since the fundamental region F has finite area.

1.3. Analyticity in string theory vs field theory. Before discussing the analytic continuation of $A_1(s,t)$, we pause to discuss briefly its physical meaning. In quantum field theory, causality is equivalent to locality and, combined with Lorentz invariance, implies the analytic dependence of the Green's functions of local fields in the external momenta. This can usually be seen through a spectral representation for the amplitudes, as in the Kählen-Lehmann representation [**7, 14**] for the off-shell two-point function

$$G(s) = \int_0^\infty dM^2 \frac{\rho(M^2)}{s - M^2 + i\varepsilon} \tag{1.20}$$

which implies that $G(s)$ is analytic off the positive real axis. In string theory, no such representation is as yet available, and this can be viewed as precisely the problem we are addressing, when we construct an analytic continuation for $A_1(s,t,u)$.

1.4. Existence and properties of the analytic continuation. In [**9, 10, 11**], Eric D'Hoker and the author established the following

THEOREM 1.1. *The partial amplitude $A_1(s,t)$ initially convergent only for*

$$\text{Re}(s) \leq 0, \quad \text{Re}(t) \leq 0, \quad \text{Re}(s+t) > -2,$$

admits an meromorphic continuation to $s \in \mathbb{C} \setminus \mathbb{R}_+$, $t \in \mathbb{C} \setminus \mathbb{R}_+$, with poles only in $u = -(s+t)$ at positive integers greater than 2. In particular, the expression (1.10) *provides then an analytic continuation for $A_1(s,t,u)$ to the cut complex plane*

$$s \in \mathbb{C} \setminus \mathbb{R}_+, \quad t \in \mathbb{C} \setminus \mathbb{R}_+, \quad u \in \mathbb{C} \setminus \mathbb{R}_+, \quad s+t+u=0. \tag{1.21}$$

We describe now the singularities along the branch cuts of the analytic continuation $A_1(s,t)$, which encode the physics of the scattering process.

1.4.1. *Double dispersion relations.* More precisely, for any positive integer N, the amplitude $A_1(s,t)$ can be expressed as

$$A_1(s,t) = \sum_{n_1+\cdots+n_4 \leq N} \sum_{|\nu_i| \leq n_i} P^{(4)}_{\{n_i\nu_i\}}(s,t) A_{\{n_i\nu_i\}}(s,t) + M_N(s,t) \tag{1.22}$$

with

- $M_N(s,t)$ is meromorphic in the half-planes $\text{Re}(s)$, $\text{Re}(t) < N$, with poles only in s, t, and $u = -(s+t)$ at positive integers greater or equal to 2;
- the coefficients $P^{(4)}_{\{n_i\nu_i\}}(s,t)$ are polynomials in s and t, indexed by integers $\{n_i\nu_i\} = (n_1, \nu_1; \ldots; n_4, \nu_4)$, $|\nu_i| \leq n_i$, which can be generated explicitly by a recursive formula;

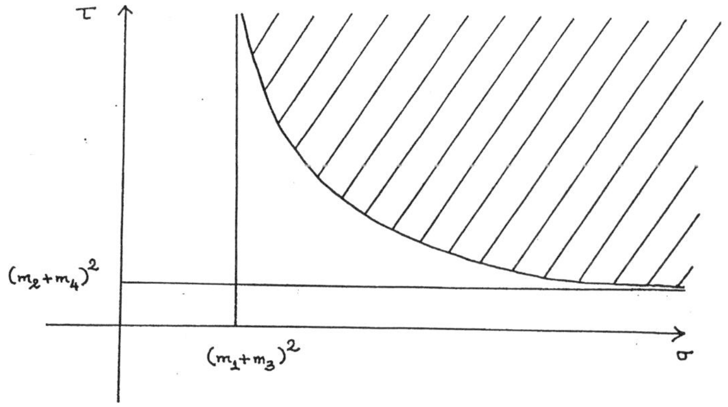

FIGURE 4. Support of the spectral density $\rho_{\{n_i\nu_i\}}$.

- $A_{\{n_i\nu_i\}}(s,t)$ are "minimal amplitudes" which admit a *double dispersion representation* of the form

$$(1.23)\quad A_{\{n_i\nu_i\}}(s,t) = \int_0^\infty\int_0^\infty d\tau d\sigma\left(\frac{\rho_{\{n_i\nu_i\}}(s,t;\sigma,\tau)}{(s-\sigma)(t-\tau)} - \Lambda_{\{n_i\nu_i\}}(s,t;\sigma,\tau)\right) + M_{\{n_i\nu_i\}}(s,t)$$

  where $M_{\{n_i\nu_i\}}(s,t)$ are meromorphic functions in the whole plane, and the term $\Lambda_{\{n_i\nu_i\}}(s,t;\sigma,\tau)$ is a subtraction term also contributing only globally meromorphic functions;
- The key expression $\rho_{\{n_i\nu_i\}}(s,t;\sigma,\tau)$ is called the *double spectral density.* For fixed σ and τ, it has simple poles in both s and t at positive even integers greater or equal to 2. If we set $m_i^2 = n_i + |\nu_i|$, then its support is given in Fig. 4. In particular, the branch cuts in s and in t of the minimal amplitude $A_{\{n_i\nu_i\}}(s,t)$ are contained respectively in $\big[(m_1+m_3)^2,\infty\big)$ and $\big[(m_2+m_4)^2,\infty\big)$.

1.4.2. *Single dispersion relations and double poles.* Recall that in string perturbation theory, the single torus diagram corresponding to $A_1(s,t,u)$ must exhibit the branch cuts, as well as the simple and double poles on top of branch cuts of the box graph and the one-particle reducible graphs of field theory (Fig. 3). The branch cuts are an immediate consequence of the double dispersion relations, and the double spectral densities $\rho_{\{n_i\nu_i\}}$ already exhibit simple poles in each of the s and t variables. To exhibit the double poles, we need to integrate out one of the σ or τ variable. The result is a *simple dispersion relation* of the form

$$(1.24)\qquad A_{\{n_i\nu_i\}}(s,t) = \int_0^\infty d\sigma\frac{\rho_{\{n_i\nu_i\}}(s,t;\sigma)}{s-\sigma}$$

where we have ignored the purely meromorphic terms, and the *simple spectral density* $\rho_{\{n_i\nu_i\}}(s,t;\sigma)$ has double poles in s at positive even integers greater or equal to 2.

1.4.3. *Explicit formulas.* The spectral densities $\rho_{\{n_i\nu_i\}}(s,t;\sigma,\tau)$, $\rho_{\{n_i\nu_i\}}(s,t;\sigma)$ can both be written down explicitly in terms of hypergeometric functions. More precisely, let $f_{n\nu}(s,\alpha)$ and $\varphi_{n\nu}(s,\beta)$ be respectively the Mellin and the Laplace transforms of the hypergeometric function $F(s/2, s/2+|\nu|, |\nu|+1; x)$, defined by

$$
\begin{aligned}
f_{n\nu}(s,\alpha) &= \frac{\Gamma(s/2+|\nu|)}{\Gamma(s/2)^2\Gamma(|\nu|+1)} \int_0^1 dx\, x^{-\alpha-1+n+|\nu|} F\left(\frac{s}{2},\frac{s}{2},1;x^2\right) \\
\varphi_{n\nu}(s,\beta)\theta(\beta) &= \frac{1}{2\pi i}\left[f_{n\nu}(s,\beta+i\varepsilon) - f_{n\nu}(s,\beta-i\varepsilon)\right].
\end{aligned} \tag{1.25}
$$

Then $f_{n\nu}(s,\alpha)$ is meromorphic everywhere, with poles at positive integers in both s and α. The function $\varphi_{n\nu}(s,\beta)$ should be viewed as a distribution. It produces entire functions of s when integrated over any finite interval in β, but leads to poles in s at positive integers when integrated over $[0,\infty)$. Then the spectral densities are given respectively by

$$
\begin{aligned}
\rho_{\{n_i\nu_i\}}(s,t;\sigma,\tau) = \int_0^\infty d\beta_1 \int_0^\infty d\beta_2 \varphi_{n_1\nu_1}(t;\beta_1)\varphi_{n_2\nu_2}(s,\beta_2) \int_0^1 du_1 \\
\times \int_0^{1-u_1} du_2 (1-u_1-u_2)^2 \\
\times \int_{(u_1\beta_1+u_2\beta_2)/(1-u_1-u_2)}^\infty \left(x - \frac{u_1\beta_1+u_2\beta_2}{1-u_1-u_2}\right)^2 \\
\times \varphi_{n_3\nu_3}(t;u_1\sigma - x)\varphi_{n_4\nu_4}(s,u_2\tau - x)
\end{aligned} \tag{1.26}
$$

and by

$$
\begin{aligned}
\rho_{\{n_i\nu_i\}}(s,t;\sigma) = \sum_{k_1,k_3=0}^\infty \frac{\Gamma(t/2+k_1)}{\Gamma(t/2)\Gamma(k_1+1)} \frac{\Gamma(t/2+k_1+|\nu_1|)}{\Gamma(t/2)\Gamma(k_1+|\nu_1|+1)} \\
\times \frac{\Gamma(t/2+k_3)}{\Gamma(t/2)\Gamma(k_3+1)} \frac{\Gamma(t/2+k_1+|\nu_3|)}{\Gamma(t/2)\Gamma(k_1+|\nu_3|+1)} \\
\times \rho_{\{n_i\nu_i\},k_1k_3}(s,t;\sigma)
\end{aligned} \tag{1.27}
$$

where

$$
\begin{aligned}
\rho_{\{n_i\nu_i\},k_1k_3}(s,t;\sigma) = \sum_{p=0}^\infty \frac{2t^p}{p!(p+3)!} \int_0^1 du \theta\left(\sigma - \frac{m_1^2}{1-u} - \frac{m_3^2}{u}\right) \\
\times \left(\sigma u(1-u) - m_3^2(1-u) - m_1^2 u\right)^{p+3} \\
\times f^{(p)}_{n_2\nu_2}(s; m_3^2 - u\sigma) f^{(p)}_{n_4\nu_4}(s; m_3^2 - u\sigma).
\end{aligned} \tag{1.28}
$$

As an example, we list the lowest order terms for the forward scattering amplitude $A_1(s,0)$, ignoring purely meromorphic contributions

$$
\begin{aligned}
\rho_{00}(s,0;\sigma) &= \frac{\sigma^3}{24}\int_0^1 du\, u^3(1-u)^3 \left[f_{00}\left(s,-\frac{1}{2}u\sigma\right)\right]^2 \\
A_{00}(s,0) &= \frac{\pi}{6}s^3\ln(-s)\int_0^1 du\, u^3(1-u)^3 f_0\left(s,-\frac{1}{2}su\right)^2.
\end{aligned} \tag{1.29}
$$

Double dispersion relations of the form (1.23) were introduced in particle theory by Mandelstam [**22, 23**].

1.4.4. *Spurious poles.* The partial amplitude $A_1(s,t)$ can have spurious poles at integers greater or equal to 2. Of course we expect only poles at even positive integers for the full amplitude $A_1(s,t,u)$, corresponding to the string spectrum. The spurious poles are due to the decomposition (1.10), and should cancel when we combine back all three terms. This is akin to the tree-level case

$$\begin{aligned} A_0(s,t,u) &= -\frac{2}{s^2}\int_{\mathbb{C}} d^2z|z|^{-2-s}|1-z|^{-s} \\ &= -\frac{4\pi}{s^2}\left[f_{00}(s,t)+f_{00}(s,u)\right] \end{aligned}$$

where $f_{00}(s,t)$ has poles only at even positive integers in t. The decomposition into partial amplitudes is similar to a splitting here of the domain of integration into $|z|\geq 1$ and $|z|\leq 1$, which cuts the pole contributing region around $z=1$ into two halves, to be treated independently.

2. The Analytic Continuation Method

2.1. Meromorphic terms for non-degenerating surfaces. We first show that for fixed τ, the integral over the emission points z_i

$$\int_{M_\tau}\prod_{i=1}^{4}\frac{d^2z_i}{\tau_2}\prod_{i<j}\exp\left(\frac{1}{2}s_{ij}G(z_i,z_j)\right) \tag{2.1}$$

can be extended to all values of the Mandelstam variables s_{ij}, with only poles at integers greater or equal to 2. To see this, we note that locally, the singularities of $\exp\big(sG(z,w)/2\big)$ are of the form $|z-w|^{-s}$. In particular, we may write

$$\int_{M_\tau} d^2z\exp\left(\frac{1}{2}sG(z,0)\right)=\int_0^\delta dr\, r^{1-s}f(r)+(\text{entire function}) \tag{2.2}$$

by breaking M_τ into a small disk of radius δ and its complement. By expanding $f(r)$ in a *finite* Taylor expansion

$$f(r)=\sum_{n=0}^{N}\frac{f^{(n)}(0)}{n!}r^n+r^{N+1}E(r)$$

the first term on the right hand side of (2.2) can be easily seen to admit a meromorphic continuation to an arbitrary half-plane $\mathrm{Re}(s)<N+3$

$$\int_0^\delta dr\, r^{1-s}f(r)=\sum_{n=0}^{N}\frac{f^{(n)}(0)}{n!}\frac{\delta^{n+2-s}}{n+2-s}+\int_0^\delta dr\, r^{N+2-s}E(r).$$

Now the integrals in (2.1) are more complicated, since the $|z_i-z_j|$'s may approach 0 simultaneously, and at different rates. However, the whole domain of integration can be broken into disjoint regions where the "scales" $|z_i-z_j|$'s can be ordered by size. By integrating out successive scales, beginning with the smallest, we can adapt the above arguments and establish the desired statement.

The meromorphicity for each τ of the integrals over emission points implies the meromorphicity of their integrals over any bounded region in τ. In particular, the region $F\cap\{\tau_2\leq 1\}$ produces only a meromorphic contribution, and we can now restrict our attention to its complement in F.

2.2. Reduction to minimal amplitudes. To deal with the region $\{|\tau_1| \leq 1/2, \tau_2 \geq 1\}$, we consider the integrand of $A_1(s,t)$ in the form (17). In this expression, the factors $R(z)$ can be safely expanded in uniformly convergent series since they are bounded away from 0 and ∞. At first sight, besides the factor $|q|^{-(su_1u_3+tu_2u_4)}$, it seems that we would need to keep the remaining 12 factors in (1.17). In fact, we need only keep

$$\prod_{i=1}^{4} \left|1 - e^{i\alpha_i}|q|^{u_i}\right|^{-s_i}. \tag{2.3}$$

To see this, we note that in this truncated fundamental region, α_4 becomes a genuine angle variable, on the same footing as α_1, α_2, α_3. The integral admits then a $u_1 \leftrightarrow u_3$ and a $u_2 \leftrightarrow u_4$ symmetry, and we may assume that $u_1 \leq u_3$, $u_2 \leq u_4$. In the region (I), where $(u_1 + u_2)\tau_2 < 1$, the factor $|q|^{-(su_1u_3+tu_2)} = \left[e^{2\pi\tau_2 u_1}\right]^{su_3}\left[e^{2\pi\tau_2 u_2}\right]^{tu_4}$ is bounded. The preceding arguments still apply to produce a meromorphic extension in all the Mandelstam variables. In the remaining region (II) $= \{(u_1+u_2)\tau_2 \geq 1\}$, $|q|^{u_i+u_j}$ is small for i and j of opposite parity, and the eight factors in (1.17)

$$\prod_{i=2k}\prod_{j=2m+1} \left|1 - e^{i(\alpha_i+\alpha_j)}|q|^{u_i+u_j}\right|^{s_i+s_j} \tag{2.4}$$

can be expanded uniformly. In this way, we arrive at the following key lemma

LEMMA 2.1. *For any positive integer N, we can write*

$$A_1(s,t) = \sum_{n_1+\cdots+n_4\leq 4N}\ \sum_{|\nu_i|\leq n_i} P_{\{n_i\nu_i\}}(s,t)A_{\{n_i\nu_i\}}(s,t) + M_N(s,t) \tag{2.5}$$

where $M_N(s,t)$ is a meromorphic function in the region $\mathrm{Re}(s), \mathrm{Re}(t) < N$, *and the amplitudes $A_{\{n_i\nu_i\}}(s,t)$ are defined by*

$$A_{\{n_i\nu_i\}}(s,t) = \int_1^\infty \frac{d\tau_2}{\tau_2^2}\int_0^{2\pi}\frac{d\alpha_i}{2\pi}\int_0^1 du_i\delta\Big(1-\sum_{i=1}^4 u_i\Big)|q|^{-su_1u_3-tu_2u_4}$$
$$\times \prod_{i=1}^{4}\left|1-e^{i\alpha_i}|q|^{u_i}\right|^{-s_i}|q|^{n_iu_i}e^{i\nu_i\alpha_i}. \tag{2.6}$$

Our next step is to recast the minimal amplitudes in terms of double dispersion relations. This can be done explicitly, first by reexpressing the $d\alpha_i$ integrals above in terms of the Laplace transforms $\varphi_{n\nu}(s,\beta)$ of hypergeometric functions, and then by arrying out the $d\tau_2$ integral, using the relation

$$\int_1^\infty \frac{d\tau_2}{\tau_2^2}e^{-\kappa\tau_2} = \frac{1}{2}\int_0^\infty dx\left(\frac{x^2}{(x+\kappa)^2} - 1 + \frac{2\kappa}{x+1}\right) + E(\kappa) \tag{2.7}$$

where $E(\kappa)$ is an entire function. (Note that this relation provides in effect an analytic continuation of the function $\kappa\ln\kappa$ from the right half-plane to the entire plane cut along the negative real axis.) A change of variables leads then to (1.17), with the double spectral density $\rho_{n\nu}(s,t;\sigma,\tau)$ given by (1.26).

This completes our description of the method of analytic continuation.

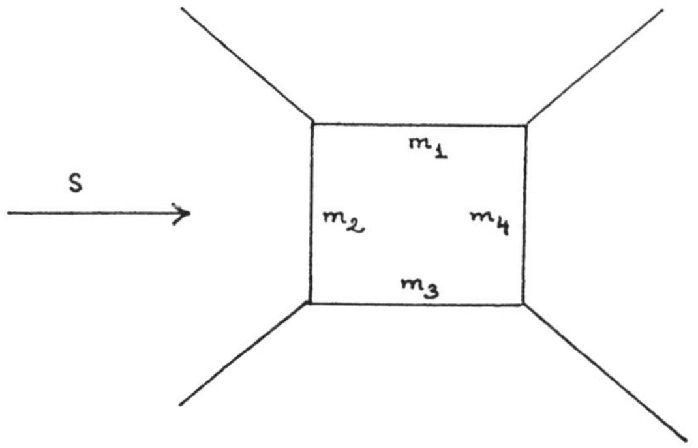

FIGURE 5. The box diagram in a ϕ^3-like theory with arbitrary mass assignments on all propagators.

2.3. Stringy effects. We can now compare with the situation in field theory. Consider now a ϕ^3-like box diagram in d space-time dimensional quantum field theory with arbitrary masses m_i for each of the propagators, and massless external on-shell states (Fig. 5). Couplings are assumed to be identical for any choice of m_i and are non-derivative ϕ^3. The box diagram (Euclidean) Feynman integral can be performed as usual after introducing Feynman parameters u_i and exponentiating the denominator. We obtain (c.f. [**20**])

$$\begin{aligned}(2.8)\quad & \int d^dk \prod_{i=1}^{4} \frac{1}{(k+p_i)^2+m_i^2} \\ &= \int_0^\infty \frac{d\tau}{\tau^{d/2-3}} \int_0^1 du_i \delta\Big(1-\sum_{i=1}^4 u_i\Big) e^{2\pi\tau\left(su_1u_3+tu_2u_4-\sum_{i=1}^4 u_i m_i^2\right)}.\end{aligned}$$

If we make the identification of the space-time dimension as $d = 10$, and the masses m_i^2 as positive even integers, in accord with the superstring spectrum, this amplitude is almost identical to the expression (2.6) for the minimal amplitudes $A_{\{n_i\nu_i\}}(s,t)$, except for two important differences:

(a) in the string case, the range in τ_2 has been truncated away from 0, a well-understood consequence of modular invariance;
(b) the string amplitudes contain the extra factor (2.3), whose role we can now appreciate.

The box graphs in field theory contain no poles, unlike the box graph in string theory which must incorporate the singularities of the field theoretic one-particle reducible graphs as well. The factors (2.3) are the factors responsible for these poles. We note that if we expanded them in a series in $|q|^{u_i}e^{i\alpha_i}$, we would get formally a series of terms resembling (3.2) (with $\{n_i\nu_i\}$ suitably shifted). This series however cannot converge everywhere, since it would produce then no poles. Mathematically, this is of course due to the fact that the expansion of (2.3) is not uniformly convergent as the $|q|^{u_i}$'s approach 1. This divergence is a key distinction between string amplitudes and an infinite superposition of point-particle amplitudes.

3. Further Developments

3.1. Decay rates of massive states and mass shifts. Some important physical information can be easily derived from the above formalism. For example, the decay width $\Gamma(2N, t)$ of a string state of mass squared $2N$ into two lower mass string states can be read off from the imaginary part of the coefficient of the double pole of the 4-point function at $s = 2N$. This implies in particular that the meromorphic terms in the analytic continuation of $A_1(s,t)$ do not play any role and

$$\Gamma(2N,t) = 2\,\mathrm{Im} \lim_{s\to 2N} (s-2N)^2 \int_0^\infty d\sigma \int_0^\infty d\tau \frac{\rho(s,t;\sigma,\tau)}{(s-\sigma+i\varepsilon)(t-\tau)}. \tag{3.1}$$

The explicit formulas at our disposal for the spectral density $\rho(s,t;\sigma,\tau)$ lead then to similar formulas for the decay widths. These formulas are given in detail in [**27**], and we shall not reproduce them here. We note only that this problem as well as the one of determining the mass shifts that we are going to consider next, has received considerable attention, with some of the earlier work listed in [**8, 26, 2, 24, 25, 21, 4, 5**].

The mass-shifts are given by the real parts of the coefficient of the double pole at $s = 2N$ of the forward-scattering amplitude $A_1(s,0)$. This means that unlike in the case of decay widths, the purely meromorphic terms cannot be ignored. For the lowest massive state, $N = 2$, we obtain

THEOREM 3.1. (a) *The coefficient of the double pole of $A_1(s,t,u)|_{t=0}$ is given by $A(-4)$, where the function $A(s)$ is the analytic continuation to the cut plane $\mathbb{C}\setminus\mathbb{R}_+$ of the integral expression*

$$A(s) = \int_F \frac{d^2\tau}{\tau_2^2} \int_{M_\tau} \frac{d^2 z}{\tau_2} \exp\left[\frac{s}{2} G(z,0)\right] \tag{3.2}$$

(b) *For any fixed positive integer N, $A(s)$ can be expanded as*

$$A(s) = s\ln s + \sum_{k=1}^{N/2-1} a_k(s) \int_{2k} \frac{d\alpha}{\alpha^2} \ln(s+\alpha) + A_N(s) \tag{3.3}$$

where the $A_N(s)$ are holomorphic in the domain $-N < \mathrm{Re}\, s < 2$, and the coefficients $a_k(s)$ are polynomials in s;

(c) *In particular $A(-4)$ is given by the following* convergent *integral*

$$\begin{aligned} 16\pi^2 A(-4) = &\int_0^1 du\, L_2\big(-4\pi u(1-u)\big) \\ &+ \int_{-1/2}^{1/2} d\tau_1 \int_{\sqrt{1-\tau_1^2}} \int_0^1 dy\, e^{4\pi\tau_2 y(1-y)} \\ &\times \int_F \frac{d^2\tau}{\tau_2^2} \int_{M_\tau} \frac{d^2 z}{\tau_2} \left[\exp\big(-2G(z,0)\big) - e^{4\pi\tau_2 y(1-y)}\right] \end{aligned} \tag{3.4}$$

where the function $L_2(s)$ is defined by

$$L_2(s) = s\ln s - \int_0^1 \frac{dx}{x^2}\big(e^{-sx-1+sx}\big) - \left(\int_1^\infty \frac{dx}{x} e^{-x} + \int_0^1 \frac{dx}{x}(e^{-x}-1) + 1\right) s + 1.$$

In principle, our method of analytic continuation allows us to track down the coefficient of the Laurent expansions of the amplitude at an arbitrary pole. This can get however quite lengthy in practice.

3.2. Toroidal, orbifold backgrounds, and heterotic strings. We have studied in detail the one-loop Type IIB superstring amplitude propagating in 10-dimensional flat Minkowski space-time. However, the above method of analytic continuation can be readily adapted to more general situations. For example, it applies practically verbatim to the scattering of 4 massless charged gauge bosons in the heterotic Spin $32/\mathbb{Z}_2$ and $E_8 \times E_8$ superstrings. The amplitude is then a function of the usual Mandelstam variables s, t, u, as well as of root lattice Mandelstam variables S, T, U taking the discrete values -8, -6, -4, -2, or 0. As before, the amplitude can be decomposed into minimal amplitudes admitting double spectral representations, and branch cuts in s and t starting at $(m_1 + m_3)^2$ and $(m_2 + m_4)^2$, with m_i^2 given now by $m_i^2 = n_i + |\nu_i + 1| - 1$, $|\nu_i| \leq n_i$.

Similarly, the case of toroidal compactifications and orbifolds has been treated in Chalmers *et al.* [**6**]. If one considers only scattering of NS-NS states which remain massless after compactification (e.g. the graviton or certain Kaluza-Klein $U(1)$ fields) the compactified amplitudes require only inserting the partition function for the zero modes and soliton sector in the integrand for the superstring (c.f. [**29, 30**]). A new difficulty does arise, which is the issue of infrared regularization and which has to be treated with some care. Otherwise, one can proceed along similar lines, and obtain an analytic continuation in terms of minimal amplitudes with branch cuts as before, but with m_i^2 this time defined by $m_i^2 = n_i + |\nu_i| + 1/2(P_L \cdot P_L + P_R \cdot P_R)$. Here (P_L, P_R) are respectively the left and right momenta parametrizing the self-dual, even integer Lorentzian lattice of the compactification.

We note that this also gives the analytic continuation of amplitudes for the heterotic string compactified on certain special points of the moduli space of Calabi-Yau backgrounds, corresponding e.g. to Gepner's models [**15**]. It may be interesting to derive as well their deformations as we move away from these special points.

3.3. Higher-loop amplitudes and singularities of the Green's functions. The program of formulating efficient Feynman rules for string theory to all orders of perturbation theory is still incomplete at the present time. We refer to [**12**] for a review of the subject, and to [**13**] for a description of some of the latest geometric machinery which has been developed to address this question. Here we only comment briefly on how a proof for the finiteness and $i\varepsilon$ prescription for higher loops may proceed, when more explicit formulas for higher-loop string scattering amplitudes become available. Again we can expect the most severe divergences to arise from when the underlying world-sheet degenerates, i.e. as one of its homology cycles is pinched to a point. It is well-known in the theory of Riemann surfaces that the degenerating behavior of the conformal structure of the world-sheet is well-modeled by that of an infinitely stretching out cylinder. Indeed the surface behaves as two essentially unchanged surfaces joined by the cylinder (c.f. Fig. 6). In particular, the singularities of the Green's functions on a world-sheet of higher

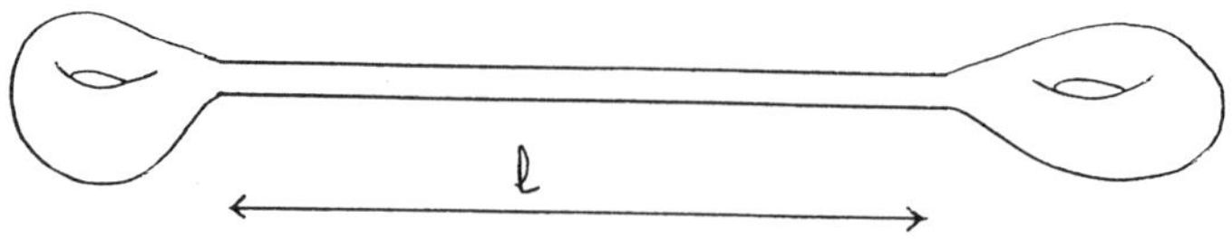

FIGURE 6. A degeneration of a Riemann surface.

genus should be similar to the Green's function with Neumann boundary conditions for of the cylinder. This can be determined explicitly

$$
\begin{aligned}
(3.5)\quad & \exp(-G_\ell(x,y;x',y')) \\
&= \exp\left(-4\pi\ell\left[\frac{1}{3} - \frac{y+y'}{2\ell}\left(1 - \frac{y+y'}{2\ell}\right) - \frac{|y-y'|}{2\ell}\left(1 - |y-y'|2\ell\right)\right]\right) \\
&\quad \times \left|1 - \exp\left(-4\pi\left(\ell - \frac{y+y'}{2} + i\frac{|x-x'|}{2}\right)\right)\right|^2 \\
&\quad \times \left|1 - \exp\left(-4\pi\left(\frac{y+y'}{2} - i\frac{|x-x'|}{2}\right)\right)\right|^2 \\
&\quad \times \left|1 - \exp\left(-4\pi\left(\frac{|y-y'|}{2} - i\frac{|x-x'|}{2}\right)\right)\right|^2 R(x,y;x',y').
\end{aligned}
$$

Here the cylinder has circumference 2π and length $\ell \gg 1$. Its coordinates are $0 \le x \le 1$, $0 \le y \le \ell$, and $R(x,y;x',y')$ is a factor uniformly bounded above and below. The singular behavior of $G(x,y;x',y')$ is almost identical to that of the Green's function on the torus, and we would expect the above method to proceed along similar lines.

Finally, we note that it may be possible, through factorization, to use the information we just obtained on the one-loop amplitude to determine the two-loop amplitude. For example, several Ansatzes satisfying the modular invariance condition for the two-loop amplitude have been proposed in [**28**]. It is conceivable that the kind of argument presented here for the one-loop amplitude can be used to lift the indeterminacy. However, the large number of intermediate states makes this impractical. Obviously, a lot of work remains ahead.

References

1. K. Amano, *A finite string loop amplitude in a finite form*, Nuclear Phys. B **328** (1989), no. 2, 510–526.
2. K. Amano and A. Tsuchiya, *Mass splittings and the finiteness problem of mass shifts in the Type* II *superstring at one-loop order*, Phys. Rev. D **39** (1989) 565–570.
3. K. Aoki, E. D'Hoker, and D. H. Phong, *Unitarity of closed superstring perturbation theory*, Nuclear Phys. B **342** (1990), no. 1, 149–230.
4. A. Berera, *Unitary string amplitudes*, Nuclear Phys. B **411** (1994), no. 1, 157–180.
5. ———, *All-order mass renormalization of string theory*, Phys. Rev. D (3) **49** (1994), no. 12, 6674–6686.
6. G. Chalmers, E. D'Hoker, and D. H. Phong, *Momentum-analyticity and finiteness of compactified string amplitudes*, in preparation.
7. G. F. Chew, *The analytic S-matrix: A basis for nuclear democracy*, W. A. Benjamin Inc., New York-Amsterdam, 1966.
8. C. B. Chiu and S. Matsuda, *Intermediate mass distribution of the dual resonance pomeron*, Nucl. Phys. B **134** (1978) 463–476.
9. E. D'Hoker and D. H. Phong, *Momentum analyticity and finiteness of the* 1*-loop superstring amplitude*, Phys. Rev. Lett.**70** (1993), no. 24, 3692–3695.
10. ———, *The box graph in superstring theory*, Nuclear Phys. B **440** (1995), no. 1-2, 24–94
11. ———, *Dispersion relations in string theory*, Teoret. Mat. Fiz. **98** (1994), no. 3, 442–455; English transl. Theoret. and Math. Phys. **98** (1994), no. 3, 306–316.
12. ———, *The geometry of string perturbation theory*, Rev. Modern Phys. **60** (1988), no. 4, 917–1065.
13. ———, *Conformal scalar fields and chiral splitting on super Riemann surfaces*, Comm. Math. Phys. **125** (1989), no. 3, 469–513

14. R. J. Eden, P. V. Landshoff, D. I. Olive, and J. C. Polkinghorne, *The analytic S-matrix*, Cambridge University Press (1966).
15. D. Gepner, *On the spectrum of 2D conformal field theories*, Nuclear Phys. B **287** (1987), no. 1, 111–130.
16. M. B. Green and J. Schwarz, *Supersymmetrical string theories*, Phys. Lett. B **109** (1982), 444–447.
17. M. Green and D. Gross (eds.), *Workshop on unified string theories* (Santa Barbara, Calif., 1985), World Scientific Publishing Co., Singapore, 1986.
18. M. B. Green, J. H. Schwarz, and L. Brink, *$N = 4$ Yang-Mills and $N = 8$ supergravity as limits of string theories*, Nucl. Phys. B **198** (1982), 474–492.
19. M. Green, J. Schwarz, and E. Witten, *Superstring theory*. II. *Loop amplitudes, anomalies and phenomenology*, Cambridge Monogr. Math. Phys., Cambridge University Press, Cambridge-New York, 1987.
20. C. Itzykson and J. B. Zuber, *Quantum field theory*, Internat. Ser.Pure Appl. Phys., McGraw-Hill International Book Co., New York, 1980.
21. N. Marcus, *Unitarity and regularized divergences in strong amplitudes*, Phys. Lett. B **219** (1989) 265–272;
22. S. Mandelstam, *Determination of the pion-nucleon scattering amplitude from dispersion relations and unitarity. General theory*, Phys. Rev. (2) **112** (1958), 1344–1360
23. ———, *Analytic properties of transition amplitudes in perturbation theory*, Phys. Rev. (2) **115** (1959), 1741–1751.
24. D. Mitchell, B. Sundborg, and N. Turok, *Decays of massive open string*, Nucl. Phys. B **335** (1990) 621-634.
25. D. Mitchell, N. Turok, R. Wilkinson, and P. Jetzer, *The decay of highly excited open strings*, Nucl. Phys. B **315** (1989), 1–24.
26. V. A. Miranski, V. P. Shelest, B. V. Struminsky, and G. M. Zinovjev, *Dual resonance model and statistical bootstrap*, Phys. Lett. B **43** (1973), 73–75.
27. J. Montag and W. I. Weisberger, *A finite representation for a superstring scattering amplitude and its low-energy limit*, Nuclear Phys. B **363** (1991), nos. 2–3, 527–542.
28. A. Morozov, *On the two-loop contribution to the superstring four-point function*, Phys. Lett. B **209** (1988), no. 4, 473–476.
29. K. S. Narain, *New heterotic string theories in uncompactified dimensions < 10*, Phys. Lett. B **169** (1986), no. 1, 41–46.
30. K. S. Narain, M.H. Sarmadi, and E. Witten, *A note on toroidal compactification of heterotic string theory*, Nuclear Phys. B **279** (1987), nos. 3–4, 369–379.
31. A. M. Polyakov, *Gauge fields and strings*, Contemp. ConceptsPhys., vol. 3, Harwood Academic Publishers, Chur, 1987.
32. J. Schwarz (ed.), *Superstrings: the first 15 years of superstring theory*, vols. 1–2, World Scientific Publishing Co., Inc., Teaneck, NJ, 1985.
33. S. T. Yau (ed.), *Mathematical aspects of string theory* (San Diego, 1986), Adv. Ser. Math. Phys., vol. 1, World Scientific, Singapore, 1987.

Department of Mathematics, Columbia University, New York, N.Y. 10027, USA
E-mail address: phong@math.columbia.edu

Centre de Recherches Mathématiques
CRM Proceedings and Lecture Notes
Volume **11**, 1997

Consequences of an Abelian Family Symmetry

Pierre Ramond

ABSTRACT. The addition of an Abelian family symmetry to the Minimal Supersymmetric Standard Model reproduces the observed hierarchies of quark and lepton masses and quark mixing angles, only if it is anomalous. Green-Schwarz compensation of its anomalies requires the electroweak mixing angle to be $\sin^2\theta_w = 3/8$ at the string scale, without any assumed GUT structure, suggesting a superstring origin for the standard model. The analysis is extended to neutrino masses and the lepton mixing matrix.

1. Introduction

The relative complexity of the standard model of strong and electroweak interactions suggests that it is the low energy manifestation of more fundamental theory. There are few hints as to the nature of this theory or value of the scale at which it becomes operative.

These questions can be studied in the context of the extension of the standard model to $N = 1$ supersymmetry [**1**] which allows for its perturbative extrapolation to near Planckian scales, where the gauge couplings [**2**] and some Yukawa couplings [**3**] appear to converge. This raises the hope that the $N = 1$ standard model at short distances is much simpler than at experimental scales, although we do not seem to have sufficient information to determine exactly the type of structure it describes, a GUT theory [**4**], or a direct descendant of superstrings. Fortunately there is more information, as not all parameters are the same at the unification scale, suggesting in fact a strong hierarchy among the masses of the quarks and the leptons, indicated by the orders of magnitude estimates [**5**]

$$\frac{m_u}{m_t} = \mathcal{O}(\lambda^8); \quad \frac{m_c}{m_t} = \mathcal{O}(\lambda^4); \tag{1.1}$$

$$\frac{m_d}{m_b} = \mathcal{O}(\lambda^4); \quad \frac{m_s}{m_b} = \mathcal{O}(\lambda^2), \tag{1.2}$$

1991 *Mathematics Subject Classification.* Primary 81T60.

This work was supported in part by the United States Department of Energy under grant DE-FG05-86-ER40272.

This is the final form of the paper.

where, following Wolfenstein's parametrization [**6**], we use the Cabibbo angle λ, as expansion parameter. The charged lepton masses also satisfy similar relations

$$\frac{m_e}{m_\tau} = \mathcal{O}(\lambda^4); \quad \frac{m_\mu}{m_\tau} = \mathcal{O}(\lambda^2). \tag{1.3}$$

The mass hierarchy appears to be geometrical in each sector. The equality

$$m_b = m_\tau,$$

known to be valid in the ultraviolet [**3**], yields the estimate

$$\frac{m_d m_s m_b}{m_e m_\mu m_\tau} = \mathcal{O}(1). \tag{1.4}$$

We assume that the mechanism which sets these orders of magnitude is an Abelian family symmetry, in the manner originally suggested by Froggatt and Nielsen [**7**].

In the following, we suggest that these orders of magnitude are determined by a gauged family Abelian symmetry. This is an old idea, some aspects of which has been revisited in the recent literature [**8, 9, 10**]. The work presented below is the result of a collaboration with P. Binétruy and S. Lavignac [**11, 12**]. There are also closely related work in the recent literature, but with different emphases [**13, 14, 15**].

Our framework is the minimal extension of the Standard Model to $N = 1$ supersymmetry, including the so-called μ term, $P = \mu H_u H_d$. The most remarkable conclusion is that to reproduce these estimates, the symmetry must be anomalous. The anomalies can be compensated by the Green-Schwarz mechanism, fixing [**16**] $\sin^2\theta_w = 3/8$, in perfect agreement with data when extrapolated to the infrared.

2. Quark Masses and Mixing Angles

The most general Abelian charge assignments to the particles of the Supersymmetric Standard Model can be written as

$$X = X_0 + X_3 + \sqrt{3X_8}, \tag{2.1}$$

where X_0 is the family-independent part, X_3 is along λ_3, and X_8 is along λ_8, the two diagonal Gell-Mann matrices of the SU(3) family space in each charge sector. In a basis where the entries correspond to the components in the family space of the fields $Q, \bar{u}, \bar{d}$, L, and $\bar{e}$, we can write the different components in the form

$$X_i = (a_i, b_i, c_i, d_i, e_i), \tag{2.2}$$

for $i = 0, 3, 8$. The Higgs doublets $H_{u,d}$ have zero X-charge because of the μ term.

The tree-level Yukawa coupling involves *only* the third family (implicitly choosing the third direction in family space),

$$y_t Q_3 \bar{u}_3 H_u + y_b Q_3 \bar{d}_3 H_d + y_\tau L_3 \bar{e}_3 H_d, \tag{2.3}$$

where the y_i's are the Yukawa couplings. This generates the relations

$$a_0 + b_0 = 2(a_8 + b_8), \quad a_0 + c_0 = 2(a_8 + c_8), \quad d_0 + e_0 = 2(d_8 + e_8).$$

The other elements of the Yukawa matrices are zero at tree-level because of X-charge conservation. Let x_{ij} be the excess X-charges at each of their entries; for the charge 2/3 Yukawa matrix they are

$$\begin{pmatrix} 3(a_8+b_8)+a_3+b_3 & 3(a_8+b_8)+a_3-b_3 & 3a_8+a_3 \\ 3(a_8+b_8)-a_3+b_3 & 3(a_8+b_8)-a_3-b_3 & 3a_8-a_3 \\ 3b_8+b_3 & 3b_8-b_3 & 0 \end{pmatrix}. \tag{2.4}$$

In the charge $-1/3$ sector, the b_i are replaced by the c_i, and in the charge -1 sector, a_i, b_i are replaced by d_i, e_i, respectively.

Introduce an electroweak singlet field θ with X-charge $-x$, to soak up the excess charge at each entry, x_{ij}, yielding an interaction of higher dimensions with no hypercharge [**7, 17**]

$$Q_i \bar{u}_j H_u \left(\frac{\theta}{M_u}\right)^{n_{ij}}, \tag{2.5}$$

where the n_{ij} are positive numbers which satisfy

$$x_{ij} - x n_{ij} = 0, \tag{2.6}$$

and M_u is some large scale. In a perturbative framework, the n_{ij} are expected to be integers. With only one field θ, not chaperoned by its vectorlike partner, invariance under supersymmetry then naturally [**8**] generates a true texture zero whenever a Yukawa matrix element has negative excess X-charge in units of $(-x)$, and non-zero entries correspond only to positive excess X-charge. Henceforth we normalize X so that $x = 1$.

We assume that the electroweak singlet θ develops a vacuum expectation value smaller than M_u, producing a *small parameter*, $\lambda_u \sim \theta/M_u$. This is what happens in many compactified superstring theories [**18**]. The n_{ij} then determine the order of magnitude of the entries in the Yukawa matrices [**7**]. It may seem that because the masses M_u, and thus the expansion parameters are in principle different in the three charge sectors, we are introducing many unknowns in our description, but most of the conclusions we can reach with this simple assumption depend only on the *existence* of these small parameters, *not on their values*. Also, since the down quark and lepton sectors share the same electroweak quantum numbers, we expect them to be the same for the charge -1 and $-1/3$ matrices.

In the following we restrict ourselves to the case where all the excess charges in each Yukawa matrix are positive. We leave to future work the study of the cases where some of the excess charges are negative, which creates a true zero in that matrix element [**8**]. The charge 2/3 Yukawa matrix is

$$Y_{uij} = \mathcal{O}(\lambda_u^{n_{ij}}), \tag{2.7}$$

normalized to the top quark mass. It is not hard to diagonalize this matrix, setting

$$Y_u = U_u D_u V_u^\dagger, \tag{2.8}$$

where

$$D_u = \mathrm{diag}\big(\mathcal{O}(\lambda_u^{3(a_8+b_8)+a_3+b_3}), \mathcal{O}(\lambda_u^{3(a_8+b_8)-a_3-b_3}), \mathcal{O}(1)\big),$$

and the unitary matrix U_u is given by

$$U_u = \begin{pmatrix} \mathcal{O}(1) & \mathcal{O}(\lambda_u^{2a_3}) & \mathcal{O}(\lambda_u^{3a_8+a_3}) \\ \mathcal{O}(\lambda_u^{2a_3}) & \mathcal{O}(1) & \mathcal{O}(\lambda_u^{3a_8-a_3}) \\ \mathcal{O}(\lambda_u^{3a_8+a_3}) & \mathcal{O}(\lambda_u^{3a_8-a_3}) & \mathcal{O}(1) \end{pmatrix}. \tag{2.9}$$

These are valid for a range of parameters such that

$$3a_8 + 3b_8 > a_3 + b_3 > 0.$$

We have a similar relation in the down quark sector, with the b_i replaced by c_i. It follows that the orders of magnitude of U_u and U_d are the same, but the expansion coefficients might be different. Let us set $\lambda_u = \lambda_d^y$, with $y > 0$. If $y > 1$, the orders of magnitude of the entries of the CKM matrix are

$$\mathcal{U}_{\text{CKM}} = \begin{pmatrix} \mathcal{O}(1) & \mathcal{O}(\lambda_d^{2a_3}) & \mathcal{O}(\lambda_d^{3a_8+a_3}) \\ \mathcal{O}(\lambda_d^{2a_3}) & \mathcal{O}(1) & \mathcal{O}(\lambda_d^{3a_8-a_3}) \\ \mathcal{O}(\lambda_d^{3a_8+a_3}) & \mathcal{O}(\lambda_d^{3a_8-a_3}) & \mathcal{O}(1) \end{pmatrix}. \tag{2.10}$$

If $y < 1$, the expansion parameter in the above is replaced by λ_u, that is its exponents all are multiplied by y. In either case the exponents satisfy the sum rule

$$n_{12} = n_{13} - n_{23}, \tag{2.11}$$

which implies that

$$\frac{V_{us}V_{cb}}{V_{ub}} = \mathcal{O}(1), \tag{2.12}$$

in agreement with data (the right hand side is ≈ 3, and the Wolfenstein parametrization ($n_{12} = 1, n_{13} = 3, n_{23} = 2$). We note the relation between our expansion parameters with the Cabibbo angle

$$\lambda \equiv V_{us} = \lambda_{u,d}^{2a_3}, \tag{2.13}$$

depending on the relative magnitudes of λ_u and λ_d. The eigenvalue order of magnitude estimates are

$$\frac{m_u}{m_t} = \mathcal{O}(\lambda_u^{3(a_8+b_8)+a_3+b_3}); \quad \frac{m_c}{m_t} = \mathcal{O}(\lambda_u^{3(a_8+b_8)-a_3-b_3});$$
$$\frac{m_d}{m_b} = \mathcal{O}(\lambda_d^{3(a_8+c_8)+a_3+c_3}); \quad \frac{m_s}{m_b} = \mathcal{O}(\lambda_d^{3(a_8+c_8)-a_3-c_3}).$$

The geometric hierarchy of the mass ratios in each quark sector suggests the further equalities

$$a_8 + b_8 = a_3 + b_3; \quad a_8 + c_8 = a_3 + c_3. \tag{2.14}$$

Agreement with experimental information on the quark mass ratios dictates the following

$$2(a_8 + c_8) = y(a_8 + b_8). \tag{2.15}$$

In addition the mixing angle relation

$$V_{us} = \sqrt{\frac{m_d}{m_s}}, \tag{2.16}$$

is satisfied provided that

$$c_3 = a_3, \tag{2.17}$$

if $y > 1$, and

$$c_3 = (2y - 1)a_3, \tag{2.18}$$

if $0 < y < 1$. We also find that

$$V_{cb} = \mathcal{O}\big(V_{us}^{(3a_8-a_3)/(2a_3)}\big), \tag{2.19}$$

from which we may deduce that

$$3a_8 = 5a_3. \tag{2.20}$$

Comparison with the data gives us six equations among seven unknown. The last unknown is y. Until we know the origins of the scales and of the expansion parameters, we cannot fix the values of λ_d and of λ_u in terms of observables. We note, for example, the interesting case $y = 2$, corresponding to $\lambda_u = \lambda_d^2$, yields $b_i = c_i$, which suggests an $SU(2)_R$ symmetry. It is quite remarkable that this simple idea is in agreement with the present data, and even predicts one successful relation among the CKM matrix elements (2.12).

3. Lepton Masses and Mixing Angles

An analysis akin to that in the previous section yields the charged lepton mass estimates

$$\frac{m_e}{m_\tau} = \mathcal{O}(\lambda_e^{3(d_8+e_8)+d_3+e_3}); \quad \frac{m_\mu}{m_\tau} = \mathcal{O}(\lambda_e^{3(d_8+e_8)-d_3-e_3}). \tag{3.1}$$

Geometric hierarchy of the charged lepton mass ratios implies that

$$d_8 + e_8 = d_3 + e_3. \tag{3.2}$$

There are no mixing angles if the neutrinos are massless. Below, we generalize the Froggatt-Nielsen analysis to massive neutrinos, without assuming any extra symmetry [**19**]. We do this by adding right-handed neutrinos to the MSSM in order to generate masses for the neutrinos via the "see-saw" mechanism [**20**].

Let us assume that the low energy chiral remnants of the primal soup come from 27 representations of E_6. This representation carries two fields with no electroweak quantum numbers. One is an SO(10) singlet, as we can see from the decomposition $27 = 16 \oplus 10 \oplus 1$. The other is an SU(5) singlet which lives in the spinor representation of $SO(10) 16 = \bar{5} \oplus 10 \oplus 1$. This same field is part of an isodoublet under the right-handed $SU(2)_R$ inside $SO(10) 16 = (2, 1, \bar{3}^c \oplus 1^c) \oplus (1, 2, 3^c \oplus 1^c)$. These two neutrino fields are not so "ino" as they are assumed to be very massive. With two fields, the Majorana mass matrix is

$$\begin{pmatrix} 0 & m_1 & m_2 \\ m_1 & \multicolumn{2}{c}{\multirow{2}{*}{$\mathcal{M}_0$}} \\ m_2 & & \end{pmatrix},$$

where $\mathcal{M}_0$ is a 2×2 symmetric matrix, and $m_{1,2}$ are the usual $\Delta I_w = 1/2$ mass entries of electroweak order. Let the eigenvalues of $\mathcal{M}_0$ be M_1 and M_2. We can go to a basis where M_0 is diagonal, in which $m_{1,2}$ are rotated into $\widehat{m}_{1,2}$, yielding the light eigenvalue

$$m_\nu = \frac{\widehat{m}_1^2 M_2 - \widehat{m}_2^2 M_1}{M_1 M_2}.$$

Thus if $M_1 < M_2$, it becomes just $\widehat{m}_1^2 / M_1$, so that it is the lighter of the singlet neutrinos that enters in the light neutrino mass. This assumes that the $\widehat{m}_i$ are of the same order of magnitude, themselves much smaller than M_1.

Thus in the following we assume only one right-handed neutrino per family, and leave the more complicated analysis to others. Assume that we have three such fields, $\overline{N}_i$, each carrying X-charge. The superpotential now contains the new interaction terms

$$L_i \overline{N}_j H_u \left(\frac{\theta}{m_\nu}\right)^{p_{ij}} + m_0 \overline{N}_i \overline{N}_j \left(\frac{\theta}{m_0}\right)^{q_{ij}}, \tag{3.3}$$

multiplied by couplings of order one, and where m_0 is some mass of the order of the GUT scale or string scale. In analogy with the quark and charged lepton sectors, we assume that $p_{33} = 0$, so that there is only the tree-level coupling for the third family. Call the X-charges of the right-handed neutrinos f_0, f_3, f_8, so that at tree-level

$$d_0 + f_0 = 2(d_8 + f_8). \tag{3.4}$$

All Yukawa couplings satisfy conservation of X, relating q_{ij} and p_{ij} to the X-charges of the fields. For three families, the 6×6 Majorana mass matrix is of the form

$$\begin{pmatrix} 0 & \mathcal{M} \\ \mathcal{M}^T & \mathcal{M}_0 \end{pmatrix}.$$

In the above $\mathcal{M}$ is the $\Delta I_w = 1/2$ mass matrix with entries not larger than the electroweak breaking scale, and $\mathcal{M}_0$ is the unrestricted $\Delta I_w = 0$ mass matrix. Assuming that the order of magnitude of the $\Delta I_w = 0$ masses is much larger than the electroweak scale, we obtain the generalized "see-saw" mechanism.

The calculation of the light neutrinos masses and mixing angles proceeds in two steps. Let U_0 be the matrix diagonalizes $\mathcal{M}_0$, that is

$$\mathcal{M}_0 = U_0 D_0 U_0^T, \tag{3.5}$$

where D_0 is diagonal. Then in terms of D_ν, the 3×3 eigenvalue matrix for the light neutrinos, and U_ν be their mixing matrix, we have

$$Y'_e \equiv U_\nu D_\nu U_\nu^T = -\mathcal{M}' \frac{1}{D_0} \mathcal{M}'^T. \tag{3.6}$$

In the "see-saw" limit, the matrices U_0 and U_ν are unitary, so that

$$\mathcal{M}' = \mathcal{M} U_0^*. \tag{3.7}$$

The orders of magnitude of the heavy neutrino mass matrix are

$$\mathcal{M}_0 = m_0 \mathcal{O} \begin{pmatrix} \lambda_0^{2(f_0+f_3+f_8)} & \lambda_0^{2(f_0+f_8)} & \lambda_0^{2f_0+f_3-f_8} \\ \lambda_0^{2(f_0+f_8)} & \lambda_0^{2(f_0-f_3+f_8)} & \lambda_0^{2f_0-f_3-f_8} \\ \lambda_0^{2f_0+f_3-f_8} & \lambda_0^{2f_0-f_3-f_8} & \lambda_0^{2(f_0-2f_8)} \end{pmatrix}.$$

Its diagonalization yields the three eigenvalues

$$M_1 = m_0 \mathcal{O}\big(\lambda_0^{2(f_0+f_3+f_8)}\big) < M_2 = m_0 \mathcal{O}\big(\lambda_0^{2(f_0-f_3+f_8)}\big) < M_3 = m_0 \mathcal{O}\big(\lambda_0^{2(f_0-2f_8)}\big).$$

We have assumed for simplicity that the charges satisfy the inequalities

$$f_0 > 2f_8, \quad 3f_8 > f_3 > 0, \tag{3.8}$$

corresponding to $M_1 < M_2 < M_3$. The diagonalizing matrix is

$$U_0 = m_0 \mathcal{O} \begin{pmatrix} 1 & \lambda_0^{2f_3} & \lambda_0^{3f_8+f_3} \\ \lambda_0^{2f_3} & 1 & \lambda_0^{3f_8-f_3} \\ \lambda_0^{3f_8+f_3} & \lambda_0^{3f_8-f_3} & 1 \end{pmatrix}.$$

The electroweak breaking mass yields the matrix

$$\mathcal{M} = m\mathcal{O}\begin{pmatrix} \lambda_\nu^{3(d_8+f_8)+d_3+f_3} & \lambda_\nu^{3(d_8+f_8)+d_3-f_3} & \lambda_\nu^{3d_8+d_3} \\ \lambda_\nu^{3(d_8+f_8)-d_3+f_3} & \lambda_\nu^{3(d_8+f_8)-d_3-f_3} & \lambda_\nu^{3d_8-d_3} \\ \lambda_\nu^{3f_8+f_3} & \lambda_\nu^{3f_8-f_3} & 1 \end{pmatrix}, \tag{3.9}$$

where λ_ν is the expansion parameter, and m is a mass of electroweak breaking size. If we let $\lambda_0 = \lambda_\nu^z$, with $z > 0$. When $z \geq 1$, we find that

$$Y_e' = \frac{m^2}{M_1}\mathcal{O}\big(\lambda_\nu^{6f_8+2f_3}\big)\mathcal{O}\begin{pmatrix} \lambda_\nu^{6d_8+2d_3} & \lambda_\nu^{6d_8} & \lambda_\nu^{3d_8+d_3} \\ \lambda_\nu^{6d_8} & \lambda_\nu^{6d_8-2d_3} & \lambda_\nu^{3d_8-d_3} \\ \lambda_\nu^{3d_8+d_3} & \lambda_\nu^{3d_8-d_3} & 1 \end{pmatrix}, \tag{3.10}$$

is the matrix whose eigenvalues yield the light neutrino masses, and their mixing angles. It is diagonalized by the unitary matrix

$$U_\nu = \mathcal{O}\begin{pmatrix} 1 & \lambda_\nu^{2d_3} & \lambda_\nu^{3d_8+d_3} \\ \lambda_\nu^{2d_3} & 1 & \lambda_\nu^{3d_8-d_3} \\ \lambda_\nu^{3d_8+d_3} & \lambda_\nu^{3d_8-d_3} & 1 \end{pmatrix}. \tag{3.11}$$

The light neutrino masses are then

$$\begin{aligned} m_{\nu_1} &= \frac{m^2}{M_1}\mathcal{O}\big(\lambda_\nu^{2(3f_8+3d_8+f_3+d_3)}\big), \quad m_{\nu_2} = \frac{m^2}{M_1}\mathcal{O}\big(\lambda_\nu^{2(3f_8+3d_8+f_3-d_3)}\big), \\ m_{\nu_3} &= \frac{m^2}{M_1}\mathcal{O}\big(\lambda_\nu^{2(3f_8+f_3)}\big). \end{aligned} \tag{3.12}$$

In order to obtain the mixing matrix which appears in the charged lepton current, we must fold this matrix with that which diagonalizes the charged lepton masses. If we let $\lambda_\nu = \lambda_e^w$, with $w > 1$, the result is

$$\mathcal{U}_\nu = \mathcal{O}\begin{pmatrix} 1 & \lambda_e^{2d_3} & \lambda_e^{3d_8+d_3} \\ \lambda_e^{2d_3} & 1 & \lambda_e^{3d_8-d_3} \\ \lambda_e^{3d_8+d_3} & \lambda_e^{3d_8-d_3} & 1 \end{pmatrix}.$$

When $0 < w < 1$, the matrix has the same form with λ_e replaced by λ_ν. It is similar to the CKM matrix. The mixing in the charged lepton current was first proposed by Maki, Nakagawa and Sakata in 1962, so we call it the MNS matrix [**21**]. We note that its elements satisfy

$$V_{e\nu_\mu}V_{\mu\nu_\tau} \sim V_{e\nu_\tau}. \tag{3.13}$$

It may be that $\lambda_e = \lambda_d$ and $\lambda_u = \lambda_\nu$, since they have the same quantum numbers, implying $w = y$. We also have the relations

$$\frac{m_{\nu_1}}{m_{\nu_2}} \approx (V_{e\nu_\mu})^w; \quad \frac{m_{\nu_2}}{m_{\nu_3}} \approx (V_{\mu\nu_\tau})^w, \tag{3.14}$$

valid only when $w > 1$. When $0 < w < 1$ the exponents in these relations is one. In this analysis, the lepton mixing matrix has the same structure as the CKM matrix. We have assumed a simple set of inequalities among the charges, to provide an example of our method. When $0 < z < 1$, the forms of the neutrino masses are the same except that $f_{3,8}$ appear multiplied by z. The mixing matrix is unchanged.

Unlike quark masses and mixing, we have little solid experimental information on the values of these parameters. The most compelling evidence for neutrino masses and mixings come from the MSW interpretation of the deficit observed in various solar neutrino fluxes. In this picture, the electron neutrino mixes with another neutrino (assumed here to be the muon neutrino) with a mixing angle θ_{12} such that

$$m_{\nu_1}^2 - m_{\nu_2}^2 \approx 7 \times 10^{-6} eV^2; \quad \sin^2 2\theta_{12} \approx 5 \times 10^{-3}. \tag{3.15}$$

The other piece of evidence comes from the deficit of muon neutrinos in the collision of cosmic rays with the atmosphere. If taken at face value, these suggest that the muon neutrinos oscillate into another species of neutrinos, say τ neutrinos, with a mixing angle θ_{23}, and masses such that

$$m_{\nu_2}^2 - m_{\nu_3}^2 \approx 2 \times 10^{-2} eV^2; \quad \sin^2 2\theta_{23} \geq .5. \tag{3.16}$$

Fitting the parameters coming from the solar neutrino data is rather easy, suggesting that

$$V_{e\nu_\mu} \sim \lambda_e^{2d_3} \sim \lambda^2,$$

together with $m_{\nu_2} \approx 1\,\text{meV}$. However it is not so easy to understand the atmospheric neutrino data. These imply

$$V_{\mu\nu_\tau} \sim \lambda_e^{3d_8 - d_3} = \mathcal{O}(1).$$

The relations (3.14) then suggest that w has to be large. For example the value $\theta_{23} \sim \frac{\pi}{9}$ for which $\sin^2 2\theta_{23} = .34$, yields $m_{\nu_2}/m_{\nu_3} \sim .01$, for $w = 4$. This gives $m_{\nu_3} \approx .1eV$, which marginally reproduces the "data", and fixes the lightest neutrino mass to $m_{\nu_1} \approx 10^{-13}eV$! The heaviest neutrino weighs one tenth of an eV, not enough to be of use for structure formation. Perhaps there are more light neutrals, coming from the extra neutral leptons in each E_6 or from end T in string compactification.

Generically, though, it is difficult to understand mixing angles of order one, as suggested by the atmospheric neutrino data. The existence of only small mixing angles in the quark sectors suggests either that the interpretation of the atmospheric neutrino data is premature, or that there is fine tuning in the neutrino matrices [**22**].

4. Anomalies

The X family symmetry is in general anomalous. The three chiral families contribute to the mixed gauge anomalies as follows

$$C_3 = 3(2a_0 + b_0 + c_0), \tag{4.1}$$

$$C_2 = 3(3a_0 + d_0), \tag{4.2}$$

$$C_1 = a_0 + 8b_0 + 2c_0 + 3d_0 + 6e_0. \tag{4.3}$$

The subscript denotes the gauge group of the Standard Model, i.e. $1 \sim$ U(1), $2 \sim$ SU(2), and $3 \sim$ SU(3). The X-charge also has a mixed gravitational anomaly, which is simply the trace of the X-charge,

$$C_g = 3(6a_0 + 3b_0 + 3c_0 + 2d_0 + e_0 + f_0) - x + C'_g, \tag{4.4}$$

where C'_g is the contribution from the particles that do not appear in the model we are discussing. One must also account for the mixed YXX anomaly, given by

$$C_{YXX} = 6(a_0^2 - 2b_0^2 + c_0^2 - d_0^2 + e_0^2) + 4A_T, \tag{4.5}$$

with the texture-dependent part given by

$$A_T = (3a_8^2 + a_3^2) - 2(3b_8^2 + b_3^2) + (3c_8^2 + c_3^2) - (3d_8^2 + d_3^2) + (3e_8^2 + e_3^2). \tag{4.6}$$

The last anomaly coefficient is that of the X-charge itself, C_X, the sum of the cubes of the X-charge. Extra particles with chiral X-charge other than those in the minimal model, will contribute to both C'_g and C_X.

These anomaly coefficients can be related to combinations of quarks and lepton masses. The reason is that the X-charge of the determinant in each charge sector is *independent* of the texture coefficients that distinguish between the two lightest families. We set

$$\det Y_u \sim y_t^3 \mathcal{O}(\lambda_u^U), \quad \det Y_d \sim y_b^3 \mathcal{O}(\lambda_d^D), \quad \det Y_l \sim y_\tau^3 \mathcal{O}(\lambda_e^E), \tag{4.7}$$

where

$$U \equiv 6(a_8 + b_8), \quad D \equiv 6(a_8 + c_8), \quad E \equiv 6(d_8 + e_8).$$

Since the down and lepton matrices have the same quantum numbers, and couple to the same Higgs, we may assume they have the same expansion parameter, $\lambda_d = \lambda_e$. In that case we can relate the products of the down quark masses to that of the leptons (assuming $y_b = y_\tau$)

$$\frac{m_d m_s m_b}{m_e m_\mu m_\tau} \sim \mathcal{O}(\lambda_d^{(D-E)}). \tag{4.8}$$

From the tree-level Yukawa couplings to the third family expressed through (2.4), we can write combinations of anomaly coefficients in terms of the family-dependent charges

$$\begin{aligned} C_1 + C_2 - \frac{8}{3}C_3 &= 12(d_8 + e_8 - a_8 - c_8) = 2(E - D), \\ C_3 &= 6(2a_8 + b_8 + c_8) = U + D. \end{aligned} \tag{4.9}$$

These allow us to relate the anomaly coefficients to the ratio of products of quark and lepton masses (4.8), (assuming $y_b = y_\tau$),

$$\frac{m_d m_s m_b}{m_e m_\mu m_\tau} \sim \mathcal{O}(\lambda_d^{-(C_1 + C_2 - 8/3C_3)/2}). \tag{4.10}$$

Compatibility with the extrapolated data requires the exponent to vanish

$$C_1 + C_2 - \frac{8}{3}C_3 = 0, \tag{4.11}$$

which expressed in other variables, reads $E = D$.

5. Green-Schwarz Cancellation of X Anomaly

If X is anomaly-free, then

$$C_1 = C_2 = C_3 = 0, \quad C_g = 0. \tag{5.1}$$

The last equation is not constraining as there are likely more fields in the theory with chiral X-charge. These are consistent with (4.10), but the vanishing of C_3 contradicts our hypothesis that all excess charges have the same sign. Indeed, using the tree-level Yukawa relations (2.4), (4.1), we see that

$$0 = C_3 = 6(a_8 + b_8) + 6(a_8 + c_8),$$

which is not consistent with our assumption that all excess charges are positive. Hence we must rely on the Green-Schwarz mechanism.

String theories naturally contain all of the ingredients we need to reproduce the Yukawa textures. They have an antisymmetric tensor Kalb-Ramond field which in four dimensions is the Nambu-Goldstone boson of an anomalous U(1) that couples like an axion through a dimension five term to the divergence of the anomalous current. Its anomalies are cancelled by the Green-Schwarz mechanism [**23**]. Under a chiral transformation, this term is capable of soaking up certain anomalies, by shifting the axion field, provided that they appear in commensurate ratios

$$\frac{C_i}{k_i} = \frac{C_X}{k_X} = \frac{C_g}{k_g}, \tag{5.2}$$

where the k_i are the Kac-Moody levels. They need to be integers only for the non-Abelian factors.

In superstring theories, this U(1) is broken spontaneously slightly below the string scale. The scale is set by the charge content of the theory [**18**]. It follows that singlets with masses protected by X can still be very massive, and not appear in the effective low-energy theory.

This chiral U(1) X-charge can fix the value of the Weinberg angle, without the use of a grand unified group, as remarked by Ibàñez [**16**]. More recently, Ibàñez and Ross [**9**] applied it to the determination of symmetric textures when the field θ is vector-like.

In superstring theories, the non-Abelian gauge groups have the same Kac-Moody levels. For Green-Schwarz cancellation, it means that

$$C_2 = C_3 \quad \text{or} \quad d_0 = b_0 + c_0 - a_0. \tag{5.3}$$

After this very generic requirement, we see that equation (4.10) reduces to

$$\frac{m_d m_s m_b}{m_e m_\mu m_\tau} \sim \mathcal{O}\big(\lambda_d^{-(C_1 - 5/3C_2)/2}\big), \tag{5.4}$$

valid whenever θ is chiral. Since the right-hand side is of order one, it means that the exponent vanishes, so that in models with an *ab initio* μ term, we *deduce* that

$$C_1 = \frac{5}{3}C_2. \tag{5.5}$$

However the gauge coupling constants at string unification scale with the anomaly coefficients, so that

$$\frac{C_1}{C_2} = \frac{g_1^2}{g_2^2}, \tag{5.6}$$

which fixes the Weinberg angle to the value

$$\sin^2\theta_w = \frac{3}{8},$$

at the string scale, the canonical GUT value, but without the excess baggage of these theories! This is a strong hint that the $N = 1$ model does indeed come from superstrings!

We note that the mixed gravitational anomaly is exactly along the anomaly-free combination of baryon minus lepton numbers, $B - L$. In fact the most general X-charge can contain an arbitrary mixture along $B - L$, but this is already taken into account by our general parametrization. In superstring models, the Green-Schwarz mechanism extends to the mixed gravitational anomaly so that

$$\frac{C_g}{C_3} = \frac{k_g}{k_3} = \eta,$$

where η is a normalization parameter; in the simplest level-one models, it is equal to 12. In general, however,

$$C_g = \eta(U + E). \tag{5.7}$$

The family independent X-charges are seen to depend only on two parameters, E, and U, assuming we know the normalization η.

Starting from very simple generic assumptions we are able to reproduce the data and even determine the Weinberg angle in terms of the ratio of quark and lepton masses. Our analysis, applied to neutrino masses, shows that it is ackward to accomodate both solar neutrino and atmospheric neutrino data.

Acknowledgements

I would like to thank Professor Vinet for inviting me to contribute to this anniversary volume. My interactions with the Centre de recherches mathématiques started in 1973, when Professor J. Patera showed me his treasure trove on exceptional groups, at a time where they were but curiosities. They are now part of the physics vocabulary, as in the heterotic string and grand unification. I wish to thank M. Booth for reading this manuscript.

References

1. C. D. Froggatt and H. B. Nielsen, *Hierarchy of quark masses, Cabibbo angles and CP violation*, Nucl. Phys. B **147** (1979), no. 3–4, 277–298.
2. H. P. Nilles, *Supersymmetry, supergravity and particle physics*, Phys. Rep. **110** (1984), no. 1–2, 1–162.
3. H. E. Haber and G. L. Kane, *The search for supersymmetry: Probing physics beyond the standard model*, Phys. Rep. **117** (1985), no. 2–4, 75–263.
4. J. C. Pati and A. Salam, *Lepton number as the fourth "color"*, Phys. Rev. D **10** (1974), 275–289.
5. H. Georgi and S. Glashow, *Unity of all elementary-particle forces*, Phys. Rev. Lett. **32** (1974), 438–441.
6. H. Fritzsch and P. Minkowski, *Unified interactions of leptons and hadrons*, Ann. Physics **93** (1975), no. 1–2, 193–266.
7. F. Gürsey, P. Ramond, and P. Sikivie, *A universal gauge theory model based on E_6*, Phys. Lett. B **60** (1976), no. 2, 177–180.
8. H. Arason, D. J. Castaño, B. E. Keszthelyi, S. Mikaelian, E. J. Piard, P. Ramond, and B. D. Wright, *Top-quark and Higgs-Boson mass bounds from a numerical study of supersymmetric grand unified theories*, Phys. Rev. Lett. **67** (1991), no. 21, 2933–2936.
9. A. Giveon, L. J. Hall, and U. Sarid, SU(5) *unification revisited*, Phys. Lett. B **271** (1991), no. 1–2, 138–144.
10. P. Ramond, R. G. Roberts, and G. G. Ross, *Stitching the Yukawa quilt*, Nuclear Phys. B **406** (1993), no. 1–2, 19–42.
11. A. Kusenko and R. Shrock, *General determination of phases in quark mass matrices*, Phys. Rev. D **50** (1994), no. 1, 30–33.
12. L. E. Ibàñez, *Computing the weak mixing angle from anomaly cancellation*, Phys. Lett. B **303** (1993), no. 1–2, 55–62.
13. L. E. Ibàñez and G. G. Ross, *Fermion masses and mixing angles from gauge symmetries*, Phys. Lett. B **332** (1994), no. 1–2, 100–110.
14. U. Amaldi, W. de Boer, and H. Furstenau, *Comparison of grand unified theories with electroweak and strong coupling constants measured a LEP*, Phys. Lett. B **260** (1991), no. 3–4, 447–455.

15. J. Ellis, S. Kelley, and D. V. Nanopoulos, *Probing the desert using gauge coupling unification*, Phys. Lett. B **260** (1991), no. 1–2, 131–137.
16. P. Langacker and M. Luo, *Implications of precision electroweak experiments for* $m_t, \rho_0, \sin^2, \theta_w$, *and grand unification*, Phys. Rev. D **44** (1991), no. 3, 817–822.
17. G. Anderson, S. Dimopoulos, L. J. Hall, S. Raby, and G. D. Starkman, *Systematic* SO(10) *operator analysis for fermion masses*, Phys. Rev. D **49** (1994), no. 7, 3660–3690.
18. M. Leurer, Y. Nir, and N. Seiberg, *Mass matrix models*, Nuclear Phys. B **398** (1993), no. 2, 319–342.
19. M. Dine, N. Seiberg, and E. Witten, *Fayet-Iliopoulos terms in string theory*, Nuclear Phys. B **289** (1987), no. 3–4, 589–598.
20. J. J. Atick, L. J. Dixon, and A. Sen, *String calculation of Fayet-Iliopoulos D-terms in arbitrary supersymmetric compactifications*, Nuclear Phys. B **292** (1987), no. 1, 109–149.
21. M. Dine, I. Ichinoise, and N. Seiberg, *F terms and D terms in string theory*, Nuclear Phys. B **293** (1987), no. 2, 253–265.
22. A. Font, L. E. Ibàñez, H. P. Nilles, and F. Quevedo, *Yukawa couplings in degenerate orbifolds: towards a realistic* $SU(3) \times SU(2) \times U(1)$ *superstring*, Phys Lett. B **210** (1988), no. 1–2, 101–112.
23. J. A. Casas, E. K. Katehou, and C. Muñoz, U(1) *charges in orbifolds: anomaly cancellation and phenomenological consequences*, Nuclear Phys. B **317** (1989), no. 1, 171–186.
24. J. A. Casas and C. Muñoz, *Three-generation* $SU(3) \times SU(2) \times U(1)_Y \times U(1)$ *orbifold models through Fayet-Iliopoulos terms*, Phys. Lett. B **209** (1988), no. 2–3, 214–220.
25. A. Font, L. E. Ibàñez, F. Quevedo, and A. Sierra, *The construction of "realistic" four-dimensional strings through orbifolds*, Nuclear Phys. B **331** (1990), no. 2, 421–474.
26. Z. Maki, M. Nakagawa, and S. Sakata, *Remarks on the unified model of elementary particles*, Progr. Theoret. Phys. **28** (1962), no. 5, 870–880.
27. N. Cabibbo, *Unitary symmetry and leptonic decays*, Phys. Rev. Lett. **10** (1963), no. 12, 531–533.
28. J. E. Kim and H. P. Nilles, *The* μ*-problem and the strong CP-problem*, Phys. Lett. B **138** (1984), no. 1–3, 150–154.
29. ———, *Gaugino condensation and the cosmological implications of the hidden sector*, Phys. Lett. B **263** (1991), no. 1, 79–85.
30. E. J. Chun, J. E. Kim, and H. P. Nilles, *A natural solution of the* μ *problem with a composite axion in the hidden sector*, Nuclear Phys. B **370** (1992), no. 1, 105–122.
31. J. A. Casas and C. Munõz, *A natural solution to the* μ *problem*, Phys. Lett. B **306** (1993), no. 3–4, 288–294.
32. G. F. Giudice and A. Masiero, *A natural solution to the* μ*-problem in supergravity theories*, Phys. Lett. B **206** (1988), no. 3, 480–484.
33. L. Wolfenstein, *Parametrization of the Kobayashi-Maskawa matrix*, Phys. Rev. Lett. **51** (1983), no. 21 1945–1947.
34. H. Dreiner, G. K. Leontaris, S. Lola, G. G. Ross, and C. Scheich, *Neutrino masses from gauge symmetries*, Nuclear Phys. B **436** (1995), no. 3, 461–473.

Institute for Fundamental Theory, Department of Physics, University of Florida, Gainesville FL 32611, USA

Centre de Recherches Mathématiques
CRM Proceedings and Lecture Notes
Volume **11**, 1997

Natural Dualities and Clones

B. A. Davey and I. G. Rosenberg

ABSTRACT. We consider algebraic dualities for the quasivarieties generated by finite algebras endowed with discrete topology. The continuous maps involved are described combinatorially and this allows us to recast the question: "Which pairs of algebras and relational system yield a duality?" in a purely relational framework. We apply this to show a nonrefinable chain of nondualizable clones on every finite universe A with $|A| > 2$.

1. Introduction, Preliminaries and Foundations

Algebraic duality grew out of two classical special cases—Pontryagin's duality for Abelian groups and Stone's duality for boolean algebras. After Priestley's duality for bounded distributive lattices it evolved into an important representation tool in universal algebra, logics and categories. In most cases the generating algebra is a finite algebra endowed with the discrete topology. We are restricting ourselves to this case and address the following question. Let $\mathfrak{P} = \langle A; \underline{P}\rangle$ be a finite algebra and and $\mathfrak{R} = \langle A; \underset{\sim}{P}\rangle$ a relational system on A. We say that the pair $\underline{P}, \underset{\sim}{P}$ yields a duality if for every subalgebra S of a power $\mathfrak{P}^J$ we can find a subset T of a power $\mathfrak{R}^J$ such that the continuous relational homomorphisms from T into $\mathfrak{R}$ form an algebra isomorphic- in a well specified way-to S. (Usually duality is formulated for the quasivariety $I\mathbb{SP}\mathfrak{P}$ which allows as well all isomorphic copies of S; clearly this extension has no impact on the essence of the problem and so for simplicity is left out). Here continuity refers to the product topology of the discrete topology on A. Since this is transparent, we express the continuity in a combinatorial way and then remold the problem. We characterize the sets T to be considered. We show that in order to prove that $\underline{P}, \underset{\sim}{P}$ yields a duality we need only to consider such subsets T of $\mathfrak{R}^J$ for sets J of arbitrarily large cardinalities. We consider briefly the special case when $\underline{P}$ is the clone of all operations commuting with a given set F of operations on A.

We apply this machinery to the clones U_h consisting of all operations on A that are either essentially at most unary or take at most h values ($h = 2, \ldots, k$

1991 *Mathematics Subject Classification*. Primary: 08C05, 08C15; Secondary: 08B99.
Part of this work was done during the second author's stay at La Trobe University in the summer of 1991. The partial financial help provided by NSERC Canada operating grant OGP0005407 is gratefully acknowledged.
This is the final form of the paper.

where $k := |A|$). It is well known that $U_1 \subset B \subset U_2 \subset \cdots \subset U_k$ is a maximal (= nonrefinable) chain from the clone U_1 of all nonessential operations to the clone U_k of all operations on $\underline{A}$ where B is the Burle clone and U_{k-1} the Słupecki clone. We show that $U_2, \ldots, U_{k-1}$ are nondualizable. As a byproduct we obtain a form of Hu's duality for primal algebras.

The scope of this paper is restricted to duality but the methods could be as well extended to full duality. In our context the following questions are intriguing but probably quite difficult:

(1) Which clones on a finite at least 3-element universe are dualizable?
(2) For a dualizable $\underline{P}$ find a preferably small $\underset{\sim}{P}$ such that $\underline{P}, \underset{\sim}{P}$ yields a duality.

1.1. Let k be a fixed positive integer. Put $\boldsymbol{k} := \{0, 1 \ldots, k-1\}$. Let I be a nonempty set. A subset ρ of $\boldsymbol{k}^I$ (i.e. a set of maps from I to $\boldsymbol{k}$) is an I-relation on $\boldsymbol{k}$. (If $I = \boldsymbol{n} = \{0, \ldots, n-1\}$ for some positive integer n, we may identify $\boldsymbol{k}^I$ with $\boldsymbol{k}^n$ and so perceive ρ as a set of n-tuples over $\boldsymbol{k}$). Each $r \in \rho$ is a *row* of ρ. For $i \in I$ the *column* $\rho^{(i)}$ of ρ is the map from ρ to $\boldsymbol{k}$ defined by setting $\rho^{(i)}(f) := f(i)$ for all $f \in \rho$. The set of all columns of ρ is denoted by ρ^T. The relation ρ is *repetition-free* if its columns are pairwise distinct. For $J \subseteq I$ set $pr_J\rho := \{r \restriction J : r \in \rho\}$ (where $r \restriction J$ denotes the restriction of r to J).

For a positive integer n an *n-ary operation* on $\boldsymbol{k}$ is a map $f\colon \boldsymbol{k}^n \to \boldsymbol{k}$. The set of all n-ary operations on $\boldsymbol{k}$ is denoted $O^{(n)}$ (for notational simplicity nullary operations are considered as constant unary operations). Set $O := \bigcup_{n=1}^{\infty} O^{(n)}$. For $n > 0$ and $n > i \geq 0$ the i-th n-ary *projection* e_i^n is defined by $e_i^n(x_1, \ldots, x_n) \approx x_i$. (The symbol $\approx$ is used to express that both sides are equal for every choice of $x_1, \ldots, x_n$ in $\boldsymbol{k}$). A *clone* is a composition closed subset of O containing all projections (for a more precise definition see [**MMT87**]). For $F \subseteq O$ the least clone containing F is denoted $\operatorname{Clo} F$. (In other words, $\operatorname{Clo} F$ is the set of all term operations of the algebra $\langle \boldsymbol{k}; F\rangle$).

For $f \in O^{(n)}$ and $g_i\colon I \to \boldsymbol{k}$ $(i = 1, \ldots, n)$ define the map $h = f[g_1, \ldots, g_n]$ from I into $\boldsymbol{k}$ by setting $h(i) := f\big(g_1(i), \ldots, g_n(i)\big)$ for all $i \in I$.

Let $\mathfrak{P} := \langle \boldsymbol{k}; \underline{P}\rangle$ be an algebra on $\boldsymbol{k}$. For a nonempty set L denote by $\operatorname{Su}(\mathfrak{P}^L)$ the set of all subuniverses of the power $\mathfrak{P}^L$. For $\rho \in \operatorname{Su}(\mathfrak{P}^L)$ let $\mathfrak{A} = \langle \rho; \underline{P}\rangle$ denote the corresponding subalgebra of $\mathfrak{P}^L$ and let

$$I := D(\rho) := \operatorname{Hom}(\mathfrak{A}, \mathfrak{P})$$

(= the set of all homomorphisms from $\mathfrak{A}$ into $\mathfrak{P}$). Observe that for every $\ell \in L$ the column $\rho^{(\ell)}$ (of the L-relation ρ on $\boldsymbol{k}$) is the ℓ-th projection on ρ, hence $\rho^{(\ell)}$ is a homomorphism from $\mathfrak{A}$ into $\mathfrak{P}$ and thus $\rho^{(\ell)} \in I$. Without loss of generality we therefore shall assume that L is a subset of I.

To every $a \in \rho$ assign the map $r_a\colon I \to \boldsymbol{k}$ defined by setting $r_a(i) = i(a)$ for each $i \in I$. Clearly $\bar{\rho} := \{r_a : a \in \rho\}$ is an I-relation on $\boldsymbol{k}$. The relations of this form are crucial in the sequel and so we characterize them right away as special subuniverses of powers of $\mathfrak{P}$.

Let $\varnothing \neq I \subseteq I'$ be sets, let $\rho \in \operatorname{Su}(\mathfrak{P}^I)$ and $\rho' \in \operatorname{Su}(\mathfrak{P}^{I'})$. The I'-relation ρ' *extends* ρ if $\rho' \restriction I = \rho$ and

$$f \restriction I = g \restriction I \Rightarrow f = g$$

holds for all $f, g \in \rho'$. Thus ρ' extends ρ whenever the restriction of every $r' \in \rho'$ to I belongs to ρ and every $r \in \rho$ has a unique extension $r^* \in \rho'$. We start with the following lemma.

LEMMA 1.1. *Let* $\rho \in \mathrm{Su}(\mathfrak{P}^I)$. *Then:*

(i) $\bar{\rho}$ *extends* ρ, *and*

(ii) $\bar{\rho}$ *is repetition-free provided* ρ *is repetition-free.* □

PROOF. Set $I' := D(\rho)$, We need:

CLAIM. $\bar{\rho} \in \mathrm{Su}(\mathfrak{P}^{I'})$.

PROOF. Let $f \in \underline{P}$ be n-ary, let $a_1, \dots, a_n \in \rho$ be arbitrary and let $h := f[r_{a_1}, \dots, r_{a_n}]$. Set $g := f[a_1, \dots, a_n]$ and notice that $h\colon I' \to \boldsymbol{k}$ and $g\colon I \to \boldsymbol{k}$. Let $\varphi \in I'$ be arbitrary. Then $\varphi \in \mathrm{Hom}(\langle \rho; \underline{P}\rangle, \mathfrak{P})$ and so

$$\begin{aligned} h(\varphi) &= f\big(r_{a_1}(\varphi), \dots, r_{a_n}(\varphi)\big) = f\big(\varphi(a_1), \dots, \varphi(a_n)\big) \\ &= \varphi\big(f[a_1, \dots, a_n]\big) = \varphi(g) = r_g(\varphi) \end{aligned}$$

proving that $h = r_g \in \bar{\rho}$ and the claim.

Now (i) follows from the claim and the definition of $\bar{\rho}$. Finally, (ii) is evident from the definition of $\bar{\rho}$. □

The next lemma is the converse of Lemma 1.1.

LEMMA 1.2. *Let* $\varnothing \neq I \subseteq I'$, $\rho \in \mathrm{Su}(\mathfrak{P}^I)$ *and let* $\rho' \in \mathrm{Su}(\mathfrak{P}^{I'})$ *extend* ρ. *Then for each* $j \in I'$ *the column* $\rho'^{(j)}$ *is a homomorphism from* $\langle \rho; \underline{P}\rangle$ *into* P.

PROOF. For every $a \in \rho$ denote by a' the element of ρ' extending uniquely a. Recall that $\varphi := \rho'^{(j)}$ is defined by setting $\varphi(a) := a'(j)$ for all $a \in \rho$. To show that $\varphi \in \mathrm{Hom}(\langle \rho, \underline{P}\rangle, \mathfrak{P})$ let $f \in \underline{P}$ be n-ary, let $a_1, \dots, a_n \in \rho$ and set $a := f[a_1, \dots, a_n]$. Taking into account that $\rho \in \mathrm{Su}(\mathfrak{P}^I)$, $\rho' \in \mathrm{Su}(\mathfrak{P}^{I'})$ and the uniqueness of extensions, we obtain $a' = f[a'_1, \dots, a'_n]$ and so

$$\begin{aligned} \varphi\big(f[a_1, \dots, a_n]\big) &= \varphi(a) = a'(j) = f[a'_1, \dots, a'_n](j) \\ &= f\big(a'_1(j), \dots, a'_n(j)\big) = f\big[\varphi(a_1), \dots, \varphi(a_n)\big]. \quad \square \end{aligned}$$

We say that $\rho \in \mathrm{Su}(\mathfrak{P}^I)$ is $\underline{P}$*-saturated* if ρ is repetition-free but every proper extension of ρ has a repeated column. We characterize the sets of the form $\bar{\rho}$.

PROPOSITION 1.3. *Let* $\sigma \in \mathrm{Su}(\mathfrak{P}^I)$. *Then* $\sigma = \bar{\rho}$ *for some* $L \subseteq I$ *and* $\rho \in \mathrm{Su}(\mathfrak{P}^L)$ *if and only if* σ *is saturated.*

PROOF. **(⇒):** Let $\sigma = \bar{\rho}$ for some $L \subseteq I$ and $\rho \in \mathrm{Su}(\mathfrak{P}^L)$. Let $J \supset I$ and let a J-relation σ' extend σ. Consider an arbitrary element j of $J \backslash I$. By Lemma 1.1 (i) the relation σ extends ρ and so σ' extends ρ. From Lemma 1.2

$$\rho'^{(j)} \in \mathrm{Hom}(\langle \rho, \underline{P}\rangle, \mathfrak{P}) = \sigma$$

and therefore $\rho'^{(j)} = \rho^{(i)} = \rho'^{(i)}$ for some $i \in I$ proving that ρ' has a repeated column.

(⇐): Let σ be saturated. Then σ is repetition-free and by Lemma 1.1 the relation $\bar{\sigma}$ is a repetition-free extension of σ. Thus $\bar{\sigma} = \sigma$ and it suffices to choose $L = I$ and $\rho = \sigma$. □

1.2. So far our approach was purely algebraic. Consider a fixed I-relation ρ on $\boldsymbol{k}$ and endow I with the following topology. The set $\rho^T = \{\rho^{(i)} : i \in I\}$ of columns of ρ is a subset of $\boldsymbol{k}^\rho$. The discrete topology on the finite set $\boldsymbol{k}$ induces the usual product topology on $\boldsymbol{k}^\rho$ whose subbase is formed by the sets $\{f \in \boldsymbol{k}^\rho : f(r) = p\}$ for all $r \in \rho$ and $p \in \boldsymbol{k}$. Clearly ρ^T inherits this topology and so we can transfer it to I by choosing the sets

$$[r : p] := \{i \in I : r(i) = p\} \tag{1.1}$$

($r \in \rho$ and $p \in \boldsymbol{k}$) as a subbase of a topology τ_ρ on I. For duality we need the continuous maps from (I, τ_ρ) into the discrete space $\boldsymbol{k}$. To describe them we need the following terminology.

A *row-block* of ρ is a nonempty intersection $B_1 \cap \dots \cap B_n$ where each B_i is a block of $\ker r_i$ for some $r_i \in \rho$. The *vicinity* $\operatorname{Vic}\rho$ of ρ is the set of all maps $f : I \to \boldsymbol{k}$ such that for each $x \in \boldsymbol{k}$ the set $f^{-1}(x)$ is the union of a set of row-blocks of ρ. Thus $f \in \boldsymbol{k}^I$ belongs to $\operatorname{Vic}\rho$ if every block of $\ker f$ is covered by the row-blocks of ρ it contains. This concept is justified by the following lemma.

LEMMA 1.4. *Let ρ be an I-relation on $\boldsymbol{k}$. The vicinity $\operatorname{Vic}\rho$ of ρ is the set of all continuous maps from the topological space (I, τ_ρ) into the discrete space $\boldsymbol{k}$.*

PROOF. Let f be continuous, let $i \in I$ and $y := f(i)$. Clearly $\{y\}$ is a neighborhood of y in the discrete topology on $\boldsymbol{k}$ and so $f(G) \subseteq \{y\}$ for some neighborhood G of i in τ_ρ. In view of (1.1) we can choose G to be of the form

$$[r_1 : p_1] \cap \dots \cap [r_n : p_n]$$

for some $r_1, \dots, r_n \in \rho$ and $p_1, \dots, p_n \in \boldsymbol{k}$. As $i \in G$, clearly $r_1(i) = p_1, \dots, r_n(i) = p_n$. For $\ell = 1, \dots, n$ denote by B_ℓ the block of $\ker r_\ell$ containing i. It is easy to see that G is the row-block $B := B_1 \cap \dots \cap B_n$ of φ. Now $f(B) = \{i\}$ means $B \subseteq f^{-1}(i)$ proving that $f \in \operatorname{Vic}\rho$.

This argument can be reversed to prove that every $f \in \operatorname{Vic}\rho$ is continuous. □

1.3. Let ρ be an I-relation on $\boldsymbol{k}$, let h be a positive integer and σ an h-ary relation on $\boldsymbol{k}$. The h-ary relation σ_ρ on I is defined by setting $(i_1, \dots, i_h) \in \sigma_\rho$ if $(r(i_1), \dots, r(i_h)) \in \sigma$ for all $r \in \rho$.

We recall the following standard terminology. A *relational system* on a set A is a set $\Sigma = \{\rho_n : n \in N\}$ of finitary relations on A. The type of Σ is the map τ from N into the positive integers assigning to each $n \in N$ the arity of ρ_n. Let ρ and ρ' be h-ary relations on A and A'. A map $f : A \to A'$ is a *relational homomorphism* from ρ into ρ' if $f(\rho) \subseteq \rho'$ (i.e. if $(f(a_1), \dots, f(a_h)) \in \rho'$ for all $(a_1, \dots, a_h) \in \rho$). Next $\operatorname{Hom}(\rho, \rho')$ denotes the set of all relational homomorphisms from ρ into ρ'. Given two relational systems

$$\Sigma = \{\rho_n : n \in N\}, \quad \Sigma' = \{\rho'_n : n \in N\}$$

of the same type, set

$$\operatorname{Hom}(\Sigma, \Sigma') := \bigcap_{n \in N} \operatorname{Hom}(\rho_n, \rho'_n).$$

1.4. Let f be an n-ary operation on $\boldsymbol{k}$ and ρ and I-relation on $\boldsymbol{k}$. We say that f *preserves* ρ if $f[r_1, \dots, r_n] \in \rho$ whenever $r_1, \dots, r_n \in \rho$; equivalently, if ρ is a subuniverse of the power $\langle \boldsymbol{k}; f\rangle^I$. Denote by $\operatorname{Pol}\rho$ the set of all $f \in O$ that preserve ρ. For a set R of relations on $\boldsymbol{k}$ set $\operatorname{Pol} R := \bigcap_{\rho\in R} \operatorname{Pol}\rho$.

Let $\underline{P} \subseteq O$ and let $\underset{\sim}{P}$ be a set of finitary relations on $\boldsymbol{k}$. The pair $\underline{P}, \underset{\sim}{P}$ is a *strict protoduality* if $\operatorname{Clo}\underline{P} = \operatorname{Pol}\underset{\sim}{P}$. Thus $\underline{P}, \underset{\sim}{P}$ is a strict protoduality if

(i) every operation of $\underline{P}$ preserves each relation from $\underset{\sim}{P}$ and
(ii) every operation on $\boldsymbol{k}$ preserving all relations from $\underset{\sim}{P}$ is a term operation of $P = \langle \boldsymbol{k}; \underline{P}\rangle$.

We assume throughout that the pair $\underline{P}, \underset{\sim}{P}$ is a strict protoduality.

Let L be a nonempty set and α a subuniverse of $\mathfrak{P}^L$. Denote by ρ the corresponding I-relation $\bar{\alpha}$ (defined in 1.1) and set

$$\underset{\sim}{P}_\rho := \{\sigma_\rho \colon \sigma \in \underset{\sim}{P}\}$$

(see 1.3). Recall from 1.2 that the topologies on I and $\boldsymbol{k}$ are τ_ρ and the discrete topology. Call ρ the *$\underline{P}, \underset{\sim}{P}$-dual* of α if every continuous map from $\operatorname{Hom}(\underset{\sim}{P}_\rho, \underset{\sim}{P})$ belongs to ρ. The strict protoduality $\underline{P}, \underset{\sim}{P}$ *yields a duality* if $\bar{\alpha}$ is the $\underline{P}, \underset{\sim}{P}$-dual of α for every subuniverse α of a power of $\mathfrak{P} = \langle \boldsymbol{k}; \underline{P}\rangle$. We have:

LEMMA 1.5. *A strict protoduality $\underline{P}, \underset{\sim}{P}$ yields a duality if and only if*

$$\operatorname{Vic}\rho \cap \operatorname{Hom}(\underset{\sim}{P}_\rho, \underset{\sim}{P}) \subseteq \rho \tag{1.2}$$

for every $\underline{P}$-saturated relation ρ on $\boldsymbol{k}$.

PROOF. Proposition 1.3 and Lemma 1.4. □

REMARK 1.6. (1.2) *is equivalent to*

$$\operatorname{Vic}\rho \cap \operatorname{Hom}(\underset{\sim}{P}_\rho, \underset{\sim}{P}) = \rho. \tag{1.3}$$

Indeed, let $r \in \rho$. As each block of $\ker r$ is trivially a row-block of ρ, we have $r \in \operatorname{Vic}\rho$. Let $\sigma \in \underset{\sim}{P}$ be h-ary and let $(i_1, \dots, i_h) \in \sigma_\rho$. Then $(r(i_1), \dots, r(i_h)) \in \sigma$ and therefore $r \in \operatorname{Hom}(\underset{\sim}{P}_\rho, \underset{\sim}{P})$ proving $\supseteq$ in (1.3).

1.5. Let ρ be a $\underline{P}$-saturated I-relation on $\boldsymbol{k}$. By its definition (see 1.2) the set $\operatorname{Vic}\rho$ is completely determined by the equivalences $\{\ker v \colon v \in \operatorname{Vic}\rho\}$. Thus an equivalence θ on I is called *ρ-vicinal* if

(i) θ has at most k blocks and
(ii) each block of θ is the union of a set of row-blocks of ρ.

For every I-relation λ on $\boldsymbol{k}$ and each equivalence θ on I set

$$\lambda^\theta := \{f \in \lambda \colon \ker f \supseteq \theta\}.$$

It is easy to see that the condition (1.2) from Lemma 1.5 can be reformulated as

$$\left(\operatorname{Hom}(\underset{\sim}{P}_\rho, \underset{\sim}{P})\right)^\theta \subseteq \rho^\theta \tag{1.4}$$

for every $\underline{P}$-saturated I-relation ρ on $\boldsymbol{k}$ and each ρ-vicinal equivalence θ on I.

Denote by n the number of blocks of θ and notice that $n \leq k$. We condense the two I-relations appearing in (1.4) into two n-ary relations. A selfmap t on I is *θ-canonical* if it maps each block B of θ onto a singleton subset of B; in other words, if $T := \operatorname{im} t$ is a set of representatives of the blocks of θ. As usual, for an h-ary relation λ on I the set

$$t \circ \lambda := \{(t(i_1), \dots, t(i_h)) \colon (i_1, \dots, i_h) \in \lambda\}$$

will be considered as an h-ary relation on $T := \operatorname{im} t$. Notice that

$$t \circ \underset{\sim}{P}_\rho := \{t \circ \sigma_\rho : \sigma \in \underset{\sim}{P}\}$$

is a relational system on T of the same type as P. Set

$$\hat{\rho} := \rho^\theta \restriction T = \{r \restriction T : r \in \rho, \ker r \supseteq \theta\}.$$

Finally denote by $\widehat{\underline{P}}$ the class of all triples (ρ, θ, t) such that ρ is a $\underline{P}$-saturated I-relation on $\boldsymbol{k}, \theta$ is a ρ-vicinal equivalence on I and t is a θ-canonical map. Notice that we list $\theta = \ker t$ only for notational convenience . The same applies to t because for a given θ, replacing t by another θ-canonical map t' has only notational impact.

We reformulate Lemma 1.5.

LEMMA 1.7. *A strict protoduality $\underline{P}, \underset{\sim}{P}$ yields a duality if and only if*

$$\operatorname{Hom}(t \circ \underset{\sim}{P}_\rho, \underset{\sim}{P}) \subseteq \hat{\rho}_t \tag{1.5}$$

holds for every $(\rho, \theta, t) \in \widehat{\underline{P}}$.

PROOF. Follows from Lemma 1.5, Subsection 1.5 and the fact that every $f \in \rho^\theta$ is constant on each block of θ. □

REMARK 1.8. From the definition of a ρ-vicinal equivalence it is immediate that the set V_ρ of ρ-vicinal equivalences is an upset (or order ideal) in the poset $(E(I), \subseteq)$ of all equivalences on I. As no chain in V_ρ has more than k elements, the set V_ρ is determined by the set M_ρ of its minimal elements (i.e. $\theta \in V_\rho$ without any proper ρ-vicinal refinement). In Lemma 1.7 we can restrict θ to a minimal ρ-vicinal equivalence. To see this, let (1.5) hold for all $(\rho, \theta, t) \in \widehat{\underline{P}}$ with $\theta \in M_\rho$ and let $(\rho, \theta', t') \in \widehat{\underline{P}}$ where $\theta' \in V_\rho \backslash M_\rho$. Choose θ so that $\theta' \supseteq \theta \in M_\rho$. Then (1.5) holds for some $(\rho, \theta, t) \in \widehat{\underline{P}}$ such that $T := \operatorname{im} t \supseteq \operatorname{im} t'$. Then $t' = v \circ t$ for a selfmap v of T satisfying $v \circ v = v$. Let $f' \in \operatorname{Hom}(t' \circ \underset{\sim}{P}_\rho, \underset{\sim}{P})$ be arbitrary. We assert that

$$f := f' \circ v \in \operatorname{Hom}(t \circ \underset{\sim}{P}_\rho, \underset{\sim}{P}).$$

Indeed, let $\sigma \in \underset{\sim}{P}$ be h-ary and let $(j_1, \dots, j_h) \in t \circ \sigma_\rho$. Then $(i_1, \dots, i_h) \in \sigma_\rho$ for some $i_\ell \in t^{-1}(j_\ell)$, $\ell = 1, \dots, h$. As $v(j_\ell) = v(t(i_\ell)) = t'(i_\ell)$, from

$$(v(j_1), \dots, v(j_h)) = (t'(i_1), \dots, t'(i_h)) \in t' \circ \sigma_\rho$$

we obtain

$$(f(j_1), \dots, f(j_h)) = (f'(v(j_1)), \dots, f'(v(j_h))) \in \sigma$$

proving the assertion. From (1.5) it follows that $f = r \restriction T$ for some $r \in \rho$ with $\ker r \supseteq \theta$. Next $f = f' \circ v$ shows that, in fact, $\ker r \supseteq \theta'$. Finally, $f' \restriction T' = r \restriction T'$ proves the required $f' \in \hat{\underline{P}}_{t'}$.

1.6. Denote by $\operatorname{Inv} \underline{P}$ the set of all finitary relations ρ on $\boldsymbol{k}$ such that $\underline{P} \subseteq \operatorname{Pol} \rho$. Observe that $\underset{\sim}{P} \subseteq \operatorname{Inv} \underline{P}$ for every strict protoduality $\underline{P}, \underset{\sim}{P}$. Following [**DW83**] we say that *brute force yields a duality for* $\underline{P}$ (or that $\underline{P}$ *is dualizable*) if $\underline{P}, \operatorname{Inv} \underline{P}$ yields a duality. The following is immediate. If some strict protoduality $\underline{P}, \underset{\sim}{P}$ yields a duality, then brute force yields a duality.

Let ρ be an I-relation on $\boldsymbol{k}$. The *local closure* $\operatorname{Loc} \rho$ of ρ consists of all $f : I \to \boldsymbol{k}$ such that for every finite subset F of I

$$f \restriction F = r \restriction F$$

for some $r \in \rho$ (i.e. f agrees with some $r \in \rho$ on F). We have:

LEMMA 1.9. *Brute force yields a duality if and only if*

$$\operatorname{Vic}\rho \cap \operatorname{Loc}\rho \subseteq \rho$$

for every $\underline{P}$-saturated relation ρ.

PROOF. Set $\underset{\sim}{P} := \operatorname{Inv}\underline{P}$ and let ρ be a $\underline{P}$-saturated I-relation on $\boldsymbol{k}$. In view of Lemma 1.5 it suffices to show that

$$\operatorname{Hom}(\underset{\sim}{P}_\rho, \underset{\sim}{P}) = \operatorname{Loc}\rho.$$

Let $f \in \operatorname{Hom}(\underset{\sim}{P}_\rho, \underset{\sim}{P})$ and let $F = \{i_1, \ldots, i_h\}$ be a finite subset of I. Since by Proposition 1.3 the relation ρ is a subuniverse of $\mathfrak{P}^I$, clearly $\sigma := \{(r(i_1), \ldots, r(i_h)) : r \in \rho\}$ is a subuniverse of $\mathfrak{P}^h$ and so $\sigma \in \operatorname{Inv}\underline{P} = \underset{\sim}{P}$. Clearly $(r(i_1), \ldots, r(i_h)) \in \sigma$ for all $r \in \rho$ and therefore $(i_1, \ldots, i_h) \in \sigma_\rho$. In view of $f \in \operatorname{Hom}(\underset{\sim}{P}_\rho, \underset{\sim}{P})$ we have $(f(i_1), \ldots, f(i_h)) \in \sigma$. By the definition of σ evidently

$$f(i_1) = r(i_1), \ldots, f(i_h) = r(i_h)$$

for some $r \in \rho$. This proves $f \restriction F = r \restriction F$ and $\operatorname{Hom}(\underset{\sim}{P}_\rho, \underset{\sim}{P}) \subseteq \operatorname{Loc}\rho$.

For the converse inclusion, let $f \in \operatorname{Loc}\rho$, let $\sigma \in \underset{\sim}{P}$ be h-ary and $(i_1, \ldots, i_h) \in \sigma_\rho$. Let $F := \{i_1, \ldots, i_h\}$. As $f \in \operatorname{Loc}\rho$, we have that $f \restriction F = r \restriction F$ for some $r \in \rho$. By the definition of σ_ρ we obtain

$$(f(i_1), \ldots, f(i_h)) = (r(i_1), \ldots, r(i_h)) \in \sigma$$

proving that $f \in \operatorname{Hom}(\underset{\sim}{P}_\rho, \underset{\sim}{P})$ and $\operatorname{Loc}\rho \subseteq \operatorname{Hom}(\underset{\sim}{P}_\rho, \underset{\sim}{P})$. □

According to the lemma and the remark preceding it, proving that for a fixed $\underline{P}$ no strict protoduality $\underline{P}, \underset{\sim}{P}$ yields a duality amounts to finding a $\underline{P}$-saturated relation ρ and an $s \in (\operatorname{Loc}\rho) \backslash \rho$ such that $\ker s$ is not ρ-vicinal. Here $\ker s$ is not ρ-vicinal if there exist a block B of $\ker s$ and $b \in B$ such that $R \not\subseteq B$ for every row-block R of ρ containing B.

2. Two Duality Properties

Let $\underline{P}, \underset{\sim}{P}$ be a strict protoduality. We show that in order to prove that $\underline{P}, \underset{\sim}{P}$ yields a duality, the validity of (1.5) need not be checked for $\underline{P}$-saturated relations of small arities.

PROPOSITION 2.1. *Let κ be an infinite cardinal. A strict protoduality $\underline{P}, \underset{\sim}{P}$ yields a duality if and only if*

$$\operatorname{Hom}(t \circ \underset{\sim}{P}_\rho, \underset{\sim}{P}) \subseteq \hat{\rho}_t$$

holds for every $(\rho, \theta, t) \in \widehat{\underline{P}}$ where ρ is an I-relation with $|I| \geq \kappa$.

PROOF. ($\Rightarrow$): Lemma 1.5.

($\Leftarrow$): Let the condition hold, let $(\rho, \theta, t) \in \widehat{\underline{P}}$ where ρ is an L-relation with $0 < |L| < \kappa$. Denote by K the set of all ordinals of cardinality less than κ. Let $I := K \times L$ and notice that $|I| = \kappa$ (due to $\aleph_0 \leq \kappa = |K| > |L|$).

For each $r \in \rho$ define $r' \in \boldsymbol{k}^I$ by setting $r'(x,y) := r(y)$ for all $x \in K$ and $y \in L$. Further construct an I-relation ι on $\boldsymbol{k}$ as follows: Fix $r_0 \in \rho$. For all $a \in K^* := k\backslash\{0\}$ define $s_a \in \boldsymbol{k}^I$ by setting

$$s_a(x,y) := \begin{cases} r_0(y) & \text{if } x = 0 \\ 1 & \text{if } x = a \\ 0 & \text{otherwise,} \end{cases}$$

and

$$\iota := \{r' \colon r \in \rho\} \cup \{s_a \colon a \in K^*\}. \tag{2.1}$$

We need the following claims.

CLAIM 1. *The $I-$relation ι has distinct columns.*

PROOF. Let $\iota^{((a,b))} = \iota^{((c,d))}$ for some (a,b), $(c,d) \in I$. From $\{r' \colon r \in \rho\} \subseteq \iota$ we obtain $\rho^{(b)} = \rho^{(d)}$. As ρ is $\underset{\sim}{P}$-saturated, it has distinct columns and so $b = d$. If $a > 0$ and $c > 0$, then from

$$s_a(a,b) = 1 = s_a(c,b)$$

we obtain the required $a = c$. Thus by symmetry we may assume $a = 0$. Suppose to the contrary that $c > 0$. Choose $e \in K\backslash\{0,c\}$. We obtain the contradiction

$$0 = s_e(c,b) = s_e(0,b) = r_0(b) = s_c(0,b) = s_c(c,b) = 1.$$

Thus $a = c = 0$ proving the claim.

Denote by ν the subuniverse of $\mathfrak{P}^I$ generated by ι. To every $s \in \boldsymbol{k}^I$ assign $s^- \in \boldsymbol{k}^L$ defined by setting $s^-(x) := s(0,x)$ for all $x \in L$. Further set $\nu^- := \{s^- \colon s \in \nu\}$ (thus ν^- is the projection $pr_{\{0\}\times L}\nu$ considered as an L-relation).

CLAIM 2. $\nu^- = \rho$.

PROOF. Clearly $\iota^- = \rho$ by the construction of ι. Now the claim follows from the fact that ρ is a subuniverse of $\mathfrak{P}^L$.

By Claim 1 the relation ν has distinct columns. Let μ be any saturated extension of ν. Clearly μ is a J-relation on $\boldsymbol{k}$ with $J \supseteq I$. For each $r \in \rho$ denote by r'' the element of μ extending r'. Notice that $r''^- = r$. Set $\alpha := \{r'' \colon r \in \rho\}$. We show that the relations ρ and α have the same columns:

CLAIM 3. $\rho^T = \alpha^T$.

PROOF. Let $j \in J$ be arbitrary. Since ν is a subuniverse of $\mathfrak{P}^I$, the column $\mu^{(j)}$ is a homomorphism from $\langle \nu, \underline{P} \rangle$ into $\mathfrak{P}$. Recall that ρ is a subuniverse of $\mathfrak{P}^L$. By its definition $\beta := \{r' \colon r \in \rho\}$ is a subuniverse of $\mathfrak{P}^I$. Here $\beta \subseteq \nu$ by (2.1) and so β is a subuniverse of $\langle \nu; \mathfrak{P} \rangle$. It follows that the restriction $\alpha^{(j)}$ of the homomorphism $\mu^{(j)}$ to β is a homomorphism from $\langle \beta; \underline{P} \rangle$ into $\mathfrak{P}$. From the definition of β it follows that $\alpha^{(j)}$ is also a homomorphism from $\langle \rho; \underline{P} \rangle$ into $\mathfrak{P}$. Now ρ is $\underline{P}$-saturated and so $\alpha^{(j)} = \rho^{(\ell)}$ for some $\ell \in L$. Thus $\alpha^T \subseteq \rho^T$. As $\alpha^- = \rho$, clearly $\rho^T \subseteq \alpha^T$ proving the claim.

Let $B_1, \ldots, B_q$ be the blocks of θ. For $i = 1, \ldots, q$ define b_i by $\{b_i\} := t(B_i)$ and set

$$B_i' := \{j \in J \colon \alpha^{(j)} = \rho^{(b)} \text{ for some } b \in B_i\}. \tag{2.2}$$

In view of Claim 3 and the fact that ρ has distinct columns, the sets $B'_1, \dots, B'_q$ partition J. Denote by θ' the equivalence on J with blocks $B'_1, \dots, B'_q$.

CLAIM 4. *The equivalence θ' is μ-vicinal.*

PROOF. Let $1 \leq i \leq q$ and let j be an arbitrary element of B'_i. By the definition $\alpha^{(j)} = \rho^{(b_i)}$. Since θ is ρ-vicinal, there exists a positive integer n and $r_1, \dots, r_n \in \rho$ such that the sets $C_\ell := r_\ell^{-1}(b_i)$ satisfy

$$b_i \in C := C_1 \cap \dots \cap C_n \subseteq B_i.$$

Set $C^*_\ell := r''^{-1}_\ell(b_i)$ and $C^* := C^*_1 \cap \dots \cap C^*_n$. In view of $\alpha^{(j)} = \rho^{(b_i)}$, we have $j \in C^* \supseteq C$ and so it remains to prove that the row-block C^* of μ is contained in B'_i. Let $c^* \in C^*$ be arbitrary. By Claim 3 we have that $\alpha^{(c^*)} = \rho^{(d)}$ for some $d \in L$. For each $\ell = 1, \dots, q$ clearly $r_\ell(b_i) = r''_\ell(c^*) = r_\ell(d)$ proving $d \in C_\ell$. Thus $d \in C \subseteq B_i$. As $\{0\} \times B_i \subseteq B'_i$ we obtain $c^* \in B'_i$ and $C^* \subseteq B'_i$. This proves that θ' is μ-vicinal and the claim.

For $i = 1, \dots, q$ define $b_i \in L$ from $\{b_i\} = t(B_i)$. Denote by μ the map sending every B'_i onto the singleton $\{b_i\}$ for all $i = 1, \dots, q$.

CLAIM 5. *$u \circ \sigma_\mu = t \circ \sigma_\rho$ for every h-ary relation σ on $\boldsymbol{k}$.*

PROOF. (1) To prove $\subseteq$, consider $(j_1, \dots, j_h) \in u \circ \sigma_\mu$. Then $(i_1, \dots, i_h) \in \sigma_\mu$ for some $i_1 \in u^{-1}(j_1), \dots, i_h \in u^{-1}(j_h)$. From $\alpha \subseteq \mu$ and the definition of σ_μ (see 1.3), we get $\big(r''(i_1), \dots, r''(i_h)\big) \in \sigma$ for all $r \in \rho$. By (2.2) and $\operatorname{im} u \subseteq L$, for every $m = 1, \dots, h$ we have that $\alpha^{(i_m)} = \rho^{((0,q_m))}$ for some $q_m \in L$ satisfying $\big(i_m, (0, q_m)\big) \in \theta'$. Consequently,

$$\big(r(q_i), \dots, r(q_h)\big) = \big(r''(i_1), \dots, r''(i_h)\big) \in \sigma$$

for every $r \in \rho$ proving that $(q_1, \dots, q_m) \in \sigma_\rho$. Finally $t(q_m) = \mu\big((0, q_m)\big) = \mu(i_m) = j_m$ due to $\big(i_m, (0, q_m)\big) \in \theta'$ and the definition of u and i_m. Together $(j_1, \dots, j_h) = \big(t(q_1), \dots, t(q_h)\big) \in t \circ \sigma_\rho$.

(2) To prove $\supseteq$ consider an arbitrary $(j_1, \dots, j_h) \in t \circ \sigma_\rho$. Then $(i_1, \dots, i_h) \in \sigma_\rho$ for some $i_m \in t^{(-1)}(j_m)$. By the definition of σ_ρ clearly

$$\big(r(i_1), \dots, r(i_h)\big) \in \sigma \tag{2.3}$$

holds for all $r \in \rho$. Recall that $\nu^- = \rho$ by Claim 2. From this and (2.3) we obtain that $\big(a\big((0, i_1)\big), \dots, a\big((0, i_h)\big)\big) \in \sigma$ holds for all $a \in \mu$. This proves $\big((0, i_1), \dots, (0, i_h)\big) \in \sigma_\mu$. Since $u\big((0, b)\big) = t(\ell)$ holds for all $\ell \in L$, we obtain the required

$$(j_1, \dots, j_h) = \big(t(i_1), \dots, t(i_h)\big) = \big(u\big((0, i_1)\big), \dots, u\big((0, i_h)\big)\big) \in u \circ \sigma_\mu.$$

This proves the claim.

Recall that $T = \operatorname{im} t = \operatorname{im} u = \{b_1, \dots, b_q\}$. Define a selfmap t' of T by setting $t'(j) := \big(0, u(j)\big)$ for all $j \in J$. Clearly $T' := \operatorname{im} t' = \{0\} \times T$. Define $v \colon T \to T'$ by setting $v(t) := (0, t)$ for all $t \in T$. For $\beta \subseteq \boldsymbol{k}^T$ set $\beta \circ v := \{b \circ v \colon b \in \beta\}$. The relation $\hat{\rho}_t$ was defined in 1.5. We need:

CLAIM 6. $\hat{\rho}_t - \hat{\mu}_{t'} \circ v$.

PROOF. First we show that $\ker r'' \supseteq \theta'$ whenever $r \in \rho$ and $\ker r \supseteq \theta$. Indeed, let $(c,d) \in \theta'$. By (2.2) clearly $\alpha^{(c)} = \rho^{(c')}$ and $\alpha^{(d)} = \rho^{(d')}$ for some $1 \le i \le q$ and $c', d' \in B_i$. From this and $\ker r \supseteq \theta$ we obtain the required

$$r''(c) = r(c') = r(d') = r''(d).$$

To prove $\subseteq$ let $a \in \hat{\rho}_t$. Then $a = r \restriction T$ for some $r \in \rho$ with $\ker r \supseteq \theta$. Set $b := r'' \restriction T'$. Here $r'' \in \mu$ and $\ker r'' \supseteq \theta'$ by the above result and so $b \in \hat{\mu}_{t'}$. Clearly $a = b \circ v$ proving $a \in \hat{\mu}_{t'} \circ v$ and $\hat{\rho}_t \subseteq \hat{\mu}_{t'} \circ v$.

For $\supseteq$ let $b \in \mu$ satisfy $\ker b \supseteq \theta'$. Define $r\colon L \to \boldsymbol{k}$ by setting $r(\ell) := b\big((0,\ell)\big)$ for all $\ell \in L$. From Claim 2 we have $r \in \rho$ and $\ker r \supseteq \theta$ follows from $\ker b \supseteq \theta'$. Thus $a := r \restriction T \in \hat{\rho}_t$ and clearly $a = b \circ v$. This proves the claim.

Recall that by its construction μ is a $\underline{P}$-saturated J-relation on $\boldsymbol{k}$ where $|J| \geq |I| = \kappa$. By Claim 4 the equivalence θ' is μ-vicinal. By our hypothesis

$$\text{(2.4)} \qquad \operatorname{Hom}(t' \circ \underset{\sim}{P}_{\mu}, \underset{\sim}{P}) \subseteq \hat{\rho}_{t'}.$$

Finally by Claim 5, (2.4) and Claim 6

$$\operatorname{Hom}(t \circ \underset{\sim}{P}_{\rho}, \underset{\sim}{P}) = \operatorname{Hom}(u \circ \underset{\sim}{P}_{\mu}, \underset{\sim}{P}) = \operatorname{Hom}(t' \circ \underset{\sim}{P}_{\mu}, \underset{\sim}{P}) \circ v \subseteq \hat{\mu}_{t'} \circ v = \hat{\rho}_t.$$

□

We consider a special type of strict protoduality. We start with the following observation.

LEMMA 2.2. *Let $(\rho, \theta, t) \in \widehat{\underline{P}}$, let $\lambda := \hat{\rho}_t$ and let σ be an h-ary relation on $\boldsymbol{k}$. Then*

$$t \circ \sigma_\rho \subseteq \sigma_\lambda.$$

PROOF. Set $T := \operatorname{im} t$. Let $(j_1, \dots, j_h) \in t \circ \sigma_\rho$. Then $(i_1, \dots, i_h) \in \sigma_\rho$ for some $i_\ell \in t^{(-1)}(j_\ell), \ell = 1, \dots, h$. Let $s \in \lambda$ be arbitrary. By the definition $s = r \restriction T$ for some $r \in \rho$ with $\ker r \supseteq \theta$. Clearly $(i_1, j_1), \dots, (i_h, j_h) \in \theta$ (because t is θ-canonical). Taking into account $\ker r \supseteq \theta$, $(i_1, \dots, i_h) \in \sigma_\rho$ and $r \in \rho$, we obtain

$$\big(s(j_1), \dots, s(j_h)\big) = \big(r(j_1), \dots, r(j_h)\big) = \big(r(i_1), \dots, r(i_h)\big) \in \sigma$$

proving $(j_1, \dots, j_h) \in \sigma_\lambda$. □

2.1. A strict protoduality $\underline{P}, \underset{\sim}{P}$ is *expedient* if $\sigma_\lambda \subseteq t \circ \sigma_\rho$ for every $(\rho, \theta, t) \in \widehat{\underline{P}}, \lambda := \hat{\rho}_t$ and each $\sigma \in \underset{\sim}{P}$. A relation λ is *$\underline{P}$-derived* if $\lambda = \hat{\rho}_t$ for some $(\rho, \theta, t) \in \widehat{\underline{P}}$. Notice that $\lambda = \hat{\rho}_t \subseteq \boldsymbol{k}^T$ (with $T := \operatorname{im} t$) whereby $|T|$ is the number h of blocks of θ. Since $h \leq k$, we can consider λ as an h-ary relation on $\boldsymbol{k}$. List the h-tuples from λ as the rows of an $n \times h$ matrix L (in any order and without repetition; i.e. $n = |\lambda|$) and denote by $\lambda^\dagger$ the set of the column vectors of L. A map $f\colon \lambda^\dagger \to \boldsymbol{k}$ is an n-ary *partial operation* on $\boldsymbol{k}$ (with *domain* $\lambda^\dagger$). As usual, f *preserves* an m-ary relation σ on $\boldsymbol{k}$ if for every $m \times n$ matrix M whose row-vectors $M_{1*}, \dots, M_{m*}$ belong to $\lambda^\dagger$ and whose column-vectors $M_{*1}, \dots, M_{*n}$ belong to σ we have $\big(f(M_{1*}), \dots, f(M_{m*})\big) \in \sigma$. Next f is a *partial projection* if there is $1 \leq i \leq n$ such that $f(a_1, \dots, a_n) = a_i$ for all $(a_1, \dots, a_n) \in \lambda^\dagger$.

We say that $\underline{P}, \underset{\sim}{P}$ satisfies the condition (C) if for every derived relation λ, each $\underset{\sim}{P}$-preserving partial operation on $\boldsymbol{k}$ with domain $\lambda^\dagger$ is a partial projection. (For a given λ with $n := |\lambda|$ there are up to $n!$ different $\lambda^\dagger s$ obtained one from another by a suitable coordinate exchange but the fact that we have fixed one of them has no effect on the condition (C).) We have:

PROPOSITION 2.3. *Let $\underline{P}, \underset{\sim}{P}$ be expedient. Then $\underline{P}, \underset{\sim}{P}$ yields a duality if and only if it satisfies the condition (C).*

PROOF. Let $(\rho, \theta, t) \in \widehat{\underline{P}}$ and $\lambda := \hat{\rho}_t$. By Lemma 2.2 and the definition of expediency we have $t \circ \sigma_\rho = \sigma_\lambda$ for all $\sigma \in \underset{\sim}{P}$ and so

$$H := \operatorname{Hom}(t \circ \underset{\sim}{P}_\rho, \underset{\sim}{P}) = \operatorname{Hom}(\underset{\sim}{P}_\lambda, \underset{\sim}{P}). \tag{2.5}$$

As mentioned in 2.1 we consider λ to be an h-ary relation on $\boldsymbol{k}$, hence ρ_λ is a relational system on $\{1, \dots, h\}$ and H is a set of maps from $\{1, \dots, h\}$ into $\boldsymbol{k}$, i.e. a subset of $\boldsymbol{k}^h$.

Let $L = [\ell_{ij}]$ be an $n \times h$ matrix whose rows are the elements of λ and let $\lambda^\dagger$ denote the set of all column vectors of L (cf. 2.1). Denote by Q the set of all n-ary partial operations on $\boldsymbol{k}$ with domain $\lambda^\dagger$. Further let $f \mapsto f^*$ denote the bijection from $\boldsymbol{k}^h$ onto Q defined by setting $f^*(L_{*\ell}) := f(\ell)$ for all $\ell = 1, \dots, h$. We need:

CLAIM. $f \in H \iff f^*$ *preserves* $\underset{\sim}{P}$.

PROOF. Let $f \in H$. To show that f^* preserves $\underset{\sim}{P}$ let $\sigma \in \underset{\sim}{P}$ be m-ary and M an $m \times n$ matrix whose rows belong to $\lambda^\dagger$ and columns to σ. Then $M_{1*} = L_{*i_1}, \dots, M_{m*} = L_{*i_m}$ for some $1 \le i_1, \dots, i_m \le h$. By the definition of σ_λ we have $(i_1, \dots, i_m) \in \sigma_\lambda$ and from $f \in H$ we obtain the required

$$\big(f^*(L_{*i_1}), \dots, f^*(L_{*i_m})\big) = \big(f(i_1), \dots, f(i_m)\big) \in \sigma.$$

Thus f preserves $\underset{\sim}{P}$. For the converse it suffices to reverse the argument. This proves the claim.

($\Rightarrow$): Assume that $\underline{P}, \underset{\sim}{P}$ yields a duality. Let $f^* \in Q$ preserve $\underset{\sim}{P}$. By the claim $f := f^{*-1} \in H$. From (2.5) and Lemma 1.7

$$H = \operatorname{Hom}(t \circ \underset{\sim}{P}_\rho, \underset{\sim}{P}) \subseteq \hat{\rho}_t = \lambda;$$

consequently $f \in \lambda$ and so $\big(f(1), \dots, f(h)\big) = L_{i*}$ for some $1 \le i \le n$. This means that $f^*(L_{*j}) = f(j) = \ell_{ij}$ for all $i = 1, \dots, h$ proving that $f*$ is the i-th partial projection on the domain $\lambda^\dagger$.

($\Leftarrow$): Let $\underline{P}, \underset{\sim}{P}$ satisfy (C). By (2.5) and Lemma 1.7 we must show that $H \subseteq \lambda$. Let $f \in H$. By the claim, f^* preserves $\underset{\sim}{P}$ and so by (C) the partial operation f^* is a partial projection. Consequently there exists $1 \le i \le n$ such that $f(j) = f^*(L_{*j}) = \ell_{ij}$ for all $j = 1, \dots, h$. Thus $f \in \lambda$.

□

2.2. For an n-ary operation f on $\boldsymbol{k}$ the $(n+1)$-ary relation

$$f^\circ := \big\{\big(x_1, \dots, x_n, f(x_1, \dots, x_n)\big) \colon x_1, \dots, x_n \in \boldsymbol{k}\big\}$$

is the *graph* (or diagram) of f. For $F \subseteq O$ set $F^\circ := \{f^\circ \colon f \in F\}$. The clone $C := \operatorname{Pol} F^\circ$, the *centralizer* of F, consists of all operations on $\boldsymbol{k}$ commuting (or permuting) with every $f \in F$ (cf. [**C65**] pp. 127–132 or [**MMT87**] p. 163). All clones of this form are known for $k = 2$ ([**P41**], [**Ha74**]) and $k = 3$ [**D77, D81**] while for $k > 3$ it is only known that there are at most countably many such clones [**BW87**].

A triple $(\rho, \theta, t) \in \widehat{\underline{P}}$ is $\underline{P}$-*taut* if

$$\theta = \cap\{\ker r \colon r \in \rho, \ker r \supseteq \theta\}.$$

We need:

LEMMA 2.4. *Let $\underline{P}, \underset{\sim}{P}$ be a strict protoduality. If $f^\circ = \sigma \in \underset{\sim}{P}$ for some $f \in O$ and a triple $(\rho, \theta, t) \in \widehat{\underline{P}}$ is $\underline{P}$-taut, then $\lambda := \hat{\rho}_t$ satisfies $\sigma_\lambda \subseteq t \circ \sigma$.*

PROOF. Let f be n-ary. Let $(i_1, \dots, i_{n+1}) \in \sigma_\lambda$ and let $\ell \in \lambda$. By the definition of σ_λ clearly $\big(\ell(i_1), \dots, \ell(i_{n+1})\big) \in \sigma = f^\circ$ and so

$$\ell(i_{n+1}) = f\big(\ell(i_1), \dots, \ell(i_n)\big). \tag{2.6}$$

According to the definition of $\lambda = \hat{\rho}_t$ the map $\ell \colon T \to \boldsymbol{k}$ extends to $\ell^* \colon I \to \boldsymbol{k}$ such that

$$\ell^* \in \rho, \quad \ker \ell^* \supseteq \theta, \ell^* \restriction T = \ell. \tag{2.7}$$

□

We need:

CLAIM 1. *Set $J := I \cup \{j\}$ where $j \notin I$ and extend every $r \in \rho$ to $r' \colon J \to \boldsymbol{k}$ by setting $r'(j) := f\big(r(i_1), \dots, r(i_n)\big)$. Then $\rho' := \{r' \colon r \in \rho\}$ is a subuniverse of $\mathfrak{P}^J$.*

PROOF. Let $g \in \underline{P}$ be p-ary and let $r_1, \dots, r_p \in \rho$. Set $r := g[r_1, \dots, r_p]$ (cf. 1.4). We must show that $r' = g[r'_1, \dots, r'_p]$. For $u = 1, \dots, p$ and $v = 1, \dots, n$ put $a_{uv} := r_u(i_v)$. In this notation

$$r'_u(j) = f\big(r_u(i_1), \dots, r_u(i_n)\big) = f(a_{u1}, \dots, a_{un}), \tag{2.8}$$

$$h := g\big(r'_1(j), \dots, r'_p(j)\big) = g\big(f(a_{11}, \dots, a_{1n}), \dots, f(a_{p1}, \dots, a_{pn})\big). \tag{2.9}$$

In view of $g \in \underline{P}$ and $f^\circ \in \underset{\sim}{P}$, the operations f and g commute, and from (2.9) and (2.8)

$$\begin{aligned} h &= f\big(g(a_{11}, \dots, a_{p1}), \dots, g(a_{1n}, \dots, a_{pn})\big) \\ &= f\big(g\big(r_1(i_1), \dots, r_p(i_1)\big), \dots, g\big(r_1(i_n), \dots, r_p(i_n)\big)\big) \\ &= f\big(r'_1(i_1), \dots, r'_p(i_n)\big) = r'(j). \end{aligned}$$

This proves $r' = g(r'_1, \dots, r'_p)$ and the claim

CLAIM 2. *$f[\rho^{(i_1)}, \dots, \rho^{(i_n)}] = \rho^{(i)}$ for some $i \in I$.*

PROOF. Follows from Claim 1 and the fact that ρ is $\underline{P}$-saturated.

CLAIM 3. *$t(i) = i_{n+1}$.*

PROOF. By (2.6), (2.7) and Claim 2 for every $r \in \rho$ with $\ker r \supseteq \theta$ we obtain

$$r(i_{n+1}) = f\big(r(i_1), \dots, r(i_n)\big) = r(i).$$

Since (ρ, θ, t) is $\underline{P}$-taut, this implies $(i_{n+1}, i) \in \theta$ leading to $t(i_{n+1}) = t(i)$. As $i_{n+1} \in T = \operatorname{im} t$, we have $t(i) = t(i_{n+1}) = i_{n+1}$. This proves the claim.

From Claim 2 we see that $(i_1, \dots, i_n, i) \in \sigma_\rho$. Finally, from $t(i_j) = i_j$ for $j = 1, \dots, n$ and Claim 3 we obtain

$$(i_1, \dots, i_{n+1}) = \big(t(i_1), \dots, t(i_n), t(i)\big) \in t \circ \sigma_\rho. \quad \square$$

Now Lemma 2.4, Subsection 2.1 and Proposition 2.3 yield:

PROPOSITION 2.5. *Let $\underline{P}, \underset{\sim}{P}$ be a strict protoduality such that*

(a) *$\underline{P} = F^\circ$ for some $F \subseteq O$ and*

(b) *each $(\rho, \theta, t) \in \widehat{\underline{P}}$ is $\underline{P}$-taut. Then*

  (i) *$\underline{P}, \underset{\sim}{P}$ is expedient and*

  (ii) *$\underline{P}, \underset{\sim}{P}$ yields a duality if and only if $\underline{P}, \underset{\sim}{P}$ satisfies (C).*

3. A Chain of Nondualizable Clones

In this section we show that the clones $U_2, \ldots, U_{k-1}$ defined below are not dualizable. This was obtained independently and via a completely different technique in [**He93**].

3.1. For $0 < h \leq k$ denote by U_h the set of all operations f on $\boldsymbol{k}$ that are either

(i) essentially unary (i.e. $f(x_1, \ldots, x_n) \approx g(x_i)$ for some $1 \leq i \leq n$ and $g \in O^{(1)}$) or

(ii) satisfy $|\operatorname{im} f| \leq h$.

It is easy to see that U_h is a clone and that $U_1 \subset U_2 \subset \cdots \subset U_k = O$. It is known [**B66, M67**] that $U_h = \operatorname{Clo}(O^{(1)} \cup \{f\})$ where $k > 1$ and f is an arbitrary essential operation with $|\operatorname{im} f| = h$. The clone U_1 is the clone of all essentially unary operations on $\boldsymbol{k}$. The clone U_{k-1}, the *Słupecki clone*, is the clone of all essentially unary or non-surjective operations on $\boldsymbol{k}$. It is the unique maximal clone (i.e. a dual atom or co-atom in the lattice of all clones on $\boldsymbol{k}$ ordered by $\subseteq$) containing $O^{(1)}$.

3.2. We describe all I-relations preserved by U_h. We denote by $E(I)$ the set of all equivalences on I. For a positive integer n, denote by $E^n(I)$ the set of all equivalences on I with at most n blocks. For $T \subseteq E^k(I)$ set $T^n := T \cap E^n(I)$ and denote by F_T the filter on $(E^k(I), \subseteq)$ generated by T (i.e. F_T consists of all $\tau \in E^k(I)$ satisfying $\tau \supseteq \tau_1 \cap \cdots \cap \tau_m$ for some positive integer m and $\tau_1, \ldots, \tau_m \in T$). For $\theta \in E(I)$ put

$$\Delta_\theta := \{f \in \boldsymbol{k}^I : \ker f \supseteq \theta\}$$

(i.e. Δ_θ consists of all $f \colon I \to \boldsymbol{k}$ constant on each block of θ). As usual, $T \subseteq E^k(I)$ is an *upset* if $\theta \in E^k(I)$ belongs to T whenever $\theta \supseteq \tau$ for some $\tau \in T$. First we characterize all I-relations preserved by U_h.

LEMMA 3.1. *Let $1 \leq h \leq k$. An I-relation λ on $\boldsymbol{k}$ is preserved by U_h if and only if*

$$\lambda = \bigcup_{\tau \in T} \Delta_\tau \tag{3.1}$$

for an upset T of $E^k(I)$ such that

$$T^h = (F_T)^h \tag{3.2}$$

PROOF. $(\Rightarrow)$: Let λ be preserved by U_h. Set $T := \{\ker \ell \colon \ell \in \lambda\}$ and consider $\tau \in T$. Since $O^{(1)} \subseteq U_h$, we have $\Delta_\tau \subseteq \lambda$ and so (3.1) holds. From $T \subseteq F_T$ we have $T^h \subseteq (F_T)^h$. To prove $\supseteq$ let $\tau \in (F_T)^h$ be arbitrary. Then $\tau \supseteq \tau_1 \cap \cdots \cap \tau_m$ for some $m > 0$ and $\tau_1, \ldots, \tau_m \in T$. Set $\tau_{m+1} := \tau$. As τ has at most h blocks, we have $\tau_1, \ldots, \tau_m \in E^k(I)$ and so we can choose $r_j \in \Delta_j$ so that $\ker r_j = \tau_j$ holds for all $j = 1, \ldots, m+1$. Define a partial m-ary operation g on $\boldsymbol{k}$ by the equations

$$g\big(r_1(i), \ldots, r_m(i)\big) = r_{m+1}(i)$$

for all $i \in I$. We claim that g is well defined. Indeed, suppose $r_\ell(i) = r_\ell(j)$ for some $i, j \in I$ and all $\ell = 1, \ldots, m$. Then, by the choice of $r_1, \ldots, r_m$, we have $(i, j) \in \tau_1 \cap \cdots \cap \tau_m \subseteq \tau_{m+1}$ and so $r_{m+1}(i) = r_{m+1}(j)$, proving the claim. As $\tau_{m+1} \in E^h(I)$, clearly $|\operatorname{im} g| \leq h$. Extend g to a (full) m-ary operation g^* on $\boldsymbol{k}$ so that $\operatorname{im} g^* = \operatorname{im} g$. Clearly $g^* \in U_h$. From this, $r_x \in \Delta_{\tau_x} \subseteq \lambda$ for all $x = 1, \ldots, m$ and $r_{m+1} = g^*[r_1, \ldots, r_m]$ we obtain $r_{m+1} \in \lambda$ and so $\Delta_{\tau_{m+1}} \subseteq \lambda$ proving $\tau_{m+1} \in T$ and (3.2).

($\Leftarrow$): Let λ satisfy (3.1) and (3.2). We show that U_h preserves λ. Let $f \in O^{(1)}$ and let $\ell \in \lambda$. Clearly $f \circ \ell \in \Delta_{\ker \ell} \subseteq \lambda$. Let $f \in U_h$ be an essential m-ary operation. Then $|\operatorname{im} f| \leq h$. Consider arbitrary $\ell_0, \ldots, \ell_{m-1} \in \lambda$. Then for every $0 \leq j < m$ we have $\ell_j \in \Delta_{\tau_j}$ for some $\tau_j \in T$. Put $\ell := f[\ell_0, \ldots, \ell_{m-1}]$. It is immediate that $|\operatorname{im} \ell| \leq |\operatorname{im} f| \leq h$ and $\tau := \ker \ell \supseteq \tau_0 \cap \cdots \cap \tau_{m-1}$. Consequently $\tau \in (F_T)^h$. From (3.2) we have $\tau \in T$ proving $\ell \in \Delta_\tau \subseteq \lambda$.

□

3.3. In the remainder of this section $1 < h \leq k$ and $\underline{P}$ satisfies $\operatorname{Clo} \underline{P} = U_h$. We first determine the saturated relations. As usual, a family B of sets has the *finite intersection property* ((FIP) for short) if the intersection of every finite nonempty subfamily of B is nonvoid. A subset T of $E(I)$ is *compact* if $\bigcap_{\tau \in T} B_\tau$ is nonempty whenever

(i) B_τ is a block of τ for every $\tau \in T$ and
(ii) $\{B_\tau \colon \tau \in T\}$ has the (FIP) (this is related to the usual compactness for any topology in which all the blocks of the equivalences from T are all closed).

We prove:

LEMMA 3.2. *An I-relation ρ is U_h-saturated if and only if*

$$\rho = \bigcup_{\tau \in T} \Delta_\tau$$

for a compact upset T of $E^k(I)$ satisfying (3.2).

PROOF. **($\Leftarrow$):** Let T be compact. Suppose to the contrary that ρ is not saturated. Denote by ρ' a saturated J-relation on $\boldsymbol{k}$ extending ρ and by T' the corresponding upset of $E^k(J)$. It is easy to see that $T' = \{\tau' \colon \tau \in T\}$ where each τ' extends τ (i.e. $\tau' \restriction I = \tau$). Fix $j \in J \backslash I$ and for each $\tau \in T$ denote by B'_τ the block of τ' containing j. Further set $B_\tau := B'_\tau \cap I$ for all $\tau \in T$ and $B := \{B_\tau \colon \tau \in T\}$. We need:

CLAIM 1. *The family B has the* (FIP).

PROOF. Suppose to the contrary that $B_S := \bigcap_{\sigma \in S} B_\sigma$ is void for some finite nonempty subset S of T. Set $B'_S := \bigcap_{\sigma \in S} B'_\sigma$ and denote by π the 2-block equivalence on J with blocks B'_S and $J \backslash B'_S$. Since $\pi \supseteq \cap\{\tau \colon \tau \in S\}$, from (3.2) we obtain $\pi \in (F_{T'})^2 = T'^2$. Define $f, g \in \boldsymbol{k}^J$ by setting $f(x) = g(x) := 0$ for all $x \in J \backslash B'_S$ and $f(x) := 0$, $g(x) := 1$ for all $x \in B'_S$. Clearly $f, g \in \Delta_\pi \subseteq \rho'$ in contradiction to the fact that ρ' extends ρ. This proves the claim.

We assume that T is compact and so $B^* := \bigcap_{\tau \in T} B_\tau$ is nonempty. As ρ has distinct columns, clearly $B^* = \{i\}$. Consider $\tau \in T$. We have $i, j \in B'_\tau$, hence $(i, j) \in \tau'$ and so $f(i) = f(j)$ for every $f \in \Delta_{\tau'}$. Taking into account $\rho' = \bigcup_{\tau \in T} \Delta_{\tau'}$, we have $f(i) = f(j)$ for all $f \in \rho'$. However, this contradicts the fact that ρ' properly extends ρ and so ρ is U_h-saturated.

($\Rightarrow$): Let T be noncompact and satisfy (3.2). Then for every $\tau \in T$ there exists a block B_τ of τ such that $B := \{B_\tau \colon \tau \in T\}$ has the (FIP) but the intersection of B is void. Choose an element j outside I, set $J := I \cup \{j\}$ and extend each $\tau \in T$ to $\tau' \in E(J)$ by joining the element j to the block B_τ.

CLAIM 2. *$T' := \{\tau' : \tau \in T\}$ satisfies* (3.2).

PROOF. Let $\xi \supseteq \tau'_0 \cap \cdots \cap \tau'_{n-1}$ hold for some $n > 0$, some $\xi \in E^k(J)$ and $\tau_0, \ldots, \tau_{n-1} \in T$. Since T satisfies (3.2), the equivalence $\sigma := \xi \restriction I$ belongs to T. As B has the (FIP), we have

$$\varnothing \neq B_{\tau_0} \cap \cdots \cap B_{\tau_{n-1}} \subseteq B_\sigma.$$

Now σ' was obtained from σ by appending the element j to B_σ. From this it follows easily that $\xi = \sigma'$ and so we have the required $\xi \in T'$. This proves the claim.

By Claim 2 and Lemma 3.1 the J-relation $\rho' := \bigcup_{\tau \in T} \Delta_{\tau'}$ is preserved by U_h. Suppose $\rho'^{(j)} = \rho'^{(\ell)}$ for some $\ell \in J$. Then clearly $(j, \ell) \in \tau$ for all $\tau \in I$. As $\bigcap_{\tau \in T} B_\tau$ is empty, we have $j \neq \ell$. Thus ρ' properly extends ρ proving that ρ is not U_h-saturated.

Now we can prove:

PROPOSITION 3.3. *For every $1 < h < k$ the clone U_h is not dualizable.*

PROOF. Put $I := \mathbb{N} = \{0, 1, \ldots\}$. For every $i \in \boldsymbol{k}$ set $B_i := \{kn + i : n \in \mathbb{N}\}$ and set $B_k := B_0$. Denote by τ and τ' the equivalences on $\mathbb{N}$ with blocks $B_0 \cup B_1, B_2, \ldots, B_{k-1}$ and $B_0, B_1, \ldots, B_{k-3}, B_{k-2} \cup B_{k-1}$, respectively. Clearly $\beta := \tau \cap \tau'$ has blocks $B_0, \ldots, B_{k-1}$. For every $n \in \mathbb{N}$ denote by τ_n the equivalence on $\mathbb{N}$ with blocks

$$B_0, \ldots, B_{i-1}, \quad B_i \backslash \{k+n\}, \quad B_{i+1} \cup \{k+n\}, \quad B_{i+2}, \ldots, B_{k-1}$$

where $i \in \boldsymbol{k}$ satisfies $n \equiv i \pmod{k}$. For each $n \in \mathbb{N}$ set $m := k + n - 1$ and denote by σ_n the equivalence on $\mathbb{N}$ with blocks

$$\{k\}, \{k+1\}, \ldots, \{m-1\}, \quad B_0 \backslash (\mathbf{m} \backslash \boldsymbol{k}), \quad B_1 \backslash (\mathbf{m} \backslash \boldsymbol{k}), \ldots, B_{k-1} \backslash (\mathbf{m} \backslash \boldsymbol{k}).$$

Finally let T consist of τ, τ', τ_n for all $n \in \mathbb{N}$ and all $\xi \in E^h(I)$ containing some σ_n.

We need the following claims:

CLAIM 1. *T satisfies* (3.2).

PROOF. Let $\xi \in (F_T)^h$. Then ξ has at most h blocks and $\xi \supseteq \cap L$ for a finite nonempty subset L of T. Since $\tau_i \supseteq \sigma_i \subseteq \beta$ for all $i \in N$, clearly $\xi \supseteq \sigma_n$ for some n proving $\xi \in T$ and $(F_T)^h \subseteq T^h$. The inclusion $\supseteq$ follows from $F_T \supseteq T$. This proves the claim.

CLAIM 2. *T is compact.*

PROOF. Suppose that a family $\{C_\tau : \tau \in T\}$ of blocks of equivalences from T has the (FIP). For every $S \subseteq T$ set

$$I(S) := \bigcap_{\sigma \in S} C_\sigma$$

and denote by $\mathcal{F}$ the family of all finite nonempty subsets of T. From the (FIP) we have $I(S) \neq \varnothing$ for every $S \in \mathcal{F}$. If, moreover, $I(S)$ is finite for some $S \in \mathcal{F}$, then again by the (FIP) we have the required $I(S) \neq \varnothing$. Thus assume $I(S)$ to be infinite for every $S \in \mathcal{F}$. Consider arbitrary $0 \leq m_1 < m_2$ and set $S_{m_1 m_2} := \{\tau_{m_1}, \tau_{m_2}\}$. The set $\{C_{\tau_{m_1}} \cap C_{\tau_{m_2}}\}$ being infinite, from the definition of τ_p we see that there exists $i \in \boldsymbol{k}$ such that for $j = 1, 2$ either $C_{m_j} = B_i$ or $C_{m_j} = B_i \backslash \{a\}$ or $C_{m_j} = B_i \cup \{a\}$ for a suitable $a \in \mathbb{N}$. From this it follows that there exists a universal $i \in \boldsymbol{k}$ such that

$$C_{\tau_p} = \begin{cases} B_i \backslash \{k+p\} & \text{if } p \equiv i \pmod{k}, \\ B_i \cup \{k+p\} & \text{if } p \equiv i-1 \pmod{k}, \\ B_i & \text{otherwise.} \end{cases}$$

Choose $j \in \boldsymbol{k}$ so that $i - 1 \not\equiv j \not\equiv i \pmod{k}$. Let $\xi \in E^h(I)$ satisfy $\xi \supseteq \sigma_n$ for some $n \in \mathbb{N}$. Now $C_\xi \mid \xi \cap C_{\tau_j} = I(\{\xi, \tau_j\})$ is infinite and so by the definition of σ_n we obtain $C_\xi \supseteq B_i \backslash \{k, \ldots, k+n\}$. A similar argument shows $C_t \supseteq B_i$ and $C_{\tau'} \supseteq B_i$. Combining these results with the definition of the equivalences from T, we see that $i \in C_\varepsilon$ for every $\varepsilon \in T$. Thus $i \in \bigcap_{\varepsilon \in T} C_\varepsilon$ and hence T is compact proving the claim.

From Claims 1 and 2 and Lemma 3.2 we obtain that

$$\rho := \bigcup_{\varepsilon \in T} \Delta_\varepsilon$$

is U_h-saturated. In view of Lemma 1.9, to prove the proposition it suffices to show:

CLAIM 3. $\operatorname{Vic} \rho \cap \operatorname{Loc} \rho \not\subseteq \rho$.

PROOF. Define $f\colon \mathbb{N} \to k$ by setting $f(x) := i$ for all $i \in \boldsymbol{k}$ and $x \in B_i$. Clearly $\ker f = \beta = \tau \cap \tau'$. Since $\tau, \tau' \in T$, evidently $f \in \operatorname{Vic} \rho$. We show that $f \in \operatorname{Loc} \rho$. Let S be a finite subset of $\mathbb{N}$. Denote by s the greatest element of S. If $s < k$ then $f \restriction S = g \restriction S$ where $\ker g = \tau_0$. Thus let $s \geq k$. Define $g\colon \mathbb{N} \to \boldsymbol{k}$ by setting $g(s+1) := \big(f(s+2)\big)$ and $g(s) := f(s)$ otherwise. It is immediate that $\ker g = \tau_{s+1-k} \in T$ and $f \restriction S = g \restriction S$. This proves $f \in \operatorname{Loc} \rho$.

We show that $f \notin \rho$. Suppose to the contrary that $f \in \rho$. Then $f \in \Delta_\varepsilon$ for some $\varepsilon \in T$. Notice that τ, τ' and all $\sigma \in E^h(\mathbb{N})$ have at most $k-1$ blocks. Since $\ker f$ has k blocks, we have $\varepsilon = \tau_n$ for some $n \in \mathbb{N}$. However, all τ_n are distinct from $\beta = \ker f$. This proves $f \notin \rho$ and the claim. □

REMARK 3.4. (1) The clone U_{k-1} is the Słupecki clone (of all essentially at most unary or nonsurjective operations). From Proposition 3.3 we obtain that U_{k-1} is not dualizable [**He93**]. It is known ([**S39, I58**], cf. [**R70**]) that U_{k-1} is a maximal clone (i.e. a coatom in the lattice of clones on $\boldsymbol{k}$) and this provides an example of a maximal nondualizable clone.

(2) We consider the case $h = k$.

Clearly U_k is the clone O of all operations on $\boldsymbol{k}$. By Lemma 3.1 an I-relation ρ on k is preserved by O if and only if

$$\rho = \bigcup_{\tau \in T} \Delta_\tau, \quad T \subseteq E(I), \quad T^k = (F_T)^k. \tag{3.3}$$

Observe that for every $\gamma \in E(I)$

$$\Delta_\gamma = \cup\{\Delta_\varepsilon \colon \varepsilon \in E^k(I) \colon \varepsilon \supseteq \gamma\}$$

and so (3.3) may be replaced by

$$\rho = \bigcup_{\gamma \in G} \Delta_\gamma \tag{3.4}$$

for a filter (i.e. an $\cap$-closet upset) G of the lattice $\big(E(I), \subseteq\big)$ (cf. [**R70**] for I finite and [**R74**] for I infinite. By Lemma 3.2 the relation ρ from (3.3) is σ-saturated if and only if T is compact. Set $\underset{\sim}{P} = \{\varnothing\}$ where $\varnothing$ is the empty (unary) relation. It is well known that $\underset{\sim}{P}$, $\varnothing$ is a strict protoduality. Now, the relation $\varnothing_\rho$ is the empty I-relation and so $\operatorname{Hom}(\underset{\sim}{P}_\rho, \underset{\sim}{P}) = \boldsymbol{k}^I$. The condition (2) from Lemma 1.5 reduces to $\operatorname{Vic}\rho \subseteq \rho$ for every ρ of the form (3.3) with a compact T. Suppose to the contrary that $\operatorname{Vic}\rho \not\subseteq \rho$. Then by (3.3) there exists a ρ-vicinal equivalence θ on I such that $\theta \notin T$. A block $\tau \in T$ is *good* if it is contained in a block of θ and *bad* otherwise. A cover of I is *good* if it consists of good blocks. The fact that θ is ρ-vicinal and (3.3) imply that I has a good cover.

We need:

CLAIM 1. *Every good cover of I is infinite.*

PROOF. Suppose to the contrary that I has a finite good cover G. For each $B \in G$ denote by τ_B any $\tau \in T$ such that B is a block of τ. Set $T' := \{\tau_B \colon B \in G\}$ and observe that $\theta \supseteq \cap T'$. From (3.3) and $\theta \in E^k(I)$ we obtain $\theta \in T$, a contradiction. This proves the claim.

From the claim it follows that every $\tau \in T$ has at least one bad block B_τ. We show that $\{B_\tau \colon \tau \in T\}$ has the (FIP). Indeed, suppose that $B_{\tau_1} \cap \cdots \cap B_{\tau_n}$ is empty for a finite subset $\{\tau_1, \ldots, \tau_n\}$ of T. Then the good blocks of $\tau_1, \ldots, \tau_n$ provide a finite cover of I. Since every $\tau \in T$ has at most k blocks, this provides a finite good cover of I and so contradicts the claim. Now T is compact and therefore there exists $i \in \bigcap_{\tau \in T} B_\tau$. However, i is contained in no good block in contradiction to the fact that I has a good cover. Thus $\operatorname{Vic}\rho \subseteq \rho$. $\square$

As usual, $\mathfrak{P} = \langle \boldsymbol{k}; \underset{\sim}{P}\rangle$ is *primal* if O is the least clone containing $\underset{\sim}{P}$, i.e. if O is the set of term operations of $\mathfrak{P}$. From the above we obtain:

PROPOSITION 3.5. [**Hu71**] *If* $\mathfrak{P} = \langle \boldsymbol{k}; \underset{\sim}{P}\rangle$ is primal, then $\underset{\sim}{P}, \varnothing$ *yields a duality.*

References

[B66] Burle, G. A., *Classes of k-valued logic which contain all functions of a single variable*, Diskret. Analiz **10** (1967), 3–7. (Russian)

[BW87] Burris, S. and Willard, R., *Finitely many primitive positive clones*, Proc. Amer. Math. Soc. **101** (1987), no. 3, 427–430.

[C65] Cohn, P. M., *Universal algebra*, 2nd ed., Math. Appl. **6**, Harper & Row 1965, D. Reidel Publ. Co., Dordrecht-Boston, Mass., 1981.

[D77] Danil'čenko, A. F., *Parametric expressibility of functions of three-valued logic*, Algebra i Logika **16** (1977), 397–416 (Russian); English transl., Algebra and Logic **16** (1977), 266–280.

[D81] ______, *On parametrical expressibility of the functions of k-valued logic*, Finite Algebra and Multiple-Valued logic (B. Csákány and I. Rosenberg, eds.), Coll. Math. Soc. János Bolyai, vol. 28, North-Holland, 1981, pp. 147–159.

[Dav93] Davey, B. A., *Duality theory on ten dollars a day*, Algebras and Orders (I. G. Rosenberg, G. Sabidussi, eds.,), NATO Adv. Sci. Inst. Ser. C Math. Phys. Sci. **389**, Kluwer Acad. Publ., Dordrecht, 1993, 71–111.

[DW83] Davey, B. A. and Werner, H., *Dualities and equivalences for varieties of algebras*, Contributions to Lattice Theory, Coll. Math. Soc. Bolyai **33**, 1983, 101–275.

[HR90] Haddad, L. and Rosenberg, I. G., *Finite clones containing all permutations* (short version), Proc. 20th Internat. Sympos. Multiple-Valued Logic (Charlotte N.C.), 1990, 248–253; Canad. J. Math. **46**, no. 5 (1994), 951–970.

[Ha74] Harnau, W., *Die Definition von Vertauschbarkeitsmengen in der k-wertigen Logik und das Maximalitätsproblem*, Z. Math. Logik Grundlag. Math. **20** (1974), no. 4, 339–352.

[He93] Heindorf, L., *Some examples of nondualizable algebras*, General Algebra and Applications (Potsdam 1992), Res. Exp. Math. **20**, Helderman, Berlin, 1993, 143–151.

[Hu71] Hu, T. K., *Stone duality for primal algebra theory*, Math. Z. **110** (1971), 180–198.

[I58] Iablonskiĭ, S. V., *Functional constructions in k-valued logics*, Trudy Mat. Inst. Steklov **51** (1958), 5–142. (Russian)

[MMT87] McKenzie, R. N., McNulty, G., and Taylor, W. F., *Algebras, lattices, varieties*, vol. I, The Wadsworth & Brooks/Cole Mathematics Series, Wadsworth & Brooks/Cole Advanced Books & Software, Monterey, Calif., 1987.

[M67] Mal'tsev, A. I., *A strengthening of the theorems of Słupecki and Iablonskiĭ*, Algebra i Logika Sem. **6** (1967), no. 3, 61–75. (Russian)

[P41] Post, E. L., *The two-valued iterative systems of mathematical logic*, Ann. of Math. Stud. **5**, Princeton Univ. Press, Princeton, N.J., 1941.

[R70] Rosenberg, I. G., *Uber die funktionale Vollständigkeit in der mehrwertigen Logiken*, Rozpravy Československ. Akademie Věd., Řada Mat. Přír. Věd. **80** (1970), no. 4, 3–93.

[R74] ———, *Universal algebras with all operations of bounded range*, Colloq. Math. **30** (1974), 177–185.

[S39] Słupecki, J., *A criterion of fullness of many-valued systems of propositional logic*, C. R. Soc. Sci. Varsovie **32** (1939), 102–109 (Polish); English transl., Studia Logica **30** (1972) 153–157.

Department of Mathematics, La Trobe University, Bundoora, 3083, Victoria, Australia.

E-mail address: B.Davey@latrobe.edu.au

Département de Mathématiques et statistique, Université de Montréal, C.P. 6128, Succ. "A", Montréal, Québec, H3C 3J7, Canada.

E-mail address: rosenb@ere.UMontreal.CA

Centre de Recherches Mathématiques
CRM Proceedings and Lecture Notes
Volume **11**, 1997

The Early Introduction of Dynamic Programming into Computational Biology

David Sankoff

This paper is dedicated to the memory of Mark Kac

1. Introduction

In 1994–1995, the DIMACS Center for Discrete Mathematics and Theoretical Computer Science (Rutgers-Princeton-AT&T-BellCore) sponsored a theme year (which has actually continued for two years) on computational biology. Among the numerous seminars and workshops was one organized by Alberto Apostolico and Raffaele Giancarlo [**1**] on the history and motivations for sequence comparison. In my participation in this event, I was led to consider some of the early interactions, at the Centre de recherches mathématiques (CRM) and elsewhere, in the field now known as computational biology. A short time earlier, I had read a paper by Walter Goad on the impact of Stanislaw Ulam in this field [**10**]. The penultimate sentence in this article was a quote from Ulam himself "I started all this", which puzzled me greatly. As far as I knew, Ulam had no impact in the early development of the field, and while he gave talks on it at least from 1971 [**33**], the distance he defined was already published and the problem he proposed had already been solved, for all intents and purposes, in the molecular biology literature [**15**] and elsewhere. I had also read a joint interview of Ulam and Mark Kac by Mitchell Feigenbaum [**7**], and this led me to reflect on this misperception on the part of Ulam (and of Goad), and to crystallize the realization that ironically, Kac, his colleague of many years had played a crucial, if indirect, role in the earliest development of the field, especially that associated with the CRM. Despite the fact that he had no personal research interest in the field, his encouragement of a number of junior mathematicians, and his role, intentional or not, in bringing researchers together, recur as important influences in several aspects of computational biology, and explain why I dedicate this article to his memory.

1991 *Mathematics Subject Classification.* Primary: 92D20; Secondary: 68Q25.
Research Supported by grants from the Natural Sciences and Engineering Research Council of Canada and the Canadian Genome Analysis and Technology program. Fellow of the Canadian Institute for Advanced Research.
This is the final form of the paper.

In presenting some of these thoughts to the DIMACS workshop, focusing on the early 1970s, my understanding of this period was broadened by comments from a number of participants, particularly Jerrold Griggs and Pavel Pevzner, and clarified by the presentation immediately following, by Michael Waterman, who concentrated on the era starting in the late 1970s.

In this article, I will address the question of the origin and propagation of certain ideas fundamental in computational biology. It is not my purpose to claim priority for any of my own work—my contributions have been largely incremental in a field endowed with many insightful and innovative mathematicians. Rather, I will draw on my recollections of the earliest phases of the field to describe how certain ideas found their ways into the vernacular of the computational biologist.

2. Dynamic Programming for Sequence Comparison

In the late summer of 1971, the sequence comparison problem was brought to my attention by Robert Cedergren, a biochemist friend in the Département de biochimie interested in studying the evolution of ribonucleic acid sequences. In discussions with him, we formulated this as the longest common subsequence problem, though at the time we called it maximum matching. I realized that the application of mathematical induction led to a simple recurrence relation for solving the problem. We consider two sequences of length m and n of terms from any alphabet. If the two sequences are $a(1), \ldots, a(m)$ and $b(1), \ldots, b(n)$, then using the notation a_i for the prefix sequence $a(1), \ldots, a(i)$ and $M(i,j)$ for the longest common subsequence of a_i and b_j,

$$M(i,j) = \text{Max}\{M(i-1,j), M(i,j-1), M(i-1,j-1) + \Delta\}$$

where

$$\Delta = 1 \quad \text{if } a(i) = b(j) \quad \text{and} \quad \Delta = 0 \quad \text{if } a(i) \neq b(j),$$

under the initial conditions $M(i,0) = M(0,j) = 0$. Then $M(m,n)$ is the length of the longest common subsequence, and all longest common subsequences may be found by a simple traceback routine on the matrix M: starting with matrix cell (m,n), we choose the predecessor cell (i,j) to be any one of the cells $(m-1,n)$, $(m,n-1)$, $(m-1,n-1)$ that satisfies

$$M(m,n) = M(m-1,n), \quad M(m,n) = M(m,n-1),$$
$$M(m,n) = M(m-1,n-1) + \Delta$$

respectively. We continue by finding a predecessor of (i,j), then its predecessor, and so on, until we reach some cell $(0,k)$ or $(k,0)$. Whenever during this procedure a predecessor to (r,s) is found that satisfies $M(r,s) = M(r-1,s-1)+1$, then $a(r) = b(s)$ forms part of the longest common subsequence.

I did not know at the time that this was discrete dynamic programming nor, I suppose, does it matter—a pertinent though preposterously unjust remark I once heard about Richard Bellman was that he had made a career out of rediscovering induction!

Enter the Polish connection. A couple of weeks at most after I found this algorithm, a colleague of mine, S. K. Zaremba, one of the original eight members of the CRM, hosted a colloquium on numerical computation of integrals in high-dimensional space, including Monte-Carlo methods—whose proceedings became the first mathematical volume arising from CRM activities [**39**]. Stanislaw

Ulam, prominent among the promoters of Monte-Carlo methods, was invited. His talk covered a wide range of subjects, among them the problem of how to calculate the minimum edit distance between two finite binary sequences, based on substitutions, insertions and deletions of single terms. He stressed in the talk and in the published version [**33**] that no algorithm was known for handling this problem. Needless to say, I was astonished since this sort of sequence comparison was uppermost in my mind, and I was in the middle of writing up my algorithm. Imagine being in an audience where Ulam formulates an important unsolved problem, which you have just cracked a few days earlier! After the talk I approached him and started to explain my result. He listened for about 30 seconds and then turned to speak to someone else, after saying my problem (maximum matching) and his (minimum edit distance) were different. I was astonished for the second time, partly because of his peremptory attitude, and partly because it was clear to me that the difference between the two problems lay only in their strategies for counting multi-term gaps (strings of terms in one sequence unmatched to any terms in the other, representing deletion or insertion of several terms in the edit calculation), but the essential nature of the two problems was the same, and that dynamic programming would work as well for one as for the other. I talked to Cedergren and he assured me that the edit distance would weight long gaps too heavily, and I promptly forgot about the incident. Recently, in the same collection as Goad's article, I read a controversial article by Gian-Carlo Rota [**18**], in which he characterizes Ulam in this period as having a very limited attention span—maybe this accounts for his dismissive reply.

I do not think Mark Kac was at the CRM during Ulam's visit, though he had spent at least six months there during 1969–1970, and had visited frequently thereafter, being a member of the advisory board. I had gotten to know him somewhat and when I wrote my article on maximum matching, I asked him to communicate it to the Proceedings of the National Academy of Sciences. He gave it to a junior colleague at Rockefeller, Peter H. Sellers, to referee (perhaps on the basis of a remote biological connection: Seller's interest in the combinatorics of organic compounds). Not only did Sellers review my paper positively, he became very interested in the problem. In the meantime Cedergren had come across the article by Needleman and Wunsch [**15**], published a year earlier, in which dynamic programming was first proposed for the comparison of protein sequences. Fortunately, I had made further progress on the problem, incorporating constraints on insertions and deletions, and refocused my PNAS article [**21**] on this. Also, the computing time of the Needleman and Wunsch algorithm was cubic function of sequence length, and my formulation of the unconstrained algorithm was quadratic—the number of steps required to compute the matrix M above is proportional to mn, or n^2 if $m = n$. Later I found out that at this time Donald Knuth knew the problem was quadratic, since he incorporated it in a technical report of open combinatorial problems that Vaclav Chvátal, D.A. Klarner and he put out in 1972 [**2**]. Pavel Pevzner suggests that Knuth knew this via his work on Young tableaux. And later I found out that speech recognition researchers had been using a quadratic dynamic programming approach for this general type of problem since the late 1960s [**34**], as documented in my book with Joseph Kruskal [**22**]. Walter Fitch once told me that when the Needleman and Wunsch article came out, he took the basic idea and programmed an algorithm so that it performed in quadratic time and did not realize that the original was cubic.

Shortly thereafter, I received a manuscript from Sellers in which he generalized the sequence comparison problem and my algorithm with constraints to the finding of optimal paths in partial orders. I made some suggestions and we worked on it together, publishing it in *Discrete Mathematics*, entitled in catchy dactylic heptameter: *Shortcuts, diversions and maximal chains in partially ordered sets* [**27**]. Some time later in early 1972 (Goad mistakenly cites 1974), Kac invited Ulam to talk at Rockefeller, where Sellers heard him pose the edit-distance version of the problem. Sellers devised a dynamic programming algorithm for it, based on maximum matching and reminiscent of my algorithm with insertion and deletion constraints, requiring cubic computing time, and sent me a copy of the relevant pages from his notebook. I pointed out how it could be done in quadratic time, as did Albert Nijenhuis, and the published version [**28**] incorporates the quadratic algorithm in a footnote, kindly giving us credit. In the above notation, using $D(i, j)$ for the minimum number of steps, insertions, deletions or substitutions, to convert a_i into b_j,

$$D(i,j) = \mathrm{Min}\{D(i-1,j)+1, D(i,j-1)+1, D(i-1,j-1)+\delta\}\cdot$$

where

$$\delta = 0 \quad \text{if } a(i) = b(j) \quad \text{and} \quad \delta = 1 \quad \text{if } a(i) \neq b(j),$$

under the initial conditions $D(i, 0) = D(0, i) = i$. Then $D(m, n)$ measures the edit distance between the two sequences, and all appropriate sets of edit steps may be found by the traceback routine in the matrix D.

The publication of this paper and similar work by Sellers [**29**] had an immediate impact, among mathematicians and biologists. They are frequently cited, though in recent years the tendency has been to realize that the important idea is the dynamic programming and not the edit distance, and appropriate credit has reverted to Needleman and Wunsch.

If the distance approach is generalized to allow a different weight $s > 0$ on substitution compared to insertion and deletion:

$$D(i,j) = \mathrm{Min}\{D(i-1,j)+1, D(i,j-1)+1, D(i-1,j-1)+s\delta\},$$

then the longest common subsequence problem and the shortest edit distance problem become essentially identical when $s \geq 2$.

A couple of years later, Wagner and Fischer [**35**] published this same algorithm in a computer science journal, oblivious of the previous computational biology work—to say nothing about the speech recognition work, and this had even greater impact. Not only was it frequently cited, but it interested a number of talented computer scientists in the problem area.

I have not even mentioned many of the other independent discoveries of the same algorithm. It did not require great mathematical power or imagination to find a good solution to the sequence comparison problem. At least a half-dozen people who chanced upon the problem, in one field or another, quickly came up with the same solution [**13**]. Ulam's characterization of it as an important unsolved problem was making a mountain out of a molehill.

Though the dynamic programming algorithm for the global alignment of two sequences was the inevitable result of a widespread need for this application—the spirit of the times, not all the early discoveries in computational biology necessarily would eventually have been made without the inspiration of their specific authors. For example, the optimal local alignment algorithm of Smith and Waterman [**31**], which is more important for applications than the global comparison method, is based on an extremely simple but non-obvious adjustment to the basic dynamic programming algorithm. Local alignment was an important problem that defied a number of attempts at solution. Sellers did find an algorithm [**5, 30**], but this was unwieldy. The Smith-Waterman algorithm was ingenious, exactly what was needed, and has had wide impact. We define

$$L(i,j) = \text{Max}\{0, L(i-1,j)-1, L(i,j-1)-1, L(i-1,j-1)+\sigma\}$$

where

$$\sigma \text{ is a constant} > 0 \quad \text{if } a(i) = b(j)$$

and

$$\sigma = -1 \quad \text{if } a(i) \neq b(j),$$

under the initial conditions $L(i,0) = L(0,i) = 0$. Then $\text{Max}_{i,j} L(i,j)$ measures the score of the optimal local alignment between the two sequences, and all appropriate sets of edit steps may be found by the traceback routine in the matrix L, starting at any cell containing this maximizing value and terminating when a cell containing zero is reached. If there are several regions of good local alignment between the two sequences, this algorithm finds all of them with a single pass of the dynamic programming followed by tracebacks from all cells containing locally maximal values of L.

3. Multiple Alignment and Phylogeny

Soon after our first applications of sequence comparison to pairs of small RNAs, Cedergren and I became interested in assessing the relative rates of the 12 possible substitution mutations among the four bases $\{A, C, G, U\}$, based on comparing small homologous RNA sequences from N species. The idea was to isolate each position in the ribonucleotide sequence, examine its evolutionary history, count the number of mutations of each kind, and then combine the data for all positions. Given the particular data available—one sequence each from a bacterium, a yeast, human, chicken and frog, there was no problem of determining the phylogenetic relationships among the species. In other words, the topology of the phylogeny was given. What was left was only the task of aligning corresponding positions in all the sequences and counting the mutations at all positions. After a little work, it was clear that dynamic programming would work here too, with one interesting difficulty. In edit distance terms:

$$D(\mathbf{i}) = \text{Min}_{\mathbf{e}\in\{0,1\}^N}\{D(\mathbf{i}-\mathbf{e}) + f(\mathbf{e} \bullet \mathbf{a}(\mathbf{i}))\}$$

where $\mathbf{i}$ is an N-vector representing a cell in the N-way matrix of sequence positions, generalizing (i,j) in the above algorithms, $\mathbf{e}$ is a vector of 0's and 1's (excluding $\mathbf{0}$), the "product" $\mathbf{e} \bullet \mathbf{a}(\mathbf{i})$ is a vector containing the i_j-th term of sequence $\mathbf{a_j}$ if $e_j = 1$, but containing Φ if $e_j = 0$. (The appropriate initial conditions are not hard to work out.)

How is f calculated? Labeling the N terminal nodes of the given phylogenetic tree with the values of $\mathbf{e} \bullet \mathbf{a}(\mathbf{i})$, the question is how to label the internal, or ancestral, nodes of the tree so as to minimize the number of edges in the tree with two differently labeled end-points. After consulting with combinatorialist colleagues at the CRM, including S. K. Zaremba, it appeared to me that this problem was non-trivial (Margaret Dayhoff [**4**], for example, was using an incorrect method at this time), but it too soon yielded to dynamic programming. Some internal node on the tree is (through biological considerations, or arbitrarily) identified as the root ρ and all tree edges are directed away from the root. Then ρ alone is ancestral to all other tree nodes. Consider any non-terminal node ν whose immediate descendants are $\nu_1, \ldots, \nu_r$. Let A be the (finite) alphabet from which the terms of all the $\mathbf{a_j}$ are drawn. The problem is solved by applying the following recurrence to any node ν whose descendants are either terminal nodes or have already had the recurrence applied to them (this is always possible in a tree structure): For all $X \in A \cup \Phi$:

$$F_\nu(X) = \mathrm{Min}_{\mathbf{Y} \in (A \cup \Phi)^r} \sum_{i=1}^{r} \left[F_{\nu_i}(Y_i) + \delta(X, Y_i)\right],$$

where

$$\delta(X,Y) = 0 \quad \text{if } X = Y, \quad \delta(X,Y) = 1 \quad \text{if } X \neq Y,$$

with the initial conditions that $F_{\nu_i}(Y_i) = 0$ if ν_i is a terminal node and Y_i is the original label assigned to that node and $F_{\nu_i}(Y_i) = \infty$ otherwise. Then

$$f(\mathbf{e} \bullet \mathbf{a}(\mathbf{i})) = \mathrm{Min}_{Y \in A \cup \Phi} F_\rho(Y).$$

The traceback part of the dynamic programming based on this algorithm reconstructs the inferred ancestral sequences at each of the nonterminal nodes. For fixed N this is still a polynomial algorithm, though the degree of the polynomial is N. Nevertheless, it can be implemented for small N and can serve as the basis for heuristic or locally optimal algorithms in the case of large N.

I rushed off a manuscript containing this algorithm to Mark Kac, again requesting him to communicate it to PNAS, but I did not receive a reply for many months. Fortunately, at the same time, I published a short paper [**24**] in Nature with Cedergren and my student Cristiane Morel, where we applied the method to RNA sequences in order to calculate the mutation frequencies that were our original interest. In retrospect, of course, it turns out that the significance of the paper was not the mutation frequencies, nor even the reconstruction of the ancestral sequences, but that this was the first formal algorithm for multiple sequence alignment. At the time, the eventual importance of multiple alignment was not realized.

During this time, a paper by John Hartigan [**12**] came out in *Biometrics* containing the essentials of the dynamic programming calculation for f. Indeed, this idea was first published by Walter Fitch [**9**]. Steve Farris [**6**] had adapted it to optimizing trees in L_1 and some time later Pascale Rousseau and I [**26**] showed how to apply it in an arbitrary metric space.

After waiting for perhaps six months for a reply from Kac about the paper (though in the interim I had seen him a few times) I finally summoned up enough courage to ask him about it. He said that he thought he had already told me that he would not accept it. The referee had reported that it did not represent enough of an advance over the $N = 2$ case, and that I should incorporate in the algorithm a way of simultaneously optimizing the tree topology! I was flabbergasted; even if complexity theory was not yet popular, I knew that finding a Steiner tree in Euclidean, L_1 or other well-known spaces was exceedingly hard and had been studied intensively for years by many prominent mathematicians. Doing it in the space of sequences under the edit distance metric was quite an order! No harm done; I was able to publish my algorithm in the *SIAM Journal on Applied Mathematics* within a year or two [**20**].

4. Secondary Structure

Continuing his interest in RNA, Cedergren suggested we explore computer methods for analyzing the secondary structure of 5S RNA. In the same way that two entirely complementary DNA strands bind together through Watson-Crick pairing in forming a double helix, different regions of a single RNA molecule can bind to each other (C ribonucleotides with G ribonucleotides and A with U) to form a structure with single-stranded loops closed by double-stranded stems. In a secondary structure, or "folding" of the sequence **a**, the two regions $a(i), \ldots, a(i+k)$ and $a(j), \ldots, a(j-k)$, where $k \geq 2$, can form a helical stem by pairing together if each $a(i+h)$ is complementary to $a(j-h)$, for $h = 0, \ldots, k$, if none of the terms is already paired to some other term, if $3 < (j-k-i-k)$ and if, in the simplest case, there are no "knots" in the structure. A knot exists if $a(r)$ is paired with $a(s)$ while $a(u)$ is paired with $a(v)$ and $r < u < s < v$.

Generally, for a given RNA molecule there are several foldings that satisfy all these conditions. A working hypothesis is that the folding that has a stable existence is the one that maximizes the energy freed when it is formed. Stem formation releases energy in an amount which depends in a known way on k, the number of terms, and on which ribonucleotides are involved. Unpaired regions have the opposite effect and require energy to be formed, to an extent which may again depend on the number of terms and on which ribonucleotides are involved, but also depends on a structural parameter R. It can be shown that every secondary structure satisfying the above conditions can be uniquely decomposed into a number of stems, at most two energetically neutral "dangling ends" containing the first or last terms of **a** if they are unpaired, and a number of R-loops, described as follows: $a(i_r), \ldots, a(k_r)$ are all unpaired, for $r = 1, \ldots, R$, $a(i_1 - 1)$ is paired to $a(k_R + 1)$, and the remaining $a(i_r - 1)$ are each paired to $a(k_{r-1} + 1)$. If $R = 1$ the loop is called a "hairpin", if $R = 2$ the loop is called an interior loop—except in the special case of a "bulge", with one null unpaired region so that $a(i_1 - 1)$ and $a(k_1 + 1)$ are paired to $a(i_2)$ and $a(i_2 - 1)$, respectively. For $R \geq 3$, the loop is called a multiple loop. Thanks to a series of experimental studies pioneered by Tinoco [**32**], estimates are available for the energetic contribution of all possible stems and loops and these can be summed to estimate the free energy of a folding.

We obtained a copy of a program by Pipas and McMahon [**17**] which essentially finds all possible stems that could be formed in a RNA sequence, and tries to piece them together to find the optimize the energy. With Annie Morin, we tried this on a sample of 5S RNA's from phylogenetically diverse organisms and found, as an early example of what is now known as "phylogenetic filtering", that only one structure recurred among the five energetically most stable solutions in almost all of the species [**25**]. This turns out to correspond to the folding which is now generally recognized for this molecule. Years later I formalized the concept of phylogenetic filtering by simultaneously solving the folding and multiple alignment problems using a weighted objective function [**19**].

While doing this research it became apparent that secondary structure stems disrupted only by bulges and other interior loops could be detected by the dynamic programming comparison of an RNA sequence with a reversed copy of itself, with negative costs for base pairing and positive costs for 2-loop formation. Trying to incorporate multiple loops into this methodology was somewhat trickier, and I devised an iterative algorithm which produced, on the first pass, only 0-branchings, namely stems alternating with 2-loops. On the second pass, it could add several such stems branching off a main stem, producing 1-branchings, and so on. Eventually, any secondary structure could be found. I gave a talk on this at the Numerical Taxonomy meetings in Kansas in the fall of 1976 and had a student, Nguyen Bach Hue, implement it the next summer in a computer program. The method turned out to be very dependent on the crude energy estimates available at the time and produced rather bizarre structures so at the end of the summer I postponed further work on the project. A year or two later I came across Michael Waterman's papers [**36, 38**] on this topic, in which, following a query of Charles Delisi, the combinatorics of folding and the same type of iterative algorithm I had been experimenting with were already rigorously developed and indeed had been announced in a 1975 Los Alamos technical report.

During this time Mark Kac invited me to give a talk at Rockefeller University and I spoke about my approach to folding. G. Piecznik, a biochemist, was present and showed a great deal of interest in the subject. I cannot recall the details of our discussion, but he may have been insisting on the possibility of a single-pass algorithm instead of the iterative approach. A year or so later the basic idea of a single-pass dynamic programming algorithm was published by Ruth Nussinov, Piecznik, Jerrold Griggs, and D. J. Kleitman [**16**]. I have since learned from Griggs that experience with dynamic programming in other contexts led him to try it on the folding problem with multiple loops. Certainly I had never thought of this solution, and did not think of it after my meeting Piecznik, nor did I have anything to teach these researchers. Nevertheless it is interesting that once again this interaction took place under the aegis of Mark Kac, and re-stimulated my interest in secondary structure [**23, 19, 8**]. Michael Zuker of the National Research Council of Canada wrote a very effective and widely disseminated program based on the Nussinov *et al.* principles [**40**], later versions of which are still in use around the world.

The dynamic programming recurrence fundamental to folding may be represented as:

$$C(p,q) = \min_{R\geq 1} \min_{s \text{ an } R-\text{loop}} \left\{ E(s) + \sum_{r=1}^{R-1} C(k_r+1, i_{r+1}-1) \right\}$$

for the partial sequence $a(p),\ldots,a(q)$, if $a(p)$ and $a(q)$ can be paired, where $p = i_1 - 1, q = k_R + 1$ for any R-loop in the above notation. For unpairable $a(p)$ and $a(q)$, $C(p,q) = \infty$. Consecutive base-pairs in stems are considered to form 2-loops with two null unpaired regions in the same way as a bulge is a 2-loop with only one unpaired region. The energy terms $E(s)$ can be looked up in tables or derived by simple calculations provided by experimentalists. There are of course initial conditions; these are easily worked out.

The recurrence as it stands may require exponential computing time, but it elegantly expresses the idea of a single-pass algorithm. Mild restrictions on the form of $E(s)$ for multiple loops lead to polynomial algorithms [**23**].

5. Conclusion

Returning to the puzzlement I described in the Introduction, it is now clear to me that though he certainly did not "start all this", Ulam *did* have some impact on the early use of dynamic programming. He did not invent the edit distance that

Goad [**10**] and Ronald Graham [**11**] associate with his name—it was published by Levenshtein years earlier [**14**]—and he exaggerated its difficulty as a mathematical problem and did not realize that dynamic programming, for example as already published by Needleman and Wunsch, represented a solution. He did, however, interest Peter Sellers, fresh from our dynamic programming collaboration, in this problem, and the resulting paper did have an impact. And he did influence a number of people at Los Alamos, including Michael Waterman, in the mathematics of sequences [**37**]. But it is clear that it was through refereeing for Mark Kac that Sellers originally learned about dynamic programming for sequence comparison and through Kac that he heard about the edit-distance problem from Ulam. Furthermore, Waterman's involvement with dynamic programming followed directly upon Seller's paper. As far as my own work is concerned, I have described Mark Kac's encouragement in connection with my first paper on sequence comparison, the work on multiple alignment, and the research on RNA secondary structure. I doubt that Mark Kac counted me among his protégés; we scarcely ever discussed the details of a mathematical problem, and he did no work in the area. One area where he probably would have been of great help was the probabilistic analysis of matching random sequences which I undertook with Vaclav Chvátal [**3**], but I never took the opportunity to discuss it with him. (Indeed, he was interested in Waterman's early work in a related area, the statistics of local alignment.) Nevertheless, his friendly encouragement, his invitations to Rockefeller, his hospitality and (I suspect) his letters of reference, constituted a crucial impetus to my work at the CRM in the early years.

Acknowledgements

Thanks to Alberto Apostolico, Martin Goldstein and Michael Waterman for reading and commenting on an earlier version, and to Erica Jen for key bibliographic leads. And to Robert Cedergren, who got me involved in this problem area, and whose collaboration over the years has been the most rewarding and enjoyable experience.

References

1. A. Apostolico and R. Giancarlo, *Sequence alignment in molecular biology*, Purdue University Computer Science Technical Report 95–075, 1995; Mathematical Support for Molecular Biology (M. Farach, F. Roberts, and M. S. Waterman, eds.), DIMACS Ser. Discrete Math. Theoret. Comput. Sci., Amer. Math. Soc., Providence. (to appear)
2. V. Chvátal, D. A. Klarner and D. E. Knuth, *Selected combinatorial research problems*, Stanford University Computer Science Technical Report 72–292 (1972), 26.
3. V. Chvátal and D. Sankoff, *Longest common subsequences of two random sequences*, J. Appl. Probability **12** (1975), 306–315.
4. M. O. Dayhoff, *Computer analysis of protein evolution*, Sci. American **221** (1969), 86–95.
5. Bruce W. Erickson and P. H. Sellers, *Recognition of patterns in genetic sequences*, Time Warps, String Edits and Macromolecules. The Theory and Practice of Sequence Comparison (D. Sankoff and J. B. Kruskal, eds.) Addison-Wesley, Reading, Mass., 1983, 55–91.
6. J. S. Farris, *Methods for computing Wagner trees*, Systematic Zoology **19** (1970), 83–92.
7. M. Feigenbaum, *Reflections of the Polish masters—an interview with Stanislaw Ulam and Mark Kac*, Los Alamos Science **3**, 1982, 54–65.
8. V. Ferretti and D. Sankoff, *A continuous analog for* RNA *folding*, Bull. Math. Bio. **51** (1989), 167–171.
9. W. M. Fitch *Towards defining the course of evolution*: *Minimum change for a specific tree topology*, Systematic Zoology **20** (1971), 406–416.

10. W. Goad, *Sequence analysis—contributions by Ulam to molecular genetics*, From Cardinals to Chaos. Reflections on the Life and Legacy of Stanislaw Ulam (N. G.Cooper, ed.), Cambridge University Press, 1989, 288–291.
11. R. L. Graham, *A similarity measure for graphs—reflections on a theme of Ulam*, From Cardinals to Chaos. Reflections on the Life and Legacy of Stanislaw Ulam (N. G.Cooper, ed.), Cambridge University Press, 1989, 114–121.
12. J. A. Hartigan, *Minimal mutation fits to a given tree*, Biometrics **29** (1973), 53–65.
13. J. B. Kruskal, *An overview of sequence comparison*, Time Warps, String Edits and Macromolecules: The theory and practice of sequence comparison (D. Sankoff and J. B. Kruskal, eds.), Addison-Wesley, Reading, Mass., 1983, 1–33.
14. V. I. Levenshtein, *Binary codes capable of correcting deletions, insertions and reversals*, Cybernetics and Control Theory **10** (1966), 707–710; Dokl. Akad. Nauk USSR **163** (1965) 845–848. (Russian)
15. S. B. Needleman and C. D. Wunsch, *A special method applicable to the search for similarities in the amino acid sequence of two proteins*, J. Molecular Biology **48** (1970), 443–453.
16. R. Nussinov, G. Piecznik, J. R. Griggs, and D. J. Kleitman, *Algorithms for loop matchings*, SIAM J. Appl. Math. **35** (1978), 68–82.
17. J. M. Pipas and J. E. McMahon, *Method for predicting* RNA *secondary structure*, Proc. Nat. Acad. Sci. U.S.A. **72** (1975), 2017–2021.
18. G.-C. Rota, *The lost café*, From Cardinals to Chaos. Reflections on the Life and Legacy of Stanislaw Ulam (N. G. Cooper, ed.), Cambridge University Press, 1989, 23–32.
19. D. Sankoff, *Simultaneous solution of the* RNA *folding, alignment and protosequence problems*, SIAM J. Appl. Math. **45** (1985), 810–825.
20. ———, *Minimal mutation trees of sequences*, SIAM J. Appl. Math. **28** (1975), 35–42.
21. ———, *Matching sequences under deletion/insertion constraints*, Proc. Nat. Acad. Sci. U.S.A. **69** (1972), 4–6.
22. D. Sankoff and J. B. Kruskal (eds.), *Time warps, string edits and macromolecules: The theory and practice of sequence comparison*, Addison-Wesley, Reading, Mass., 1983.
23. D. Sankoff, R. J. Cedergren, J. B. Kruskal, and S. Mainville, *Fast algorithms to predict* RNA *secondary structures containing multiple loops.* Time Warps, String Edits and Macromolecules. The Theory and Practice of Sequence Comparison (D. Sankoff and J.B. Kruskal, eds.), Addison-Wesley, Reading, Mass., 1983, 93–120.
24. D. Sankoff, C. Morel, and R. J. Cedergren, *Evolution of* 5S RNA *and the non-randomness of base replacement*, Nat. New Biol. **245** (1973), 232–234.
25. D. Sankoff, A.-M. Morin and R.J. Cedergren, *The evolution of* 5S RNA *secondary structures*, Canad. J. Biochem. **56** (1978), 440–443.
26. D. Sankoff and P. Rousseau, *Locating the vertices of a Steiner tree in an arbitrary metric space*, Math. Programming **9** (1975), 240–246; Steiner Minimal Trees (F. K.Hwang and D. Richards, eds.), Adv. Discrete Math. Comput. Sci., vol. 4–5, Hadronic Press, Palm Arbor, FL.
27. D. Sankoff and P. H. Sellers, *Shortcuts, diversions and maximal chains in partially ordered sets*, Discrete Math. **4** (1973), 287–293.
28. P. H. Sellers, *An algorithm for the distance between two finite sequences*, J. Combin. Theory Ser. A **16** (1974), 253–258.
29. ———, *On the theory and computation of evolutionary distance*, SIAM J. Appl. Math. **26** (1974), 787–793.
30. ———, *The theory and computation of evolutionary distances: Pattern recognition*, J. Algebras **1** (1980), 359–373.
31. T. F. Smith and M. S. Waterman, *Identification of common molecular subsequences*, J. Molecular Biology **147** (1981), 195–197.
32. I. Tinoco, P. N. Borer, B. Dengler, M. D. Levine, O. C. Uhlenbeck, D. M. Crothers, and J. Gralla, *Improved estimation of secondary structure in ribonucleic acids*, Natur. New Biol. **246** (1973), 40–41.
33. S. M. Ulam, *Some combinatorial problems studied experimentally on computing machines*, Applications of Number Theory to Numerical Analysis (S. K. Zaremba, ed.) (Proc. Sympos., Univ. Montréal, Montréal, Qué., 1971), Academic Press, New York, 1972, 1–10.
34. T. K. Vintsyuk, *Speech discrimination by dynamic programming*, Kybernetika **4** (1968), 81–88. (Russian); Engl. Transl.: Cybernetics **4**, 52–57.

35. R. A. Wagner and M. J. Fischer, *The string-to-string correction problem*, J. Assoc. Comput. Mach. **21** (1974) 168–173.
36. M. S. Waterman, *Secondary structure of single-stranded nucleic acids*, Studies in Foundations and Combinatorics, Adv. in Math. Suppl. Stud., vol. 1, Academic Press, New York-London, 1978, pp. 167–212.
37. ———, *Introduction to computational biology*, Chapman and Hall, London, 1995, xiv.
38. M. S. Waterman and T. F. Smith, RNA *secondary structure: A complete mathematical analysis*, Math. Biosci. **42** (1978), 257–266.
39. S. K. Zaremba (ed.), *Applications of number theory to numerical analysis*, (Proc. Sympos., Univ. Montréal, Montréal, Qué., 1971), Academic Press, New York, 1972.
40. M. Zuker and P. Stiegler, *Optimal folding of large* RNA *sequences using thermodynamics and auxiliary information*, Nucleic Acids Research **9** (1981), 133–148.

Centre de recherches mathématiques, Université de Montréal, C.P. 6128 succursale centre-ville, Montréal, Québec H3C 3J7

E-mail address: sankoff@crm.umontreal.ca

Centre de Recherches Mathématiques
CRM Proceedings and Lecture Notes
Volume **11**, 1997

On Character Generators

N. Hambli and R. T. Sharp

ABSTRACT. The fact that incompatibilities between fundamental representation states are all or degree 2 is exploited to derive character generators more simply. As illustrative examples of the method we treat compact simple algebras of rank 1 (A_1) and rank 2 (A_2, B_2, G_2).

1. Introduction

Character generators for the classical algebras A_l, B_l, C_l, D_l have been given by several authors [**1, 6, 7, 10**]; Gaskell's method [**3**] also deals with the exceptional algebras G_2, F_4, E_6, E_7, E_8. See also Refs. [**2, 4, 5**].

The usual form of the character generator $X(A;\mu)$ depends on l (the algebra rank) dummy variables A_i which carry representation variables λ_i as exponents and on l dummies μ_j which carry weight components ω_j as exponents. $X(A;\mu)$ is a rational function which can be expanded in powers of the dummy variables:

$$X(A;\mu) = \sum_{\lambda,\omega} A^{\lambda}\mu^{\omega}C_{\lambda\omega}. \tag{1.1}$$

Here $A^{\lambda}\mu^{\omega}$ means $\prod_{i=1}^{l} A_i^{\lambda_i} \prod_{j=1}^{l} \mu_j^{\omega_j}$; $C_{\lambda\omega}$ is the multiplicity of the weight $[\omega]$ in the irreducible representation (IR) (λ); thus $\sum_{\omega}\mu^{\omega}C_{\lambda\omega}$ is the character of the IR (λ). The form of X which we consider has no negative terms in its expansion and hence no cancellations between terms. We will shortly say a little more about the form of X.

According to Cartan the IR $(\lambda) = (\lambda_1, \lambda_2, \dots, \lambda_l)$ is the completely stretched part (representation labels additive) of the direct product of λ_i copies of the i-th fundamental IR $(\lambda_j) = \delta_{ij}$ $(i = 1, \dots, l)$. Thus the basis states of the fundamental IR's form an integrity basis for states of the general IR (λ), which correspond to products of powers of them, the degree in states of the i-th fundamental IR being λ_i. Actually the states just described contain admixtures of states belonging to lower representations, which should be projected out to get the actual states of the IR (λ); for many purposes, e.g., counting weights or calculating generator matrix elements, they can be used as they stand, without projecting out lower IR's. Since we know the integrity basis, all we need are the compatibility rules to determine all

---

1991 *Mathematics Subject Classification*. Primary: 20C15; Secondary: 20C35.
This is the final form of the paper.

allowed products, i.e., to determine the character generator X which can be written as a sum of terms of the form $(\prod_j \eta_j)\prod_i(1-\eta_i)^{-1}$, where the η_i, η_j separately are distinct fundamental IR states, with only mutually compatible ones appearing in the same term. A desirable form of X has, for a given set of compatible denominators η_i, only one term, whose numerator is a product of a subset of the η_i appearing in the denominator; the role of the numerator is to avoid multiple counting of states. In Gaskell's version [**3**] of the character generator, a state belonging to an interior Weyl orbit of a fundamental IR is always incompatible with itself, and appears only in the numerator of a term in X; in our view such a form of X is less desirable than that described just earlier. In all the character generators appearing in this paper interior fundamental IR states are on the same footing as exterior states and appear in denominator factors as shown above.

It was shown recently [**2**] that all elementary incompatibilities are of degree 2; i.e., two states may be incompatible, or a state may be incompatible with itself. To find all incompatibilities one has only to examine IR's in which a single label has the value 2 ($\lambda_j = 2\delta_{ij}$) and IR's in which 2 labels have the value 1 ($\lambda_k = \delta_{ij} + \delta_{jk}$).

In Section 2 we deal with A_1 and A_2 and in Sections 3, 4 respectively we treat the other rank 2 algebras B_2, G_2. We complete this section with some general considerations about the number of denominator factors and about the number of incompatible pairs.

Let the dimension of the i-th fundamental IR be d_i. Then the number of degree 2 monomials is $1/2d_i(d_i+1)$; and let the dimension of the IR with $\lambda_j = 2\delta_{ij}$ be $d_i^{(2)}$. Then there are $1/2d_i(d_i+1)-d_i^{(2)}$ unwanted (i.e., belonging to unstretched IR's) states. This is the number of incompatibilities between pairs of states in the i-th fundamental IR, or of single states with themselves.

Similarly, with two nonzero labels λ_i and λ_j, $i \neq j$, we are led to consider the $d_i d_j$ products of two states, one from the i-th fundamental IR and one from the j-th. Let d_{ij} be the dimension of the stretched (wanted) IR with $\lambda_k = \delta_{ik} + \delta_{jk}$. Among such states $d_i d_j - d_{ij}$ pairs are incompatible.

It is straightforward to compute the number of denominator factors in the character generator of an algebra, including the case of degenerate IR's (one or more labels zero).

The number of denominator factors is just the number of labels, IR labels and internal labels required for states of all IR's. But there are l IR labels and, according to Racah [**8**], $1/2(r-l)$ internal labels. So the character generator terms have $1/2(r+l)$ denominator factors; r here is the order of the algebra (number of generators). For degenerate IR's, with specified labels zero the number of denominator factors is the number of nonzero IR labels plus the number of internal labels required by those degenerate IR's. The number of internal labels is given by Seligman and Sharp [**9**]; it is the number of internal labels of the algebra minus the number of internal labels of the subalgebra whose Dynkin diagram is that with nodes corresponding to the zero labels.

It can be checked that the counting of incompatibilities and of denominator factors for the algebras to be treated here complies with the counting done in this section.

2. The Algebras A_1 and A_2

In this section and the following two, the square of a fundamental IR means the symmetric or polynomial part of the direct product of two copies.

A_1 has only one fundamental IR (1) with $d = 2$. We find

$$(1)_3^2 \supset (2)_3. \tag{2.1}$$

On the right side of (2.1) and of similar equations which follow the first IR is the stretched or wanted part; any other IR's are unstretched and unwanted. Subscripts give dimensions. The number of incompatible pairs is the dimension of the unwanted IR's, zero for A_1 by (2.1). We can take the two states as η and ξ with $\omega = \pm 1$. Since there are no incompatibilities the character generator is

$$X(A,\mu) = \left[(1 - A\mu)(1 = A\mu^{-1}\right]^{-1} = \left[(1-\eta)(1-\xi)\right]^{-1}.$$

The number of denominator factors is $1/2(r + l) = 1/2(3 + 1) = 2$.

A_2 has two fundamental IR's $(1, 0)$ and $(0, 1)$, each with dimension $d_1 = d_2 = 3$. We find

$$\begin{aligned} (1,0)_6^2 &\supset (2,0)_6 \\ (0,1)_6^2 &\supset (0,2)_6 \\ (1,0)\times(0,1)_9 &\supset (1,1)_8 + (0,0)_1. \end{aligned} \tag{2.2}$$

We see that there is one incompatibility, between a $(1, 0)$ state and a $(0, 1)$ state. We can find the character generator analytically by substituting the SU(2) character generator $X(B;\mu) = \left[(1-B\mu)(1-B\mu^{-1})\right]^{-1}$ into the SU(3) $\supset$ SU(2)$\times$U(1) branching rules generating function $G(A_1, A_2; B, Y) = \left[(1 - A_1BY^{1/3})(1 - A_1Y^{-2/3})(1 - A_2BY^{-1/3})(1 - A_2Y^{2/3})\right]^{-1}$, i.e., evaluating

$$X(B;\mu)G(A_1, A_2; B^{-1}, Y)|_{B^0}. \tag{2.3}$$

Here $|_{B^0}$ is an instruction to retain the terms of degree zero in B; the dummy Y carries as its exponent the hypercharge y corresponding to the vertical weight component. The result of evaluating (2.3) is

$$\begin{aligned} X(A_1, A_2, \mu, Y) &= \left[(1-\eta)(1-\zeta)(1-\eta^*)(1-\zeta^*)\right]^{-1} \\ &\quad \times \left[(1-\xi)^{-1} + \xi^*(1-\xi^*)^{-1}\right] \end{aligned} \tag{2.4}$$

where $\eta = A_1\mu Y^{1/3}$, $\xi = A_1\mu^{-1}Y^{1/3}$, $\zeta = A_1 Y^{-2/3}$ can be interpreted as the three states of the fundamental IR $(1, 0)$. The starred variables are the same with A_1 replaced by A_2 and the exponents of μ, Y reversed in sign; they are the states of $(0, 1)$. So the incompatible pair is $\xi\xi^*$.

But it is much simpler to evaluate $(1, 0) \times (0, 1)$. We get the same result, that ξ, ξ^* are incompatible. An alternative, equivalent, conclusion is that $\eta\eta^*$ are incompatible, a result that could have been obtained by an alternative evaluation of (2.3) leading to the result (2.4) with η and ξ, and η^* and ξ^*, interchanged. The number of denominator factors is $1/2(r + l) = 1/2(8 + 2) = 5$. For a degenerate character generator, with $\lambda_1 = 0$ or $\lambda_2 = 0$ set $\eta^* = \xi^* = \zeta^* = 0$ or $\eta = \xi = \zeta = 0$. It has 3 denominator factors.

3. The Algebra B_2

We discuss the character generator of B_2 using the subgroup $A_1 \times A_1$. The $B_2 \supset A_1 \times A_1$ branching rules generating function is

$$F(A_1, A_2; B_1, B_2) = \left[(1 - A_1)(1 - A_1B_1B_2)(1 - A_2B_1)(1 - A_2B_2)\right]^{-1}. \tag{3.1}$$

Substituting into this the $A_1 \times A_1$ character generator $X(B_1, B_2; \mu_1, \mu_2) = \left[(1 - B_1\mu_1)(1 - B_1\mu_1^{-1})(1 - B_2\mu_2)(1 - B_2\mu_2^{-1})\right]^{-1}$, i.e. evaluating

$$F(A_1, A_2; B_1, B_2)X(B_1^{-1}, B_2^{-1}; \mu_1, \mu_2)|_{B_1^0 B_2^0}$$

gives the B_2 character generator

$$X(A; \mu) = \frac{1}{\beta^*\gamma\delta\varepsilon^*}\left[\frac{1}{\alpha\beta} + \frac{\alpha^*}{\beta\alpha^*} + \frac{\delta^*}{\alpha^*\delta^*} + \frac{\varepsilon}{\delta^*\varepsilon} + \frac{\alpha\varepsilon}{\alpha\varepsilon}\right]. \tag{3.2}$$

We have adopted the space-saving convention that a denominator variable denotes 1 minus that variable; the variables on the right side of (3.2) are the states of the fundamental IR's $(1, 0)$ and $(0, 1)$, whose dimensions are respectively, 5 and 4. Thus,

$$\alpha = A_1\mu_1\mu_2, \quad \beta = A_1\mu_1\mu_2^{-1}, \quad \gamma = A_1, \quad \delta = A_2\mu_1, \quad \varepsilon = A_2\mu_2.$$

The starred variables are the same with the weights, i.e., the exponents of μ_1, μ_2 reversed. As a check we notice that

$$(1, 0)^2_{15} \supset (2, 0)_{14} + (0, 0)_1$$
$$(0, 1)^2_{10} \supset (0, 2)_{10}$$
$$(1, 0) \times (0, 1)_{20} \supset (1, 1)_{16} + (0, 1)_4.$$

This is in agreement with the incompatibilities implied by (3.2). Incompatible pairs are $\alpha\alpha^*$, $\alpha\delta^*$, $\alpha^*\varepsilon$, $\beta\delta^*$, $\beta\varepsilon$ i.e., one, $\alpha\alpha^*$, in $(1, 0)^2$ and the other 4 in $(1, 0) \times (0, 1)$. Also the number of denominator factors in (3.2) is $6 = 1/2(r + l) = 1/2(10 + 2)$. In the degenerate IR $(\lambda_1, 0)$ (set δ, δ^*, ε, ε^* equal to 0 in (3.2)) there are 4. In the degenerate IR $(0, \lambda_2)$ (set α, β, α^*, β^*, γ equal to 0 but keep 1 from the first term $1/\alpha\beta$) there are also 4.

As in the case of A_2 there are other equivalent versions of $X(A; \mu)$ which can be obtained both from the analytic evaluation and from the direct products $(1, 0) \times (0, 1)$, $(1, 0)^2$ and $(0, 1)^2$.

4. The Algebra G_2

The dimensions of the two fundamental IR's are $d_1 = 14$, $d_2 = 7$. We find

$$(1, 0)^2_{105} \supset (2, 0)_{77} + (0, 2)_{27} + (0, 0)_1$$
$$(0, 1)^2_{28} \supset (0, 2)_{27} + (0, 0)_1$$
$$(1, 0) \times (0, 1)_{98} \supset (1, 1)_{64} + (0, 2)_{27} + (0, 1)_7.$$

Among polynomials of degree 2 in $(1, 0)$ there are 28 incompatibilities; in $(0, 1)$ there is 1; in $(1, 0) \times (0, 1)$ there are 34. The incompatible pairs are tabulated in Refs. [**2, 5**]. Knowing them, it is straightforward to write down the character generator; it can be checked that it has $1/2(14{+}2) = 8$ denominator factors and that for the degenerate versions there are 6 denominator factors Ref. [**2**, equations (2.7), (2.11)]. The incompatibilities were found analytically in Refs. [**2, 5**] by substituting the SU(3) character generator into the $G_2 \supset$ SU(3) branching rules generating function. But they can be found much more simply by considering the direct products $(1,0)^2$, $(0,1)^2$, $(1,0)\times(0,1)$.

5. Conclusions

We plan to consider the rank 3 algebras A_3, B_3, C_3 in a similar way. In the case of B_3 the subalgebra D_3 can be used. It is sometimes stated that the subalgebra must have the same rank as the algebra in order to get weights in an l-dimensional space. However it may be that a lower dimensional subalgebra, e.g., G_2 in B_3 can be used with the method followed in this paper, i.e., by finding incompatibilities between B_3 fundamental basis states. We hope to report later on this possibility (we plan to determine $B_3 \supset G_2$ states and generator matrix elements between them).

We thank Sophie Pelland for help in the early stages of the work.

References

1. K. Baclawski, *Character generators for unitary and symplectic groups*, J. Math. Phys. **24** (1983), no. 7, 1688–1694.
2. L. Farell, C. S. Lam, and R. T. Sharp, *G_2 generator matrix elements for degenerate representations in an* SU(3) *basis*, J. Phys. A **27** (1994), no. 8, 2761–2771.
3. R. W. Gaskell, *Character generators for compact semisimple Lie groups*, J. Math. Phys. **24** (1983), no. 10, 2379–2386.
4. R. Gaskell and R. T. Sharp, *Generating functions for G_2 characters and subgroup branching rules*, J. Math. Phys. **22** (1981), no. 12, 2736–2739.
5. N. Hambli and R. T. Sharp, *Generator matrix elements for $G_2 \supset$* SU(3). II. *Generic representations*, J. Phys. A **28** (1995), no. 9, 2581–2588.
6. R. C. King and N. G. I. El-Sharkaway, *Standard Young tableaux and character generators of classical Lie groups*, J. Phys. A **17** (1984), no. 1, 19–45.
7. J. Patera and R. T. Sharp, *Generating functions for characters of group representations and their applications*, Lecture Notes in Phys. **94** (1979) 175–183.
8. G. Racah, *Group theory and spectroscopy*, Ergeb. Exakt. Naturwiss., Band **37** (1965), 28–84.
9. T. H. Seligman and R. T. Sharp, *Internal labels of degenerate representations*, J. Math. Phys. **24** (1983), 769–771.
10. R. P. Stanley, *The character generator of* SU(n), J. Math. Phys. **21** (1980), no. 9, 2321–2326.

Institut des Hautes Études Scientifiques, 35 Route des Chartres, Bures-sur-Yvette, 91440, Paris.
E-mail address: hambli@ihes.fr

Department of Physics, McGill University, 3600 rue University, Montréal, Québec H3A 2T8.
E-mail address: sharp@physics.mcgill.ca

Centre de Recherches Mathématiques
CRM Proceedings and Lecture Notes
Volume **11**, 1997

Quantum Transport in Solids

Carolyne M. Van Vliet

ABSTRACT. Several aspects of quantum transport in solids are discussed. After a brief introduction, we discuss the quantum field description for the degenerate electron gas, with electron confinement in $3-\nu$ dimensions, in Section 2. Next we deal with the basic many-body description, developing generalized and Markovian master equations from the von Neumann equation in some detail, Section 3. From these results, using the occupation number formalism of second quantization, the quantum Boltzmann equations for the diagonal occupancy operator $\langle n_\zeta \rangle_t$, as well as for the nondiagonal operator $\langle c^\dagger_{\zeta_1} c_{\zeta_2} \rangle_t$, are stated in Section 4, with references to the earlier quite laborious computations (not carried out here); also the Wigner formalism is briefly considered. Finally, in the last two sections we deal with applications to the integral quantum Hall effect and with recent theories of fluctuations in mesoscopic conductors.

1. Introduction

Transport of electricity and heat forms one of the oldest branches of nonequilibrium statistical mechanics, with roots in the developments of the previous century. On the many-body level, the basic transport equation in Γ-space or phase space for the phase-density function $\rho(p, q, t)$ was developed by Liouville in 1838 [**28**]; on the one-particle level, permissible if interparticle interactions are weak, The Boltzmann transport equation (1871 [**4**]; see also [**5**]) describes the rate of change of the occupation function in μ-space $f(\boldsymbol{v}, \boldsymbol{r}, t)$ and the ensuing, dissipative irreversible behaviour as contained in Boltzmann's "H-theorem".[1] Early in the present century both approaches were endowed with quantum mechanical meaning. Replacing Poisson's bracket with the commutor bracket, see Dirac [**13**], the Liouville equation goes over into the von Neumann equation; for solids with a periodic structure and wave functions satisfying Floquet's theorem [**49**], the transport equation for the occupation number function $f(\boldsymbol{k}, \boldsymbol{r}, t)$ obtains a form very similar to Boltzmann's original equation, referred to as the semi-quantal BTE, with $\boldsymbol{F}/m$ in the streaming term due to the external field now replaced by $\boldsymbol{F}/\hbar$, where $\boldsymbol{F}$ is still the Lorentz force $q(\boldsymbol{E} + \boldsymbol{v}_k \times \boldsymbol{B})$, $\boldsymbol{v}_{\boldsymbol{k}}$ being the Bloch velocity $(1/\hbar)\boldsymbol{\nabla}_{\boldsymbol{k}}\varepsilon(\boldsymbol{k})$.

Obviously, the many-body description has a much wider scope than the semiclassical BTE approach. Mathematically the many-body description is in principle

1991 *Mathematics Subject Classification.* Primary: 82C10; Secondary: 82C70.
This is the final form of the paper.
[1] H in reality denotes capital eta, for entropy.

less complex, since linear operator algebra applies to the evolution of the density operator ρ and other system operators $A^H(t)$—in Heisenberg form—being elements of Liouville space (Fano [**14**]). In the late fifties new, closed formulas were developed for generalized conductivities and susceptibilities, now usually referred to as the Kubo-Green relations [**25, 15**]. To evaluate the response functions or correlation functions occurring in these relations, as a rule one must apply diagrammatic perturbation theory. The validity of Kubo's linear response theory (LRT) has been a wide subject of discussion, see van Kampen [**21**], Cohen [**12**], and Van Vliet [**40**].

A major new development was undertaken at the Centre de recherches mathématiques by Van Vliet and coworkers in the late seventies and eighties. The objective was to obtain useful *convergent* correlation expressions and *irreversible* kinetic equations on the many-body level, by applying Zwanzig's projection operators for diagonal and nondiagonal Liouville space [**50**], together with Leon Van Hove's perturbation procedure to all orders [**19**]. Thus, starting from the von Neumann equation, we obtained new Kubo-like expressions and an inhomogeneous Markovian master equation (ME) which accounts for all interactions and streaming due to external fields, as well as quantum interference effects; the developments are laid down in a series of papers in the Journal of Mathematical Physics [**47, 48, 9, 38**]. From the many-body ME it was now possible to obtain quantum Boltzmann equations (QBE) for the occupancies of one-body states $|\zeta)$, both for the evolution of the diagonal part, $\boldsymbol{n}_\zeta = c_\zeta^\dagger c_\zeta$, and the nondiagonal part, $c_{\zeta_1}^\dagger c_{\zeta_2}$, $\zeta_1 \neq \zeta_2$, where $c^\dagger$ and c are creation and annihilation operators, respectively. It is important to note here that the states $|\zeta)$ can be both extended ($\zeta \to \boldsymbol{k}$) and localized (e.g. Landau states in a magnetic field), so that ponderomotive current as well as collisional current conductivities can be obtained. Thus the QBE goes far beyond the semiquantal BTE described above; in fact, it is the first general quantum form which replaces Boltzmann's 1871 kinetic equation. Many applications, including the quantum Hall effect and magneto-phonon effects, were treated by Vasilopoulos [**35, 36**], Charbonneau [**8**], Chaubey [**10, 11**], and Van Vliet [**45**]. Higher order kinetic equations (ME-hierarchy) were developed by Vasilopoulos and Van Vliet [**37, 42**].

At the time of the development of the ME and QBE formalisms, the results were in a sense "futuristic"; localized states and quantum interference effects mainly occurred in a magnetic field or in hopping conductivity. However, in the late eighties and nineties quantum confinement involving localized states or subbands was realized in many devices: mesoscopic conductors, ballistic conductors,double resonant barrier diodes, etc. For such conductors, often of reduced dimensionality (2D, 1D) the quantum approach is imperative. Thus we begin this survey with a generalized quantum field description of the current density in Fermion systems.

2. Quantum Field Description for the Current Density

Although both ponderomotive and collisional currents are amenable to a quantum field description [**46**], we restrict ourselves here to ponderomotive conductance, which implies that in at least one direction the states must be extended, represented by a crystal momentum in the first Brillouin zone, $\boldsymbol{k} = \{k_i\}$, $i = 1 \ldots \nu$, $\nu \geq 1$; in the other directions quantum confinement will occur when the dimensions are less than or comparable to the De Broglie wavelength, thus involving a set of quantum numbers $\{\alpha\}$; finally spin is denoted by σ. The full one-particle states are

$|\zeta) = |\boldsymbol{k}, \alpha, \sigma)$, while their wave mechanical realization implies the position functions $u_\zeta(\boldsymbol{q}) \equiv (\boldsymbol{q}|\zeta)$, $\boldsymbol{q} = (q_1 \dots q_\nu)$. Characterizing further the quantum field of all Fermion particles by the Schrödinger field operators $\Psi(\boldsymbol{q})$ and $\Psi^\dagger(\boldsymbol{q})$, we have the usual expansion

$$\text{(1)} \qquad \begin{cases} \Psi(\boldsymbol{q}) = \sum_\zeta u_\zeta(\boldsymbol{q}) c_\zeta, \\ \Psi^\dagger(\boldsymbol{q}) = \sum_{\zeta'} u^*_{\zeta'}(\boldsymbol{q}) c^\dagger_{\zeta'}. \end{cases}$$

The current density is accordingly

$$\text{(2)} \qquad \boldsymbol{J}(\boldsymbol{q}) = \int d^\nu q \Psi^\dagger(\boldsymbol{q}) \boldsymbol{j}(\boldsymbol{q}; \boldsymbol{q}') \Psi(\boldsymbol{q}),$$

where $\boldsymbol{j}$ is the Hermitized one-particle operator

$$\text{(3)} \qquad \boldsymbol{j}(\boldsymbol{q}; \boldsymbol{q}') = \frac{q}{2m} \big[\boldsymbol{p}\delta(\boldsymbol{q} - \boldsymbol{q}') + \delta(\boldsymbol{q} - \boldsymbol{q}')\boldsymbol{p}\big].$$

(scalar $q = \pm e$ represents the charge, not to be confused with the position vector $\boldsymbol{q}$). With $\boldsymbol{p} = (\hbar/i)\boldsymbol{\nabla}_{\boldsymbol{q}}$ substitution of (3) into (2), using Green's theorem for the first term, we obtain the well-known expression

$$\text{(4)} \qquad \boldsymbol{J}(\boldsymbol{q}) = \frac{q\hbar}{2mi} \big\{ \big[-\boldsymbol{\nabla}_{\boldsymbol{q}} \Psi^\dagger(\boldsymbol{q})\big] \Psi(\boldsymbol{q}) + \Psi^\dagger(\boldsymbol{q}) \boldsymbol{\nabla}_{\boldsymbol{q}} \Psi(\boldsymbol{q}) \big\};$$

or, employing (1), alternatively

$$\text{(5)} \qquad \boldsymbol{J}(\boldsymbol{q}) = \frac{q\hbar}{2mi} \sum_{\zeta\zeta'} c^\dagger_{\zeta'} c_\zeta \big[u^*_{\zeta'}(\boldsymbol{q}) \boldsymbol{\nabla}_{\boldsymbol{q}} u_\zeta(\boldsymbol{q}) - u_\zeta(\boldsymbol{q}) \boldsymbol{\nabla}_{\boldsymbol{q}} u^*_{\zeta'}(\boldsymbol{q}) \big].$$

Let $\rho(t)$ be the density operator subject to initial $\rho(0)$. Then for the ensemble average conditional to $\rho(0)$ we have

$$\begin{aligned} \text{(6)} \qquad \big\langle \boldsymbol{J}(\boldsymbol{q}) \big\rangle_t &= \operatorname{Tr}\big[\boldsymbol{J}(\boldsymbol{q}) \rho(t)\big] \\ &= \frac{q\hbar}{2mi} \sum_{\alpha\alpha'\sigma\sigma'} \sum_{\boldsymbol{k}\boldsymbol{k}'} \langle c^\dagger_{\boldsymbol{k}'} c_{\boldsymbol{k}} \rangle_t \big[u^*_{\boldsymbol{k}'}(\boldsymbol{q}) \boldsymbol{\nabla}_{\boldsymbol{q}} u_{\boldsymbol{k}}(\boldsymbol{q}) - u_{\boldsymbol{k}}(\boldsymbol{q}) \boldsymbol{\nabla}_{\boldsymbol{q}} u^*_{\boldsymbol{k}'}(\boldsymbol{q}) \big] \end{aligned}$$

where we suppressed the quantum numbers α and σ in the states $u(\boldsymbol{q})$ and in c, $c^\dagger$. For normalized plane waves[2] $\boldsymbol{\nabla}_{\boldsymbol{q}} u_{\boldsymbol{k}}(\boldsymbol{q}) = i\boldsymbol{k} u_{\boldsymbol{k}}(\boldsymbol{q})$. Thus (6) yields

$$\text{(7)} \qquad \langle \boldsymbol{J}(\boldsymbol{q}) \rangle_t = \frac{q\hbar}{2m} \sum_{\alpha\alpha'\sigma\sigma'} \sum_{\boldsymbol{k}\boldsymbol{k}'} \langle c^\dagger_{\boldsymbol{k}'} c_{\boldsymbol{k}} \rangle_t (\boldsymbol{k} + \boldsymbol{k}') \big[u^*_{\boldsymbol{k}'}(\boldsymbol{q}) u_{\boldsymbol{k}}(\boldsymbol{q}) \big].$$

We now introduce the one-particle Wigner form, being the Weyl transform of the one-particle density function, see [**17**]. So we have, denoting by Ω the ν-dimensional volume of the system[3]

$$\text{(8)} \qquad \rho^1_{\alpha\sigma}(\boldsymbol{k}, \boldsymbol{q}, t) = \frac{\Omega}{(2\pi)^\nu h^\nu} \int d^\nu u e^{i\boldsymbol{u}\cdot\boldsymbol{q}} \langle c^\dagger_{\boldsymbol{k}-1/2\boldsymbol{u}} c_{\boldsymbol{k}+1/2\boldsymbol{u}} \rangle_t,$$

---

[2]For Bloch functions Fourier series on the reciprocal lattice should be used.

[3]We note that in (8) and (9) we deal with ensemble averages, i.e., $\langle c^\dagger c \rangle = \operatorname{Tr}\big[\rho(t) c^\dagger c\big] = \operatorname{Tr}\big[\rho^H c^\dagger(t) c(t)\big]$ where ρ^H, $c^\dagger(t)$, and $c(t)$ are Heisenberg operators and $\rho^H \equiv \rho(0)$. Analogous to (8) and (9) one can also introduce the non-averaged Heisenberg operator transform pair:

$$\text{(8')} \qquad \hat{\rho}^1_{\alpha\sigma}(\boldsymbol{k}, \boldsymbol{q}, t) = \frac{\Omega}{(2\pi)^\nu h^\nu} \int d^\nu u e^{i\boldsymbol{u}\cdot\boldsymbol{q}} c^\dagger_{\boldsymbol{k}-1/2\boldsymbol{u}}(t) c_{\boldsymbol{k}+1/2\boldsymbol{u}}(t),$$

$$\text{(9')} \qquad c^\dagger_{\boldsymbol{k}-1/2\boldsymbol{u}}(t) c_{\boldsymbol{k}+1/2\boldsymbol{u}}(t) = \frac{h^\nu}{\Omega} \int d^\nu q e^{-i\boldsymbol{u}\cdot\boldsymbol{q}} \hat{\rho}^1_{\alpha\sigma}(\boldsymbol{k}, \boldsymbol{q}, t).$$

with inversion

(9) $$\langle c^{\dagger}_{\boldsymbol{k}-1/2\boldsymbol{u}} c_{\boldsymbol{k}+1/2\boldsymbol{u}}\rangle_t = \frac{h^\nu}{\Omega}\int d^\nu q e^{-i\boldsymbol{u}\cdot\boldsymbol{q}} \rho^1_{\alpha\sigma}(\boldsymbol{k},\boldsymbol{q},t).$$

In (7) we replace the summation variables by $\boldsymbol{K}$ and $\boldsymbol{K}'$ and we make the transformation

(10) $$\begin{cases} \boldsymbol{k} - \frac{1}{2}\boldsymbol{u} = \boldsymbol{K}' \\ \boldsymbol{k} + \frac{1}{2}\boldsymbol{u} = \boldsymbol{K} \end{cases} \quad \text{or} \quad \begin{cases} \boldsymbol{k} = (\boldsymbol{K}+\boldsymbol{K}')/2 \\ \boldsymbol{u} = \boldsymbol{K} - \boldsymbol{K}' \end{cases}$$

Let $u_{\boldsymbol{k}\alpha\sigma}(\boldsymbol{q}) = \chi_\alpha(\sigma,\bar{\boldsymbol{q}})\Omega^{-1/2}e^{i\boldsymbol{k}\cdot\boldsymbol{q}}$, where $\bar{\boldsymbol{q}}$ denotes the ignorable coordinates of the $3-\nu$ quasi dimensions. From (7) and (9) we then have

(11) $$\langle \boldsymbol{J}(\boldsymbol{q})\rangle_t = \frac{q\hbar}{m}\sum_{\alpha\alpha'\sigma\sigma'}\sum_{\boldsymbol{k},\boldsymbol{u}}\frac{\boldsymbol{k}h^\nu}{\Omega^2}\chi^*_{\alpha'}(\sigma',\bar{\boldsymbol{q}})\chi_\alpha(\sigma,\bar{\boldsymbol{q}})\int d^\nu \boldsymbol{q}' e^{i\boldsymbol{u}\cdot(\boldsymbol{q}-\boldsymbol{q}')}\rho^1_{\alpha\sigma}(\boldsymbol{k},\boldsymbol{q}',t).$$

Writing further $\sum_{\boldsymbol{u}} \to [\Omega/(2\pi)^\nu]\int d^\nu u$ and noting

(12) $$\int d^\nu u e^{i\boldsymbol{u}\cdot(\boldsymbol{q}-\boldsymbol{q}')} = (2\pi)^\nu\delta(\boldsymbol{q}-\boldsymbol{q}'),$$

while averaging over the ignorable coordinates $\bar{\boldsymbol{q}}$ and all spin diads implies $\delta_{\alpha\alpha'}\delta_{\sigma\sigma'}$, Eq. (11) results in

(13) $$\langle \boldsymbol{J}(\boldsymbol{q})\rangle_t = \frac{q\hbar}{m\Omega}\sum_{\alpha\sigma}\sum_{\boldsymbol{k}}\boldsymbol{k}h^\nu\rho^1_{\alpha\sigma}(\boldsymbol{k},\boldsymbol{q},t),$$

reminescent of the semiquantal Boltzmann expression, with additional summation over subbands and spin. Usually the limit $\lim_{h\to 0}\rho^1(\boldsymbol{k},\boldsymbol{q},t)$ is called the "Wigner distribution" (see e.g., Groenewold [**16**]). However, noting that h^ν is a microcell in the one-particle phase space, we introduce the "Wigner occupation function"

(14) $$g(\boldsymbol{k},\boldsymbol{q},t) = h^\nu\tilde{\rho}^1(\boldsymbol{k},\boldsymbol{q},t),$$

where $\tilde{\rho}(\boldsymbol{k},\boldsymbol{q},t)$ is coarse-grained over a minimum support volume $\Delta\omega \equiv \Delta^\nu p\Delta^\nu q$ of magnitude h^ν, i.e.,

(15) $$\tilde{\rho}(\boldsymbol{k},\boldsymbol{q},t) = (1/h^\nu)\iint_{\Delta\omega}\rho^1(\boldsymbol{k},\boldsymbol{r},t)d^\nu p d^\nu q.$$

Note that g as defined here is a *number*: it is the "occupancy of a state $\boldsymbol{k}$ centered on $\Delta^\nu k$ in a neighbourhood $\Delta^\nu q$ of $\boldsymbol{q}$ (for $\nu = 3$ this agrees with the formulation of Ziman [4]). Finally from (13) and the above, replacing $\rho^1_{\alpha\sigma}$ by $\tilde{\rho}^1_{\alpha\sigma}$,

(16) $$\langle \boldsymbol{J}(\boldsymbol{q})\rangle_t = \frac{q\hbar}{m\Omega}\sum_{\alpha\sigma}\sum_{\boldsymbol{k}}\boldsymbol{k}g_{\alpha\sigma}(\boldsymbol{k},\boldsymbol{q},t).$$

We still note that the approximation associated with coarse graining is quite permissible, since for the directions in which extended states exist the dimensions must be large with respect to the appropriate localization lengths.

In Section 4 be shall further discuss the transport equation for the Wigner occupation function, after we first derive the master equations on the many-body level.

---

[4] Ref. [**50**], Section 7.3.

3. Diagonal and Nondiagonal Master Equations

The system Hamiltonian (without applied external fields) will be denoted by $H = H^0 + \lambda V$. Here H^0 is the largest sub-Hamiltonian which in the relevant description (e.g., electron-phonon gas) can be diagonalized and λV is an interparticle interaction potential, amenable to perturbation techniques. It is well-known that the evolution operator $\exp\big[i(H^0 + \lambda V)\big]$, with $[H^0, V] \neq 0$, can be expressed in a power series in λ^n of time-ordered integrals, see e.g. Messiah [**31**]; this route was followed in LRTI, Ref. [**47**]. A far more elegant presentation is obtained through the use of projection operators $\mathcal{P}$ and $\mathcal{Q} = 1 - \mathcal{P}$ of the Liouville space, introduced by Zwanzig and employed in LRTII, Ref. [**48**]. There are some essential difficulties, however, because of the possible non-validity of the Dirac notation, as will be discussed now.

3.1. Preliminaries. Let the state space pertaining to the sub-dynamics of H^0 be $\mathcal{H}$, with the eigenstates of H^0 spanning this space being $\big\{|\gamma\rangle\big\}$; we assume that a wave mechanical realization can be assigned, implying functions $\gamma\big(\{q_k\}\big) \equiv \big\langle\{q_k\}|\gamma\big\rangle$ as customary [**13**]. For γ we assume periodic boundary conditions over the volume of the system, as usual in solid state physics. We also assume that $\mathcal{H}$ is embedded in a bigger space $\mathcal{S}$ in which the scalar product is still defined as in $\mathcal{H}$, but without the restriction of periodic b.c.; this is necessary since many physical operators, i.e. functions of position, carry the new state $K|\gamma\rangle$ outside $\mathcal{H}$. Thus, let K be an arbitrary operator of $\mathcal{S}$. Using the closure property for the states $\big\{|\gamma\rangle\big\}$ we can generally write in Dirac notation

$$(17)\quad K = \sum_{\gamma\gamma'} |\gamma\rangle\langle\gamma|K|\gamma'\rangle\langle\gamma'| = \sum_{\gamma} |\gamma\rangle\langle\gamma|K|\gamma\rangle\langle\gamma| + \sum_{\gamma\neq\gamma'} |\gamma\rangle\langle\gamma|K|\gamma'\rangle\langle\gamma'| = \mathcal{P}K + (1-\mathcal{P})K$$

where $\mathcal{P}$ and $1 - \mathcal{P}$ are projection operators acting in the Liouville space $\mathcal{H} \otimes \overline{\mathcal{H}}$, where $\overline{\mathcal{H}}$ is the dual space of $\mathcal{H}$. More precisely we define:

$$(17a)\qquad K_d \equiv \mathcal{P}K = \sum_{\gamma} |\gamma\rangle\langle\gamma|K|\gamma\rangle\langle\gamma| = \sum_{\gamma} |\gamma\rangle\langle\gamma|\,\langle\gamma|K|\gamma\rangle,$$

$$(17b)\qquad K_{nd} \equiv (1-\mathcal{P})K\left(= \sum_{\gamma\neq\gamma'} |\gamma\rangle\langle\gamma'|\,\langle\gamma|K|\gamma'\rangle\right);$$

the ( ) of the far right-hand side are explained below. We note that $\mathcal{P}$ has the form $\sum_\gamma (P_\gamma \rightarrow \leftarrow P_\gamma)$ where P_γ is the projector $|\gamma\rangle\langle\gamma|$; as shown by Fano [**14**], all superoperators in Liouville space have the form $\sum_{\alpha\beta}(A_\alpha \rightarrow \leftarrow B_\beta)$. Other examples are the identity operator:

$$(18)\qquad \mathcal{I} \equiv 1 = \sum_{\gamma\gamma'} |\gamma\rangle\langle\gamma| \rightarrow \leftarrow |\gamma'\rangle\langle\gamma'|$$

and the Liouville operator based on the commutator bracket,

$$(19)\qquad \mathcal{L}K = \frac{1}{\hbar}[H, K] \quad \text{or} \quad \mathcal{L} = \frac{1}{\hbar}[H \rightarrow \leftarrow I - I \rightarrow \leftarrow H].$$

While the ordinary operators K, being *elements* of the Liouville space, have the regular diadic representation $K_{\alpha\beta}$, superoperators have a tetradic representation; this will not be further discussed here (see Appendix A of Ref. [**48**]).

To return to (17), note $\mathcal{P}$ and $Q = 1 - \mathcal{P}$ satisfy the usual rules for projection operators, $\mathcal{P}^2 = \mathcal{P}$, $\mathcal{P}Q = 0$. Clearly also the diagonal part of any operator K commutes with H^0, i.e.,

(20) $$[\mathcal{P}K, H^0] = 0 \quad \text{or} \quad (\mathcal{P}K)H^0 = H^0\mathcal{P}K.$$

Note that $\mathcal{P}$ acts on everything to its right, except when restricted by parentheses. In addition to the Liouville operator $\mathcal{L}$ we need the subsystem Liouvillian $\mathcal{L}^0$, defined similarly

(21) $$\mathcal{L}^0 K = \frac{1}{\hbar}[H^0, K].$$

The following theorems are easily proven by series expansion:

THEOREM 1. *The evolution operator in Liouville space satisfies*

(22) $$e^{i\mathcal{L}_0 t} K = U^\dagger(t) K U(t), \quad U(t) = e^{-iH^0 t/\hbar}.$$

THEOREM 2. *We have the following identity involving diagonal parts:*

(23) $$e^{\sum_\gamma |\gamma\rangle\langle\gamma| f(\gamma)} = \sum_\gamma |\gamma\rangle\langle\gamma| e^{f(\gamma)}.$$

THEOREM 3. *For nondiagonal parts it holds that*

(24) $$e^{-it(1-\mathcal{P})\mathcal{L}}(1-\mathcal{P})K = (1-\mathcal{P})e^{-it\mathcal{L}(1-\mathcal{P})}K.$$

THEOREM 4. *For* $\lambda \to 0$ *one has*

(25) $$e^{-it(1-\mathcal{P})\mathcal{L}} K = \left[e^{-it\mathcal{L}^0} + O(\lambda)\right] K,$$

providing

(26) $$\mathcal{P}[K, H^0] = 0.$$

The proof is simple since

(27) $$\begin{aligned} e^{-it(1-\mathcal{P})\mathcal{L}} K &= e^{-it(1-\mathcal{P})\mathcal{L}^0} K + O(\lambda) \\ &= \sum_{n=0}^{\infty} \left[\frac{(-it\mathcal{L}^0 + it\mathcal{P}\mathcal{L}^0)^n}{n!} \right] K + O(\lambda). \end{aligned}$$

But $\mathcal{P}\mathcal{L}^0 K = 0$ *if* (26) *is satisfied.*

At this moment it may surprise the reader that (26) is *not* a consequence of (20). The problem involves here *the validity of the Dirac notation*, which assumes that there is no boundary term arising from Green's Theorem, which in turn implies that no transformations occur which carry the states outside the space $\mathcal{H}$ determined by H^0 and the boundary conditions. To appreciate this, we revert to the standard algebraic notation for the inner product in a Hilbert space [**33**]; we have if $|\phi\rangle$ and $|\psi\rangle$ are arbitrary states,

(28) $$\langle\phi|\{K|\psi\rangle\} \equiv (K\psi, \phi) \quad \text{and} \quad \{\langle\phi|K\}|\psi\rangle \equiv (\psi, K^\dagger\phi),$$

where in the first operation K acts to the right in $\mathcal{H}$, while in the second operation K acts to the left, i.e., in the dual space $\overline{\mathcal{H}}$. Green's theorem reads

(29) $$(K\psi, \phi) - (\psi, K^\dagger\phi) = Q_K(\psi, \phi),$$

where Q_K is the bilinear concomitant; it vanishes iff ψ satisfies the b.c. and ϕ the adjoint b.c. which, in view of the fact that K is self adjoint—not just Hermitian—must be identical to those of the elements of $\mathcal{H}$. Thus, when dealing with a transformation $K_1K_2,\ldots,K_n\gamma$, we find that $Q(K_1,\ldots,K_n\gamma,\gamma)=0$ iff $K_1,\ldots,K_n\gamma$ satisfies the assumed periodic b.c. We conclude that the *the standard Dirac projector algebra is valid whenever the operators $K_1,\ldots K_n$ are invariant against translation over the dimensions of the system.* To avoid ambiguity we adopt the consensus that in the Dirac bilinear form $\langle\phi|K|\psi\rangle$ the operator K acts to the right, unless otherwise indicated by curly brackets. As to our formulas of departure, (17), (17a), and (17b), *we note that* (17a) *presents no problems,* since $\langle\gamma|K|\gamma\rangle$ involves the same states and clearly $Q_K(\gamma,\gamma)\equiv 0$; on the other hand, (17b) is ambiguous, depending on whether in $\langle\gamma|K|\gamma'\rangle$ we operate to the left or to the right. For this reason we prefer to define K_{nd} as $K_{nd}\equiv K-K_d=(1-\mathcal{P})K$, *omitting the far rhs of* (17b). Further the following rules apply. Let L_d be a diagonal operator; then we have

(30a) $$\mathcal{P}(K_1\ldots K_nL_d)=(\mathcal{P}K_1\ldots K_n)L_d=L_d\mathcal{P}(K_1\ldots K_n),$$

(30b) $$\mathcal{P}(L_dK_1\ldots K_n)=L_d\mathcal{P}(K_1\ldots K_n)+\sum_\gamma|\gamma\rangle\langle\gamma|Q_{L_d}(K_1\ldots K_n\gamma,\gamma).$$

Or, for the commutator, we obtain

THEOREM 5.

(30) $$\mathcal{P}[K_1\ldots K_n,L_d]\ =-\sum_\gamma|\gamma\rangle\langle\gamma|Q_{L_d}(K_1\ldots K_n\gamma,\gamma).$$

In particular for the commutator (26):

(31) $$\mathcal{P}[K,H^0]=-\sum_\gamma|\gamma\rangle\langle\gamma|Q_{H^0}(K\gamma,\gamma).$$

The explicit proof is easy. Employing (17a) we write

$$\begin{aligned}\mathcal{P}[K,H^0]&=\sum_\gamma|\gamma\rangle\langle\gamma|(\langle\gamma|KH^0|\gamma\rangle-\langle\gamma|H^0K|\gamma\rangle)\\&=\sum_\gamma|\gamma\rangle\langle\gamma|(\langle\gamma|KH^0|\gamma\rangle-\{\langle\gamma|H^0\}K|\gamma\rangle-Q_{H^0}(K\gamma,\gamma))\\&=\sum_\gamma|\gamma\rangle\langle\gamma|(\langle\gamma|K|\gamma\rangle\mathcal{E}_\gamma-\mathcal{E}_\gamma\langle\gamma|K|\gamma\rangle-Q_{H^0}(K\gamma,\gamma))\\&=-\sum_\gamma|\gamma\rangle\langle\gamma|Q_{H^0}(K\gamma,\gamma),\end{aligned}$$

in which we used (29). Though for suitable operators K the bilinear concomitant can be obtained, it is usually easier to first evaluate the commutator, and to find afterwards its diagonal part. The problems raised here are essential for the derivation of the ME; improper use of the Dirac notation leads to fallicious results, see also Appendix B of Ref. [**48**].

3.2. The derivation. As before the system Hamiltonian is $H=H^0+\lambda V$, where it will now be stipulated that V involves no diagonal part (if it has such a part it should be added to H^0). In addition, we assume that an external field $F(t)$, with Hamiltonian—$AF(t)$, is switched on at $t=0$; here A is a system operator. E.g., if $F(t)$ represents the force $q\boldsymbol{E}(t)\bullet$ where $\boldsymbol{E}$ is an electric field, $A=\sum_i(\boldsymbol{r}_i-\boldsymbol{r}_{i0})$, $\boldsymbol{r}_i$

being the position coordinates. The von Neumann equation for the density operator now reads

$$\frac{\partial \rho}{\partial t} + i\mathcal{L}\rho = \frac{\partial \rho}{\partial t} + \left(\frac{i}{\hbar}\right)[H, \rho] = \left(\frac{i}{\hbar}\right)u(t)F(t)[A, \rho] \tag{32}$$

where $u(t)$ is the Heaviside function (unit step function). Writing $\rho = \rho_d + \rho_{nd}$ we find by application of the projection operators $\mathcal{P}$ and $1 - \mathcal{P}$ to (32)

$$\frac{\partial \rho_d}{\partial t} + i\mathcal{P}\mathcal{L}\rho_d + i\mathcal{P}\mathcal{L}\rho_{nd} = \left(\frac{i}{\hbar}\right)F(t)\mathcal{P}[A, \rho_d] + \left(\frac{i}{\hbar}\right)\mathcal{P}[A, \rho_{nd}] \tag{33}$$

$$\begin{aligned} \frac{\partial \rho_{nd}}{\partial t} &+ i(1-\mathcal{P})\mathcal{L}\rho_d + i(1-\mathcal{P})\mathcal{L}\rho_{nd} \\ &= \left(\frac{i}{\hbar}\right)F(t)(1-\mathcal{P})[A, \rho_d] + \left(\frac{i}{\hbar}\right)F(t)(1-\mathcal{P})[A, \rho_{nd}]. \end{aligned} \tag{34}$$

The second term on the left in (33) is zero, since $\mathcal{L}^0\rho_d = 0$ and $\mathcal{P}[V, \rho_d] = 0$, given that V, being usually the sum of all two-particle interactions, $\sum_{i,j} v(\boldsymbol{r}_i - \boldsymbol{R}_j)$, is invariant against translation of all particle coordinates $\{q\}$. Eqs. (33) and (34) now simplify to

$$\frac{\partial \rho_d}{\partial t} + i\mathcal{P}\mathcal{L}\rho_{nd} = \left(\frac{i}{\hbar}\right)F(t)\mathcal{P}[A, \rho], \tag{35}$$

$$\frac{\partial \rho_{nd}}{\partial t} + i\mathcal{L}\rho_d + i(1-\mathcal{P})\rho_{nd} = \left(\frac{i}{\hbar}\right)F(t)(1-\mathcal{P})[A, \rho]. \tag{36}$$

We first formally solve (36). Using the Green's operator

$$\mathcal{G}(t, t') = u(t - t')e^{-i(t-t')(1-\mathcal{P})\mathcal{L}} \tag{37}$$

we obtain

$$\begin{aligned} \rho_{nd}(t') = {} & \mathcal{G}(t, 0)\rho_{nd}(0) \\ &+ \int_0^t dt' G(t - t')\left\{-i\mathcal{L}\rho_d(t') + \left(\frac{i}{\hbar}\right)F(t')(1-\mathcal{P})\big[A, \rho(t')\big]\right\}. \end{aligned} \tag{38}$$

Substitution into (35) yields the integro-differential equation

$$\begin{aligned} \frac{\partial \rho_d}{\partial t} + \mathcal{P}\mathcal{L}\int_0^t dt'\mathcal{G}(t-t')\mathcal{L}\rho_d(t') = {} & \left(\frac{i}{\hbar}\right)F(t)\mathcal{P}\big[A, \rho(t)\big] \\ &+ \left(\frac{1}{\hbar}\right)\mathcal{P}\mathcal{L}\int_0^t dt' F(t')\mathcal{G}(t-t')(1-\mathcal{P})\big[A, \rho(t')\big] - i\mathcal{P}\mathcal{L}\mathcal{G}(t, 0)\rho_{nd}(0). \end{aligned} \tag{39}$$

Equations (38) and (39) are still fully equivalent to the original von Neumann equation, except that a fixed direction of time, $t \geq 0$, has been introduced.

We now make the standard simplification of linear response theory in that in the field terms $\rho(t)$ is replaced by $\rho(0)$ for which we assume the canonical (or grand canonical) distribution,

$$\rho(0) = \rho_{\mathrm{eq}} = \frac{e^{-\beta H}}{z}, \quad z = \mathrm{Tr}\, e^{-\beta H}, \quad \beta = \frac{1}{kT}. \tag{40}$$

In addition, we omit $\rho_{nd}(0)$, justifiable in the limit $\lambda \to 0$, see below. The generalized ME's are now:

$$(41)\quad \frac{\partial \rho_d}{\partial t} + \mathcal{P}\mathcal{L}\int_0^t dt'\mathcal{G}(t-t')\mathcal{L}\rho_d(t') = \left(\frac{i}{\hbar}\right)F(t)\mathcal{P}[A,\rho_{\text{eq}}] + \left(\frac{1}{\hbar}\right)\mathcal{P}\mathcal{L}\int_0^t dt' F(t')\mathcal{G}(t-t')(1-\mathcal{P})[A,\rho_{\text{eq}}]$$

$$(42)\quad \rho_{nd}(t) = \int_0^t dt'\mathcal{G}(t-t')\{-i\mathcal{L}\rho_d(t') + (i/\hbar)F(t')(1-\mathcal{P})[A,\rho_{\text{eq}}]\}.$$

The convolution integrals show the memory effects, involving the entire history of interactions and fields for $t' \le t$.

As first shown by Zwanzig [**50**], the Van Hove limit decouples these equations and reveals the asymptotic Markovian behaviour. More precisely, *we assume that t is large compared to the time of randomizing transitions $\gamma \to \gamma'$ caused by λV, while $\lambda \to 0$ but $(\lambda^2 t)^n$ is finite for any power $n \ge 1$.* For $\mathcal{L}\rho_d(t')$ we have $(\lambda/\hbar)\big[V,\rho_d(t')\big]$ since $\mathcal{L}^0\rho_d = 0$. For $\mathcal{G}(t-t')\rho_d(t')$ we use Theorem 4. We find

$$(43)\quad \mathcal{G}(t-t')\mathcal{L}\rho_d(t') = \left(\frac{\lambda}{\hbar}\right)\big[\mathcal{G}_0(t-t') + O(\lambda)\big]\big[V,\rho_d(t')\big]$$

where

$$(44)\quad \mathcal{G}_0(t) = u(t)e^{-i\mathcal{L}^0 t}.$$

Operating on (43) with $\mathcal{P}\mathcal{L} = \mathcal{P}\mathcal{L}^0 + \lambda\mathcal{P}\mathcal{L}^1$, we find that $\mathcal{P}\mathcal{L}^0$ does not contribute since $\mathcal{L}^0(\dots)$ leaves no diagonal part; consequently, using (17a) and (22) we have for the memory term of (41) in the weak interaction limit $\lambda \to 0$,

$$(45)\quad \mathcal{P}\mathcal{L}\int_0^t dt'\mathcal{G}(t-t')\mathcal{L}\rho_d(t') \to \sum_\gamma |\gamma\rangle\langle\gamma|\left(\frac{\lambda^2}{\hbar^2}\right)\int_0^t dt'\langle\gamma|\big[V, e^{-iH^0(t-t')/\hbar}[V,\rho_d(t')]e^{iH^0(t-t')/\hbar}\big]|\gamma\rangle.$$

With the ensemble probability $p(\gamma,t) = \langle\gamma|\rho_d(t)|\gamma\rangle$ one easily finds for the matrix element under the integral of (45):

$$(46)\quad \langle\gamma|[V,\dots]|\gamma\rangle = \sum_{\gamma''}\{\langle\gamma|V|\gamma''\rangle e^{-i\epsilon_{\gamma''}(t-t')/\hbar}\big[\langle\gamma''|V|\gamma\rangle p(\gamma,t') - p(\gamma'',t')\langle\gamma''|V|\gamma\rangle\big] \times e^{i\epsilon_\gamma(t-t')/\hbar} - e^{-i\epsilon_\gamma(t-t')/\hbar}\big[\langle\gamma|V|\gamma''\rangle p(\gamma'',t') - p(\gamma,t')\langle\gamma|V|\gamma''\rangle\big]e^{i\epsilon_{\gamma''}(t-t')/\hbar}\langle\gamma''|V|\gamma\rangle\}$$
$$= 2\sum_{\gamma''}\big|\langle\gamma|V|\gamma''\rangle\big|^2\cos\left[\frac{(\epsilon_{\gamma''}-\epsilon_\gamma)(t-t')}{\hbar}\right]\big[p(\gamma,t') - p(\gamma'',t')\big].$$

Denoting (45) by $-(\partial\rho/\partial t)_{\text{mem}}$ we now have

$$(47)\quad \left(\frac{\partial \rho_d}{\partial t}\right)_{\text{mem}} = -\sum_{\gamma\gamma''}|\gamma\rangle\langle\gamma|\left(\frac{2\lambda^2}{\hbar^2}\right)|\langle\gamma|V|\gamma''\rangle|^2 \times \int_0^t dt'\cos\left[\frac{(\epsilon_{\gamma''}-\epsilon_\gamma)(t-t')}{\hbar}\right]\big[p(\gamma,t') - p(\gamma'',t')\big].$$

This has the Laplace transform[5] employing the convolution theorem,

$$\Psi(s) = -\sum_{\gamma\gamma''} |\gamma\rangle\langle\gamma| \left(\frac{2\lambda^2}{\hbar^2}\right) |\langle\gamma|V|\gamma''\rangle|^2 \times \frac{s}{s^2 + (\epsilon_{\gamma''} - \epsilon_\gamma)^2/\hbar^2} \left[P(\gamma, s) - P(\gamma'', s)\right]. \tag{48}$$

For $|s| \ll (\epsilon_{\gamma''} - \epsilon_\gamma)/\hbar$ (corresponding to $t \gg \tau_t \equiv \hbar/\delta\epsilon$ in the time domain, τ_t being the transition time) we have the well-known limit

$$\lim_{s\to 0^+} \frac{1}{(\epsilon_{\gamma''} - \epsilon_\gamma)/\hbar \pm is} = \mp\pi i\hbar\delta(\epsilon_{\gamma''} - \epsilon_\gamma) + \mathbb{P}\frac{1}{\epsilon_{\gamma''} - \epsilon_\gamma} \tag{49}$$

where $\mathbb{P}$ denotes the Cauchy principal value. Hence,

$$\begin{aligned} &\lim_{s\to 0^+} \frac{s}{s^2 + (\epsilon_{\gamma''} - \epsilon_\gamma)^2/\hbar^2} \\ &\quad = \lim_{s\to 0^+} \frac{1}{2i}\left[\frac{1}{(\epsilon_{\gamma''} - \epsilon_\gamma)/\hbar - is} - \frac{1}{(\epsilon_{\gamma''} - \epsilon_\gamma)/\hbar + is}\right] = \pi\hbar\delta(\epsilon_{\gamma''} - \epsilon_\gamma). \end{aligned} \tag{50}$$

The inverse transform becomes

$$\left(\frac{\partial\rho_d}{\partial t}\right)_{\text{mem}} = -\Lambda_d\rho_d, \tag{51}$$

where in a natural fashion the master operator emerged:

$$\Lambda_d K = -\sum_{\gamma\gamma''} |\gamma\rangle\langle\gamma| \left[W_{\gamma''\gamma}\langle\gamma''|K|\gamma''\rangle - W_{\gamma\gamma''}\langle\gamma|K|\gamma\rangle\right], \tag{52}$$

with $W_{\gamma\gamma''}$ being the transition probability of Fermi's golden rule

$$W_{\gamma\gamma''} = \left(\frac{2\pi\lambda^2}{\hbar}\right) |\langle\gamma|V|\gamma''\rangle|^2 \delta(\epsilon_\gamma - \epsilon_{\gamma''}) = W_{\gamma''\gamma}. \tag{53}$$

The equality of the first and third member expresses the property of microscopic reversibility, though the memory term (51) exhibits macroscopic irreversibility as evident from the positive entropy production associated with the Markovian terms (51) [**47**]. It is important to note here that Λ_d destroys the nondiagonal part, while $\mathfrak{L}^0$ destroys a diagonal part:

$$\Lambda_d K = \Lambda_d K_d, \quad \mathfrak{L}^0 K = \mathfrak{L}^0 K_{nd}. \tag{54}$$

Next we must evaluate the field terms on the rhs of Eq. (41). The following assumptions are made:

(i) A commutes with λV. This is the case for all conductivity problems for which $A = \sum_i (\boldsymbol{r}_i - \boldsymbol{r}_{i0})$ is a positions operator. It is also true for most spin relaxation phenomena if $F(t)$ refers to a magnetic field [**47, 48**].

(ii) $\dot{A}$ is translational invariant. For conductivity problems $\dot{A}$ represents the total momentum, which is related by the translation operator T_ℓ via $T_\ell = \exp(i\boldsymbol{P}\cdot\ell/\hbar)$; clearly $\dot{A}$ and T_ℓ commute.

---

[5] Laplace transforms could have been used from the beginning [**40**] using resolvent expansions, but we believe the physics of the memory effects are more clearly displayed in the time-domain.

For ρ_{eq} we assume (40) and we use Kubo's lemma

$$[A, e^{-\beta H}] = -\int_0^\beta d\beta' e^{-\beta H}\left[A^H(-i\hbar\beta'), H\right] \tag{55}$$
$$= -i\hbar \int_0^\beta d\beta' e^{-\beta H}\dot{A}^H(-i\hbar\beta'),$$

where $A^H(t)$ is the Heisenberg operator; in the second equality we used the Heisenberg equation of motion for $t = -i\hbar\beta'$. For $\lambda \to 0$, $\dot{A}^H \to \dot{A}^I$ where the latter is the interaction-picture operator[6]

$$\dot{A}^I(-i\hbar\beta') = e^{\hbar\beta'\mathcal{L}^0}\dot{A} = e^{\beta' H^0}\dot{A}e^{-\beta' H^0}. \tag{56}$$

Because of (ii) the $\mathcal{P}$ operator of the first field term can be carried through, leaving

$$e^{\hbar\beta'\mathcal{L}^0}(\dot{A})_d = e^{\beta' H^0}(\dot{A})_d e^{-\beta' H^0} = (\dot{A})_d, \tag{57}$$

where the parentheses indicate that differentiation must precede diagonal projection. Hence,

$$\mathcal{P}\left[\frac{A, e^{-\beta H}}{z}\right] \to -i\hbar\beta\rho_{\text{eq}}(\dot{A})_d. \tag{58}$$

The field term now becomes:

$$\text{first field term} \;= \beta F(t)\rho_{\text{eq}}(\dot{A})_d. \tag{59}$$

Finally, we turn to the second field term on the rhs of (41). It contains both the memory effect of the collisions and the external field; it will turn out to represent the "collisional current", predicted by Titeica, but first obtained in Refs. [**48**] and [**9**]. The $\mathcal{PL}$ in front of the terms is split into $\mathcal{PL}^0 + \lambda\mathcal{PL}^1$, while $\mathcal{G}$ is replaced by $\mathcal{G}^0 + O(\lambda)$ as before. Clearly $\mathcal{PL}^0\ldots$ yields zero. We therefore arrive for the second field term at, using $\mathcal{L}_0^1\bullet = \hbar^{-1}[V, \bullet]$ and using once more Kubo's lemma,

$$-\left(\frac{\lambda}{\hbar}\right)^2 \mathcal{P}\Big[V, \int_0^t \mathrm{d}t' F(t')(e^{-i\mathcal{L}^0(t-t')} + 0(\lambda)) \tag{60}$$
$$\times(1-\mathcal{P})\int_0^\beta d\beta' e^{-\beta H}\frac{[A^H(-i\hbar\beta'), H]}{z}\Big].$$

We examine separately the effects of the "1" and of the "$-\mathcal{P}$". For the first we write,

$$\left[A^H(-i\hbar\beta), H\right] \simeq \left[A^H(0), H^0\right] + \lambda[A, V] = [A, H^0] \tag{61}$$

since $[A, V] = 0$ by assumption (i). The commutator $[A, H^0]$ is basically unbounded; we may assume, however, that it averages to zero due to the rapidly oscillating superoperator $\exp\left[-i\mathcal{L}^0(t-t')\right]$ (the dismissal of this term needs further scrutiny).

[6]Instead of replacing $\dot{A}^H$ by $\dot{A}^I$, a series in λ^n can be given, of which we retained only the term with $n = 0$; higher order terms do not contribute to the first field term since it does not involve asymptotic time. In the second field term, effects of order $\lambda^n t$, with $n = 3, 4\ldots$, will occur. These are of no concern in the "Van Hove limit proper"; see Loss [**29**] for extensions.

For the other contribution we have

$$(62)\qquad -\mathcal{P}\int_0^\beta d\beta'\left(\frac{e^{-\beta H^0}}{z^0}\right)(\hbar i)\dot{A}^H(-i\hbar\beta')$$
$$= -\int_0^\beta d\beta'\left(\frac{e^{-\beta H^0}}{z^0}\right)(\hbar i)\big(\dot{A}^H(-ih\beta)\big)_d$$
$$\to -\int_0^\beta d\beta'\rho_{\rm eq}\big[A_d^I(-i\hbar\beta'),\lambda V\big] = \beta\lambda\rho_{\rm eq}[V, A_d].$$

We note here that we first evaluated the commutator $\big[A^H(-i\hbar\beta'), H\big]$ via the Heisenberg equation, prior to carrying through the $\mathcal{P}$, after which we reverted back to the commutator form, replacing A^H by A^I; analogous to (57) we noted $A_d^I(t) \equiv A_d$ for all t. Substituting (62) into (60) we obtain, using the explicit form (17a) for the opening projector $\mathcal{P}$:

$$(63)\quad \text{second field term} = -\left(\frac{\lambda^2}{\hbar^2}\right)\beta\rho_{\rm eq}\sum_\gamma |\gamma\rangle\langle\gamma|\,\langle\gamma|\Big[V, \int_0^t dt' F(t') \times e^{-i\mathcal{L}^0(t-t')}[V, A_d]\Big]|\gamma\rangle.^7$$

We note that (63) has a structure entirely similar to (45). Thus, for its Laplace transform we obtain

$$(64)\quad \chi(s) = -\sum_{\gamma\gamma''}|\gamma\rangle\langle\gamma|\left(\frac{2\lambda^2}{\hbar^2}\right)\beta\rho_{\rm eq}\big|\langle\gamma|V|\gamma''\rangle\big|^2\hat{F}(s) \times \frac{s}{s^2+(\epsilon_{\gamma''}-\epsilon_\gamma)^2/\hbar^2}\big[\langle\gamma|A_d|\gamma\rangle - \langle\gamma''|A_d|\gamma''\rangle\big].$$

With (50) and the statements (52) and (53) one finds upon inverse transformation:

$$(65)\qquad \text{second field term} = -\beta F(t)\rho_{\rm eq}\Lambda_d A_d.$$

We observe that in both memory terms (51) and (65) Λ_d acts on a *diagonal* operator, viz., ρ_d and A_d, respectively; no use is made of (54), Λ_d being principally an operator in diagonal Liouville space [**47**]. Collecting terms, the entire diagonal master equation now becomes

$$(66)\qquad \boxed{\frac{\partial\rho_d}{\partial t} + \Lambda_d\rho_d(t) = \beta F(t)\rho_{\rm eq}\big[(\dot{A})_d - \Lambda_d A_d\big].}$$

The [ ] on the rhs will also be referred to as the "reduced current operator", i.e., after the Van Hove limit; we have

$$(67)\qquad J_{A,d} \equiv (\dot{A}^R)_d = (\dot{A})_d - \Lambda_d A_d.$$

The first term denotes ponderomotive current while the second term (not existing in the full dynamics of H) is called collisional current.

Rests to return to the nondiagonal result (42). The first term is found to be of order λ, so it will be dismissed. For the second term we use Kubo's lemma. We

[7] In contrast to the treatment of Ref. [**48**] the operator $\rho_{\rm eq}$ has been taken outside the commutator since in the ensuing transitions the energy is conserved.

find

$$\rho_{nd}(t) = \left(\frac{i}{\hbar}\right)\int_0^t dt' F(t')\mathcal{G}(t-t')(1-\mathcal{P})[A, \rho_{\text{eq}}] \tag{68}$$
$$= \int_0^t dt' F(t') e^{-i\mathcal{L}^0(t-t')} \int_0^\beta d\beta' \rho_{\text{eq}} e^{\mathcal{L}^0\hbar\beta'}(\dot{A})_{nd} + O(\lambda).$$

The reduced nondiagonal operators show no relaxation and we have

$$\mathcal{J}_{A,nd} \equiv (\dot{A}^R)_{nd} = (\dot{A})_{nd}, \tag{69}$$

the extreme rhs being the Schrödinger operator. Differentiating (68) to the time we obtain

$$\boxed{\frac{\partial \rho_{nd}}{\partial t} + i\mathcal{L}^0 \rho_{nd}(t) = F(t)\rho_{\text{eq}} \int_0^\beta d\beta' e^{\mathcal{L}^0\hbar B'}(\dot{A})_{nd}.} \tag{70}$$

Finally, we can combine (66) and (70) upon employing (54) and (57); they then read

$$\frac{\partial \rho_d}{\partial t} + \Lambda_d \rho(t) = F(t)\rho_{\text{eq}} \int_0^\beta d\beta' e^{\hbar\beta'\mathcal{L}^0}(\dot{A}^R)_d, \tag{71}$$

$$\frac{\partial \rho_{nd}}{\partial t} + i\mathcal{L}^0 \rho(t) = F(t)\rho_{\text{eq}} \int_0^\beta d\beta' e^{\hbar\beta'\mathcal{L}^0}(\dot{A}^R)_{nd}, \tag{72}$$

Or, after adding,

$$\boxed{\frac{\partial \rho}{\partial t} + (\Lambda_d + i\mathcal{L}^0)\rho(t) = F(t)\rho_{\text{eq}} \int_0^\beta d\beta' e^{\hbar\beta'\mathcal{L}^0}\dot{A}^R,} \tag{73}$$

where we have for the total current associated with A:

$$\boxed{\mathcal{J}_A^R \equiv \dot{A}^R = \dot{A} - \Lambda_d A.} \tag{74}$$

Comparing with the original von Neuman equation we see that the Van Hove limit has provided damping to $\rho(t)$ and the Heisenberg operators $A^H(t)$ since $\|\Lambda_d\|$ is semi-positive definite. In particular we established

$$\lim_{\lambda,t} e^{\pm i\mathcal{L}t} = e^{(-\Lambda_d \pm i\mathcal{L}^0)t}. \tag{75}$$

For the Heisenberg operators the evolution involves the plus sign; thus

$$\lim_{\lambda,t} A^H(t) = A^R(t) = e^{(-\Lambda_d + i\mathcal{L}^0)t} A. \tag{76}$$

Upon differentiation one obtains

$$\dot{A}^R(0) \equiv \dot{A}^R = -\Lambda_d A_d + \dot{A}_d + \dot{A}_{nd}, \tag{77}$$

in agreement with (74).

4. The Quantum Boltzmann Equations and the Wigner Form

In order to obtain one-particle equations (or higher order equations of the ME hierarchy) we must resort to the occupation-number representation for the states $|\gamma\rangle$. We thus have in the Fock space of all particles

(78) $$|\gamma\rangle = \big|\{n_\zeta\}, \{N_\eta\}\big\rangle = \big|\{n_\zeta\}\big\rangle \otimes \big|\{N_\eta\}\big\rangle,$$

where n_ζ and N_η are occupancy numbers, being the eigenvalues of the operators $\boldsymbol{n}_\zeta = c_\zeta^\dagger c_\zeta$ and $\boldsymbol{N}_\eta = a_\eta^\dagger a_\eta$ of Fermion states $|\zeta)$ and Boson states $|\eta)$, respectively. The Hamiltonian H^0 is now

(79) $$H^0 = \sum_\zeta \boldsymbol{n}_\zeta \epsilon_\zeta + \sum_n \boldsymbol{N}_\eta E_\eta$$

while the general binary interaction is

(80) $$\lambda V = \sum_{\zeta'\zeta''\eta'\eta''} (\zeta''\eta''|\lambda v|\zeta'\eta') c_{\zeta''}^\dagger a_{\eta''}^\dagger a_{\eta'} c_{\zeta'}.$$

We shall also be concerned with electron-phonon interaction, see Ref. [**9**]. In that case

(81) $$\lambda V = (i) \sum_{\zeta'\zeta''\boldsymbol{q}'} \mathcal{F}(\boldsymbol{q}') \big[c_{\zeta''}^\dagger c_{\zeta'} a_{\boldsymbol{q}'} (\zeta''|e^{i\boldsymbol{q}'\cdot\boldsymbol{r}}|\zeta') \pm c_{\zeta''}^\dagger c_{\zeta'} a_{\boldsymbol{q}'}^\dagger (\zeta''|e^{-i\boldsymbol{q}'\cdot\boldsymbol{r}}|\zeta')\big],$$

where the two terms represent phonon absorption and emission. $\mathcal{F}(q)$—as well as the sign and the presence of the factor i—depends on the type of phonons (acoustical, optical, polar optical, etc) involved. Both for (80) and (81) we assume that the Bosons (phonons) remain in thermal equilibrium, given by the Bose-Einstein distribution.

To obtain the diagonal quantum Boltzmann equation (diagonal QBE), we multiply the diagonal ME (66) by $\boldsymbol{n}_\zeta$ and take the trace. We thus obtain

(82) $$\frac{\partial \langle n_\zeta \rangle_t}{\partial t} + \operatorname{Tr}\{\boldsymbol{n}_\zeta \Lambda_d \rho_d(t)\} = \beta F(t) \operatorname{Tr}\{\rho_{\mathrm{eq}} \boldsymbol{n}_\zeta [-\Lambda_d A_d + (\dot{A})_d]\}$$

where we noticed that ρ_{eq} and $\boldsymbol{n}_\zeta$ commute. A general lemma, derived in Ref. [**48**] permits us to switch the positions of operators preceding and following Λ_d in the trace. We therefore arrive at

(83) $$\frac{\partial \langle n_\zeta \rangle_t}{\partial t} + \operatorname{Tr}\{\rho_d(t) \Lambda_d \boldsymbol{n}_\zeta\} = \beta F(t) \operatorname{Tr}\{\rho_{\mathrm{eq}} \boldsymbol{n}_\zeta (-\Lambda_d A_d)\} + \beta F(t) \operatorname{Tr}\{\rho_{\mathrm{eq}} \boldsymbol{n}_\zeta (\dot{A})_d\}.$$

The evaluation of the second term, stemming from the interactions, is again the most crucial step. Essential here is the computation of $W_{\gamma\bar{\gamma}}$, employing the interaction forms (80) or (81) and finding the "connected states" based on the action of the creation and annihilation operators. For the latter we employ the results

(84a) $$a_\eta^\dagger \big|\{N_\eta\}\big\rangle = \sqrt{N_\eta + 1} |N_1 \ldots N_\eta + 1 \ldots\rangle,$$

(84b) $$a_\eta \big|\{N_\eta\}\big\rangle = \sqrt{N_\eta} |N_1 \ldots N_\eta - 1 \ldots\rangle,$$

(84c) $$c_\eta^\dagger \big|\{n_\zeta\}\big\rangle = (-1)^\alpha \sqrt{1 - n_\zeta} |n_1 n_2 \ldots 1 - n_\zeta \ldots\rangle,$$

(84d) $$c_\zeta \big|\{n_\zeta\}\big\rangle = (-1)^\beta \sqrt{n_\zeta} |n_1 n_2 \ldots 1 - n_\zeta \ldots\rangle.$$

While the Fermion occupancies are 0 or 1, yet the radicals of (84c) and (84d) are necessary in nonequilibrium statistical mechanics (Ref. [**48**], Appendix D). For a given set $\zeta'\zeta''\eta'\eta''$ of the series (80) the connected state (i.e. with non-vanishing matrix element) is denoted by $\bar{\gamma}_{\zeta'\zeta''\eta'\eta''}$; one finds from Eqs. (84):

$$\bar{n}_\zeta = n_\zeta(1 - \delta_{\zeta\zeta'} - \delta_{\zeta\zeta''}) + (1 - n_\zeta)(\delta_{\zeta\zeta'} + \delta_{\zeta\zeta''}) \tag{85}$$

and

$$\bar{N}_\eta = N_\eta - \delta_{\eta\eta'} + \delta_{\eta\eta''}. \tag{86}$$

The operator $\Lambda_d \boldsymbol{n}_{\zeta^0}$, involving $n_{\zeta^0} - \bar{n}_{\zeta^0}$, is now easily evaluated, cf. Ref. [**9**], Eqs. (2.17)–(2.22). Because the interaction matrix elements are squared in $W_{\gamma\bar{\gamma}}$, the result becomes of second degree in the occupation numbers $\{N_\eta\}$ and $\{\eta_\zeta\}$—although on the many-body level the kinetic equations are linear! A straightforward, though laborious computation yields for the Boson average of $\Lambda_d \boldsymbol{n}_{\zeta^0}$

$$\langle \Lambda_d \boldsymbol{n}_{\zeta^0} \rangle_b = \sum_{\{n_\zeta\}} |\{n_\zeta\}\rangle\langle\{n_\zeta\}| \mathcal{B}_{\zeta^0} n_{\zeta^0}, \tag{87}$$

where $\mathcal{B}_\zeta$ is the Boltzmann collision operator:

$$\mathcal{B}_\zeta n_\zeta = \sum_{\bar{\zeta}} \{ w_{\zeta\bar{\zeta}} n_\zeta (1 - n_{\bar{\zeta}}) - w_{\bar{\zeta}\zeta} n_{\bar{\zeta}} (1 - n_\zeta) \}. \tag{88}$$

The Fermion transition rates are given by

$$w_{\zeta''\zeta'} = \sum_{\eta'\eta''} Q(\zeta''\eta''; \zeta', \eta') \big\langle N_{\eta''}(1 + N_{\eta'}) \big\rangle_{\mathrm{eq}} \tag{89}$$

with

$$Q(\zeta''\eta''; \zeta'\eta') = \left(\frac{2\pi\lambda^2}{\hbar}\right) \big|(\zeta''\eta''|v|\zeta'\eta')\big|^2 \delta(\epsilon_{\zeta''} - \epsilon_{\zeta'} + E_{\eta''} - E_{\eta'}). \tag{90}$$

For electron-phonon interactions the Boltzmann collision operator retains the form (88); however, the Fermion transition rates, based on (81), are given by

$$w_{\zeta''\zeta'} = \sum_{\boldsymbol{q}} [Q(\zeta'', \boldsymbol{q} \to \zeta')\langle N_{\boldsymbol{q}}\rangle_{\mathrm{eq}} + Q(\zeta'' \to \zeta', \boldsymbol{q})(1 + \langle N_{\boldsymbol{q}}\rangle_{\mathrm{eq}}] \tag{91}$$

with

$$Q(\zeta'', \boldsymbol{q} \to \zeta') = \left(\frac{2\pi}{\hbar}\right) |\mathcal{F}(\boldsymbol{q})|^2 \big|(\zeta'|e^{i\boldsymbol{q}\cdot\boldsymbol{r}}|\zeta'')\big|^2 \delta(\epsilon_{\zeta''} - \epsilon_{\zeta'} + E_{\boldsymbol{q}}), \tag{92a}$$

$$Q(\zeta'' \to \zeta', \boldsymbol{q}) = \left(\frac{2\pi}{\hbar}\right) |\mathcal{F}(\boldsymbol{q})|^2 \big|(\zeta'|e^{-i\boldsymbol{q}\cdot\boldsymbol{r}}|\zeta'')\big|^2 \delta(\epsilon_{\zeta''} - \epsilon_{\zeta'} - E_{\boldsymbol{q}}). \tag{92b}$$

For both cases the second term of (83), obtained from (87) by taking the trace in the Fermion Fock-subspace yields $\langle \mathcal{B}_\zeta n_\zeta \rangle_t \simeq \mathcal{B}_\zeta \langle n_\zeta \rangle_t$ where we used the truncation rule that in the n-particle equation averages involving $(n+1)$ particles can be factured, see Ref. [**48**], Eq. (8.1). For the second term on the rhs of (83) we write $(\dot{A})_d = \sum_{\zeta'} \boldsymbol{n}_{\zeta'} (\zeta'|\dot{a}|\zeta')$ where $\dot{a}$ is the one-particle velocity operator. We thus must evaluate the average

$$\sum_{\zeta'} \langle n_\zeta n_{\zeta'} \rangle_{\rm eq} (\zeta'|\dot{a}|\zeta') \tag{93}$$
$$= \sum_{\zeta' \neq \zeta} \langle n_\zeta \rangle_{\rm eq} \langle n_{\zeta'} \rangle_{\rm eq} (\zeta'|\dot{a}|\zeta') + \langle n_\zeta^2 \rangle_{\rm eq} (\zeta|\dot{a}|\zeta)$$
$$= \langle n_\zeta \rangle_{\rm eq} \sum_{\zeta'} \langle n_{\zeta'} \rangle_{\rm eq} (\zeta'|\dot{a}|\zeta') + \left[\langle n_\zeta^2 \rangle_{\rm eq} - \langle n_\zeta \rangle_{\rm eq}^2\right] (\zeta|\dot{a}|\zeta).$$

The first sum is zero, since $\langle (\dot{A})_d \rangle_{\rm eq} = 0$. We are left with

$$2^{\rm nd} \text{ term rhs } = \beta F(t) \langle \eta_\zeta \rangle_{\rm eq} (1 - \langle n_\zeta \rangle_{\rm eq}) (\zeta|\dot{a}|\zeta). \tag{94}$$

For the first term on the rhs we write $A_d = \sum_{\zeta'} \boldsymbol{n}_{\zeta'} (\zeta'|a|\zeta)$; this term, then, involves the average

$$\sum_{\zeta'} \langle n_\zeta \langle -\Lambda_d \boldsymbol{n}_{\zeta'} \rangle_b (\zeta'|a|\zeta') \rangle_{\rm eq} = - \sum_{\zeta'} \langle n_\zeta \mathcal{B}_{\zeta'} n_{\zeta'} \rangle_{\rm eq} (\zeta'|a|\zeta'). \tag{95}$$

For a computation, see Ref. [**48**], Eqs. (8.43)ff. The result is

$$1^{\rm st} \text{ term rhs } = -\beta F(t) \langle n_\zeta \rangle_{\rm eq} (1 - \langle n_\zeta \rangle_{\rm eq}) \tag{96}$$
$$\times \sum_{\zeta'} \left\{ \left[(\zeta|a|\zeta) - (\zeta'|a|\zeta')\right] \left[w_{\zeta\zeta'} (1 - \langle n_{\zeta'} \rangle_{\rm eq}) + w_{\zeta'\zeta} \langle n_{\zeta'} \rangle_{\rm eq}\right] \right\}.$$

Thus the diagonal QBE is obtained:

$$\frac{\partial \langle n_\zeta \rangle_t}{\partial t} - \beta F(t) \langle n_\zeta \rangle_{\rm eq} (1 - \langle n_\zeta \rangle_{\rm eq}) (\zeta|\dot{a}|\zeta) - \beta F(t) \langle n_\zeta \rangle_{\rm eq} (1 - \langle n_\zeta \rangle_{\rm eq}) \tag{97}$$
$$\times \sum_{\zeta'} \left\{ \left[(\zeta'|a|\zeta') - (\zeta|a|\zeta)\right] \left[w_{\zeta\zeta'} (1 - \langle n_{\zeta'} \rangle_{\rm eq}) + w_{\zeta'\zeta} \langle n_{\zeta'} \rangle_{\rm eq}\right] \right\}$$
$$= -\mathcal{B}_\zeta \langle n_\zeta \rangle_t.$$

We note that there are two streaming terms associated with the external field; the second one, dealing with the position matrix element, has no classical analog.

By multiplying the nondiagonal ME by $c_{\zeta_1}^\dagger c_{\zeta_2}$ and taking the trace, the nondiagonal QBE can be obtained. We refer to Ref. [**9**] for details. The result is, with $\zeta_1 \neq \zeta_2$:

$$\frac{\partial \langle c_{\zeta_1}^\dagger c_{\zeta_2} \rangle_t}{\partial t} - F(t) \frac{1 - \exp[-\beta(\epsilon_{\zeta_1} - \epsilon_{\zeta_2})]}{\epsilon_{\zeta_1} - \epsilon_{\zeta_2}} \langle n_{\zeta_2} \rangle_{\rm eq} (1 - \langle n_{\zeta_1} \rangle_{\rm eq}) \tag{98}$$
$$\times (\zeta_2|\dot{a}|\zeta_1) = -\left(\frac{i}{\hbar}\right) (\epsilon_{\zeta_2} - \epsilon_{\zeta_1}) \langle c_{\zeta_1}^\dagger c_{\zeta_2} \rangle_t.$$

Note that the nondiagonal QBE does not depend on the interaction potential λV; it describes universal quantum-interference effects.

Finally, we can combine both equations. (Alternatively, we could have started from the full ME(73).) The complete QBE reads (ζ_1 and ζ_2 arbitrary):

$$
\begin{aligned}
(99)\quad & \frac{\partial\langle c^\dagger_{\zeta_1} c_{\zeta_2}\rangle_t}{\partial t} - F(t)\frac{1-\exp[-\beta(\epsilon_{\zeta_1}-\epsilon_{\zeta_2})]}{\epsilon_{\zeta_1}-\epsilon_{\zeta_2}}\langle n_{\zeta_2}\rangle_{\rm eq}\big(1-\langle n_{\zeta_1}\rangle_{\rm eq}\big)(\zeta_2|\dot{a}|\zeta_1) \\
& - \beta F(t)\langle n_{\zeta_1}\rangle_{\rm eq}\big(1-\langle n_{\zeta_1}\rangle_{\rm eq}\big) \\
& \times \sum_{\zeta'}\Big\{\big[(\zeta'|a|\zeta')-(\zeta_1|a|\zeta_1)\big]\big[w_{\zeta_1\zeta'}\big(1-\langle n_{\zeta'}\rangle_{\rm eq}\big)+w_{\zeta'\zeta_1}\langle n_{\zeta_1}\rangle_{\rm eq}\big]\Big\}\delta_{\zeta_1\zeta_2} \\
& = \sum_{\zeta'}\big[w_{\zeta'\zeta_1}\langle n_{\zeta'}\rangle_t\big(1-\langle n_{\zeta_1}\rangle_t\big) - w_{\zeta_1\zeta'}\langle n_{\zeta_1}\rangle_t\big(1-\langle n_{\zeta'}\rangle_t\big)\big]\delta_{\zeta_1\zeta_2} \\
& \quad - \left(\frac{i}{\hbar}\right)(\epsilon_{\zeta_2}-\epsilon_{\zeta_1})\langle c^\dagger_{\zeta_1}c_{\zeta_2}\rangle_t .
\end{aligned}
$$

This transport equation comprises all quantum effects, providing we do not go too far from the equilibrium state, as evidenced by the equilibrium averages in the streaming terms which stem from the *linear* response assumption in the ME. (A quantum transport equation valid far from equilibrium undoubtedly exists, but—in general form—has so far been illusive.) Even so, a great many applications of (99) have been published in the last decade.

We now state the forms for the electrical current operator and relevant conductivity expressions [**45**]. From (74) we have the *many-body form* ($q = -e$)

$$
(100)\qquad \boldsymbol{J} = -\frac{e}{\Omega}\left[-\Lambda_d\sum_i(\boldsymbol{r}_i-\boldsymbol{r}_{i,\rm eq})+\sum_i \boldsymbol{v}_i\right].
$$

Employing the occupation formalism, we also have the *second quantization form*

$$
(101)\qquad \boldsymbol{J} = -\frac{e}{\Omega}\sum_{\zeta'\zeta''}\big[-\Lambda_d\boldsymbol{n}_{\zeta'}(\zeta'|\boldsymbol{r}-r_{\rm eq}|\zeta')\delta_{\zeta'\zeta''}+c^\dagger_{\zeta''}c_{\zeta'}(\zeta''|\boldsymbol{v}|\zeta')\big].
$$

Finally, with Λ_d representing the effect of binary interactions, we have the Boltzmann form for $\mathrm{Tr}(\rho\boldsymbol{J})$:

$$
(102)\qquad \langle\boldsymbol{J}\rangle_t = -\frac{e}{\Omega}\sum_{\zeta'\zeta''}\big[-\mathcal{B}_{\zeta'}\langle n_{\zeta'}\rangle_t(\zeta'|\boldsymbol{r}-\boldsymbol{r}_{\rm eq}|\zeta')\delta_{\zeta'\zeta''}+\langle c^\dagger_{\zeta''}c_{\zeta'}\rangle_t(\zeta''|\boldsymbol{v}|\zeta')\big].
$$

As to the conductivities, there are three types, viz. $\sigma^d_{\mu\nu}$ (coll.), $\sigma^d_{\mu\nu}$ (pond.), and $\sigma^{nd}_{\mu\nu}$ (pond.); here ν and μ refer to Cartesian components (field in direction ν, current in direction μ).

The collisonal conductivity is important for magneto-phonon effects. It arises from the motion of the centre of the Landau orbits. It is most easily obtained for frequency $\omega = 0$. Then in the stationary regime, when $(\zeta|\boldsymbol{v}|\zeta) = 0$ but $(\zeta|\boldsymbol{r} - \boldsymbol{r}_{\rm eq}|\zeta) \equiv \boldsymbol{R}_\zeta \neq 0$ and $F = -e\boldsymbol{E}\bullet$, (97) yields

$$
\begin{aligned}
(103)\quad & \beta e E_\nu\langle n_\zeta\rangle_{\rm eq}\big(1-\langle n_\zeta\rangle_{\rm eq}\big) \\
& \times\Big\{\sum_{\zeta'}(R_{\nu\zeta'}-R_{\nu\zeta})\big[w_{\zeta\zeta'}\big(1-\langle n_{\zeta'}\rangle_{\rm eq}\big)+w_{\zeta'\zeta}\langle n_{\zeta'}\rangle_{\rm eq}\big]\Big\} \\
& \qquad = -\mathcal{B}_\zeta\langle n_\zeta\rangle_{\rm stat}.
\end{aligned}
$$

Substitution into (108) gives

$$(104)\quad \langle \mathcal{J}_{u,\mathrm{coll}}\rangle_{\mathrm{stat}} = -\frac{\beta e^2}{\Omega} E_\nu \sum_{\zeta,\sigma} \langle n_\zeta\rangle_{\mathrm{eq}}\big(1-\langle n_\zeta\rangle_{\mathrm{eq}}\big) \sum_{\zeta'} (R_{\nu\zeta'} - R_{\nu\zeta}) R_{\mu\zeta} \times \big[w_{\zeta\zeta'}(1-\langle n_{\zeta'}\rangle_{\mathrm{eq}}) + w_{\zeta'\zeta}\langle n_{\zeta'}\rangle_{\mathrm{eq}}\big].$$

Note we added a summation over spin. Using detailed balance

$$(105)\quad w_{\zeta\zeta'}\langle n_\zeta\rangle_{\mathrm{eq}}\big(1-\langle n_{\zeta'}\rangle_{\mathrm{eq}} = w_{\zeta'\zeta}\langle n_{\zeta'}\rangle_{\mathrm{eq}}\big(1-\langle n_\zeta\rangle_{\mathrm{eq}}\big)$$

one easily finds

$$(106)\quad \sigma_{\mu\nu}(\mathrm{coll}, \omega\to 0) = -\frac{2\beta e^2}{\Omega}\sum_{\zeta\zeta'} (R_{\nu\zeta'} - R_{\nu\zeta}) R_{\mu\zeta} w_{\zeta\zeta'}\langle n_\zeta\rangle_{\mathrm{eq}}\big(1-\langle n_{\zeta'}\rangle_{\mathrm{eq}}\big).$$

If $\mu = \nu = x$ we find, by interchanging the two indices ζ, ζ' and taking half the sum,

$$(107)\quad \sigma_{xx}(\mathrm{coll}, \omega\to 0) = \frac{\beta e^2}{\Omega}\sum_{\zeta\zeta'} (R_{x\zeta'} - R_{x\zeta})^2 w_{\zeta\zeta'}\langle n_\zeta\rangle_{\mathrm{eq}}\big(1-\langle n_{\zeta'}\rangle_{\mathrm{eq}}\big).$$

This result was first obtained by Argyres and Roth in a rather obscure final section of their paper [**2**]. For elastic collisions $\big(1-\langle n_{\zeta'}\rangle_{\mathrm{eq}}\big)$ is to be replaced by $(1-\langle n_\zeta\rangle_{\mathrm{eq}})$; the result so found was obtained with density matrix methods by Adams and Holstein [**1**].

When we deal with extended states such as Bloch states, $(\zeta|\boldsymbol{r}-\boldsymbol{r}_{\mathrm{eq}}|\zeta) = 0$ but $(\zeta|\boldsymbol{v}|\zeta) \neq 0$, a formal solution of the diagonal QBE exists if $\mathcal{B}$ is linear, like for elastic collisions or inelastic collisions in a nondegenerate electron gas. We have

$$(108)\quad \langle n_\zeta\rangle_t = \langle n_\zeta\rangle_{\mathrm{eq}} - e\int_0^t dt' E_\nu(t') e^{-\mathcal{B}_\zeta(t-t')} \frac{\partial\langle n_\zeta\rangle_{\mathrm{eq}}}{\partial\epsilon_\zeta}(\zeta|v_\nu|\zeta).$$

Hence,

$$(109)\quad \langle \mathcal{J}^d_\mu\rangle_t = -\frac{e^2}{\Omega}\int_0^t dt' E_\nu(t') e^{-\mathcal{B}_\zeta(t-t')}\sum_\zeta \frac{\partial\langle n_\zeta\rangle_{\mathrm{eq}}}{\partial\epsilon_\zeta}(\zeta|v_\nu|\zeta)(\zeta|v_\mu|\zeta),$$

which is of the form

$$(110)\quad \langle \mathcal{J}_\mu\rangle_t = \int_0^t dt' \phi_{\mu\nu}(t-t') E_\nu(t'),$$

where ϕ is the response function. Its Fourier-Laplace transform is the conductivity. With $\mathcal{B}_\zeta \to 1/\tau_\zeta$ being the reciprocal relaxation time we find, writing $(\zeta|v_\mu|\zeta) \equiv v_{\mu\zeta}$;

$$(111)\quad \sigma^d_{\mu\nu}(\mathrm{pond.}\, i\omega) = -\frac{e^2}{\Omega}\sum_\zeta \frac{\partial\langle n_\zeta\rangle_{\mathrm{eq}}}{\partial\epsilon_\zeta}\frac{\tau(\epsilon_\zeta)}{1+i\omega\tau(\epsilon_\zeta)} v_{\nu\zeta}v_{\mu\zeta}.$$

Finally, we consider the nondiagonal conductivity, which is of paramount importance for the ordinary and quantum Hall effect. The solution of (98) is

$$(112)\quad \langle c^\dagger_{\zeta_1} c_{\zeta_2}\rangle_t = e\int_0^t dt' E_\nu(t') e^{-i\Delta(t-t')/\hbar}\frac{1-e^{-\beta\Delta}}{\Delta}\langle n_{\zeta_2}\rangle_{\mathrm{eq}} \times \big(1-\langle n_{\zeta_1}\rangle_{\mathrm{eq}}\big)(\zeta_2|v_\nu|\zeta_1),$$

where $\Delta = \epsilon_{\zeta_2} - \epsilon_{\zeta_1}$. This yields

$$(113) \qquad \langle \mathcal{J}_\mu^{nd}\rangle_t = \sum_{\zeta_1\zeta_2} \text{(above)} \times \frac{e}{\Omega}(\zeta_1|v_\mu|\zeta_2).$$

This is again of the response from (110). Taking the Fourier-Laplace transform of $\phi_{\mu\nu}(t)$ yields

$$(114) \quad \sigma_{\mu\nu}^{nd}(\text{pond.}, \omega \to 0) = \frac{e^2\hbar i}{\Omega} \sum_{\zeta\zeta',\sigma\sigma'}{}' \langle n_\zeta\rangle_{\text{eq}}(1 - \langle n_{\zeta'}\rangle_{\text{eq}}) \times \frac{1 - \exp[\beta(\epsilon_\zeta - \epsilon_{\zeta'})]}{(\epsilon_\zeta - \epsilon_{\zeta'})^2}(\zeta|v_\nu|\zeta')(\zeta'|v_\mu|\zeta).$$

An application will follow in Section 5.

Lastly, we return to the Wigner form. We consider extended states in ν dimensions and electron confinement in 3-ν dimensions, as in Section 2, so $\zeta = (\boldsymbol{k}, \alpha, \sigma)$ with $\boldsymbol{k} = (k_1 \dots k_\nu)$. For simplicity the indices α and σ will be suppressed. The full QBE now reads

$$(115) \quad \frac{\partial}{\partial t}\langle c^\dagger_{\boldsymbol{k}'} c_{\boldsymbol{k}}\rangle_t - q\boldsymbol{E}(t)\langle n_{\boldsymbol{k}}\rangle_{\text{eq}}\big[1 - \langle n_{\boldsymbol{k}'}\rangle_{\text{eq}}\big] \bullet \left[\frac{1 - e^{-\beta(\epsilon_{\boldsymbol{k}'} - \epsilon_{\boldsymbol{k}})}}{\epsilon_{\boldsymbol{k}'} - \epsilon_{\boldsymbol{k}}}\right](\boldsymbol{k}'|\boldsymbol{v}|\boldsymbol{k}) - \left(\frac{i}{\hbar}\right)(\epsilon_{\boldsymbol{k}'} - \epsilon_{\boldsymbol{k}})\langle c^\dagger_{\boldsymbol{k}'} c_{\boldsymbol{k}}\rangle_t = -\mathcal{B}_{\boldsymbol{k}}\langle n_{\boldsymbol{k}}\rangle_t \delta_{\boldsymbol{k}\boldsymbol{k}'}.$$

For plane waves or Bloch states we have[8]

$$(116) \qquad (\boldsymbol{k}'|\boldsymbol{v}|\boldsymbol{k}) = \hbar^{-1}\boldsymbol{\nabla}_{\boldsymbol{k}}\epsilon_{\boldsymbol{k}}\delta_{\boldsymbol{k}'\boldsymbol{k}} = \boldsymbol{v}_{\boldsymbol{k}}\delta_{\boldsymbol{k}'\boldsymbol{k}},$$

$$(117) \qquad \langle n_{\boldsymbol{k}}\rangle_{\text{eq}}\big(1 - \langle n_{\boldsymbol{k}}\rangle_{\text{eq}}\big) = -\frac{1}{\beta}\frac{\partial\langle n_{\boldsymbol{k}}\rangle_{\text{eq}}}{\partial\epsilon_{\boldsymbol{k}}},$$

$$(118) \qquad \lim_{\boldsymbol{k}'\to\boldsymbol{k}} \frac{1 - \exp\{-\beta(\epsilon_{\boldsymbol{k}'} - \epsilon_{\boldsymbol{k}})\}}{\epsilon_{\boldsymbol{k}'} - \epsilon_{\boldsymbol{k}}}\langle n_{\boldsymbol{k}}\rangle_{\text{eq}}\big(1 - \langle n_{\boldsymbol{k}'}\rangle_{\text{eq}}\big)\boldsymbol{\nabla}_{\boldsymbol{k}}\epsilon_{\boldsymbol{k}} = -\frac{\partial\langle n_{\boldsymbol{k}}\rangle_{\text{eq}}}{\partial\epsilon_k}\boldsymbol{\nabla}_{\boldsymbol{k}}\epsilon_{\boldsymbol{k}} = -\boldsymbol{\nabla}_{\boldsymbol{k}}\langle n_{\boldsymbol{k}}\rangle_{\text{eq}} \to -\nabla_{\boldsymbol{k}}\langle n_{\boldsymbol{k}}\rangle_t;$$

the force term thus becomes $\big[q\boldsymbol{E}(t)/\hbar\big] \bullet \boldsymbol{\nabla}_{\boldsymbol{k}}\langle n_{\boldsymbol{k}}\rangle_t$.[9]

In (115) we now set $\boldsymbol{k}' \to \boldsymbol{k} - \frac{1}{2}\boldsymbol{u}$, $\boldsymbol{k} \to \boldsymbol{k} + \frac{1}{2}\boldsymbol{u}$, multiply by $\big[\Omega/(2\pi)^\nu\big]\exp(i\boldsymbol{u}\cdot\boldsymbol{q})$, and integrate over $\boldsymbol{u}$; this gives

$$(119) \quad \frac{\partial}{\partial t}\frac{\Omega}{(2\pi)^\nu}\int d^\nu u e^{i\boldsymbol{u}\cdot\boldsymbol{q}}\langle c^\dagger_{\boldsymbol{k}-1/2\boldsymbol{u}} c_{\boldsymbol{k}+1/2\boldsymbol{u}}\rangle_t + \frac{q\boldsymbol{E}(t)}{\hbar}\frac{\Omega}{(2\pi)^\nu}\int d^\nu u e^{i\boldsymbol{u}\cdot\boldsymbol{q}} \cdot \boldsymbol{\nabla}_{\boldsymbol{k}}\langle c^\dagger_{\boldsymbol{k}-1/2\boldsymbol{u}} c_{\boldsymbol{k}+1/2\boldsymbol{u}}\rangle_t\delta_{\boldsymbol{u},0} + \frac{i\Omega}{(2\pi)^\nu\hbar}\int d^\nu u e^{i\boldsymbol{u}\cdot\boldsymbol{q}}(\epsilon_{\boldsymbol{k}-1/2\boldsymbol{u}} - \epsilon_{\boldsymbol{k}+1/2\boldsymbol{u}}) \times \langle c^\dagger_{\boldsymbol{k}-1/2\boldsymbol{u}} c_{\boldsymbol{k}+1/2\boldsymbol{u}}\rangle_t = -\frac{\Omega}{(2\pi)^\nu}\int d^\nu u e^{i\boldsymbol{u}\cdot\boldsymbol{q}}\mathcal{B}_{\boldsymbol{k}}\langle n_{\boldsymbol{k}}\rangle_t\delta_{\boldsymbol{u},0}.$$

Employing the one-particle Wigner density $\rho^1(\boldsymbol{k}, \boldsymbol{q}, t)$, see Eqs. (8) and (9), and the Wigner occupation function as defined in (14) and (15), we note that the

[8]The nabla operator $\boldsymbol{\nabla}_{\boldsymbol{k}}$ is the gradient in ν-dimensional $\boldsymbol{k}$-space.

[9]The last transition in (118) is the obvious nonequilibrium generalization of the streaming term. Its rigorous proof must await a non-linearized response generalization of the ME.

first term simply yields the differential $\partial g(\boldsymbol{k},\boldsymbol{q},t)/\partial t$. The third term is likewise straightforward. We note

(120) $$\epsilon_{\boldsymbol{k}-1/2u}-\epsilon_{\boldsymbol{k}+1/2u}\simeq-(\boldsymbol{\nabla}_{\boldsymbol{k}}\epsilon_k)\cdot\boldsymbol{u}=-\hbar\boldsymbol{v}_{\boldsymbol{k}}\cdot\boldsymbol{u}.$$

Using (9) and (apply $\boldsymbol{\nabla}_{\boldsymbol{q}'}$ to (12))

(121) $$\int d^\nu u\boldsymbol{u}e^{i\boldsymbol{u}\cdot(\boldsymbol{q}-\boldsymbol{q}')}=(2\pi)^\nu i\boldsymbol{\nabla}_{\boldsymbol{q}'}\delta(\boldsymbol{q}-\boldsymbol{q}')$$

we obtain:

(122) $$\begin{aligned}\text{third term }&=\frac{i}{(2\pi)^\nu}\int d^\nu q'h^\nu\rho^1(\boldsymbol{k},\boldsymbol{q}',t)\int d^\nu u\boldsymbol{v}_{\boldsymbol{k}}\cdot\boldsymbol{u}e^{i\boldsymbol{u}\cdot(\boldsymbol{q}-\boldsymbol{q}')}\\&\overset{\text{c.gr.}}{=}-\boldsymbol{v}_{\boldsymbol{k}}\cdot\int d^\nu q'g(\boldsymbol{k},\boldsymbol{q}',t)\boldsymbol{\nabla}_{\boldsymbol{q}'}\delta(\boldsymbol{q}-\boldsymbol{q}')=\boldsymbol{v}_{\boldsymbol{k}}\cdot\boldsymbol{\nabla}_{\boldsymbol{q}}g(\boldsymbol{k},\boldsymbol{q},t).\end{aligned}$$

Note that the gradient term stems from the nondiagonal part of the ME!

The second term offers some problems. Strictly it yields

(123) $$\frac{1}{(2\pi)^\nu\hbar}q\boldsymbol{E}(t)\cdot\boldsymbol{\nabla}_{\boldsymbol{k}}\int d^\nu ue^{i\boldsymbol{u}\cdot\boldsymbol{q}}\int d^\nu q'h^\nu\rho^1(\boldsymbol{k},\boldsymbol{q}',t)e^{-i\boldsymbol{u}\cdot\boldsymbol{q}'}\delta_{\boldsymbol{u},0}.$$

If $\boldsymbol{u}=0$, i.e., the momentum is entirely sharp, the rhs of (123) involves integration over all positions. If on the other hand $\boldsymbol{u}$ has a large leeway, (12) implies an exact position. Since neither can be true some coarse graining must be carried out judiciously. We will give an extension $|\Delta u|^\nu$ in $\boldsymbol{u}$-space. We thus write

(124) $$\int d^\nu ue^{i\boldsymbol{u}\cdot(\boldsymbol{q}-\boldsymbol{q}')}\delta_{\boldsymbol{u},0}\to\int_{|\Delta u|^\nu}d^\nu ue^{i\boldsymbol{u}\cdot(\boldsymbol{q}-\boldsymbol{q}')}=\prod_i^\nu\frac{\sin[\Delta u_i(q_i'-q_i)/2]}{(q_i'-q_i)/2}$$

where we integrated over the intervals $(-\Delta u_i/2,\Delta u_i/2)$. The rhs has its maximum $\prod_i\Delta u_i$ for $q_i'=q_i$; the width in the direction u_i is

(125) $$\frac{1}{\Delta u_i}\int_{-\infty}^{\infty}\frac{\sin[\Delta u_i(q_i'-q_i)/2]}{(q_i'-q_i)/2}dq_i'=\frac{2\pi}{\Delta u_i}.$$

We may thus replace the rhs of (124) by a function in $\boldsymbol{q}'$ space which has a magnitude $\prod_i\Delta u_i$ within a box of volume $(2\pi)^\nu/\prod_i\Delta u_i$, centered on $\boldsymbol{q}$, and which is zero elsewhere. We write $|\Delta u|^\nu\equiv\prod_i\Delta u_i$ and now carry out the $\boldsymbol{q}'$ integration indicated in (123), to obtain

(126) $$(123)\;\to\frac{1}{(2\pi)^\nu\hbar}q\boldsymbol{E}(t)\cdot\boldsymbol{\nabla}_{\boldsymbol{k}}\left[|\Delta u|^\nu\int_{(2\pi)^\nu/|\Delta u|^\nu}d^\nu q'\rho^1(\boldsymbol{k}',\boldsymbol{q}',t)\right].$$

Since as in Section 2 the coarse graining in phase space involves a microcell h^ν, the minimum support in $\boldsymbol{q}'$-space for given $|\Delta p|^\nu=\left[\hbar|\Delta u|\right]^\nu$ amounts to

(127) $$|\Delta q'|^\nu=\frac{h^\nu}{[\hbar|\Delta u|]^\nu}=\frac{(2\pi)^\nu}{|\Delta u|^\nu}.$$

Hence, the [ ] under the $\boldsymbol{\nabla}_{\boldsymbol{k}}$ operator represents exactly the Wigner occupation function as defined previously in Eqs. (14) and (15). Therefore we obtained:

(128) $$\text{second term }=\left[\frac{q\boldsymbol{E}(t)}{\hbar}\right]\cdot\boldsymbol{\nabla}_{\boldsymbol{k}}g(\boldsymbol{k},\boldsymbol{q},t).$$

Finally we must consider the collision term, i.e., the rhs of (119). As in (88) we have a part that is linear in $\langle n_k\rangle_t$ and a part that is quadratic involving $\langle n_{\boldsymbol{k}}\rangle_t\langle n_{\boldsymbol{k}'}\rangle_t$. The linear part is treated entirely analogous to the streaming term we discussed.

For the quadratic terms we refer to Ref. [**48**], Eqs. (8.32′)ff. The final result for (119) is

$$\frac{\partial g(\boldsymbol{k},\boldsymbol{q},t)}{\partial t} + \boldsymbol{v}_{\boldsymbol{k}} \cdot \boldsymbol{\nabla}_{\boldsymbol{q}} g(\boldsymbol{k},\boldsymbol{q},t) + \left(\frac{q\boldsymbol{E}(t)}{\hbar}\right) \cdot \boldsymbol{\nabla}_{\boldsymbol{k}} g(\boldsymbol{k},\boldsymbol{q},t) = \sum_{\boldsymbol{k}'} \{ w_{\boldsymbol{k}'\boldsymbol{k}} g(\boldsymbol{k}',\boldsymbol{q},t)[1-g(\boldsymbol{k},q,t)] - w_{\boldsymbol{k}\boldsymbol{k}'} g(\boldsymbol{k},\boldsymbol{q},t)[1-g(\boldsymbol{k}',\boldsymbol{q},t)]\}. \tag{129}$$

Thus the semiquantal BTE *can be used in the framework of the Wigner occupation function for extensive states.* The reader is reminded, however, that properly we must consider $g_{\alpha\sigma}(\boldsymbol{k},\boldsymbol{q},t)$; i.e., full quantum effects including localization and spin must still be accounted for. *While the Wigner formalism is useful for inhomogeneous systems, in homogeneous systems a direct solution with the* QBE *is to be preferred.* Examples will be provided in the next two sections.

5. Application of the QBE to the Quantum Hall effect

In the presence of a strong magnetic field there is no $\boldsymbol{k}$-space, except along $\boldsymbol{B}$. In the plane quantization leads to Landau states. Previously, transverse magneto resistance and the Hall effect were treated with density-matrix methods [**2, 1, 20**]. Since there are no diagonal matrix elements of the momentum $\boldsymbol{p}$ for Landau states, it is at once clear that a proper treatment should be based on the nondiagonal QBE. The author showed around 1980 that the latter yields both the ordinary and the oscillating Hall effect; later, Vasilopoulos showed that the 2D quantum Hall effect follows from the n.d. QBE in minute detail. We follow here his treatment [**35, 36**].

A 2D electron gas is easily realized in the inversion layer of MOSFETs or the quantum well of a $GaAs - Ga_xAl_{1-x}As$ heterojunction. Let z be the direction of confinement (i.e. along the well width) and xy the directions of free motion (i.e. along the hetero junction plane). With the $\boldsymbol{B}$-field along z, the electron orbits in the xy plane according to the classical Lorentz force picture. Quantum mechanically the Hamiltonian for one electron of effective mass m^* is

$$h^0 = \frac{(\boldsymbol{p}+e\boldsymbol{A})^2}{2m^*}, \quad (\boldsymbol{A} = 0, Bx, 0) \tag{130}$$

where $\boldsymbol{A}$ is the vector potential in the Landau gauge. The eigenstates are

$$|\zeta) = |N, k_y) \quad \text{or} \quad \zeta(x,y) = \phi_N(x + l^2 k_y)\frac{e^{ik_y y}}{L_y^{1/2}}, \tag{131}$$

where $x_0 \equiv l^2 k_y$ is the centre of the orbit, $l^2 = \hbar/m^*\omega_0$ is its radius squared and $\omega_0 = eB/m^*$ is the cyclotron frequency. The ϕ_N are harmonic oscillator eigenfunctions—basically Hermite polynomials—and L_x, L_y are the dimensions in the plane of area $\mathcal{A} = L_x L_y$. The eigenvalues depend only on the quantum number N,

$$\epsilon_N = (N+1)\hbar\omega_0. \tag{132}$$

The relevant matrix elements are found in any textbook on quantum mechanics; they only exist next to the diagonal,

$$(133a)\qquad (\zeta|v_x|\zeta') = i\left(\frac{\hbar\omega_0}{2m^*}\right)^{1/2}\left[-(N+1)^{1/2}\delta_{N',N+1} + N^{1/2}\delta_{N',N-1}\right]\delta_{k_y,k_y'}$$

$$(133b)\qquad (\zeta|v_y|\zeta') = \left(\frac{\hbar\omega_0}{2m^*}\right)^{1/2}\left[\left(N+\frac{1}{2}\right)\delta_{N',N+1} + N^{1/2}\delta_{N',N-1}\right]\delta_{k_y,k_y'}.$$

For the Hall effect we need the conductivity σ_{xy}, for which (114) is applicable, with Ω replaced by $\mathcal{A}$. From Eqs. (133):

$$(134)\quad (\zeta|v_x|\zeta')(\zeta'|v_y|\zeta) = \left(\frac{i\hbar\omega_0}{2m^*}\right)\left[-(N+1)^{1/2}(N')^{1/2}\delta_{N',N+1} + (N)^{1/2}(N'+1)^{1/2}\delta_{N',N-1}\right]\delta_{k_y,k_y'}.$$

For the allowed transitions we have

$$(135)\qquad \epsilon_{\zeta'} - \epsilon_\zeta = \pm\hbar\omega_0,$$

where the $\pm$ signs hold for $N' = N \pm 1$, respectively.

From (114) and (134) we now find, changing $N \to N+1$ in the second summation,

$$(136)\ \sigma_{yx} \equiv \sigma_{yx}^{nd}(0) = \frac{e}{2B\mathcal{A}}\sum_{k_y} \times \sum_{N=0,1\ldots}(N+1)\{\langle n_N\rangle_{\text{eq}}(1-\langle n_{N+1}\rangle_{\text{eq}}) \times (1-e^{-\beta\hbar\omega_0}) - \langle n_{N+1}\rangle_{\text{eq}}(1-\langle n_N\rangle_{\text{eq}})(1-e^{\beta\hbar\omega_0})\},$$

(we suppressed the index k_y; thus $n_N \equiv n_{Nk_y}$, etc.). For the k_y summation we write

$$(137)\qquad \sum_{k_y} \to \frac{L_y}{2\pi}\int_{-L_x/2l^2}^{L_x/2l^2} dk_y = \frac{\mathcal{A}}{2\pi l^2}.$$

We further set $\langle n_N\rangle_{\text{eq}} \equiv f_N$; from the equilibrium Fermi-Dirac function one easily shows

$$(138)\qquad f_N(1-f_{N+1})(1-e^{-\beta\hbar\omega_0}) = -(f_{N+1})(1-f_N)(1-e^{\beta\hbar\omega_0}).$$

Thus (136) yields noting $l^2 = \hbar/m^*\omega_0 = h/2\pi eB$ and using N' for the summation index,

$$(139)\qquad \sigma_{yx} = \frac{e^2}{h}\sum_{N'}(N'+1)f_{N'}(1-f_{N'+1})(1-e^{-\beta\hbar\omega_0}).$$

We now consider the zero-temperature limit ($\beta \to \infty$) when the Fermi level is pinned in a gap between the broadened Landau levels (see Fig. 1 of Ref. [**35**]),

$$(140)\qquad \epsilon_N < \epsilon_F < \epsilon_{N+1}.$$

For $N' < N$, N' is at most $N-1$ or $N'+1$ is at most N. Thus

$$(141)\qquad \epsilon_{N'+1} \le \epsilon_N < \epsilon_F \to f_{N'+1} = 1.$$

These terms therefore do not contribute in (139). Alternatively, for $N' > N$, N' is at least $N+1$, so that

$$\epsilon_{N'} \geq \epsilon_{N+1} > \epsilon_F \rightarrow f_{N'} = 0. \tag{142}$$

Clearly these terms do not contribute either. The only contribution of (139), given the condition (140), stems from $N' = N$. Whence

$$\lim_{T \to 0} \sigma_{yx} = (N+1)\left(\frac{e^2}{h}\right), \quad N = 0, 1, 2, \ldots. \tag{143}$$

This is the famous quantum Hall expression, first obtained and observed by von Klitzing, Dorda, and Pepper [**24**] (Nobel prize von Klitzing 1981).

For finite temperature we can expand in $\delta = e^{-\beta\hbar\omega_0}$. Setting further $b = (\epsilon_F - \epsilon_N)/\hbar\omega_0 < 1$, we then find

$$\sigma_{yx} = (N+1)\left(\frac{e^2}{h}\right)(1 - \delta^b - \delta^{1-b} + \delta^{2b} + \delta^{2(1-b)} - \ldots). \tag{143'}$$

It was found [**35**] that the result agrees with the observed plateaus to at least the fifth decimal place! The correctness of the QBE is abundantly illustrated by this striking example. For other notable examples, e.g., the Aharonov-Bohm effect, we refer to the literature.

6. Fluctuations and Noise Spectra in Mesoscopic Conductors

Mesoscopic conductors, which have a length L equal to or smaller than the phase-breaking or inelastic scattering mean free path, have enjoyed great attention since the early eighties. Fluctuations of the current have been extensively studied since the work of Büttiker et al. in 1988 [**6**]; see also Landauer and Martin [**27**] and the papers by Reggiani et al. [**32, 26**]. Thermal equilibrium noise was also studied by VanVliet and Sreenivasan [**44**], whereas very recently (1996) we succeeded in obtaining the nonequilibrium noise for all frequencies, up to terahertz [**41**]. Clearly, space lacks to deal with this exciting new field; we restrict ourselves therefore to some aspects of the equilibrium noise in a 1D quantum wire, using the Wigner formalism as set forth in this paper, confirming, *grosso motto*, the results of Reggiani and Kuhn.

We consider a mesoscopic conductor with contacts A and B, at $-L/2$ and $L/2$, respectively. The contacts are of macroscopic size and in quasi-equilibrium, establishing a temperature T in the system composed of these reservoirs and the mesoscopic conductor. Carriers are emitted and absorbed by both reservoirs; in thermal equilibrium the flow $A \rightarrow B$ balances the flow $B \rightarrow A$. A nonequilibrium state is established when the electro-chemical potentials (or quasi-Fermi levels) are different. In the Reggiani-Kuhn model it is assumed that contact B is grounded, so $\epsilon_{FB} = \epsilon_F$, while a negative voltage $-V$ is applied to contact A, yielding $\epsilon_{FA} = \epsilon_F + eV$[10]; net current flow in the nonequilibrium state is from B to A corresponding to electron flow from A to B. For a picture, see Figure 1.

The local current is as in Eq. (4); however, since we want to describe current fluctuations rather than average behaviour we must employ Heisenberg operators,

---

[10]The reader is reminded that the electron band diagram and the concepts of ϵ_F and $\epsilon(\boldsymbol{k})$ always refer to electrons; whence $\epsilon_{FA} = \epsilon_F + eV$, regardless of whether the current is carried by electrons ($q - -e$) or holes ($q - +e$).

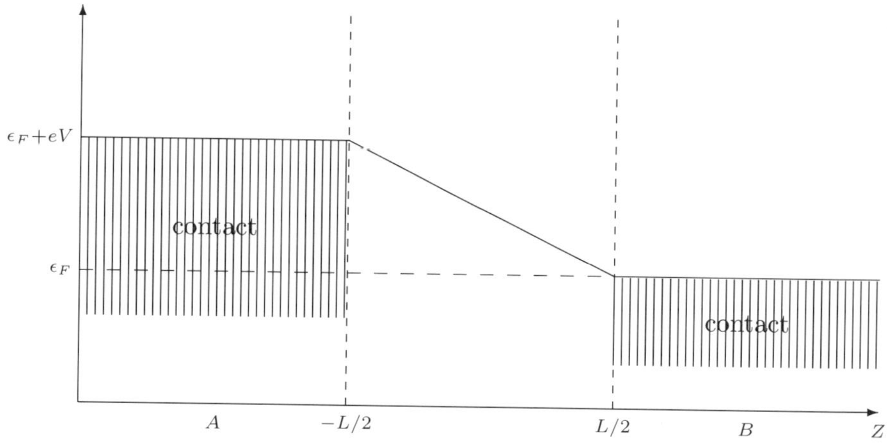

FIGURE 1. Mesoscopic conductor with contact reservoirs

so we now write,

(144) $$i(z,t) = \frac{q\hbar}{2m^*i}\sum_{\alpha\sigma}\left\{\left[-\frac{\partial}{\partial z}\Psi^\dagger_{\alpha\sigma}(z,t)\right]\Psi_{\alpha\sigma}(z,t) + \Psi^\dagger_{\alpha\sigma}(z,t)\frac{\partial}{\partial z}\Psi_{\alpha\sigma}(z,t)\right\}.$$

Expanding the field operators in the basis $u_{k\alpha\sigma}(\boldsymbol{q})$ we find analogous to (7)

(145) $$i(z,t) = \frac{q\hbar}{2m^*}\sum_{\alpha\alpha'\sigma\sigma'}\sum_{kk'} c^\dagger_{k'\alpha'\sigma'}(t)c_{k\alpha\sigma}(t)(k'+k)\big[u^*_{k'\alpha'\sigma'}(z)u_{k\alpha\sigma}(z)\big].$$

Here m^* is the effective mass and k the 1D crystal momentum. Following a similar treatment as in Section 2 we find analogous to (16)

(146) $$i(z,t) = \frac{q\hbar}{m^*L}\sum_{\alpha\sigma}\sum_{k} k\hat{g}_{\alpha\sigma}(k,z,t)$$

where $\hat{g}_{\alpha\sigma}$ is the non-ensemble-averaged Wigner occupation function based on (8′) and (9′). For a single system of the ensemble, $\hat{g}_{\alpha\sigma}(k,z,t)$ is a stochastic distribution, randomized by the interactions λV, in accordance with the ergodic theorem [**22**, **7**].[11]

We now consider a subensemble with initial condition $\hat{g}_{\alpha\sigma}(k,z,0) \equiv g_{0,\alpha\sigma}(k,z)$. Clearly from a conditional ensemble average we recover the Wigner occupation function of Section 2, i.e.,

(147) $$\big\langle\hat{g}_{\alpha\sigma}(k,z,t)\big\rangle_{\text{cond}} = \text{Tr}\big[\rho g_{\alpha\sigma}(k,z,t)\big] = g_{\alpha\sigma}(k,z,t).$$

Thus, $\big\langle\hat{g}_{\alpha\sigma}(k,z,t)\big\rangle_{\text{cond}}$ follows the phenomenological equation, *in casu* the BTE, as expected from Onsager's hypothesis.

Fluctuations are defined as deviations from the stationary values as usual,

(148) $$\Delta g_{\alpha\sigma}(k,z,t) = \hat{g}_{\alpha\sigma}(k,z,t) - \lim_{t\to\infty}\hat{g}_{\alpha\sigma}(k,z,t).$$

[11] In this respect we are reminded that the ordinary distribution of classical statistical mechanics is in essence also a stochastic quantity; an ensemble average over a microcanonical ensemble yields "the" Boltzmann distribution, while the covariance of the fluctuations is easily obtained with the Darwin-Fowler method [**39**].

By Bayes' theorem we have for the two-point correlation function[12]

(149) $\langle \Delta g_{\alpha\sigma}(k,z,t)\Delta g_{\alpha'\sigma'}(k,z,0)\rangle = \langle\langle \Delta g_{\alpha\sigma}(k,z,t)\rangle_{\text{cond}}\Delta g_{0,\alpha\sigma}\rangle$.[13]

Let further $G_{\alpha\sigma}(k,z,s)$ be the Laplace transform of $\langle \Delta g_{\alpha\sigma}(k,z,t)\rangle_{\text{cond}}$. We then have, using a relaxation-time model for the collision operator (exact for elastic collisions [**43**]),

(150) $$\left[s + \frac{1}{\tau_c(k)} + \frac{\hbar k}{m^*}\frac{\partial}{\partial z} + \frac{qE}{\hbar}\frac{\partial}{\partial k}\right] G_{\alpha\sigma}(k,z,s) = \Delta g_{0,\alpha\sigma}(k,z).$$

The formal solution is

(150') $$G_{\alpha\sigma}(k,z,s) = \left[s + \frac{1}{\tau_c(k')} + \frac{\hbar k}{m^*}\frac{\partial}{\partial z} + \frac{qE}{\hbar}\frac{\partial}{\partial k}\right]^{-1} \Delta g_{0,\alpha\sigma}(k,z)$$

with inversion

(151) $$\begin{aligned}\langle \Delta g_{\alpha\sigma}(k,z,t)\rangle_{\text{cond}} &= \left\{\exp\left(-t\left[\frac{1}{\tau_c(k')} + \frac{\hbar k}{m^*}\frac{\partial}{\partial z} + \frac{qE}{\hbar}\frac{\partial}{\partial k}\right]\right)\right\} \\ &\quad \times \Delta g_{0,\alpha\sigma}(k,z) \\ &= e^{-t/\tau_c(k')}\Delta g_{0,\alpha\sigma}\left(k - \frac{qE}{\hbar}t, z - \frac{\hbar k}{m^*}t\right),\end{aligned}$$

where we used the identity $\left[\exp(\alpha\partial/\partial u)\right] f(u) = f(u+\alpha)$, verifiable by Taylor expansion. Introducing the stationary covariance function

(152) $$\begin{aligned}\Gamma_{\alpha\sigma}(kz;k'z') &= \langle \Delta g_{\alpha\sigma}(k,z,t)\Delta g_{\alpha\sigma}(k',z',t)\rangle \\ &= \langle \Delta g_{0,\alpha\sigma}(k,z)\Delta g_{0,\alpha\sigma}(k'z')\rangle,\end{aligned}$$

we obtain

(153) $$\langle \Delta g_{\alpha\sigma}(k,z,t)\Delta g_{\alpha\sigma}(k,z,0)\rangle = e^{-t/\tau_c(k')}\Gamma\left(k - \frac{qE}{\hbar}t, z - \frac{\hbar k}{m^*}t; k'z'\right).$$

We remark that the trajectories in phase space entailed in the above Γ differ from those of Ref. [**32, 26**], whose results for $E \neq 0$ are incorrect. Only for equilibrium situations is the covariance function simple; then from the grand canonical ensemble,

(154) $$\Gamma_{\alpha\sigma}(kz;k'z') = f_{\text{eq},\alpha\sigma}(k,z)\left[1 - f_{\text{eq},\alpha\sigma}(k',z')\right]\delta_{kk'}\delta_{zz'}$$

where $f_{\text{eq},\alpha\sigma}$ follows from Fermi statistics for a quantum well with subbands [**34**]. For nonequilibrium we need a "double Pauli principle for counting transfers $A \rightleftarrows B$, see Landauer and Martin [**27**] and Van Vliet [**41**].

Returning to the current (146) and its space-average we have for the fluctuations in an equilibrium situation ($E = 0$):

(155) $$\langle \Delta i(z,t)\rangle_{\text{cond}} = \frac{q\hbar}{m^*\pi}\sum_\alpha \int_{-\infty}^{\infty} dk k \langle \Delta g_\alpha(k,z,t)\rangle_{\text{cond}}$$

---

[12] Since $W(A,B) = P(A \mid B)W(B)$, where P denotes the probability of finding A for given B, we have

$$\langle AB\rangle = \sum_{AB} ABW(A,B) = \sum_B BW(B)\sum_A AP(A \mid B) = \langle B\langle A\rangle_{\text{cond}B}\rangle.$$

[13] As in Section 2, $\delta_{\alpha\alpha'}\delta_{\sigma\sigma'}$ is implied.

(we summed over the spin and used the density of states $Z(k) = 2\pi/L$),

$$(156)\qquad \langle \Delta I(t)\rangle_{\text{cond}} = \frac{q\hbar}{\pi m^* L}\sum_{\alpha}\int_{-L/2}^{L/2} dz \int_{-\infty}^{\infty} dkk\langle \Delta g_\alpha(k,z,t)\rangle_{\text{cond}}.$$

Employing (153) and (154), replacing $\delta_{kk'}\delta_{zz'} \to \pi\delta(k-k')\delta(z-z')$ this finally yields for the current correlation function

$$(157)\qquad C_I(t) = \langle \Delta I(t)\Delta I(0)\rangle = \langle\langle \Delta I(t)\rangle_{\text{cond}}\Delta I(0)\rangle$$
$$= \frac{1}{\pi}\left(\frac{e\hbar}{h^* L}\right)^2 \sum_\alpha \int_{-L/2}^{L/2}\int_{-L/2}^{L/2} dzdz' \int_{-\infty}^{\infty}\int_{\infty}^{\infty} dkdk'kk'$$
$$\times\, e^{-\tau_c(k')} f_{\text{eq},\alpha}(k)\big[1 - f_{\text{eq},\alpha}(k')\big]\delta(k-k')\delta\Big(z - \frac{\hbar k}{m^*}t - z'\Big)$$
$$= \frac{1}{\pi}\left(\frac{e\hbar}{m^* L}\right)^2 \int_{-L/2}^{L/2} dz \int_{m^*/[(-1/2)L-z]\hbar t}^{m^*/[(1/2)L-z)]\hbar t} dkk^2 e^{-t/\tau_c(\epsilon_k)} f_{\text{eq},\alpha}(\epsilon_{k,\alpha})$$
$$\big[1 - f_{\text{eq}}(\epsilon_{k,\alpha})\big].$$

Finally, consider only one band with $\epsilon_k = \hbar^2k^2/2m^*$. Changing the order of the integrals over dz and $dk(\hbar t/m^*)$, using inversion symmetry $\epsilon_{-k} = \epsilon_k$, and applying the Wiener-Khintchine theorem [**23**], the spectral density is found to be

$$(158)\quad S_I(\omega) = \frac{8e^2}{hL}\sqrt{\frac{2}{m^*}}\int_{-\infty}^{\infty} dte^{-i\omega t}\int_0^{m^*L^2/2t^2} d\epsilon\sqrt{\epsilon}e^{-t/\tau_c(\epsilon)}\left(1 - t\sqrt{\frac{2\epsilon}{m^*L^*}}\right)$$
$$\times\, f_{\text{eq}}(\epsilon)\big[1 - f_{\text{eq}}(\epsilon)\big],$$

which expression agrees with Kuhn and Reggiani, except that the collision term is brought under the integral. Numerical solutions for many conditions are found in Refs. [**32, 26**] and [**44**].

Perhaps one of the most striking examples is the asymptotic regime $t \ll \tau_{\text{tr}} = L/v_F$ (v_F is the Fermi velocity) in a ballistic degenerate electron gas near $T = 0$, for which $f_{\text{eq}}(1 - f_{\text{eq}}) \to k_BT\partial f_{\text{eq}}/\partial\epsilon_F \simeq k_BT\delta(\epsilon - \epsilon_F)$. The spectrum is found to be

$$(159)\qquad S_I(\omega) = 4kT\left(\frac{2e^2}{h}\right)\frac{\sin^2(\omega\tau_{\text{tr}}/2)}{(\omega\tau_{\text{tr}}/2)^2},$$

showing resonances when ω is a multiple of the reciprocal transit time. We note that this is entirely similar to the spectrum for ambipolar sweep-out in germanium samples in 1958 by Hill and Van Vliet [**18**]. Indeed, "there is nothing new under the sum" (Solomon in Eccl. 1:9). Yet many more exotic computations can be expected for the transport and fluctuations in the nanometer reduced-dimensional structures generated by modern technology. Very detailed computations for "many-channel" mesoscopic conductors are found in the recent Ph.D. thesis of Benamira [**3**] showing likewise the power of present-day quantum nonequilibrium statistical mechanical theory.

Acknowledgments

I am greatly indebted to many colleagues, collaborators, and former graduate students, whose criticisms and keen understanding helped to shape my ideas. Foremost in my mind are the late Aisenstadt-chair titulaires, Marc Kac and Sybren R.

de Groot; further, N. G. Van Kampen, Ch. G. Van Weert, E. G. D. Cohen, L. de Sobrino, M. Charbonneau, Th. Samson, and P. Vasilopoulos. Also, throughout my years at the CRM, these projects have been sponsored by grants from NSERC (#A–9522) and par la subvention au CRM du fonds FCAR pour l'aide et le soutien à la recherche; their support has been greatly appreciated.

References

1. E. N. Adams and T. D. Holstein, *Quantum theory of transverse galvano-magnetic phenomena*, J. Phys. Chem. Solids **10** (1959), 254.
2. P. N. Argyres and L. M. Roth, *Theory of electrical conduction in high magnetic fields*, J. Phys. Chem. Solids **12** (1958), 189.
3. F. Benamira, *Termes de corrections à la limite de Van Hove, Conductance mésoscopique et interférences quantiques*, thèse de Doctorat, Université de Montréal, 1996.
4. L. Boltzmann, *Analytischer Beweis des zweiten hauptsatzes der mechanischen wärmetheorie aus den Sätzen über das Gleichgewicht der lebendigen Kraft*, Wiener Berichte **63** (1871), 712.
5. ______, *Vorlesungen über gas theorie*, J. A.Barth, Leipzig (1896, 1898); Lectures on Gas Theory, Univ. of California Press, 1964.
6. M. Büttiker, *Scattering theory of thermal and excess noise in open conductors*, Phys. Rev. Lett. **65** (1990), 2901.
7. P. Caldirola (ed.), *Ergodic theories*, Proc. Intern. School of Physics "Enrico Fermi", vol. XIV, Academic Press, N.Y., 1961.
8. M. Charbonneau and K. M. Van Vliet, *Conductance et magnétorésistance transversale à partir de l'équation de Boltzmann quantique*, Phys. Status Solidi (b) **101** (1980), 509.
9. M. Charbonneau, K. M. van Vliet, and P. Vasilopoulos, *Linear response theory revisited* III. *One-body response formulas and generalized Boltzmann equations*, J. Math. Phys. **23** (1982), 318–336.
10. M. P. Chaubey and C. M. Van Vliet, *Transverse magnetoconductivity of quasi-twodimensional semiconductor layers in the presence of phonon scattering*, Phys. Rev. B **33**, 5617 (1986).
11. ______, *Theory of cyclotron resonance of a quasi-twodimensional gas in a quantum well*, Phys. Rev. B **34** (1986), 3932.
12. E. G. D. Cohen, *On two objections by van Kampen*, Statist. Mech. Meeting in honour of van Kampen 60th birthday, Utrecht, 1981.
13. P. A. M. Dirac, *The principles of quantum mechanics*, 4th ed., The Clarendon Press, Oxford, 1958.
14. U. Fano, *Description of states in quantum mechanics by density matrix and operator techniques*, Rev. Modern Phys. **29** (1957), 74; *Liouville representation of quantum mechanics with application to relaxation processes*, Lectures on the Many-Body Problem (E. R. Cajeniello, ed.), Academic Press, N.Y., 1964.
15. M. S. Green, *Markoff random processes and the statistical mechanics of time-dependent phenomena*, J. Chem. Phys. **20** (1952), 1281; II *Irreversible processes in fluids*, J. Chem. Phys. **22** (1954), 398.
16. H. J. Groenewold, *On the principle of elementary quantum mechanics*, Physica **12** (1946), 405.
17. S. R. de Groot, *La transformation de Weyl et la fonction de Wigner; une forme alternative de la mécanique quantique*, Les Presses de l'Université de Montréal, 1975.
18. J. E. Hill and K. M. Van Vliet, *Ambipolar transport of carrier-density fluctuations in Germanium*, Physica **24** (1958), 709.
19. L. van Hove, *Quantum mechanical perturbations giving rise to a statistical transport equation*, Physica **21** (1955), 517–540 ; ibid, *The approach to equilibrium in quantum statistics; a perturbation treatment to general order*, Physica **23** (1957), 441.
20. A. H. Kahn and H. P. R. Frederikse, *Oscillatory behavior of magnetic susceptibility and electronic conductivity*, Solid State Physics, vol. 9 (F. Seitz and D. Turnbull, eds.), Academic Press, N.Y., 1959, p. 259.
21. N. G. van Kampen, *The case against linear response theory* (Festschrift Prof. H. Wergeland's 60th birthday), Physica Norveg. **5** (1971), 279.
22. A. I. Khinchin, *Mathematical foundations of statistical mechanics*, Dover, N.Y., 1949.

23. A. Khintchine, *Korrelationstheorie der stationären stochastischen Prozesse*, Math. Ann. **109** (1934), 604.
24. K. von Klitzing, G. Dorda, and M. Pepper, *New method for high accuracy determination of the fine-structure constant based on quantized Hall resistance*, Phys. Rev. Lett. **45** (1980), 494.
25. R. Kubo, *Statistical-mechanical theory of irreversible processes*. I, J. Phys. Soc. Japan **34** (1957), 570 .
26. T. Kuhn, L. Reggiani, and L. Varani, *Correlation functions and quantized noise in mesoscopic systems*, Superlattices and Microstructures **11** (1992), 205.
27. R. Landauer and Th. Martin, *Equilibrium and shot noise in mesoscopic systems*, Physica B **175** (1991), 167; ibid, Physica B **183** (1992), 288.
28. J. Liouville, *Note sur la théorie de variation de constantes arbitraires*, J. Math. Pures Appl. (9) **3** (1838), 342.
29. D. Loss, *Correction terms to the $\lambda^2 t$ limit of van Hove by the Liouville-operator method*. I, *General perturbation treatment*, Physica A **139** (1986), 305; ibid, II. *Evaluation of the Kubo-formula of the electrical conductivity*, Physica A **139** (1986), 526.
30. D. Loss and A. Thellung, *The electrical conductivity for inhomogeneous electric fields by the Liouville-operator method*, Physica A **144** (1987), 17.
31. A. Messiah, *Mécanique quantique*, vol. 2, chap. XVII, Dunod, Paris, 1964.
32. L. Reggiani and T. Kuhn, *Noise in small and ultrasmall geometries*, Proc. Nato Adv. Study Inst. on Granular Nano-Electronics, Il Ciocco, Italy, 1990
33. R. Riesz and B. Sz. Nagy, *Leçons d'analyse fonctionnelle*, Gauthier-Villars, Paris, 1965.
34. M. Shur, *GaAs devices and circuits*, Plenum Press, N.Y., 1986.
35. P. Vasilopoulos, *Finite temperature aspects of the quantum Hall effect: A Boltzmann-equation approach*, Phys. Rev. B **32** (1985) , 771.
36. ———, *Magnetophonon oscillations in quasi-twodimensional quantum wells*, Phys. Rev. B **33** (1986), 8587.
37. P. Vasilopoulos and C. M. Van Vliet, *Master hierarchy of kinetic equations for binary interactions in the adiabatic approximation*, Canad. J. Phys. **61** (1983), 102.
38. ———, *Linear response theory revisited* IV. *Applications*, J. Math. Phys. **25**, 1391–1403 (1984); ibid, J. Math. Phys. **25** (1985), 3358.
39. C. M. Van Vliet, *Statistical thermodynamics*. I, *Equilibrium theory*, Part B, CRM-1457, notes de cours (1987).
40. ———, *On van Kampen's objections against linear response theory*, J. Statist. Phys. **53** (1988), 49.
41. ———, *Gigahertz nonequilibrium noise in mesoscopic conductors*, Proc. Asia Pacific Microwave Conf., Delhi, India, 1996.
42. ———, *Master hierarchy of kinetic equations for binary collisions*. II, *Addenda*, Canad. J. Phys. (1996). (submitted)
43. ———, *Statistical thermodynamics* II, *Non-equilibrium theory*, Part E, CRM-1618, Notes de cours (1989).
44. C. M. Van Vliet and R. Sreenivasan, *Quantum Transport in the transition from mesoscopic to ballistic devices*, Physica Status Solidi (b) (1996), (to appear).
45. C. M. Van Vliet and P. Vasilopoulos, *The quantum Boltzmann equations and some applications to transport phenomena*, J. Phys. Chem. Solids **49** (1988), 639.
46. C. M. Van Vliet, Ch. G. van Weert, and A. H. Marshak, *Electrical current through extended and localized states from the quantum Boltzmann equation*, Physica A **134** (1985), 249.
47. K. M. Van Vliet, *Linear response theory revisted* I. *The many-body Van Hove limit*, J. Math. Phys. **19** (1978), 1345-1370.
48. K. M. Van Vliet, *Linear response theory revisited* II, *The master equation approach*, J. Math. Phys. **20** (1979) , 2573–2595.
49. J. Ziman, *Electrons and phonons*, The Clarendon Press, Oxford, 1960.
50. R. Zwanzig, *Statistical mechanics of irreversibility*, Lectures in Theoretical Physics, vol. III, (W. E. Britten, B. W. Downs and J. Downs, eds.), Boulder, Colorado, 1960, Interscience, N.Y. (1961), 116–141.

Centre de recherches mathématiques, Université de Montréal, C.P. 6128, succursale centre-ville, Montréal, Québec, H3C 3J7, Canada.
E-mail address: vanvliet@eng.fiu.edu

Centre de Recherches Mathématiques
CRM Proceedings and Lecture Notes
Volume **11**, 1997

Families of Transitive Primitive Maximal Simple Lie Subalgebras of diff_n

L. Michel and P. Winternitz

ABSTRACT. Transitive primitive Lie algebras, corresponding to pairs (L, L_0), where L is a classical complex simple Lie algebra and L_0 is a maximal parabolic subalgebra of L, are constructed as subalgebras of diff_n for all finite values of n. All such realizations of $\text{sl}(N,\mathbb{C})$ are shown to be maximal in diff_n. Mutual inclusions involving realizations of orthogonal and symplectic Lie algebras are pointed out.

1. Introduction

The purpose of this article is expressed in its title, namely we construct, in a systematic manner, certain infinite series of finite dimensional classical simple Lie algebras that are maximal subalgebras of diff_n, the algebra of vector fields on $\mathbb{C}^n$. In other words we construct certain Lie algebras that are realized by vector fields of the form:

$$\widehat{X}_i = \sum_{\mu=1}^{n} f_i^\mu(z)\frac{\partial}{\partial z^\mu}, \quad 1 \le i \le r < \infty, \quad z \in \mathbb{C}^n, \quad f_i \in \mathbb{C}^n. \tag{1.1}$$

We are thus addressing two related problems going back to Sophus Lie [**11, 12, 13**]. They can be stated as follow:

PROBLEM A. Classify all finite dimensional subalgebras of diff_n into conjugacy classes under the action of the group Diff_n of local diffeomorphisms of $\mathbb{C}^n$ and construct a representative of each class.

PROBLEM B. Construct all dynamical systems

$$\frac{dz_\mu}{dt} = \eta^\mu(\vec{z}, t), \quad t \in \mathbb{R}, \quad \vec{z} \in \mathbb{C}^n, \quad 1 \le \mu \le n, \tag{1.2}$$

allowing a fundamental set of solutions, i.e. having a superposition formula

$$\vec{z}(t) = \vec{F}\big(\vec{z}_1(t), \dots, \vec{z}_m(t), c_1, \dots, c_n\big), \tag{1.3}$$

expressing the general solution $\vec{z}(t)$ in terms of a finite number m of particular solutions $\vec{z}_i(t)$ and n significant constants $c_1, \dots, c_n$, specifying the initial conditions.

1991 *Mathematics Subject Classification.* Primary: 17B66; Secondary: 34A26.
This is the final form of the paper.

The direct relation between Problems A and B is expressed in a theorem, proven by Sophus Lie [**13**]. Namely, a necessary and sufficient condition for a fundamental set of solutions to exist is that the right hand side of (1.2) should lie in a finite dimensional Lie algebra, i.e. we must have (with $1 \le k, \ell, m \le r < \infty$):

$$\eta^\mu(z,t) = \sum_{k=1}^{r} Z_k(t)\xi_k^\mu(z), \quad \widehat{X}_k = \sum_{\mu=1}^{n} \xi_k^\mu(z)\partial_{z^\mu}, \quad [\widehat{X}_k, \widehat{X}_\ell] = \sum_m f_{k\ell}^m \widehat{X}_m. \tag{1.4}$$

Lie's study of differential equations [**11**] led him to give a realization of all classes of finite dimensional subalgebras of diff_n for $n = 1$ and 2. He summed up the results, when studying the case $n = 3$, in Abtheilung II of a book with Engel [**12**].

For $n = 1$ the result is very simple: the only finite dimensional subalgebras of diff_1 are sl(2) and its subalgebras, aff_1, and $\mathbb{C}$, of dimensions 3,2, and 1, respectively. For $n = 2$ the situation is already quite different. The algebra diff_2 has infinitely many different, mutually nonconjugate finite dimensional subalgebras and their dimension r can be arbitrary. Lie classified all of them into a not too large number of different types.

Among the infinite series of subalgebras of diff_2 we mention as a type those that are contained in the Abelian infinite dimensional algebra $A_\infty^{(1)}$ consisting of all vector fields $f(x)\partial_y$ (we denote temporarily x, y the two variables of diff_2, instead of z_1, z_2) where $f(x)$ are all (e.g. analytic) functions of x. Notice that $A_\infty^{(1)}$ has finite dimensional subalgebras of any dimension; none of them is maximal: indeed we can always continue indefinitely any increasing chain of $A_\infty^{(1)}$ subalgebras $a_1 \subset a_2 \subset a_3 \subset \ldots$.

Another type corresponds to an interesting infinite series of subalgebras of diff_2; it contains the symmetry algebras of the equations $y^{(m)}(x) = 0$ for $m \ge 3$. Let us denote them by W_m; their dimension is $\dim W_m = m + 4$ and a basis can be chosen to be (with $0 \le k \le m-1$):

$$\begin{gathered} \widehat{X}_1 = \partial_x, \quad \widehat{X}_2 = x\partial_x + \frac{m-1}{2} y\partial_y, \\ \widehat{X}_3 = x^2\partial_x + (m-1)xy\partial_y, \quad \widehat{X}_4 = y\partial_y, \quad \widehat{Y}_k = x^k\partial_y. \end{gathered} \tag{1.5}$$

Each subalgebra W_m is actually a maximal finite dimensional subalgebra of diff_2, which has only two more maximal finite dimensional subalgebras: sl(3) and sl(2) $\oplus$ sl(2) (see Section 3 and [**10**] for more details).

The number of different types of finite subalgebras of diff_n increases rapidly with n and it seems hopeless to try and proceed beyond $n = 3$ by dimension. We address a more limited problem, namely the following one:

PROBLEM C. Classify all maximal simple or semisimple finite dimensional subalgebras of diff_n into conjugacy classes under the action of Diff_n and choose a representative of each class.

This problem is also out of reach of the present mathematical methods although we know that for each value of n the number of solutions is finite. There is a natural grouping of these algebras into infinite families. Those we will construct are defined by the title whose technical meaning will be explained in Section 3.

Lie himself already knew two families of maximal simple finite dimensional subalgebras of diff_n (each with one representative for each n). One family contains O_{n+2} realized in Eq. (5.23) as the Lie algebra of all local conformal transformations;

the other family contains $\mathrm{sl}(n+1)$ realized as the Lie algebra of all local projective transformations of $\mathbb{C}^n$. This series was used in the context of Problem B to construct and solve systems of projective Riccati equations [**1, 2**].

It was pointed out [**15, 16**] in the context of Problem B that it is important to first construct "indecomposable" systems of ordinary differential equations with superposition formulas. These are systems of n equations from which it is not possible to split of $n_1 < n$ equations that themselves have a superposition formula. Decomposability occurs if the variables z can be split into two subsets

$$z = (x_1, x_2, \ldots, x_{n_1}, y_1, y_2, \ldots, y_{n_2}), \quad n_1 + n_2 = n, \tag{1.6}$$

such that all infinitesimal operators $\widehat{X}_k$ satisfy:

$$\widehat{X}_k = \sum_{a=1}^{n_1} \eta_k^a(x)\partial_{x^a} + \sum_{b=1}^{n_2} \xi_k^b(x,y)\partial_{y^b}, \tag{1.7}$$

i.e. the coefficients of ∂_{x^a} depend only on the variables x.

It was shown [**15**] that the system of equations (1.2) is indecomposable, if and only if the algebra (1.4) is transitive, primitive and effective, i.e. it is defined by a pair of Lie algebras $L_0 \subset L$ where L_0 is maximal subalgebra of L and does not contain a proper ideal of L. Ref. [**15**] was devoted to constructing indecomposable systems of ordinary differential equations with superposition formulas. This has a significant relation to Problem C, formulated above.

From Eq. (1.7) we see why constructing the transitive primitive Lie algebras is a crucial step towards solving Problem C. Indeed, assume that we have already introduced the coordinates (1.6) and that no further subset of the coordinates $(x_1, \ldots x_{n_1})$ can be split off. Then the "truncated" vector fields

$$\widehat{X}_k^{\top} = \sum_{a=1}^{n_1} \eta_k^a(x)\partial_{x^a}, \tag{1.8}$$

realize the same Lie algebra L as do the fields (1.7), but in $n_1 < n$ dimensions. In this lower dimensional case we have a transitive primitive Lie algebra (L, L_1), where $L_0 \subset L_1 \subset L$ and L_1 is maximal in L. This provides a method for constructing the imprimitive Lie algebras, once the primitive ones are found. Indeed, we start from the primitive ones in dimension n_1, then add the second sum in Eq. (1.7) for the chosen value of n_2. The coefficients $\eta_k^a(x)$ are to be determined from the commutation relations: a lower dimensional and simpler task.

In [**15**] families of classical simple Lie algebras were constructed as subalgebras of diff_n. The present paper can be viewed as a continuation of Ref. [**15**], in that we integrate the results into a systematic solution of Problem C. Moreover we address a question, not posed earlier [**15**], namely that of maximality of the algebra within diff_n and possible mutual inclusions and equivalences among various finite simple subalgebras of diff_n.

The problem of finding all maximal simple subalgebras of diff_n is of considerable mathematical interest as it stands. As far as physical motivation is concerned, it comes from several directions. One is the above mentioned study of ordinary differential equations with superposition formulas (Problem B). Another is the fact that in studies of quantum mechanical systems with degeneracies, Lie algebras usually occur as algebras of differential operators. They may of course have a more complicated form than given in (1.1), i.e. involve nonderivative terms and higher derivatives. Solving Problem C is however definitely an important ingredient in the study of algebras of quantum mechanical operators.

Throughout this article we consider all Lie algebras and Lie groups over $\mathbb{C}$.

2. The Lie Algebra diff_n of Vector Fields on $\mathbb{C}^n$ and Some of its Subalgebras

The aim of this section is to establish the notations and recall some definitions and some basic properties of diff_n.

A point $z \in \mathbb{C}^n$ is given by the set of its n coordinates z_i, $1 \le i \le n$. We denote by T_a the tangent plane to $\mathbb{C}^n$ at the point a. All functions we shall consider are analytic. We distinguish between the maps (e.g. φ) from $\mathbb{C}^n$ to itself and those from $\mathbb{C}^n$ to $\mathbb{C}$ (e.g. f):

$$\mathbb{C}^n \xrightarrow{\varphi} \mathbb{C}^n, \quad \mathbb{C}^n \xrightarrow{f} \mathbb{C}. \tag{2.1}$$

The coordinates of $\varphi(z)$ are denoted by $\varphi_i(z)$. The differential $d_a\varphi$ of the map φ at a point a is a linear operator:

$$T_a \xrightarrow{d_a\varphi} T_{\varphi(a)} : \frac{\partial \varphi_i}{\partial z_j}(a), \tag{2.2}$$

whose matrix is called the Jacobian of φ at a. If $d_a\varphi$ is invertible, that is if $\det(\partial_i\varphi/\partial z_j)(a) \neq 0$, then φ is called a local diffeomorphism at a. We have a diffeomorphisms on $\mathbb{C}^n$ (respectively on the open set $\mathcal{O} \subset \mathbb{C}^n$) when $d_a\varphi$ is invertible at every point of $\mathbb{C}^n$ (respectively of $\mathcal{O}$). We emphasize that in this paper we generally consider *local diffeomorphism*, i.e. defined on the neighbourhood of a regular point. The diffeomorphisms on the same domain form a local group. The action of the diffeomorphism φ on the function f is

$$\varphi \cdot f = f \circ \varphi^{-1} \iff (\varphi \cdot f)(z) = f(\varphi^{-1}(z)); \quad \varphi(\varphi^{-1}(z)) = z. \tag{2.3}$$

By multiplication (of their values) the complex valued functions on $\mathbb{C}^n$ form an associative and commutative algebra. The derivations of that algebra are called vector fields and are denoted by $\hat{v} \equiv \sum_i v_i(z)\partial_i$ where ∂_i is short for $\partial/\partial z_i$:

$$\hat{v}f \equiv \sum_i v_i\partial_i f, \quad \hat{v}(fg) = (\hat{v}f)g + f\hat{v}g. \tag{2.4}$$

One shows that the Lie algebra formed by these derivations:

$$[\hat{v}, \hat{w}] = \sum_{ij}(v_i\partial_i w_j - w_i\partial_i v_j)\partial_j, \tag{2.5}$$

is diff_n, the Lie algebra of the group of diffeomorphisms. A linear combination of vector fields $\sum_i \lambda_i \hat{v}^{(i)}$, $\lambda_i \in \mathbb{C}$ is a vector field. An m-dimensional vector space V_m of vector fields defines an m-dimensional Lie algebra when $\hat{v}, \hat{w} \in V_m \implies [\hat{v}, \hat{w}] \in V_m$; for example:

$$\mathcal{A}_m^{(1)} = \{z_1^{k-1}\partial_2\}, \quad 1 \le k \le m, \tag{2.6}$$

is an m-dimensional Abelian Lie algebra, which is a finite dimensional (nonmaximal) subalgebra of $\mathcal{A}_\infty^{(1)}$ introduced in Section 1. Notice that if $\hat{v}$ is a vector field and f a function, $f\hat{v}$ is a vector field.

Consider a set of m vector fields $\hat{v}^{(k)} = \sum_i v_i^{(k)}\partial_i$, $1 \le k \le m$, depending on n variables $1 \le i \le n$.

DEFINITION 2.1. *The* functional rank r_f *of the set of* m *vector fields* $v^{(k)}$ *is equal to the rank of the matrix of the coefficients of these fields:*

$$r_f = rank\ \{v_i^{(k)}\}, \quad 1 \le i \le n,\ 1 \le k \le m. \tag{2.7}$$

The functional rank can be defined at a point a*, in a neighbourhood of* a*, on some open domain* $\mathcal{O} \subset \mathbb{C}^n$*, or on* $\mathbb{C}^n$ *itself. For a given* n*, the functional rank satisfies* $r_f \le n$*. For instance, the* m *dimensional Lie algebra* $\mathcal{A}_m^{(1)}$ *defined in* (2.6) *has functional rank* 1 (*that explains the upper index*).

The action $\varphi \cdot \hat{v}$ of a diffeomorphism $y = \varphi(z)$ on a vector field $\hat{v}$ is given explicitly by

$$\varphi \cdot \Big(\sum_i v_i \partial_{y_i}\Big) = \sum_{ij} \Big(v_i(\varphi^{-1}(y)) \frac{\partial \varphi_j}{\partial z_i}\Big) \partial_{y_j}. \tag{2.8}$$

One can check the properties:

$$\begin{gathered} \varphi \cdot (\lambda_1 \hat{v}^{(1)} + \lambda_2 \hat{v}^{(2)}) = \lambda_1 \varphi \cdot \hat{v}^{(1)} + \lambda_2 \varphi \cdot \hat{v}^{(2)}, \quad \varphi \cdot (f\hat{v}) = (\varphi \cdot f)(\varphi \cdot \hat{v}), \\ \varphi \cdot [\hat{v}, \hat{w}] = [\varphi \cdot \hat{v}, \varphi \cdot \hat{w}], \end{gathered} \tag{2.9}$$

and prove from them that a diffeomorphism transforms a subalgebra of diff_n into an isomorphic subalgebra with same dimension and same functional rank.

It is well know that every nonsingular vector field can be transformed by a diffeomorphism into ∂_{z_1}. In other words all dimension 1 Abelian subalgebras are equivalent. This can be extended to:

LEMMA 2.1. *The* k*-dimensional Abelian subalgebras of functional rank* k, $k \le n$*, form one equivalence class* $[\mathcal{A}_k^{(k)}]$ *under the diffeomorphisms of* Diff_n.

For instance they can be transformed into $\{\partial_{z_1}, \partial_{z_2}, \dots, \partial_{z_k}\}$ or into the set $\{z_1\partial_{z_1}, z_2\partial_{z_2}, \dots, z_k\partial_{z_k}\}$ when $z_i \ne 0$, $1 \le i \le k$.

Let $\mathcal{L}$ be a Lie algebra and $\mathcal{A}, \mathcal{B}$ two vector subspaces of $\mathcal{L}$. We denote by $[\mathcal{A}, \mathcal{B}]$ the vector space:

$$[\mathcal{A}, \mathcal{B}] = \big\{[a, b], a \in \mathcal{A}, b \in \mathcal{B}\big\}. \tag{2.10}$$

For example: $[\mathcal{B}, \mathcal{B}] \subseteq \mathcal{B}$ means: $\mathcal{B}$ is a subalgebra of $\mathcal{L}$ and $[\mathcal{K}, \mathcal{L}] \subseteq \mathcal{K}$ means: $\mathcal{K}$ is an ideal of $\mathcal{L}$. Let us recall the following definitions:

DEFINITION 2.2. $C_{\mathcal{L}}(\mathcal{B})$, *the* centralizer *of* $\mathcal{B}$ *in* $\mathcal{L}$*, is the largest subalgebra* $\mathcal{H} \subseteq \mathcal{L}$ *such that* $[\mathcal{B}, \mathcal{H}] = 0$*; e.g.* $C_{\mathcal{L}}(\mathcal{L})$ *is the center of* $\mathcal{L}$.

DEFINITION 2.3. $N_{\mathcal{L}}(\mathcal{B})$, *the* normalizer *of* $\mathcal{B}$ *in* $\mathcal{L}$*, is the largest subalgebra* $\mathcal{H} \subseteq \mathcal{L}$ *such that* $\mathcal{B}$ *is an ideal of* $\mathcal{H}$.

If a family of $\mathcal{L}$ subalgebras forms one orbit for the group G, then the same is true of its centralizers and its normalizers.

For an Abelian subalgebra $\mathcal{A} \subseteq \mathcal{L}$, one has $\mathcal{A} \subseteq C_{\mathcal{L}}(\mathcal{A})$; when the equality holds, $\mathcal{A}$ is a *maximal Abelian subalgebra* of $\mathcal{L}$. Obviously the n-dimensional, functional rank n subalgebra $\{\partial_k,\ 1 \le k \le n\}$ is a maximal Abelian subalgebra of diff_n; this extends to all its conjugates, hence:

LEMMA 2.2. *Every* n*-dimensional, functional rank* n *algebra* $\mathcal{A}_n^{(n)}$ *is a maximal Abelian subalgebra of* diff_n.

Similarly:

$$
\begin{gathered}
1 \leq k \leq n; \\
C_{\mathrm{diff}_n}(\mathcal{A}_k^{(k)}) = \mathcal{A}_k^{(k)} \oplus \mathrm{diff}_{n-k} \iff C_{\mathrm{diff}_n}(\mathrm{diff}_{n-k}) = \mathrm{diff}_k\,.
\end{gathered} \tag{2.11}
$$

In this paper we shall restrict the qualification of simple and semisimple to *finite dimensional* Lie algebras. By definition a simple Lie algebra has no nontrivial ideals (i.e. different from itself and from 0) and a semisimple Lie algebra is a direct sum of simple Lie algebras. We have also to recall the existence and properties of some of their remarkable subalgebras.

On the N dimensional vector space of a Lie algebra $\mathcal{L}$ there is a natural linear representation, called the *adjoint representation* and denoted by ad$\mathcal{L}$. The representant of $a \in \mathcal{L}$ is the linear operator traditionally written as

$$
a \mapsto \mathrm{ad}(a), \quad \forall x \in \mathcal{L}, \quad \mathrm{ad}(a)x = [a, x]. \tag{2.12}
$$

The Killing form of a finite dimensional Lie algebra is defined by:

$$
(a, b) = \mathrm{tr}\,\mathrm{ad}(a)\mathrm{ad}(b). \tag{2.13}
$$

We have

$$
\forall x \in \mathcal{L}, \quad \big([x, a], b) = (a, [x, b]\big). \tag{2.13'}
$$

This property of the Killing form is the infinitesimal expression of its invariance under the group G. For any semisimple algebra $\mathcal{S}$, its adjoint representation is faithful and its Killing form is nondegenerate.

DEFINITION 2.4. *A* Cartan subalgebra *of a semisimple Lie algebra $\mathcal{S}$ is a maximal Abelian subalgebra $\mathcal{H}$ consisting of nonnilpotent elements (i.e. elements h such that there is no positive integer m such that* $\mathrm{ad}(h)^m = 0$*). All Cartan subalgebras form one orbit of $\mathcal{S}$ (over $\mathbb{C}$). Equivalently, a Cartan subalgebra is a maximal Abelian selfnormalizing subalgebra of $\mathcal{S}$. The rank of $\mathcal{S}$ is the dimension of its Cartan subalgebra*: $r(\mathcal{S}) = \dim \mathcal{H}$.

There exists a set of common eigenvectors of all $\mathrm{ad}(h)$; they span the space $\mathcal{L}$. Let $\{e_\alpha\}$ be a set of $(\dim \mathcal{S} - r(\mathcal{S}))$ eigenvectors forming a basis of the eigenspace with nonvanishing eigenvalues (i.e. complementary to $\mathcal{H}$). Each e_α defines a linear form on $\mathcal{H}$ that we can write (using the Killing form) as $h \mapsto (h, r_\alpha)$. We have

$$
\mathrm{ad}(h)e_\alpha \equiv [h, e_\alpha] = (h, r_\alpha)e_\alpha. \tag{2.14}
$$

These elements $r_\alpha \in \mathcal{H}$ are called the *roots* of $\mathcal{H}$. It is easy to prove that $-r\alpha$ (that we shall also denote by $r_{-\alpha}$) is also a root. By choosing a normalization of the e_α's we have

$$
[e_\alpha, e_{-\alpha}] = r_\alpha. \tag{2.15}
$$

The last two equations show that in $\mathcal{S}$ each root r_α, or each eigenvector e_α, defines a unique $\mathrm{sl}(2)_\alpha$ subalgebra with basis $e_\alpha, e_{-\alpha}, r_\alpha$.

Let us choose in the Cartan subalgebra H a hyperplane which contains no roots. We denote by R_+ the set of roots on one side of the hyperplane and R_- the roots on the other side. Each set $R_\pm$ contains $\big(d(\mathcal{S}) - r(\mathcal{S})\big)/2$ roots ($d(\mathcal{S})$ is short for $\dim(\mathcal{S})$). To each set $R_\pm$ corresponds a set of eigenvectors which generate a subspace $\mathcal{N}_\pm$ of $\mathcal{S}$. Each space contains only nilpotent elements and forms a maximal nilpotent subalgebra of $\mathcal{S}$. Their normalizers $\mathcal{B}_\pm = \mathcal{H} \oplus \mathcal{N}_\pm$ are maximal solvable subalgebras of $\mathcal{S}$. Such algebras are called Borel subalgebras. A complex semi-simple Lie algebra $\mathcal{S}$ has precisely one Borel subalgebra up to a conjugacy in $\mathcal{S}$.

We can prove now a theorem directly related to our problem:

THEOREM 2.1. *The functional rank $r_f(\mathcal{H})$ of a Cartan subalgebra of a semisimple algebra $\mathcal{S} \subset \mathrm{diff}_n$ is equal to rank $(\mathcal{S})$.*

PROOF. Consider a semisimple Lie algebra $\mathcal{S}$ of rank r. We choose a root $r \in \mathcal{H}$, a Cartan subalgebra. We take a basis in $\mathcal{H}$: $r, h_1, \dots, h_{r-1}$, such that $(r, h_i) = 0$, $1 \le i \le r-1$. We need to consider only the $\mathrm{sl}_2(e_\pm, r)$ subalgebra corresponding to the chosen root. From equation (2.14), (2.15) we have the partial set of commutation relations in $\mathcal{L}$

$$1 \le i, \quad j \le r-1,$$
$$[h_i, h_j] = [r, h_i] = [h_i, e_\pm] = 0, \quad [r, e_\pm] = \pm(r,r)e_\pm, [e_+, e_-] = r. \tag{2.16}$$

Assume that $\mathcal{S}$ is represented by a subalgebra of diff_n and that the vector fields $\widehat{R}$, $\widehat{H}_i$, representing basis elements of $\mathcal{H}$ are functionally dependent i.e.,

$$\widehat{R} = \sum_k \varphi_k \widehat{H}_k, \text{ with } [\widehat{H}_i, \widehat{H}_j] = 0, \tag{2.17}$$

where the $\varphi(x)$'s are functions (and not constants!). Then

$$0 = [\widehat{R}, \widehat{H}_i] = -\sum_k (\widehat{H}_i \varphi_k)\widehat{H}_k = 0. \tag{2.18}$$

The other commutation relations are represented by:

$$[\widehat{H}_i, \widehat{E}_\pm] = 0, \quad \pm(r,r)\widehat{E}_\pm = [\widehat{R}, \widehat{E}\pm] = -\sum_k (\widehat{E}_\pm \varphi_k)\widehat{H}_k, \tag{2.19}$$

and the last commutation relation in (2.16) yields

$$\widehat{R} = [\widehat{E}_+, \widehat{E}_-] = -(r,r)^{-2} \sum_{k,\ell} \big[(\widehat{E}_+ \varphi_k)\widehat{H}_k, (\widehat{E}_- \varphi_\ell)\widehat{H}_\ell\big]. \tag{2.20}$$

With a possible change of labeling of mute indices and the further use of the first relation in (2.19), we obtain

$$(r,r)^2 \widehat{R} = \sum_{k,\ell} \big((\widehat{E}_- \varphi_k)(\widehat{E}_+ \widehat{H}_k \varphi_\ell) - (\widehat{E}_+ \varphi_k)(\widehat{E}_- \widehat{H}_k \varphi_\ell)\big)\hat{H}_\ell. \tag{2.21}$$

Equation (2.18) shows that this expression vanishes, so $\widehat{R} = 0$, which is a contradiction. Hence Eq. (2.17) is contradictory. This ends the proof of the theorem. □

COROLLARY 2.1. *Every semisimple subalgebra of* diff_n *has a rank $r \le n$.*

We shall show at the end of this section that the equality $r = n$ can be reached for $\mathrm{sl}(n+1)$. In the conclusion of the paper we show that it is not reached for any other simple Lie subalgebra of diff_n.

For convenience, let us relate the Cartan notation for the classical simple Lie algebras of rank r to the standard notation used for the special linear $\mathrm{sl}(n)$, orthogonal $o(n)$, symplectic $\mathrm{sp}(n)$ algebras in dimension n:

$$\begin{aligned} &r \ge 1, \quad A_r = \mathrm{sl}(r+1); \qquad r \ge 2, \quad B_r = o(2r+1); \\ &r \ge 3, \quad C_r = \mathrm{sp}(2r); \qquad r \ge 4, \quad D_r = o(2r). \end{aligned} \tag{2.22}$$

TABLE 1. Lower bounds on n for which the Lie algebra L can be realized as a subalgebra of diff_n. The rank of L is denoted r.

L	$\mathrm{sl}(N)$	$o(N); N \neq 4, 6$	$o(6)$	$\mathrm{sp}(2N)$	G_2	F_4	E_6	E_7	E_8
n	$N-1$	$N-2$	3	$2N-1$	5	15	16	27	57
r	$N-1$	$[\frac{N}{2}]$	3	N	2	4	6	7	8

There exist the following isomorphisms:

$$o(3) \sim \mathrm{sl}(2) \sim \mathrm{sp}(2); \quad o(4) \sim \mathrm{sl}(2) \oplus \mathrm{sl}(2); \quad o(5) \sim \mathrm{sp}(4); \quad o(6) \sim \mathrm{sl}(4). \tag{2.23}$$

Notice that $O(2)$ is a 1-dimensional Abelian Lie algebra and $o(4)$ is not simple. Among the simple Lie algebras over $\mathbb{C}$, the only ones which have a simple subalgebra of the same rank are:

$$\begin{gathered} r \geq 3, \quad o(2r+1) \supset o(2r); \\ G_2 \supset \mathrm{sl}(3), \quad F_4 \supset o(8), \quad E_7 \supset \mathrm{sl}(8), \quad E_8 \supset \mathrm{sl}(9), \quad E_8 \supset o(16). \end{gathered} \tag{2.24}$$

There is another limitation for the minimum value of n such that the simple Lie algebra G can be realised a a subalgebra of diff_n. Indeed in this realization it acts locally on a manifold of dimension n and the minimum n value corresponds to a (locally) transitive action. This can occur only if L has a subalgebra L_0 of codimension $n = \dim L - \dim L_0$. For simple Lie algebras L over the field $\mathbb{C}$ the algebra L_0 of maximal dimension is always one of the maximal parabolic subalgebras of L.

In Table 1 we give a list of the minimal values of n of the homogeneous spaces G/G_0, together with the rank r of L, for all complex simple Lie algebras. From the Table we see that for $\mathrm{sl}(n)$ (including $o(3) \sim \mathrm{sl}(2)$, $o(6) \sim \mathrm{sl}(4)$) we have $n = r$. In all other cases we have $n > r$.

The dimension n is obtained according to the formula $n = \dim L - \dim P$ where P is the highest dimensional maximal parabolic subalgebra of L. The dimension of a maximal parabolic subalgebra is obtained as $\dim P = r + n_1 + n_+$ where n_+ is the number of positive roots of L, n_1 the number of positive roots of a regular maximal simple subalgebra S of L and r is the rank of L. The numbers n_1 and n_+ can be read off from tables [**4, 5**].

We end this section by giving for every n the realization of the simple Lie algebra $\mathrm{sl}(n+1)$ of rank n as a subalgebra of diff_n.

We recall first that the $n \times n$ complex matrices form the Lie algebra $gl(n, \mathbb{C})$ when the Lie bracket of two elements is their commutator $[a, b] = ab - ba$. Let us denote by e_{ij}, $i \neq j$ the matrices whose all elements are vanishing except the one of the i-th line and j-th column which is equal to 1. In this basis for the vector space of $n \times n$ matrices the Lie algebra law is:

$$[e_{ij}, e_{k\ell}] = \delta_{jk} e_{i\ell} - \delta_{i\ell} e_{kj}. \tag{2.25}$$

The multiples of I_n form the center of $\mathrm{gl}(n)$. So the algebra $\mathrm{sl}(n)$ is faithfully represented by the $n \times n$ traceless matrices. The diagonal matrices represent a Cartan subalgebra $\mathcal{H}$: indeed they are nonnilpotent matrices and they form an (n-1)-dimensional maximal Abelian subalgebra. Eq. (2.25) shows that the $e_{k\ell}$'s, $k \neq \ell$ are the eigenvectors of $\mathrm{ad}(h)$, $h \in \mathcal{H}$ and the corresponding root is $e_{kk} - e_{\ell\ell}$.

The upper (respectively lower) triangular matrices form the Borel subalgebra $\mathcal{B}_+$ (respectively $\mathcal{B}_-$). The antisymmetrical matrices represent the subalgebra $o(n)$ which is a simple algebra when $n = 3$ and $n \geq 5$. Notice that for $n > 2$ this representation of $o(n)$ is irreducible. The identity $-[a,b]^\top = [-a^\top, -b^\top]$ shows that $a \mapsto -a^\top$ represents an automorphism of $\mathrm{sl}(n)$ which reduces to the identity on the subalgebra $o(n)$. So this automorphism of $\mathrm{sl}(n)$ is an outer automorphism for $n > 2$ and an inner one for $n = 2$ (conjugation by the Pauli matrix σ_2).

From the commutation rule of vector fields:

$$[z_i\partial_j, z_k\partial_\ell] = \delta_{jk} z_i \partial_\ell - \delta_{i\ell} z_k \partial_j, \tag{2.26}$$

and the comparison with (2.25) one sees that the map $e_{ij} \mapsto z_i\partial_j$ gives a realization of the Lie algebra $\mathrm{gl}(n)$ as subalgebra of diff_n. This realization can be extended to the rank n simple algebra $\mathrm{sl}(n+1)$, with $1 \leq i,j,k,\ell \leq n$:

$$\begin{gathered} \mathcal{D} = \sum_i z_i\partial_i, \quad e_{ij} \mapsto z_i\partial_j, \\ e_{n+1,j} \mapsto \partial_j, \quad e_{i,n+1} \mapsto -z_i\mathcal{D}, \quad e_{n+1,n+1} \mapsto -\mathcal{D}. \end{gathered} \tag{2.27}$$

This example was known to Lie for any n. It shows that the equality can be reached in Corollary 2.1.

We have shown above that $a \mapsto -a^\top$ is an outer automorphism of any $\mathrm{sl}(m)$ algebra. That automorphism transforms into itself the $\mathrm{sl}(n)$ of (2.26) as a subalgebra of $\mathrm{sl}(n+1)$ realized in (2.27) and it exchanges the two n-dimensional Abelian subalgebras $\mathcal{A}_n^n(\partial_j)$, $\mathcal{A}_n^1(z_i\hat{d})$, For $n > 1$, these two algebra are not conjugate in diff_n since their functional ranks are respectively n and 1.

It is easy to verify that the normalizer of $\mathcal{A}_n^{(n)}(\partial_i)$ is an affine algebra:

$$\mathrm{aff}_n^- \sim N_{\mathrm{diff}_n}(\mathcal{A}_n^{(n)}) = \mathcal{A}_n^{(n)} \rtimes \mathrm{gl}(n), \quad gl(n) = \{z_i\partial_j\}. \tag{2.28}$$

Since the n-dimensional, functional rank n Abelian subalgebras of diff_n form a unique orbit of the group Diff_n, that is also true of their normalizers in diff_n. We denote this orbit of affine algebras in dimension n by $[\mathrm{aff}_n^-]$. while the other orbit of n dimensional affine algebras is that of

$$\mathrm{aff}_n^+ = \mathcal{A}_n^{(1)}(z_i\hat{d}) \rtimes \mathrm{gl}(n), \tag{2.29}$$

i.e. that of the normalizer of $\mathcal{A}_n^{(1)}(z_i\mathcal{D})$. The $\mathrm{sl}(n+1)$ algebra is generated by the two subalgebras $\mathrm{aff}_n^\pm$ satisfying $\mathrm{aff}_n^+ \cap \mathrm{aff}_n^- = \mathrm{gl}(n)$. Since each type of affine algebras forms one orbit of Diff_n that is also true of the algebras $\mathrm{sl}(n+1) \subset \mathrm{diff}_n$.

3. Construction of the Homogeneous Spaces and Algebras of Vector Fields

The method to be applied in this article is closely related to one used earlier to construct systems of nonlinear ordinary differential equations with superposition formulas [**15**].

We are constructing finite dimensional subalgebras L of diff_n so that elements of L have the form (1.1). The algebra can be integrated to obtain a local Lie group G of local diffeomorphisms that act on a manifold $M \sim \mathbb{C}^n$. We make the restriction that G acts regularly, that is that it can be stratified into orbits of orbit type G/G_0 where G_0 is a Lie subgroup of G. Its Lie algebra L_0 is realized by vector fields vanishing at the origin. Abusing notation, we shall assume that this stratification has already been performed and we shall consider M to be a homogeneous space

$$M \sim G/G_0. \tag{3.1}$$

Let us assume that G_0 is not a maximal subgroup of G; let G_1 be a maximal subgroup of G containing G_0. The orbit g/g_0 is a disjoint union of imprimitivity cells which are permuted by the action of G. There exists a G-equivariant surjective map σ between the orbits: $G/G_0 \xrightarrow{\sigma} G/G_1$ which projects the imprimitivity cells of G/G_0 on the points of G/G_1. Each cell can be considered as an orbit G_1/G_0. In the case of Lie groups, the imprimitivity cells are leaves of a foliation. Here we restrict all considerations to local ones; G_0 is the stabilizer (= isotropy group) of the origin of coordinates. Transitive group actions have been studied in a geometric context [**6, 7, 8, 9**]. In particular a decomposition of the type indicated in eqs. (1.6)–(1.7) was shown to correspond to the existence of an invariant foliation on M. Locally this occurs if the pair of Lie algebras $L_0 \subset L$ does *not* define a transitive primitive and effective Lie algebra. We recall their definition:

DEFINITION 3.1. *The pair of Lie algebras $L_0 \subset L$ defines a transitive primitive and effective Lie algebra if L_0 is a maximal subalgebra of L and does not contain a proper ideal of L.*

DEFINITION 3.2. *The transitive primitive effective pair of Lie algebras is nonlinear if there exists a nontrivial subalgebra $L_1 \subset L_0$ such that $[L_1, L] \subseteq L_0$.*

DEFINITION 3.3. *The transitive primitive effective nonlinear pair of Lie algebras is* irreducible *if the only subspaces S satisfying $L_0 \subseteq S \subseteq L$ and $[L_0, S] \subseteq S$ are $S = L_0$ and $S = L$. Otherwise the pair $L_0 \subset L$ is* reducible.

DEFINITION 3.4. *A parabolic subalgebra of a simple Lie algebra L is any subalgebra containing a Borel subalgebra (the unique, up to a conjugation, maximal solvable subalgebra). A maximal parabolic subalgebra is not properly contained in any other subalgebra of L.*

DEFINITION 3.5. *A reductive Lie algebra is a direct sum of simple and Abelian Lie algebras containing at least one simple Lie algebra (but not necessarily any Abelian ones).*

All transitive primitive pairs of Lie algebras over $\mathbb{C}$ have been classified [**6, 7, 8, 9**]. Let us state, without proof, the classification theorem, summing up results due to these authors.

THEOREM 3.1. *Precisely five types of transitive primitive effective pairs $L_0 \subset L$ of Lie algebras over $\mathbb{C}$ exist. They are distinguished by the nature of L_0 and L.*

1. *L is simple and L_0 is a maximal parabolic irreducible subalgebra.*
2. *L is simple and L_0 is a maximal parabolic irreducible subalgebra.*
3. *L is simple and L_0 is a maximal reductive subalgebra.*
4. *L is semi-simple and has the form $L = K \oplus K$, $L_0 = K_D$, where K is simple and L_0, isomorphic to K, is the diagonal subalgebra of $K \oplus K$.*
5. *L is an affine Lie algebra, either* aff_n *or an affine subalgebra of it. Then L_0 is reductive and acts faithfully and irreducibly on an Abelian ideal A:*

$$L = A \rtimes L_0, \quad [A, A] = 0, \quad [A, L_0] = A, \quad [L_0, L_0] \subseteq L_0. \tag{3.2}$$

With L simple, the types 1,2,3 are effective. This is also true of the other types when they are realized as subalgebras of diff_n and all are transitive. So from now on we will use the shorter expression "primitive Lie algebras".

REMARK 1. A primitive Lie algebra has no center. In Case 5, that is implied by the faithful action of L_0 on A; for the other case it is implied by the definition of simple or semisimple Lie algebras.

The list of the primitive subalgebras of diff_2 and diff_3 has been given by S. Lie in [**12**]. There are 3 of them for dimension $n = 2$ (Theorem 6, p. 71): sl(3), type 1, maximal finite dimensional subalgebra; aff_2^- and it subalgebra saff_2^-, both of type 5 and nonmaximal. Notice that $\text{sl}(2) \oplus \text{sl}(2) \equiv o(4)$ and the infinite family of subalgebras W_m, $m \geq 3$ (given in (1.5)) are maximal finite dimensional subalgebras but they are not primitive. For $n = 3$ (Chap. 7, Theorem 9, p. 139), there are 8 primitive subalgebras but only 2 of them are maximal: they are of type 1 (sl(4) and $o(5)$); the other primitive subalgebras are: 1 of type 2 (another $o(5) \subset \text{sl}(4) \sim o(6)$), one of type 4 ($\text{sl}(2) \oplus \text{sl}(2) \subset \text{sl}(4) \sim o(6)$), four of type 5: aff_3^- and three of its subalgebras. The primitive algebra ([2] p. 134) is interesting; its two factors $K \sim \text{sl}(2)$ are realized by (with $\hat{d} = x\partial_x + y\partial_y + z\partial_z$):

$$\hat{a}_1 = \partial_y + x\partial_z, \quad \hat{a}_2 = y\partial_y + z\partial_z, \quad \hat{a}_3 = -z\partial_x + y\hat{d}, \tag{3.3}$$

$$\hat{b}_1 = \partial_x + y\partial_z, \quad \hat{b}_2 = x\partial_x + z\partial_z, \quad \hat{b}_3 = -z\partial_y + x\hat{d}. \tag{3.3'}$$

Notice that the functional rank of each sl(2) is $r_f = 3$ for $xy - z \neq 0$.

More generally the minimum value of n for the realization of the primitive algebra (L, L_0) as a subalgebra of diff_n is

$$n = \dim M = \dim L - \dim L_0, \tag{3.4}$$

where M is the homogeneous space of the transitive L action. Of course we are interested only in the minimal n realizations.

REMARK 2. In the Case 5 of Theorem 3.1 we have, $n = \dim M = \dim A$. As we have shown as the end of Section 2, there are only two realizations of aff_n in diff_n and both are subalgebras of the projective $\text{sl}(n+1)$. So Case 5 does not yield maximal subalgebras.

LEMMA 3.1. *Let (L, L_0) be a transitive primitive Lie algebra realized as a subalgebra of* diff_n *(with $n = \dim L - \dim L_0$). Let K be another finite dimensional Lie algebra, satisfying $L \subset K \subset$* diff_n*. Then there exists an algebra $K_0 \subset K$ such that (K, K_0) is a transitive primitive Lie algebra and we have $L_0 = L \cap K_0$, $n = \dim K - \dim K_0$.*

PROOF. Consider the groups G_0 and G, corresponding to the Lie algebras L_0 and L. Locally in the neighbourhood of the origin in $\mathbb{C}^n$, we have $M \sim G_0/G$, the manifold on which G acts transitively. We denote $K_0 \subset K$ the subalgebra of K realized by vector fields vanishing at the origin, and H_0, H the Lie groups corresponding to K_0 and K. Since G acts transitively on M and $G \subset H$, H also acts transitively. Hence we must also have $M \sim H/H_0$. If (K, K_0) did not define a primitive Lie algebra, then the local action of H on M would allow an invariant foliation. Then so would any subgroup of H, in particular G. Then (L, L_0) would not be primitive either and we obtain a contradiction.

In other words, if (L, L_0) is primitive and L is properly contained in $K \subset \text{diff}_n$, the (K, K_0) is also primitive. □

COROLLARY 3.1. *Assume that* (L, L_0) *is a primitive algebra realized as a subalgebra of* diff_n *with* n *satisfying* (3.4); *then the centralizer* $C_{\mathrm{diff}_n}(L)$ *is trivial.*

Assume the contrary. Since L has no center (Remark 1) $L \cap C_{\mathrm{diff}_n}(L) = 0$. Let $\hat{a}$ be a vector field of the centralizer; $L' = L \oplus \mathbb{C}\hat{a}$ is an algebra containing L. From the preceding lemma, it is primitive and from Theorem 3.1 it is not. End of proof per absurdum.

COROLLARY 3.2. $\mathrm{sl}(n+1)$ *is a maximal finite subalgebra of* diff_n.

Let us assume the contrary: there exists H such that $\mathrm{sl}(n+1) \subset H \subset \mathrm{diff}_n$. From the preceding lemma, H is primitive. From Remark 2 and the end of Section 2, Case 5 of Theorem 3.1 is excluded. So H is semisimple and from Corollary 2.1, rank $H =$ rank $\mathrm{sl}(n+1) = n$. Case 4 of Theorem 3.1 is excluded: the rank of such an H has to be larger than n. From (2.24) there are 3 values $n = 2, 7, 8$ for which there exists a simple H of rank n containing $\mathrm{sl}(n+1)$, but Table 1 shows that these simple algebras cannot be realised in diff_n. So cases 1,2,3 of Theorem 3.1 are also ruled out and the corollary is proven.

LEMMA 3.2. *Assume that* (L, L_0) *is a primitive Lie algebra realized as a subalgebra of* diff_n *with* n *satisfying* (3.4), *and that* L *is a proper subalgebra* $L \subset H \subset \mathrm{diff}_n$ *of the finite dimensional* **maximal** *subalgebra* H *of* diff_n. *Then* H *is simple.*

PROOF. From the previous lemma we know that H corresponds to a primitive algebra. Since it is maximal we know from Remark 2, that it cannot belong to type 5. We prove in three steps that it does not belong to type 4, i.e. $H = S \oplus S$, S simple. We have $L_0 \subset L \subset H \subset \mathrm{diff}_n$. First assume that L is simple (type 1,2,3). If L were a subalgebra of one of the factors S of H the other factor S would be in the centralizer of L; that contradicts the Corollary 3.1. So L has to be a subalgebra of the diagonal $S^d \subset S \oplus S$. Then $n = \dim M = \dim S \geq \dim L > \dim M$ which is self contradictory. Now assume that L is semisimple, i.e. $L = S' \oplus S'$ of type 4, then $n = \dim M = \dim S = \dim S'$ which implies $S = S'$, hence $H = L$. Finally, if L is of type 5, we have shown $L \subseteq \mathrm{aff}_n \subset \mathrm{sl}(n+1)$ and we have shown in Section 2 that aff_n is a maximal subalgebra of $\mathrm{sl}(n+1)$ and Corollary 3.1 has shown that this simple algebra is maximal in diff_n; That concludes the proof of the lemma. □

This article is devoted to classical simple Lie algebras, so we will only encounter the first three types of transitive primitive Lie algebras listed in Theorem 3.1. The coordinates in the first two cases are easy to construct. For both of them the homogeneous space M is a Grassmanian of r-planes. If the group G acting on M has an invariant metric (i.e. G is $O(N)$ or $\mathrm{SP}(2N)$), then we shall deal with a Grassmanian of null planes.

(a) If the maximal parabolic subalgebra L_0 is irreducible, then coordinates can be so chosen that the coefficients of the vector fields (1.1) are second order polynomials.
(b) If L_0 is a maximal parabolic reducible subalgebra,then the coefficients of the vector fields can again be chosen to be polynomials but their degree may be higher than 2. For the classical simple Lie algebras and for g_2 [**3**] the degree of the polynomials is at most 4.

(c) If L_0 is a maximal reductive subalgebra, the coefficients of the vector fields are not necessarily polynomials (in any coordinate system). The construction of the homogeneous spaces M is less uniform than for the case of maximal parabolic subalgebras.

The following sections are devoted to the specific construction of the homogeneous spaces M, corrcsponding to the two first cases of Theorem 3.1. We construct convenient coordinate patches in the neighbourhood of the origin. Then we obtain realizations of $\mathrm{sl}(N)$ in Section 4, of $o(N)$ in Section 5, of $\mathrm{sp}(2N)$ in Section 6. In Section 7 we find the list of those which may be, up to an equivalence, subalgebras of others. Excluding them, we obtain several large families of nonequivalent and independent realizations of these classical simple Lie algebras by vector fields.

4. The Homogeneous Spaces $\mathbf{M} = \mathrm{SL}(\mathbf{N}, \mathbb{C})/\,\mathrm{P}(\mathbf{r}, \mathbf{s})$

In order to present explicitly the complete list of maximal parabolic subalgebras of $\mathrm{sl}(N)$, we realize this Lie algebra of the Lie group $\mathrm{SL}(N)$ by the matrices

$$W = \left\{ \begin{pmatrix} C & A \\ -D & -B \end{pmatrix} \right\}, \quad r + s = N, \quad 1 \le s \le r \le N-1, \tag{4.1}$$

$$C \in \mathbb{C}^{r\times r}, \quad A \in \mathbb{C}^{r\times s}, \quad D \in \mathbb{C}^{s\times r}, \quad B \in \mathbb{C}^{s\times s}, \quad \mathrm{tr}\, B = \mathrm{tr}\, C, \tag{4.2}$$

where A, B, C, D represent all possible matrices satisfying conditions (4.2).

For each different possible set of values of r, s, the elements of W with $A = 0$ form a maximal parabolic subalgebra $L_0 \sim p(r,s) \subset \mathrm{sl}(N)$; they are all irreducible. For $N > 2$ there are two inequivalent N-dimensional representations of $\mathrm{sl}(N)$ which are transformed into each other by the outer automorphism $W \leftrightarrow -W^\top$. We notice that this outer automorphism exchanges r and s in the decomposition of W. However, these two inequivalent representations have the same image; since the construction which follows depends only on the representation image, the subalgebras $p(r,s)$ and $p(s,r)$ lead to equivalent results; that justifies the convention $s \le r$ made in (4.1). We realize the homogeneous space

$$M = \mathrm{SL}(N)/\,\mathrm{P}(r,s), \quad \dim M = rs, \tag{4.3}$$

as the Grassmannian of r-planes in $\mathbb{C}^N$.

Following Ref. [**15**] we first introduce (redundant) homogeneous coordinates as the components of a matrix in $\mathbb{C}^{N\times s}$in which the action of $G = GL(N)$ is linear:

$$\begin{pmatrix} X' \\ Y' \end{pmatrix} = \begin{pmatrix} G_{11} & G_{12} \\ G_{21} & G_{22} \end{pmatrix} \begin{pmatrix} X \\ Y \end{pmatrix}, \quad \begin{pmatrix} X \\ Y \end{pmatrix} \sim \begin{pmatrix} XG_0 \\ YG_0 \end{pmatrix}, \quad G_0 = GL(s). \tag{4.4}$$

The origin is the point $(0, I)^\top$. In the neighbourhood of the origin we remove the redundancy inherent in homogeneous coordinates by introducing (complex) affine coordinates $Z = XY^{-1}$ in which the action of $\mathrm{SL}(N)$ is a fractional linear one:

$$Z = XY^{-1}, \quad Z' = (G_{11}Z + G_{12})(G_{21}Z + G_{22})^{-1}. \tag{4.5}$$

The corresponding infinitesimal action is represented by vector fields $\hat{L}$:

$$\hat{L} = \{\mathrm{tr}(A + ZB + CZ + ZDZ)\nabla^\top\}, \text{ where } (\nabla)_{a\alpha} = \partial_{a\alpha}, \tag{4.6}$$

and where A, B, C, D are the matrices figuring in (4.1)–(4.2). Eq. (4.6) is a realization of

$$\mathrm{sl}(N) \subset \mathrm{diff}_n, \quad N = r + s, \quad 1 \le s \le r \le N-1, \quad n = rs. \tag{4.7}$$

As basis is given by the vector fields

$$(4.8)\qquad \hat{a}_{a\alpha} = \partial_{z_{a\alpha}}, \quad \hat{b}_{\alpha\beta} = z_{c\alpha}\partial_{z_{c\beta}}, \quad \hat{c}_{ab} = z_{b\gamma}\partial_{z_{a\gamma}}, \quad \hat{d}_{\alpha a} = z_{c\alpha}z_{a\gamma}\partial_{z_{c\gamma}}.$$

There is a linear relation among the fields $\hat{b}$'s and $\hat{c}$'s:

$$(4.9)\qquad \sum_a \hat{c}_{aa} = \sum_{a,\alpha} z_{a\alpha}\partial_{z_{a\alpha}} = \sum_\alpha \hat{b}_{\alpha\alpha},$$

which is related to the last equality of (4.2).

The sets of vector fields denoted $\hat{a}$, $\hat{b}$, $\hat{c}$, $\hat{d}$ in (4.8) generate four Lie subalgebras of $\mathrm{sl}(N)$ that we denote $\hat{A}$, $\widehat{B}$, $\widehat{C}$, $\widehat{D}$, respectively. Evidently $\hat{A}$ is Abelian; this is also the case of $\widehat{D}$. Moreover

$$(4.10)\qquad [\hat{A}, \widehat{B}] \subset \hat{A}, \quad [\hat{A}, \widehat{C}] \subset \hat{A}, \quad [\widehat{D}, \widehat{B}] \subset \widehat{D}, \quad [\widehat{D}, \widehat{C}] \subset \widehat{D}.$$

We also check that $\widehat{B} \cup \widehat{C}$ is also a subalgebra; from (2.26) we know that it is the subalgebra $\mathrm{sl}(r) \oplus \mathrm{sl}(s) \oplus \mathbb{C}$.

The Abelian Lie algebra $\hat{A}$ has functional rank n. Since $s \le r$ (see (4.7), we note that the Abelian algebra $\widehat{D}$ is of functional rank s^2; indeed all its $n = rs$ basic vector fields are different linear combinations with coefficients linear in $z_{a\gamma}$ of s^2 different fields $z_{c\alpha}\partial_{z_{c\gamma}}$ that are linearly independent at each generic point $z_{c\alpha}$. To summarize:

$$(4.11)\qquad \hat{A} \subset [\mathcal{A}_n^{(n)}], \qquad \text{functional rank } (\widehat{D}) = s^2 \le n.$$

The equality holds in the last expression only when $r = s = N/2$.

When $s = 1$, $N = n + 1$: we obtain the realization of $\mathrm{sl}(n+1)$ in diff_n already given in (2.27). When n is not prime, it can be written in different manners as the product of two integers. We then obtain the realization of more than one $\mathrm{sl}(N)$ Lie algebra in diff_n. We need to study if they are essentially distinct or if one of them, $\mathrm{sl}(N_1)$ can be conjugate to a subalgebra of another one, $\mathrm{sl}(N_2)$. That will be done in Section 7.

5. The Homogeneous Spaces $\mathbf{M} = \mathbf{O(N)}/\,\mathrm{P}(\lambda, \mu, \lambda)$

Following Ref. [**15**] we realize the Lie algebra $o(N)$ of the orthogonal group $O(N)$ by the set of matrices W with the division into blocks

$$(5.1)\quad W = \left\{ \begin{pmatrix} A & -B^\top & C \\ D & E & B \\ F & -D^\top & -A^\top \end{pmatrix} \right\} \quad C^\top = -C; \quad E^\top = -E; \quad F^\top = -F,$$

where A, B, C, D, E, F represent all possible matrices satisfying conditions (5.1) with the dimensions:

$$(5.2)\qquad A, C, F \in \mathbb{C}^{\lambda\times\lambda}, \quad B, D \in \mathbb{C}^{\mu\times\lambda}, \quad E \in \mathbb{C}^{\mu\times\mu}.$$

The invariant quadratic form is

$$(5.3)\quad K_N = \begin{pmatrix} 0 & 0 & I_\lambda \\ 0 & I_\mu & 0 \\ I_\lambda & 0 & 0 \end{pmatrix}, \quad WK_N + K_N W^\top = 0, \quad 1 \le \lambda \le \left[\frac{N}{2}\right],$$
$$0 \le \mu \le N-2, \quad 2\lambda + \mu = N.$$

As we shall see, to study the maximality of parabolic algebras, it is easier to use an equivalent matrix representation of $o(N)$, corresponding to the invariant quadratic form represented by the matrix K'_N with 1's along the second diagonal and 0's elsewhere; i.e.

$$m > 0,\ (K'_m)_{ij} = \delta_{i,m+1-j};\quad K'_N = \begin{pmatrix} 0 & 0 & K'_\lambda \\ 0 & K'_\mu & 0 \\ K'_\lambda & 0 & 0 \end{pmatrix}. \tag{5.4}$$

Then the set W' of matrices of $o(N)$ are the $N \times N$ matrices antisymmetrical with respect to the second diagonal; indeed

$$W'K' + K'W'^{\top} = 0 \iff W'^{\top} = -K'W'K'. \tag{5.5}$$

The block division of W' corresponding to λ, μ with $2\lambda + \mu = N$ is:

$$W' = \begin{pmatrix} A' & B'' & C' \\ D' & E' & B' \\ F' & D'' & A'' \end{pmatrix}, \tag{5.6}$$

where the block matrices A', B', C', D', E', F' have the dimensions of the corresponding unprimed matrices given in (5.2) and satisfy the conditions:

$$\begin{aligned} A'' &= -K'A'^{\top}K', & B'' &= -K'B'^{\top}K', & D'' &= -K'D'^{\top}K', \\ C' &= -K'C'^{\top}K', & E' &= -K'E'^{\top}K', & F' &= -K'F'^{\top}K'. \end{aligned} \tag{5.7}$$

The parabolic subalgebras, candidates for maximality, $L_0 = p(\lambda, \mu, \lambda)$ are obtained by setting $B' = 0 = B''^{\top}$, $C' = 0$ in (5.7) (correspondingly $B = 0$, $C = 0$ in (5.1). We recall the dimension of $o(N)$ and that of the homogeneous space $M = O(N)/P(\lambda, \mu, \lambda)$ as functions of λ and μ:

$$\dim o(N) = \binom{N}{2} = \lambda(2\lambda + 2\mu - 1) + \binom{\mu}{2};\quad \dim(M) = n = \lambda\mu + \binom{\lambda}{2}. \tag{5.8}$$

Let us compare two different parabolic subalgebras of $o(N)$ for N fixed, namely $p_2 \equiv p(\lambda+1, \mu-2, \lambda+1)$ and $p_1 \equiv p(\lambda, \mu, \lambda)$. The matrix W' of (5.6) for p_2 is obtained from that of p_1 by adding λ new elements to the last column of the old matrix A' and eliminating $\mu - 2$ elements from the first row of the old matrix E'. We hence have $\dim p_2 - \dim p_1 = \lambda - \mu + 2$. Now consider the case $\mu = 2$. We obtain $p(\lambda, 2, \lambda) \subset p(\lambda+1, 0, \lambda+1)$ since in this case, and only this case, the smaller parabolic subalgebra is obtained from the larger one by setting $a_{1,\lambda+1}, \ldots, a_\lambda$ equal to zero, *without* adding any new elements in the matrix E. Indeed, we have $e_{11}(\text{new}) = a_{\lambda+1,\lambda+1}(\text{old})$, $e_{12}(\text{new}) = 0$ (because of the second relation in (5.7)).

When the exceptional case $\mu = 2$ is excluded one has the lemma:

LEMMA 5.1. *The parabolic subalgebras $p(\lambda, \mu, \lambda)$ with $\mu \neq 2$ are maximal subalgebras of $o(N)$.*

However, most of the general formulas we will obtain are also valid for $\mu = 2$. When this is not true, we shall state it explicitly.

The matrices W's with $C = 0$, $B \neq 0$ form a subspace $\mathcal{S}$ satisfying:

$$L_0 = p(\lambda, \mu, \lambda),\quad L_0 \subseteq \mathcal{S} \subseteq o(N),\quad [L_0, \mathcal{S}] \subseteq \mathcal{S},\quad [\mathcal{S}, \mathcal{S}] = o(N). \tag{5.9}$$

From Definition 3.3, the pair $p(\lambda, \mu, \lambda) \subset o(N)$ determines a transitive primitive *reducible* Lie algebra unless we have

(a) $S = o(N)$, i.e. $\lambda = 1$ so $C = -C^{\top} = 0$;
(b) $S = L_0$, i.e. $\mu = 0$ so $N = 2\lambda$, $B = D = 0$, $E = 0$.

In these two exceptional cases (L, L_0) is an irreducible transitive primitive subalgebra.

In all cases one can realize the homogeneous space $O(N)/P(\lambda,\mu,\lambda)$ as a Grassmannian of isotropic planes in $\mathbb{C}^N$. Following Ref. [**15**], Eqs. (3.72)–(3.73) we first introduce homogeneous coordinates

$$(5.10)\qquad U_1, U_3 \in \mathbb{C}^{\lambda\times\lambda},\quad U_2 \in \mathbb{C}^{\mu\times\lambda},\quad U = \begin{pmatrix} U_1 \\ U_2 \\ U_3 \end{pmatrix},\quad U^\top K U = 0,$$

which satisfy

$$(5.11)\qquad \forall g \in GL(\lambda),\quad U' = \begin{pmatrix} U_1 g \\ U_2 g \\ U_3 g \end{pmatrix} \sim U,$$

i.e. U' and U represent the same point of M. To remove the redundancy represented by the arbitrary matrix g, we introduced affine coordinates (in the neighbourhood of the origin with $U_3 = I$, $U_1 = 0$, $U_2 = 0$),

$$(5.12)\qquad Y' = U_1 U_3^{-1},\quad Z = U_2 U_3^{-1}.$$

In these coordinates the last equality of (5.10) becomes

$$(5.13)\qquad 2X + Z^\top Z = 0 \text{ with } X = \frac{1}{2}(Y' + Y'^\top),\quad Y = \frac{1}{2}(Y' - Y'^\top).$$

That means that X, the symmetric part of Y' does not represent independent coordinates; so we will eliminate X in the following computations. The coordinates on M are thus provided by the matrix elements of the matrices:

$$(5.14)\qquad Z \in \mathbb{C}^{\mu\times\lambda},\quad Y = -Y^\top \in \mathbb{C}^{\lambda\times\lambda}.$$

Introducing the symbolic matrices:

$$(5.15)\qquad \begin{gathered} 1 \le a, b \le \mu,\quad 1 \le \alpha, \beta \le \lambda, \\ (\nabla_Y)_{\alpha\beta} = \partial_{y_{\alpha\beta}},\quad y_{\beta\alpha} = -y_{\alpha\beta},\quad (\nabla_Z)_{a\alpha} = \partial_{z_{a\alpha}}, \end{gathered}$$

we obtain directly the formula corresponding to (3.64) of Ref. [**15**], that is the realization of $o(N)$ as subalgebra of diff_n:

$$(5.16)\qquad \begin{aligned} \hat{L} = \operatorname{tr}\Big(& C + AY + YA^\top + \frac{1}{2}(-B^\top Z + Z^\top B) - YFY \\ & + \frac{1}{2}(YD^\top Z + Z^\top DY) + \frac{1}{4}Z^\top(DZ^\top - ZD^\top)Z \\ & - \frac{1}{4}Z^\top ZFZ^\top Z\Big)\nabla_Y^\top \\ + \operatorname{tr}\Big(& B + DY + EZ + ZA^\top - \frac{1}{2}DZ^\top Z \\ & + ZD^\top Z - ZFY + \frac{1}{2}ZFZ^\top Z\Big)\nabla_Z^\top, \end{aligned}$$

where the matrices A, B, C, D, E, F are arbitrary complex matrices satisfying the conditions (5.1) and (5.2). That symbolic notation simply means that a basis for the representation of $o(N)$ in diff_n is given by the following vector fields (summation over identical indices is assumed):

$$\hat{c}_{\alpha\beta} = \partial_{y_{\alpha\beta}}, \quad \hat{a}_{\alpha\beta} = 2y_{\beta\mu}\partial_{y_{\alpha\mu}} + z_{a\beta}\partial_{z_{a\alpha}}, \tag{5.17}$$

$$\hat{f}_{\alpha\beta} = y_{\alpha\mu}y_{\beta\nu}\partial_{y_{\mu\nu}} - \frac{1}{4}z_{a\mu}z_{a\alpha}z_{b\beta}z_{b\nu}\partial_{y_{\mu\nu}} - z_{a\alpha}y_{\beta\mu}\partial_{z_{a\mu}} + \frac{1}{2}z_{a\alpha}z_{b\beta}z_{b\mu}\partial_{z_{a\mu}}, \tag{5.18}$$

$$\hat{d}_{a\alpha} = z_{a\mu}y_{\alpha\nu}\partial_{y_{\mu\nu}} + z_{a\mu}z_{b\alpha}z_{b\nu}\partial_{y_{\mu\nu}} + y_{\alpha\mu}\partial_{z_{a\mu}} + z_{b\alpha}z_{a\mu}\partial_{z_{b\mu}} - \frac{1}{2}z_{b\alpha}z_{b\mu}\partial_{z_{a\mu}}, \tag{5.19}$$

$$\hat{b}_{a\alpha} = -z_{a\beta}\partial_{y_{\alpha\beta}} + \partial_{z_{a\alpha}}, \quad \hat{e}_{ab} = z_{b\mu}\partial_{z_{a\mu}} - z_{a\mu}\partial_{z_{b\mu}}. \tag{5.20}$$

As we did in the previous section, we can denote by $\hat{A}, \widehat{B}, \widehat{C}, \dots$ the vector spaces generated by the vector fields $\hat{a}, \hat{b}, \hat{c}, \dots$. Some of them are subalgebras, e.g; $\hat{A}, \widehat{C}, \widehat{E}$ but not $\widehat{B}$. The sum of those which are of degree 0 or 1 in the variables $y_{\alpha\beta}$, $z_{a\alpha}$ form the subalgebra:

$$\begin{gathered} \widehat{C} \oplus \widehat{B} \oplus \hat{A} \oplus \widehat{E}, \quad [\widehat{C}, \widehat{C}] = [\widehat{C}, \widehat{B}] = [\widehat{C}, \widehat{E}] = 0, \quad [\widehat{E}\hat{A}] = 0, \\ [\widehat{C}, \hat{A}] \subset \widehat{C}, \quad [\widehat{B}, \widehat{B}] \subset \widehat{C}, \quad [\widehat{B}, \widehat{E}] \subset \widehat{B}, \\ [\widehat{B}, \hat{A}] \subset \widehat{B}, \quad [\widehat{E}, \widehat{E}] \subset \widehat{E}, \quad [\hat{A}, \hat{A}] \subset \hat{A}. \end{gathered} \tag{5.21}$$

The centralizer in diff_n of the subalgebra $\widehat{C} \oplus \widehat{B}$ is:

$$C_{\text{diff}_n}(\widehat{C} \oplus \widehat{B}) = \widehat{C}. \tag{5.22}$$

So the centralizer of $\hat{L} = o(n)$ is trivial (verification of Corollary 3.1).

We study now the two cases of transitive primitive irreducible Lie algebras we have already announced and also a reducible case, for which the vector fields have quadratic coefficients.

CASE 1 ($\lambda = 1$, so $n = \mu$, $N = \mu + 2$). Then $C = 0$ from (5.1), $Y = 0$ from (5.14): there are no y variables and Z is a one column matrix; so we denote simply by z_i the $n = \mu = N - 2$ variables (see (5.8)). Then we obtain, as a particular case of (5.17)–(5.20), the basic vector fields ($1 \le i, j \le n = \mu$):

$$\begin{gathered} b_i = \partial_{z_i}, \quad a = -\sum_j z_j\partial_{z_j} = \mathcal{D}, \quad e_{ij} = z_j\partial_{z_i} - z_i\partial_{z_j}, \\ d_i = z_i\mathcal{D} - \frac{1}{2}\Big(\sum_j z_j^2\Big)\partial_{z_i} \end{gathered} \tag{5.23}$$

defines the realization of the algebra $o(n+2) \subset \text{diff}_n$, which is the conformal algebra on M. Lie found it first for $o(5)$ ($n = 3$) and quoted Liouville for recognizing it as the conformal algebra. Later, in **[12]**, he wrote it for arbitrary n. There is a grading of the conformal algebras by the degree of the polynomial fields:

$$\mathcal{L} = L^{(0)} \oplus L^{(1)} \oplus L^{(2)}, \quad L^{(0)} = \widehat{B}, \quad L^{(1)} = \hat{A} \oplus \widehat{E}, \quad L^{(2)} = \widehat{D}, \tag{5.24}$$

with (summation on identical indices is implied)

$$\widehat{B} = B_i\hat{b}_i, \quad \hat{A} = A\hat{a}, \quad \widehat{E} = E_{ij}\hat{e}_{ij}, \quad \widehat{D} = D_i\hat{d}_i. \tag{5.24'}$$

Notice that $\hat{A}$ represents the dilations on $M = \mathbb{C}^n$, $\widehat{E}$ the algebra $o(n)$ and $L^{(0)} = \widehat{B}$, $L^{(2)} = \widehat{D}$ are two n-dimensional Abelian algebras of functional rank n, so they are maximal Abelian subalgebras of diff_n (i.e. of the class $[\mathcal{A}_n^n]$). They are respectively ideals of $L^{(0)} \oplus L^{(1)}$, $L^{(2)} \oplus L^{(1)}$, which are both isomorphic to the *similitude algebra* in dimension n (i.e. the Euclidean algebra with the dilations). The subalgebras $L^{(0)}$ and $L^{(2)}$ generate the conformal algebra. To summarize:

$$\widehat{B} \sim \widehat{D} \sim A_n^{(n)}, \quad \widehat{E} \sim o(n), \quad \hat{A} = \mathbb{C}\mathcal{D} \sim \mathcal{A}_1. \tag{5.25}$$

Remark that $\dim \mathrm{sl}(n+1) \geq \dim o(n+2)$ with the equality holding for $n = 1$. In this case:

- **$n = 1$:** The conformal algebra $o(3)$ is identical to $\mathrm{sl}(2)$, the projective Lie algebra on the line: $o(3) \sim \mathrm{sl}(2) = \{\partial_z, z\partial_z, z^2\partial_z\}$.
- **$n > 1$:** The conformal algebra cannot be a subalgebra of the projective algebra. Indeed, from (2.27) and (5.25) we note that their intersection is the similitude algebra in n dimensions. Only linear transformations preserve this intersection. Then they also preserve the functional rank, r_f, of the n-dimensional Abelian subalgebras formed by the quadratic fields: these r_f's are respectively n and 1, so they are different for $n > 1$.

Let us discuss some low dimensional cases.

- **$n = 2$:** We have the conformal algebra $o(4)$ of the plane. Notice that $\hat{a}$ and $\hat{e}$ are 1 dimensional subalgebras. The six basic fields can be combined into $\hat{h}_\pm = -\hat{a} \pm i\hat{e}$, $\hat{b}_\pm = \hat{b}_1 \pm i\hat{b}_2$, $\hat{d}_\pm = \hat{d}_1 \pm i\hat{d}_2$ satisfying $[\hat{h}_\pm, \hat{d}_\pm] = 2\hat{d}_\pm$, $[\hat{h}_\pm, \hat{b}_\pm] = -2\hat{b}_\pm$, $[\hat{d}_\pm, \hat{b}_\pm] = 2\hat{h}_\pm$ and any field with $+$ index commutes with any with—index. This corresponds to the known isomorphism: $o(4) \sim \mathrm{sl}(2) \oplus \mathrm{sl}(2)$; indeed $o(4)$ is semi-simple but not simple.
- **$n = 3$:** This case was studied by Lie who showed that $o(5)$ is a maximal algebra and quoted Liouville for recognizing it as the conformal algebra. Later in his book [**12**], the conformal algebra $o(n+2)$ series was given for arbitrary n.
- **$n = 4$:** There is another primitive Lie algebra in this case: $S = \big(\mathrm{sl}(4), p(2,2)\big)$: its Abelian subgroup formed by the quadratic fields has maximal functional rank $r_f = n = 4$. We verify that the conformal algebra $Q = o(6)$ is identical (in agreement with the known isomorphism recalled in (2.23)). The two algebras S, Q have the same constant fields: $L_0 \sim [\mathcal{A}_4^{(4)}]$. Their subalgebras $L^{(2)}$ belong to the same class $[\mathcal{A}_4^{(4)}]$ and can be transformed into each other by a linear transformation that we do not compute explicitly. We only remark that the linear fields in S form the algebra $\widehat{B} \oplus \widehat{C}$, $[\widehat{B}, \widehat{B}] \subset \widehat{B}$, $[\widehat{C}, \widehat{C}] \subset \widehat{C}$, $[\widehat{B}, \widehat{C}] = 0$, with the two 2×2 matrices B, C satisfying $\operatorname{tr} B = \operatorname{tr} C$; this algebra is isomorphic to $\mathrm{sl}(2) \oplus \mathrm{sl}(2) \oplus \mathcal{A}_1 \sim o(4) \oplus \mathcal{A}_1$. So the subalgebras $L^{(0)} \oplus L^{(1)}$ of S and Q are isomorphic to the similitude algebra in dimension 4.

CASE 2 ($\mu = 0$). Then $B = D = E = 0$, $Z = 0$. The only variables are

$$y_{\alpha\beta} = -y_{\beta\alpha}, \quad n = \frac{\lambda(\lambda-1)}{2}, \quad N = 2\lambda, \quad \lambda \geq 3. \tag{5.26}$$

We have added the last condition because for $\lambda \leq 2$, $o(2\lambda)$ is not primitive and not simple. The realization of $o(2\lambda)$ is given by:

$$\begin{gathered} o(2\lambda)\colon C_{\alpha\beta}\partial_{y_{\alpha\beta}} + 2A_{\alpha\mu}y_{\mu\beta}\partial_{y_{\alpha\beta}} + y_{\alpha\mu}F_{\mu\nu}y_{\nu\beta}\partial_{y_{\alpha\beta}}, \\ C = -C^\top, \; F = -F^\top, \end{gathered} \tag{5.27}$$

or, with matrix notation:

$$o(2\lambda) : \mathrm{tr}(C + 2AY - YFY)\nabla_Y. \tag{5.27'}$$

We first show that for $\lambda = 3, 4$ we have already obtained these algebras.

$\lambda = 3 = n$: Any 3×3 antisymmetric matrix Y can be written as $Y_{ij} = \varepsilon_{ijk} y_k$ were ε_{ijk} is the completely antisymmetric representation of the permutation of the three indices and the components of y_k (which transform as those of a vector for SL_3) are the three independent matrix elements of Y. Then

$$\begin{aligned} -(YFY)_{ij} &= -\sum_{abrst} \varepsilon_{iar}\varepsilon_{abs}\varepsilon_{bjt} y_r f_s y_t \\ &= \sum_{abrst} (\delta_{ib}\delta_{rs} - \delta_{is}\delta_{br})\varepsilon_{bjt} y_r f_s y_t = (y, f) Y_{ij}. \end{aligned} \tag{5.28}$$

Hence $-\mathrm{tr}\, YFY\nabla^\top = (f, y)(y_j \partial_{y_j}) = f_k y_k \mathcal{D}$. Thus we have shown that the quadratic terms $L^{(2)}$ of (5.27) are identical to those of (2.27). The $L^{(0)} = \{\partial_{y_i}\}$ terms are also identical. Since $L^{(0)}$ and $L^{(2)}$ generate the whole algebra we have proven the identity between the two algebras; that is a verification of the well known isomorphism $\mathrm{sl}(4) \sim o(6)$.

$\lambda = 4$, $n = 6$: We show that the obtained algebra $\mathrm{o}(8)$ is the conformal one. Indeed an explicit computation of the quadratic terms yields:

$$1 \le i \le 6, \quad L^{(2)} = \{y_j \mathcal{D} - \mu(y)\partial_{y_j}\}, \text{ with} \quad \mu(y) = y_1 y_4 + y_2 y_5 + y_3 y_6 = \frac{1}{2}\left(y, \begin{pmatrix} 0 & I_3 \\ I_3 & 0 \end{pmatrix} y\right). \tag{5.29}$$

The linear transformation of variables

$$S = \frac{1+i}{\sqrt{2}} \begin{pmatrix} I_3 & -iI_3 \\ -iI_3 & I_3 \end{pmatrix} \text{ satisfies } S^\top = S, \quad SS^\top = \begin{pmatrix} 0 & I_3 \\ I_3 & 0 \end{pmatrix}, \tag{5.30}$$

so it transforms the quadratic form I_6 (which appears in the last quadratic form of (5.23) for $n = 6$) into $SS^\top$ and therefore $1/2(y, y)$ into $\mu(y)$. So the conformal algebra for $n = 6$ is identified with the O_8 algebra defined by (5.27) with $n = 6$. We recall the triality principle of E. Cartan. The outer automorphism group of O_8 is $\mathcal{S}_3$ (the permutation group of the three branches of the Coxeter diagram of D_4); it permutes the 3 inequivalent (the vector one and the two spinors) 8-dimensional linear irreducible representations of O_8. This leads to three inequivalent monomorphism $o_8 \to \mathrm{diff}_6$; however, their images are equivalent subalgebras of diff_6.

CASE 3 ($\mu = 1$). Then $N = 2\lambda + 1$ and $n = \lambda(\lambda + 1)/2$. That shows that the dimension n is identical to that of the previous family (that with $\mu = 0$) with λ replaced by $\lambda' = \lambda + 1$. The N of $O(N)$ in this case satisfies $N = N' - 1$, where $o(N')$ was obtained for the same n, but $\mu = 0$ (the previous family). We shall show that this $o(2\lambda+1)$ algebra is a subalgebra of $o(2\lambda')$ obtained in the previous family with $\mu = 0$, $\lambda' = \lambda + 1$.

With $\mu = 1$, in the general equation (5.1) we have to introduce the following modifications: $E = 0$, $B^\top$ and $D^\top$ become one column matrices that we replace by the "vectors" b, d. Then (5.1) reads:

$$W_{\lambda,1,\lambda} = \begin{pmatrix} A & -b & C \\ d^\top & 0 & b^\top \\ F & -d & -A^\top \end{pmatrix}, \quad \text{with } C^\top = -C, \quad F^\top = -F. \tag{5.31}$$

It is straightforward to write the corresponding realization of $o(2\lambda+1)$ as a particular case of the equations (5.16). Indeed, we replace $Z^\top$ and $\nabla^\top$ by the λ component column vectors z and ∂; for instance $-B^\top Z$ becomes a rank one $\lambda \times \lambda$ matrix that we denote by $bz^\top$ while the matrix $DZ^\top$ becomes the number ($= 1 \times 1$ matrix) $d^\top z = (d, z)$:

$$\begin{aligned} \operatorname{tr}\Big(C + AY + YA^\top + \frac{1}{2}(bz^\top - zb^\top) - YFY + \frac{1}{2}(Y(dz^\top) - (zd^\top)Y)\Big)\nabla_Y^\top \\ + \Big(b^\top + z^\top A^\top + d^\top Y + (z^\top d)z^\top - \frac{1}{2}(d^\top z)z^\top - z^\top FY, \partial_z \Big), \end{aligned} \tag{5.32}$$

where we have $(ab^\top) \in \mathbb{C}^{\lambda\times\lambda}$, $(a^\top b) \in \mathbb{C}$.

LEMMA 5.2. *The algebra $o(2\lambda+1)$ constructed in* (5.32) *is not maximal in* diff_n *but is a subalgebra of $o(2\lambda+2)$ constructed in* (5.27):

$$\begin{gathered} o(2\lambda+1) \subset o(2\lambda+2) \subset \mathrm{diff}_n, \quad n = \frac{\lambda(\lambda+1)}{2} \\ O(2\lambda+1)/\mathrm{P}(\lambda,1,\lambda) \sim O(2\lambda+2)/\mathrm{P}(\lambda+1,0,\lambda+1). \end{gathered} \tag{5.33}$$

PROOF. We consider the algebra $o(2\lambda+2)$ realized in the space $M \sim O(2\lambda+2)/\mathrm{P}(\lambda+1,0,\lambda+1)$ in the realization (5.27). We relabel all the $(\lambda+1)\times(\lambda+1)$ matrices involved as follows:

$$\tilde{A} = \begin{pmatrix} A & a \\ r^\top & \alpha \end{pmatrix}, \quad \widetilde{C} = \begin{pmatrix} C & c \\ -c^\top & 0 \end{pmatrix}, \quad \widetilde{F} = \begin{pmatrix} F & f \\ -f^\top & 0 \end{pmatrix}, \tag{5.34}$$

$$\widetilde{Y} = \begin{pmatrix} Y & z \\ -z^\top & 0 \end{pmatrix}, \quad \widetilde{\nabla} Y = \begin{pmatrix} \nabla_Y & \partial_z \\ -\partial_z^\top & 0 \end{pmatrix}, \tag{5.35}$$

$$\begin{gathered} A, C, F, Y, \nabla_Y \in \mathbb{C}^{\lambda\times\lambda}, \quad a, c, f, z, \partial_z \in \mathbb{C}^{\lambda\times 1}, \\ C + C^\top = 0, \quad F + F^\top = 0, \quad Y + Y^\top = 0. \end{gathered} \tag{5.36}$$

Rewriting Eq. (5.27) for $o(2\lambda+2)$, using the notations (5.34)–(5.36) we see that (5.32) is a subcase of the obtained equations with the identification

$$2a = c = -b, \quad 2f = -r = d, \quad \alpha = 0. \tag{5.37}$$

This proves the assertions in Eq. (5.33). □

Finally, let us consider an example when the algebra $\{L, L_0\}$ is not primitive, namely $L \sim o(N, \mathbb{C})$, $L_0 \sim p(\lambda, 2, \lambda)$, $N = 2\lambda + 2$. More specifically, we consider $N = 6$, $\lambda = \mu = 2$. The construction (5.16) is valid, just as in the primitive cases. However, for $\mu = 2$ we can transform from the coordinates $\{y_{\alpha\beta}, z_{a\alpha}\}$ to new coordinates (u, v) in which the invariant foliation becomes manifest. Indeed, consider the case of the algebra $o(6)$ with coordinates $\{y_{12} \equiv y, z_{11}, z_{12}, z_{21}, z_{22}\}$ and put

$$\begin{gathered} u_1 = y + \frac{i}{2}(z_{11}z_{22} - z_{12}z_{21}), \quad u_2 = z_{21} + iz_{11}, \\ u_3 = z_{12} - iz_{22}, \quad w_1 = z_{12} + iz_{22}, \quad w_2 = z_{21} - iz_{11}. \end{gathered} \tag{5.38}$$

The basis vector fields (5.17), ... , (5.20) (for $N = 6$, $\lambda = \mu = 2$) are transformed into

$$\hat{c}_{12} \equiv \hat{c} = \partial_{u_1}, \quad \hat{b}_{11} + i\hat{b}_{21} = 2i\partial_{u_2}, \quad b_{22} - ib_{12} = -2i\partial_{u_3},$$
$$\hat{b}_{11} - i\hat{b}_{21} = -u_3\partial_{u_1} - 2i\partial_{w_2}, \quad \hat{b}_{22} + i\hat{b}_{12} = u_2\partial_{u_1} + 2i\partial_{w_1},$$
$$\hat{a}_{11} = u_1\partial_{u_1} + u_2\partial_{u_2} + w_2\partial_{w_2}, \quad \hat{a}_{12} = i(u_3\partial_{u_2} - w_1\partial_{w_2}),$$
$$\hat{a}_{21} = i(-u_2\partial_{u_3} + w_2\partial_{w_1}), \quad \hat{a}_{22} = u_1\partial_{u_1} + u_3\partial_{u_3} + w_1\partial_{w_1},$$
$$\hat{e}_{12} \equiv \hat{e} = i(u_2\partial_{u_2} + u_3\partial_{u_3} - w_1\partial_{w_1} - w_2\partial_{w_2}),$$

$$\begin{aligned} \hat{d}_{11} + i\hat{d}_{21} &= 2u_1\partial_{u_3} + iw_2(w_1\partial_{w_1} + w_2\partial_{w_2}), \\ -\hat{d}_{22} + i\hat{d}_{12} &= 2u_1\partial_{u_2} + iw_1(w_1\partial_{w_1} + w_2\partial_{w_2}), \\ i\hat{d}_{11} + \hat{d}_{21} &= u_2(u_1\partial_{u_1} + u_2\partial_{u_2} + u_3\partial_{u_3}) + [2iu_1 - u_2w_1 - u_3w_2]\partial_{w_1}, \\ -i\hat{d}_{22} + \hat{d}_{12} &= u_3(u_1\partial_{u_1} + u_2\partial_{u_2} + u_3\partial_{u_3}) + [2iu_1 - u_2w_1 - u_3w_3]\partial_{w_2}, \\ \hat{f}_{12} \equiv \hat{f} &= -u_1(u_1\partial_{u_1} + u_2\partial_{u_2} + u_3\partial_{u_3}) + w_1\left[u_1 - \frac{i}{2}(w_1u_2 + w_2u_3)\right]\partial_{w_1} \\ &\quad + w_2\left[u_1 - \frac{i}{2}(w_1u_2 + w_2u_3)\right]\partial_{w_2}. \end{aligned} \tag{5.39}$$

The point of the above exercise is that in Eq. (5.39) the coefficients of ∂_{u_1}, ∂_{u_2} and ∂_{u_3} depend only on (u_1, u_2, u_3). Thus, we could have started out from the transitive primitive case $o(6) \supset p(3, 0, 3)$, constructed the $n = 3$ realization (with $y_{12} = u_3$, $y_{23} = u_1$, $y_{31} = u_2$) and then extended it to the higher dimension $n = 5$ by adding further coordinates w_1, w_2, labeling leaves in an invariant foliation. This corresponds to the procedure described in the Introduction for constructing nonprimitive algebras from primitive ones.

SUMMARY OF THE RESULTS OF THIS SECTION. We have constructed a double series, $\lambda \geq 1$, $\mu \geq 0$ of algebras $o(N)$, $N = 2\lambda + \mu$, as subalgebras of diff_n, $n = \lambda\mu + \lambda(\lambda - 1)/2$. The series $\lambda = 1$, which constructs the conformal algebras $o(n + 2)$ (with $n = \mu$), was known to S. Lie. For $n = 1$ and $n = 4$ they have been already obtained in Section 4, due to the respective isomorphisms $\mathrm{sl}(2) \sim o(3)$ and $\mathrm{sl}(4) \sim o(6)$; for $n = 2$, $o(4)$ is not a simple algebra (it is maximal in diff_2). The series $\mu = 0$ starts at $\lambda = 3$; the corresponding $o(6)$ is identical to the projective $\mathrm{sl}(4)$ and for $\lambda = 4$ the corresponding $o(8)$ is identical to the conformal $o(8)$. The series $\mu = 1$ gives only subalgebras of the series $\mu = 0$ and the series $\mu = 2$ does not yield primitive algebras. The vector fields in general have polynomial coefficients of order up to four. They are of second order if the primitive algebra is irreducible ($\lambda = 1$, or $\mu = 0$) or if it is reducible, but not maximal, ($\mu = 1$) and contained in a larger irreducible subalgebra $\big(o(2\lambda + 1) \subset o(2\lambda + 2) \subset \mathrm{diff}_n\big)$. □

6. The Homogeneous Spaces $\mathbf{M} = \mathrm{Sp}(\mathbf{2N})/\,\mathrm{P}(\lambda, \mathbf{2}\mu, \lambda)$

The symplectic group $\mathrm{Sp}(2N)$ is the subgroup of $\mathrm{SL}(2N)$ which leaves invariant the quadratic form

$$J_N = \begin{pmatrix} 0 & I_N \\ -I_N & 0 \end{pmatrix} \text{ satisfying } J_N^\top = -J_N. \tag{6.1}$$

In order to introduce maximal parabolic subalgebras, we will use a more refined block decomposition $\lambda, 2\mu, \lambda$:

$$N = \lambda + \mu, \quad J_N = \begin{pmatrix} 0 & 0 & I_\lambda \\ 0 & J_\mu & 0 \\ -I_\lambda & 0 & 0 \end{pmatrix}, \quad J_\mu^\top = -J_\mu, \tag{6.2}$$

which leads to the realization the Lie algebra sp$(2N)$ of the symplectic group by the set of matrices W:

$$W = \left\{ \begin{pmatrix} A & B^\top & C \\ D & E & J_\mu B \\ F & -D^\top J_\mu & -A^\top \end{pmatrix} \right\} \; C^\top = C; \quad F^\top = F; \quad E^\top = J_\mu E J_\mu, \tag{6.3}$$

where A, B, C, D, E, F represent all possible matrices with the dimensions:

$$A, C, F \in \mathbb{C}^{\lambda\times\lambda}, \quad B, D \in \mathbb{C}^{2\mu\times\lambda}, \quad E \in \mathbb{C}^{2\mu\times 2\mu}, \tag{6.4}$$

satisfying conditions given in (6.3).

As in the previous section, following Ref. [**15**] we realize the homogeneous space $M = \mathrm{SP}(2N)/\mathrm{P}(\lambda, 2\mu, \lambda)$ as a Grassmannian of isotropic planes in $\mathbb{C}^{2N}$ (where $\mathrm{P}(\lambda, 2\mu, \lambda)$ is the maximal parabolic subgroup corresponding to the maximal parabolic subalgebra obtained by putting $B = 0$, $C = 0$ in (6.3)). To eliminate the redundancy of homogeneous coordinates we transform to affine coordinates namely elements of the matrices

$$Z \in \mathbb{C}^{2\mu\times\lambda}, \quad Y' \in \mathbb{C}^{\lambda\times\lambda}, \tag{6.5}$$

verifying the relation:

$$2X = Z^\top J_\mu Z \text{ with } X = \frac{1}{2}(Y' - Y'^\top), \quad Y = \frac{1}{2}(Y' + Y'^\top). \tag{6.6}$$

That means that X, the antisymmetric part of Y, does not represent independent coordinates; so we shall eliminate it. That yields for $n = \dim M$:

$$n = 2\lambda\mu + \frac{1}{2}\lambda(\lambda+1), \quad 1 \le \lambda, \quad 0 \le \mu, \quad N = \lambda + \mu. \tag{6.7}$$

Introducing the symbolic matrices:

$$\begin{gathered} 1 \le a,\ b \le 2\mu, \ 1 \le \alpha, \beta \le \lambda, \\ (\nabla_Y)_{\alpha\beta} = \partial_{y_{\alpha\beta}}, \quad y_{\beta\alpha} = y_{\alpha\beta}, \quad (\nabla_Z)_{a\alpha} = \partial_{z_{a\alpha}}. \end{gathered} \tag{6.8}$$

We obtain the formula corresponding to (3.76) of Ref. [**15**], that is the realization of sp$(2N)$ as subalgebra of diff$_n$:

$$\begin{aligned} \hat{L} = \operatorname{tr}\Big(& C + AY + YA^\top + \frac{1}{2}(B^\top Z + Z^\top B) - YFY \\ & + \frac{1}{2}(YD^\top J_\mu Z - Z^\top J_\mu DY) \\ & \quad + \frac{1}{4}Z^\top J_\mu (DZ^\top + ZD^\top) J_\mu Z - \frac{1}{4} Z^\top J_\mu Z F Z^\top J_\mu Z \Big) \nabla_Y^\top \\ & + \operatorname{tr}\Big(J_\mu B + DY + EZ + ZA^\top + \frac{1}{2} DZ^\top J_\mu Z + ZD^\top J_\mu Z \\ & \qquad - ZFY - \frac{1}{2} ZFZ^\top J_\mu Z \Big) \nabla_Z^\top, \end{aligned} \tag{6.9}$$

where the matrices A, B, C, D, E, F are arbitrary complex matrices satisfying the conditions (6.3)–(6.4)).

Let us now study the two cases when the quartic vector fields in Eq. (6.9) reduce to quadratic ones.

CASE 1 ($\mu = 0$). Then $B = D = E = 0$, $Z = 0$ and this corresponds to a transitive primitive irreducible Lie algebra. The only variables are

$$y_{\alpha\beta} = y_{\beta\alpha}, \quad n = \frac{1}{4}\lambda(\lambda+1), \quad 2N = 2\lambda. \tag{6.10}$$

The corresponding realization of $\text{sp}(2\lambda)$ is given by:

$$\text{sp}(2\lambda)\colon \operatorname{tr}(C + AY + YA^\top - YFY)\nabla_Y, \quad C = C^\top, \quad F = F^\top. \tag{6.11}$$

We notice that

$$r_f(\operatorname{tr} YFY\nabla_Y^\top) = n. \tag{6.12}$$

For $\lambda = 1$, $n = 1$ we simply verify the known isomorphism $\text{sl}(2) = \text{sp}(2)$. For $\lambda = 2$, $n = 3$; since $\text{sp}(4) \sim o(5)$, we have obtained one of the two $o(5)$'s found in Section 5 and known to Lie. From the value of the functional rank of the quadratic fields, this realization of $\text{sp}(4)$ is the conformal $o(5)$.

There is another series of realizations of symplectic Lie algebras with polynomial vector fields of degree ≤ 2; it corresponds to $\lambda = 1$. Remark that it is obtained (for $\mu > 0$) from reducible primitive transitive algebras:

CASE 2 ($\lambda = 1$). $2N = 2\mu + 2$, $n = 2\mu + 1$, so $2N = n + 1$. The elements of the W matrix can be decomposed into 3 numbers: $(\alpha, \gamma, \varphi)$, 4 vectors with μ components (b, b', d, d' represented by a μ lines, 1 column matrix), and three $\mu \times \mu$ matrices: (R, $S = S^\top$, $T = T^\top$). Explicitly:

$$W = \begin{pmatrix} \alpha & b^\top & b'^\top & \gamma \\ d & R & S & b' \\ d' & T & -R^\top & -b \\ \varphi & d'^\top & -d^\top & -\alpha. \end{pmatrix}. \tag{6.13}$$

The $n = 2\mu + 1$ variables are η, z, z'. We denote by $(b.d) = b^\top d$ the scalar product of two vectors. The derivative operators are ∂_η and two symbolic vectors: $\nabla_z, \nabla_{z'}$. With this structure for W one finds that the algebra $\hat{L}$ of (6.9), when decomposed according to the degree of the vector fields, has only polynomials of degree $0, 1, 2$ (the vanishing of higher degree terms is due to the vanishing of $Z^\top J_\mu Z = 2X$). The set of quadratic terms is:

$$(-\varphi\eta - (d' \cdot z) + (d \cdot z'))\mathcal{D}, \text{ with } \mathcal{D} = \eta\partial_\eta + z \cdot \nabla_z + z' \cdot \nabla_{z'}, \tag{6.14}$$

where $\mathcal{D}$ is the dilation operator. This set of vector fields has functional rank $r_f = 1$. So the realization in diff_n of this algebra $\text{sp}(2\mu+2) \equiv \text{sp}(n+1)$ contains only linear combinations with constant coefficients of the basis vector fields of $\text{sl}(n+1)$. We have arrived at the following result.

LEMMA 6.1. *The* $\text{sp}(n+1)$ *algebra* (n *odd*) *constructed in* (6.14) *as a subalgebra of* diff_n *is not maximal. It is contained in the projective* $\text{sl}(n+1)$ *realized in* (2.27) *for the same* n:

$$\begin{gathered} \text{sp}(n+1) \subset \text{sl}(n+1) \subset \text{diff}_n, \\ \text{SP}(n+1)/\,\text{P}(1n-11) \sim \text{SL}(n+1)/P(n,1). \end{gathered} \tag{6.15}$$

As in the case of $o(N)$ we see that the coefficients of vector fields are at most second order polynomials, if the parabolic subalgebra is irreducible ($\mu = 0$) or if it is reducible, but the algebra is not maximal. Then it is contained in an irreducible one.

The other symplectic algebras described by (6.9), with $\lambda \neq 1$, $\mu \neq 0$ are represented by polynomial fields of degree up to 4. The smallest values of $2N$ and n for these algebras are for $\lambda = 2$, $\mu = 1$; they define $\mathrm{sp}(6) \subset \mathrm{diff}_7$.

7. Summary and Conclusions

We have explicitly constructed all transitive primitive Lie algebras (L, L_0), where L is a classical complex simple Lie algebra and L_0 is one of its maximal parabolic subalgebras. They have all been realized as subalgebras of diff_n for some value of n. We thus have a list of all such subalgebras of diff_n for all values of n.

The dimension of the corresponding homogeneous spaces are:

$$\begin{aligned} &\mathrm{SL}(N)/\mathrm{P}(r,s)\colon n = rs, && N = r+s,\\ &O(N)/\mathrm{P}(\lambda,\mu,\lambda)\colon n = \lambda\left(\mu + \frac{\lambda-1}{2}\right), && N = 2\lambda+\mu, \quad \mu \neq 2,\\ &\mathrm{Sp}(2N)/\mathrm{P}(\lambda,2\mu,\lambda)\colon n = \lambda\left(2\mu + \frac{\lambda+1}{2}\right), && N = \lambda+\mu. \end{aligned} \tag{7.1}$$

All such pairs (L, L_0) are summed up in Table 2 for $1 \leq n \leq 20$.

We have already seen that some seemingly different realizations are actually equivalent under local diffeomorphisms. Moreover, not all of the constructed subalgebras of diff_n are maximal.

Let us discuss the question of mutual inclusions amongst the constructed algebras somewhat further.

The following lemma is of use in this analysis.

LEMMA 7.1. *Let L and S be two classical simple complex Lie algebras and let us have $L \subset S$. Then L can be a maximal subalgebra of S in only two cases:*

1. *If L has an irreducible representation of dimension N where S is $\mathrm{sl}(N)$, $o(N)$, or $\mathrm{sp}(N)$ (N even in the last case).*
2. *If S is $o(N)$ and L is $o(N-1)$.*

PROOF. The simple subalgebra $L \subset S$ can be imbedded in the defining representation of S (and any other representation of S), either reducibly, or irreducibly. If it is imbedded irreducibly, it leaves no nontrivial subspace of the representation space invariant. In this case L is maximal in S if it has no centralizer in S.

If L is imbedded in S reducibly, it leaves a nontrivial subspace V of the representation space $\mathbb{C}^N$ invariant. If S is $\mathrm{sl}(N)$ the invariant subspace are completely characterized by their dimension. Moreover, all maximal reducibly imbedded subalgebras are parabolic ones. If S is $o(N)$, or $\mathrm{sp}(N)$ then any invariant subspace is characterized by its dimension and the degree of its degeneracy, i.e. the number of zero length vectors in an orthogonal basis (with respect to the corresponding orthogonal, or symplectic invariant form). If the invariant subspace is degenerate, the corresponding maximal subalgebra will again be parabolic.

TABLE 2. Primitive subalgebras (L, L_0) of diff_n for $1 \leq n \leq 20$ with L simple classical and L_0 maximal parabolic

	$\mathrm{sl}(N)/\mathrm{P}(r,s)$			$o(N)/\mathrm{P}(\lambda,\mu,\lambda)$			$\mathrm{Sp}(2N)/\mathrm{P}(\lambda,2\mu,\lambda)$		
n	N	r	s	N	λ	μ	N	λ	μ
1	2	1	1	3	1	1	1	1	0
2	3	2	1						
3	4	3	1	6	3	0	2	1	1
				5	1	3	2	2	0
				5	2	1			
4	5	4	1	6	1	4			
	4	2	2						
5	6	5	1	7	1	5	3	1	2
6	7	6	1	8	1	6	3	3	0
	5	3	2	8	4	0			
				7	3	1			
7	8	7	1	9	1	7	4	1	3
				7	2	3	3	2	1
8	9	8	1	10	1	8			
	6	4	2						
9	10	9	1	11	1	9	5	1	4
	6	3	3	8	2	4			
10	11	10	1	12	1	10	4	4	0
	7	5	2	10	5	0			
				9	4	1			
11	12	11	1	13	1	11	6	1	5
				9	2	5	4	2	2
12	13	12	1	14	1	12	4	3	1
	8	6	2	9	3	3			
	7	4	3						
13	14	13	1	15	1	13	7	1	6
				10	2	6			
14	15	14	1	16	1	14			
15	16	15	1	17	1	15	8	1	7
	8	5	3	12	6	0	5	2	3
				11	5	1	5	5	0
				11	2	7			
				10	3	4			
16	17	16	1	18	1	16			
	10	8	2						
	8	4	4						
17	18	17	1	19	1	17	9	1	8
				12	2	8			
18	19	18	1	20	1	18	5	3	2
	11	9	2	11	4	3	5	4	1
	9	6	3	11	3	5			
19	20	19	1	21	1	19	10	1	9
				13	2	9	6	2	4
20	21	10	1	22	1	20			
	12	10	2						
	9	5	4						

Let $V \subset \mathbb{C}^N$ be nondegenerate. Then the orthogonal complement $V^\perp$ is also invariant. The maximal subalgebra leaving V and $V^\perp$ invariant is in general semisimple, namely $o(N_1) \oplus o(N_2)$, $N_1 + N_2 = N$ for $S \sim o(N)$, or $\mathrm{sp}(N_1) \oplus \mathrm{sp}(N_2)$ for $S \sim \mathrm{sp}(N)$ (N_1, N_2 and N even).

The sole exception occurs for $N_1 = N-1$, $N_2 = 1$ when the reducibly imbedded maximal subalgebra of $o(N)$ is $o(N-1)$. The reason for the exception is that the complement $o(1)$ is the Lie algebra consisting of only the null element.

This completes the proof of Lemma 7.1. □

COROLLARY 7.1. *Let (L, L_0) and (S, S_0) be two transitive primitive Lie algebras with*

$$n = \dim L - dim L_0 = \dim S - \dim S_0, \quad \dim L < \dim S. \tag{7.2}$$

We can have $L \subset S$ only if L has an irreducible linear representation of dimension N, the dimension of the defining representation of S, or if we have $L \sim o(N-1)$, $S \sim o(N)$.

PROOF. Any subalgebra of S is contained in at least one maximal subalgebra of L. The maximal subalgebra must be primitive, if L is to be primitive. According to Lemma 7.1 this is only possible in the two cases covered by the corollary. □

Lemma 7.1, together with dimensional considerations, help us to establish, or in the contrary, to rule out many possible mutual inclusions. Other cases need a detailed analysis of the form of the vector fields. This we postpone to a future publication.

We have seen that for all Lie algebras considered in this article the vector fields have polynomial coefficients. The order of these polynomials for the primitive Lie algebras is at most four.

The polynomials are at most quadratic in the following cases.

1. L_0 is a maximal parabolic irreducible subalgebra of L. This covers the cases $\big[(\mathrm{sl}(N), p(r,s)\big]$, $\big[o(2,k), p(k,0,k)\big]$, $\big[o(N), p(1,N-2,1)\big]$, $\big[\mathrm{sp}(2N), p(N,o,N)\big]$.

2. L_0 is a maximal parabolic reducible subalgebra of L, L is not maximal in diff_n, but we have

$$L \subset S \subset \mathrm{diff}_n, \quad \dim L - \dim L_0 = \dim S - \dim S_0, \tag{7.3}$$

and S_0 is a maximal parabolic irreducible subalgebra of S. The only such cases correspond to the spaces

$$\begin{aligned} O(2k+1)/\,\mathrm{P}(k,1,k) &\sim O(2k+1)/\,\mathrm{P}(k+1,0,k+1), \\ \mathrm{Sp}(2N)/\,\mathrm{P}(1,2N-2,1) &\sim \mathrm{SL}(2N)/\,\mathrm{P}(2N-1,1). \end{aligned} \tag{7.4}$$

Notice that both cases of Lemma 7.1 and its corollary are represented here. Indeed $o(2k+1)$ is imbedded reducibly in $o(2k)$, $\mathrm{sp}(2N)$ is imbedded irreducibly in $\mathrm{sl}(2N)$.

The situation is particularly simple for the $\mathrm{sl}(N)$ subalgebras of diff_n.

THEOREM 7.1. *The primitive Lie algebras $\mathrm{sl}(N)$ constructed in Section 4 are maximal among the simple subalgebras of diff_n for all values of n.*

PROOF. Let $\mathrm{sl}(N') \subset \mathrm{sl}(N) \subset \mathrm{diff}_n$, $N' = \lambda' + \mu' \le N = \lambda + \mu$ and $\lambda'\mu' = \lambda\mu = n$. We can exclude the case $N' = N$ because that implies $\lambda' = \lambda$ and $\mu' = \mu$, i.e. the two algebra would be identical. So n must be the product of two different pairs of integers; that requires $n \ge 4$. From Section 4 we know that $N \le n+1$. From the theory of irreducible representations of the $\mathrm{sl}(N)$'s, it is known that N is the smallest dimension of a nontrivial irreducible representation and that the next smallest dimension greater than N for an irreducible representation is the antisymmetric tensor square of the defining ($= N$-dimensional) representation; its dimension is $N(N-1)/2$. So we must have $d = N'(N'-1)/2 \leq N \leq n+1$. Using $n = \lambda'\mu'$, an explicit computation gives:

$$(7.5) \quad d = \frac{1}{2}(\lambda'+\mu')(\lambda'+\mu'-1) = \frac{1}{2}\lambda'(\lambda'-1) + \frac{1}{2}\mu'(\mu'-1) + n \leq n+1$$
$$\iff \lambda'(\lambda'-1) + \mu'(\mu'-1) \leq 2.$$

The inequality is incompatible with $n = \lambda'\mu' \geq 4$. □

The smallest orthogonal representation of $\mathrm{sl}(N')$ is the adjoint representation, of dimension $d = N'^2 - 1$ except for $\mathrm{sl}(4) \sim o(6)$ (special case that we have already studied). The largest value of N for $o(N) \subset \mathrm{diff}_n$ is $n+2$ (except for $n = 3$: it is $N = 6$; case that we have studied). So we have to study when the inequality $d \leq n+2$ can be satisfied with $N' = \lambda' + \nu'$ and $n = \lambda'\mu'$. We obtain

$$(7.6) \qquad d = \lambda'^2 + \mu'^2 + 2n - 1 \leq n+2 \iff \lambda'^2 + \mu'^2 \leq 3 - n \iff n \leq 1.$$

That shows that no $\mathrm{sl}(N')$ of Section 4 is a subalgebra of $o(N)$ of Section 5 except for the two isomorphisms already quoted.

The smallest symplectic linear representation of $\mathrm{sl}(N')$ has a larger dimension than the adjoint one except for $\mathrm{sl}(6) \subset \mathrm{sp}(20)$. Relation (7.6) applies a fortiori in the general case and the special case is directly ruled out.

Let us now run through the low dimensional cases of diff_n following Table 2 and identify all maximal subalgebras, all equivalences and all inclusions. The notation $L/L_0 \sim L'/L'_0$ means that the two realizations are equivalent $L/L_0 \subset S/S_0$ means that $L \subset S$ and the realization of L is not maximal in diff_n.

$n = 1$ $\quad \mathrm{sl}(2)/p(1,1) \sim o(3)/p(1,1,1) \sim \mathrm{sp}(2)/p(1,0,1)$
Maximal in diff_n.

$n = 2$ $\quad \mathrm{sl}(3)/p(2,1)$
Maximal in diff_2.

$n = 3$ $\quad \mathrm{sl}(4)/p(3,1) \sim o(6)/p(3,0,3)$
$o(5)/p(1,3,1) \sim \mathrm{sp}(4)/p(2,0,2)$
$o(5)/p(2,1,2) \sim \mathrm{sp}(4)/p(1,2,1) \subset \mathrm{sl}(4)/p(3,1)$
The first two are maximal in diff_3.

$n = 4$ $\quad \mathrm{sl}(5)/p(4,1)$
$\mathrm{sl}(4)/p(2,2) \sim o(6)/p(1,4,1)$
Both are maximal in diff_4.

$n = 5$ $\quad \mathrm{sl}(6)/p(5,1)$
$o(7)/p(1,5,1)$
$\mathrm{sp}(6)/p(1,4,1) \subset \mathrm{sl}(6)/p(5,1)$
The first two are maximal.

$n = 6$ $\quad \mathrm{sl}(7)/p(6,1)$
$\mathrm{sl}(5)/p(3,2)$
$o(8)/p(1,6,1) \sim o(8)/p(4,0,4)$
$\mathrm{sp}(6)/p(3,0,3)$
$o(7)/p(3,1,3) \subset o(8)/p(1,6,1)$

The first four are maximal in diff_6.

$n = 7$ $\quad$ $\mathrm{sl}(8)/p(7,1)$

$o(9)/p(1,7,1)$

$\mathrm{sp}(6)/p(2,2,2)$

$\mathrm{sp}(8)/p(1,6,1) \subset \mathrm{sl}(8)/p(7,1)$

The first three are maximal. The $\mathrm{sp}(6)/p(2,2,2)$ case is the first primitive one with quartic coefficients.

$n = 8$ $\quad$ All three algebras in Table 2 are maximal.

$n = 9$ $\quad$ $\mathrm{sl}(10)/p(1,9)$

$o(11)/p(1,9,1)$

$o(8)/p(2,4,2)$

$\mathrm{sl}(6)/p(3,3)$

$\mathrm{sp}(10)/p(1,4,1) \subset \mathrm{sl}\,\mathrm{sl}(10)/p(9,1)$

The first four are maximal.

$n = 10$ $\quad$ All algebras are maximal except $o(9)/p(4,1,4) \subset o(10)/p(5,0,5)$.

$n = 11$ $\quad$ All algebras are maximal except $\mathrm{sp}(12)/p(1,10,1) \subset \mathrm{sl}(12)/p(11,1)$.

$n = 12, 13, 14$ $\quad$ All algebras are maximal.

$n = 15$ $\quad$ $\mathrm{sl}(16)/p(15,1)$

$\mathrm{sl}(8)/p(5,3)$

$o(17)/p(1,15,1)$

$o(12)/p(6,0,6)$

$\mathrm{sp}(10)/p(5,0,5)$

$\mathrm{sp}(10)/p(2,6,2)$

$o(11)/p(5,1,5) \subset o(12)/p(6,0,6)$

$\mathrm{sp}(16)/p(1,14,1) \subset \mathrm{sl}(16)/p(15,1)$

The first six are maximal; $\mathrm{sp}(10)/p(2,6,2)$ has quartic coefficients.

For $n = 15$ we encounter the first case when our simple criteria based on dimensions and ranks of algebras, dimensions of representations and on the order of the polynomials involved, are not sufficient to decide upon mutual inclusions. Indeed, consider the pair $o(10)/p(3,4,3)$ and $o(11)/p(2,7,2)$. As we know, $o(10)$ can be imbedded reducibly in $o(11)$, without having a centralizer. For both algebras, the coefficients of the vector fields involve quartic polynomials. A different study is needed to decide whether this $o(10)$ can be transformed into the $o(10)$ contained reducibly in $o(11)$.

This problem is typical of the ones that we are postponing to a future article and that occur for higher values of n. For a given value of n we can have two different realizations of $o(N)$, corresponding to two different pairs (λ_1, μ_1) and (λ_2, μ_2). Similarly, we can have two different realizations of $\mathrm{sp}(2N)$. Pairs of algebras $o(N)$, $o(N+1)$ occur for many values of n. An analysis of inclusions and equivalences among such algebras is in preparation.

Acknowledgements

We thank Dr. J. Patera for many helpful discussions. The authors thank each others institutions for hospitality during mutual visits. The research of P. W. was partly supported by research grants from NSERC of Canada and FCAR du Québec.

References

1. R. L. Anderson, *A nonlinear superposition principle admitted by coupled Riccati equations of the projective type*, Lett. Math. Phys. **4** (1980), no. 1, 1–7
2. R. L. Anderson, J. Harnad, and P. Winternitz, *Systems of ordinary differential equations with nonlinear superposition principles*, Phys. D **4** (1982), 164–182.
3. J. Beckers, V. Hussin, and P. Winternitz, *Nonlinear equations with superposition formulas and the exceptional group G_2. I. Complex and real forms of g_2 and their maximal subalgebras*, J. Math. Phys. **27** (1986), no. 9, 2217–2227; II. *Classification of the equations*, J. Math. Phys. **28** (1987), no. 3, 520–529.
4. M. R. Bremner, R. V. Moody, and J. Patera, *Tables of dominant weight multiplicities for representations of simple Lie algebras*, Marcel Dekker, New York, 1985.
5. J. F. Cornwell, *Group theory in Physics*, vol. II, Academic Press, London, 1984.
6. M. Demazure, *Classification des algèbres de Lie filtrées*, Séminaire Bourbaki, vol. 1966–1967, 19e année, W. A. Benjamin Inc., New York-Amsterdam, 1968, pp. 326-01–326-11.
7. M. Golubitsky, *Primitive actions and maximal subgroups of Lie groups*, J. Differential Geom. **7** (1972), 175–191.
8. V. W. Guillemin and S. Sternberg, *An algebraic model of transitive differential geometry*, Bull. Amer. Math. Soc. (N.S.) **70** (1964), 16–47.
9. S. Kobayashi and T. Nagano, *On filtered Lie algebras and geometric structures*. I, J. Math. Mech. **13** (1964), 875–907; II, J. Math. Mech. **14** (1965), 513–521.
10. J. Krause and L. Michel, *Classification of symmetries of ordinary differential equations*, Lecture Notes in Physics, vol. 382, Springer-Verlag, Berlin, 1991, pp. 251–262.
11. S. Lie, *Klassifikation und Integration von gewöhnlichen Differentialgleichungen zwischen x, y die eine Gruppe von Transformationens gestatten*. I, ..., IV, Sophus Lie Gessamunelte Abhandlungen, vol. 5, X, XI, XIV, XVI, Teubner, Leipzig, 1924.
12. ———, *Theorie der Transformationsgruppen, Dritter Abschnitt, Abteilun*. I, Unter Mitwirkung von Pr. F. Engel, Teubner, Leipzig, 1893.
13. ———, *Vorlesungen über continuirliche Gruppen mit geometrischen und anderen Anwendungen*, Bearbeitet und herausgegeben von Dr. G. Scheffers, Teubner, Leipzig, 1893.
14. T. Ochiai, *Classification of the finite nonlinear primitive Lie algebras*, Trans. Amer. Math. Soc. **124** (1966), 313–322.
15. S. Shnider and P. Winternitz, *Classification of systems of nonlinear ordinary differential equations with superposition principles*, J. Math. Phys. **25** (1984), no. 11, 3155–3165.
16. ———, *Nonlinear equations with superposition principles and the theory of transitive primitive Lie algebras*, Lett. Math. Phys. **8** (1984), no. 1, 69–78.
17. B. Yu. Veisfeiler, *A certain property of semisimple algebraic groups*, Funksional. Anal. i Prilozhen. **2** (1968) no. 3, 84–85. (Russian)

Institut des Hautes Etudes Scientifique, 91440 Bures-sur-Yvette, France
E-mail address: `michel@ihes.fr`

Centre de recherches mathématiques, Université de Montréal, C.P. 6128, succ. Centre-ville, Montréal, Québec, H3C 3J7, Canada
E-mail address: `wintern@CRM.UMontreal.CA`

Centre de Recherches Mathématiques
CRM Proceedings and Lecture Notes
Volume **11**, 1997

Shape Boundary Derivative for an Elastic Membrane

Fabrice Desaint and Jean-Paul Zolésio

ABSTRACT. We characterize the first variation of the displacement field u of an elastic membrane under a static loading F with respect to the variation of the compact manifold Γ which represents the membrane position without loading. That first variation is the "Shape Boundary Derivative" u'_Γ introduced in [**12**], [**13**] and then in the book [**10**] which has the property to be only dependant on the normal component of the vector field V moving the manifold. That shape boundary derivative appears at Theorem 4.1 as solving itself a membrane problem with coefficients depending explicitely on the second and third fundamental forms of the manifold.

1. Introduction

Shape derivative has been introduced in [**12**], [**13**] as the intrinsic concept for "domain derivative $\partial/\partial\Omega y(\Omega)$" for the solution $y(\Omega) \in H^m(\Omega)$ of classical boundary value problems, see also [**14**] for further developments. In 1 this concept is extended to boundary functions $z(\partial\Omega) \in H^k(\partial\Omega)$. The two main properties of the shape derivative are that it depends only on the normal component of the vector field which moves the domain and it switches with usual derivatives, roughly speaking if $y'(\Omega, V)$ denotes the shape derivative we have $y'(\Omega, V) = \tilde{y}'(\Omega, V(0).n)$ and $\nabla\big(y'(\Omega, V)\big) = (\nabla y)'(\Omega, V)$.

Let us emphasis here the fact that we definitively not deal with derivatives or generalized gradients of scalar functionals, but with derivatives of functions ranging in a Banach space which moves with the variable. For example if $y(\Omega)$ denotes the solution of the classical Laplace-Dirichlet boundary value problem in Ω, then y is a mapping defined from a family of open sets into a linear fiber bundle.

In a previous work [**6**], [**7**], [**8**] we characterized the shape derivative of the solution of the Laplace-Beltrami problem both on a compact manifold (without boundary) and on an open set in the previous manifold with Dirichlet and Neuman conditions on the (relative) boundary.

Our objective is to reach that characterization for the solutions of elastic shell systems expressed in intrinsic geometry in [**4**], [**3**]. We present here a first step in that direction as we solve that question for a membrane-like problem. The considered modeling of the membrane had been introduced in [**1**] as a "poor modeling"

1991 *Mathematics Subject Classification.* Primary: 73K15; Secondary: 73O62.
This is the final form of the paper.

which led to numerical experiments matching experiments, see [9]. In some sense that membrane problem is simpler than the shell system as it only involves one vectorial tangential boundary value problem. On the mathematical view point it has an "incompressible" difficulty lying on the fact that the kernel of the tangential deformation tensor $\varepsilon_\Gamma(u)$ of the displacement vector u is not reduced to zero and is not known. The introduction of Dirichlet boundary condition would not avoid that point and then the solution is not unique as it is for clamped shell system. We bypass here that difficulty in adding a "small" zero order term $a\vec{u}$ in the membrane equation, which forces the kernel to be zero in all cases. Then for simplicity we deal here with a compact manifold but the addition of Dirichlet boundary conditions would be similar to the scalar work in [8], [7] and would not give extra difficulties.

2. Shape Boundary Derivative

We use notations of the book [10]. The compact manifold Γ in $\mathbb{R}^N$ is oriented by a unitary vector field n. In fact we assume Γ to be the boundary of a smooth bounded domain Ω and n to be the out going normal field. The perturbated manifolds Γ_t are built from Γ by the flow transformation $T_t(V)$ of a smooth non autonomeous vector field $V \in C^0([0,\tau[, C^k(R^N, R^N))$. For a function $u_t \in H^1(\Gamma_t, R^N)$ the material derivative, if it exists, is

DEFINITION 2.1. $\dot{u} = \lim_{t\searrow 0}\{u_t \circ T_t - u/t\}$, the limit being a strong one in $H^1(\Gamma)$.

The shape boundary derivative can be defined (cf. [14]) via the material derivative as:

DEFINITION 2.2. $u'_\Gamma = \dot{u}_{|\Gamma} - D_\Gamma u.V \in L^2(\Gamma, R^N)$.

For a vector $\Phi \in H^2(\Gamma_t; \mathbb{R}^N)$, we introduce the shape boundary derivative of the transposed jacobian matrix ${}^*D_{\Gamma_t}\Phi$:

DEFINITION 2.3. The shape boundary derivative of ${}^*D_{\Gamma_t}\Phi$ is defined via the shape boundary derivatives of the vectors $(\nabla_{\Gamma_t}\Phi_i)$:

$$({}^*D_{\Gamma_t}\Phi)'_\Gamma = \big((\nabla_{\Gamma_t}\Phi_1)'_\Gamma, (\nabla_{\Gamma_t}\Phi_2)'_\Gamma, \ldots, (\nabla_{\Gamma_t}\Phi_N)'_\Gamma\big), \tag{2.1}$$
$$= \big((\dot{\widehat{\nabla_\Gamma\Phi_1}}) - D_\Gamma(\nabla_\Gamma\Phi_1).V, \ldots, (\dot{\widehat{\nabla_\Gamma\Phi_N}}) - D_\Gamma(\nabla_\Gamma\Phi_N).V\big).$$

3. The Membrane Problem

The characterization of u is done via the minimization of the total energy of the system. The internal energy E stored in the "membrane" is assumed to be given by: $E = 1/2 \int_\Gamma [au.u + A\varepsilon_\Gamma(u)..\varepsilon_\Gamma(u)]\, d\Gamma$ where $A \in L^\infty(\Gamma)$ with $A(x) \geq q > 0$ a.e. $x \in \Gamma$ and a is a positive small number. This modeling of the membrane has been introduced in [9] and led to numerical experiments matching the experimental data, see [1].

We study the existence of the solution (or displacement u) to the following vectorial membrane problem:

$$-\vec{\mathrm{div}}_\Gamma(A\varepsilon_\Gamma(u)) + HA\varepsilon_\Gamma(u).n + au = \vec{f}(\Gamma) \text{ on } \Gamma, \tag{3.1}$$

where $f(\Gamma)$ is a force which acts on the membrane and creates a displacement field u. The problem is well posed in $H^1(\Gamma; \mathbb{R}^N)$, without any boundary conditions. The right hand side f is the restriction to Γ of a function $F \in H^{1/2} + \varepsilon(D)$. Then f, the trace of F on Γ is well defined in $L^2(\Gamma;\mathbb{R}^N)$. If a was taken equal to zero f would have to verify a compatibility condition in order to get the existence and the unicity of u:

Let us denote by $\mathcal{N}_\Gamma$ the kernel of ε_Γ in $H^1(\Gamma;\mathbb{R}^N)$. The solution would be searched in the space $H^1(\Gamma;\mathbb{R}^N)/\mathcal{N}_\Gamma$. And the data f should be such that $\int_\Gamma fe\,d\Gamma = 0$ for any e in $\mathcal{N}_\Gamma$. When we will movc the domain, that condition will not be satisfied for perturbed domains as we shall assume f to be the restriction to Γ of a given F.

From [**3**], we have the tangential Korn's inequality:

THEOREM 3.1. *For any vector u in $H^1(\Gamma;\mathbb{R}^N)$, with $\Gamma \subset \Omega$, Ω of class C^3, there exists a constant $C_1(\Gamma)$ such that:*

$$\int_\Gamma D_\Gamma(u)..D_\Gamma(u)\,d\Gamma \le C_1(\Gamma)\left[\|u\|^2_{L^2(\Gamma)} + \|\varepsilon_\Gamma(u)\|^2_{L^2(\Gamma)}\right]. \tag{3.2}$$

THEOREM 3.2. *Let be given $F \in H^{1/2} + \varepsilon(D)$ and define $f(\Gamma)$ as $F_{|\Gamma}$ (the restriction—or trace—on Γ) then there exists a unique $u \in H^1(\Gamma;\mathbb{R}^N)$ solution of* (3.1).

PROOF. We consider the functional

$$\mathcal{E}(\varphi) = \int_\Gamma \left\{\frac{1}{2}A\varepsilon_\Gamma(\varphi)..\varepsilon_\Gamma(\varphi) + \frac{1}{2}a\varphi.\varphi - f\varphi\right\} d\Gamma, \tag{3.3}$$

and the variational problem

$$\mathrm{Min}_{\{\varphi\in H^1(\Gamma;\mathbb{R}^N)\}}\,\mathcal{E}(\varphi). \tag{3.4}$$

From Korn's inequality it derives that $\mathcal{E}$ is coercive on $H^1(\Gamma;\mathbb{R}^N)$. □

4. Shape Derivative of the Solution

Let D a hold all of $\mathbb{R}^N$ contain Ω an open set of $\mathbb{R}^N, \Gamma = \partial\Omega$ its boundary of class C^2 . The problem is then equivalent to the minimization of the elastic energy $\mathcal{E}(u)$, $\varepsilon_\Gamma = 1/2(D_\Gamma + {}^*D_\Gamma)$ is the tangential deformation tensor.

DEFINITION 4.1. B is a $N \times N$ matrix, then we denote by S its symmetrized part $S(B) = 1/2\{B +^* B\}$.

THEOREM 4.1. *Let u_t be the solution of the problem* (3.1) *defined on the perturbed manifold Γ_t with a right hand side f, the trace of some $\vec{F}$ in $H^{3/2+\varepsilon}(D)$ and $A \in W^{1,\infty}(D)$, $A \ge q > 0$ a.e. The shape boundary derivative u'_Γ of u_t exists in $L^2(\Gamma)$ and is the unique solution of the following equation:*

$$\begin{aligned} -\,\vec{\mathrm{div}}_\Gamma\big(A\varepsilon_\Gamma(u'_\Gamma)\big) + HA\varepsilon_\Gamma(u'_\Gamma).n + a\,u'_\Gamma \\ = \vec{\mathrm{div}}_\Gamma(G) - HG.n + 2v\left(\frac{\partial F}{\partial n} + Hf\right), \end{aligned} \tag{4.1}$$

where $v = \langle V(0), n\rangle$ being the normal component of the field,

$$G = A\{Hv\varepsilon_\Gamma(u) + S(D_\Gamma uM(v)) + \varepsilon_\Gamma(u)^*M(v)\} + \frac{\partial A}{\partial n}v\varepsilon_\Gamma(u),$$

$$M(v) = \nabla_\Gamma(v).^*n - vD^2b,$$

H is the mean curvature on Γ, D^2b is the second fundamental form of the manifod Γ, see [**4**] .

REMARK 4.1. As ${}^*M(v).n = 0$ and $M(v).n = \nabla_\Gamma(v)$, the term $G.n$ simplifies for:

$$G.n = \Big\{ HAv\varepsilon_\Gamma(u).n + AS\big(D_\Gamma u M(v)\big).n \Big\} + \frac{\partial A}{\partial n} v\varepsilon_\Gamma(u).n.$$

The existence of the material derivative of u can be classiquely derived by the use of the implicit function theorem which would also provide for the linearity of that material derivative onto the vector field $V(0)$. From the Definition 2.2 the existence of the shape boundary derivative in L^2 is then derived and we shall use its property concerning derivative of boundary integral at Lemma 4.4.

In the proof, for the sake of simplicity, we drop the coefficient a but the kernel $\mathcal{N}_\Gamma = 0$.

PROPERTY 4.1. Let E, P, Q be any $N \times N$ matrices, the following identity holds:

$$E..{}^*P = {}^*E..P, \quad E..(P.Q) = ({}^*P.E)..Q = (E.{}^*Q)..P. \tag{4.2}$$

Moreover if $E = {}^*E$ then

$$E..S(P) = E..P = E..{}^*P.$$

LEMMA 4.1. *Let $Q = (q_{ij})_{i,j}$, $Q \in H^1(\Gamma, R^{N^2})$ and $\Psi \in \mathcal{D}(\mathbb{R}^N)$. We have the following formula of integration by parts:*

$$\int_\Gamma Q..D_\Gamma(\Psi)\, d\Gamma = \int_\Gamma \langle -\vec{\operatorname{div}}_\Gamma Q, \Psi\rangle\, d\Gamma + \int_\Gamma H\langle \Psi, Q.n\rangle\, d\Gamma, \tag{4.3}$$

where H is the mean curvature on Γ, $\vec{\operatorname{div}}_\Gamma(Q)$ denotes the vector whose i^{th} component is built with the divergence of the ith line of the matrix Q and where the second integral should be understood as the duality bilinear form beetwen $H^{-1}(\Gamma, R^N)$ and $H^1(\Gamma, R^N)$.

PROOF. Let Γ be the boundary of a smooth domain Ω in $\mathbb{R}^N$, E being any vector in $\mathbb{R}^N$ such that its restriction to Γ is well defined, the classical result of integration by parts on Γ is given by (cf. [**10**]): $\int_\Gamma -\operatorname{div}_\Gamma(\vec{E}).\psi\, d\Gamma = \int_\Gamma \{\vec{E}.\nabla_\Gamma\Psi - H\vec{E}.n\Psi\}\, d\Gamma$ so, as ${}^*D_\Gamma\Psi = (\nabla_\Gamma\Psi_1, \dots, \nabla_\Gamma\Psi_N)$, we can write: $\int_\Gamma Q..D_\Gamma(\Psi)\, d\Gamma = \int_\Gamma q_{ij}(\nabla_\Gamma\Psi_i)_j$ and we have $\int_\Gamma \vec{Q}_{i,.}.\nabla_\Gamma\Psi_i\, d\Gamma = \int_\Gamma\{-\operatorname{div}_\Gamma\vec{Q}_{i,.}.\Psi_i + H\vec{Q}_{i,.}.n\Psi\}\, d\Gamma$, for i fixed, where $\vec{Q}_{i,.}$ denotes a vector built from the ith line of the matrix Q. $\big(\vec{\operatorname{div}}_\Gamma\big(A\varepsilon_\Gamma(u)\big)\big)_i = \operatorname{div}_\Gamma\big(A\varepsilon_\Gamma(u)\big)_{i,.}$. Then we conclude that the above equality holds for all index i which can be written in matrix form as: $\int_\Gamma q_{ij}(\nabla_\Gamma\Psi_i)_j\, d\Gamma = \int_\Gamma\{-\vec{\operatorname{div}}_\Gamma Q.\Psi + HQ.n\Psi\}\, d\Gamma$.
□

Then we get the weak formulation of the problem (3.1) as:

LEMMA 4.2. $\forall\tilde{u} \in H^1(\Gamma)$, $\forall A \in L^\infty(\Gamma)$, *we have:*

$$\int_\Gamma\{A(x)\varepsilon_\Gamma(u)..\varepsilon_\Gamma(\tilde{u})\}\, d\Gamma = \int_\Gamma \langle\{-\vec{\operatorname{div}}_\Gamma\big(A\varepsilon_\Gamma(u)\big) + HA\varepsilon_\Gamma(u).n\}, \tilde{u}\rangle\, d\Gamma. \tag{4.4}$$

PROOF. The matrix $\varepsilon_\Gamma(.)$ is symmetric then we have $\int_\Gamma \varepsilon_\Gamma(\tilde{u})..{}^*D_\Gamma(u)\, d\Gamma = 1/2\int_\Gamma \varepsilon_\Gamma(\tilde{u})..{}^*D_\Gamma(u)\, d\Gamma + 1/2\int_\Gamma \varepsilon_\Gamma(\tilde{u})..D_\Gamma(u)\, d\Gamma = \int_\Gamma \varepsilon_\Gamma(\tilde{u})..\varepsilon_\Gamma(u)\, d\Gamma$ from Property 4.1 and the result derives from Lemma 4.1.
□

In order to characterize the equation whose u'_Γ is the solution, we shall use test functions defined over $\mathbb{R}^N$, $\Phi \in \mathcal{D}(\mathbb{R}^N)$ with the property

$$\frac{\partial}{\partial n}\Phi = 0 \text{ on } \Gamma,$$

and we shall set $\varphi \in H^1(\Gamma;\mathbb{R}^N) : \varphi = \Phi_{|\Gamma}$; for example Φ could be taken in the special form $\varphi \circ p$ in a neighbourhood of Γ. For such elements we have :

LEMMA 4.3. *Let $\Phi \in H^{3/2}(\mathbb{R}^N)$, with $\partial/\partial n\Phi = 0$ on Γ,*

$$\varepsilon(\Phi)_{|\Gamma} = \varepsilon_\Gamma(\Phi)_{|\Gamma} = \varepsilon_\Gamma(\varphi),$$

where $\varphi = \Phi|_\Gamma$ lies in $H^1(\Gamma;\mathbb{R}^N)$.

The weak formulation of the problem (3.1) over Γ_t is $u_t \in H^1(\Gamma_t,\mathbb{R}^N)$ such that, $\forall\Phi \in \mathcal{D}(\mathbb{R}^N)$, $\int_{\Gamma_t}\langle -\vec{\operatorname{div}}_{\Gamma_t}(A\varepsilon_{\Gamma_t}(u_t)) + HA\varepsilon_{\Gamma_t}(u_t).n_t, \Phi\rangle\, d\Gamma_t = \int_{\Gamma_t}\langle f,\Phi\rangle\, d\Gamma_t$ but using the symmetry that appears in the Lemma 4.2 we can exchange u_t and Φ in the above equation, then we get:

$$\int_{\Gamma_t}\big\langle -\vec{\operatorname{div}}_{\Gamma_t}\big(A\varepsilon_{\Gamma_t}(\Phi)\big) + HA\varepsilon_{\Gamma_t}(\Phi).n_t, u_t\big\rangle\, d\Gamma_t = \int_{\Gamma_t}\langle f,\Phi\rangle\, d\Gamma_t \tag{4.5}$$

In order to differentiate (4.5) with respect to the parameter t we recall the

LEMMA 4.4. (*see* [**10**]) *Let $\Gamma_t = \partial\Omega_t$, $\lambda_t \in H^1(\Omega_t)$, and denote by λ' the shape derivative at Ω, i.e. $\langle\lambda'(\Omega,V) = \dot\lambda - \langle\nabla\lambda, V(0)\rangle$ and λ'_Γ the shape boundary derivative at Γ of $\lambda_t|_{\Gamma_t}$. Then we have*

$$\lambda'_\Gamma(\Gamma,V) = \big(\lambda'(\Omega V)\big)\big|_\Gamma + \frac{\partial}{\partial n}\lambda \text{ on } \Gamma.$$

Moreover

$$\frac{\partial}{\partial t}\Big\{\int_{\Gamma_t}\lambda_t\, d\Gamma_t\Big\}_{t=0} = \int_\Gamma(\lambda'_\Gamma + H\lambda V.n)\, d\Gamma, \tag{4.6}$$
$$= \int_\Gamma\Big\{\lambda'(\Omega,V) + \Big(\frac{\partial\lambda}{\partial n} + H\lambda\Big)\langle V(0),n\rangle\Big\}\, d\Gamma.$$

Where $\lambda = (\lambda_0)|_\Gamma \in H^1(\Gamma)$.

Then the derivative of (4.5) is : $\forall\Phi \in \mathcal{D}(\mathbb{R}^N)$,

$$\begin{aligned}\int_\Gamma\{\langle[-\vec{\operatorname{div}}_{\Gamma_t}(A\varepsilon_{\Gamma_t}(\Phi)) + H_tA\varepsilon_{\Gamma_t}(\Phi).n_t]'_\Gamma, u\rangle\\ + \langle(-\vec{\operatorname{div}}_\Gamma(A\varepsilon_\Gamma(\Phi)) + HA\varepsilon_\Gamma(\Phi).n), u'_\Gamma\rangle\\ + H\langle(-\vec{\operatorname{div}}_\Gamma(A\varepsilon_\Gamma(\Phi)) + HA\varepsilon_\Gamma(\Phi).n), u\rangle V.n\}\, d\Gamma\\ = \int_\Gamma\{\langle f'_\Gamma,\Phi\rangle + \langle f,\Phi'_\Gamma\rangle + H\langle f,\Phi\rangle V.n\}\, d\Gamma.\end{aligned}$$

The two last terms of the left hand side are easy to integrate in order to "liberate" the test function Φ and then to characterize the equation satisfied by u'_Γ. The difficulties are to express the term $\big[-\vec{\operatorname{div}}_{\Gamma_t}\big(A\varepsilon_{\Gamma_t}(\Phi)\big) + H_tA\varepsilon_{\Gamma_t}(\Phi).n_t\big]'_\Gamma$ and then in the integration by parts we will have to do in order to "liberate" the test function Φ. In order to do that we work using a transposition method shifting the problem on the test functions. We must first solve that question for test functions which are smooth and not depending on the parameter t. In that direction we need the three following results

4.1. Shape boundary derivatives of first and second order operators.

PROPOSITION 4.1. *Let $\Psi \in \mathcal{D}(\mathbb{R}^N)$ such that $\partial/\partial n \Psi = 0$ on Γ. We denote $\psi = \Psi|_\Gamma$. Then we have:*

$$(\Psi)'_\Gamma - \frac{\partial}{\partial n}\Psi = 0, \tag{4.7}$$

$$(\nabla_{\Gamma_t}\Psi)'_\Gamma = -D^2 b.\nabla_\Gamma\psi\langle V(0), n\rangle + \nabla_\Gamma\psi.\nabla_\Gamma\big(\langle V(0), n\rangle\big)\vec{n}. \tag{4.8}$$

PROPOSITION 4.2. *Let $\Phi \in \mathcal{D}(\mathbb{R}^N)$. with $\partial/\partial n\Phi = 0$ We have:*

$$\big(\varepsilon_{\Gamma_t}(\Phi)\big)'_\Gamma = S\big(D_\Gamma\varphi.M(v)\big), \tag{4.9}$$

where $M(v) = \big[\nabla_\Gamma(v)^* n - vD^2 b\big]$ and S is the symmetrized part of the matrix $D_\Gamma\varphi.M(v)$ with $v = \big\langle V(0), n\big\rangle$.

So we need the expression of $\big[-\vec{\operatorname{div}}_{\Gamma_t}\big(A\varepsilon_{\Gamma_t}(\Phi)\big) + H_t A\varepsilon_{\Gamma_t}(\Phi).n_t\big]'_\Gamma$

PROPOSITION 4.3. *Let $\Phi \in \mathcal{D}(\mathbb{R}^N)$, setting $v = \big\langle V(0), n\big\rangle$ and:*

$$M(v) = \nabla_\Gamma(v).^* n - vD^2 b.$$

We have:

$$\begin{aligned}
&\big[-\vec{\operatorname{div}}_{\Gamma_t}\big(A\varepsilon_{\Gamma_t}(\Phi)\big) + H_t A\varepsilon_{\Gamma_t}(\Phi).n_t\big]'_\Gamma \\
&\quad = \big[\vec{\operatorname{div}}_\Gamma\big(A\varepsilon_\Gamma(\Phi)\big) - HA\varepsilon_\Gamma(\Phi).n\big]Hv - \vec{\operatorname{div}}_\Gamma\big(HAv\varepsilon_\Gamma(\Phi)\big) \\
&\qquad + H^2 Av\varepsilon_\Gamma(\Phi).n - \vec{\operatorname{div}}_\Gamma\big(A\varepsilon_\Gamma(\Phi).^* M(v)\big) \\
&\qquad + HA\big(\varepsilon_\Gamma(\Phi).^* M(v)\big).n - \vec{\operatorname{div}}_\Gamma\big(AS\big(D_\Gamma\Phi.M(v)\big) + A'_\Gamma\varepsilon_\Gamma(\Phi)\big) \\
&\qquad + \big\{HAS\big(D_\Gamma\Phi.M(v)\big) + HA'_\Gamma\varepsilon_\Gamma(\Phi)\big\}.n.
\end{aligned}$$

PROOF. Applying the Lemma 4.4 to the expression (4.4) we get: $\forall\Psi \in \mathcal{D}(\mathbb{R}^N)$,

$$\begin{aligned}
\int_\Gamma \{\langle\big[-\vec{\operatorname{div}}_{\Gamma_t}\big(A\varepsilon_{\Gamma_t}(\Phi)\big) + H_t A\varepsilon_{\Gamma_t}(\Phi).n_t\big]'_\Gamma, \Psi\rangle& \\
+ \big\langle -\vec{\operatorname{div}}_\Gamma\big(A\varepsilon_\Gamma(\Phi)\big) + HA\varepsilon_\Gamma(\Phi).n, \Psi'_\Gamma\big\rangle& \\
+ H\big\langle -\vec{\operatorname{div}}_\Gamma\big(A\varepsilon_\Gamma(\Phi)\big) + HA\varepsilon_\Gamma(\Phi).n, \Psi\big\rangle V.n\big\}\, d\Gamma& \\
= \int_\Gamma \big\{\big(A.\varepsilon_{\Gamma_t}(\Phi)\big)'_\Gamma..\varepsilon_\Gamma(\Psi) + A\varepsilon_\Gamma(\Phi)..\big(\varepsilon_{\Gamma_t}(\Psi)\big)'_\Gamma& \\
+ HA.\varepsilon_\Gamma(\Phi)..\varepsilon_\Gamma(\Psi)V.n\big\}\, d\Gamma&
\end{aligned} \tag{4.10}$$

at this step we need to express the shape boundary derivative of $\varepsilon_{\Gamma_t}(\Phi)$ and also an integration by parts in order to derive the expression of $\big[-\vec{\operatorname{div}}_{\Gamma_t}\big(A\varepsilon_{\Gamma_t}(\Phi)\big) + H_t A\varepsilon_{\Gamma_t}(\Phi).n_t\big]'_\Gamma$ by identification of the terms of both right and left hand sides of equality (4.10). We also have the very usefull property concerning the shape boundary derivative of the matrix $D_\Gamma u$:

The shape boundary derivative of $D_\Gamma u$ is related to the shape boundary derivative of ${}^*D_\Gamma u$ in the following way:

$$(D_{\Gamma_t} u)'_\Gamma = {}^*\big\{({}^*D_{\Gamma_t} u)'_\Gamma\big\}.$$

□

LEMMA 4.5. *We have the following developments:*

(i) $\nabla(V.n) = ^*DV.n + ^*Dn.V.$
(ii) $\nabla_\Gamma(V.n) = ^*D_\Gamma V.n + ^*D_\Gamma n.V.$

LEMMA 4.6. *Let* Jac *denote the boundary Jacobian, then we have*

$$\frac{\partial}{\partial t}[\mathrm{Jac}^{-1}]_{t=0}. = -\operatorname{div}_\Gamma V(0)$$

LEMMA 4.7. *Let u be a vector defined on the manifold Γ, $u \in H^1(\Gamma, \mathbb{R}^N)$, then*

$$\begin{aligned}(u \circ T_t^{-1})'_\Gamma &= (\widehat{u \circ \dot{T}_t^{-1}})|_\Gamma - D_\Gamma u.V,\\ &= -D_\Gamma u.V.\end{aligned}$$

LEMMA 4.8.

$$\|(^*DT_t^{-1} \circ T_t^{-1}).n \circ T_t^{-1}\|'_\Gamma = -\langle \nabla(\langle V(0), n\rangle), n\rangle_{\mathbb{R}^N}.$$

PROOF. The transport of the normal is given by:

$$n_t \circ T_t = \frac{^*DT_t^{-1}.n}{\|^*DT_t^{-1}.n\|_{\mathbb{R}^N}} \iff n_t = \frac{^*DT_t^{-1} \circ T_t^{-1}.n \circ T_t^{-1}}{\|^*DT_t^{-1} \circ T_t^{-1}.n \circ T_t^{-1}\|_{\mathbb{R}^N}},$$

then we get

$$\|(^*DT_t^{-1} \circ T_t^{-1}).n \circ T_t^{-1}\| = \langle(^*DT_t^{-1} \circ T_t^{-1}).n \circ T_t^{-1}, n_t\rangle_{\mathbb{R}^N},$$

and

$$\begin{aligned}\|(^*DT_t^{-1} \circ T_t^{-1}).n \circ T_t^{-1}\|'_\Gamma &= \frac{1}{2}\big(\|(^*DT_t^{-1} \circ T_t^{-1}).n \circ T_t^{-1}\|^2\big)'_\Gamma,\\ &= -\langle ^*DV.n, n\rangle - \langle D_\Gamma n.V, n\rangle,\end{aligned}$$

but $\nabla\langle V, n\rangle = {}^*DV.n + {}^*D(\nabla b).V$ (as ∇b is an extension of the normal n) then $\|(^*DT_t^{-1} \circ T_t^{-1}).n \circ T_t^{-1}\|'_\Gamma = -\langle\nabla(V.n), n\rangle_{\mathbb{R}^N} + \langle ^*D(\nabla b).V, n\rangle_{\mathbb{R}^N} - \langle D_\Gamma n.V, n\rangle_{\mathbb{R}^N}$. But from [4], we have $^*D(\nabla b) = D_\Gamma n = D^2 b$ a symmetric matrix and $n \in \operatorname{Ker} D^2 b$. □

LEMMA 4.9.

$$\big[(^*DT_t^{-1} \circ T_t^{-1}).u \circ T_t^{-1}\big]'_\Gamma = DV.u - D_\Gamma u.V.$$

LEMMA 4.10.

$$n'_\Gamma = -\nabla_\Gamma(V.n) = -(^*D_\Gamma V.n + {}^*D_\Gamma n.V).$$

PROOF. The definition for the shape boundary derivative leads to

$$\begin{aligned}n'_\Gamma &= \dot{n_t}|_\Gamma - D_\Gamma n.V = \partial_t(n_t \circ T_t)_{|t=0} - D_\Gamma n.V,\\ &= \partial_t\big(\{^*DT_t^{-1}.n\} \times \{\|^*DT_t^{-1}.n\|_{\mathbb{R}^N}\}^{-1}\big)|_{t=0} - D_\Gamma n.V,\end{aligned}$$

but

$$\partial_t\|^*DT_t^{-1}.n\|_{\mathbb{R}^N_{t=0}} = \frac{1}{2}\partial_t\|^*DT_t^{-1}.n\|_{\mathbb{R}^N{}^2_{t=0}} = -\langle ^*DV.n, n\rangle,$$

then

$$\begin{aligned}n'_\Gamma &= -{}^*DV.n + <^* DV.n, n> n - D_\Gamma n.V,\\ n'_\Gamma &= -(^*D_\Gamma V.n + ^* D_\Gamma n.V),\\ n'_\Gamma &= -\nabla_\Gamma(V.n).\end{aligned}$$

□

PROOF OF PROPOSITION 4.2. We have $\varepsilon_{\Gamma_t}(\Phi) = 1/2\{D_{\Gamma_t}\Phi + {}^*D_{\Gamma_t}\Phi\}$ with ${}^*D_{\Gamma_t}\Phi = (\nabla_{\Gamma_t}\Phi_1, \ldots, \nabla_{\Gamma_t}\Phi_N)$ and then using the definition of the shape boundary derivative for a vector and the result of Proposition 3.2, the jth column of the matrix $({}^*D_{\Gamma_t}\Phi)'_\Gamma$ will be equal to: $(\nabla_{\Gamma_t}\Phi_j)'_\Gamma = -D^2b.\nabla_\Gamma\varphi_j V.n + \nabla_\Gamma\varphi_j.\nabla_\Gamma(V.n)\vec{n}$.

From this expression, we get:

$$\begin{aligned}({}^*D_{\Gamma_t}\Phi)'_\Gamma &= -D^2b.{}^*D_\Gamma\varphi V.n + n.{}^*\big[D_\Gamma\varphi\nabla_\Gamma(V.n)\big],\\ &= -D^2b.{}^*D_\Gamma\varphi V.n + n^*\nabla_\Gamma(V.n)^*D_\Gamma\varphi,\end{aligned}$$

and using the Proposition 4.1, we first deduce the expression of $(D_{\Gamma_t}\Phi)'_\Gamma$ and then the expression of $\big(\varepsilon_{\Gamma_t}(\Phi)\big)'_\Gamma$ is the following:

$$\begin{aligned}\big(\varepsilon_{\Gamma_t}(\Phi)\big)'_\Gamma &= \frac{1}{2}\{(D_{\Gamma_t}\Phi)'_\Gamma + ({}^*D_{\Gamma_t}\Phi)'_\Gamma\}\\ &= -\frac{1}{2}(D^2b.{}^*D_\Gamma\varphi + D_\Gamma\varphi.D^2b)V.n + \frac{1}{2}D_\Gamma\varphi.\nabla_\Gamma(V.n).{}^*n\\ &\qquad\qquad + \frac{1}{2}n^*\big(D_\Gamma\varphi.\nabla_\Gamma(V.n)\big)\\ &= -S\big(D_\Gamma\varphi.D^2bV.n\big) + S\big(D_\Gamma\varphi.\nabla_\Gamma(V.n).{}^*n\big)\\ &= S\big(D_\Gamma\varphi.\nabla_\Gamma(V.n).{}^*n - D_\Gamma\varphi.D^2bV.n\big).\end{aligned}$$

At this step, we are able to express the terms $\big(A\varepsilon_{\Gamma_t}(\Phi)\big)'_\Gamma$ and $\big(\varepsilon_{\Gamma_t}(\Psi)\big)'_\Gamma$.

As we went over this difficulty, we will have then to integrate by parts the result in order to isolate the function Ψ in the equation (4.10) and to conclude with the expression of

$$\big[-\vec{\operatorname{div}}_{\Gamma_t}\big(A\varepsilon_{\Gamma_t}(\Phi)\big) + H_tA\varepsilon_{\Gamma_t}(\Phi).n_t\big]'_\Gamma,$$

given in Proposition 4.3. □

REMARK 4.2. The Property 4.1

1) Allows us to integrate the term $\int_\Gamma E..S\big(D_\Gamma\Psi.M(V.n)\big)\,d\Gamma$.

Using the Proposition 4.2 and the Properties 4.1, we can write:

$$\int_\Gamma E..S\big(D_\Gamma\Psi.M(V.n)\big)\,d\Gamma = \frac{1}{2}\int_\Gamma \big(M(V.n).E\big)..{}^*D_\Gamma\Psi\,d\Gamma + \frac{1}{2}\int_\Gamma \big(E.{}^*M(V.n)\big)..D_\Gamma\Psi\,d\Gamma, \tag{4.11}$$

these two terms being equal.

As $\Psi'_\Gamma = 0$, Ψ is independant on t, the equation (4.11) can be re-written as: $\forall\Phi \in \mathcal{D}(\mathbb{R}^N)$, $\forall\Psi \in \mathcal{D}(\mathbb{R}^N)$,

$$\begin{aligned}&\int_\Gamma \big\langle\big[-\vec{\operatorname{div}}_{\Gamma_t}\big(A\varepsilon_{\Gamma_t}(\Phi)\big) + H_tA\varepsilon_{\Gamma_t}(\Phi).n_t\big]'_\Gamma, \Psi\big\rangle\,d\Gamma\\ &\quad= \int_\Gamma \Big\{A'_\Gamma.\varepsilon_\Gamma(\Phi)..\varepsilon_\Gamma(\Psi) + A\big(\varepsilon_{\Gamma_t}(\Phi)\big)'_\Gamma..\varepsilon_\Gamma(\Psi)\\ &\qquad + A\varepsilon_\Gamma(\Phi)..\big(\varepsilon_{\Gamma_t}(\Psi)\big)'_\Gamma + HA\varepsilon_\Gamma(\Phi)..\varepsilon_{\Gamma_t}(\Psi)V.n\\ &\qquad + HV.n\big\langle\big[\vec{\operatorname{div}}_\Gamma\big(A\varepsilon_\Gamma(\Phi)\big) - HA\varepsilon_\Gamma(\Phi).n\big], \Psi\big\rangle\Big\}\,d\Gamma\end{aligned} \tag{4.12}$$

that, for convenience, we shall write as follows:

$$(4.13)\quad \int_\Gamma \langle [-\vec{\mathrm{div}}_{\Gamma_t}(A\varepsilon_{\Gamma_t}(\Phi)) + H_t A\varepsilon_{\Gamma_t}(\Phi).n_t]'_\Gamma, \Psi\rangle \, d\Gamma$$
$$= \int_\Gamma \{T_1 + T_2 + T_3 + T_4 + HV.n\langle [\vec{\mathrm{div}}_\Gamma(A\varepsilon_\Gamma(\Phi)) - HA\varepsilon_\Gamma(\Phi).n], \Psi\rangle\} \, d\Gamma,$$

where the terms $T_1, \dots, T_4$ defined by:

$$T_1 = A'_\Gamma.\varepsilon_\Gamma(\Phi)..\varepsilon_\Gamma(\Psi,), \qquad T_2 = A(\varepsilon_{\Gamma_t}(\Phi))'_\Gamma..\varepsilon_\Gamma(\Psi),$$
$$T_3 = A\varepsilon_\Gamma(\Phi)..(\varepsilon_{\Gamma_t}(\Psi))'_\Gamma, \qquad T_4 = AH\varepsilon_\Gamma(\Phi)..\varepsilon_\Gamma(\Psi)V.n,$$

have to be integrated by parts in order to isolate Ψ and to deduce the result of Proposition 4.3.

From Lemma 4.1, it easily comes that:

$$(4.14)\quad \int_\Gamma T_1 \, d\Gamma = \int_\Gamma \langle [-\vec{\mathrm{div}}_\Gamma(A'_\Gamma\varepsilon_\Gamma(\Phi)) + HA'_\Gamma\varepsilon_\Gamma(\Phi).n], \Psi\rangle\} \, d\Gamma$$

and

$$(4.15)\quad \int_\Gamma T_4 \, d\Gamma = \int_\Gamma \langle [-\vec{\mathrm{div}}_\Gamma(HAV.n\varepsilon_\Gamma(\Phi)) + H^2AV.n\varepsilon_\Gamma(\Phi).n], \Psi\rangle \, d\Gamma.$$

Concerning the terms T_2 and T_3, we use the Proposition 4.3 where $(\varepsilon_{\Gamma_t}(\Phi))'_\Gamma$ is given and whose expression is similar to the one of $(\varepsilon_{\Gamma_t}(\Psi))'_\Gamma$:

$$(4.16)\quad \int_\Gamma T_2 \, d\Gamma = \int_\Gamma AS(D_\Gamma\Phi.[\nabla_\Gamma(V.n)^* n - V.nD^2 b])..\varepsilon_\Gamma(\Psi) \, d\Gamma$$
$$= \frac{1}{2}\int_\Gamma AS(D_\Gamma\Phi.[M(V.n)])..^* D_\Gamma(\Psi) \, d\Gamma$$
$$+ \frac{1}{2}\int_\Gamma AS(D_\Gamma\Phi.[M(V.n)])..D_\Gamma(\Psi) \, d\Gamma$$
$$= \int_\Gamma \langle -\vec{\mathrm{div}}_\Gamma(AS(D_\Gamma\Phi.[M(V.n)]))$$
$$+ HAS(D_\Gamma\Phi.[M(V.n)]).n, \Psi\rangle \, d\Gamma$$

and for T_3 we get:

$$\int_\Gamma T_3 \, d\Gamma = \int_\Gamma A\varepsilon_\Gamma(\Phi)..S(D_\Gamma\Psi.[\nabla_\Gamma(V.n)^* n - V.nD^2 b]) \, d\Gamma.$$

Using Lemma 4.2 we have

$$\int_\Gamma T_3 \, d\Gamma = \int_\Gamma \frac{1}{2} A\varepsilon_\Gamma(\Phi)..D_\Gamma\Psi.[\nabla_\Gamma(V.n)^* n - V.nD^2 b] \, d\Gamma$$
$$+ \int_\Gamma \frac{1}{2} A\varepsilon_\Gamma(\Phi)..^*[\nabla_\Gamma(V.n)^* n]^* D_\Gamma\Psi \, d\Gamma$$
$$- \int_\Gamma \frac{1}{2} A\varepsilon_\Gamma(\Phi)..V.nD^2 b^* D_\Gamma\Psi \, d\Gamma$$
$$= \int_\Gamma \frac{1}{2} AM(V.n)\varepsilon_\Gamma(\Phi)..^* D_\Gamma\Psi \, d\Gamma + \int_\Gamma \frac{1}{2} A\varepsilon_\Gamma(\Phi).^* M(V.n)..D_\Gamma\Psi \, d\Gamma$$
$$= \int_\Gamma A(\varepsilon_\Gamma(\Phi).^* M(V.n))..D_\Gamma\Psi \, d\Gamma$$

and then from Lemma 4.1 we finally obtain

$$\text{(4.17)}\quad \int_\Gamma T_3\, d\Gamma = \int_\Gamma \langle -\vec{\operatorname{div}}_\Gamma\big(A\varepsilon_\Gamma(\Phi).^*M(V.n)\big), \Psi\rangle\, d\Gamma + \int_\Gamma AH\langle \Psi, \varepsilon_\Gamma(\Phi).^*M(V.n).n\rangle\, d\Gamma.$$

All the required integrations by parts to isolate the test function Ψ associated with Φ are done. Then regrouping equations (19), (20), (21) and (22) we get:

$$\begin{aligned}
\int_\Gamma \langle [-\vec{\operatorname{div}}_{\Gamma_t}\big(A\varepsilon_{\Gamma_t}(\Phi)\big) + H_t A\varepsilon_{\Gamma_t}(\Phi).n_t]'_\Gamma, \Psi\rangle\, d\Gamma
&= \int_\Gamma \langle -\vec{\operatorname{div}}_\Gamma\big(A'_\Gamma\varepsilon_\Gamma(\Phi)\big) + HA'_\Gamma\varepsilon_\Gamma(\Phi).n, \Psi\rangle\, d\Gamma \\
&\quad + \int_\Gamma \langle -\vec{\operatorname{div}}_\Gamma\big(AS(D_\Gamma(\Phi).M(V.n))\big) + HAS\big(D_\Gamma(\Phi).M(V.n)\big).n, \Psi\rangle\, d\Gamma \\
&\quad + \int_\Gamma \langle -\vec{\operatorname{div}}_\Gamma\big(A\varepsilon_\Gamma(\Phi).^*M(V.n)\big) + HA\varepsilon_\Gamma(\Phi).^*M(V.n).n, \Psi\rangle\, d\Gamma \\
&\quad + \int_\Gamma \langle -\vec{\operatorname{div}}_\Gamma\big(HAV.n\varepsilon_\Gamma(\Phi)\big) + H^2AV.n\varepsilon_\Gamma(\Phi).n, \Psi\rangle\, d\Gamma \\
&\quad + \int_\Gamma \langle HV.n\{\vec{\operatorname{div}}_\Gamma\big(A\varepsilon_\Gamma(\Phi)\big) - HA\varepsilon_\Gamma(\Phi).n\}, \Psi\rangle\, d\Gamma
\end{aligned}$$

whose strong formulation is the following:

$$\begin{aligned}
\big[-\vec{\operatorname{div}}_{\Gamma_t}\big(A\varepsilon_{\Gamma_t}(\Phi)\big) + H_tA\varepsilon_{\Gamma_t}(\Phi).n_t\big]'_\Gamma
&= \big[\vec{\operatorname{div}}_\Gamma\big(A\varepsilon_\Gamma(\Phi)\big) - HA\varepsilon_\Gamma(\Phi).n\big]HV.n - \vec{\operatorname{div}}_\Gamma\big(HAV.n\varepsilon_\Gamma(\Phi)\big) \\
&\quad + H^2AV.n\varepsilon_\Gamma(\Phi).n - \vec{\operatorname{div}}_\Gamma\big(A\varepsilon_\Gamma(\Phi).^*M(V.n)\big) \\
&\quad + HA\big(\varepsilon_\Gamma(\Phi).^*M(V.n)\big).n - \vec{\operatorname{div}}_\Gamma\big(AS(D_\Gamma\Phi.M(V.n))\big) \\
&\quad + \big(A'_\Gamma\varepsilon_\Gamma(\Phi)\big) + \big\{HAS\big(D_\Gamma\Phi.M(V.n)\big) + HA'_\Gamma\varepsilon_\Gamma(\Phi)\big\}.n
\end{aligned}$$

but some terms can be brought together and some others can be simplified.

Then finally we get for the shape boundary derivative,

$$\begin{aligned}
-\vec{\operatorname{div}}_\Gamma\big(A\varepsilon_\Gamma(u'_\Gamma)\big) + HA\varepsilon_\Gamma(u'_\Gamma).n
&= \vec{\operatorname{div}}_\Gamma\big(HAV.n\varepsilon_\Gamma(u)\big) - H^2AV.n\varepsilon_\Gamma(u).n \\
&\quad + \vec{\operatorname{div}}_\Gamma\big(AS(D_\Gamma(u).M(V.n))\big) - HAS\big(D_\Gamma(u).M(V.n)\big).n \\
&\quad + \vec{\operatorname{div}}_\Gamma\big(A'_\Gamma\varepsilon_\Gamma(u)\big) - HA'_\Gamma\varepsilon_\Gamma(u).n + \vec{\operatorname{div}}_\Gamma\big(A\varepsilon_\Gamma(u)^*M(V.n)\big) \\
&\quad - HA\big(\varepsilon_\Gamma(u)^*M(V.n)\big).n + 2\frac{\partial F}{\partial n}V.n + HfV.n
\end{aligned}$$

we could rearrange some terms in order to make appear the deviatoric part of the curvature tensor $(D^2b - H/2I_d)$ as in the scalar situation of [4] but this does not make simpler the expression.

REMARK 4.3. It is easy to verify that $\big(\varepsilon_\Gamma(u)^*M(V.n)\big).n = 0$ as long as $n \in \operatorname{Ker} D^2b$, using the expression of $M(V.n)$. The same way we have $D_\Gamma u.n = 0$ as long as $^*D_\Gamma E = (\nabla_\Gamma E_1, \dots, \nabla_\Gamma E_N)$, for $E \in H^1(\Gamma; \mathbb{R}^N)$.

From this expression, it is easy to "shape differentiate" the additive term "a u" that appears in the problem (3.1) and then to obtain the result of Theorem 4.1 where the shape boundary derivative u'_Γ of u_t, solution of the problem (3.1) on the perturbed manifold Γ_t, only depends on $V(0).n$.

References

1. N. Clariond, J.-P. Zolésio, *Hydroelastic behavior of an elastic membrane*, Boundary Control and Variation (J.-P. Zolésio ed.), Lecture Note in Pure and Appl. Math. **163**, New-York, 1994, pp. 73–93.
2. M. C. Delfour and J.-P. Zolésio, *Shape analysis via oriented distance functions*, J. Func. Anal. **123** (1994), no. 1, 129–201.
3. ______, *Intrinsic differential geometry and theory of thin/shallow shells*, Publication of the Scuola Normale Superiore, Pisa, 1997.
4. ______, *A boundary differential equation for thin shells*, J. Differential Equations **119** (1995), no. 2, 426–449; *Tangential differential equations for dynamical thin/shallow shells*, J. Differential Equations **128** (1996), no. 1, 125–167.
5. ______, *Velocity method and Lagrangian formulation for the computation of the shape Hessian*, SIAM J. Control Optim. **29** (1991), no. 6, 1414–1442.
6. F. Desaint, *Dérivées par rapport au domaine en géométrie intrinsèque: Application aux équations de coques*, thèse de doctorat, Université de Nice, Sophia Antipolis, 1996.
7. F. Desaint and J.-P. Zolésio, *Shape derivative for Laplace–Beltrami equation*, PDE's Method in Control and Shape Analysis (G. Da Prato, J. P. Zolésio, eds.), Lecture Notes in Pure and Appl. Math., Marcel Dekker, 1997.
8. ______, *Dérivation de forme dans les problèmes elliptiques tangentiels*, C. R. Acad. Sci. Paris Sér. I, vol. 283 (1996), pp. 781–786.
9. J. P. Marmorat, *Static and dynamic behavior of a fluid-shell system* (G. Da Prato, J. P. Zolésio eds.), Lecture Notes in Pure and Appl. Math., Marcel Dekker, 1996.
10. J. Sokolowski and J.-P. Zolésio, *Introduction to shape optimization. Shape sensitivity analysis*, Comput. Math. **16**, Springer Verlag, 1992.
11. R. Temam, *Mathematical problems in elasticity*, Gauthier-Villars, Paris, 1985.
12. ______, *Identification de domaines par déformation* , thèse de doctorat-d'état, Université de Nice, France, 1979.
13. ______, *The material derivative (or speed method) for shape optimization*, Optimization of Distributed Parameter Structures (E. J. Haug and J. Céa, eds.), vol. II, Sijhoff and Nordhoff, Alphen aan den Rijn, 1981, pp. 1089–1151.
14. ______, *Introduction to shape optimization problems and free boundary problems*, Shape Optimization and Free Boundaries (M. C. Delfour, G. Sabidussi, eds.), Nato Adv. Sci. Inst. Ser. C Math. Phys. Sci. **380**, Kluwer Acad. Publ., Cordrecht (1992), 397–458.

Université Montpellier II, Université des Sciences et Technologies du Languedoc, Place Eugene Bataillon, 34060 Montpellier, France

E-mail address: desaint@icm.edu.pl

CNRS-Institut Non-Linéaire de Nice, Sophia Antipolis 06560 Valbonne, France

E-mail address: Jean-Paul.Zolesio@sophia.inria.fr